MATTI

THE DESIGN OF MODERN MICROWAVE OSCILLATORS FOR WIRELESS APPLICATIONS

THE DESIGN OF MODERN MICROWAVE OSCILLATORS FOR WIRELESS APPLICATIONS

Theory and Optimization

Ulrich L. Rohde, Ajay Kumar Poddar, and Georg Böck*

Synergy Microwave Corporation, Paterson, NJ

*Technische Universität Berlin

JOHN WILEY & SONS, INC.

This book is printed on acid-free paper. ∞

Copyright © 2005 by John Wiley & Sons, Inc. All rights reserved.

Published by John Wiley & Sons, Inc., Hoboken, New Jersey.
Published simultaneously in Canada.

No part of this publication may be reproduced, stored in a retrieval system, or transmitted in any form or by any means, electronic, mechanical, photocopying, recording, scanning, or otherwise, except as permitted under Section 107 or 108 of the 1976 United States Copyright Act, without either the prior written permission of the Publisher, or authorization through payment of the appropriate per-copy fee to the Copyright Clearance Center, Inc., 222 Rosewood Drive, Danvers, MA 01923, 978-750-8400, fax 978-646-8600, or on the web at www.copyright.com. Requests to the Publisher for permission should be addressed to the Permissions Department, John Wiley & Sons, Inc., 111 River Street, Hoboken, NJ 07030, (201) 748-6011, fax (201) 748-6008.

Limit of Liability/Disclaimer of Warranty: While the publisher and author have used their best efforts in preparing this book, they make no representations or warranties with respect to the accuracy or completeness of the contents of this book and specifically disclaim any implied warranties of merchantability or fitness for a particular purpose. No warranty may be created or extended by sales representatives or written sales materials. The advice and strategies contained herein may not be suitable for your situation. You should consult with a professional where appropriate. Neither the publisher nor author shall be liable for any loss of profit or any other commercial damages, including but not limited to special, incidental, consequential, or other damages.

For general information on our other products and services please contact our Customer Care Department within the U.S. at 877-762-2974, outside the U.S. at 317-572-3993 or fax 317-572-4002.

Wiley also publishes its books in a variety of electronic formats. Some content that appears in print, however, may not be available in electronic format.

Library of Congress Cataloging-in-Publication Data:

Rohde, Ulrich L.
    The design of modern microwave oscillators for wireless applications theory and optimization / by Ulrich L. Rohde, Ajay Kumar Poddar, Georg Böck.
    p. cm.
Includes bibliographical references and index.
ISBN 0-471-72342-8 (cloth)
    1. Oscillators, Microwave. 2. Wireless communication systems--Equipment and supplies. I. Poddar, Ajay Kumar, 1967- II. Böck, Georg, 1951- III. Title.

TK7872.O7R643 2005
621.381′323--dc22

Printed in the United States of America

10 9 8 7 6 5 4 3 2

CONTENTS

Foreword		ix
Preface		xiii
Biographies		xv
1	**Introduction**	1
	1.1 Organization	5
2	**General Comments on Oscillators**	9
	2.1 Sinusoidal Oscillators	10
	2.2 Phase Noise Effects	11
	2.3 Specifications of Oscillators and VCOs	13
	2.4 History of Microwave Oscillators	17
	2.5 Three Approaches to Designing Microwave Oscillators	18
	2.6 Colpitts Oscillator, Grounded Base Oscillator, and Meissen Oscillator	21
	2.7 Three-Reactance Oscillators Using Y-Parameters: An Introduction	28
	2.8 Voltage-Controlled Oscillators (VCOs)	32
3	**Transistor Models**	37
	3.1 Introduction	37
	3.2 Bipolar Transistors	40
	3.3 Field-Effect Transistors (FETs)	47
	3.4 Tuning Diodes	56
4	**Large-Signal S-Parameters**	61
	4.1 Definition	61
	4.2 Large-Signal S-Parameter Measurements	63
5	**Resonator Choices**	71
	5.1 LC Resonators	71

	5.2	Microstrip Resonators	72
	5.3	Ceramic Resonators	79
	5.4	Dielectric Resonators	81
	5.5	YIG-Based Resonators	83

6 General Theory of Oscillators — 87

- 6.1 Oscillator Equations — 87
- 6.2 Large-Signal Oscillator Design — 94

7 Noise in Oscillators — 123

- 7.1 Linear Approach to the Calculation of Oscillator Phase Noise — 123
- 7.2 The Lee and Hajimiri Noise Model — 137
- 7.3 Nonlinear Approach to the Calculation of Oscillator Phase Noise — 139
- 7.4 Phase Noise Measurements — 148
- 7.5 Support Circuits — 153

8 Calculation and Optimization of Phase Noise in Oscillators — 159

- 8.1 Introduction — 159
- 8.2 Oscillator Configurations — 159
- 8.3 Oscillator Phase Noise Model for the Synthesis Procedure — 159
- 8.4 Phase Noise Analysis Based on the Negative Resistance Model — 162
- 8.5 Phase Noise Analysis Based on the Feedback Model — 185
- 8.6 2400 MHz MOSFET-Based Push–Pull Oscillator — 199
- 8.7 Phase Noise, Biasing, and Temperature Effects — 210

9 Validation Circuits — 233

- 9.1 1000 MHz CRO — 233
- 9.2 4100 MHz Oscillator with Transmission Line Resonators — 237
- 9.3 2000 MHz GaAs FET-Based Oscillator — 241
- 9.4 77 GHz SiGe Oscillator — 242
- 9.5 900–1800 MHz Half-Butterfly Resonator-Based Oscillator — 245

10 Systems of Coupled Oscillators — 247

- 10.1 Mutually Coupled Oscillators Using the Classical Pendulum Analogy — 247
- 10.2 Phase Condition for Mutually Locked (Synchronized) Coupled Oscillators — 254

10.3	Dynamics of Coupled Oscillators	257
10.4	Dynamics of N-Coupled (Synchronized) Oscillators	263
10.5	Oscillator Noise	266
10.6	Noise Analysis of the Uncoupled Oscillator	271
10.7	Noise Analysis of Mutually Coupled (Synchronized) Oscillators	276
10.8	Noise Analysis of N-Coupled (Synchronized) Oscillators	282
10.9	N-Push Coupled Mode (Synchronized) Oscillators	300
10.10	Ultra-Low-Noise Wideband Oscillators	315

11 Validation Circuits for Wideband Coupled Resonator VCOs — **341**

11.1	300–1100 MHz Coupled Resonator Oscillator	341
11.2	1000–2000/2000–4000 MHz Push–Push Oscillator	346
11.3	1500–3000/3000–6000 MHz Dual Coupled Resonator Oscillator	355
11.4	1000–2000/2000–4000 MHz Hybrid Tuned VCO	361

References — **367**

Appendix A	Design of an Oscillator Using Large-Signal S-Parameters	381
Appendix B	Example of a Large-Signal Design Based on Bessel Functions	389
Appendix C	Design Example of Best Phase Noise and Good Output Power	397
Appendix D	A Complete Analytical Approach for Designing Efficient Microwave FET and Bipolar Oscillators	407
Appendix E	CAD Solution for Calculating Phase Noise in Oscillators	437
Appendix F	General Noise Presentation	457
Appendix G	Calculation of Noise Properties of Bipolar Transistors and FETs	471
Appendix H	Noise Analysis of the N-Coupled Oscillator Coupled Through Different Coupling Topologies	509

Index — **517**

We believe that the abbreviations used in this book are common knowledge for the intended audience, and therefore require no explanation.

FOREWORD

Wireless and mobile communications, as well as the development of RF test and measurement equipment, have been some of the most powerful drivers of RF technology during the last few years. Oscillators belong to the key elements of such systems—both analog and digital. As a consequence of the progress in technology and design methodology, continuously improved performance of oscillators has been observed.

Although many technical and scientific works have been published with respect to oscillator development, there has been no comprehensive work covering all the important aspects of oscillator development ranging from fundamentals, device and board technology, supply noise, analysis methods, design, and optimization methodologies to practical design of various types of single and coupled oscillators. In addition, most articles concentrate on classic design strategies based on measurements, simulation, and optimization of output power and phase noise (in this sequence) and not on a systematic composition of the whole design procedure, starting with phase noise and output power requirements and leading to optimum performance of all relevant oscillator features.

The purpose of this book, which is based on practical and theoretical research and decades of work, is to fill this gap. There is a need for a deep understanding of all fundamental mechanisms affecting oscillator performance. This work is a tutorial introduction for engineers just entering the exciting field of oscillator design, as well as a reference book for senior engineers and engineering managers that will enable them to evaluate new trends and developments. All the necessary mathematics and practical design information for many oscillator types are given in the text. A wealth of additional information can be found in the modern and substantial referenced literature.

The book is organized in 11 chapters. Chapter 1 summarizes the historical evolution of harmonic oscillators and analysis methods. Chapters 2 to 5 deal with the most important building blocks. Following a discussion of the basic theory on semiconductor devices and large-signal parameters, the most modern devices are chosen as examples. A whole chapter is dedicated to the various types of resonators, an important subject for microstrip applications.

Chapter 6 concentrates on the fundamentals of oscillator design, starting with the derivation of the oscillating condition under the assumption of a simplified linear network. Next, the more realistic nonlinear case is analyzed. The startup and steady-state conditions are derived and discussed. The time-domain behavior of voltage and current in a bipolar transistor as an active device in the oscillator is evaluated. The relationship between current conducting angle, phase noise, and element values of the external capacitors is derived, and two practical design examples at different frequencies are given. The chapter is illustrated with numerous figures. In summary, this is a good overview of the large-signal operation of the oscillator and the correlation between conducting angle and oscillator properties.

Chapter 7 addresses the fundamentals of oscillator noise. Derivation and application of the extended Leeson formula is shown extensively. Although Leeson's formula describes phase noise which is correct in principle, the problem is to find the correct parameters for a quantitative correct description of the practical case. Therefore, locating the correct parameters of Leeson's formula is the focus of this chapter. Formulas for practical application are given. A tutorial treatise on phase noise measurements closes the chapter.

Chapter 8 concentrates on analysis and optimization of phase noise in oscillators. Following the simple Leeson formula, the analysis is based on two models: the negative resistance model and the feedback model. The concept of a time-dependent, "noisy" negative resistance was originally introduced by Kurokawa [82] for synchronized oscillators. It is extended here to real oscillators with real noise sources and allows for a complete bias-dependent noise calculation. In the second approach, the oscillator is considered as a closed loop. A thorough study of the influence of different transistor noise sources on overall oscillator phase noise is provided by applying this approach. Both methods are applicable in practical oscillator design. They are different in approach but lead to the same results, and agree with simulation using advanced harmonic balance simulators as well as measurements. The chapter is rounded out by the description of a typical oscillator design cycle and the validation of the previously described theory.

In Chapter 9, five design examples are given for the design of practical oscillators and for extensive validation of the circuit synthesis described above. The phase noise performance is compared with measurements and simulated results from Microwave Harmonica from Ansoft Corporation. The power of the novel synthesis procedure is thus proven.

Chapter 10 addresses the topic of coupled oscillators. For the first time, it is shown that the design of ultra-low-phase noise voltage-controlled oscillators (VCOs) with a wide tuning range (octave band) is possible using this concept. A complete and correct noise analysis of coupled oscillators is carried out, and the relative improvement in phase noise with respect to the single free-running oscillator is discussed. Design criteria and the complete practical design procedure are given in detail, as well as numerous design examples.

In Chapter 11, a variety of validation circuits for wideband coupled resonator VCOs are given. This is followed by the references and appendices.

This book will be a powerful reference tool for the analysis and practical design of single and coupled low-noise microwave oscillators. It has evolved from our interest in and desire to understand the fundamental mechanisms contributing to phase noise and to build world-class oscillators with extremely low phase noise over a wide tuning range.

I hope this work, which is the result of several dissertations sponsored at the Technische Universität Berlin, will become the standard of excellence as a tool in oscillator design.

The early leader in phase noise analysis, Professor David Leeson of Stanford University, comments that "The analysis and optimization methods described in this book represent helpful contributions to the art of low noise oscillators, a subject of great interest to me over the past forty years. The extension to multiple resonators and oscillators promises the realization of new types of wideband low noise oscillators."

PROF. DR.-ING GEORG BÖCK,
Technische Universität, Berlin

PREFACE

Microwave oscillators, because of their importance in communication systems, have been a subject of interest for many decades. Initially, when engineers worked at lower frequencies, frequency generation was done with vacuum tube oscillators. Later, field-effect transistors and bipolar transistors were used. These were free-running, and used air variable capacitors for tuning and banks of inductors to select the frequency range. It was always very difficult to find the proper temperature compensation of the oscillator to improve long-term frequency stability.

Due to the high Q of these resonators and the large voltage swing across them, phase noise was never an issue. When synthesizers were introduced, there still remained the problem of phase noise outside the loop bandwidth, which determined the performance far off the carrier. These phase-locked loop systems also worked at increasingly high frequencies. Systems today, operating at 100 GHz using yttrium ion garnet (YIG) oscillators, are fully synthesized. The YIG oscillators have extraordinarily good phase noise, but they are bulky, expensive, and require complicated power supplies. To overcome this problem, at least in the SHF range to about 10 GHz, we at Synergy Microwave Corporation have developed a new class of microwave oscillators which can be tuned over a very wide frequency range, yet have good phase noise, are low in cost, and are small in size. This work, which was partially sponsored and triggered by two government contracts from DARPA and the U.S. Army, has also produced a variety of patent applications.

The contents of this book are based on this work and fall into two major categories. Chapters 1–9 and Appendices A–G deal with the classic single resonator oscillator, focusing on both nonlinear analysis and phase noise. These have been written by Ulrich L. Rohde. The second category, Chapters 10 and 11 and Appendix H, were written by Ajay Kumar Poddar.

We believe that this book is unique because it is the only complete and thorough analysis of microwave oscillators examining all necessary aspects of their design, with specific emphasis on their operating conditions and phase noise. Our concept of multiple coupled oscillators in printed circuit design has opened the door to a huge improvement in performance. Much relevant and useful information is presented in the literature quoted in this book or covered by our key patent applications. To cover all those details would make the book much too long, and since this information is available to the public, we encourage readers to do the research themselves. To support this book, we recommend the following books, which deal with related topics:

Microwave Circuit Design Using Linear and Nonlinear Techniques, second edition, by George Vendelin, Anthony M. Pavio, and Ulrich L. Rohde, John Wiley & Sons, April, 2005.

RF/Microwave Circuit Design for Wireless Applications by Ulrich L. Rohde and David P. Newkirk, John Wiley & Sons, 2000.

Microwave and Wireless Synthesizers: Theory and Design by Ulrich L. Rohde, John Wiley & Sons, 1997.

Professor Georg Böck from Technische Universität Berlin, whom we have asked to write the Foreword, has helped us stay on track and not get lost in too much mathematics. He also reviewed the manuscript.

Patent Applications

1. U.S. Application No. 60/493075, Tunable Frequency, Low Phase Noise and Low Thermal Drift Oscillator.
2. U.S. Application Nos. 60/501371 and 60/501790, Wideband Tunable, Low Noise and Power Efficient Coupled Resonator/Coupled Oscillator Based Octave-Band VCO.
3. U.S. Application Nos. 60/527957 and 60/528670, Uniform and User Defined Thermal Drift Low Noise Voltage-Controlled Oscillator.
4. U.S. Application No. 60/563481, Integrated Ultra Low Noise Microwave Wideband Push-Push VCO.
5. U.S. Application No. 60/564173, An Ultra Low Noise Wideband VCO Employing Evanescent Mode Coupled Resonator.
6. U.S. Application No. 60/589090, Ultra Low Phase Noise, Low Cost, Low Thermal Drift and Tunable Frequency Ceramic Resonator-Based Oscillator.
7. U.S. Application No. 60/601823, Ultra Low Noise, Hybrid-Tuned and Power Efficient Wideband VCO.
8. U.S. Application No. 60/605791, Visual Inspect Able Surface Mount Device Solder Pads with Improved Mechanical Performance.

ULRICH L. ROHDE
AJAY KUMAR PODDAR

Spring 2005

BIOGRAPHIES

Ulrich L. Rohde, Ph.D., Dr.-Ing., IEEE Fellow, is a professor of microwave and RF technology at the Technische Universität Cottbus, Germany; Chairman of Synergy Microwave Corp., Paterson, NJ; partner of Rohde & Schwarz, Munich, Germany; and a member of the Innovations for High Performance Microelectronics (IHP) Scientific Advisory Council, Frankfurt Oder, Germany. He was previously President of Compact Software, Inc., and a member of the Board of Directors of Ansoft Corporation. He has published 6 books and more than 60 scientific papers. His main interests are communications systems and circuits, specifically low-noise oscillators and high-performance mixers and synthesizers.

Ajay Kumar Poddar, Dr.-Ing., IEEE Senior member, is a Senior Design Engineer for Synergy Microwave Corp., Paterson, NJ; previously, he was a Senior Scientist for the Defense Research and Development Organization (DRDO) (1991–2001), ARDE, Pune, India. He has published more than 20 scientific papers. His main interests are communications systems and circuits, specifically (RF-MEMS) and microwave and millimeterwave oscillators.

Georg Böck, Dr.-Ing., IEEE Senior Member, is a full professor of microwave engineering at the Technische Universität Berlin, Germany, and head of the microwave department. His main areas of research are characterization, modeling, and design of microwave passive and semiconductor devices, MICs, and MMICs up to the millimeterwave range. He has published more than 100 scientific papers. His current interests are MEMS, nonlinear and noise modeling, especially for mixer and oscillator design, as both discrete and ICs.

1 Introduction

The need for oscillators has existed for a long time. The first time it became an important issue was when Maxwell's equations were to be experimentally proven. Heinrich Hertz made the first known oscillator. He used a dipole as the resonator and a spark gap generator as the oscillator circuit, as shown in Figure 1-1.

The spark gap oscillator changes AC or DC power into a spark, which is energy rich and wideband. The dipole then takes the energy at the resonant frequency and radiates it. Other discharges such as lightning, with short pulse duration, generate resonant frequency (RF) power ranging from a few tens of kilohertz to hundreds of megahertz.

Figures 1-2 and 1-3 show additional examples of early oscillators. The pictures in this chapter are taken from [1].

Today, oscillators are used in test and measurement equipment and communication equipment. Given the large number of two-way radios and handies (cell phones) in use, they are the largest group of users. In this monograph, high-performance and high-volume applications are considered, but not the mass-market applications. We will consider external resonators rather than monolithic resonators because thus far, high-quality phase noise requirements have only been met using external resonators.

In these applications, the oscillators have to meet a variety of specifications, which affect the quality of the operation of the system. Two important features are the cleanliness of the oscillator (low phase noise) and its freedom from spurious signals and noise. While the oscillator is almost always used as a voltage-controlled oscillator (VCO) in a frequency synthesizer system, its free-running noise performance outside the loop is still extremely important and is determined solely by the oscillator.

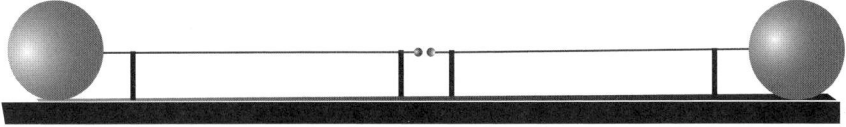

Figure 1-1 Original dipole made by Heinrich Hertz in 1887 using balls at the end to form a capacitive load (Deutsches Museum, Munich).

The Design of Modern Microwave Oscillators for Wireless Applications: Theory and Optimization, by Ulrich L. Rohde, Ajay Kumar Poddar, Georg Böck
Copyright © 2005 John Wiley & Sons, Inc.

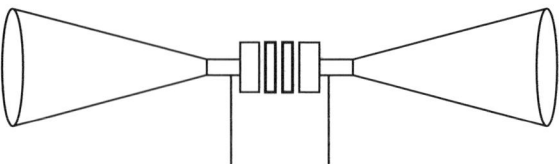

Figure 1-2 Dipole formed by two conical resonators with spark gap (1914).

An oscillator is a circuit that consists of an amplifier and a resonator. The feedback circuit takes a portion of the energy from the output of the amplifier and feeds it into the resonator to compensate for its losses. The amplitude of the oscillator depends on the DC input power and the circuit itself. A small portion of the energy is used to sustain oscillation. Most of the RF power is available to be withdrawn at the output to be further amplified and used, depending on the application. The frequency of the oscillator is largely determined by the resonator.

The classic papers deal with the maximum output power and noise properties of an oscillator as they were first measured and then optimized by trial and error. Even this did not always provide the best possible answer. The purpose of this work is to give, for the first time, a new, simple, complete, efficient way to "synthesize" the design of a high-performance, low-noise oscillator. A general solution will be discussed and validation for the popular Colpitts/Clapp oscillator will be found. This approach will be valid for all types of oscillators [2–9].

An intensive literature search has been done to cover all the relevant previously published discussions.

The first paper, concerning the noise properties of an oscillator, was done by Leeson [70] in 1966. This classic paper is still an extraordinarily good design guide. The advantage of this approach is that it is easy to understand and leads to a good approximation of the phase noise. The drawback is that the values for the flicker noise contribution, which is a necessary input to the equation, the RF output power, the loaded Q, and the noise factor of the amplifier under large-signal

Figure 1-3 Dipole oscillator after Ludenia placed in a parabolic mirror to increase efficiency (1929).

conditions, are not known. Other classic papers, such as that of Kurokawa [82], indicate where the operating point for the best phase noise lies, but the value of the phase noise as such is not known [17]. The next breakthrough in oscillator noise analysis came from Rizzoli et al. [77–79]. It is based on a noise correlation matrix and incorporates the various noise sources from the active device. Commercial simulation programs use a fixed topology for the transistor models. Available are the Gummel-Poon bipolar transistor model, an heterojunction bipolar transistor (HBT) model, and the various gallium arsenide field-effect transistor (FET)-based models, as well as metal oxide semiconductor (MOS) and JFET models. The implementation of the noise sources for these semiconductor devices is shown in the user's manual of the simulator, specifically the model library.

The latest approach is a general noise theory for arbitrary circuits, as shown by Lee et al. [64–67]. Their noise model is based on the time-varying properties of the current waveform of the oscillator, and the phase noise analysis is based on the effect of noise impulse on a periodic signal.

If an impulse is injected into the tuned circuit at the peak of the signal, it will cause maximum amplitude modulation and no phase modulation. If an impulse is injected at the zero crossing of the signal, there will be no amplitude modulation but maximum phase modulation. If noise impulses are injected between zero crossing and the peak, there will be components of both phase and amplitude modulation. Based on this theory and the intention to obtain the best phase noise, a special technique has to be adopted to make sure that any noise impulse occurs at the peak of the output voltage rather than at any other point. Lee and Hajimiri introduced an impulse sensitivity function (ISF), which is different for each oscillator topology. This ISF is a dimensionless function periodic in 2π. It has the largest value when maximum phase modulation occurs and has the smallest value when only amplitude modulation occurs.

This approach appears to be purely mathematical and lacks practicality. The calculation of the ISF is tedious and depends upon the oscillator topology. The flicker noise conversion is not clearly defined. Also, there is no general mathematical equation that can be written for the phase noise in terms of components of the circuit, which can be differentiated to obtain both maximum power and best noise performance. Recent publications by Tom Lee have shown that the noise analysis for a given topology can be expressed and gives good results once all the data are known, but it does not lead to exact design rules [107]. Similar to the Leeson equation, it suffers from the fact that the actual noise performance of the device, the loaded Q, and the output power are not known a priori. As a matter of fact, some of the published oscillators by Lee and Hajimiri could be "optimized." This means that the published oscillator circuit did not have the best possible phase noise. By using the optimizer of a commercial harmonic-balance program, the phase noise could be improved significantly. Of course, a good direct synthesis procedure would have given the correct answer immediately.

The oscillators considered in this work are based on commercially available silicon bipolar transistors and silicon germanium transistors. As most designers and companies do not have elaborate and expensive equipment for parameter extraction (to obtain accurate nonlinear models), the design process relies on using available data from the manufacturer, as well as measurements of large-signal S-parameters using a network analyzer. Modern microwave transistors are very well characterized by the manufacturer up to approximately 6 GHz. Noise data, as well as SPICE-type Gummel-Poon model data sets, are available.

At present, a number of topologies exist that can be used for designing oscillators. For the purpose of validating the general synthesis procedure, initially, a simple transistor circuit (Colpitts/Clapp oscillator) is used as the basis of discussion [2–8]. In Chapter 2, the history of microwave oscillators shows that several circuit configurations are possible, and by rotating the circuit, others can be obtained. A particularly useful derivative of the Colpitts oscillator is the Meissen oscillator, which in VCO configurations has a better phase noise. All other oscillator configurations, including VCOs, are a derivative of this basic Colpitts oscillator. The only exception to this is an oscillator with an inductor in the base. While it can also be obtained from rotating the grounding point, its basic advantage is a wider tuning range at the expense of phase noise. All mathematical derivations also apply to the other configurations.

Using various resonators, we will show that a coupled resonator, rather than a single resonator, vastly improves the phase noise. The oscillator itself can be described as a one-port device supplying a negative resistance to the tuned circuit, which is ideal to determine the best feedback network. Alternatively, it can be described as a two-port device using a resonator and an amplifier, and allows us to calculate the complete noise analysis. It will be shown that both cases provide the same answer. The second case gives more insight into the phase noise calculation. For the first time, this new mathematical approach will show a step-by-step procedure using large-signal conditions for designing an oscillator with good output power (high efficiency) and phase noise. As a third case, the values for P_{out}, Q_l, and F required for the Leeson equation will be numerically determined. All three cases give excellent agreement with the oscillator built under these test conditions and its measurements. Starting with the simple oscillator, a more complex circuit, including all the parasitics, will be used to show the general validity of this approach (Appendix C).

Due to the use of multiple rather than single resonators (since the resonators are less than quarter wave long, coupled inductors should be used), this book has been broken down into two sections (1–9). One concerns the single-resonator case, which is used to show the systematic approach of noise optimization in general, while Chapters 10 and 11 are covering multiple resonators, coupled inductors, and coupled oscillators.

Any successful design for microwave oscillators mandates, besides building and measuring it, the use and validation with a microwave harmonic-balance simulator. In the harmonic-balance analysis method, there are two techniques in use to convert between the time-domain nonlinear model and the frequency-domain evaluation of the harmonic currents of the linear network. One technique is the almost periodic discrete Fourier transform technique (APDFT), and the other is the multi-dimensional fast Fourier transform technique (MFFT) using quasi-analytic or analytic derivatives to evaluate the Jacobian matrix. The first, which has a somewhat random sampling approach, has a typical dynamic range of 75–80 dB, while the second offers a dynamic range exceeding 180 dB.

In mixer designs and intermodulation analysis, which includes the calculation of noise in oscillator circuits, it is important to be able to accurately predict a small signal in the presence of a large signal. To reliably predict this, the dynamic range (the ratio of a large signal to a minimally detectable small signal) needs to be more than 175 dB. The APDFT technique was found to have a dynamic range of 75 dB, while the MFFT, with analytically calculated derivatives, was found to have a final dynamic range of 190 dB. Given the fact that in noise calculations for oscillators a noise floor of -174 dBm/Hz is the lower reference and the reference level can be as high as $+20$ dBm, a dynamic range of up to 190 dB is required. Therefore, a numerically stable approach is definitely required [10–13].

1.1 ORGANIZATION

This work is organized into 11 chapters.

Chapter 1, the Introduction, describes the purpose of the work and defines the problem.

Chapter 2 defines the oscillator, its application, and parameters. In addition, the history of microwave oscillators is briefly discussed and various types of oscillators are introduced.

Chapter 3 describes the various transistor models and gives insight into their parameters. For better understanding, examples of current models are shown.

Chapter 4 develops the concept of large-signal S parameters. The transistor models shown in Chapter 3 are mostly provided in linear form; the large-signal conditions have to be determined from the SPICE-type time-domain signal parameters. A good way of describing a transistor under large-signal conditions involves the use of "large-signal S-parameters," which are introduced here. Examples of measured S-parameters are shown [18–53].

Chapter 5 discusses resonators used for the frequency-selective circuit of the oscillator. The popular resonators are shown and resulting Q factors are discussed.

Chapter 6 presents a comprehensive treatment of the oscillator. Initially, the linear theory is shown, which explains the design strategy. Two types of oscillator configurations are relevant. One is the parallel type and the other is the series type. For both cases, a numerical design is shown. The more precise design method of an oscillator is an approach which considers large-signal conditions. Therefore, the start-up conditions are described, followed by the steady-state behavior. Under large-signal conditions, the time-domain behavior has to be considered, as the collector (or drain) current now consists of a DC component and harmonically related RF currents. In order to describe this, a normalized drive level is introduced which determines the conducting angle. As the conducting angle becomes narrower, the efficiency increases and the noise improves. However, there is a wide range over which the output power is constant but the noise varies widely. Finding the optimum condition is the objective of Chapter 8. This chapter is supported by Appendix A.

Chapter 7 provides a detailed discussion of noise in oscillators, both linear and nonlinear. For the linear case, the Leeson model is derived, which is used as the best case model; aproaches to calculate it are presented. It contains the loaded Q, the noise factor, and the output power. These three variables determine the phase noise of an oscillator. The linear example is now useful because these three values are practically unknown. An accurate calculation based on large-signal S-parameters, specifically S_{21}, is possible for the first time. Finally, a phase noise test setup is shown which is used to validate this large-signal noise theory. This chapter is supported by Appendix B.

Chapter 8 is the key contribution of this monograph. Chapter 6 provides good insight into the large-signal operation of the oscillator, including its optimization for power, and discusses some phase noise results under these conditions. As mentioned, the normalized drive level x can vary over a broad range, with output powers only changing a few decibels, while the phase noise changes drastically. A change in a few decibels of the output power drastically changes the efficiency, but the goal here is to find the best phase noise condition. After showing that reducing the conducting angle and proving that the phase noise improves, the actual noise calculation and termination of the feedback capacitance is shown. Three cases are considered here:

1. First is the Leeson equation, which contains a need for output power, operating noise figure, and testing of the loaded Q for its validity. To do this, the exact calculations for the output power, the loaded Q, and the resulting noise factor are presented. An example shows that the accuracy of this approach is limited, however, since an ideal transistor without parasitics is assumed.
2. The second approach calculates the noise contribution of a time-varying negative resistance that cancels the losses. It will be shown that this is a time average value, that the noise calculations can be further improved, and that the optimum feedback conditions are found.
3. The third and final approach is based on the loop approach and considers all noise contributions. Therefore, it is the most accurate way to determine the oscillator's performance. A graphical differentiation of the phase noise equation shows a point to obtain the best phase noise. The phase noise increases on either side of this optimal point.

Appendix C shows a complete approach to the design.

Chapter 9 shows five selected microwave oscillators for validation purposes which provide state-of-the-art phase noise. Their design was based upon the optimization shown in Chapter 8. Bipolar transistors and gallium arsenide (GaAs) FETs are used. Measured data were available for the 1000 and 4100 MHz oscillators with bipolar transistors and for the 2000 MHz GaAs FET oscillator. The ceramic resonator-based oscillator shows a measured phase noise of 125 dBc/Hz at 10 kHz, 145 dBc/Hz at 100 kHz, and 160 dBc/Hz or better at 1 MHz. The 4.1 GHz oscillator shows a phase noise above 89 dBc/Hz at 10 kHz, 113 dBc/Hz at 100 kHz, and 130 dBc/Hz at 1 MHz, and these results are in excellent agreement with the prediction. Due to its high flicker corner frequency, the GaAs FET may cause modeling problems. The fourth example is a 77 GHz oscillator based on published information, in which the design, the simulation, and the measurements agree well. The fifth, and last, example is based on a wide microstrip resonator, which requires electromagnetic tools for simulation. This case shows the design items necessary for such an oscillator.

The previous chapters deal with single-resonator design. In Chapter 10, we will introduce a system of coupled oscillators and a multiple-resonator approach. This gives far better results than a single oscillator, but the level of effort is also much higher. This chapter describes both the coupling and locking mechanisms, as well as the noise analysis applicable for these cases. Also of particular interest is the wideband VCO, which is described at the end of this chapter. In particular, the design criteria for this oscillator are covered in great detail, including the influence of the components.

Chapter 11 shows validation circuits for multiple-coupled resonator-based and multiple-coupled oscillators; specifically, the case of the VCOs is shown. Both design information and results are given. This hybrid-tuned ultra-low-noise wideband VCO has set the state of the art and is a result of many techniques covered in this book.

The book ends with Appendices A to H, which contain three oscillator designs. The appendices are very important because they apply all the design rules presented in this monograph. The chosen circuits were also used for verification purposes.

Appendix A describes the oscillator design using large-signal S-parameters for optimum power. The unique approach used here also shows that an inductor, instead of a feedback capacitor, may be needed to make the design for a given transistor possible. This has not been shown in the literature before.

Appendix B describes the Bessel functions in a large-signal design for best output power. Consistent with Appendix A, a detailed numerical approach is given so that the step-by-step procedure can be easily followed. Again, the design is a typical application for a high-performance oscillator. In this case, the output power was the priority and the phase noise was allowed to degrade.

Appendix C combines all the technologies discussed in this monograph to show the optimum design of oscillators. It starts with the specific requirements for output power and phase noise. It further assumes a real transistor with its parasitics considered. It shows the schematic first, which is the optimum choice for this application. The Bessel function approach is used to determine the operating point and the bias point. The design calculation shows that the key equation (8-94) in Chapter 8, despite its simplification, gives an accurate answer for the phase noise. The result is consistent with the predicted phase noise, predicted by the equation derived here, the predicted values using Ansoft's harmonic balanced simulator designer, and the measured phase noise.

8 INTRODUCTION

Appendix D discusses a complete analytical approach for designing good microwave FET and bipolar oscillators. The approach for FET and bipolar transistors, due to their impedances, is different.

Appendix E offers the CAD solution for calculating phase noise in oscillators.

Appendix F is a general noise presentation to better describe the way noise is handled.

Appendix G shows the calculation of noise properties of both bipolar transistors and FETs, including temperature effects.

Appendix H shows a thorough noise analysis of the N-coupled oscillator coupled through the arbitrary coupling network.

2 General Comments on Oscillators

An oscillator consists of an amplifier and a resonant element, as well as a feedback circuit. In many cases the intention is to build a selective amplifier, but an oscillator ends up being built because of internal or external feedback either in the active device or as part of the external circuit. An amplifier is an electrical circuit with a defined input and output impedance which increases the level of the input signal to a predetermined value at the output. The energy required for this is taken from the DC power supply connected to the amplifier. The amplifier impedances can vary from several ohms to several Meg ohms, but for high-frequency application, it is standard to build amplifiers with 50 Ω real input and output impedance. The active circuit responsible for the gain can be a bipolar transistor, a FET, or a combination of both, or a gain block like a wideband amplifier offered by several companies. These are typically a combination of Darlington stages with RF feedback. Wideband amplifiers can cover frequencies such as a few hundred kilohertz to over 10,000 MHz. An oscillator built with an amplifier and a tuned circuit transforms DC energy into RF energy. It does this at a desired frequency at an acceptable power-added efficiency (RF power out/DC power in). The efficiency of a low-noise oscillator varies, depending upon frequencies and configurations between 10% and 70%. In most cases, the efficiency is a secondary problem, while the primary task is to have a signal frequency output which is stable, free of spurious signals (clean), with low phase noise, and of sufficient level [12].

The term *stability* refers to both short-term and long-term stability, and the oscillator should be clean in the sense that it does not pick up unwanted signals and noise in the circuit. Various noise sources contribute to oscillator noise. These include the loss of the resonator, the noise sources inside the transistor, noise (hum) modulated on the power supply, and noise contributions from the tuning diode(s).

This work will focus on phase noise optimization at a given and reasonable DC efficiency. The investigated oscillators are VCOs with a sinusoidal voltage output and are produced by adding a tuning diode. Most systems cannot tolerate high harmonics from the oscillator, as these cause unwanted mixing products.

The Design of Modern Microwave Oscillators for Wireless Applications: Theory and Optimization,
by Ulrich L. Rohde, Ajay Kumar Poddar, Georg Böck
Copyright © 2005 John Wiley & Sons, Inc.

2.1 SINUSOIDAL OSCILLATORS

All amplifier-based oscillators are inherently nonlinear. Although the nonlinearity results in some distortion of the signal, linear analysis techniques can normally be used for the initial analysis and design of oscillators. Figure 2-1 shows, in block diagram form, the linear model of an oscillator. It contains an amplifier with a frequency-dependent forward loop gain $G(j\omega)$ and a frequency-dependent feedback network $H(j\omega)$.

The output voltage is given by

$$V_o = \frac{V_{in} G(j\omega)}{1 + G(j\omega)H(j\omega)} \quad (2\text{-}1)$$

For an oscillator, the output V_o is nonzero even if the input signal $V_{in} = 0$. This can only be possible if the forward loop gain is infinite (which is not practical) or if the denominator is

$$1 + G(j\omega)H(j\omega) = 0 \quad (2\text{-}2)$$

at some frequency ω_o. This leads to the well-known condition for oscillation (the *Nyquist criterion*), where at some frequency ω_0

$$G(j\omega_o)H(j\omega_0) = -1 \quad (2\text{-}3)$$

That is, the magnitude of the open-loop transfer function is equal to 1

$$|G(j\omega_0)H(j\omega_0)| = 1 \quad (2\text{-}4)$$

and the phase shift is 180°

$$\arg[G(j\omega_0)H(j\omega_0)] = 180° \quad (2\text{-}5)$$

This can be more simply expressed as follows: if in a negative-feedback system the open-loop gain has a total phase shift of 180° at some frequency ω_0, the system will oscillate at that frequency provided that the open-loop gain is unity. If the gain is

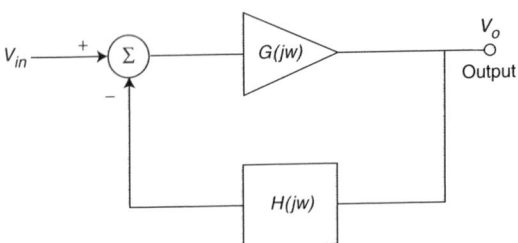

Figure 2-1 Block diagram of an oscillator showing forward and feedback loop components.

less than unity at the frequency where the phase shift is 180°, the system will be stable, whereas if the gain is greater than unity, the system will be unstable.

This statement is not correct for some complicated systems, but it is correct for those transfer functions normally encountered in oscillator design. The conditions for stability are also known as the *Barkhausen criterion* [188], which states that if the closed-loop transfer function is

$$\frac{V_o}{V_{in}} = \frac{\mu}{1 - \mu\beta} \tag{2-6}$$

where μ is the forward voltage gain and β is the feedback voltage gain, the system will oscillate provided that $\mu\beta = 1$. This is equivalent to the Nyquist criterion, the difference being that the transfer function is written for a loop with positive feedback. Both versions state that the total phase shift around the loop must be 360° at the frequency of oscillation and the magnitude of the open-loop gain must be unity at that frequency.

2.2 PHASE NOISE EFFECTS

A noisy oscillator causes interference at adjacent channels, a phenomenon called *blocking* or *reciprocal mixing*. Figure 2-2 shows how phase noise affects the signal of an ideal oscillator.

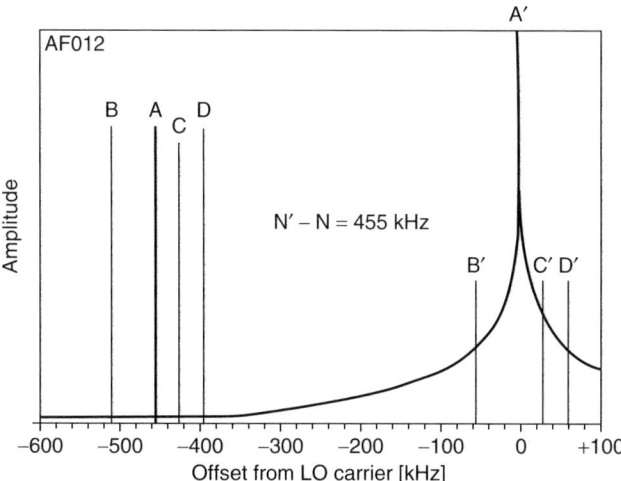

Figure 2-2 Reciprocal mixing occurs when incoming signals mix energy from an oscillator's sidebands to the IF. In this example, the oscillator is tuned so that its carrier, at A', heterodynes the desired signal, A, to the 455 kHz as intended. At the same time, the undesired signals B, C, and D mix with the oscillator noise-sideband energy at B', C', and D', respectively, to the IF. Depending on the levels of the interfering signals and the noise-sideband energy, the result may be a significant rise in the receiver noise floor.

12 GENERAL COMMENTS ON OSCILLATORS

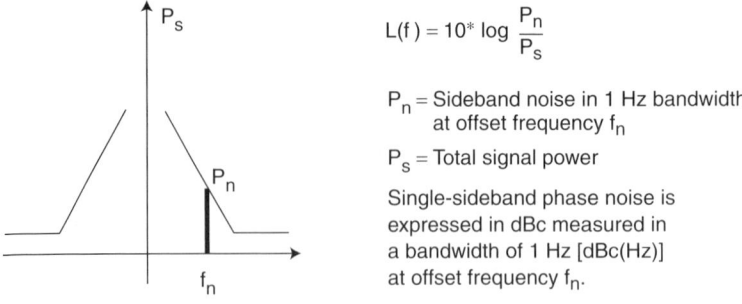

Figure 2-3 Phase noise calculation.

The spectral density, or phase noise, is measured in dBc (dB below the carrier) in a bandwidth of 1 Hz at an offset frequency f_n. The phase noise, therefore, is related to the output power. The noise power and the curve shown in Figure 2-3 can have different shapes based on the noise sources, as seen in Figure 2-8.

If the oscillator is configured to be a VCO, the phase noise inside the loop bandwidth (hopefully) improves. Outside the loop bandwidth, the phase noise is determined solely by the resonator of the oscillator, as seen in Figure 2-4.

The maximum condition, or the best phase noise number, is $10 \times \log(P_{output}/kT)$ at room temperature calculating from kT (-174 dBm/Hz) to the output power typically between 0 dBm and 30 dBm. The tuned circuit is responsible for most of the filtering. This phenomenon was first observed by Leeson in 1966 [70] and has been the basis of all linear-based assumptions. Later, it will be shown that his approach, with some additional terms, forms a useful but not always scientifically accurate method of characterizing the oscillator.

Again, if a strong signal is fed to the receiving system, it will mix with the noise bands of the oscillator and produce a noise signal at one or more adjacent channels. This effect desensitizes or blocks the channel or one or more adjacent channels. *Reciprocal mixing* is a descriptive term, as it shows that the phase noise of the oscillator at a given space is being mixed as an unwanted effect to the desired channel.

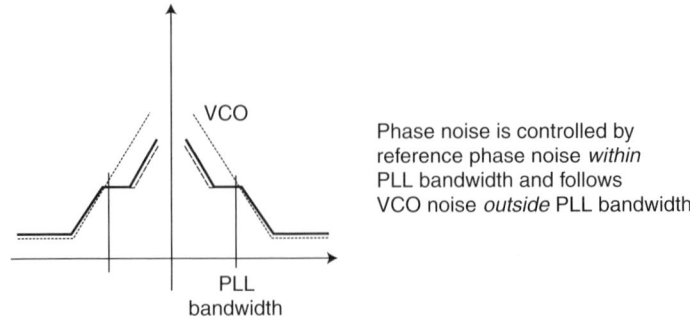

Figure 2-4 Phase noise of an oscillator controlled by a phase-locked loop.

2.3 SPECIFICATIONS OF OSCILLATORS AND VCOs

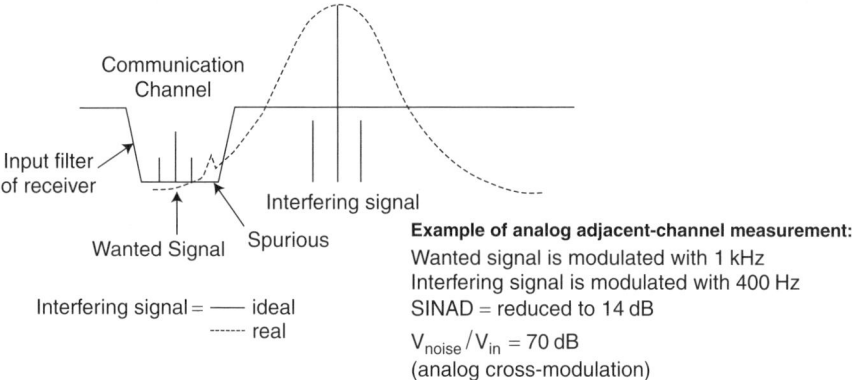

Figure 2-5 Principle of selectivity measurement for analog receivers.

Figures 2-5 and 2-6 show the mechanism of reciprocal mixing for both an analog and a digital receiver. In the case of the analog receiver, the phase noise of the oscillator and its spurious signals create interfering signals, while in the case of the digital receiver desensitization occurs [14].

2.3 SPECIFICATIONS OF OSCILLATORS AND VCOs

The properties of an oscillator can be described in a set of parameters. Following is a discussion of the important and relevant parameters for oscillators.

2.3.1 Frequency Range

The output frequency of a VCO can vary over a wide range. The frequency range is determined by the architecture of the oscillator, particularly its tuning circuit. A standard VCO has a frequency range typically less than 2:1—for example, 925–1650 MHz.

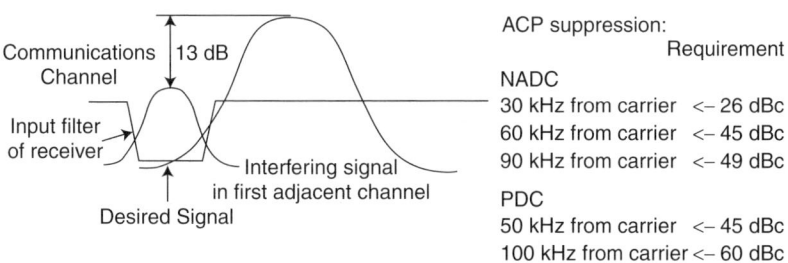

Figure 2-6 Principle of selectivity measurement for digital receivers.

2.3.2 Phase Noise

Unfortunately, oscillators do not generate perfect signals. The various noise sources in and outside of the transistor modulate the VCO, resulting in energy or spectral distribution on both sides of the carrier. This occurs via modulation and frequency conversion. The noise or, better, AM and FM noise is expressed as the ratio of output power divided by the noise power relative to a 1 Hz bandwidth measured at an offset of the carrier. Figure 2-7 shows a typical measured phase noise plot of a high Q oscillator using a ceramic resonator.

The stability or phase noise of an oscillator can be determined in the time and frequency domains. Phase noise is a short-term phenomenon and has various components. Figure 2-8a shows the stability in the time domain. For an oscillator or VCO, this is rarely relevant because the oscillator is used in a PLL system. The phase noise characteristics are more important, and Figure 2-8b shows the various contributions. These contributions will be analyzed in Chapter 3.

2.3.3 Output Power

The output power is measured at the designated output port of the oscillator/VCO. Practical designs require one or more isolation stages between the oscillator and the output. The VCO output power can vary as much as ± 2 dB over the tuning range. A typical output level is 0 to $+10$ dBm.

2.3.4 Harmonic Suppression

The oscillator/VCO has a typical harmonic suppression of more than 15 dB. For high-performance applications, a low-pass filter at the output will reduce the

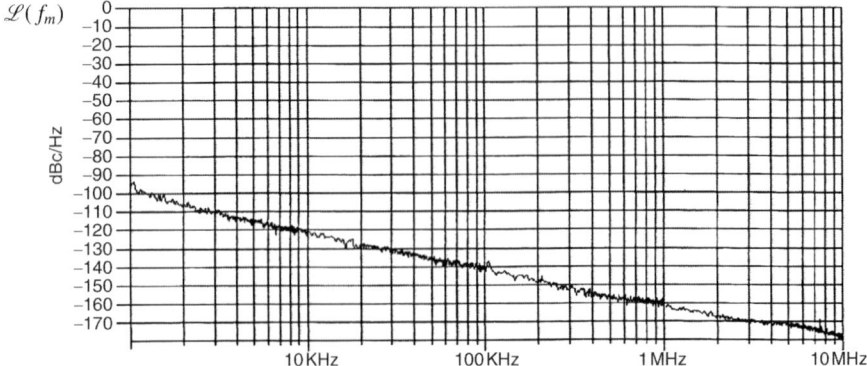

Figure 2-7 Measured phase noise of an 880-MHz resonator-based oscillator with a small tuning range.

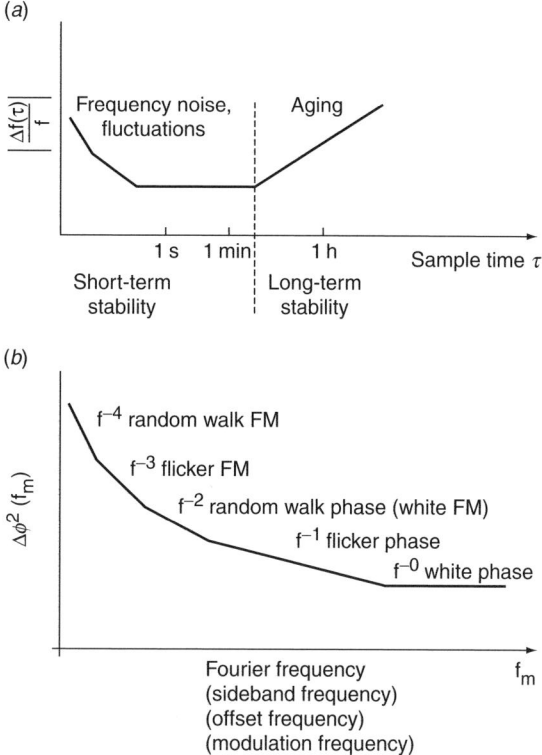

Figure 2-8 Characterization of a noise sideband in the time and frequency domains and its contributions: (*a*) time domain and (*b*) frequency domain. Note that two different effects are considered, such as aging in (*a*) and phase noise in (*b*).

harmonic contents to a desired level. Figure 2-9 shows a typical output power plot of a VCO.

2.3.5 Output Power as a Function of Temperature

All active circuits vary in performance as a function of temperature. The output power of an oscillator over a temperature range should vary less than a specified value, such as 1 dB.

2.3.6 Spurious Response

Spurious outputs are signals found around the carrier of an oscillator which are not harmonically related. A good, clean oscillator needs to have a spurious-free range of 90 dB, but this requirement makes it expensive. Oscillators typically have no spurious frequencies other than possibly 60 Hz and 120 Hz pickup. The digital electronics in a synthesizer generate many signals and, when modulated on the VCO, are responsible for these unwanted output products.

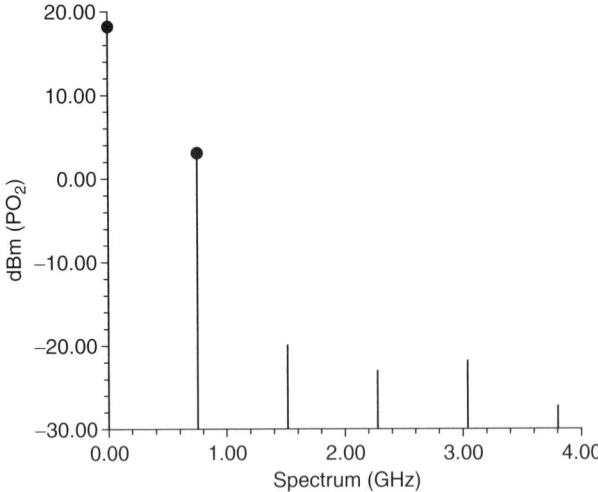

Figure 2-9 Predicted harmonics at the output of a microwave oscillator.

2.3.7 Frequency Pushing

Frequency pushing characterizes the degree to which an oscillator's frequency is affected by its supply voltage. For example, a sudden current surge caused by activating a transceiver's RF power amplifier may produce a spike on the VCO's DC power supply and a consequent frequency jump. Frequency pushing is specified in frequency/voltage form and is tested by varying the VCO's DC supply voltage (typically ± 1 V) with its tuning voltage held constant. Frequency pushing must be minimized, especially in cases where power stages are close to the VCO unit and short pulses may affect the output frequency. Poor isolation can make phase locking impossible.

2.3.8 Sensitivity to Load Changes

To keep manufacturing costs down, many wireless applications use a VCO alone, without the buffering action of a high reverse-isolation amplifier stage. In such applications, frequency pulling, the change of frequency resulting from partially reactive loads, is an important oscillator characteristic. Pulling is commonly specified in terms of the frequency shift that occurs when the oscillator is connected to a load that exhibits a nonunity VSWR (such as 1.75, usually referenced to 50 Ω), compared to the frequency that results with a unity-VSWR load (usually 50 Ω).

2.3.9 Post-tuning Drift

After a voltage step is applied to the tuning diode input, the oscillator frequency may continue to change until it settles to a final value. Post-tuning drift is one of the parameters that limits the bandwidth of the VCO input and the tuning speed.

2.3.10 Tuning Characteristic

This specification shows the relationship, depicted as a graph, between the VCO operating frequency and the tuning voltage applied. Ideally, the correspondence between operating frequency and tuning voltage is linear.

2.3.11 Tuning Linearity

For stable synthesizers, a linear deviation of frequency versus tuning voltage is desirable. It is also important to make sure that there are no breaks in the tuning range—for example, that the oscillator does not stop operating with a tuning voltage of 0 V.

2.3.12 Tuning Sensitivity, Tuning Performance

This datum, typically expressed in megahertz per volt (MHz/V), characterizes how much the frequency of a VCO changes per unit of tuning voltage change.

2.3.13 Tuning Speed

This characteristic is defined as the time necessary for the VCO to reach 90% of its final frequency upon the application of a tuning voltage step. Tuning speed depends on the internal components between the input pin and the tuning diode, including, among other things, the capacitance present at the input port. The input port's parasitic elements, as well as the tuning diode, determine the VCO's maximum possible modulation bandwidth.

2.3.14 Power Consumption

This characteristic conveys the DC power, usually specified in milliwatts and sometimes qualified by operating voltage, required by the oscillator to function properly.

2.4 HISTORY OF MICROWAVE OSCILLATORS

Early microwave oscillators were based on electron tubes, and great efforts were made to obtain gain and power at high frequencies. Starting with simple glass triodes (lighthouse tubes) and coaxial ceramic triodes, a large number of circuits designed to obtain reasonable performance were built. After using the Lecher lines (quarter-wavelength U-shaped parallel wires, shorted at the end, with a few centimeters' spacing), the next step was the use of coaxial systems, which became mechanically very difficult and expensive. At higher frequencies, cavities dominated the application, and many publications dealt with the various resonant modes. For special applications such as microwave ovens and radar applications, magnetrons and reflex klystrons were developed.

18 GENERAL COMMENTS ON OSCILLATORS

Today, the good understanding of the planar structures, such as microstrip, stripline, and coplanar waveguide, have been instrumental in extending the practical frequency range up to 100 GHz and higher.

Early transistors followed the same trend. Siemens at one time produced a coaxial microwave transistor, Model TV44, and Motorola offered similar devices. Today, microwave transistors, when packaged, are also in microstrip form or are sold as bare die, which can be connected via bond wires to the circuit. These bond wires exhibit parasitic effects and can be utilized as part of the actual circuit. The highest form of integration is radio frequency integrated circuits (RFICs), either in gallium arsenide (GaAs) or silicon germanium (SiGe) technology.

The SiGe circuits are typically more broadband because of lower impedances, and GaAs FETs are fairly high impedance at the input. From an application point of view, in oscillators, SiGe seems to be winning. From a practical design standpoint both transistor types can be considered a black box with a set of S parameters, which are bias and frequency dependent.

We will see that the transistor operates in large-signal condition and that, historically, FETs have been used to demonstrate that there is little change in parameters from small- to large-signal operation. Bipolar transistors have much more pronounced changes.

Early pioneers invented a variety of oscillator circuits which are named after them. Figure 2-10, shows a set of schematics applicable for both bipolar transistors and FETs. The ones using magnetic coupling are not useful for microwave applications.

The Colpitts and Clapp oscillator configurations had an air variable capacitor as a tuning element. As the technology developed, they were replaced by tuning diodes.

2.5 THREE APPROACHES TO DESIGNING MICROWAVE OSCILLATORS

The oscillator configurations discussed in Chapter 1 were shown for historical reasons. As mentioned, the Colpitts oscillator and its cousin, the Clapp oscillator, have found their way into modern microwave applications. The basic concept is that the capacitive feedback arrangement generates a negative resistance across the tuned circuit which compensates for the losses of that circuit.

These types of oscillators are called *one-port oscillators* because the transistor and the feedback circuit can be substituted with a negative resistance [185, 186]. The approach to the design of these oscillators is called a *one-port approach*. For stable operation, the feedback capacitance values have to be carefully selected, as will be shown in the next chapter. As gain varies with frequency, it limits the tuning range, which also affects the phase noise and the output power. Figure 2-11 shows the Colpitts-type oscillator. Tuning this circuit will alter the conditions for optimal phase noise and requires additional tunable elements to achieve optimal performance.

2.5 THREE APPROACHES TO DESIGNING MICROWAVE OSCILLATORS 19

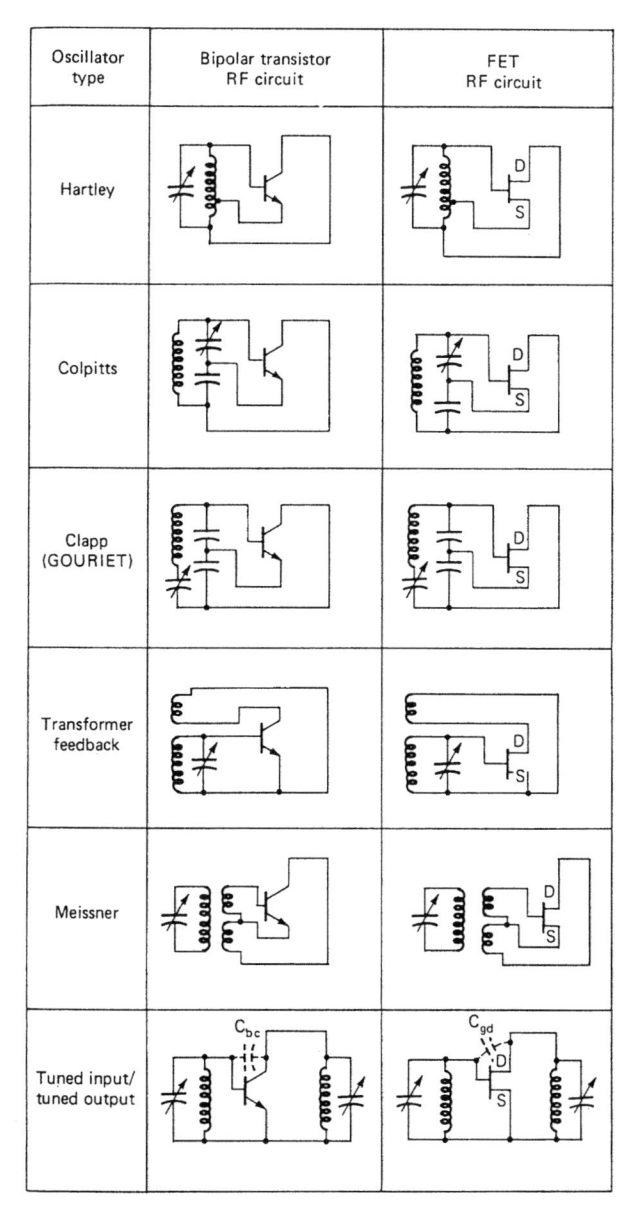

Oscillator type	Bipolar transistor RF circuit	FET RF circuit
Hartley		
Colpitts		
Clapp (GOURIET)		
Transformer feedback		
Meissner		
Tuned input/ tuned output		

Figure 2-10 Six different configurations which can be based on either bipolar transistors or FETs. Some modern microwave oscillators are built around the Colpitts and Clapp oscillator circuits [15].

A second design approach is the *two-port approach*. Now the transistor is used as a two-terminal device, with the third terminal at ground and its tuned circuit as a feedback element that determines the frequency, as shown in Figure 2-12.

As the transistor, or black box in question, has a frequency-dependent Y_{21} (or S_{21}), an additional element is needed to compensate for the phase shift generated inside

20 GENERAL COMMENTS ON OSCILLATORS

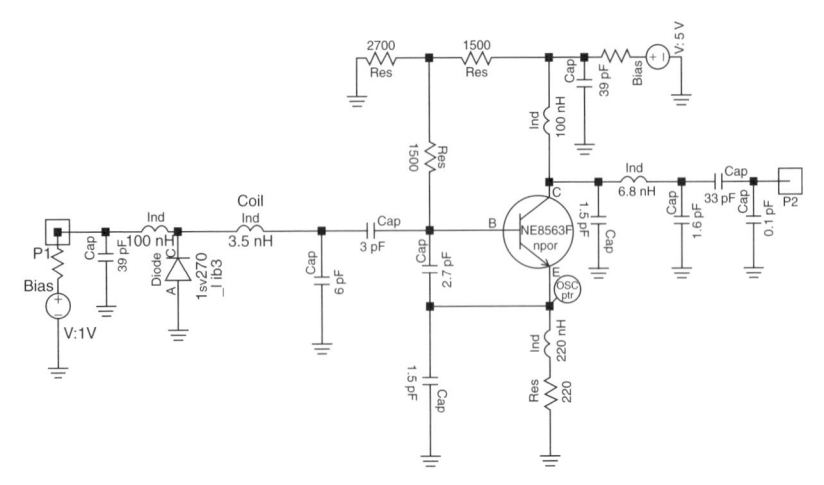

Figure 2-11 Schematic of a voltage-controlled Colpitts-type oscillator. The Colpitts circuit is recognized by the capacitive feedback network composed of the capacitors connected between the base and emitter and between the emitter and ground. This will be further explained in Section 2.6.

the black box. These types of oscillators therefore have a gain block, a phase shift/ matching circuit, and a resonator. In this configuration, the resonators are typically used in a Π configuration and in series resonant mode. A typical candidate for this is a surface acoustical wave (SAW) resonator, shown in Figure 2-12. The resonator is used in a series resonant mode.

The third design approach, which is typically used for microwave frequencies greater than 4 GHz, is shown in Figure 2-13. The base of the device (three-port,

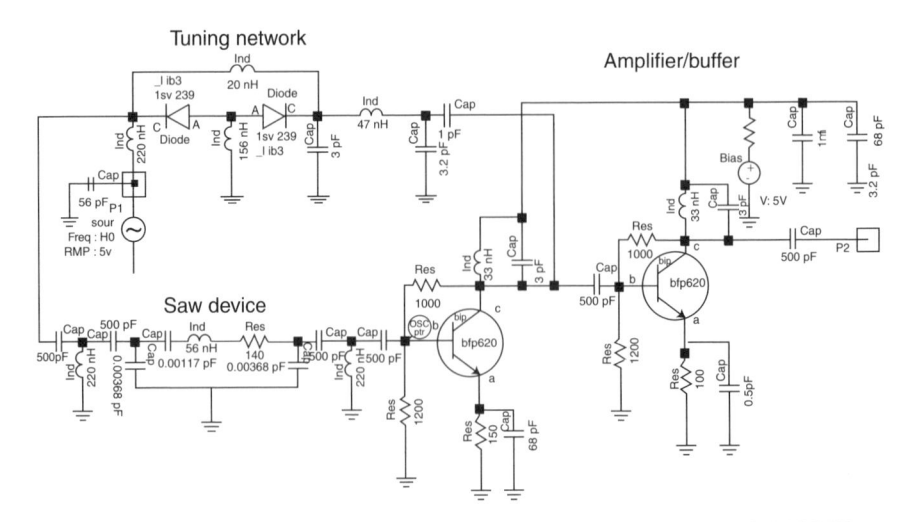

Figure 2-12 Circuit diagram of a two-port low-noise SAW oscillator at 622.08 MHz.

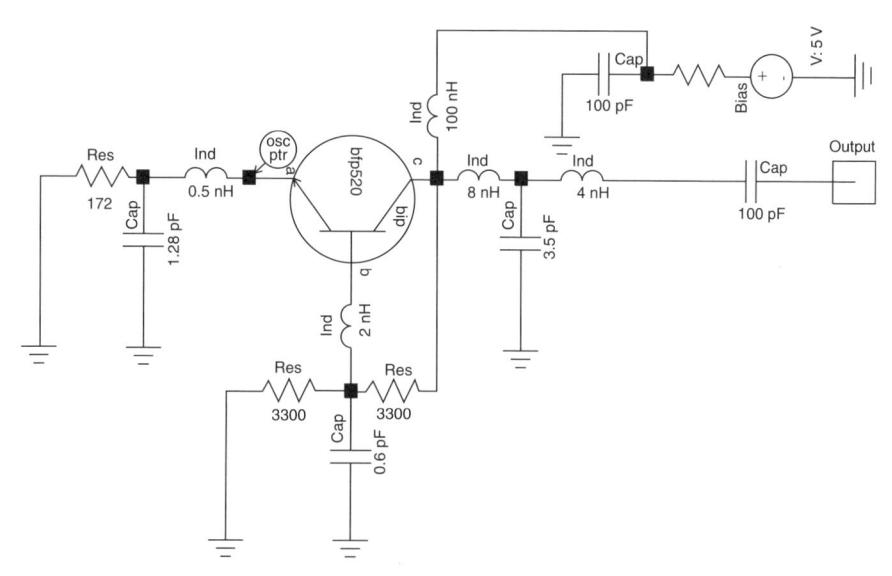

Figure 2-13 Circuit diagram of a 4 GHz oscillator using series feedback with the Infineon transistor BFP520.

three-terminal device) is floating aboveground via a base (or gate) inductor. The feedback occurs due to the inductor and the capacitive portion of Y_{22} (or S_{22}). This circuit type was first explained by [2]. This type of feedback generates a negative input and output resistance for the circuit, which, as an example, can match the 50 Ω load [16, 17].

2.6 COLPITTS OSCILLATOR, GROUNDED BASE OSCILLATOR, AND MEISSEN OSCILLATOR

There are three types of oscillators which can be translated from one to the other, starting with the Colpitts oscillator. The output power there is frequently taken at the collector/drain, but for phase noise improvement, inductive coupling is preferred. A buffer stage following this oscillator is recommended. The second type of oscillator is a grounded base type in which the RF ground is at the base. This oscillator also requires a buffer stage. It uses the same bias point and has about the same phase noise as the original Colpitts oscillator. The Colpitts oscillator shows a more pronounced $1/Q^2$ noise. The grounded base oscillator is slightly better. The third type of oscillator is the Meissen oscillator. This oscillator, with the same circuit values, gives the best phase noise. As a matter of fact, the phase noise is much better than that of the other two oscillators. It appears that the loading on the tuned circuit is much less than that of the previous two circuits. Figure 2-10 shows some of these configurations. The Meissen oscillator shown here uses a single inductor rather than three.

Figure 2-14 shows the implementation of a Colpitts oscillator which is recognized by its capacitive voltage divider (consisting of C_1 and C_2). The two PNP

22   GENERAL COMMENTS ON OSCILLATORS

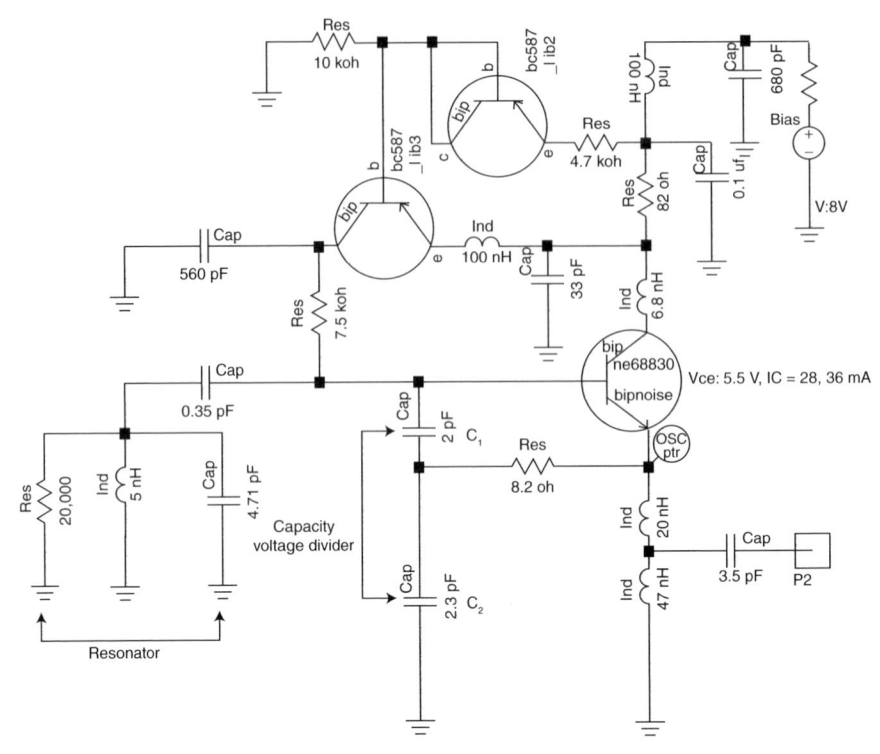

Figure 2-14   Schematic of a 1 GHz Colpitts oscillator with a DC feedback stabilization circuit.

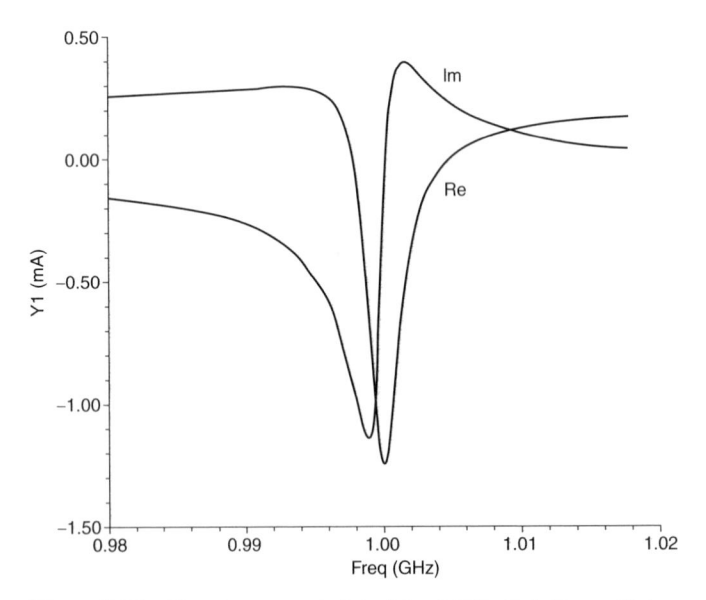

Figure 2-15   Negative current plot of the 1 GHz Colpitts oscillator.

2.6 COLPITTS, GROUNDED BASE, MEISSEN OSCILLATORS

transistors above the oscillator transistor provide a constant current source and maintain the bias condition, in this case 5.5 V and 28.36 mA. The emitter of the transistor is connected to the voltage divider by a 8.2 Ω resistor, and the tuned-circuit consists of the 5 nH and the 4.71 pF capacitor. The 20 KΩ resistor determines the Q of the resonator. As a novelty, the energy is taken out at the emitter using an inductive divider. This suppresses harmonics and improves the phase noise.

A linear analysis can be used to determine the oscillation condition. Figure 2-15 shows the reactive and real currents at the emitter test point. The condition for best phase noise is where the reactance passes through zero (resonance condition) and the most negative current occurs.

The resonator is similar to a ceramic resonator in its performance, and we can expect a high Q. This can be validated by analyzing the resulting phase noise. Figure 2-16 shows the predicted phase noise of this oscillator.

The Colpitts oscillator can be rotated to become a grounded base circuit. This is shown in Figure 2-17. The tuned circuit is loosely coupled, as in the Colpitts oscillator, this time to the connector. The feedback capacitors (C_1 and C_2), which are the same as in the Colpitts oscillator, are connected from the collector to the emitter and from the emitter to the ground. The base is grounded via a 100 pF capacitor. The output from the oscillator is taken from the collector via a coupling capacitor (C_0).

Figure 2-18 shows the resulting output and its harmonic contents. Figure 2-19 shows the resulting phase noise, which is similar to that of the Colpitts oscillator presented in Figure 2-16.

The oscillators shown in Figures 2-14 and 2-17 both have a resonator that is grounded at one side. Another possibility derived from the vacuum tube era is the

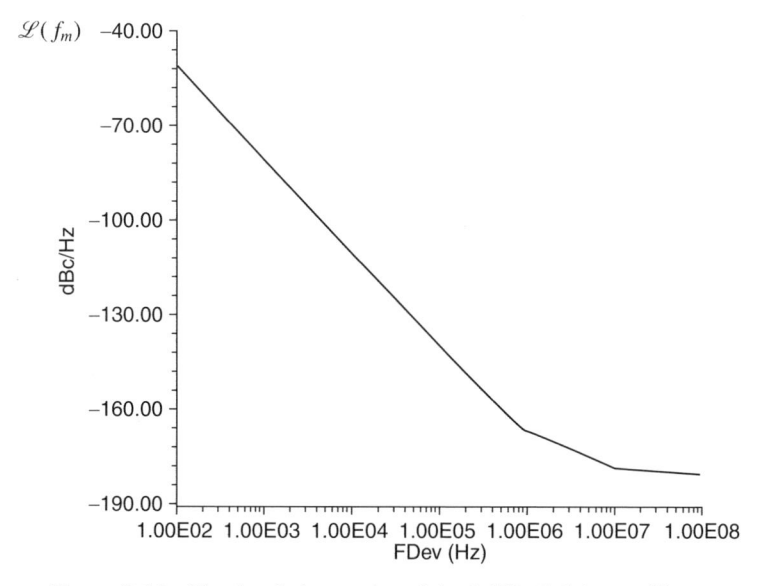

Figure 2-16 Simulated phase noise of the 1 GHz Colpitts oscillator.

24 GENERAL COMMENTS ON OSCILLATORS

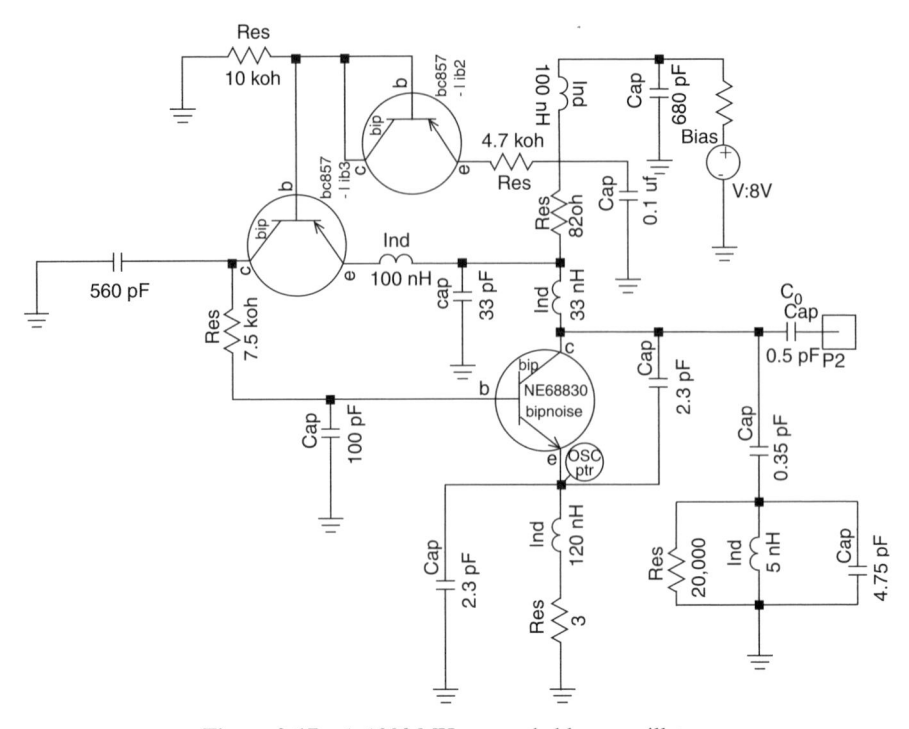

Figure 2-17 A 1000 MHz grounded base oscillator.

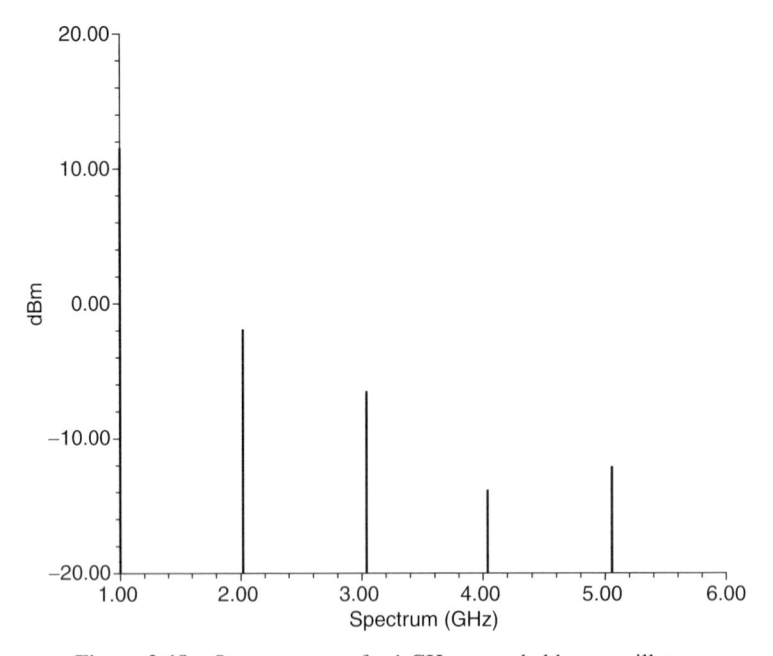

Figure 2-18 Output power of a 1 GHz grounded base oscillator.

2.6 COLPITTS, GROUNDED BASE, MEISSEN OSCILLATORS 25

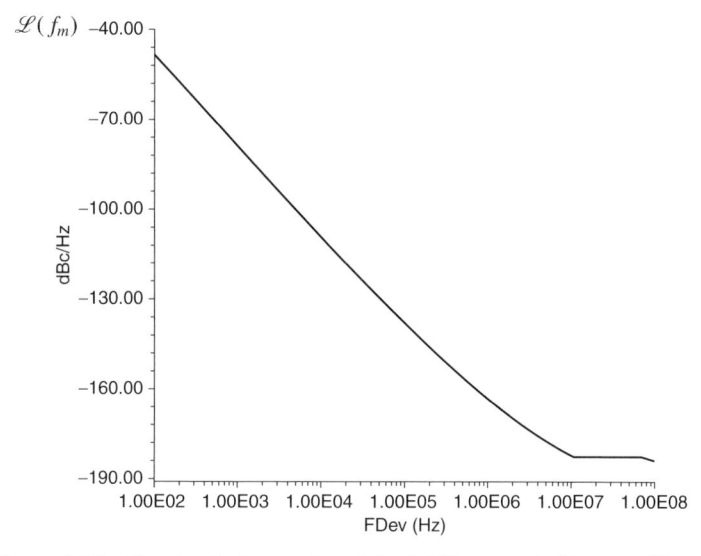

Figure 2-19 Simulated phase noise of the 1 GHz grounded base oscillator.

Meissen oscillator. It has a parallel-tuned circuit between the base and the collector, and the feedback capacitors are connected from base to ground and from collector to ground. The Meissen oscillator shown in Figure 2-20 is derived from a basic Colpitts oscillator and uses identical feedback values and coupling capacitors.

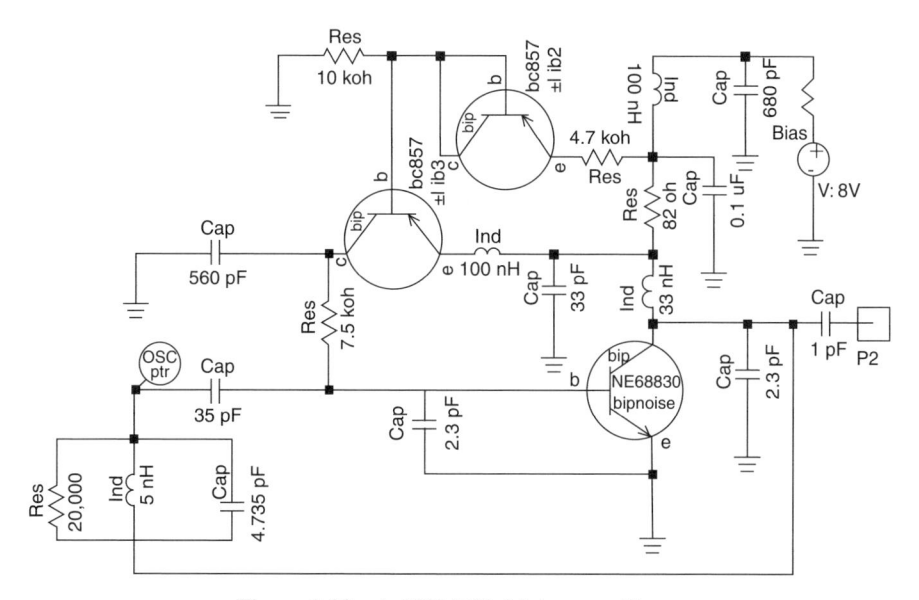

Figure 2-20 A 1000 MHz Meissen oscillator.

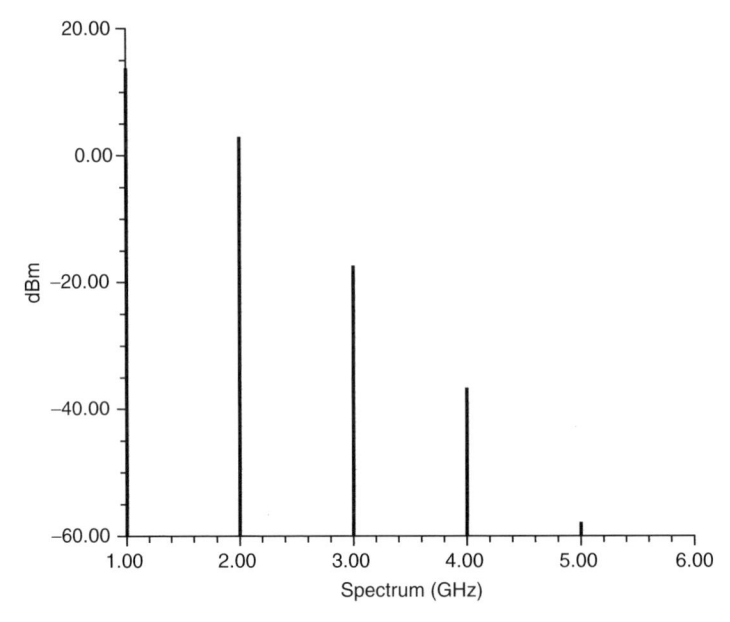

Figure 2-21 Simulation of the 1 GHz Meissen oscillator output power and its harmonics.

Analyzing the output power as shown in Figure 2-21 indicates a higher level, as before. Analyzing the phase noise, however, provides an interesting result; the phase noise is now better than in the previous two examples. To build such a symmetrical oscillator is not easy, but it is highly rewarding because the phase noise can be

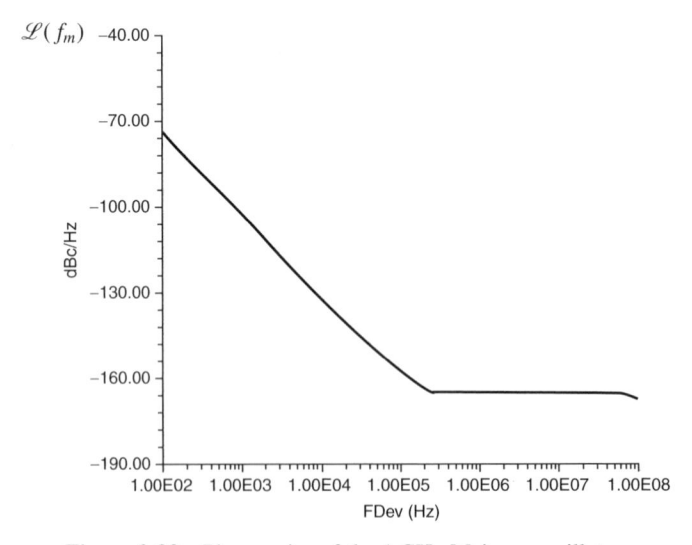

Figure 2-22 Phase noise of the 1 GHz Meissen oscillator.

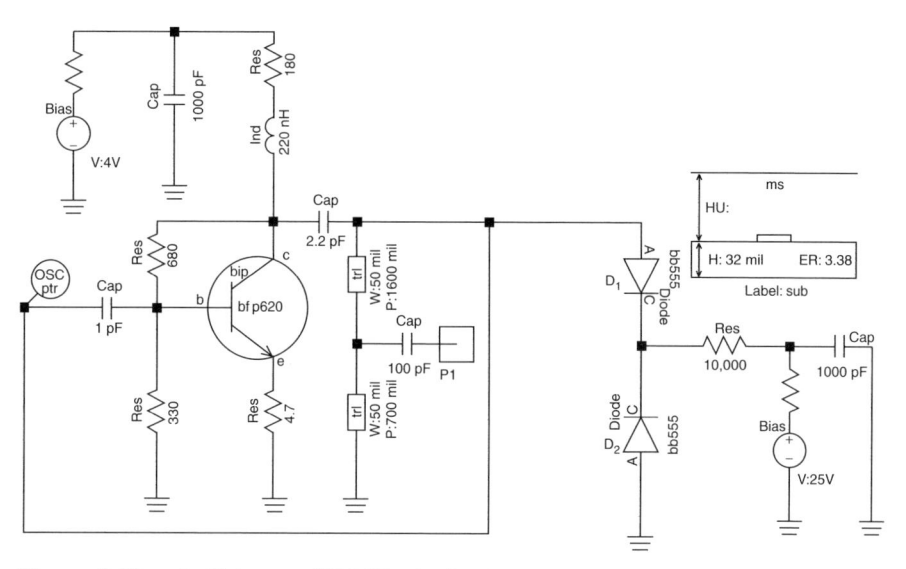

2.6 COLPITTS, GROUNDED BASE, MEISSEN OSCILLATORS

Figure 2-22a A high-pass 400 MHz feedback oscillator using transmission lines as resonators.

improved. This type of oscillator will be used later as a transition from single to push-push VCOs. In the Meissen oscillator, the loading for the tuned circuit is reduced, which results in the better phase noise. This is confirmed by the results presented in Figure 2-22.

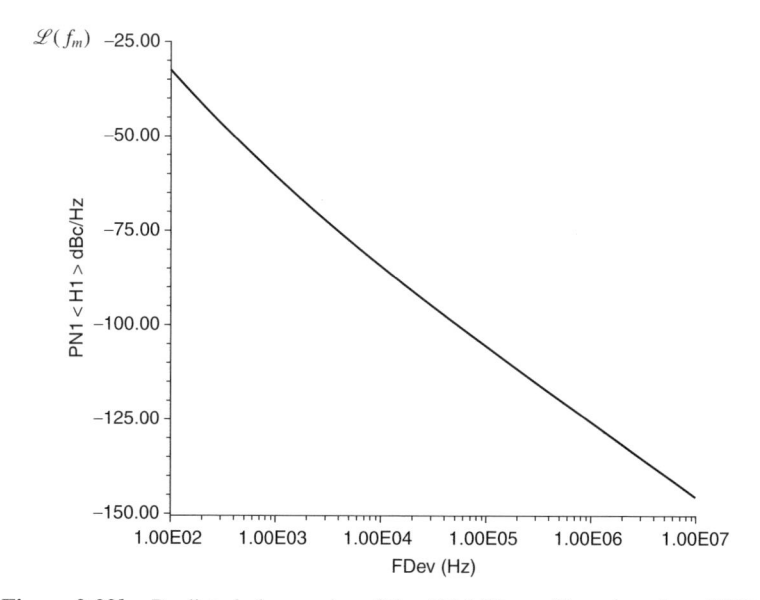

Figure 2-22b Predicted phase noise of the 400 MHz oscillator based on [120].

28 GENERAL COMMENTS ON OSCILLATORS

The circuits shown so far are those of three-reactance oscillators. An alternative to this low-pass configuration is a high-pass configuration. In [120] a suggested high-pass configuration is shown. Figure 2-22a shows the circuit diagram. The oscillator has a tuned circuit connected between the collector and emitter, which is the high-pass filter. Instead of a lumped inductor, two transmission lines are used. The main inductance is provided by the upper transmission line, and the second one is used for a tap for the output. The tuning diodes (D_1 and D_2) incorporated in the circuit, as shown in Figure 2-22a, provide for a small tuning range to correct manufacturing tolerances. This oscillator operates at 400 MHz. Figure 2-22b shows the predicted phase noise of the oscillator based on Figure 2-22a. It should be noted that this is not a particularly good phase noise, mostly due to the low Q resonator. This high-pass filter provides a phase shift of 180°, a necessary condition for oscillation.

The following section starts an introduction to the mathematics of the oscillator.

2.7 THREE-REACTANCE OSCILLATORS USING Y-PARAMETERS: AN INTRODUCTION

Although the block diagram formulation of the stability criteria shown earlier is the easiest to express mathematically, it is frequently not the easiest to apply since it is often difficult to identify the forward loop gain $G(j\omega)$ and the feedback ratio $H(j\omega)$ in electronic systems. A direct analysis of the circuit equations is frequently simpler than the block diagram interpretation (particularly for single-stage amplifiers). Figure 2-23 shows a generalized circuit for an electronic amplifier.

Figure 2-23 shows a simplified Colpitts oscillator, which is derived from Figure 2-11. It has a capacitive feedback network, C_1 and C_2, and a tuned circuit, which is built from the inductor L and the capacitor C and coupled to the transistor circuit via a coupling capacitor C_C. Figure 2-23 can be redrawn by putting the tuned

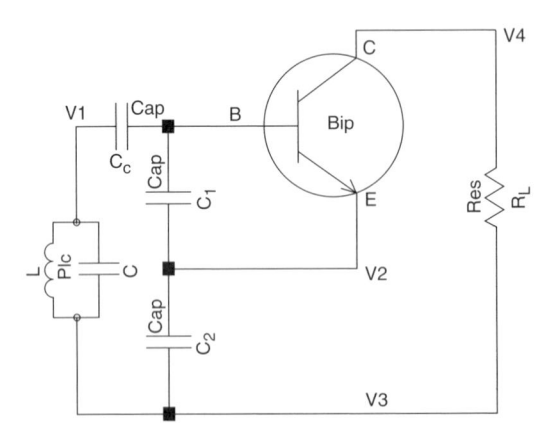

Figure 2-23 General topology of the Colpitts oscillator.

2.7 THREE-REACTANCE OSCILLATORS USING Y-PARAMETERS 29

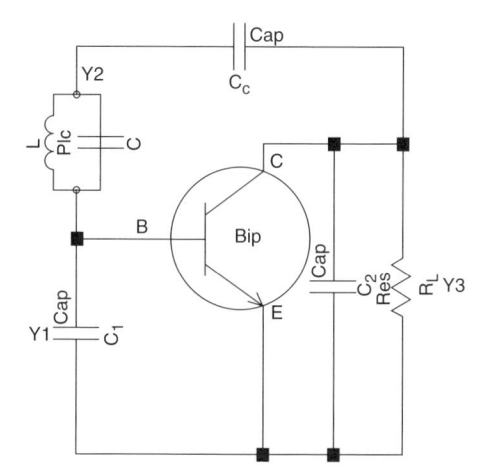

Figure 2-24 Meissen oscillator.

circuit with its coupling capacitor between the base and collector. The feedback capacitor C_1 is now in parallel to the base emitter junction, and the feedback capacitor C_2 is in parallel to the collector emitter connection. We also assume a load resistor, marked R_L, in parallel with C_2. This is shown in Figure 2-24.

For the purpose of having a general arrangement, we now introduce the admittance Y_1, Y_2, and Y_3. For this early introduction, we also assume that the transistor is ideal, that is, we assume that $Y_{12} = 0$ and $\text{Im}(Y_{11}, Y_{22}) = 0$. Later we will reformulate without this assumption.

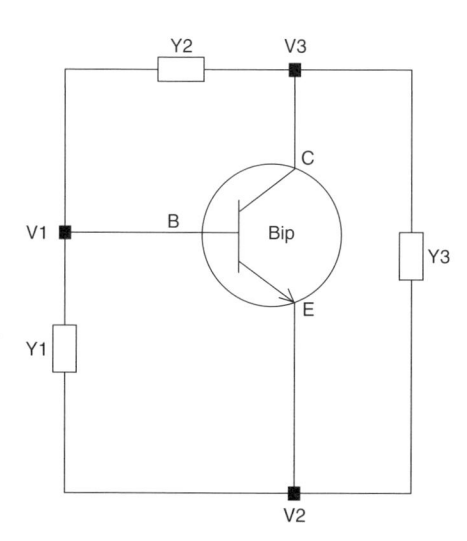

Figure 2-25 Equivalent circuit of the feedback oscillator.

30 GENERAL COMMENTS ON OSCILLATORS

Figure 2-25 shows the feedback circuit with three parallel admittances. Y_2 can either be a reactance or a more complex circuit such as a resonance circuit with a capacitance in series. In the case of a crystal oscillator, Y_2 is a series resonant circuit with a parallel capacitor, which comes from the crystal holder. The voltages V_1, V_2, and V_3 are measured relative to ground. The circuit is assumed floating.

It can be shown that

$$\begin{bmatrix} (Y_1 + Y_3 + Y_i) & -Y_3 \\ -(Y_3 - Y_{21}) & (Y_2 + Y_3) \end{bmatrix} \begin{bmatrix} V_1 \\ V_2 \end{bmatrix} = 0 \tag{2-7}$$

In this equation $Y_i = Y_{11}$ and Y_{21} are the Y-parameters of the transistor.

In order for this circuit to oscillate, it must satisfy the matrix condition $[Y][V] = 0$ for a nonzero value of $[V]$ (output power). Assuming that the feedback circuit has lossless components, then

$$\begin{aligned} (Y_2 + Y_3)(Y_1 + Y_3 + Y_i) - Y_3(Y_3 - Y_{21}) &= 0 \\ Y_1 Y_2 + Y_1 Y_3 + Y_2 Y_3 + Y_i Y_2 + Y_i Y_3 + Y_3 Y_{21} &= 0 \end{aligned} \tag{2-8}$$

If the feedback network has ideal reactive components and Y_{11} = real, then

$$Y_i = G_i, \quad Y_1 = jB_1, \quad Y_2 = jB_2, \quad Y_3 = jB_3 \tag{2-9}$$

and the equation above is further simplified to

$$\begin{aligned} -B_1 B_2 - B_1 B_3 - B_2 B_3 + jB_2 G_i + jB_3 G_i + jB_3 Y_{21} &= 0 \\ -(B_1 B_2 + B_2 B_3 + B_1 B_3) + j(B_2 G_i + B_3 G_i + B_3 Y_{21}) &= 0 \end{aligned} \tag{2-10}$$

From the equation above, the real and imaginary parts will need to be zero separately to satisfy det $[Y] = 0$.

$$\begin{aligned} B_1 B_2 + B_2 B_3 + B_1 B_3 = 0 &\Longrightarrow \frac{1}{B_1} + \frac{1}{B_2} + \frac{1}{B_3} = 0 \\ Y_{21} B_3 + G_i B_3 + G_i B_2 = 0 &\Longrightarrow \frac{1}{B_3} + \left(1 + \frac{Y_{21}}{G_i}\right) \frac{1}{B_2} = 0 \end{aligned} \tag{2-11}$$

If we convert susceptance to reactance and let

$$X_1 = \frac{1}{B_1}, \quad X_2 = \frac{1}{B_2}, \quad X_3 = \frac{1}{B_3} \tag{2-12}$$

and

$$X_1 + X_2 + X_3 = 0 \Longrightarrow X_3 + \left(1 + \frac{Y_{21}}{G_i}\right) X_2 = 0 \tag{2-13}$$

2.7 THREE-REACTANCE OSCILLATORS USING Y-PARAMETERS

then

$$\frac{Y_{21}}{G_i} = \frac{X_1}{X_2} \Longrightarrow \frac{Y_{21}}{Y_{11}} \tag{2-14}$$

Feedback conditions:

$$\frac{C_2}{C_1} = \frac{Y_{21}}{Y_{11}} \tag{2-15}$$

If Y_{21} and G_i are positive (simplified transistor model), it is implied that X_1 and X_2 have the same sign and, therefore, either capacitors or inductors. $X_1 + X_2 + X_3 = 0$ implies that X_3 must be opposite in sign from X_1 and X_2 and, therefore, the opposite type of component.

$$X_1 + X_2 + X_3 \Longrightarrow \frac{1}{\omega C_1} + \frac{1}{\omega C_2} - \omega L_3 = 0 \rightarrow \text{Colpitts oscillator}$$

$$\omega = \sqrt{\frac{1}{L_3}\left(\frac{1}{C_1} + \frac{1}{C_2}\right)} \quad \text{angular frequency, resonant condition} \tag{2-16}$$

For complex value of Y_3 (lossy inductor):

$$Y_3 = G_3 + jB_3 \Longrightarrow Y_3 = \frac{1}{Z_3}$$

$$Z_3 = R + j\omega L_3$$

$$\frac{1}{j\omega C_1} + \frac{1}{j\omega C_2} + \frac{G_i R}{j\omega C_1} + j\omega L_3 = 0 \tag{2-17}$$

$$\omega = \sqrt{\frac{1}{L_3}\left(\frac{1}{C_1} + \frac{1}{C_2} + \frac{G_i R}{C_1}\right)} \Longrightarrow \omega = \sqrt{\frac{1}{L_3}\left(\frac{1}{C_1'} + \frac{1}{C_2}\right)}$$

where

$$C_1' = \frac{C_1}{1 + G_i R} \tag{2-18}$$

and

$$\frac{R}{G_i} = \left(\frac{(1 + (Y_{21}/G_i))}{\omega^2 C_1 C_2}\right) - \frac{L_3}{C_1} \tag{2-19}$$

32 GENERAL COMMENTS ON OSCILLATORS

For steady oscillation the following condition has to be satisfied:

$$R_{Loss} < G_i \left[\frac{\left(1 + \frac{Y_{21}}{G_i}\right)}{\omega^2 C_1 C_2} - \frac{L_3}{C_1} \right] \quad (2\text{-}20)$$

Since (Y_{21}/G_i) is the frequency-dependent current gain β:

$$R_{Loss} < \left| \text{real}\left(Y_{11} \left[\frac{(1+\beta)}{\omega^2 C_1 C_2} - \frac{L_3}{C_1} \right] \right) \right| \quad (2\text{-}21)$$

A similar approach is found in [15].

2.8 VOLTAGE-CONTROLLED OSCILLATORS (VCOs)

The VCO is an oscillator in which the reactance(s), which determine the frequency of the oscillator, have been made variable. A typical LC oscillator consists of an inductor and a capacitor. The inductor can be made switchable. A perfect example is a binary-coded set of inductors which are switched by pin diodes or other types of switches and one or more tuning diodes. In some cases, a whole array of tuning diodes can be used. Figure 2-26 shows an example of an oscillator in which both inductors and capacitors are selected with diode switches and the emitter feedback can also be selected as needed. This approach can be found up to very high frequencies and is applied to integrated circuits. Figure 2-27 shows a 200–400 MHz wideband VCO. It is derived from the Colpitts oscillator shown in

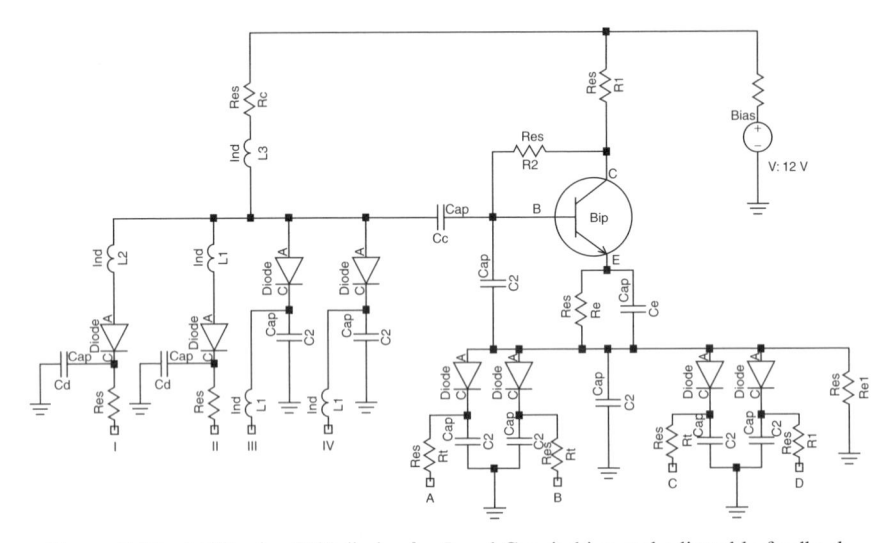

Figure 2-26 VCO using PIN diodes for L and C switching and adjustable feedback.

2.8 VOLTAGE-CONTROLLED OSCILLATORS 33

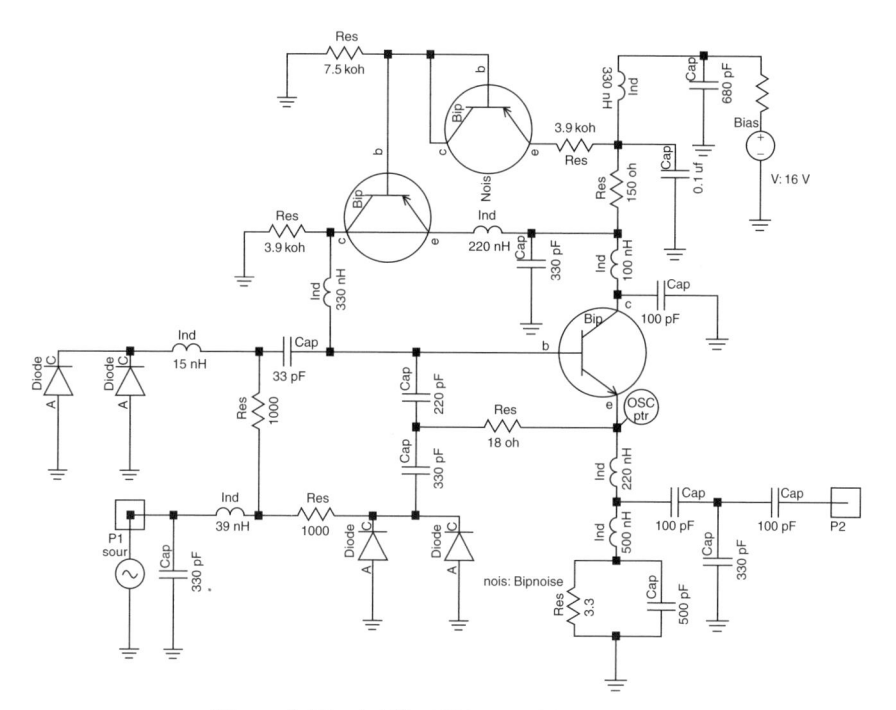

Figure 2-27 A 200–400 MHz wideband VCO.

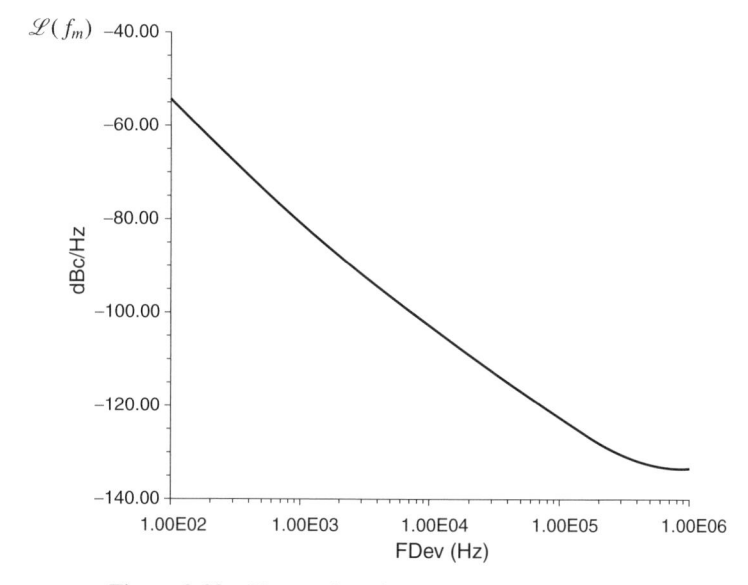

Figure 2-28 Phase noise of a dynamically tuned VCO.

34 GENERAL COMMENTS ON OSCILLATORS

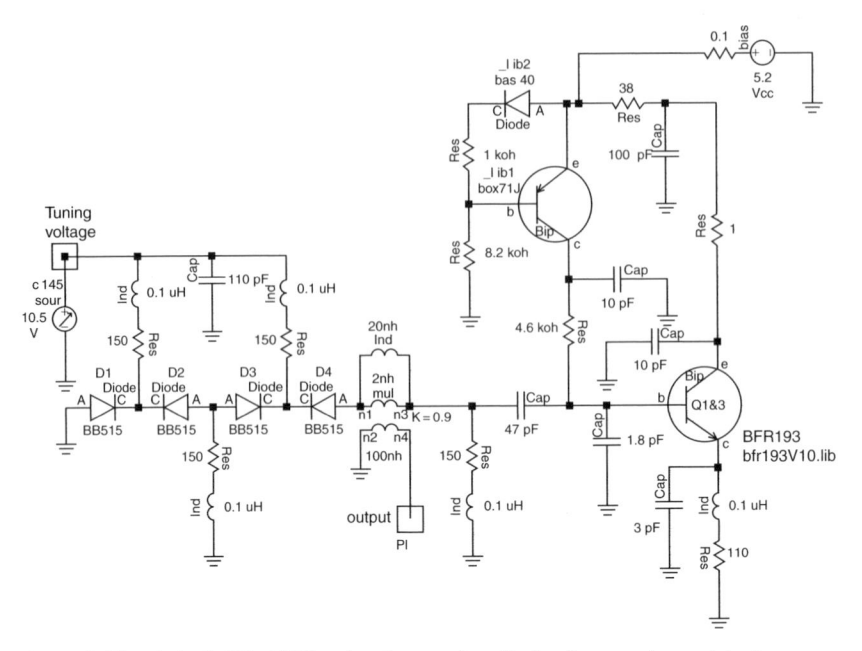

Figure 2-29 A 1–2 GHz VCO using four tuning diodes for an enhanced tuning range.

Figure 2-14; however, the tuned circuit now operates in series mode, which means that the values of the feedback capacitors between base and emitter and from emitter to ground are different. The capacitor from the emitter connection to ground is now made variable and tracks the main tuning diodes on the left. Due to the influence of

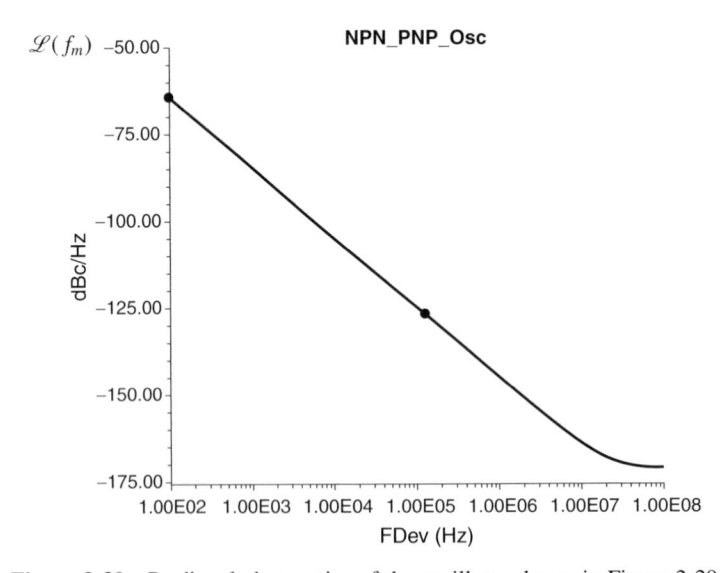

Figure 2-30 Predicted phase noise of the oscillator shown in Figure 2-29.

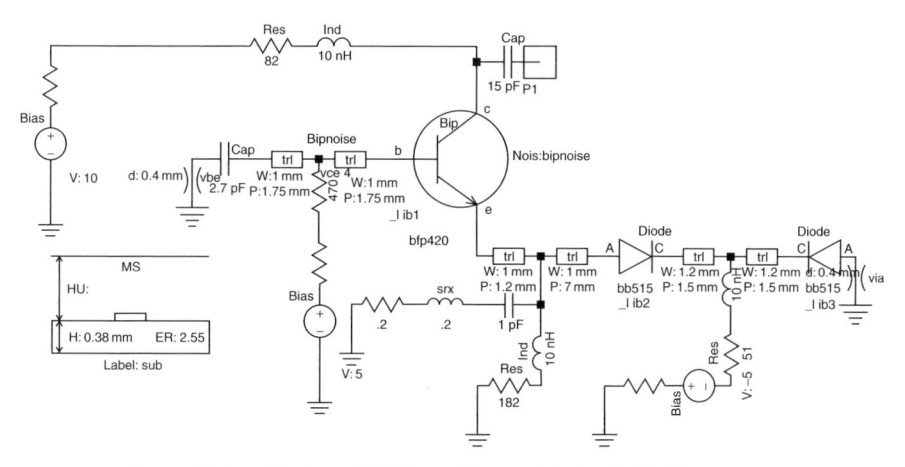

Figure 2-31 A 3.4 to 4.6 GHz oscillator with the BFP420 transistor.

the tuning diodes (see Chapter 7, Leeson's model), the phase noise is now largely determined by the tuning diode. The power output is also done using two inductors, and there is a particular harmonic cancellation emitter feedback for which Synergy Microwave Corporation has applied for patents (patent numbers: 60/493075, 60/501790, 60/528670, 60/563481, 60/564173). The resulting phase noise, simulated in Figure 2-28, agrees well with the measured data and is determined at 300 MHz. The term *dynamically tuned* refers to the fact that the feedback values are tuned at the same time that the frequency-determining element is tuned. This VCO is a typical example used to illustrate the principle, but its performance is not optimized.

Another good example of the complexity of a VCO is presented in Figure 2-29. This oscillator is used in the Rohde and Schwarz portable receiver Model EB-100 and is used for battery operation. The oscillator operates at 1000 MHz, and the output is provided via inductive coupling to the resonator. The main inductor is the 20 nH value, and the 2 µH/100 nH transformer was chosen for output simulation. The phase noise of this oscillator can be seen in Figure 2-30.

Another voltage-controlled oscillator, intended to be used in battery-operated receivers, is shown in Figure 2-31.

3 Transistor Models

3.1 INTRODUCTION

For the design of oscillators we are looking at members of bipolar and FET families. In bipolar transistors, conventional microwave silicon transistors are manufactured with an f_T up to 25 GHz, while the more advanced SiGe transistors take over from this frequency range. Today, SiGe transistors are available up to 100 GHz if used as part of an RFIC. Their cousins, the heterojunction bipolar transistors (HBTs), based on GaAs technology, can achieve similar cutoff frequencies, but this technology is much more expensive for medium to large integrated circuits. HBTs also have a higher flicker noise corner frequency. SiGe transistors have a much lower flicker noise corner frequency and lower breakdown voltages (typically 2–3 V). However, because of the losses of the transmission line in practical circuits, there is not much difference between HBT and SiGe oscillator noise, as f_T is the same.

There is a similar competing situation between BiCMOS transistors implemented in 0.12 μm technology and GaAs FETs, specifically p-HEMTs. The GaAs FETs have well-established performance with good models, and the BiCMOS transistors are currently being investigated to determine what models are best. Also, there is the $1/f$ noise problem specifically, more so with GaAs FETs than with MOS transistors. The 0.12 μm technology is somewhat impractical because of poor modeling. This means that most CAD predictions do not translate into a good design.

For the purpose of this discussion, it should be assumed that designers have the ability to do their own parameter extraction to obtain an accurate model, or receive these data from the transistor manufacturer or as part of the foundry service.

While the bipolar transistor models tend to be more physics-based, the FET models are mostly based on analytic equations, which are generated using curve-fitting techniques.

There are two types of models:

1. The models which describe DC and RF behavior are SPICE-type models, which means that they can be incorporated in a frequency/time domain simulator and give reasonable agreement with measured data in both the DC and RF areas.
2. Linear RF microwave models based on equivalent circuits.

The Design of Modern Microwave Oscillators for Wireless Applications: Theory and Optimization, by Ulrich L. Rohde, Ajay Kumar Poddar, Georg Böck
Copyright © 2005 John Wiley & Sons, Inc.

38　TRANSISTOR MODELS

In the bipolar modeling world, the Ebers-Moll equations have been used to generate the Gummel-Poon model (transport model) and its various subtle, modified (sometimes) proprietary implementations. Only in SPICE, particularly in Berkley-SPICE and H-SPICE, do we find these models that the industry has agreed upon [18, 19]. Two other important and popular SPICE programs are P-SPICE and RF Spectre from CADENCE Design Systems [20, 21]. As an extension to this modeling effort, the University of Delft/Philips has introduced a complex model consisting of an NPN and a complementary PNP transistor for better RF modeling [22–24]. This model claims better simulation results for intermodulation distortion products under large-signal conditions. For the purpose of noise calculation of oscillators, one needs to know that the modulation indices are very small and the resulting noise currents and voltages are very small compared to the RF currents/voltages. The standard models, which are continuous and the third derivative to voltage and current exists, are sufficient.

For members of the FET family, the development of models was based on junction FETs (JFETs) and MOS transistors as implemented in early forms of SPICE. JFETs are available only up to 1000 MHz and have lost importance in the RF and microwave area.

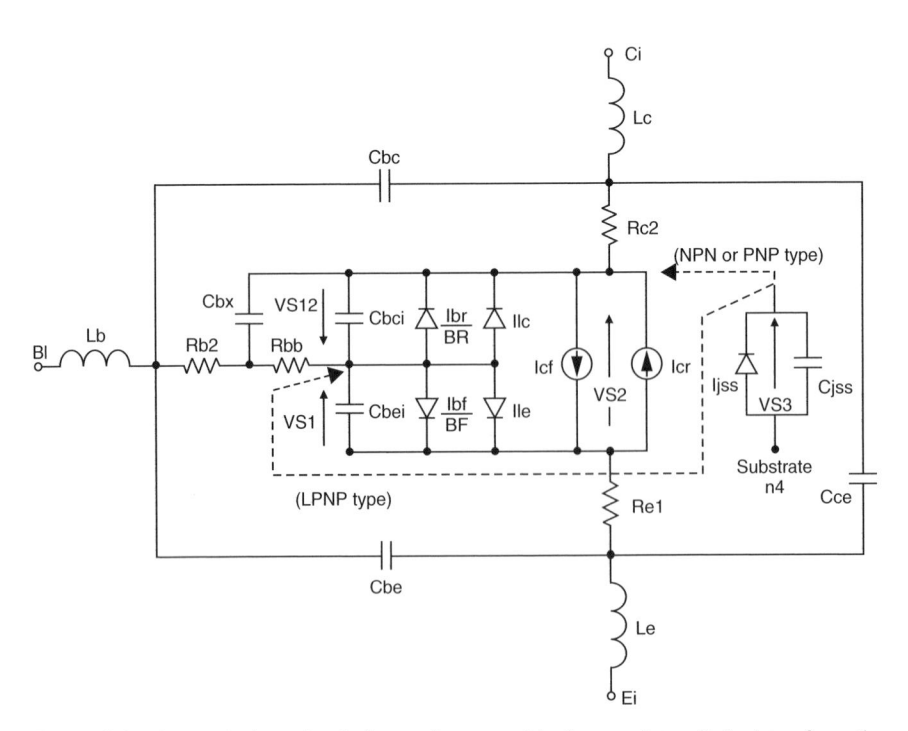

Figure 3-1 An equivalent circuit for a microwave bipolar transistor. It deviates from the SPICE implementation in having two base-spreading resistors.

3.1 INTRODUCTION

The first models for GaAs transistors were the Curtice-Ettenberg models, particularly the quadratic and cubic models. They were developed from the MOS model by adding a diode at the gate and removing the MOS capacitor at the input. Today, there are many GaAs models believed by individual researchers to have big advantages.

These models are used to describe the behavior of the transistors over a wide frequency, temperature, and bias range. Their accuracy varies. The most important factors are the input parameters, which are obtained from a process called *parameter extraction*. Given the long-term use of CAD tools, it appears that a majority of the discrepancies between measured and modeled results can be explained by a lack of accuracy stemming from parameter extraction. There is very little software available for reliable parameter extraction. Many companies have written their own software and assembled the necessary test equipment to obtain those parameters. The most popular software tools are from Silvaco and Agilent.

A successful method for generating model parameters has been the extraction of DC and RF parameters using DC IV data and S-parameter data sets under various bias and frequency conditions. The model equations are curve fit to the data. To more accurately extract parameters that affect the noise characteristics of the

Transistor Chip Data:

IS	=	15	aA	BF	=	235	–	NF	=	1	–
VAF	=	25	V	IKF	=	0.4	A	ISE	=	25	fA
NE	=	2	–	BR	=	1.5	–	NR	=	1	–
VAR	=	2	V	IKR	=	0.01	A	ISC	=	20	fA
NC	=	2	–	RB	=	11	Ω	IRB	=	–	A
RBM	=	7.5	Ω	RE	=	0.6		RC	=	7.6	Ω
CJE	=	235	fF	VJE	=	0.958	V	MJE	=	0.335	–
TF	=	1.7	ps	XTF	=	10	–	VTF	=	5	V
ITF	=	0.7	A	PTF	=	50	deg	CJC	=	93	fF
VJC	=	0.661	V	MJC	=	0.236	–	XCJC	=	1	–
TR	=	50	ns	CJS	=	0	fF	VJS	=	0.75	V
MJS	=	0.333	–	XTB	=	-0.25	–	EG	=	1.11	eV
XTI	=	0.035	0	FC	=	0.5	–	TNOM		298	K

Package Equivalent Circuit:

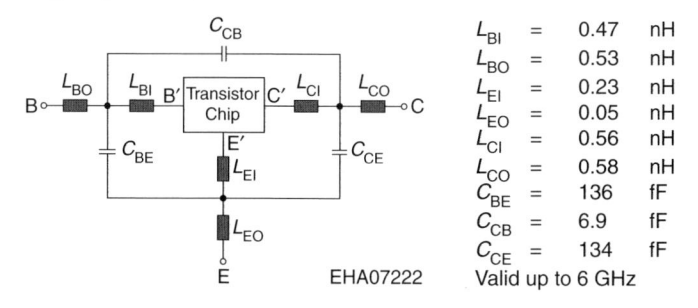

L_{BI}	= 0.47 nH
L_{BO}	= 0.53 nH
L_{EI}	= 0.23 nH
L_{EO}	= 0.05 nH
L_{CI}	= 0.56 nH
L_{CO}	= 0.58 nH
C_{BE}	= 136 fF
C_{CB}	= 6.9 fF
C_{CE}	= 134 fF

EHA07222 Valid up to 6 GHz

Figure 3-2 SPICE parameters and package equivalent circuit of the Infineon transistor BFP520.

40 TRANSISTOR MODELS

model, it has also been proposed to include noise data as part of the parameter extraction procedure [25, 26].

When experimenting with these parameter extraction programs, under certain conditions input parameters for the nonlinear models are generated which no longer have any practical meaning. This means that they are not realizable in the manufacturing process, but in a given frequency range, they may give the right S-parameters. The inclusion of noise parameters solves this problem to a large extent [27, 28].

The following discussion presents the information obtained for bipolar transistors and their modeling. The synthesis approach is applicable to all three terminal devices, including not only bipolar transistors, but all FETs such as JFETs, GaAs FETs, and MOSFET transistors.

3.2 BIPOLAR TRANSISTORS

The bipolar transistor has been known and used for many decades. Many scientists have explained its behavior, and probably the best analysis in the DC/RF area is

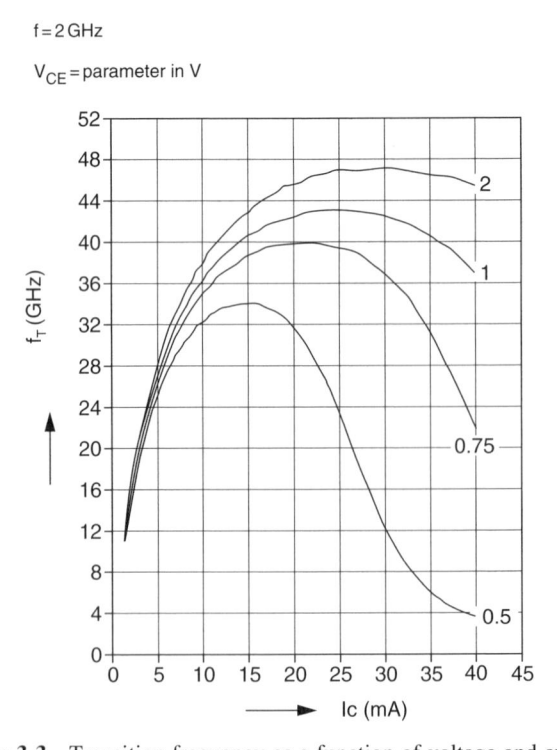

Figure 3-3 Transition frequency as a function of voltage and current.

3.2 BIPOLAR TRANSISTORS

summarized in [29]. This summary is based largely on the Infineon transistor BFP520 but is also applicable to other transistors.

The first thing we need to do is look at the model used to calculate the DC and RF performance of a microwave transistor. There are subtle differences between the standard SPICE implementation and the one suited for higher frequencies. Figure 3-1 shows a modification that was necessary for greater accuracy by introducing an additional base-spreading resistor, R_{b2}.

Obtaining the input parameters is a major issue. For the purpose of this work, the Infineon transistor BFP520 was chosen. It is well documented and has a high enough operating frequency. Figures 3-2 to 3-5 and Table 3-1 reproduce the manufacturer's data.

BFP520 NPN Silicon RF Transistor

- For the highest-gain, low-noise amplifier at 1.8 GHz and 2 mA/2V
  Outstanding $G_{ms} = 23$ dB
  Noise figure $F = 0.95$ dB
- For oscillators up to 15 GHz
- Transition frequency $f_T = 45$ GHz
- Gold metalization for high reliability
- SIEGET 45-Line
  45 GHZ f_T–Line

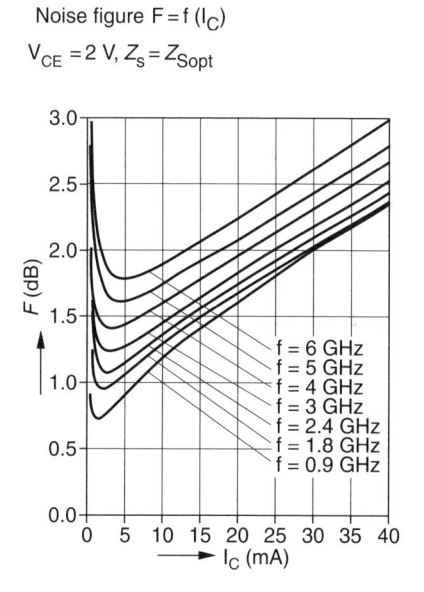

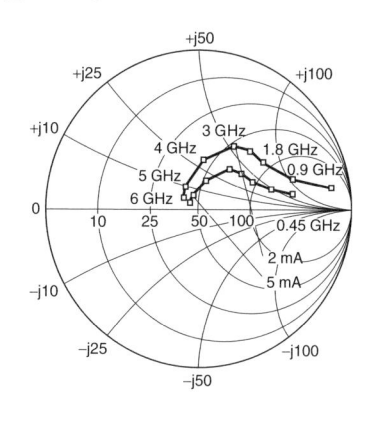

Figure 3-4 Noise figure and source impedance for the best noise figure as a function of current and frequency of the Infineon transistor BFP520.

SIEGET®45 BFP520

Electrical Characteristics at $T_A = 25°C$ unless otherwise specified.

Parameter	Symbol	min.	typ.	max.	Unit
AC characteristics (verified by random sampling)					
Transition frequency $I_C = 30$ mA, $V_{CE} = 2$ V, $f = 2$ GHz	f_T	–	45	–	GHz
Collector-base capacitance $V_{CB} = 2$ V, $f = 1$ MHz	C_{cb}	–	0.06	–	pF
Collector-emitter capacitance $V_{CE} = 2$ V, $f = 1$ MHz	C_{ce}	–	0.3	–	
Emitter-base capacitance $V_{EB} = 0.5$ V, $f = 1$ MHz	C_{eb}	–	0.35	–	
Noise figure $I_C = 2$ mA, $V_{CE} = 2$ V, $Z_S = Z_{Sopt}$, $f = 1.8$ GHz	F	–	0.95	–	dB
Power gain, maximum stable [1] $I_C = 20$ mA, $V_{CE} = 2$ V, $Z_S = Z_{Sopt}$, $Z_L = Z_{Lopt}$, $f = 1.8$ GHz	G_{ms}	–	23	–	
Insertion power gain $I_C = 20$ mA, $V_{CE} = 2$ V, $f = 1.8$ GHz, $Z_S = Z_L = 50\Omega$	$\|S_{21}\|^2$	–	21	–	dB
Third order intercept point at output $V_{CE} = 2$ V, $f = 1.8$ GHz, $Z_S = Z_{Sopt}$, $Z_L = Z_{Lopt}$, $I_C = 20$ mA $I_C = 7$ mA	IP_3	– –	25 17	– –	dBm
1 dB compression point $V_{CE} = 2$ V, $f = 1.8$ GHz, $Z_S = Z_{Sopt}$, $Z_L = Z_{Lopt}$, $I_C = 20$ mA $I_C = 7$ mA	P_{-1dB}	– –	12 5	– –	

[1] $G_{ms} = |S_{21}/S_{12}|$.

Figure 3-5 Some technical data of the Infineon transistor BFP520. *Source*: Infineon Technologies.

TABLE 3-1 Common Emitter S-Parameters

f	S_{11}		S_{21}		S_{12}		S_{22}	
GHz	MAG	ANG	MAG	ANG	MAG	ANG	MAG	ANG
$V_{CE} = 2$ V, $I_C = 20$ mA								
0.01	0.7244	−0.7	32.273	178.6	0.0007	69.4	0.9052	1.2
0.1	0.7251	−8.4	31.637	171.4	0.0041	92.8	0.9363	−4.4
0.5	0.6368	−40.7	27.293	140.7	0.0194	75.9	0.8523	−26.7
1	0.4768	−73.6	19.601	113.5	0.0351	66.5	0.6496	−46.1
2	0.2816	−123.8	11.021	84.9	0.0057	56.3	0.3818	−64.6
3	0.2251	−166.1	7.481	67.6	0.0788	49.2	0.2407	−73.6
4	0.2552	156.2	5.636	53.1	0.0994	41.5	0.1544	−95.3
5	0.3207	133.6	4.488	39.7	0.1177	32.9	0.0951	−128.9
6	0.3675	118.7	3.683	27.5	0.1343	24.7	0.0545	177.6

Note: ANG = angle; MAG = magnitude.

Let Δf be the bandwidth (usually normalized to a 1 Hz bandwidth). The noise generators introduced in the intrinsic device are shown below and have mean square values of:

$$\langle i_{bn}^2 \rangle = 2qI_B \Delta f + KF \frac{I_B^{AF}}{f^{FCP}} \Delta f$$

$$\langle i_{cn}^2 \rangle = 2qI_C \Delta f$$

$$\langle i_{Rbb}^2 \rangle = \frac{4kT}{R_{bb}} \Delta f$$

$$\langle i_{Re1}^2 \rangle = \frac{4kT}{RE1} \Delta f$$

$$\langle i_{Rc2}^2 \rangle = \frac{4kT}{RC2} \Delta f$$

$$I_B = \frac{I_{bf}}{BF} + I_{le}$$

$$I_C = I_{cf} - I_{cr}$$

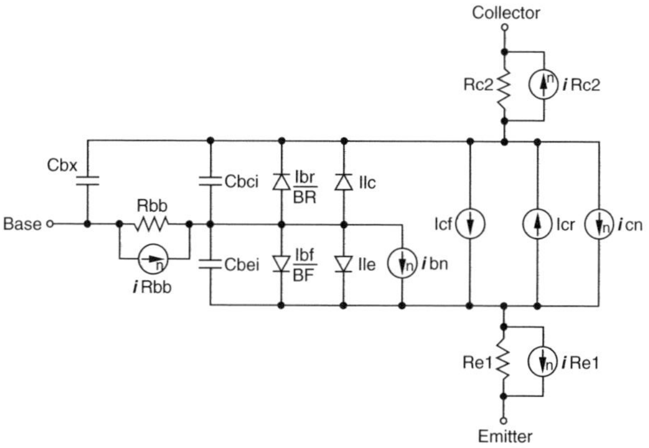

Figure 3-6 Noise model of the bipolar transistor (not showing extrinsic parasitics). Current sources with "n" are noise sources.

Figure 3-2 shows measured data provided by Infineon to further characterize the transistor.

These parameters are fit to the Berkley-SPICE 2G.6 syntax. In the case of the base-spreading resistor, it should be separated into two equal terms to fit the improved model shown above. The feedback capacitor should also be split into two terms. The equivalent circuit of the package is valid up to 6 GHz and assumes that the two emitter leads are tied together.

NPN Silicon Germanium RF Transistor

· High-gain, low-noise RF transistor

· Outstanding noise figure F = 0.7 dB at 1.8 GHz

Outstanding noise figure F = 1.3 dB at 6 GHz

· Maximum stable gain

G_{ms} = 21.5 dB at 1.8 GHz

G_{ma} = 11 dB at 6 GHz

· Gold metallization for extra high reliability

· 70 GHz f_T - silicon germanium technology

Figure 3-7 Highlights of the BFP620.

NPN Silicon Germanium RF Transistor

- For high-power amplifiers
- Ideal for low phase noise oscillators
- Maxim. available gain G_{ma} = 21 dB at 1.8 GHz
  Noise figure F = 0.9 dB at 1.8 GHz
- Gold metallization for high reliability
- 70 GHz f_T - silicon germanium technology

Figure 3-8 Highlights of the BFP650.

Transitor chip data: **BFP620**

IS	=	0.22	fA	BF	=	425	–	NF	=	1.025	–
VAF	=	1000	V	IKF	=	0.25	A	ISE	=	21	fA
NE	=	2	–	BR	=	50	–	NR	=	1	–
VAR	=	2	V	IKR	=	10	mA	ISC	=	18	pA
NC	=	2	–	RB	=	3.129	Ω	IRB	=	1.522	mA
RBM	=	2.707	Ω	RE	=	0.6	–	RC	=	2.364	Ω
CJE	=	250.7	fF	VJE	=	0.75	V	MJE	=	0.3	–
TF	=	1.43	ps	XTF	=	10	–	VTF	=	1.5	V
ITF	=	2.4	A	PTF	=	0	deg	CJC	=	124.9	fF
VJC	=	0.6	V	MJC	=	0.5	–	XCJC	=	1	–
TR	=	0.2	ns	CJS	=	128.1	fF	VJS	=	0.52	V
MJS	=	0.5	–	NK	=	-1.42	–	EG	=	1.078	eV
XTI	=	3	–	FC	=	0.8	–	TNOM		298	K
AF	=	2	–	KF	=	7.291E-11					
TITF1		-0.0065	–	TITF2		1.0E-5					

All parameters are ready to use; no scalling is necessary.

Package equivalent circuit:

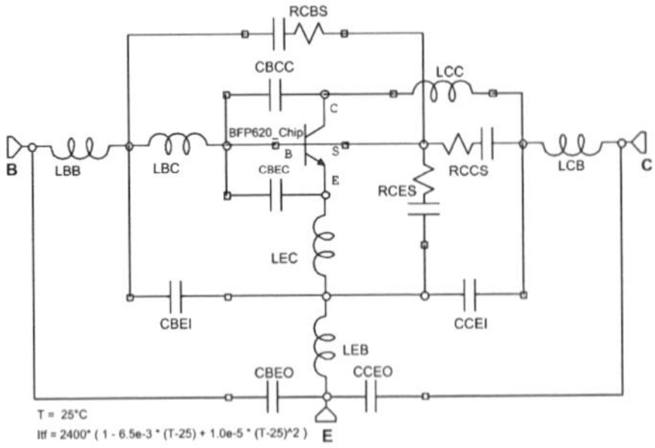

LBC	=	60	pH
LCC	=	50	pH
LEC	=	15	pH
LBB	=	764.5	pH
LCB	=	725.4	pH
LEB	=	259.6	pH
CBEC	=	98.4	fF
CBCC	=	55.9	fF
CES	=	140	fF
CBS	=	54	fF
CCS	=	50	fF
CCEO	=	106.5	fF
CBEO	=	106.7	fF
CCEI	=	132.4	fF
CBEI	=	99.6	fF
RBS	=	1200	Ω
RCS	=	1200	Ω
RES	=	300	Ω

Valid up to 6GHz

T = 25°C
Itf = 2400 * (1 - 6.5e-3 * (T-25) + 1.0e-5 * (T-25)^2)

Figure 3-9 SPICE parameters and package equivalent circuit of the BFP620.

3.2 BIPOLAR TRANSISTORS 45

Transitor chip data: BFP 650

IS	=	0.61	fA	BF = 450	-	NF	=	1.025	-
VAF	=	1000	V	IKF = 0.47	A	ISE	=	62	fA
NE	=	2	-	BR = 42	-	NR	=	1	-
VAR	=	2	V	IKR = 18	mA	ISC	=	700	fA
NC	=	1.8	-	RB = 1.036	Ω	IRB	=	4.548	mA
RBM	=	0.895	Ω	RE = 0.2	-	RC	=	1.006	Ω
CJE	=	682.5	fF	VJE = 0.8	V	MJE	=	0.3	-
TF	=	1.9	ps	XTF = 10	-	VTF	=	1.5	V
ITF	=	1.25	A	PTF = 0	deg	CJC	=	204.6	fF
VJC	=	0.6	V	MJC = 0.5	-	XCJC	=	1	-
TR	=	0.2	ns	CJS = 294.9	fF	VJS	=	0.6	V
MJS	=	0.27	-	NK = -1.42	-	EG	=	1.078	eV
XTI	=	3	-	FC = 0.8	-	TNOM		298	K
AF	=	2	-	KF = 2.441E-11					
TITF1		-0.0065	-	TITF2 1.0E-5					

All parameters are ready to use, no scalling is necessery. Extracted on behalf of Infineon Technologies AG by: Institut für Mobil- und Satellitentechnik (IMST)

Package equivalent circuit:

LBC	=	50	pH
LCC	=	50	pH
LEC	=	4	pH
LBB	=	554.6	pH
LCB	=	606.9	pH
LEB	=	138.7	pH
CBEC	=	327.6	fF
CBCC	=	171.4	fF
CES	=	490	fF
CBS	=	120	fF
CCS	=	135	fF
CBCO	=	7.5	fF
CCEO	=	112.6	fF
CBEO	=	121.5	fF
CCEI	=	5.7	fF
CBEI	=	6.9	Ω
RBS	=	710	Ω
RCS	=	710	Ω
RES	=	140	Ω

Valid up to 6 GHz

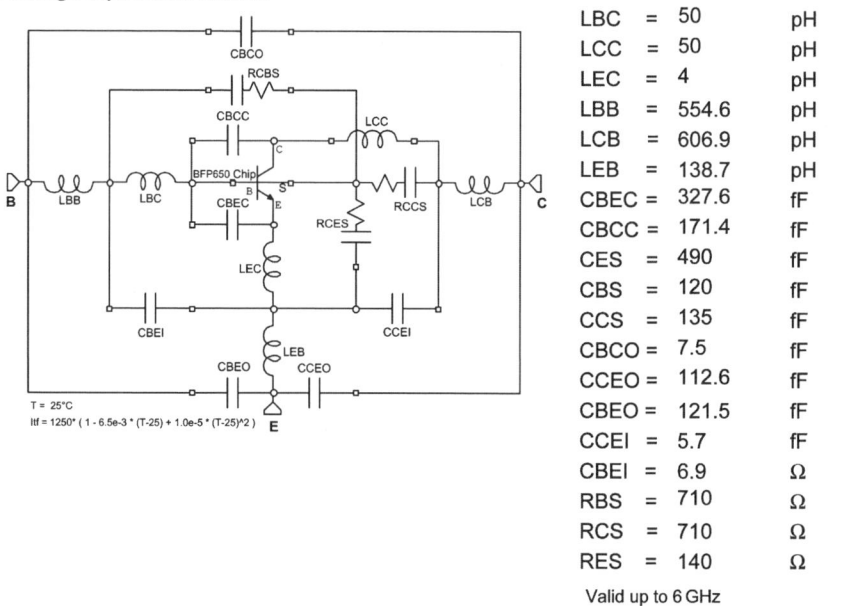

T = 25°C
Itf = 1250* (1 - 6.5e-3 * (T-25) + 1.0e-5 * (T-25)^2)

Figure 3-10 SPICE parameters and package equivalent circuit of the BFP650.

The Infineon data set also contains noise parameters and *S*-parameters (see Table 3-1). However, these are less relevant because the large-signal conditions will modulate the intrinsic nonlinear capacitances and other elements, so a large-signal noise model is needed. The following equivalent circuit in Figure 3-6 can be used to calculate the noise performance. Noise is mostly assumed to be

46 TRANSISTOR MODELS

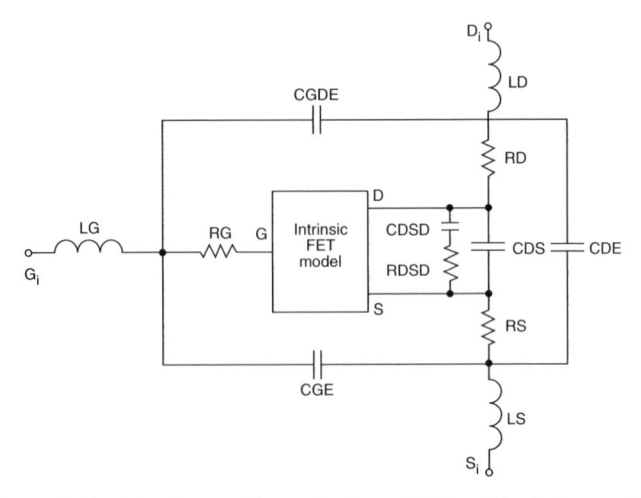

Figure 3-11 Topology of the extrinsic model for all intrinsic models.

derived in a linear mode; specifically, the flicker corner frequency under a large-signal condition will change significantly [30]. Information about SiGe transistors with much higher operating frequencies is presented in [89–98].

In Chapter 6, Section 6.2, the complete complex noise theory will be addressed.

This chapter does not address any GaAs HBTs, as they are not obtainable in packaged form. These transistors are typically part of an integrated circuit, and do not behave totally differently than the SiGe transistors if properly modeled with the help of an adequate parameter extraction program. During the validation of this

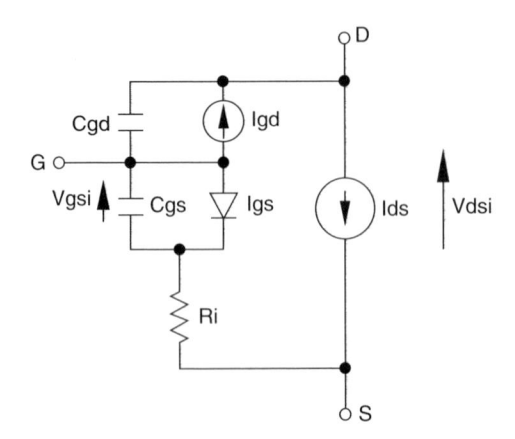

Figure 3-12 Topology of the Materka-Kacprzak intrinsic model.

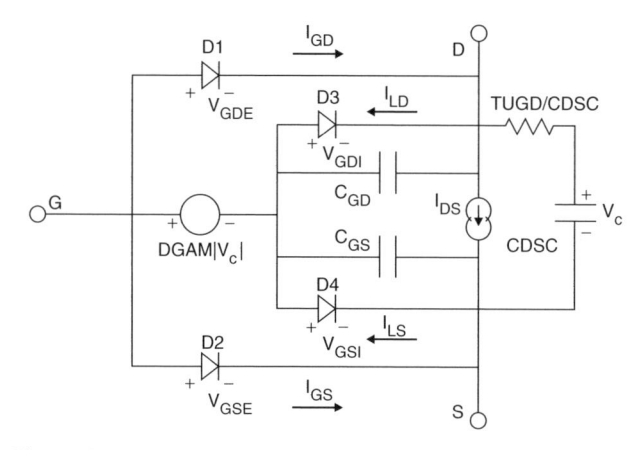

Figure 3-13 Topology of the Triquint (TOM-3) intrinsic model.

new synthesis approach, the BFP620 from Infineon is also being used to extend the frequency range. This family of SiGe transistors is currently being extended to incorporate other members with different power levels. For completeness, the following large-signal parameters of both the BFP620 and BFP650 transistors are added. The BFP650 is specifically designed for oscillators [31–37].

3.3 FIELD-EFFECT TRANSISTORS (FETs)

For RF applications, three types of transistors can be used. The first is a junction FET, which has been shown to be useful up to 1 GHz. Its fairly high input capacitance of about 1 pF and large feedback capacitance of about 0.1 pF limits its use. It is mentioned here only for completeness. The other two FETs of importance are members of the GaAs FET family and the family of BiCMOS transistors. Recent advances in technology have push the BiCMOS process close to 100 GHz if the BiCMOS transistor is built on SiGe technology.

3.3.1 GaAs FETs

For the purpose of modeling GaAs FET, many models are available; the list is even longer if company or university internal models are added. The following models are popular with CAD tools:

- Chalmers (Angelov) Model
- Curtice-Ettenberg Cubic Model
- Curtice Quadratic Model
- IAF (Berroth) Model
- ITT PFET and TFET Models

Channel Current Model Keywords

Keywords	Description	Unit	Default
IDSS	Drain saturation current for Vgs = 0	amp	0.1
VP0	Pinch-off voltage for Vds = 0	volt	-2.0
GAMA	Voltage slope parameter of pinch-off voltage	/volt	0.0
E	Constant part of power law parameter		2.0
KE	Dependence of power law on Vgs	/volt	0.0
SL	Slope of the Vgs = 0 drain characteristic in the linear region	amp/volt	0.15
KG	Drain dependence on Vgs in the linear region	/volt	0.0
SS	Slope of the drain characteristic in the saturated region	amp/volt	0.0
T	Channel transit-time delay	sec	0.0
DLVL	Select diode breakdown model (default = diode model)		1
IG0	Diode saturation current	amp	0
AFAG	Slope factor of forward diode current	/volt	38.696
IB0	Breakdown saturation current	amp	0
AFAB	Slope factor of breakdown current	/volt	0
VBC	Breakdown voltage	volt	inf.
DLVL	Select Raytheon enhanced breakdown model by setting DLVL = 2		
GMAX	Breakdown conductance	amp/volt	0
K1D	Fitting parameter	/volt	0
K2D	Fitting parameter	volt	0
K3D	Fitting parameter	$volt^2$	0
R10	Intrinsic channel resistance for Vgs = 0	ohm	0.0
KR	Slope factor of intrinsic channel resistance	/volt	0.0

Note: The flicker noise parameter of the Materka model is KFN so as not to conflict with the KF parameter in the capacitance model.

Figure 3-14 Materka FET keywords.

- Modified Materka-Kacprzak Model
- Physics-Based MESFET Model
- Raytheon (Statz) Model
- TriQuint (TriQuint Own Model, TOM-1, TOM-2, and TOM-3) Model

Their properties and different model equations can be found in the CAD user's manual. As an example, the Ansoft Designer's reference manual gives all the details and references on how the model was implemented. It is difficult to obtain a reliable parameter extractor for the models. Probably the most popular foundry for commercial application is TriQuint. The model for which good parameters can be extracted with reasonable effort is the Materka model. One of the more

Intrinsic Model Keywords

Keyword	Description	Unit	Default
BETA	Transconductance coefficient	amp/μm/volt^Q	0.1
VT0	Threshold voltage	volt	-2.0
GAMA	Threshold-shifting parameter	/volt	0.0
Q	Power law parameter		2.0
LAMB	Slope of drain characteristic in the saturated region	/volt	0.0
ALFA	Slope of drain characteristic in the linear region	/volt	2.0
T	Channel transit-time delay	sec	0.0
VBI	Built-in barrier potential	volt	1.0
KAPP	Knee-function parameter		2.0
VST	Subthreshold slope	volt	1.0
MST	Subthreshold slope drain parameter	/volt	0.0
IS	Forward gate diode saturation current	amp/μm	1.0e-14
N	Forward gate diode ideality factor		1.0
ILK	Leakage diode current parameter	amp/μm	0
PLK	Leakage diode potential parameter	volt	1.0
TUGD	Time constant of the drain conductance dispersion feedback network		1.0e-6
CDSC	Capacitance of the drain conductance dispersion feedback network		1.0e-12
DGAM	Drain conductance dispersion parameter		0.0
QGQH	Charge parameter		-2.0e-16
QGSH	Charge parameter		1.0e-16
QGDH	Charge parameter		0.0
QGI0	Charge parameter		1.0e-6
QGQL	Charge parameter		5.0e-16
QGAG	Charge parameter		1.0
QGAD	Charge parameter		1.0
QGCL	Charge parameter		2.0e-16
QGGB	Charge parameter		100.0
QGG0	Charge parameter		0.0

Figure 3-15 Materka FET keywords.

flexible programs is the SCOUT program from Ansoft. It requires a DC/IV curve tracer and a network analyzer which operates up to 40 GHz.

For the purpose of validating the theory of this research, it was important to select a transistor which had been fully characterized; see Figures 3-16 to 3-19. The Infineon (Siemens) transistor CFY30 was chosen for convenience.

The parameter extraction was done at Ansoft/Compact Software and has been validated by building some amplifier circuits around them. The noise parameters vary with voltage and current. Since only one set of measurements has been taken, the best way to obtain other noise parameters at different voltages, currents, and frequencies is to use an appropriate model in the CAD tool. The Ansoft circuit simulator has a scalable model for noise for FETs which was published and has shown excellent results under various conditions. It is based on the RPC values

50 TRANSISTOR MODELS

$I_D = 15$ mA $\quad V_{DS} = 3.5$ V $\quad Z_0 = 50\ \Omega$

f	F_{min}	G_a	Γ_{opt}		R_n	N	$F_{50\Omega}$	$G(F_{50\ \Omega})$
GHz	dB	dB	MAG	ANG	Ω	-	dB	dB
2	1.0	15.5	0.72	27	49	0.17	2.9	10.0
4	1.4	11.5	0.64	61	29	0.17	2.7	9.3
6	2.0	8.9	0.46	101	19	0.30	2.8	7.5
8	2.5	7.1	0.31	153	9	0.31	2.8	6.4
10	3.0	5.8	0.34	-133	14	0.38	3.4	4.2
12	3.5	5.0	0.41	-93	28	0.42	4.1	2.9

CFY30

* Low noise ($F_{min} = 1.4$ dB @ 4 GHz)
* High gain (11.5 dB typ. @ 4 GHz)
* For oscillators up to 12 GHz
* For amplifiers up to 6 GHz
* Ion implanted planar structure
* Chip all gold metallization
* Chip nitride passivation

Figure 3-16 Typical common source noise parameters of the CFY30.

Electrical characteristics at $T_A = 25°C$ unless otherwise specified

Characteristics	Symbol	Min	Typ	Max	Unit
Drain-source saturation current $V_{DS} = 3.5$ V, $V_{DS} = 0$ V	I_{DSS}	20	50	80	mA
Pinch-off voltage $V_{DS} = 3.5$ V $I_D = 1$ mA	$V_{GS(P)}$	-0.5	-1.3	-4.0	V
Transconductance $V_{DS} = 3.5$ V $I_D = 15$ mA	g_m	20	30	-	mS
Gate leakage current $V_{DS} = 3.5$ V $I_D = 15$ mA	I_G	-	0.1	2	μA
Noise figure $V_{DS} = 3.5$ V $I_D = 15$ mA $f = 4$ GHz $\qquad\qquad\qquad\qquad\qquad f = 6$ GHz	F	- -	1.4 2.0	1.6 -	dB
Associated gain $V_{DS} = 3.5$ V $I_D = 15$ mA $f = 4$ GHz $\qquad\qquad\qquad\qquad\qquad f = 6$ GHz	G_a	10 -	11.5 8.9	- -	dB
Maximum available gain $V_{DS} = 3.5$ V $I_D = 15$ mA $f = 6$ GHz	MAG	-	11.2	-	dB
Maximum stable gain $V_{DS} = 3.5$ V $I_D = 15$ mA $f = 4$ GHz	MSG	-	14.4	-	dB
Power output at 1 dB compression $V_{DS} = 4$ V $I_D = 30$ mA $f = 6$ GHz	P_{1dB}	-	16	-	dBm

Figure 3-17 Some technical data of the CFY30. *Source*: Infineon Technologies.

FIELD-EFFECT TRANSISTORS 51

$I_D = 30$ mA $U_D = 3.5$ V $Z_o = 50\ \Omega$

f	S_{11}		S_{21}		S_{12}		S_{22}	
GHz	Mag	Ang	Mag	Ang	Mag	Ang	Mag	Ang
0.1	1.00	-2	3.23	178	0.002	85	0.71	-1
0.4	1.00	-8	3.21	171	0.009	79	0.70	-6
0.8	0.99	-16	3.19	162	0.017	73	0.69	-11
1.2	0.97	-24	3.18	153	0.025	70	0.67	-16
1.6	0.95	-32	3.17	143	0.034	65	0.66	-21
2.0	0.92	-40	3.17	135	0.042	61	0.65	-26
2.4	0.90	-48	3.17	127	0.051	56	0.63	-31
2.8	0.87	-58	3.17	119	0.059	50	0.61	-36
3.2	0.83	-68	3.16	109	0.067	45	0.58	-42
3.6	0.79	-79	3.12	99	0.073	40	0.55	-48
4.0	0.75	-91	3.08	88	0.079	34	0.52	-54
4.4	0.71	-102	3.04	78	0.084	28	0.50	-60
4.8	0.67	-114	3.00	68	0.089	21	0.47	-66
5.2	0.63	-126	2.95	58	0.092	15	0.43	-73
5.6	0.60	-138	2.87	49	0.094	10	0.38	-81
6.0	0.57	-150	2.77	40	0.096	4	0.34	-89
6.4	0.54	-162	2.68	31	0.097	-1	0.30	-99
6.8	0.52	-174	2.58	22	0.098	-6	0.27	-109
7.2	0.51	173	2.50	14	0.099	-11	0.24	-121
7.6	0.50	160	2.43	5	0.099	-16	0.21	-134
8.0	0.50	147	2.36	-4	0.099	-20	0.18	-148
8.4	0.51	135	2.26	-13	0.099	-24	0.16	-164
8.8	0.52	125	2.15	-22	0.099	-29	0.16	176
9.2	0.54	115	2.04	-30	0.099	-33	0.17	158
9.6	0.55	107	1.93	-39	0.099	-37	0.19	142
10.0	0.57	99	1.82	-47	0.099	-41	0.22	128
10.4	0.59	91	1.71	-54	0.100	-44	0.25	118
10.8	0.60	85	1.60	-62	0.101	-47	0.27	109
11.2	0.61	79	1.51	-69	0.102	-49	0.30	100
11.6	0.62	73	1.44	-75	0.103	-52	0.32	92
12.0	0.62	68	1.38	-82	0.104	-55	0.34	85

Figure 3-18 Typical common source S-parameters of the CFY30.

recommended by Pucel. It turns out that these values are scalable. The Pospieszalski model is a subset of this approach, as it uses only the case for $S_{12} = 0$ [38–49].

3.3.2 MOS Transistors (BiCMOS)

Modern RF and microwave integrated circuits are based increasingly on MOS technology. The reasons are low cost, low power consumption, and higher integration

TABLE 3-2 SPICE Parameters CFY30

```
MODEL CFY30 FET (
+IDSS = .4425-0I      VPO  = -.1413E+01    GAMA = -.7250E-01    E    =  .1581E+01
+KE   = -.1169E-01    SL   =  .1399E+00    KG   =  .3045E+00    MGS  =  .5000E+00
+MGD  =  .5000E+00    FCC  =  .8000E+00    T    =  .5234E-11    SS   =  .4856E-03
+IG0  =  .9350E-12    AFAG =  .1790E+02    IBO  =  .1270E-06    AFAE =  .1120E+01
+VBC  = -.9000E-01    R10  =  .1229E+02    KR   = -.9807E-01    C10  =  .3583E-12
+K1   =  .8877E+00    C1S  =  .0000E+00    CFO  =  .2034E-13    KF   =  .5000E+00
+RG   =  .6887E+01    RD   =  .8653E+01    RS   =  .4053E+00    LG   =  .8197E-09
+LD   =  .8812E-09    LS   =  .7181E-10    CDS  =  .1129E-12    CDSD =  .1000E-04
+RDSD =  .2967E+03    CGE  =  .2257E-13    CDE  =  .4703E-13    CGSP =  .5359E-13
+CDSP =  .2827E-13    ZGT  =  .5000E+02    LGT  =  .2317E-02    ZDT  =  .5000E+02
+LDT  =  .2389E-02    CGDP =  .8908E-17    LST  =  .5000E+02    LST  =  .2606E-03
+CGDE =  .2987E-13    VDMX = 10 )
```

FIELD-EFFECT TRANSISTORS 53

Output characteristics $I_D = f(V_{DS})$

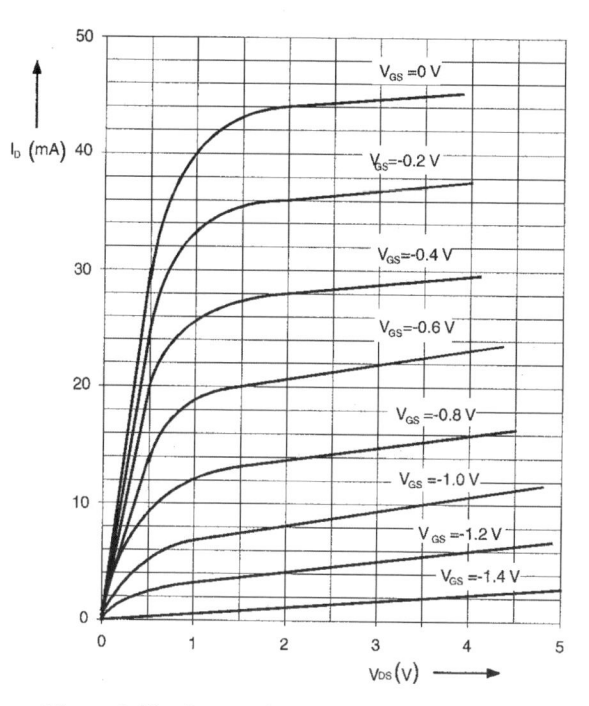

Figure 3-19 Output characteristics of the CFY30.

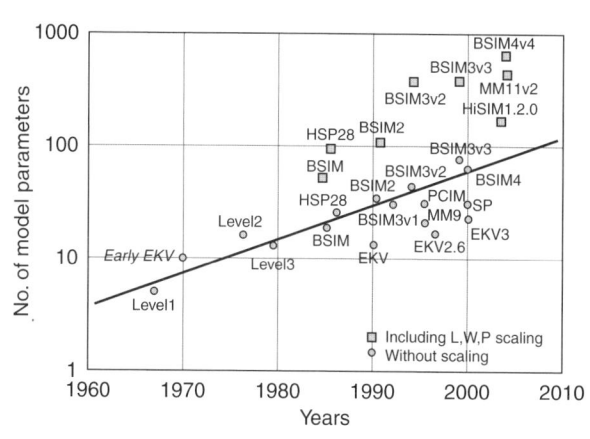

- Number of DC model parameters vs. the year of the introduction of the model
  Most recent versions of the EKV, HiSIM, MM11 and SP models are included
- Significant growth of the parameter number that includes geometry (W/L) scaling

Figure 3-20 Overview of the MOS models developed since 1960 and the number of required parameters [180]. From C. Enz, MOS Transistor Modeling for RF IC Design, June 2004.

density. Similarly, as with the bipolar transistor and GaAs FET transistors, the transistors are being described using a model and model parameters [50].

For larger transistors, Level 1, 2, and 3 MOS models, as implemented in the simulators, have been used. For submicron technology, the MOS BISIM Model 3v3 is currently the industry standard. For microwave and RF application, a variety of other models, which are more or less complex, have been developed. Figure 3-20 shows an overview of the currently available MOS models; the EKV3 model looks extremely attractive. The abbreviation EKV is derived from the names of the authors: Enz, Krummenacher, and Vittoz [180–182]. The model extractors for this are Aurora, IC-CAP, and UTMOS.

One of the difficulties in using MOS transistors is that small devices are only available as part of an IC design. For the purpose of evaluating the RF performance of these models, Motorola produced a small lateral defused MOS (LDMOS) device for us. For validation purposes, an RF power amplifier was built with these devices, in addition to comparing DC/IV curves and S-parameters. The other difficulty is that foundries make their manuals available only after signing a nondisclosure

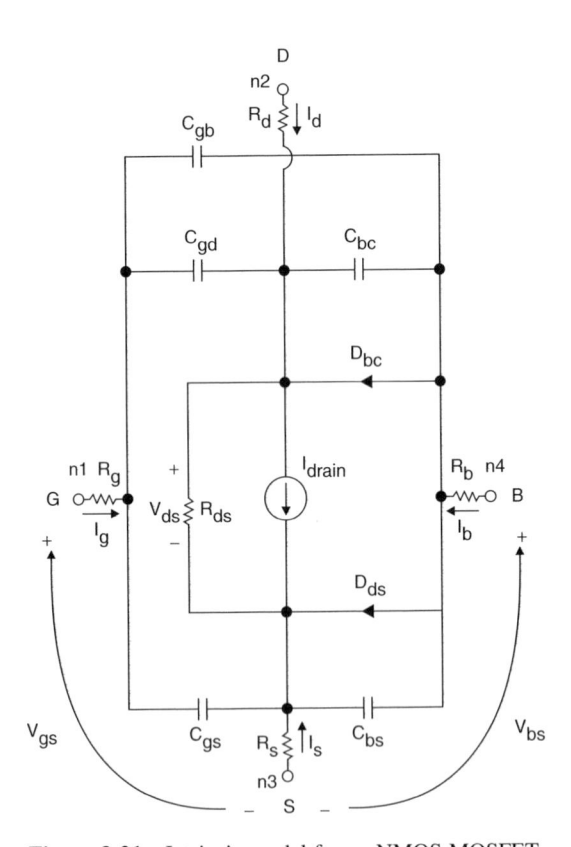

Figure 3-21 Intrinsic model for an NMOS MOSFET.

agreement, which prevents us from documenting some of the other data. These manuals can cost up to $5,000.

The model shown in Figure 3-21 is the popular MOSFET model. It has been implemented in this configuration and in most of the simulators (ADS from Agilent, Designer from Ansoft, and Microwave Office from AWR, to name a few). Addressing the model Level 3, Section 3.3.3 gives the parameters extracted for the LDMOS device, which works up to 3 GHz and was used for a 2 GHz oscillator.

3.3.3 MOS Model Level 3

L = 1 um W = 1.5 mm CBD = 0.863E-12 CGD0 = 166E-12 CGS0 = 246E-12
GAMA = 0.211 + IS = 6.53E-16 KAPA = 0.809 MJ = 0.536 NSUB = 1E15
PB = 0.71 PBSW = 0.71 + PHI = 0.579 RD = 39 RS = 0.1 THET = 0.588
TOX = 4E-8 U0 = 835 VMAX = 3.38E5 + VT0 = 2.78 XQC = 0.41.

Let Δf be the bandwidth (normalized to 1 Hz). The noise generators introduced in the intrinsic device are shown below and have mean square values of:

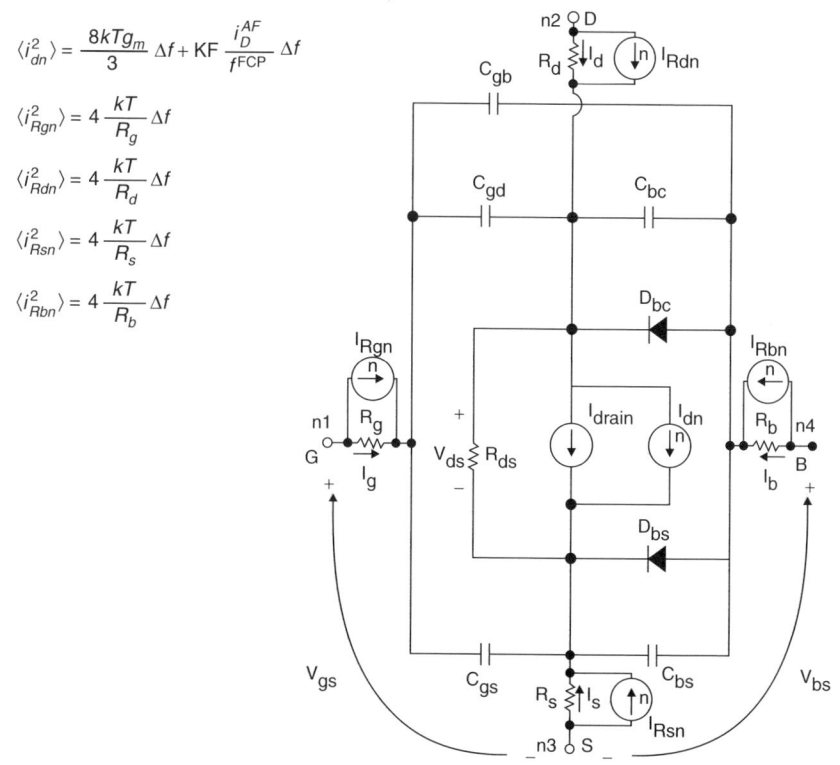

$$\langle i_{dn}^2 \rangle = \frac{8kTg_m}{3} \Delta f + KF \frac{i_D^{AF}}{f^{FCP}} \Delta f$$

$$\langle i_{Rgn}^2 \rangle = 4 \frac{kT}{R_g} \Delta f$$

$$\langle i_{Rdn}^2 \rangle = 4 \frac{kT}{R_d} \Delta f$$

$$\langle i_{Rsn}^2 \rangle = 4 \frac{kT}{R_s} \Delta f$$

$$\langle i_{Rbn}^2 \rangle = 4 \frac{kT}{R_b} \Delta f$$

Figure 3-22 Noise model of the MOSFET transistor (not showing extrinsic parasitics). Current sources with "n" are noise sources.

56   TRANSISTOR MODELS

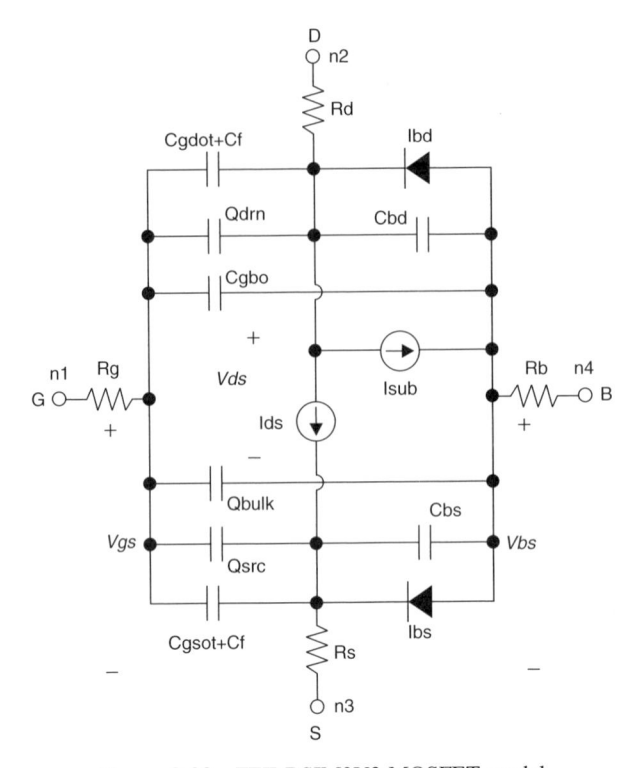

Figure 3-23   FET-BSIM3V3 MOSFET model.

These data were obtained for the Motorola LDMOS device.

The noise calculation for the Level 3 model is described in Figure 3-22.

For the RFIC application, the BISIM 3V3 is the model of choice [49–51]. Figures 3-23 and 3-24 show the MOSFET model representation.

The next more advanced foundry process is the 0.12 μm or even the 0.09 μm technology. The drawback of these technologies is that the breakdown voltages are now very small, and the actual voltage swing across the transistor is limited to 1 V or less.

3.4 TUNING DIODES

Varactors or voltage-controlled capacitors are used as tuning elements (as variable capacitors) for the (wideband) VCO circuit and are mostly the Q factor limitation of the resonating network. Typically, the inductor is the second element that determines the Q. Tuning diodes are either abrupt or hyperabrupt [195–205]. The abrupt junction is made with a linearly doped PN junction, typically has a

Let Δf be the bandwidth (normalized to 1 Hz). The noise generators introduced in the intrinsic device are shown below and have mean square values of:

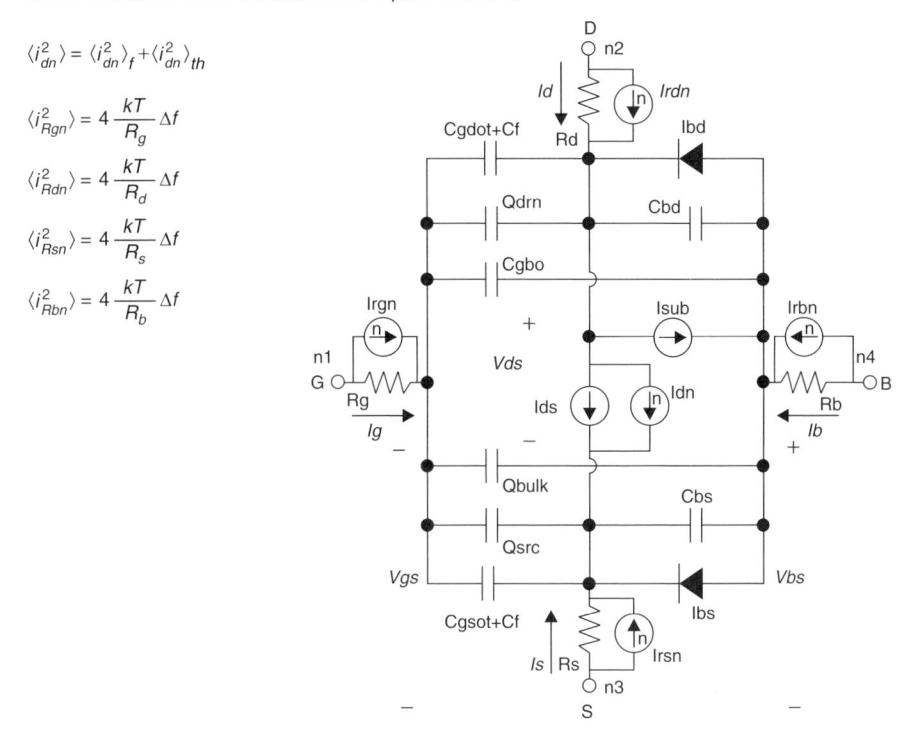

$$\langle i_{dn}^2 \rangle = \langle i_{dn}^2 \rangle_f + \langle i_{dn}^2 \rangle_{th}$$

$$\langle i_{Rgn}^2 \rangle = 4\frac{kT}{R_g}\Delta f$$

$$\langle i_{Rdn}^2 \rangle = 4\frac{kT}{R_d}\Delta f$$

$$\langle i_{Rsn}^2 \rangle = 4\frac{kT}{R_s}\Delta f$$

$$\langle i_{Rbn}^2 \rangle = 4\frac{kT}{R_b}\Delta f$$

Figure 3-24 Noise model of the MOSFET transistor (not showing extrinsic parasitics). Current sources with "n" are noise sources.

capacitance change of 4:1 or less over the specified range of reverse bias, and is available with maximum reverse bias voltages between 5 V and 60 V. The higher-voltage devices are advantageous, as they lower the voltage gain (MHz/V) of the oscillation, but they require a large supply voltage.

The hyperabrupt junction has a nonlinear-doped PN junction that increases the capacitance change versus reverse bias on the order of 10:1 or more. The disadvantages of a hyperabrupt junction are its higher series resistance and its lower value of Q. The equivalent model of a tuning diode, as shown in Figure 3-25, consists of a lead inductor, a shunt capacitor, a semiconductor PN junction, which is shown as a capacitor, and a loss resistor. Tuning diodes or voltage variable capacitors are manufactured in Si or GaAs technology; the GaAs process offers lower capacitance for the same resistance due to the higher electron mobility than the Si process. Hence, the Q of a GaAs varactor is better than that of a silicon diode, but the flicker noise of a varactor made of GaAs is high, and therefore the phase noise deteriorates.

TABLE 3-3 Extracted Parameters for a 0.25 μm Technology Used for RFICs

```
*NMOS model (0.25 μm, typical case)
.model nmos_fin nmos
+ level = 7                  + dvt0 = 2.55099             + pscbe2 = 3.71537e-08
+ mobmod = 2                 + dvt1 = 0.675305            + pvag = 2.59543
+ capmod = 2                 + dvt2 = -0.214288           + lint = 2.246420e-08
+ tox = 5.860790e-09         + dvt0w = 0                  + wr = 0.901078
+ cdsc = -0.001              + dvt1w = 5.3e+06            + wint = 1.042471e-08
+ cdscb = -0.000479266       + dvt2w = -0.032             + dwg = -1.270008e-09
+ cdscd = 6.24328e-07        + drout = 2.46755e-21        + dwb = -2.32044e-09
+ cit = 0                    + dsub = 0.823533            + b0 = 4.82337e-08
+ nfactor = 1.0568           + vth0 = 6.192220e-01        + b1 = 1e-07
+ xj = 1.8e-07               + ua = 0                     + alpha0 = 4.63789e-07
+ vsat = 82740               + ua1 = -9.86571e-10         + beta0 = 16.9841
+ at = 30422.1               + ub = 4.26165e-18           + cgsl = 2.01914e-10
+ a0 = 0.728222              + ub1 = -4.21823e-18         + cgdl = 2.01914e-10
+ ags = 0.340953             + uc = -6.28585e-10          + ckappa = 0.6
+ a1 = 0                     + uc1 = -4.7773e-11          + cf = 1.183270e-10
+ a2 = 1                     + u0 = 392.443               + clc = 0
+ keta = 0                   + ute = -2.15213             + cle = 0.6
+ nch = 3.96176e+17          + voff = -0.0993023          + dwc = 1.85529e-08
+ k1 = 0.91929               + delta = 0                  + dlc = 1.90182e-08
+ kt1 = -0.288637            + rsh = 6.000000e+00         + cgso = 4.100000e-11
+ kt1l = 7.2664e-09          + rdsw = 93.0861             + cgdo = 4.100000e-11
+ kt2 = -0.0405178           + prwg = 0.135659            + xpart = 0.5
+ k2 = -0.155928             + prwb = -0.517536           + js = 2.129500e-08
+ k3 = -1.30218              + eta0 = 0                   + pb = 7.048520e-01
+ k3b = 0.96651              + etab = -0.110514           + mj = 3.945980e-01
+ w0 = 0                     + pclm = 6.74271             + pbsw = 4.402380e-01
+ nlx = 1e-08                + pdiblc1 = 0                + mjsw = 2.355570e-01
                             + pdiblc2 = 0.01             + cj = 1.086670e-03
                             + pdiblcb = 0                + cjsw = 4.774750e-11
                             + pscbe1 = 1e+08
```

Note: The parameters can be extracted by using IC-CAP from Agilent.

3.4 TUNING DIODES   59

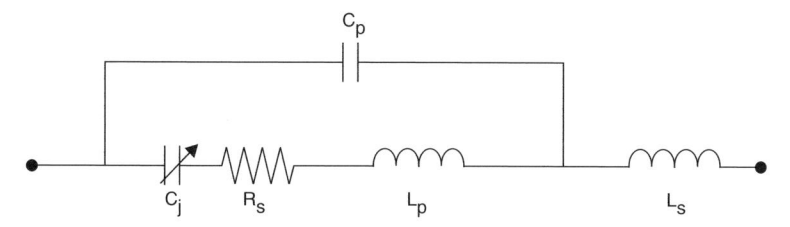

Figure 3-25 Equivalent circuit of a tuning diode.

TABLE 3-4 SPICE and Package Parameters of the Toshiba 1SV280 Varactor Diode

Parameters*	1SV280	Parameters*	1SV280	Package	1SV280
IS	5.381E−16	GC1	−1.669E−3	L_S	5E−10
N	1.037	GC2	1.303E4	L_p	1E−9
BV	15	GC3	−9.742E−6	C_p	8E−13
IBV	1E−6	V_J	3.272		
R_S	0.44	M	0.9812		
CJ0	6.89E−12	IMAX	10E−3		

*Berkley-SPICE parameters.

For this work, the Si-abrupt tuning diode, 1SV280 from Toshiba, is chosen. This diode has a very low series resistance, is low package parasitic, and a tuning range of 1 to 15 V. Table 3-4 shows the SPICE parameters of the 1SV280 for the purpose of further characterizing the tuning diode. Figures 3-26a and 3-26b show the measured plots of series resistance R_S and capacitance versus reverse bias voltage. Figure 3-27 shows the plot of the Q versus the reverse voltage for frequencies of 2, 3.5, and 4.5 GHz.

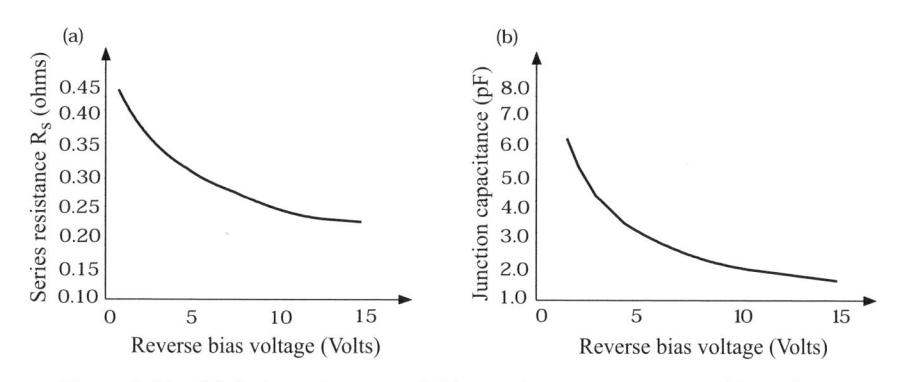

Figure 3-26 (*a*) Series resistance and (*b*) capacitance versus reverse bias voltage.

60   TRANSISTOR MODELS

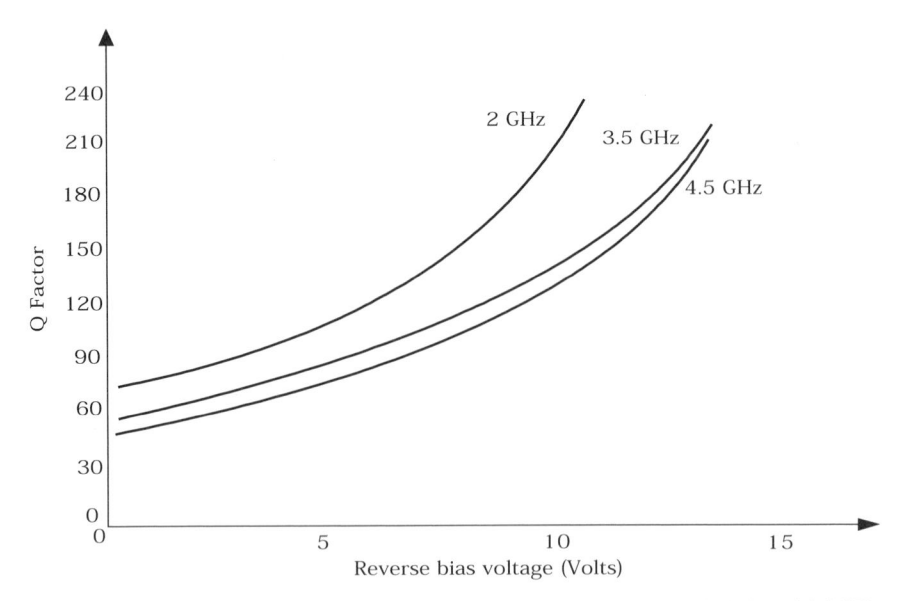

Figure 3-27 Q factor versus reverse voltage for the Toshiba 1SV280 at 2, 3.5, and 4.5 GHz.

From Figures 3-26 and 3-27, it can be seen that the tuning diode shows a lower Q factor at low reverse bias voltages. The capacitance and resistance are then at maximum values.

4 Large-Signal S-Parameters

The description of linear, active, or passive two-ports can be explained in various forms. In the early days, Z-parameters were commonly used; they were then replaced by the Y-parameters. Z-parameters are assumed to be open-ended measurements, and Y-parameters are short-circuit measurements relative to the output or input, depending on the parameter. In reality, however, the open circuit condition does not work at high frequencies because it becomes capacitive and results in erroneous measurements. The short-circuit measurements also suffer from nonideal conditions, as most "shorts" become inductive. Most RF and microwave circuits, because of the availability of $50\,\Omega$ coaxial cables, are now using $50\,\Omega$ impedances. Component manufacturers are able to produce $50\,\Omega$ termination resistors which maintain their $50\,\Omega$ real impedance up to tens of gigahertz (40 GHz). The $50\,\Omega$ system has become a de facto standard. While the Z- and Y-parameter measurements were based on voltage and currents at the input and output, the S-parameters refer to forward and reflected power.

4.1 DEFINITION

For low-frequency applications, one can safely assume that the connecting cable from the source to the device under test or from the device under test to the load plays no significant role. The wavelength of the signal at the input and output is very large compared to the physical length of the cable. At higher frequencies, such as microwave frequencies, this is no longer true. Therefore, a measuring principle was established to consider the incoming and outgoing power at the input and output. The following is a mathematical explanation of the S-parameters. It follows the definitions of [14] as outlined by Hewlett-Packard.

$$b_1 = S_{11}a_1 + S_{12}a_2 \qquad (4\text{-}1)$$

$$b_2 = S_{21}a_1 + S_{22}a_2 \qquad (4\text{-}2)$$

The Design of Modern Microwave Oscillators for Wireless Applications: Theory and Optimization, by Ulrich L. Rohde, Ajay Kumar Poddar, Georg Böck
Copyright © 2005 John Wiley & Sons, Inc.

or, in matrix form,

$$\begin{bmatrix} b_1 \\ b_2 \end{bmatrix} = \begin{bmatrix} S_{11} & S_{12} \\ S_{21} & S_{22} \end{bmatrix} \begin{bmatrix} a_1 \\ a_2 \end{bmatrix} \quad (4\text{-}3)$$

where, referring to Figure 2-78:

a_1 = (incoming signal at Port 1)
b_1 = (outgoing signal at Port 1)
a_2 = (incoming signal at Port 2)
b_2 = (outgoing signal at Port 2)
E_1, E_2 = electrical stimuli at Port 1, Port 2

From Figure 4-1 and defining linear equations, for $E_2 = 0$, then $a_2 = 0$, and (skipping through numerous rigorous steps):

$$|S_{11}| = \frac{b_1}{a_1}$$

$$= \left[\frac{\text{outgoing input power}}{\text{incoming input power}} \right]^{1/2} \quad (4\text{-}4)$$

$$= \frac{\text{reflected voltage}}{\text{incident voltage}}$$

$$= \text{input reflection coefficient}$$

$$|S_{21}| = \frac{b_2}{a_1}$$

$$= \left[\frac{\text{outgoing output power}}{\text{incoming input power}} \right]^{1/2} \quad (4\text{-}5)$$

$$= \left[\frac{\text{output power}}{\text{available input power}} \right]^{1/2}$$

$$= [\text{foward transducer gain}]^{1/2}$$

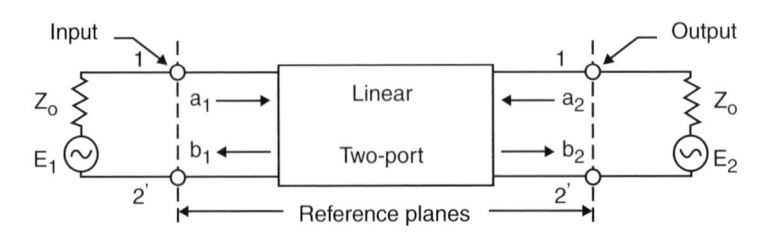

Figure 4-1 Two-port S-parameter definition.

or, more precisely in the case of S_{21}:

$$\text{Forward transducer gain} = G_{TF} = |S_{21}|^2 \tag{4-6}$$

$$Z_i = Z_o \tag{4-7}$$

Similarly at Port 2 for $E_1 = 0$; then $a_1 = 0$ and

$$S_{22} = \frac{b_2}{a_2} = \text{output reflection coefficient} \tag{4-8}$$

$$S_{12} = \frac{b_1}{a_2} = [\text{reverse transducer gain}]^{1/2} \tag{4-9}$$

$$G_{TR} = |S_{12}|^2 \tag{4-10}$$

Since many measurement systems display S-parameter magnitudes in decibels, the following relationships are particularly useful [54–57]:

$$\begin{aligned}|S_{11}|_{dB} &= 10 \log|S_{11}|^2 \\ &= 20 \log|S_{11}| \end{aligned} \tag{4-11}$$

$$|S_{22}|_{dB} = 20 \log|S_{22}| \tag{4-12}$$

$$\begin{aligned}|S_{21}|_{dB} &= 10 \log|S_{21}|^2 \\ &= 20 \log|S_{21}| \\ &= 10 \log|G_{TF}| = |G_{TF}|_{dB} \end{aligned} \tag{4-13}$$

$$\begin{aligned}|S_{12}|_{dB} &= 10 \log|S_{12}|^2 \\ &= 20 \log|S_{12}| \\ &= 10 \log|G_{TR}| = |G_{TR}|_{dB} \end{aligned} \tag{4-14}$$

4.2 LARGE-SIGNAL S-PARAMETER MEASUREMENTS

Assume that S_{11} and S_{21} are functions only of incident power at Port 1 and S_{22} and S_{12} are functions only of incident power at Port 2. Note: the plus (+) sign indicates the forward wave (voltage) and the minus (−) sign refers to the reflected wave (voltage).

$$S_{11} = S_{11}(|V_1^+|) \qquad S_{12} = S_{12}(|V_2^+|) \tag{4-15}$$

$$S_{21} = S_{21}(|V_1^+|) \qquad S_{22} = S_{22}(|V_2^+|) \tag{4-16}$$

The relationship between the traveling waves now becomes

$$V_1^- = S_{11}(V_1^+)V_1^+ + S_{12}(V_2^+)V_2^+ \qquad (4\text{-}17)$$
$$V_2^- = S_{21}(V_1^+)V_1^+ + S_{22}(V_2^+)V_2^+ \qquad (4\text{-}18)$$

Measurement is possible if V_1^+ is set to zero:

$$S_{12}(V_2^+) = \frac{V_1^-(V_2^+)}{V_2^+} \qquad (4\text{-}19)$$

Check the assumption by simultaneous application of V_1^+ and V_2^+:

$$\begin{bmatrix} V_1^- \\ V_2^- \end{bmatrix} = \begin{bmatrix} F_1(V_1^+, V_2^+) \\ F_2(V_1^+, V_2^+) \end{bmatrix} \qquad (4\text{-}20)$$

If harmonics are neglected, a general decomposition is

$$\begin{bmatrix} V_1^-(V_1^+, V_2^+) \\ V_2^-(V_1^+, V_2^+) \end{bmatrix} = \begin{bmatrix} S_{11}(V_1^+, V_2^+) & S_{12}(V_1^+, V_2^+) \\ S_{21}(V_1^+, V_2^+) & S_{22}(V_1^+, V_2^+) \end{bmatrix} \begin{bmatrix} V_1^+ \\ V_2^+ \end{bmatrix} \qquad (4\text{-}21)$$

As mentioned before, these measurements are initially done under small-signal conditions, with the RF power increasing up to 0 dBm or larger as needed. Small-signal conditions mean power levels of approximately −40 dBm. The network analyzers used to measure these S-parameters have bias tees built in and have 90 dB dynamic range. For the measurement of S-parameters of transistors, a much smaller dynamic range is sufficient.

Figure 4-2 Test fixture to measure large-signal S-parameters. A proper de-embedding has been done [27].

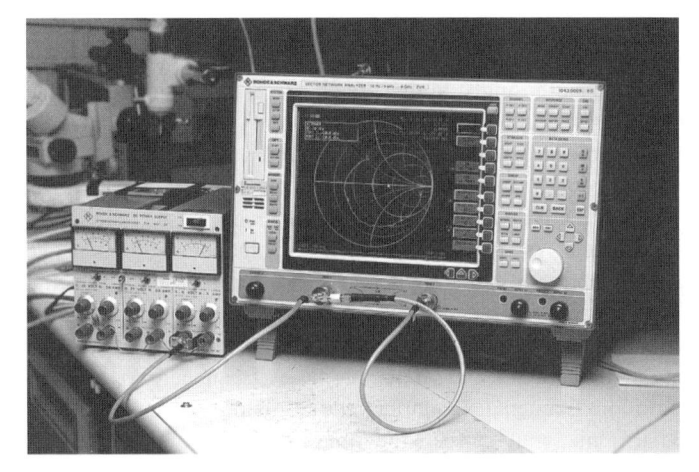

Figure 4-3 Rohde & Schwarz 3 GHz network analyzer used to measure large-signal S-parameters at different drive levels.

If the output signal from the signal generator is increased in power, it essentially has no impact on passive devices until a level of several hundred watts is reached, where intermodulation distortion products can be created due to dissimilar alloys. However, active devices, depending on the DC bias point, can only tolerate relatively low RF levels to remain in the linear region.

In the case of the oscillator, there is a large RF signal, voltage and current, imposed on the DC voltage/current. Assuming an RF output power from 0 to 10 dBm, and assuming 10–15 dB gain in the transistor, the RF power level driving the emitter/source or base/gate terminal is somewhere in the vicinity of −15 dBm.

An RF drive of −15 dBm will change the input and output impedance of the transistor even if the transistor operates at fairly large DC currents.

It is important to note that the input and output impedances of FETs are much less RF voltage-dependent or power-dependent than those of the bipolar transistor. The generation of large-signal S-parameters is, therefore, much more important for bipolar transistors than for FETs.

Figure 4-2 shows the test fixture used to measure the large-signal S-parameters for the device under test (DUT). The test fixture was calibrated to provide 50 Ω to the transistor leads. The test setup shown in Figure 4-3 consists of a DC power supply and a network analyzer for combined S-parameter measurements. The Rohde & Schwarz ZVR network analyzer, as shown in Figure 4-3, was chosen because its output power can be changed from +10 to −60 dBm. This feature is necessary to perform these measurements. The actual test system shown in this figure is very simple but, unfortunately, very expensive.

Currents and voltages follow Kirchof's law in a linear system. A linear system implies that there is a linear relationship between currents and voltages. All transistors, when driven at higher levels, show nonlinear characteristics. The FET shows a square law characteristic, while the bipolar transistor has an exponential transfer characteristic. The definition of S-parameters in a large-signal environment is ambiguous compared to that of small-signal S-parameters. When an active device is driven with an increasingly higher level, the output current consists of a DC current and RF currents, the fundamental frequency, and its harmonics. When the drive level is increased, the harmonic content rapidly increases. S_{12}, mostly defined by the feedback capacitance, now reflects harmonics back to the input. If these measurements are done in a 50 Ω system, which has no reactive components, then we have an ideal system for termination. In practical applications, however, the output is a tuned circuit or matching network which is frequency selective. Depending on the type of circuit, it typically presents either a short-circuit or an open circuit for the harmonic. For example, suppose that the matching network has a resonant condition at the fundamental and second harmonic frequencies or at the fundamental and third harmonic frequencies (quarterwave resonator). Then a high voltage occurs at the third harmonic, which affects the input impedance and, therefore, S_{11} (Miller effect).

This indicates that S-parameters measured under large-signal conditions in an ideal 50 Ω system may not correctly predict device behavior when used in a non-50 Ω environment. A method called *load pulling*, which includes fundamental harmonics, has been developed to deal with this issue [58–62, 68–172].

TABLE 4-1   Frequency-Dependent S-Parameters (S-Parameters at −20 dBm)

Frequency	S_{11}		S_{12}		S_{21}		S_{22}	
1.000E+08	0.78	−17.15	29.57	−160.6	0.01	69.66	0.96	−7.63
1.500E+08	0.74	−19.95	30.87	−175.17	0.01	73.05	0.94	10.27
2.000E+08	0.71	−23.01	30.87	174.87	0.01	73.61	0.92	12.8
2.500E+08	0.69	−26.34	30.43	167.17	0.01	73.11	0.9	−15.25
3.000E+08	0.66	−29.8	29.8	160.76	0.01	72.13	0.87	−17.61
3.500E+08	0.64	−33.28	29.08	155.2	0.01	70.91	0.85	−19.92
4.000E+08	0.61	−36.73	28.3	150.22	0.01	69.59	0.83	−22.16
4.500E+08	0.59	−40.1	27.5	145.68	0.02	68.24	0.81	−24.33
5.000E+08	0.56	−43.36	26.68	141.5	0.02	66.91	0.78	−26.44
5.500E+08	0.53	−46.47	25.85	137.62	0.02	65.66	0.76	−28.44
6.000E+08	0.51	−49.42	25.02	134	0.02	64.51	0.73	−30.33
6.500E+08	0.48	−52.19	24.18	130.62	0.02	63.5	0.7	−32.07
7.000E+08	0.46	−54.78	23.35	127.46	0.02	62.63	0.68	−33.64
7.500E+08	0.44	−57.2	22.54	124.52	0.02	61.9	0.65	−35.04
8.000E+08	0.42	−59.44	21.74	121.76	0.02	61.3	0.63	−36.26
8.500E+08	0.39	−61.53	20.98	119.19	0.02	60.82	0.6	−37.31
9.000E+08	0.38	−63.48	20.24	116.77	0.03	60.43	0.58	−38.2
9.500E+08	0.36	−65.29	19.53	114.51	0.03	60.13	0.56	−38.95
1.000E+09	0.34	−66.99	18.85	112.38	0.03	59.88	0.54	−39.57
1.500E+09	0.22	−80.06	13.7	96.21	0.04	58.66	0.41	−41.5
2.000E+09	0.14	−91.02	10.61	85.03	0.04	57.04	0.33	−40.51
2.500E+09	0.09	−105.04	8.64	76	0.05	54.51	0.29	−39.1
3.000E+09	0.06	−129.69	7.27	68.07	0.06	51.33	0.25	−37.7

4.2 LARGE-SIGNAL S-PARAMETER MEASUREMENTS

In the case of an oscillator, however, there is only one high Q resonator which suppresses the harmonics of the fundamental frequency (short-circuit). In this limited case, the S-parameters stemming from a 50 Ω system are useful. The following tables show two sets of measurements generated from the Infineon transistor BFP520 at different drive levels.

Since the oscillator will be in quasi-large-signal operation, we will need the large-signal S-parameters as a starting condition for the large-signal design (output power, harmonics, and others). The S-parameters generated from this will be converted into Y-parameters, defined under large-signal conditions, and then used for calculating the large-signal behavior. We will use the symbol Y^+ to designate large-signal Y-parameters. Tables 4-1 and 4-2 show the large-signal S-parameters for -20 and -10 dBm. However, in some cases, the analysis starts at small-signal conditions. All derivations have been verified with MATHCAD, and the original text input has been used. Therefore, in some cases the Y^+ marker has *not* been used. The use of the MATHCAD equation set allows for error-free reuse of the equations.

The following four plots, Figures 4-4 to 4-7, show S_{11}, S_{12}, S_{21}, and S_{22} measured from 50 to 3000 MHz with driving levels from -20 to $+5$ dBm. The DC operation conditions were 2 V and 20 mA, as shown in Figure 4-2.

TABLE 4-2 Frequency-Dependent S-Parameters (S-Parameters at −10 dBm)

Frequency	S_{11}		S_{12}		S_{21}		S_{22}	
1.00E+08	0.81	−12.8	14.53	179.18	0.02	39.17	0.55	−20.62
1.50E+08	0.79	−14.26	14.51	170.01	0.02	51.38	0.6	−24.42
2.00E+08	0.77	−16.05	14.46	163.78	0.02	57.11	0.65	−27.11
2.50E+08	0.76	−17.94	14.4	158.86	0.03	60.47	0.67	−28.33
3.00E+08	0.74	−19.85	14.31	154.78	0.03	62.9	0.69	−28.28
3.50E+08	0.73	−21.74	14.21	151.32	0.03	64.83	0.7	−27.33
4.00E+08	0.72	−23.62	14.1	148.32	0.03	66.46	0.71	−25.99
4.50E+08	0.71	−25.51	13.99	145.65	0.03	67.72	0.73	−24.6
5.00E+08	0.7	−27.42	13.88	143.19	0.03	68.57	0.74	−23.39
5.50E+08	0.68	−29.37	13.76	140.87	0.03	68.99	0.76	−22.5
6.00E+08	0.67	−31.38	13.65	138.62	0.04	68.98	0.77	−21.93
6.50E+08	0.66	−33.45	13.54	136.4	0.04	68.59	0.77	−21.68
7.00E+08	0.64	−35.56	13.42	134.2	0.04	67.95	0.78	−21.68
7.50E+08	0.63	−37.71	13.31	132	0.04	67.2	0.78	−21.89
8.00E+08	0.61	−39.88	13.19	129.83	0.04	66.31	0.77	−22.25
8.50E+08	0.59	−42.06	13.07	127.7	0.04	65.37	0.77	−22.62
9.00E+08	0.58	−44.23	12.95	125.6	0.04	64.48	0.76	−23.26
9.50E+08	0.56	−46.4	12.82	123.57	0.04	63.69	0.76	−24.04
1.00E+09	0.54	−48.55	12.69	121.6	0.04	62.82	0.75	−24.71
1.50E+09	0.37	−70.76	11.35	104.37	0.05	52.76	0.67	−33.77
2.00E+09	0.21	−91.19	9.99	88.64	0.05	46.68	0.48	−43.79
2.50E+09	0.12	−107.22	8.43	77.36	0.06	49.37	0.33	−43.13
3.00E+09	0.07	−130.38	7.18	68.7	0.06	48.69	0.27	−40.46

68 LARGE-SIGNAL *S*-PARAMETERS

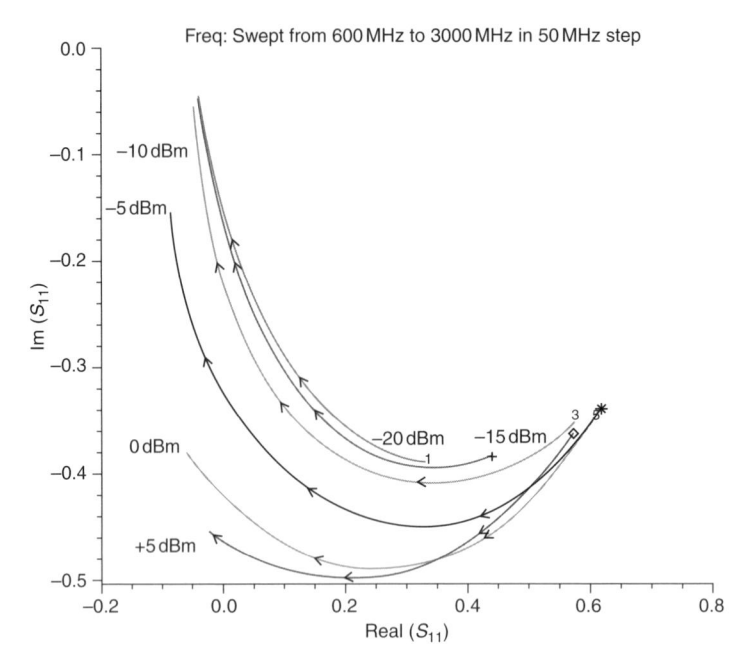

Figure 4-4 Measured large-signal S_{11} of the BFP520.

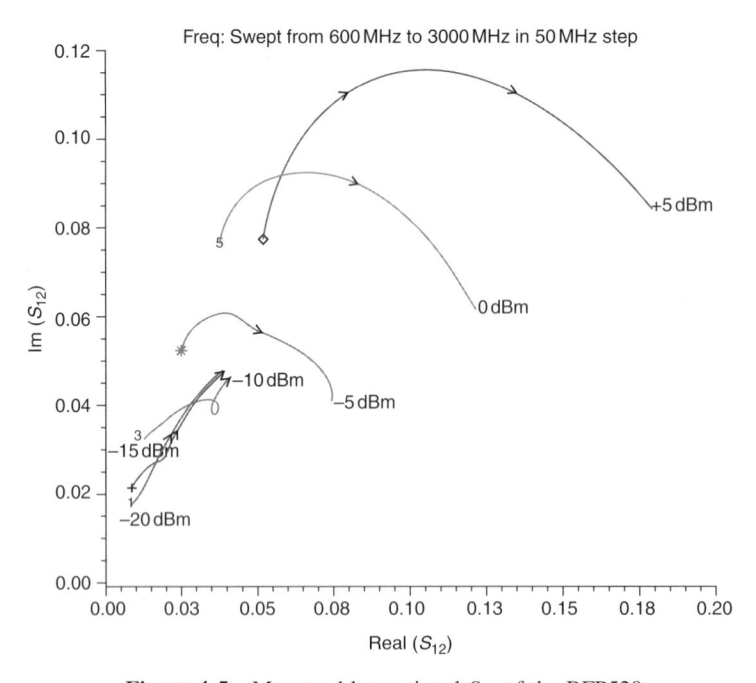

Figure 4-5 Measured large-signal S_{12} of the BFP520.

4.2 LARGE-SIGNAL S-PARAMETER MEASUREMENTS 69

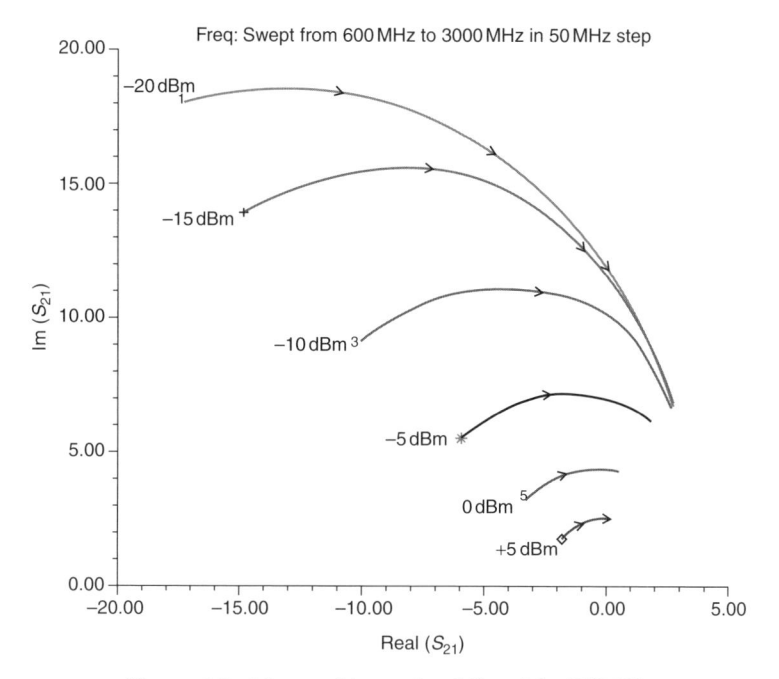

Figure 4-6 Measured large-signal S_{21} of the BFP520.

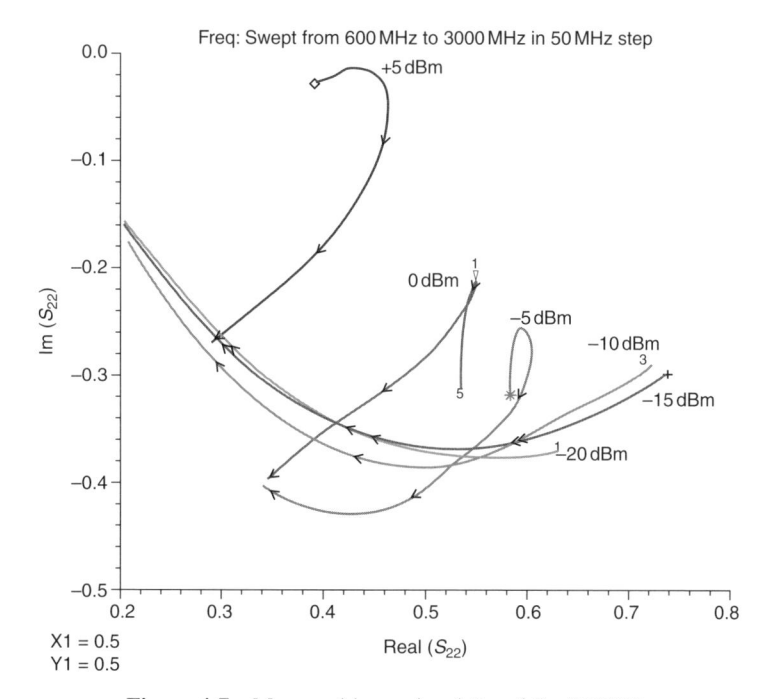

Figure 4-7 Measured large-signal S_{22} of the BFP520.

5 Resonator Choices

The following resonators are found in various microwave oscillators.

5.1 LC RESONATORS

Figure 5-1 shows the circuit diagram of the resonator; the coupling is represented by a lumped capacitor. The lumped resonator consists of a lossy 2 pF capacitor and a

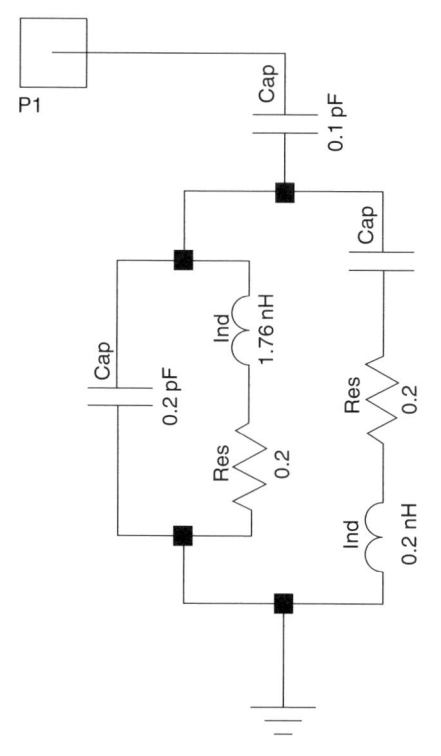

Figure 5-1 The circuit diagram of a parallel tuned circuit with lossy components and parasitics loosely coupled to the input.

The Design of Modern Microwave Oscillators for Wireless Applications: Theory and Optimization, by Ulrich L. Rohde, Ajay Kumar Poddar, Georg Böck
Copyright © 2005 John Wiley & Sons, Inc.

72 RESONATOR CHOICES

lossy 1.76 nH inductor with a 0.2 pF parasitic capacitor. The capacitor has a lead inductor of 0.2 nH and 0.2 Ω loss. The inductor has the same loss resistor. To measure the operating Q, this combination is attached to a network analyzer, which determines S_{11}. The operating Q is calculated by dividing the center frequency by the 3 dB bandwidth of S_{11}. The quality factor Q is defined as stored energy/dissipated energy. If there is no energy loss, Q is infinite. Figure 5-1 shows the circuit diagram of the resonator and the coupling.

To determine the operating Q of the circuit, let us calculate the Q of the individual branches representing the resonator. The total Q of the circuit can be calculated by combining the two individual Q values following the equation:

$$Q = \frac{Q_1 \times Q_2}{Q_1 + Q_2} \tag{5-1}$$

$$Q_1 = 2 \times \pi \times 2.4 \text{ GHz} \times 1.76 \text{ nH}/0.2\,\Omega = 133,\ Q_2 = 165.$$

$$Q = 73$$

The low Q is due to the 0.2 Ω loss resistor. It should be possible to reduce this by more than a factor of 2.

5.2 MICROSTRIP RESONATORS

5.2.1 Distributed/Lumped Resonators

The same parallel tuned circuit can be generated by using a printed transmission line instead of the lumped inductor, maintaining the same capacitance. This is shown in Figure 5-2. Since the transmission line has losses due to the material, these losses need to be considered. It is not practical to calculate these by hand; instead one should use a CAD program which does this accurately [115, 116]. These references describe how to derive the Q factor from S_{11} measurements.

The Q can be determined from the 3 dB bandwidth ($\Delta f/f_0$) shown in Figure 5-3 and was determined to be 240. This is also valid if the Y- or Z-parameters are used. This is a typical value for a microstrip resonator. Values up to 300 are possible if the appropriate layout and material are used.

5.2.2 Integrated Resonators

The same circuit can be generated not only using printed circuit board material, but also in GaAs or in silicon. Figure 5-4 shows the schematic of a parallel tuned circuit using a rectangular inductor and an interdigital capacitor. The ground connection is achieved using a via. At 2.4 GHz, the number of turns and the size of the inductor would be significant. The same applies to the capacitor. This arrangement should be reserved for much higher frequencies, that is, above 5 GHz. The inductor losses,

5.2 MICROSTRIP RESONATORS 73

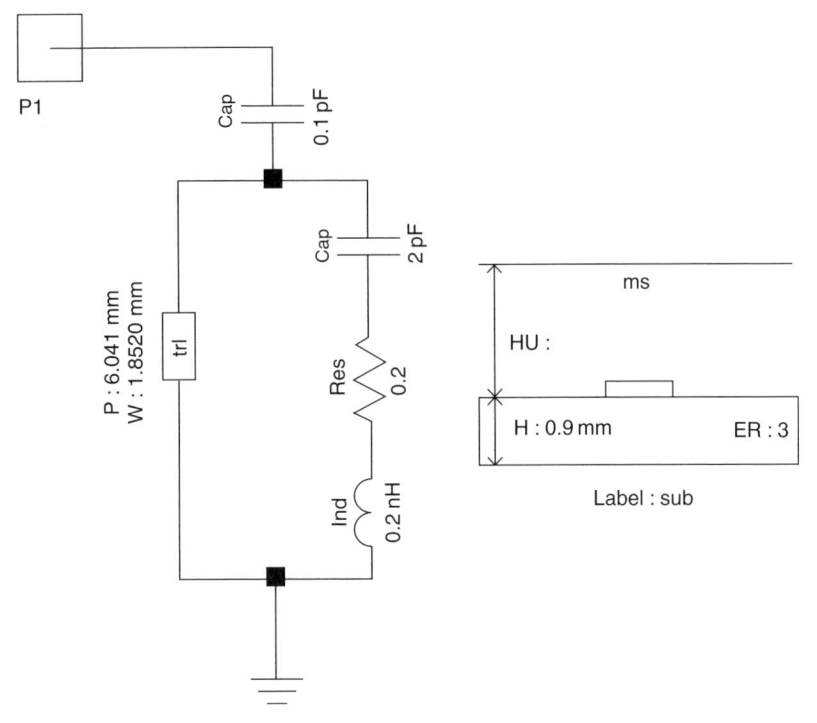

Figure 5-2 A 2.4 GHz resonator using both lumped and distributed components. The physical dimensions are given.

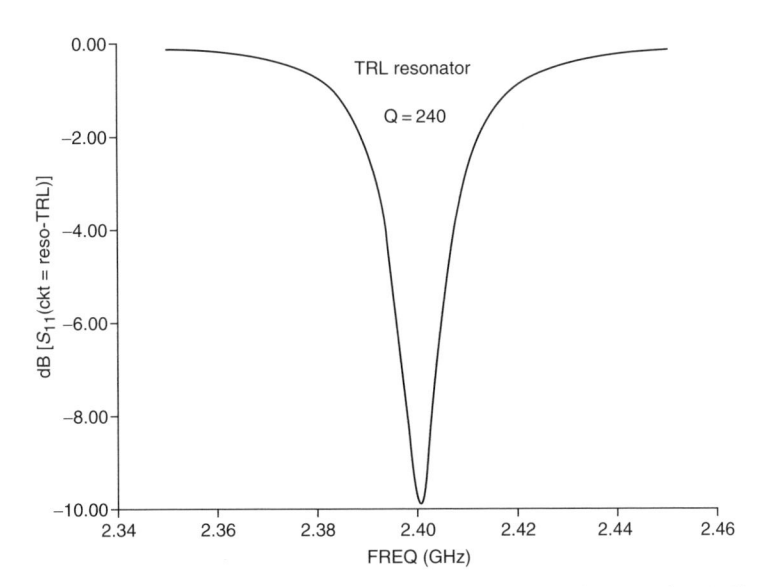

Figure 5-3 Simulated reflection coefficient S_{11} used to determine the operating Q. Since this material has fairly high losses, an operating Q of only 240 was achieved.

74 RESONATOR CHOICES

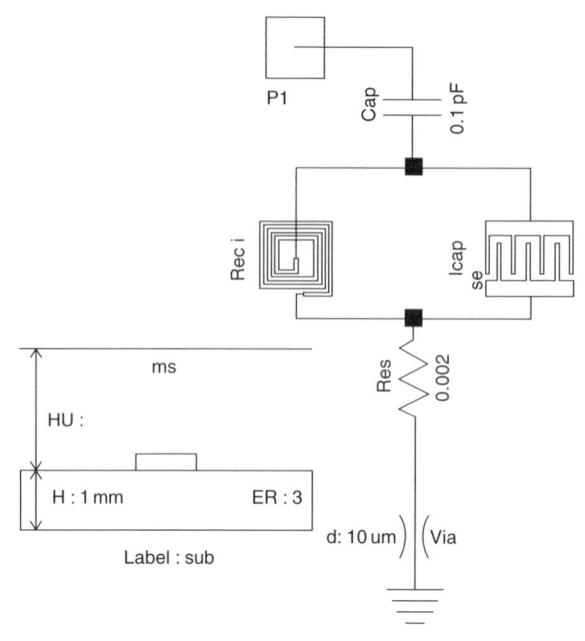

Figure 5-4 Parallel tuned circuit using a rectangular inductor (a spiral inductor could also be used) and an interdigital capacitor. If implemented on GaAs on silicon, it exhibits low Q.

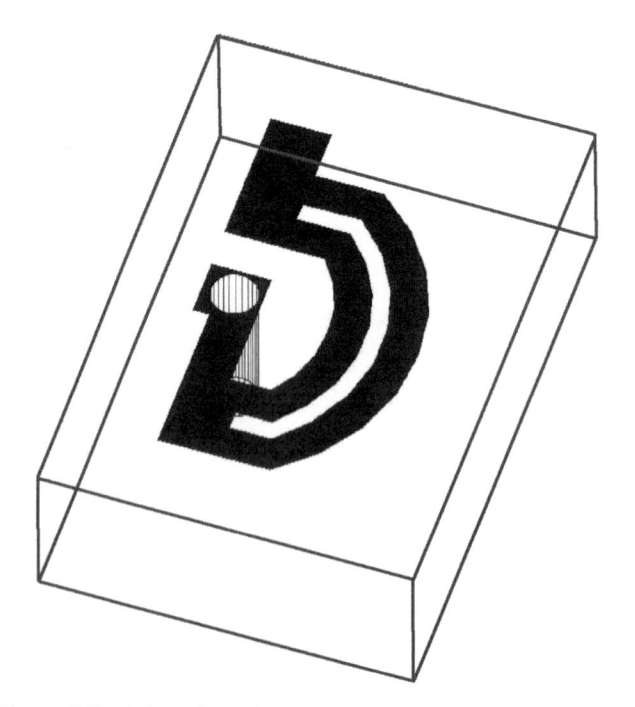

Figure 5-5 A 3-D view of the coupled resonator (patent pending).

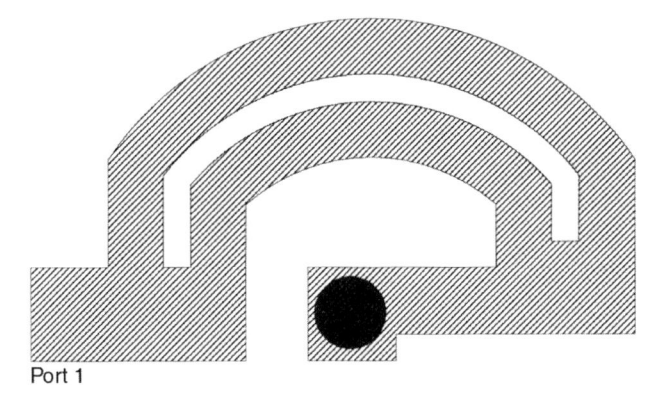

Figure 5-6 A 2-D view of the coupled microstrip resonator.

both in GaAs and silicon, are substantial, and this case is shown only for completeness. Where possible, an external resonator should be used.

Referring to integrated resonators, a high-Q resonator consisting of two coupled inductors has been developed by us. Figure 5-5 shows a three-dimensional array where the two coupled resonators are easily identifiable. One side of the resonator is connected to ground through a via.

Throughout the book, we are referring to coupled resonators for this arrangement, but purists could argue rightfully that these are coupled inductors. In reality, we are dealing with coupled microstrips which behave differently than coupled inductors.

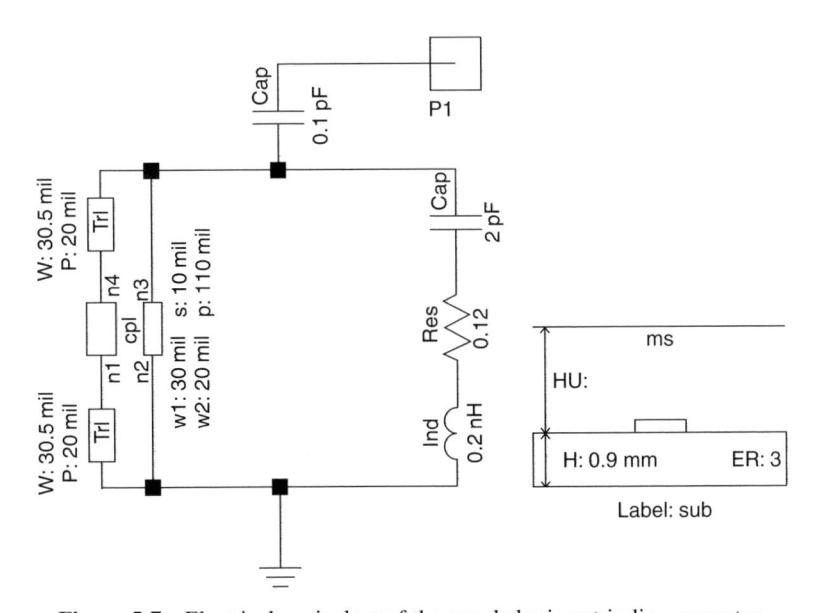

Figure 5-7 Electrical equivalent of the coupled microstrip line resonator.

76 RESONATOR CHOICES

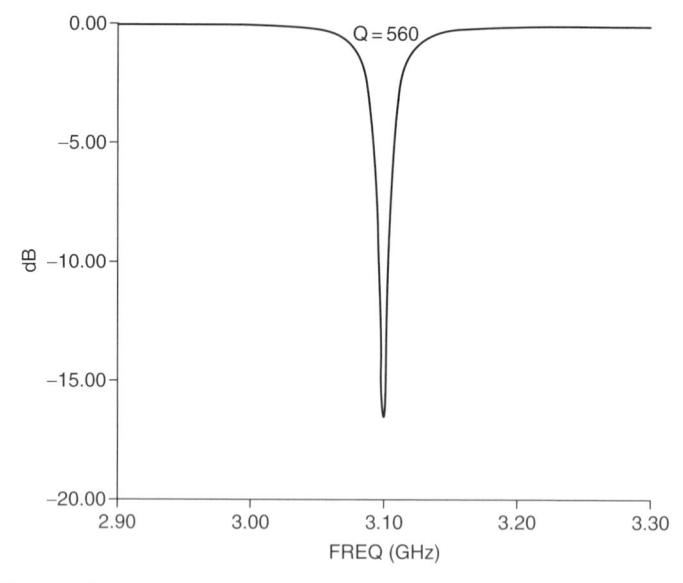

Figure 5-8 Frequency response of the coupled microstrip line resonator.

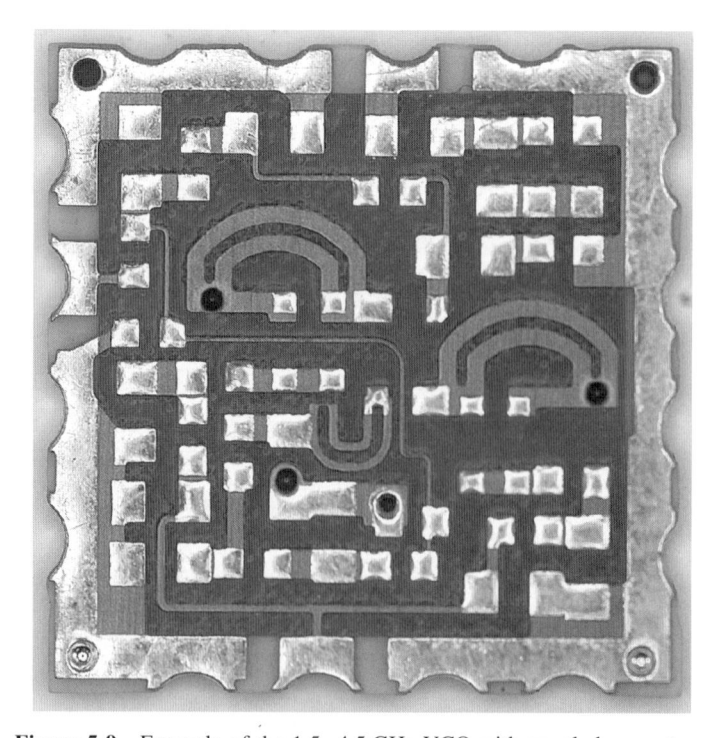

Figure 5-9 Example of the 1.5–4.5 GHz VCO with coupled resonators.

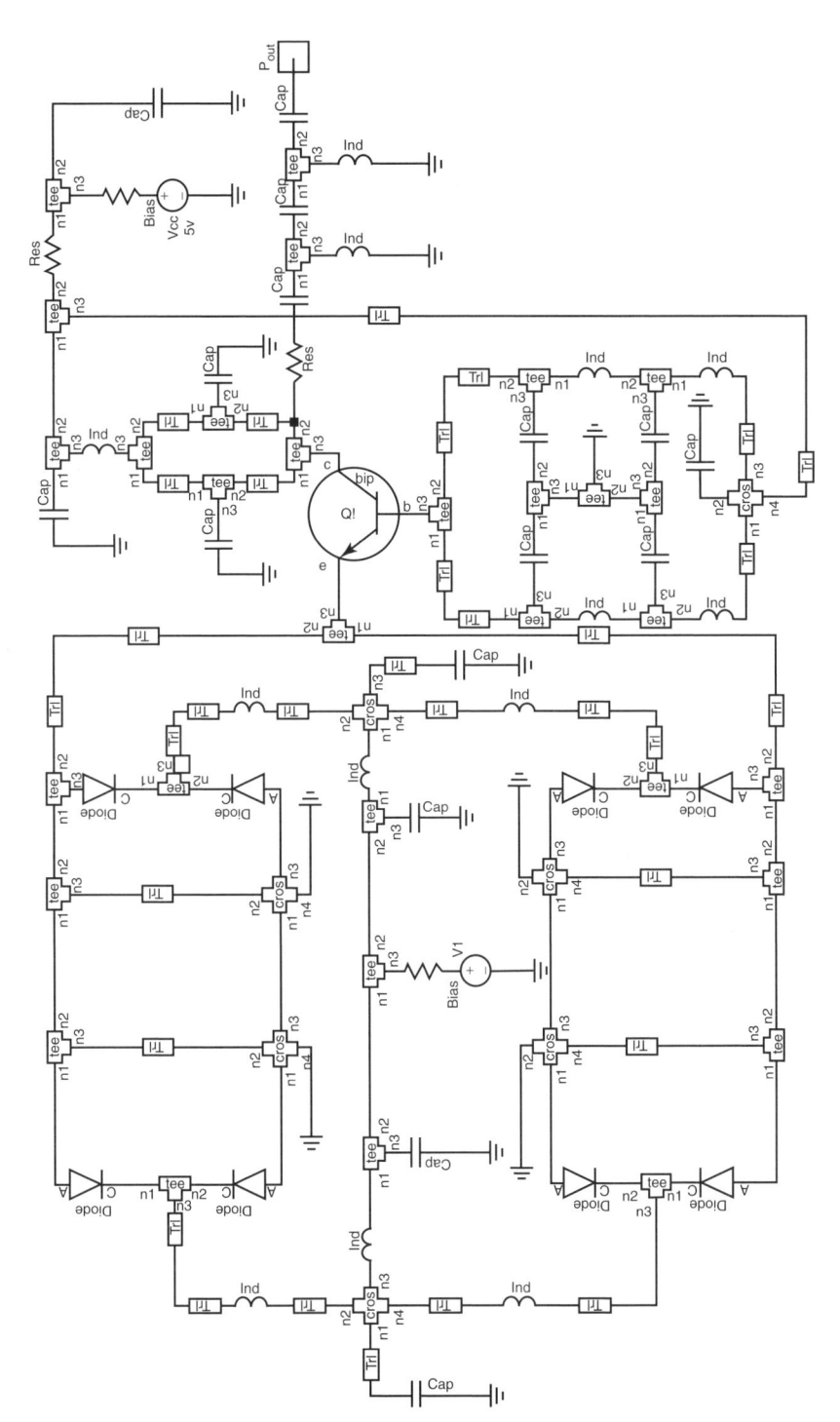

Figure 5-10 Schematic of the 1.5–4.5 GHz oscillator shown in Figure 5-9.

We have validated this experimentally and the exact explanation for this behavior would require a lengthy treatment. Figure 5-109 from [14] gives a detailed insight to this. The reason why it is frequently called coupled transmission lines or inductors is the fact that the elements themselves are not resonant and much shorter than quarterwave length. This is applicable for two or more coupled transmission lines. The resulting Q can be as much as 10 times as the Q of a single-tuned circuit. The rate of change of phase versus frequency in the configuration is much faster and is responsible for the higher Q.

The 3-D plot can be reduced to a 2-D plot, as shown in Figure 5-6, which gives further details about the resonator. The resonator analysis was done using Ansoft Designer, specifically the 2.5-D simulator Ensemble.

Taking the S-parameters obtained from the structure, a more conventional resonator analysis can be performed. Figure 5-7 shows the electrical equivalent circuit of the coupled microstrip line resonator.

Finally, the S_{11} resonant curve is analyzed. The curve seen in Figure 5-8 shows a coupled microstripline resonator response, and the resulting Q is determined to be 560. This structure and application are covered by U.S. patent application nos. 60/564173 and 60/501371.

This type of resonator, as shown in Figure 5-5, plays a major role in the design of oscillators described in Chapter 11. Figure 5-9 shows the actual circuit of a coupled resonator oscillator as it is built on a multilayer printed circuit board. It is a 1.5–4.5 GHz ultra-wideband, ultra-low-noise circuit. This was achieved using our

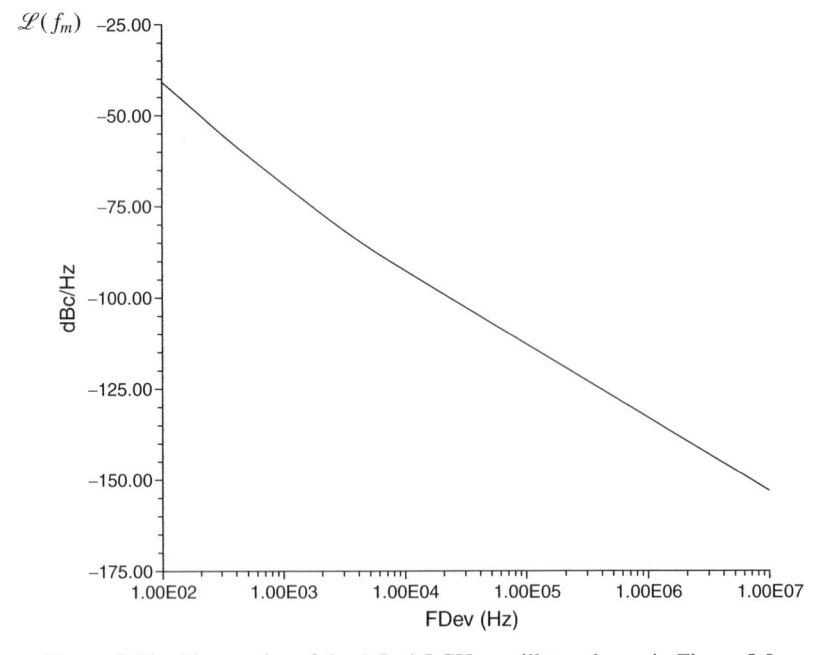

Figure 5-11 Phase noise of the 1.5–4.5 GHz oscillator shown in Figure 5-9.

patented approach, U.S. copyright registration VAU-603984. Besides the coupled filter, which determines the resonant frequency, there is an additional resonator used for noise filtering. Figure 5-10 shows the schematic of the ultra-wideband oscillator, and Figure 5-11 shows the achievable phase noise for this oscillator.

5.3 CERAMIC RESONATORS

A very popular resonator is the ceramic resonator, which is based on a quarter-wavelength arrangement. The ceramic is silver plated and can be modeled as a quarter-wave section of a transverse electromagnetic (TEM) transmission line (cable) with a high relative dielectric constant ranging from 33 to 88. The two 1 pF capacitors shown in Figure 5-12 load the resonator to achieve a resonance frequency of 2.4 GHz. This resonator has an operating Q of about 400, which is much higher than that of the previous cases. By varying the capacitors, one can alter the frequency. These are commercial parts, which are made by a variety of companies. Most of the high-performance 800 MHz to 2.4 GHz oscillators use these resonators.

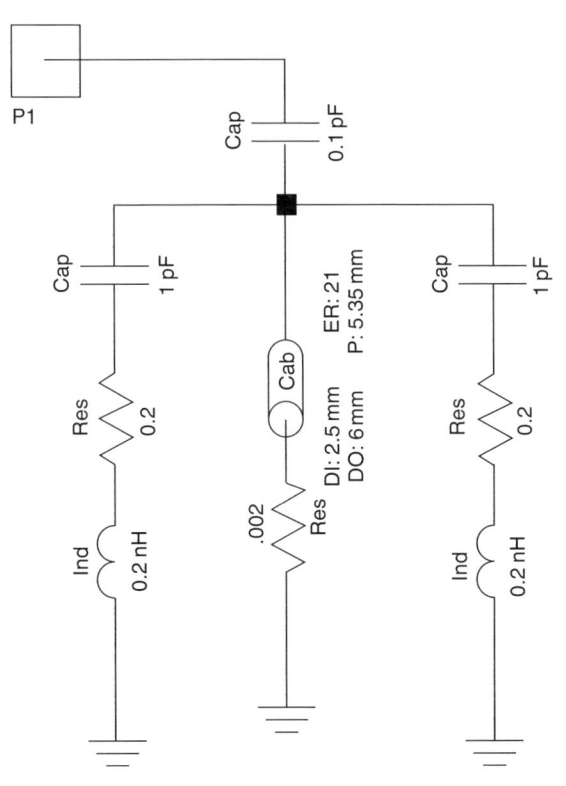

Figure 5-12 A high-Q resonator based on a quarter-wavelength ceramic resonator. Operating Q's of up to 500 are easily achievable.

80 RESONATOR CHOICES

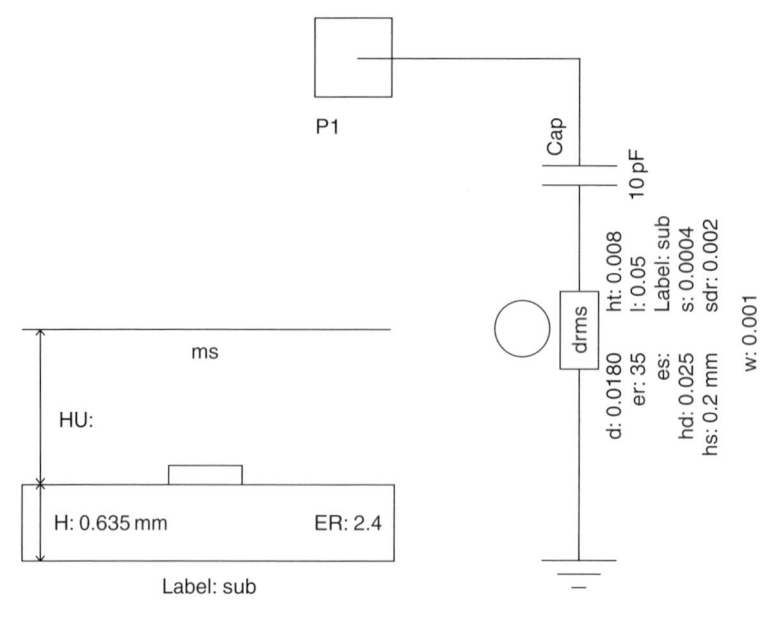

Figure 5-13 A stop-band-type resonant arrangement using a dielectric resonator.

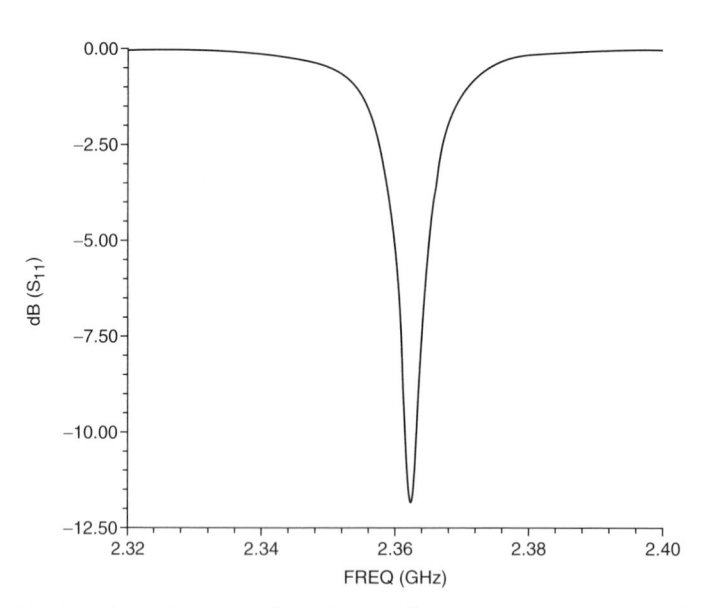

Figure 5-14 Simulated S_{11} in decibels for a dielectric resonator. This can be used to determine the 3 dB bandwidth and the Q.

5.4 DIELECTRIC RESONATORS

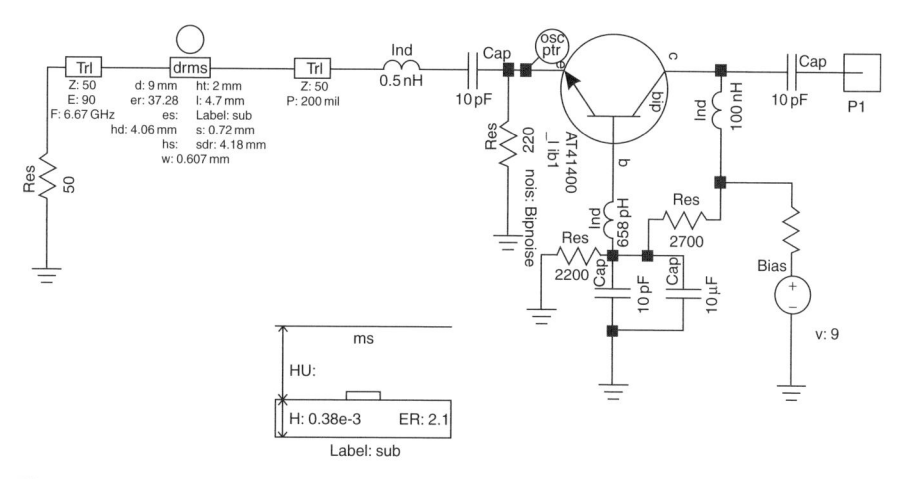

Figure 5-15 A 6.6 GHz dielectric resonator. The dimensions of the microstrip T/L, within which the design is accomplished, are provided. The transistor used is the AT41400.

5.4 DIELECTRIC RESONATORS

One of the highest-Q resonators is the dielectric resonator. This is a resonant structure coupled to a transmission line. Its physical dimensions and dielectric constant determine the operating frequency. These resonators can have a Q of several

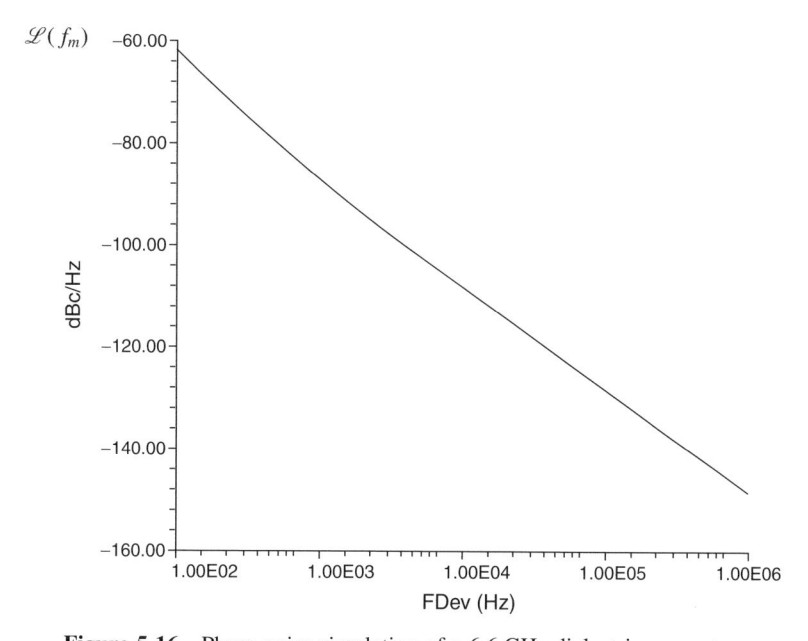

Figure 5-16 Phase noise simulation of a 6.6 GHz dielectric resonator.

thousand. By using the dielectric resonator/transmission line combination as part of an oscillator, very low phase noise oscillators can be obtained. The drawbacks of this device are its temperature coefficient and the fact that it is tuned by using a mechanical post, which is a few millimeters above the resonator. This is used for coarse tuning. Its equivalent circuit is a high-Q parallel circuit replacing the resonator. It is used as a stop-band filter, as shown in Figure 5-13.

Figure 5-14 shows the simulated 3 dB bandwidth based on the S_{11}. The Q, of course, depends on the coupling to the transmission line and other factors, such as

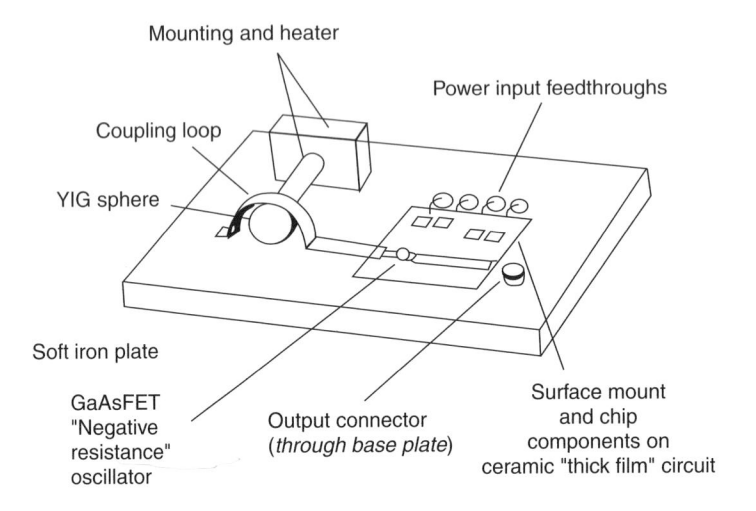

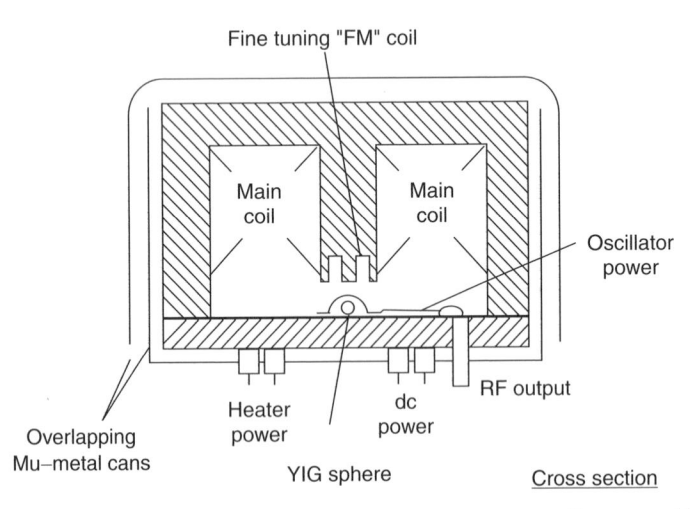

Figure 5-17 The YIG sphere serves as the resonator in the sweep oscillators used in many spectrum analyzers.

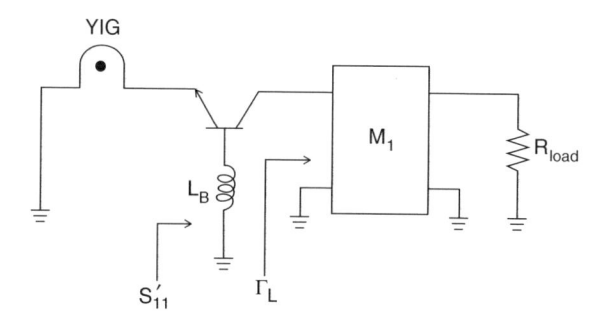

Figure 5-18 Actual circuit diagram for the YIG-tuned oscillator depicted in Figure 5-17.

the ratio of the diameter to the height of the resonator. These types of resonators can be used up to 20 GHz and above, and have become quite popular as a point of reference for low-noise designs.

A typical application of a dielectric resonator in a 6.6 GHz oscillator is shown in Figure 5-15.

The resulting phase noise is shown in Figure 5-16.

Feedback is provided through the base inductor of a 650 pH inductor. This is a series-tuned circuit, as described. The phase noise shows several breakpoints which indicate the flicker corner frequency, $1/Q^2$ area, and other contributions.

5.5 YIG-BASED RESONATORS

For wideband electrically tunable oscillators, we use either a yttrium ion garnet (YIG) or a varactor resonator. The YIG resonator is a high-Q, ferrite sphere of

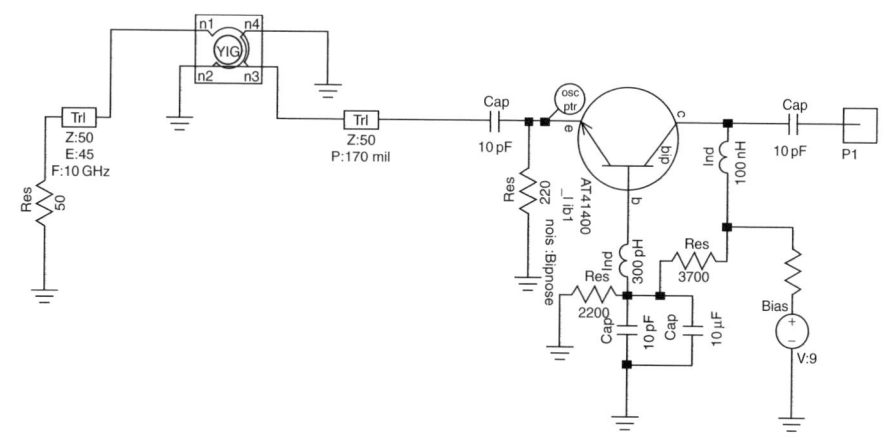

Figure 5-19 Schematic of a 10 GHz YIG oscillator, which is a typical application for the YIG resonator.

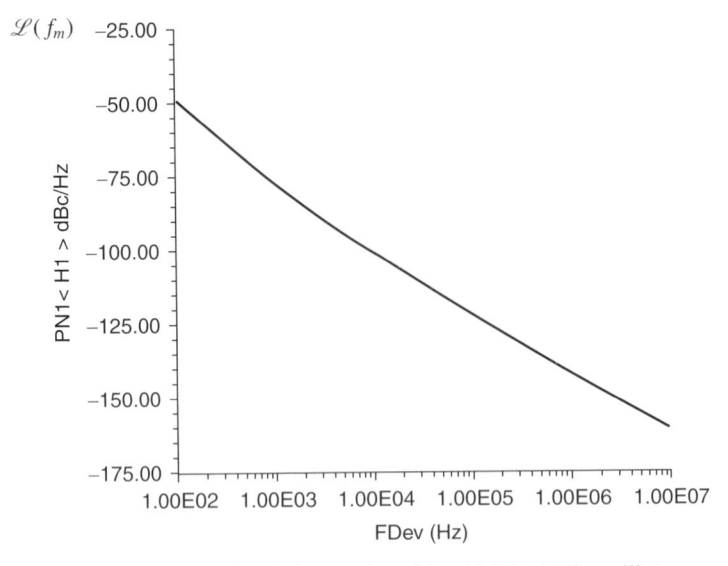

Figure 5-20 Simulated phase noise of the 10 GHz YIG oscillator.

YIG, $Y_2Fe_2(FeO_4)_3$, that can be tuned over a wide band by varying the biasing DC magnetic field. Its high performance and convenient size for applications in microwave integrated circuits make it an excellent choice in a large number of applications, such as filters, multipliers, discriminators, limiters, and oscillators. A YIG resonator makes use of the ferrimagnetic resonance, which, depending on the

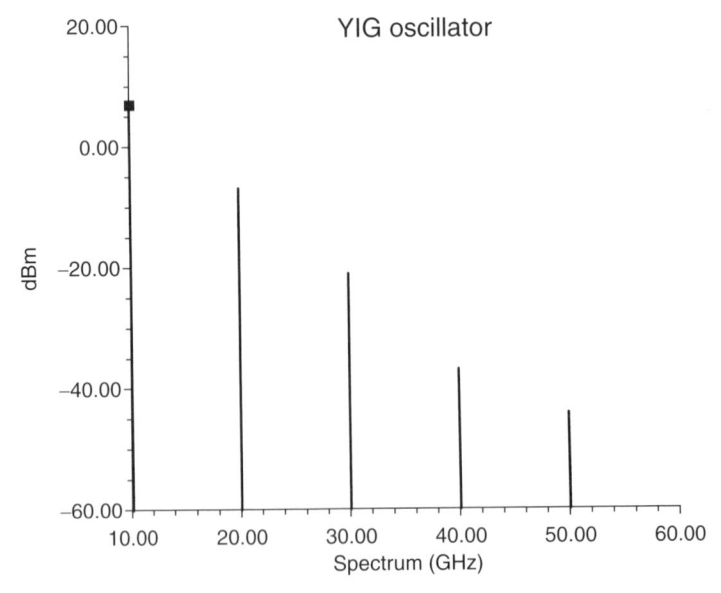

Figure 5-21 Predicted output power and harmonic contents of the 10 GHz YIG oscillator.

material composition, size, and applied field, can be achieved from 500 MHz to 50 GHz. An unloaded Q greater than 1000 is usually achieved with typical YIG material.

Figure 5-17 shows the mechanical drawing of a YIG oscillator assembly. The drawing is somewhat simplified, and the actual construction is more difficult. Its actual circuit diagram is shown in Figure 5-18.

YIG oscillators are manufactured by fewer than five companies worldwide. To build them requires specific skills. Very little information is published about their internal workings.

Figure 5-19 shows the schematic of a 10 GHz YIG oscillator using Ansoft's Designer program. The circuit uses a physical model for the YIG resonator. The users' manual contains the default values for 10 GHz. Figures 5-20 and 5-21 show the predicted phase noise for this oscillator and the output power and harmonic contents.

6 General Theory of Oscillators

6.1 OSCILLATOR EQUATIONS

The following section describes the linear theory required for understanding and designing oscillators to be optimized for a specific frequency of oscillation and output power. This is followed by an analysis of the oscillator as a nonlinear system using a set of nonlinear equations after the start-up conditions in the time domain have been explored.

6.1.1 Linear Theory

The following is a general Y matrix approach used to describe the feedback requirements for an oscillator. Deviating from the standard approach, this will be done by S-parameters, which will then be converted into Y-parameters for easy calculations. In the nonlinear analysis in Section 6.2, we will distinguish between the start-up condition and the sustaining condition. The linear theory does not allow us to do this. For the final calculation in the linear theory, the large-signal S-parameters converted to Y-parameters will be used rather than the small-signal values.

6.1.2 Calculation of the Oscillating Condition

6.1.2.1 Considering Parasitics In the practical case, the device parasitics and loss resistance of the resonator play an important role in the oscillator's design. Figure 6-2 incorporates the base lead inductance L_p and the package capacitance C_p.

The expression of input impedance is given as

$$Z_{IN}|_{package} = -\left[\frac{Y_{21}}{\omega^2(C_1+C_p)C_2}\frac{1}{(1+\omega^2 Y_{21}^2 L_p^2)}\right]$$

$$-j\left[\frac{C_1+C_p+C_2}{\omega(C_1+C_p)C_2} - \frac{\omega Y_{21} L_p}{(1+\omega^2 Y_{21}^2 L_p^2)}\frac{Y_{21}}{\omega(C_1+C_p)C_2}\right]$$

(6-1)

The Design of Modern Microwave Oscillators for Wireless Applications: Theory and Optimization, by Ulrich L. Rohde, Ajay Kumar Poddar, Georg Böck
Copyright © 2005 John Wiley & Sons, Inc.

88   GENERAL THEORY OF OSCILLATORS

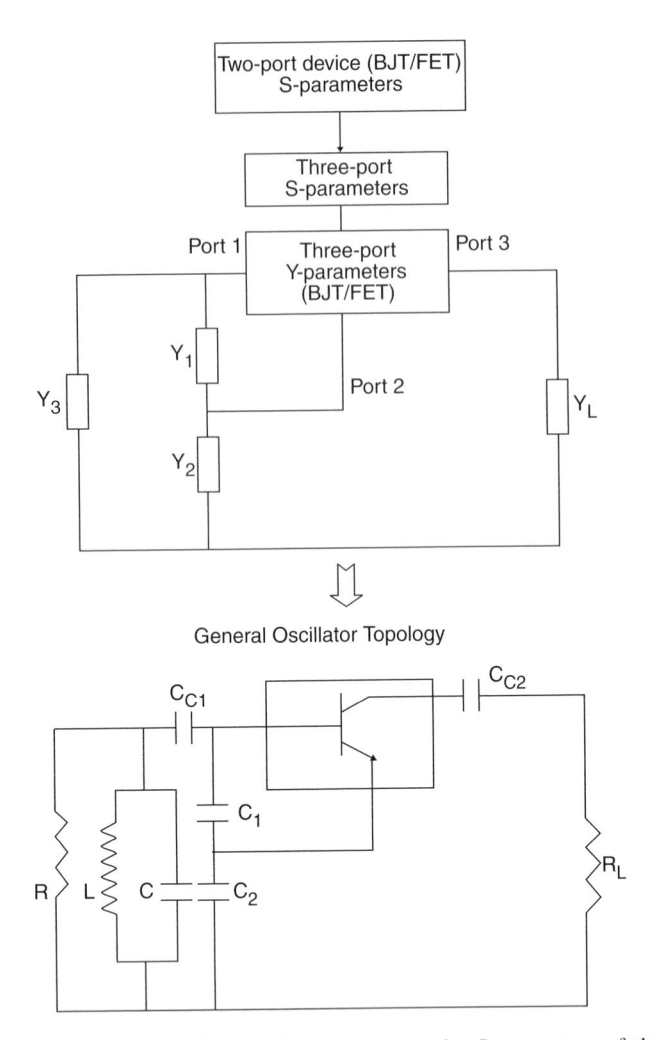

Figure 6-1 A flow diagram showing how to convert the S-parameters of the measured device to a three-port configuration for the transistor. The elements for the Colpitts oscillator, as an example, are then added.

$$Z_{IN}|_{without\ package} = -\left[\frac{Y_{21}}{\omega^2 C_1 C_2}\right] - j\left[\frac{C_1 + C_2}{\omega C_1 C_2}\right] \qquad (6\text{-}2)$$

where L_p is the base lead inductance of the bipolar transistor and C_p is base emitter package capacitance. All further circuits are based on this model.

From the expression above, it is obvious that the base lead inductance makes the input capacitance appear larger and the negative resistance appear smaller.

6.1 OSCILLATOR EQUATIONS

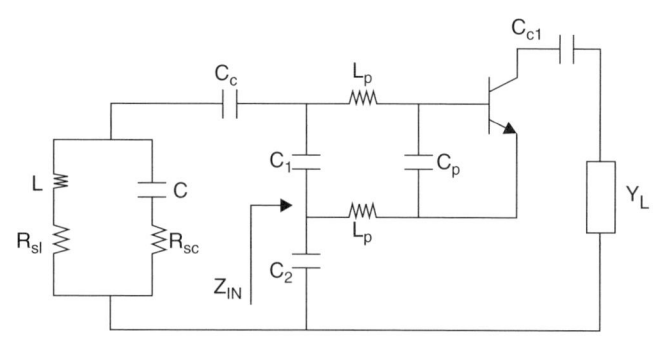

Figure 6-2 Colpitts oscillator with base lead inductances and package capacitance. C_C is neglected. The equivalent circuit of the intrinsic transistor is shown in Figure 8-12.

The equivalent negative resistance and capacitance can be defined as

$$R_{NEQ} = \frac{R_N}{(1 + \omega^2 Y_{21}^2 L_p^2)} \quad (6\text{-}3)$$

$$\frac{1}{C_{EQ}} = \left\{ \left[\frac{1}{((C_1 + C_p)C_2)/(C_1 + C_2 + C_p)} \right] - \left[\frac{\omega^2 Y_{21} L_p}{(1 + \omega^2 Y_{21}^2 L_p^2)} \right] \cdot \left[\frac{Y_{21}}{\omega(C_1 + C_p)C_2} \right] \right\} \quad (6\text{-}4)$$

$$R_N = -\frac{Y_{21}}{\omega^2 C_1 C_2} \quad (6\text{-}5)$$

where

R_N = negative resistance without lead inductance and package capacitance
R_{NEQ} = negative resistance with base lead inductance and package capacitance
C_{EQ} = equivalent capacitance with base lead inductance and package capacitance

At resonance:

$$j\left[\frac{\omega L}{1 - \omega^2 LC} - \frac{1}{\omega C_C} \right] - j\left[\frac{C_1 + C_p + C_2}{\omega(C_1 + C_p)C_2} - \frac{\omega Y_{21} L_p}{(1 + \omega^2 Y_{21}^2 L_p^2)} \frac{Y_{21}}{\omega(C_1 + C_p)C_2} \right] = 0 \quad (6\text{-}6)$$

$$\Rightarrow \left[\frac{\omega L}{1 - \omega^2 LC} - \frac{1}{wC_C} \right] = \left[\frac{C_1 + C_p + C_2}{\omega(C_1 + C_p)C_2} - \frac{Y_{21}}{\omega(C_1 + C_p)C_2} \frac{\omega Y_{21} L_p}{(1 + \omega^2 Y_{21}^2 L_p^2)} \right] \quad (6\text{-}7)$$

$$\Rightarrow \left[\frac{\omega^2 LC_C - (1 - \omega^2 LC)}{\omega C_C (1 - \omega^2 LC)} \right] = \left[\frac{(1 + \omega^2 Y_{21}^2 L_p^2)(C_1 + C_p + C_2) - \omega L_p Y_{21}^2}{\omega(C_1 + C_p)C_2 (1 + \omega^2 Y_{21}^2 L_p^2)} \right] \quad (6\text{-}8)$$

GENERAL THEORY OF OSCILLATORS

The expression above can be rewritten in terms of a determinant as

$$\text{Det}\begin{vmatrix} [\omega^2 LC_C - (1-\omega^2 LC)] & [(1+\omega^2 Y_{21}^2 L_p^2)(C_1+C_p+C_2)-\omega L_p Y_{21}^2] \\ [\omega C_C(1-\omega^2 LC)] & [\omega(C_1+C_p)C_2(1+\omega^2 Y_{21}^2 L_p^2)] \end{vmatrix} = 0 \quad (6\text{-}9)$$

and the resonance condition is given as $K_1 - K_2 \to 0$ where

$$K_1 = [\omega^2 LC_C - (1-\omega^2 LC)][\omega(C_1+C_p)C_2(1+\omega^2 Y_{21}^2 L_p^2)] \quad (6\text{-}10)$$

$$K_2 = [\omega C_C(1-\omega^2 LC)][(1+\omega^2 Y_{21}^2 L_p^2)(C_1+C_p+C_2)-\omega L_p Y_{21}^2] \quad (6\text{-}11)$$

K_1 and K_2 are expressed in terms of the polynomial as

$$K_1 = K_{11}\omega^5 + K_{12}\omega^4 + K_{13}\omega^3 + K_{14}\omega^2 + K_{15}\omega \quad (6\text{-}12)$$

$$K_2 = K_{21}\omega^5 + K_{22}\omega^4 + K_{23}\omega^3 + K_{24}\omega^2 + K_{25}\omega \quad (6\text{-}13)$$

where the coefficients of the polynomial K_1 are

$$K_{11} = [Y_{21}^2 LL_p^2 C_2(C_1 C_C + C_C C_p + CC_1 + CC_p)] \quad (6\text{-}14)$$

$$K_{12} = 0 \quad (6\text{-}15)$$

$$K_{13} = [LC_2(C_1 C_C + C_p C_C + C_1 C + CC_p) - Y_{21}^2 L_p^2 C_2(C_1+C_p)] \quad (6\text{-}16)$$

$$K_{14} = 0 \quad (6\text{-}17)$$

$$K_{15} = -[C_2(C_1+C_p)] \quad (6\text{-}18)$$

and the coefficients of the polynomial K_2 are

$$K_{21} = -[(C_1+C_p+C_2)Y_{21}^2 L_p^2 LCC_C] \quad (6\text{-}19)$$

$$K_{22} = [L_p Y_{21}^2 LCC_C] \quad (6\text{-}20)$$

$$K_{23} = [(C_1+C_p+C_2)C_C(Y_{21}^2 L_p^2 - LC)] \quad (6\text{-}21)$$

$$K_{24} = -[L_p Y_{21}^2 C_C] \quad (6\text{-}22)$$

$$K_{25} = [(C_1+C_p+C_2)C_C] \quad (6\text{-}23)$$

Now $K_1 - K_2$ can be further simplified as

$$[K_{11}\omega^5 + K_{13}\omega^3 + K_{15}\omega] - [K_{21}\omega^5 + K_{22}\omega^4 + K_{23}\omega^3 + K_{24}\omega^2 + K_{25}\omega] = 0 \quad (6\text{-}24)$$

$$[(K_{11} - K_{21})\omega^5 - K_{22}\omega^4 + (K_{13} - K_{23})\omega^3 - K_{24}\omega^2 + (K_{15} - K_{25})\omega] = 0 \quad (6\text{-}25)$$

$$K_1 - K_2 \Longrightarrow A_{11}\omega^5 + A_{12}\omega^4 + A_{13}\omega^3 + A_{14}\omega^2 + A_{15}\omega \quad (6\text{-}26)$$

$$f(\omega, A_{11}, A_{12}, A_{13}, A_{14}, A_{15}) = A_{11}\omega^5 + A_{12}\omega^4 + A_{13}\omega^3 + A_{14}\omega^2 + A_{15}\omega = 0 \quad (6\text{-}27)$$

where

$$A_{11} = (K_{11} - K_{21}) = [Y_{21}^2 L L_p^2 C_2 (C_1 C_C + C_C C_p + C C_1 + C C_p)]$$
$$+ [Y_{21}^2 L_p^2 L C C_C (C_1 + C_p + C_2)] \quad (6\text{-}28)$$

$$A_{12} = (K_{12} - K_{22}) = [0 - L_p Y_{21}^2 L C C_C] \quad (6\text{-}29)$$

$$A_{13} = (K_{13} - K_{23}) = [L C_2 (C_1 C_C + C_p C_C + C_1 C + C C_p) - Y_{21}^2 L_p^2 C_2 (C_1 + C_p)]$$
$$- [(C_1 + C_p + C_2) C_C (Y_{21}^2 L_p^2 - L C)] \quad (6\text{-}30)$$

$$A_{14} = (K_{14} - K_{24}) = L_p Y_{21}^2 C_C \quad (6\text{-}31)$$

$$A_{15} = (K_{15} - K_{25}) = -[C_2(C_1 + C_p) + (C_1 + C_p + C_2) C_C] \quad (6\text{-}32)$$

The polynomial function $f(w, A_{11}, A_{12}, A_{13}, A_{14}, A_{15})$ will have five possible solutions, which can be solved with the help of MathCAD.
For $L_p \to 0$:

$$\omega_0 = \sqrt{\frac{C_2 C_1 + C_1 C_C + C_2 C_C}{L[C_1 C_2 C_C + C_1 C_2 C + C_1 C C_C + C_2 C C_C]}} \quad (6\text{-}33)$$

$$\omega_0 = \sqrt{\frac{1}{L[C + (C_1 C_2 C_C / (C_1 C_2 + C_1 C_C + C_2 C_C))]}} \quad (6\text{-}34)$$

C_C is a coupling capacitor used for separating the bias circuit. Its value is normally small but similar to C_1 and C_2, typically 0.2 to 2 pF.
Rewriting the polynomial equation as ($C_C \to \infty$)

$$\omega_0 = \sqrt{\frac{1}{L[(C_1 C_2 / (C_1 + C_2)) + C]}} \quad (6\text{-}35)$$

6.1.3 Parallel Feedback Oscillator

This example uses the large-signal Y-parameters of a transistor derived from the large-signal S-parameters (measured or simulated).

GENERAL THEORY OF OSCILLATORS

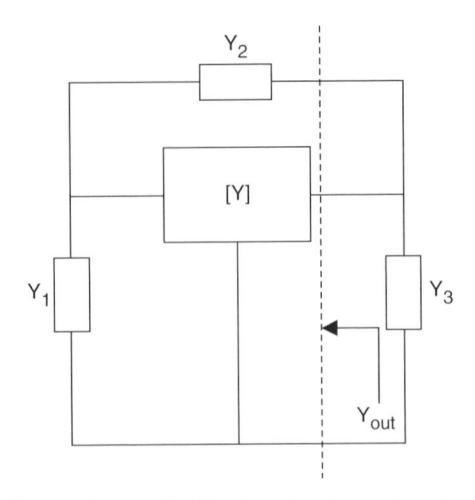

Figure 6-3 Parallel feedback oscillator topology.

The steady-state oscillation condition for the parallel feedback oscillators given in Figure 6-3 is shown as

$$Y_{out} + Y_3 \Longrightarrow 0 \qquad (6\text{-}36)$$

The steady-state stationary condition can be expressed as

$$Det \begin{bmatrix} Y_{11} + Y_1 + Y_2 & Y_{12} - Y_2 \\ Y_{21} - Y_2 & Y_{22} + Y_2 + Y_3 \end{bmatrix} = 0 \qquad (6\text{-}37)$$

$$Y_3 = -[Y_{22} + Y_2] + \frac{[Y_{12} - Y_2][Y_{21} - Y_2]}{[Y_{11} + Y_1 + Y_2]} \qquad (6\text{-}38)$$

where Y_{ij} $(i,j = 1,2)$ are Y-parameters of the hybrid bipolar/FET model.

As shown in Figure 6-3, the active two-port network, together with the feedback elements Y_1 and Y_2, are considered to be a one-port negative resistance oscillator circuit. The output admittance Y_{out} can be given as

$$Y_{out} = -Y_3 \Longrightarrow [Y_{22} + Y_2] - \frac{[Y_{12} - Y_2][Y_{21} - Y_2]}{[Y_{11} + Y_1 + Y_2]} \qquad (6\text{-}39)$$

According to the optimum criterion [177, 179], the optimum values of feedback susceptance B_1 and B_2, at which the negative value of Re[Y_{out}] is maximum, are determined by solving the following differential equation:

$$\frac{\partial \text{Re}[Y_{out}]}{\partial B_1} = 0 \quad \text{and} \quad \frac{\partial \text{Re}[Y_{out}]}{\partial B_2} = 0 \qquad (6\text{-}40)$$

The solution of the above differential equation will give the optimum values of output admittance Y^*_{out} and feedback susceptance B^*_1 and B^*_2, which can be expressed in terms of the two-port Y-parameters of the active device (BJT/FET) as

$$B^*_1 = -\left\{B_{11} + \left[\frac{B_{12}+B_{21}}{2}\right] + \left[\frac{G_{21}-G_{12}}{B_{21}-B_{12}}\right]\left[\frac{G_{12}+G_{21}}{2} + G_{11}\right]\right\} \quad (6\text{-}41)$$

$$B^*_2 = \left[\frac{B_{12}+B_{21}}{2}\right] + \left[\frac{(G_{12}+G_{21})(G_{21}-G_{12})}{2(B_{21}-B_{12})}\right] \quad (6\text{-}42)$$

The optimum values of the real and imaginary parts of the output admittance are

$$Y^*_{out} = [G^*_{out} + jB^*_{out}] \quad (6\text{-}43)$$

where G^*_{out} and B^*_{out} are given as

$$G^*_{out} = G_{22} - \left[\frac{(G_{12}+G_{21})^2(B_{21}-B_{12})^2}{4G_{11}}\right] \quad (6\text{-}44)$$

$$B^*_{out} = B_{22} + \left[\frac{G_{21}-G_{12}}{B_{21}-B_{12}}\right] - \left[\frac{(G_{12}+G_{21})}{2} + G_{22} - G^*_{out}\right] + \left[\frac{B_{21}+B_{12}}{2}\right]$$

$$(6\text{-}45)$$

Thus, in the steady-state stationary oscillation mode, the general condition for oscillation is given as

$$|Y^*_{out} + Y_3| \Longrightarrow 0 \quad \text{and} \quad [G^*_{out} + G_L] + [B^*_{out} + B^*_3]; \quad \text{finally, we get}$$
$$G^*_{out} + G_L = 0 \quad \text{and} \quad B^*_{out} + B^*_3 = 0 \quad (6\text{-}46)$$

where $Y_3 = [G_L + jB_3]$, $G_L = 1/R_L$, and R_L is the load resistance.

The output power for the given load $Y_3 = [G_L + jB_3]$ is given as $P_{out} = (1/2)V^2_{out}G_L$, where V_{out} is the voltage across the load. We can now introduce the voltage feedback factor m and phase Φ_n, which can be expressed in terms of the transistor Y-parameters as

$$m = \frac{\sqrt{(G_{12}+G_{21}-2G_2)^2 + (B_{21}-B_{12})^2}}{2(G_{12}+G_{21}-G_2)} \quad (6\text{-}47)$$

$$\Phi_n = \tan^{-1}\frac{B_{21}-B_{12}}{G_{12}+G_{21}-2G_2} \quad (6\text{-}48)$$

An example is shown in Appendix A and explains the use of the two terms.

6.2 LARGE-SIGNAL OSCILLATOR DESIGN

Traditionally, oscillators have been designed using small-signal parameters and establishing a negative resistance using a feedback network to compensate for the losses. This approach allows for the determination of the startup condition, but not for the time it takes to reach steady-state oscillation, nor the steady-state operating conditions, nor the output power. Using a nonlinear approach described here, a design rule for maximum power will be given. Some applications give higher priority to the DC efficiency rather than the best phase noise, and the following steps will demonstrate a procedure to accomplish this. Noise in oscillators and the optimization of oscillators for best noise will be discussed in Chapter 7.

6.2.1 Startup Condition

The oscillator is an autonomous circuit. The noise present in the active device or power supply turn-on transients leads to the initial oscillation buildup. Linear analysis is useful only for analyzing oscillation startup. It loses validity as the oscillation amplitude continues to grow and the nonlinearity of the circuit becomes important. Nonlinear analysis must be used to predict the oscillation amplitude and spectral purity of the oscillator output signal. As a basic requirement for producing a self-sustained near-sinusoidal oscillation, an oscillator must have a pair of complex-conjugate poles on the imaginary axis, that is, in the right half of the s-plane with $\alpha > 0$.

$$P(p_1, p_2) = \alpha \pm j\beta \qquad (6\text{-}49)$$

While this requirement does not guarantee an oscillation with a well-defined steady state (squeaking), it is a necessary condition for any oscillator. When subjected to excitation due to a power supply turn-on transient or noise associated with the oscillator circuit, the right half plane poles in the equation above produce a sinusoidal signal with an exponentially growing envelope given as

$$v(t) = V_0 \exp(\alpha t) \cos(\beta t) \qquad (6\text{-}50)$$

$$v(t)|_{t=0} \rightarrow V_o \qquad (6\text{-}51)$$

V_o is determined by the initial conditions. $v(t)$ is eventually limited by the associated nonlinearity of the oscillator circuit.

Using either a feedback model approach or a negative resistance model, one can perform the linear analysis of the oscillator. Depending on the oscillator topology, one approach is preferred over the other. The condition of oscillation buildup and steady-state oscillation will be discussed using both approaches.

Figure 6-4 shows, in block diagram form, the necessary components of an oscillator. It contains an amplifier with the frequency-dependent forward amplifier gain block, $G(j\omega)$, and a frequency-dependent feedback network, $H(j\omega)$.

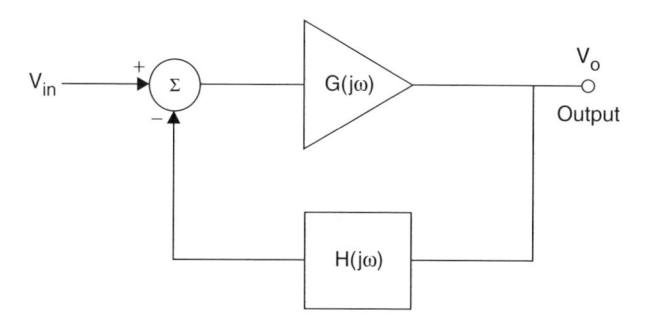

Figure 6-4 Block diagram of a basic feedback model oscillator.

When oscillation starts up, the signal level at the input of the amplifier is very small, and the amplitude dependence of the forward amplifier gain can be initially neglected until it reaches saturation.

The closed loop transfer function (TF) and output voltage $V_0(\omega)$ are given by

$$\text{TF} = \frac{V_0(\omega)}{V_{in}(\omega)} = \frac{G(j\omega)}{1 + G(j\omega)H(j\omega)} \qquad (6\text{-}52)$$

$$V_0(\omega) = \left[\frac{G(j\omega)}{1 + G(j\omega)H(j\omega)}\right] V_{in}(\omega) \qquad (6\text{-}53)$$

For an oscillator, the output voltage V_0 is nonzero even if the input signal $V_i = 0$. This is possible only if the forward loop gain is infinite (which is not practical) or if the denominator $1 + G(j\omega)H(j\omega) = 0$ at some frequency ω_0, that is, if the loop gain is equal to unity for some values of the complex frequency $s = j\omega$.

This leads to the well-known condition for the *Barkhausen criteria*, and can be mathematically expressed as

$$|G(j\omega_0)H(j\omega_0)| = 1 \qquad (6\text{-}54)$$

and

$$Arg[G(j\omega_0)H(j\omega_0)] = 2n\pi \quad \text{where } n = 0, 1, 2, \ldots \qquad (6\text{-}55)$$

When the Barkhausen criteria are met, the two conjugate poles of the overall transfer function are located on the imaginary axis of the s-plane. Any departure from that position will lead to an increase or a decrease in the amplitude of the oscillator output signal in the time domain, which is shown in Figure 6-5.

In practice, the equilibrium point cannot be reached instantaneously without violating some physical laws. As an example, high-Q oscillators take longer than low-Q types to achieve full amplitude. The oscillator output sine wave cannot start at full amplitude instantaneously after the power supply is turned on. The design of the circuit must be such that at startup, the poles are located in the right

96   GENERAL THEORY OF OSCILLATORS

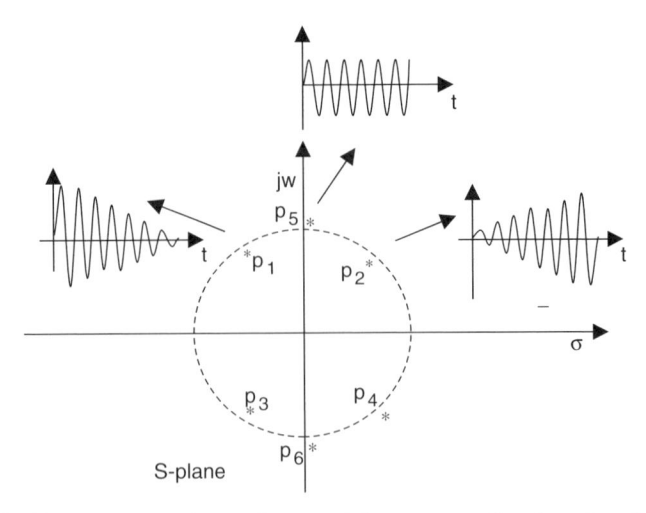

Figure 6-5 Frequency domain root locus and the corresponding time domain response.

half plane, but not too far from the *Y*-axis. However, the component tolerances and the nonlinearities of the amplifier will play a role. This oscillation is achievable with a small-signal loop gain greater than unity. As the output signal builds up, at least one parameter of the loop gain must change its value in such a way that the two complex-conjugate poles migrate in the direction of the *Y*-axis. The parameter must then reach that axis for the desired steady-state amplitude value at a given oscillator frequency.

Figure 6-6 shows the general schematic diagram of a one-port negative resistance model. The oscillator circuit is separated into a one-port active circuit, which is non-linear time variant (NLTV) and a one-port frequency-determining circuit, which is linear time invariant (LTIV).

The frequency-determining circuit or resonator sets the oscillation frequency and is amplitude independent.

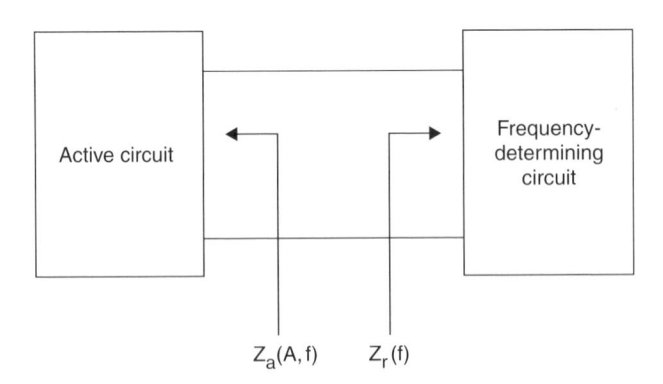

Figure 6-6 Schematic diagram of a one-port negative resistance model.

Assuming that the steady-state current at the active circuit is almost sinusoidal, the input impedance $Z_a(A, f)$ can be expressed in terms of a negative resistance and reactance as

$$Z_a(A, f) = R_a(A, f) + jX_a(A, f) \tag{6-56}$$

A is the amplitude of the steady-state current into the active oscillator circuit and f is the resonant frequency. $R_a(A, f)$ and $X_a(A, f)$ are the real and imaginary parts of the active circuit and depend on amplitude and frequency.

Since the frequency-determining circuit is amplitude independent, it can be represented as

$$Z_r(f) = R_r(f) + jX_r(f) \tag{6-57}$$

$Z_r(f)$ is the input impedance of the frequency-determining circuit. $R_r(f)$ and $X_r(f)$ are the loss resistance and reactance associated with the resonator/frequency-determining circuit.

To support the oscillator buildup, $R_a(A, f) < 0$ is required so that the total loss associated with the frequency-determining circuit can be compensated for. Oscillation will start building up if the product of the input reflection coefficient, $\Gamma_r(f_0)$, looking into the frequency-determining circuit and the input reflection coefficient, $\Gamma_a(A_0, f_0)$, of the active part of the oscillator circuit is unity at $A = A_0$ and $f = f_0$.

The steady-state oscillation condition can be expressed as

$$\Gamma_a(A, f)\Gamma_r(f)|_{f=f_0} \Longrightarrow \Gamma_a(A_0, f_0)\Gamma_r(f_0) = 1 \tag{6-58}$$

Figure 6-7 shows the input reflection coefficient $\Gamma_a(A_0, f_0)$ and $\Gamma_r(f_0)$, which can be represented in terms of the input impedance and the characteristic

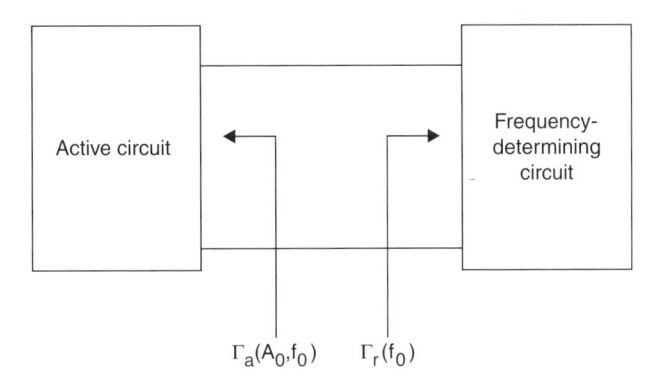

Figure 6-7 Schematic diagram of a one-port negative reflection model.

GENERAL THEORY OF OSCILLATORS

impedance Z_0 as

$$\Gamma_a(A_0, f_0) = \frac{Z_a(A_0, f_0) - Z_0}{Z_a(A_0, f_0) + Z_0} \qquad (6\text{-}59)$$

and

$$\Gamma_r(f_0) = \frac{Z_r(f_0) - Z_0}{Z_r(f_0) + Z_0} \qquad (6\text{-}60)$$

$$\Gamma_a(A_0, f_0)\Gamma_r(f_0) = 1 \Longrightarrow \left[\frac{Z_a(A_0, f_0) - Z_0}{Z_a(A_0, f_0) + Z_0}\right]\left[\frac{Z_r(f_0) - Z_0}{Z_r(f_0) + Z_0}\right] = 1 \qquad (6\text{-}61)$$

$$[Z_a(A_0, f_0) - Z_0][Z_r(f_0) - Z_0] - [Z_a(A_0, f_0) + Z_0][Z_r(f_0) + Z_0] = 0 \qquad (6\text{-}62)$$

$$\Longrightarrow Z_a(A_0, f_0) + Z_r(f_0) = 0 \qquad (6\text{-}63)$$

The characteristic equation $Z_a(A_0, f_0) + Z_r(f_0) = 0$ can be written as

$$R_a(A_0, f_0) + R_r(f_0) = 0 \qquad (6\text{-}64)$$

and

$$X_a(A_0, f_0) + X_r(f_0) = 0 \qquad (6\text{-}65)$$

To support the oscillator buildup, $R_a(A, f) < 0$ is required so that the total loss associated with $R_r(f_0)$ of the frequency-determining circuit can be compensated for. This means that the one-port circuit is unstable for the frequency range $f_1 < f < f_2$ where $R_a(A, f)|_{f_1 < f < f_2} < 0 \Rightarrow |R_a(A, f)|_{f_1 < f < f_2} > |R_r(f)|$. At the startup oscillation, when the signal amplitude is very small, the amplitude dependence of $R_a(A, f)$ is negligible and the oscillation buildup conditions can be given as

$$[R_a(f) + R_r(f)]_{f=f_x} \Longrightarrow R_a(f_x) + R_r(f_x) = \leq 0 \qquad (6\text{-}66)$$

requires slight negative resistance

and

$$[X_a(f) + X_r(f)]_{f=f_x} \Longrightarrow X_a(f_x) + X_r(f_x) = 0 \qquad (6\text{-}67)$$

f_x denotes the resonance frequency at which the total reactive component equals zero. The conditions above are necessary but not sufficient for oscillation buildup, particularly in the case when multiple frequencies exist to support the above conditions.

To guarantee oscillation buildup, the following conditions at the given frequency need to be met:

$$\frac{\partial}{\partial f}[X_a(f) + X_r(f)]_{f=f_x} > 0 \quad (6\text{-}68)$$

$$R_a(f_x) + R_r(f_x) < 0 \quad (6\text{-}69)$$

$$X_a(f_x) + X_r(f_x) = 0 \quad (6\text{-}70)$$

Alternatively, for a parallel admittance topology

$$Y_a(f_x) + Y_r(f_x) = 0 \quad (6\text{-}71)$$

$$G_a(f_x) + G_r(f_x) < 0 \quad (6\text{-}72)$$

$$B_a(f_x) + B_r(f_x) = 0 \quad (6\text{-}73)$$

$$\frac{\partial}{\partial f}[B_a(f) + B_r(f)]_{f=f_x} > 0 \quad (6\text{-}74)$$

Figure 6-8 shows the start and steady-state oscillation conditions.

To demonstrate transient behavior, a ceramic resonator–based oscillator was designed and simulated, as shown in Figure 6-9. This transistor circuit, shown in Figure 9-2, is used several times in this work. The transient analysis function of Ansoft Designer was used to show the DC bias shift and the settling of the amplitude of the oscillation that occurs after 60 nS. The transient analysis of microwave circuits, because of the distributed elements, is a major problem for most CAD microwave simulators. The voltage shown is sampled at the emitter of the Colpitts oscillator.

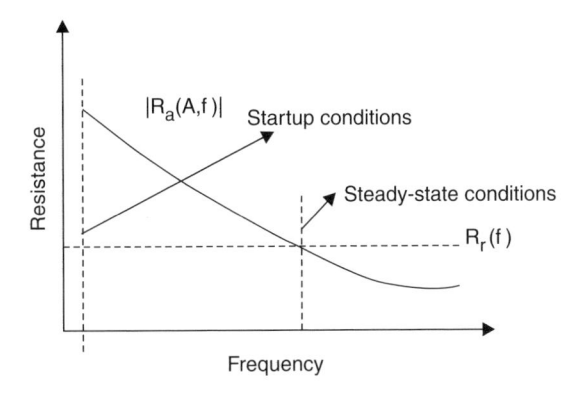

Figure 6-8 Plot of start and steady-state oscillation conditions.

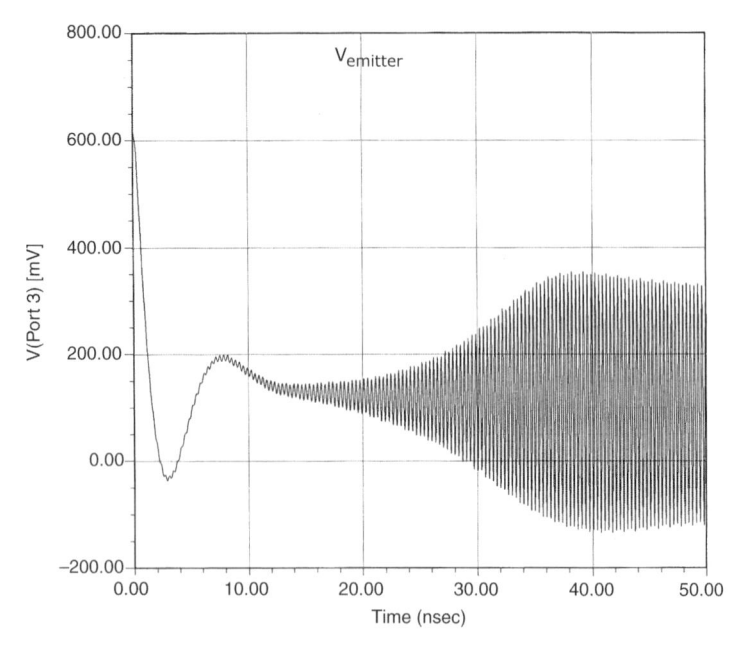

Figure 6-9 Transient simulation of the ceramic resonator–based high-Q oscillator shown in Figure 9-2. The voltage displayed is taken from the emitter.

6.2.2 Steady-State Behavior

As discussed earlier, if the closed-loop voltage gain has a pair of complex-conjugate poles in the right half of the *s*-plane, close to the imaginary axis, then due to the ever-present noise voltage generated in the circuit or power-on transient, a growing, near-sinusoidal voltage appears. As the oscillation amplitude grows, the amplitude-limiting capabilities, due to the change in transconductance from the small-signal g_m to the large-signal G_m of the amplifier, produce a change in the location of the poles. The changes are such that the complex-conjugate poles move towards the imaginary axis. At some value of the oscillation amplitude, the poles reach the imaginary axis, giving the steady-state oscillation.

$$|G(j\omega_o)H(j\omega_o)| = 1 \qquad (6\text{-}75)$$

In the case of the negative resistance model, the oscillation will continue to build up as long as $R_a(A,f)|_{f_1<f<f_2} < 0$; for the active circuit or $\Rightarrow |R_a(A,f)|_{f_1<f<f_2} > |R_r(f)|$, resonant circuit, see Figure 6-6.

The frequency of oscillation is determined by $R_a(A_0, f_0) + R_r(f_0) = 0$. $X_a(A_0, f_0) + X_r(f_0) = 0$ might not be stable because $Z_a(A, f)$ is frequency and amplitude dependent. To guarantee stable oscillation, the following conditions must be satisfied: The first term of equation (6-76) is larger than the second term because the derivative of the negative resistance must be larger than or equal to the derivative

6.2 LARGE-SIGNAL OSCILLATOR DESIGN

of the loss resistance. This can be rewritten in the form of equation (6-77):

$$\frac{\partial}{\partial A}[R_a(A)|_{A=A_0}] * \frac{\partial}{\partial w}[X_r(f)|_{f=f_0}] - \frac{\partial}{\partial A}[X_a(A)|_{A=A_0}] * \frac{\partial}{\partial w}[R_r(f)|_{f=f_0}] > 0 \quad (6\text{-}76)$$

$$\frac{\partial}{\partial A}[R_a(A)|_{A=A_0}] * \frac{\partial}{\partial w}[X_r(f)|_{f=f_0}] > \frac{\partial}{\partial A}[X_a(A)|_{A=A_0}] * \frac{\partial}{\partial w}[R_r(f)|_{f=f_0}] \quad (6\text{-}77)$$

In the case of an LC resonant circuit, $R_r(f)$ is constant and equation (6-77) can be simplified to

$$\frac{\partial}{\partial A}[R_a(A)|_{A=A_0}] * \frac{\partial}{\partial w}[X_r(f)|_{f=f_0}] > 0 \quad (6\text{-}78)$$

Alternatively, for a parallel tuned circuit, the steady-state oscillation condition is given as

$$G_a(f_0) + G_r(f_0) = 0 \quad (6\text{-}79)$$
$$B_a(f_0) + B_r(f_0) = 0 \quad (6\text{-}80)$$

$$\frac{\partial}{\partial A}[G_a(A)|_{A=A_0}] * \frac{\partial}{\partial w}[B_r(f)|_{f=f_0}] > 0 \quad (6\text{-}81)$$

6.2.3 Time-Domain Behavior

The large-signal transfer characteristic affecting the current and voltage of an active device in an oscillator circuit is nonlinear. It limits the amplitude of the oscillation and produces harmonic content in the output signal. The resonant circuit and resulting phase shift set the oscillation frequency. The nonlinear exponential relationship between the voltage and current of a bipolar transistor is given as

$$i(t) = I_s e^{(qv(t)/kT)} \quad (6\text{-}82)$$

I_s is device saturation current, $v(t)$ is the voltage drive applied across the junction, k is Boltzman's constant, q is the electronic charge, and T is the temperature of the device in Kelvin. The bipolar case is mathematically more complex than the FET case. For the FET, a similar set of equations exist which can be derived. Since most RFICs now use SiGe bipolar transistors, the bipolar case has been selected.

The voltage $v(t)$ across the base-emitter junction consists of a DC component and a driven signal voltage $V_1 \cos(wt)$. It can be expressed as

$$v(t) = V_{dc} + V_1 \cos(wt) \quad (6\text{-}83)$$

As the driven voltage $V_1 \cos(wt)$ increases and develops enough amplitude across the base-emitter junction, the resulting current is a periodic series of pulses whose amplitude depends on the nonlinear characteristics of the device and is given as

$$i_e(t) = I_s e^{(qv(t))/(kT)} \tag{6-84}$$

$$i_e(t) = I_s e^{(qV_{dc})/(kT)} e^{(qV_1 \cos(wt))/(kT)} \tag{6-85}$$

$$i_e(t) = I_s e^{(qV_{dc})/(kT)} e^{x \cos(wt)} \tag{6-86}$$

Assuming $I_c \approx I_e$ ($\beta > 10$)

$$x = \frac{V_1}{(kT/q)} = \frac{qV_1}{kT} \tag{6-87}$$

$i_e(t)$ is the emitter current and x is the drive level, which is normalized to kT/q. From the Fourier series expansion, $e^{x \cos(wt)}$ is expressed as

$$e^{x \cos(wt)} = \sum_n a_n(x) \cos(nwt) \tag{6-88}$$

$a_n(x)$ is a Fourier coefficient and is given as

$$a_0(x)|_{n=0} = \frac{1}{2\pi} \int_0^{2\pi} e^{x \cos(wt)} d(wt) = I_0(x) \tag{6-89}$$

$$a_n(x)|_{n>0} = \frac{1}{2\pi} \int_0^{2\pi} e^{x \cos(wt)} \cos(nwt) d(wt) = I_n(x) \tag{6-90}$$

$$e^{x \cos(wt)} = \sum_n a_n(x) \cos(nwt) = I_0(x) + \sum_1^\infty I_n(x) \cos(nwt) \tag{6-91}$$

$I_n(x)$ is the modified Bessel function.

$$\text{As } x \to 0 \Longrightarrow I_n(x) \to \frac{(x/2)^n}{n!} \tag{6-92}$$

$I_0(x)$ are monotonic functions having positive values for $x \geq 0$ and $n \geq 0$; $I_0(0)$ is unity, whereas all higher-order functions start at zero.

The short current pulses are generated from the growing large-signal drive level across the base-emitter junction, which leads to strong harmonic generation. The

6.2 LARGE-SIGNAL OSCILLATOR DESIGN

emitter current represented above can be expressed in terms of harmonics as

$$i_e(t) = I_s e^{(qV_{dc})/(kT)} I_0(x) \left[1 + 2 \sum_{1}^{\infty} \frac{I_n(x)}{I_0(x)} \cos(nwt) \right] \quad (6\text{-}93)$$

$$I_{dc} = I_s e^{(qV_{dc})/(kT)} I_0(x) \quad (6\text{-}94)$$

$$V_{dc} = \frac{kT}{q} \ln\left[\frac{I_{dc}}{I_s I_0(x)}\right] \Rightarrow \frac{kT}{q} \ln\left[\frac{I_{dc}}{I_s}\right] + \frac{kT}{q} \ln\left[\frac{1}{I_0(x)}\right] \quad (6\text{-}95)$$

I_s = collector saturation current.

$$V_{dc} = V_{dcQ} - \frac{kT}{q} \ln I_0(x) \quad (6\text{-}96)$$

$$i_e(t) = I_{dc}\left[1 + 2 \sum_{1}^{\infty} \frac{I_n(x)}{I_0(x)} \cos(nwt) \right] \quad (6\text{-}97)$$

V_{dcQ} and I_{dc} are the operating DC bias voltage and the DC value of the emitter current. Furthermore, the Fourier transform of $i_e(t)$, a current pulse or series of pulses in the time domain, yields a number of frequency harmonics common in oscillator circuit designs using nonlinear devices.

The peak amplitude of the output current, the harmonic content defined as $[I_N(x)/I_1(x)]$, and the DC offset voltage are calculated analytically in terms of the drive level, as shown in Table 6-1. This table provides good insight into the non-linearities involved in the oscillator design.

TABLE 6-1 For T = 300 K, Data are Generated at a Different Drive Level

Drive Level [x]	Drive Voltage $[(kT/q)*x]$mV	Offset Coefficient $\ln[I_0(x)]$	DC Offset $(kT/q)[\ln I_0(x)]$ (mV)	Fundamental Current $2[I_1(x)/I_0(x)]$	Second Harmonic $[I_2(x)/I_1(x)]$
0.00	0.000	0.000	0.000	0.000	0.000
0.50	13.00	0.062	1.612	0.485	0.124
1.00	26.00	0.236	6.136	0.893	0.240
2.00	52.00	0.823	21.398	1.396	0.433
3.00	78.00	1.585	41.210	1.620	0.568
4.00	104.00	2.425	63.050	1.737	0.658
5.00	130.00	3.305	85.800	1.787	0.719
6.00	156.00	4.208	206.180	1.825	0.762
7.00	182.00	5.127	330.980	1.851	0.794
8.00	208.00	6.058	459.600	1.870	0.819
9.00	234.00	6.997	181.922	1.885	0.835
10.00	260.00	7.943	206.518	1.897	0.854
15.00	390.00	12.736	331.136	1.932	0.902
20.00	520.00	17.590	457.340	1.949	0.926

As shown in the table, the peak current $2[I_1(x)/I_0(x)]$ in the fifth column approaches $1.897 I_{dc}$ for a drive level ratio $x = 10$.
For T = 300 K,

$$\frac{kT}{q} = 26\,\text{mV} \qquad (6\text{-}98)$$

and

$$V_1 = 260\,\text{mV} \quad \text{for } x = 10 \qquad (6\text{-}99)$$

The second harmonic distortion [63] $I_2(x)/I_1(x)$ is 85% for a normalized drive level of $x = 10$, and the corresponding DC offset is 205.518 mV. When referring to the amplitude, x is always meant as normalized to kT/q. Figure 6-10 is generated with the help of Math-CAD, and shows the plot of the normalized fundamental and second harmonic current with respect to the drive level.

Notice that as the drive level x increases, the fundamental $2I_1(x)/I_0(x)$ and the harmonic $I_2(x)/I_1(x)$ increase monotonically. Figure 6-11 shows the plot of the coefficient of offset $[\ln I_0(x)]$ with respect to drive level x so that the DC offset voltage can be calculated at different temperatures by simply multiplying the factor kT/q [61].

At T = 300 K the DC voltage shift is $-26[\ln I_0(x)]$ mV

$$\text{for } x = 10 \qquad (6\text{-}100)$$

$$V_{dc} = V_{dcQ} - \frac{kT}{q} \ln I_0(x) \qquad (6\text{-}101)$$

$$V_{dc\ offset} = \frac{kT}{q} \ln I_0(x) = 206\,\text{mV} \qquad (6\text{-}102)$$

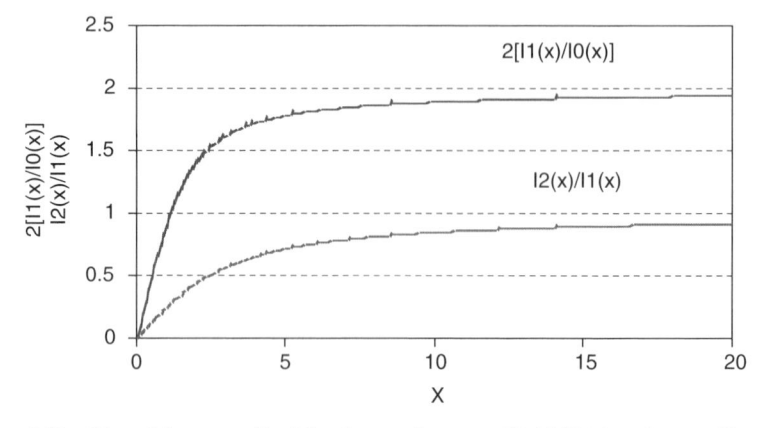

Figure 6-10 Plot of the normalized fundamental current $2I_1(x)/I_0(x)$ and second harmonic $I_2(x)/I_1(x)$ with respect to drive level x.

6.2 LARGE-SIGNAL OSCILLATOR DESIGN

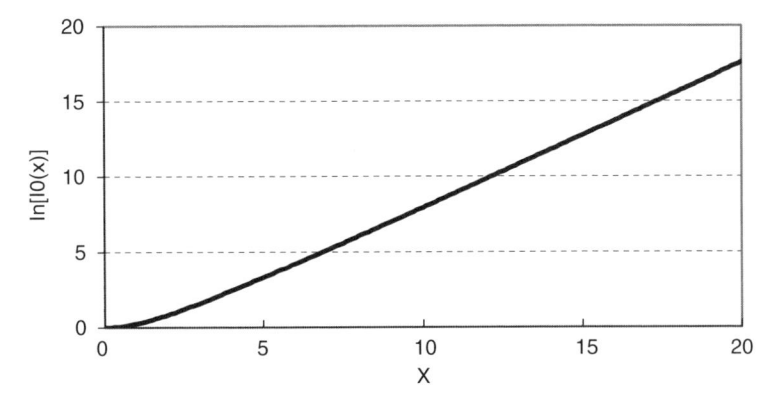

Figure 6-11 Plot of [ln $I_0(x)$] versus drive level x.

V_{dcQ} and $V_{dc\,offset}$ are the operating bias points and DC offsets due to an increase in the drive level. The DC voltage shift at $x = 10$ is 206 mV. Figure 6-12 shows the shape of the output current with respect to the drive level and demonstrates that as the drive level increases, the output current pulse width becomes shorter and the peak current amplitude becomes greater.

$$i_e(t)|_{x=10} \rightarrow 0 \quad \text{for conduction angle} \geq 60° \quad (6\text{-}103)$$
$$i_e(t)|_{x=5} \rightarrow 0 \quad \text{for conduction angle} \geq 90° \quad (6\text{-}104)$$
$$i_e(t)|_{x=2} \rightarrow 0 \quad \text{for conduction angle} > 180° \quad (6\text{-}105)$$

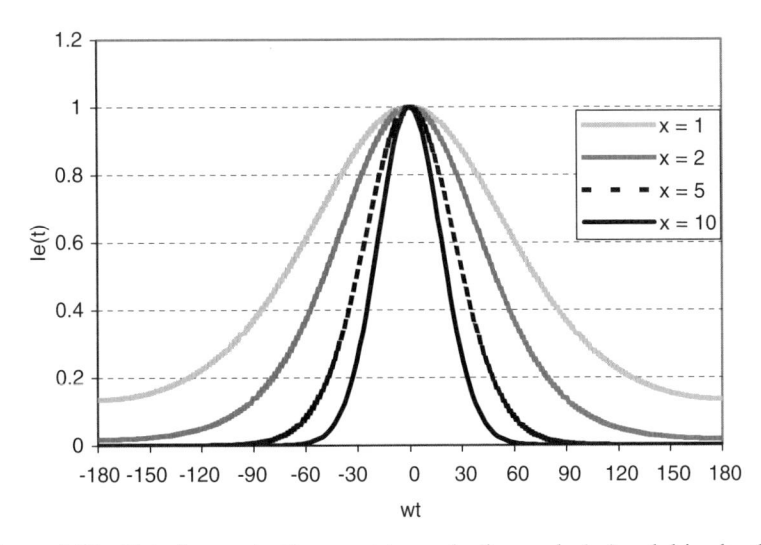

Figure 6-12 Plot of current with respect to conduction angle (wt) and drive level x.

106   GENERAL THEORY OF OSCILLATORS

The harmonic content trade-off is an important consideration in reducing the noise content by using shorter current pulses [64–67].

The designer has limited control over the physical noise sources in a transistor. He can only control the device selection and the operating bias point. However, knowing the bias level, the designer can substantially improve the oscillator phase noise by reducing the duty cycle of the current pulses, in turn reducing the conduction angle of the current.

From equations (6-103) to (6-105) it can be seen that the emitter current $i_e(t) = I_s e^{(qV_{dc}/kT)} e^{x\cos(wt)}$ is proportional to $e^{x\cos(wt)}$ for any fixed drive value of x. The output current is normalized with respect to e^x for the purpose of plotting the graph. Figure 6-13 shows the dependence of the conduction angle with respect to the drive level over one cycle of the input drive signal $v_1 = V_1 \cos(wt)$.

$$i_e(t) = I_s e^{(qV_{dc}/kT)} e^{(qV_1 \cos(wt)/kT)} \qquad (6\text{-}106)$$

$$v_1 = V_1 \cos(wt), \quad x = \frac{qV_1}{kT} \Longrightarrow i_e(t) = I_s e^{(qV_{dc}/kT)} e^{x\cos(wt)} \qquad (6\text{-}107)$$

$$i_e(t) \propto \frac{e^{x\cos(wt)}}{e^x} \qquad (6\text{-}108)$$

Note that for small values of x, the output current is almost a cosine function, as expected. However, as drive level x increases, the output current becomes pulse-like and most of the cycle is nonconducting; there will be only negligible current during the interval between current pulses. Therefore, aside from thermal noise, the noise sources that depend on the transistor on-current, such as shot noise, partition noise, and $1/f$ noise, exist only during the conducting angle of the output current pulses. The operation of the oscillator will cause the current pulse to be centered on the negative peak of the output voltage because of the 180° phase shift between base and collector. If the current pulse, and consequently the noise pulse,

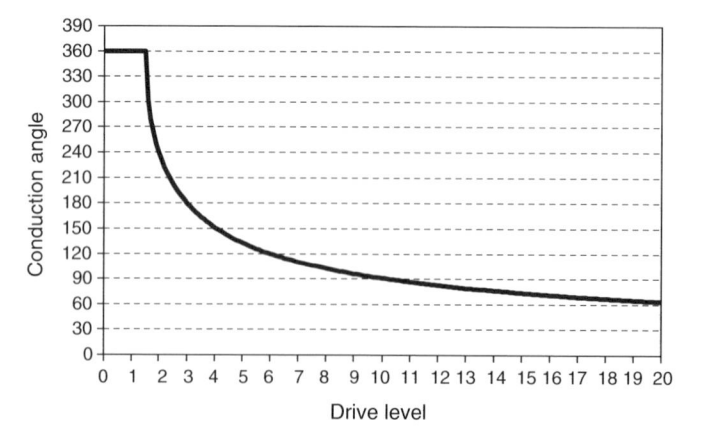

Figure 6-13   Plot of conduction angle versus drive level.

are wide, the current pulse will have a component which contributes a substantial amount of phase noise. If the drive level is increased, the current/noise pulses will become narrower, and therefore will have less PM noise contribution than wider conduction angle pulses. This is seen in Figure 6-20.

Due to the exponential nature of $i_e(t)$, it is not possible to define a conduction angle for these pulses in a conventional sense. However, we may define a special conduction angle as the angular portion of the cycle for which $e^{x\cos(wt)}/e^x \geq 0.05$, for which a solution for φ is derived:

$$\frac{e^{x\cos(wt)}}{e^x} = 0.05 \tag{6-109}$$

$$\phi = wt \tag{6-110}$$

$$\frac{e^{x\cos\varphi}}{e^x} = 0.05 \Longrightarrow \varphi = \cos^{-1}\left[1 + \frac{\ln(0.05)}{x}\right] \tag{6-111}$$

The plot of the conduction angle 2φ versus the drive level is shown in Figure 6-13. When the drive level increases above $x = 2$, the overall current wave shapes rapidly change from co-sinusoidal to impulse-shaped and cause a DC bias shift. This effectively aids the signal by shutting off the base-emitter junction for a good portion of the cycle and thereby makes the conduction angle of the output current narrower. This analysis is valid for the intrinsic transistor. In practice, all parasitics need to be considered.

Figure 6-14 is the oscillator circuit for the calculation of the startup and sustained condition.

The bipolar transistor is represented by a current source and an input conductance at the emitter for easier analysis of the reactance transformation. For easier calculation of the capacitive transformation factor n, the oscillator circuit is rearranged as shown in Figure 6-15 [68].

αI_e and Y_{21} are the current source and large-signal transconductance of the device given by the ratio of the fundamental frequency component of the current to the fundamental frequency of the drive voltage:

$$Y_{21} = \frac{I_{1peak}}{V_{1peak}}\bigg|_{fundamental\ frequency} \tag{6-112}$$

$$I_1|_{n=1} = I_{dc}\left[1 + 2\sum_1^\infty \frac{I_1(x)}{I_0(x)}\cos(wt)\right] \Longrightarrow I_{1peak} = 2I_{dc}\frac{I_1(x)}{I_0(x)} \tag{6-113}$$

108 GENERAL THEORY OF OSCILLATORS

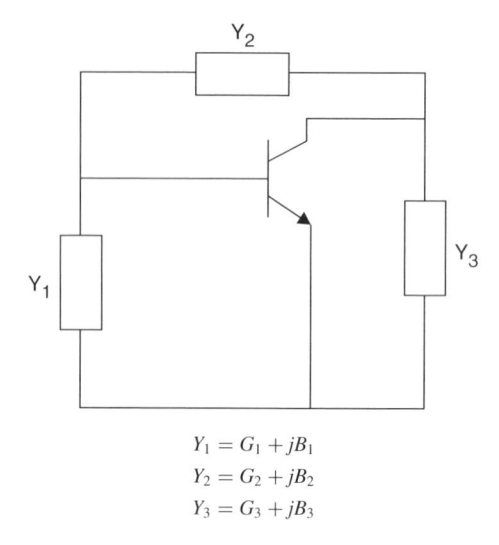

[Figure: General oscillator topology with transistor, Y_1, Y_2, Y_3]

$$Y_1 = G_1 + jB_1$$
$$Y_2 = G_2 + jB_2$$
$$Y_3 = G_3 + jB_3$$

Figure 6-14 General topology of an oscillator.

$x = $ normalized drive level from (6-87).

$$V_1|_{peak} = \frac{kT}{q} x \tag{6-114}$$

$$Y_{21}|_{large\ signal} = G_m(x) \tag{6-115}$$

$$Y_{21}|_{small\ signal} = \frac{I_{dc}}{kT/q}\bigg| = g_m \tag{6-116}$$

$$Y_{21}|_{large\ signal} = G_m(x)$$
$$= \frac{qI_{dc}}{kTx}\left[\frac{2I_1(x)}{I_0(x)}\right]_{n=1} = \frac{g_m}{x}\left[\frac{2I_1(x)}{I_0(x)}\right]_{n=1} \tag{6-117}$$

$$\frac{[Y_{21}|_{large\ signal}]_{n=1}}{[Y_{21}|_{small\ signal}]_{n=1}} = \frac{G_m(x)}{g_m} \Longrightarrow \frac{2I_1(x)}{xI_0(x)} \tag{6-118}$$

$$|Y_{21}|_{small\ signal} > |Y_{21}|_{large\ signal} \Longrightarrow g_m > G_m(x) \tag{6-119}$$

[Figure: Equivalent oscillator circuit with Collector, Emitter, Base labels, αI_e, Y_2, Y_3, Y_1, Y_{21}, V_{be}]

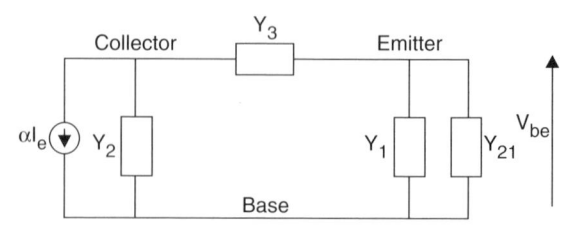

Figure 6-15 Equivalent oscillator circuit for the analysis of the transformed conductance seen by the current source.

TABLE 6-2 Plot of $G_m(x)/g_m = 2[I_1(x)/xI_0(x)]$ versus Drive Level $= x$

Drive Level: x	$G_m(x)/g_m = 2[I_1(x)/xI_0(x)]$
0.00	1
0.50	0.970
1.00	0.893
2.00	0.698
3.00	0.540
4.00	0.432
5.00	0.357
6.00	0.304
7.00	0.264
8.00	0.233
9.00	0.209
10.00	0.190
15.00	0.129
20.00	0.0975

The ratio of the large-signal and small-signal transconductance as a function of drive level is given in Table 6-2, and its graph is plotted in Figure 6-16.

The voltage division for transformation purposes is computed with respect to V_{cb} (voltage from collector to base) because the current source is connected from the collector to the base. The quality factor of the tuned circuit is assumed to be reasonably high for the calculation of the impedance transformation. Finally, the current source, which is connected from collector to base, will see a total conductance

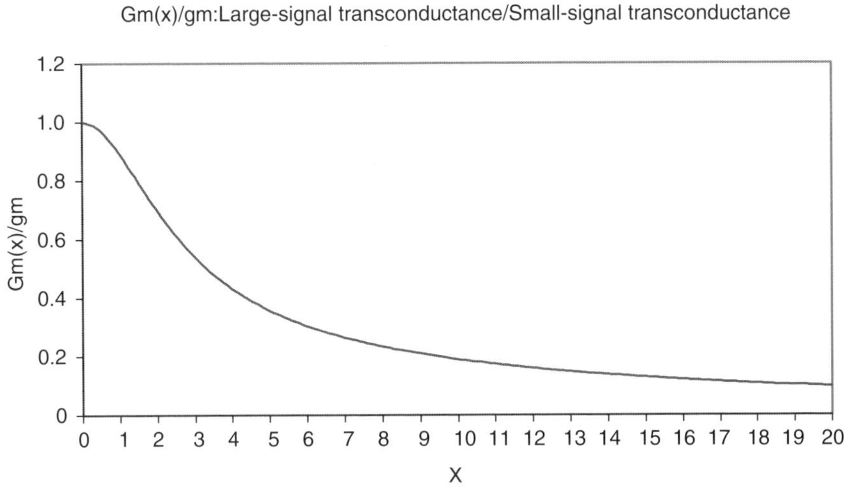

Figure 6-16 Plot of $G_m(x)/g_m = 2[I_1(x)/xI_0(x)]$ versus drive level $= x$.

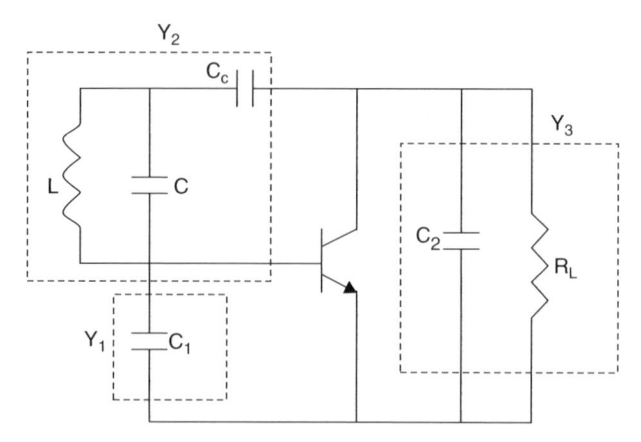

Figure 6-17 Oscillator circuit with the passive components Y_1, Y_2, and Y_3. The equivalent circuit is shown in Figure 8-12.

G_{total}. The oscillator circuit with passive component parameters is shown in Figure 6-17, where

$$Y_1 = G_1 + jB_1 \Longrightarrow j\omega C_1 \quad \text{for } G_1 = 0$$

$$Y_2 = G_2 + jB_2 \Longrightarrow G_2 + j\left[\frac{(\omega^2 LC - 1)\omega C_c}{\omega^2 L(C_c + C) - 1}\right] \quad (6\text{-}120)$$

G_2 = loss parameter/load conductance of the resonator connected parallel to the resonator components C_1, C_2, and L, respectively.

$$Y_3 = G_3 + jB_3 \Longrightarrow G_3 + j\omega C_2$$

G_3 = conductance of the bias resistor placed across C_2, $1/R_L$ in Figure 6-17.

The large-signal transconductances Y_{21} and G_1 are transformed into the current source through the voltage divider V_{eb}/V_{cb}. The voltage V_{eb} must be added to V_{ce} to calculate the transformation ratio, which can be written as

$$\frac{V_{eb}}{V_{cb}} = \frac{C_2}{C_1 + C_2} = \frac{1}{n} \quad (6\text{-}121)$$

and

$$\frac{V_{ce}}{V_{cb}} = \frac{C_1}{C_1 + C_2} = \frac{n-1}{n} \quad (6\text{-}122)$$

The conductance G_2 is already in parallel with the current source, so it remains unchanged. The factor n represents the ratio of the collector-base voltage to the

6.2 LARGE-SIGNAL OSCILLATOR DESIGN

emitter-base voltage at the oscillator resonance frequency.

$$G_1 \rightarrow \frac{G_1}{n^2} \tag{6-123}$$

$$Y_{21} \rightarrow \frac{Y_{21}}{n^2} \Longrightarrow \frac{G_m}{n^2} \tag{6-124}$$

$$G_3 \rightarrow \left[\frac{n-1}{n}\right]^2 G_3 \tag{6-125}$$

G_2 remains constant.

The transformed conductance is proportional to the square of the voltage ratios given in equations (6-121) and (6-122), producing a total conductance as seen by the current source at resonance as

$$G_{total} = G_2 + \frac{G_m + G_1}{n^2} + \left[\frac{n-1}{n}\right]^2 G_3 \tag{6-126}$$

For sustained oscillation, the closed-loop gain at resonance is given as

$$\left[\frac{\left(\frac{V_{be} Y_{21} \alpha}{nG_{total}}\right)}{V_{be}}\right] = 1 \Longrightarrow nG_{total} = Y_{21}\alpha \tag{6-127}$$

$$\frac{Y_{21}}{nG_{total}} = \frac{1}{\alpha} \Longrightarrow \frac{Y_{21}}{nG_{total}} > 1 \tag{6-128}$$

α is assumed to be 0.98, and variation in the value of α does not greatly influence the expression above. Rearranging the device conductance and circuit conductance, the general oscillator equation, after multiplying equation (6-126) with n on both sides, is written as

$$nG_{total} = n\left[G_2 + \frac{Y_{21} + G_1}{n^2} + \left(\frac{n-1}{n}\right)^2 G_3\right] \tag{6-129}$$

$$Y_{21}\alpha = n\left[G_2 + \frac{Y_{21} + G_1}{n^2} + \left(\frac{n-1}{n}\right)^2 G_3\right] \Rightarrow \left[\frac{-(1-n\alpha)}{n^2}\right]Y_{21}$$

$$= \left[G_2 + \frac{G_1}{n^2} + \left(\frac{n-1}{n}\right)^2 G_3\right] \tag{6-130}$$

$$n^2(G_2 + G_3) - n(2G_3 + Y_{21}\alpha) + (G_1 + G_3 + Y_{21}) = 0 \qquad (6\text{-}131)$$

$$n = \frac{(2G_3 + Y_{21}\alpha) \pm \sqrt{(2G_3 + Y_{21}\alpha)^2 - 4(G_2 + G_3)(G_1 + G_3 + Y_{21})}}{2(G_2 + G_3)} \qquad (6\text{-}132)$$

$$n_1 = \frac{(2G_3 + Y_{21}\alpha)}{2(G_2 + G_3)} + \frac{\sqrt{(2G_3 + Y_{21}\alpha)^2 - 4(G_2 + G_3)(G_1 + G_3 + Y_{21})}}{2(G_2 + G_3)} \qquad (6\text{-}133)$$

$$n_2 = \frac{(2G_3 + Y_{21}\alpha)}{2(G_2 + G_3)} - \frac{\sqrt{(2G_3 + Y_{21}\alpha)^2 - 4(G_2 + G_3)(G_1 + G_3 + Y_{21})}}{2(G_2 + G_3)} \qquad (6\text{-}134)$$

From the quadratic equation above, the value of the factor n can be calculated, and thereby an estimation of the capacitance can be done a priori. To ensure higher loop gain, n_1 is selected from $n[n_1, n_2]$.

Once the value of n is fixed, the ratio of the capacitance is calculated as

$$\frac{C_2}{C_1 + C_2} = \frac{1}{n} \qquad (6\text{-}135)$$

$$C_2 = \frac{C_1}{n-1} \Longrightarrow \frac{C_1}{C_2} = n - 1 \qquad (6\text{-}136)$$

If G_3 and G_1 are zero, then the above quadratic equation is reduced to

$$n^2 G_2 - n Y_{21}\alpha + Y_{21} = 0 \qquad (6\text{-}137)$$

$$Y_{21} \cong \frac{n^2}{1-n} G_2 \Longrightarrow Y_{21} = \left[\frac{n^2}{1-n}\right]\frac{1}{R_P} \qquad (6\text{-}138)$$

$$\frac{Y_{21} R_P}{n} = \frac{n}{1-n} \qquad (6\text{-}139)$$

$$R_P = \frac{1}{G_2}$$

$$\frac{Y_{21} R_P}{n} \longrightarrow \text{loop gain} \qquad (6\text{-}140)$$

$$\text{loop gain} \frac{Y_{21} R_P}{n} \longrightarrow 1 \qquad (6\text{-}141)$$

From equations (6-135) and (6-138)

$$Y_{21} \Longrightarrow G_m(x) = \frac{1}{R_P} \frac{[C_1 + C_2]^2}{C_1 C_2} \qquad (6\text{-}142)$$

For relatively optimum phase noise, the drive level has to be adjusted in such a way that the output current pulse conducts for a short period without appreciably

6.2 LARGE-SIGNAL OSCILLATOR DESIGN

increasing the harmonic content. Chapter 8 will show the absolute best phase noise operating point.

From equation (6-117)

$$Y_{21}|_{large\ signal} = G_m(x) = \frac{qI_{dc}}{kTx}\left[\frac{2I_1(x)}{I_0(x)}\right]_{n=1} = \frac{g_m}{x}\left[\frac{2I_1(x)}{I_0(x)}\right]_{n=1} \quad (6\text{-}143)$$

From equation (6-142)

$$G_m(x) = \frac{1}{R_P}\frac{[C_1+C_2]^2}{C_1C_2} \quad (6\text{-}144)$$

From equations (6-143) and (6-144)

$$\frac{g_m}{x}\left[\frac{2I_1(x)}{I_0(x)}\right]_{n=1} = \frac{1}{R_P}\frac{[C_1+C_2]^2}{C_1C_2} = \frac{1}{R_P}\frac{C_1}{C_2}\left[1+\frac{C_2}{C_1}\right]^2 \quad (6\text{-}145)$$

$$\frac{C_1}{C_2}\left[1+\frac{C_2}{C_1}\right]^2 = \frac{R_P g_m}{x}\left[\frac{2I_1(x)}{I_0(x)}\right]_{n=1} \quad (6\text{-}146)$$

From equation (6-119)

$$\frac{R_P g_m}{x}\left[\frac{2I_1(x)}{I_0(x)}\right]_{n=1} \leq \frac{C_1}{C_2}\left[1+\frac{C_2}{C_1}\right]^2 \leq R_P g_m \quad (6\text{-}147)$$

$$x = \frac{g_m[2I_1(x)/I_0(x)]_{n=1}}{((1/R_P)(C_1/C_2))(1+(C_2/C_1))^2} \quad (6\text{-}148)$$

The value of $[(2I_1(x)/I_0)(x)]_{n=1}$ increases monotonically as drive level x increases, and for large values of x and $C_2 < C_1$, $n > 1$, the dependency of x can be expressed as

$$x = \frac{R_P G_m C_2}{C_1} \quad (6\text{-}149)$$

For large drive levels, $x \propto C_2$, and the corresponding conduction angle of output current is given as

$$\varphi = \cos^{-1}\left[1+\frac{\ln(0.05)}{x}\right] \Longrightarrow \varphi \approx \cos^{-1}\left[1-\frac{3}{x}\right] \quad (6\text{-}150)$$

$$\varphi = \cos^{-1}\left[1 - \frac{C_1}{3R_P G_m C_2}\right] \quad (6\text{-}151)$$

$$\varphi \propto \frac{1}{C_2} \quad (6\text{-}152)$$

$$x \propto C_2 \quad (6\text{-}153)$$

Normally, the value of C_1 is kept fixed to avoid loading by the transistor. By increasing the value of C_2, the conduction angle can be reduced, thereby shortening the output current pulse. Any change in designed frequency, due to the variation of C_2, can be compensated for by changing the value of the resonator inductance without much change in the value of drive level x.

The following discussion shows an example for a 100 MHz and a 1 GHz oscillator circuit for different normalized drive levels x. This is provided to give some insight into the relationship between the drive level, the current pulse, and the phase noise.

Figure 6-18 shows the circuit diagram of a 100 MHz Colpitts oscillator with a load of 500 Ω. The value of 100 MHz is selected because transistor parasitics do not play a major role at such a low frequency. For this example, an NEC 85630 transistor has been selected. The emitter to ground capacitor determines the normalized drive level x. As drive level x produces narrow pulses, the phase noise improves. This can be seen in Figure 6-19.

The collector current plotted in Figure 6-20 becomes narrower as the normalized drive level x moves towards $x = 20$. At the same time, the base-emitter voltage swing increases. This is plotted in Figure 6-21.

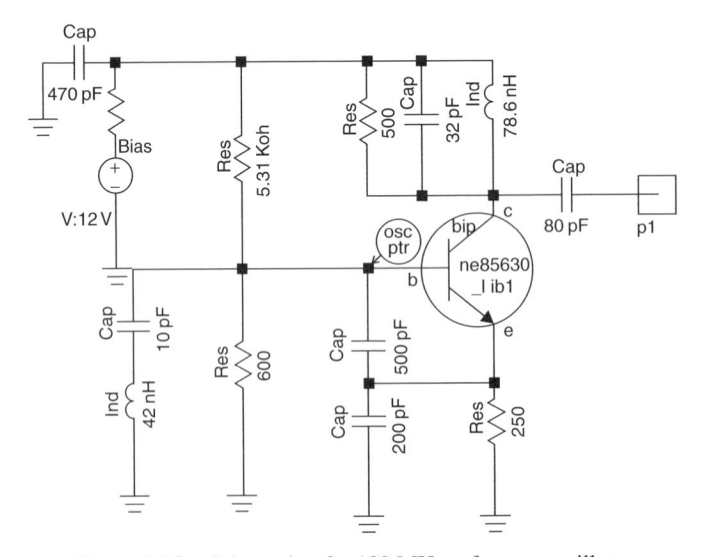

Figure 6-18 Schematic of a 100 MHz reference oscillator.

6.2 LARGE-SIGNAL OSCILLATOR DESIGN 115

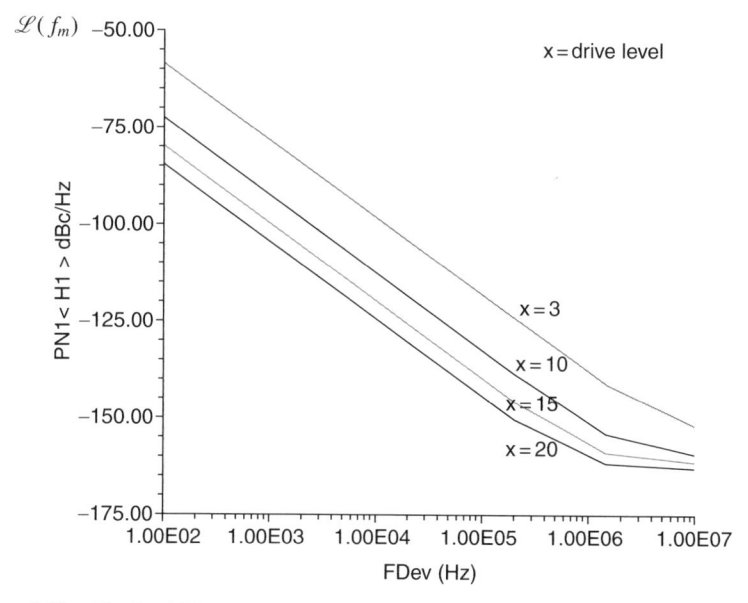

Figure 6-19 Single-sideband phase noise as a function of the normalized drive level x.

Now, moving to the 1000 MHz oscillator and using the BFP520, which has a much higher cutoff frequency, we will evaluate the same conditions. Figure 6-22 shows the 1000 MHz oscillator, and Figure 6-23 shows the collector pulses as a function of the normalized drive.

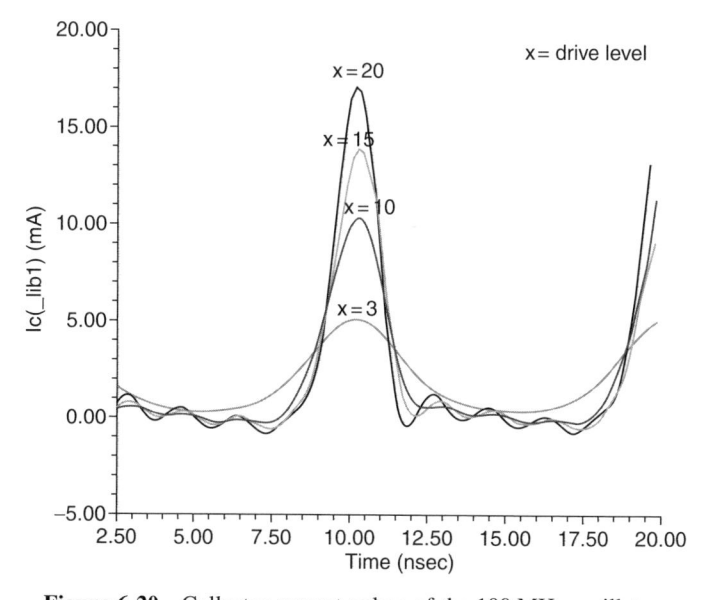

Figure 6-20 Collector current pulses of the 100 MHz oscillator.

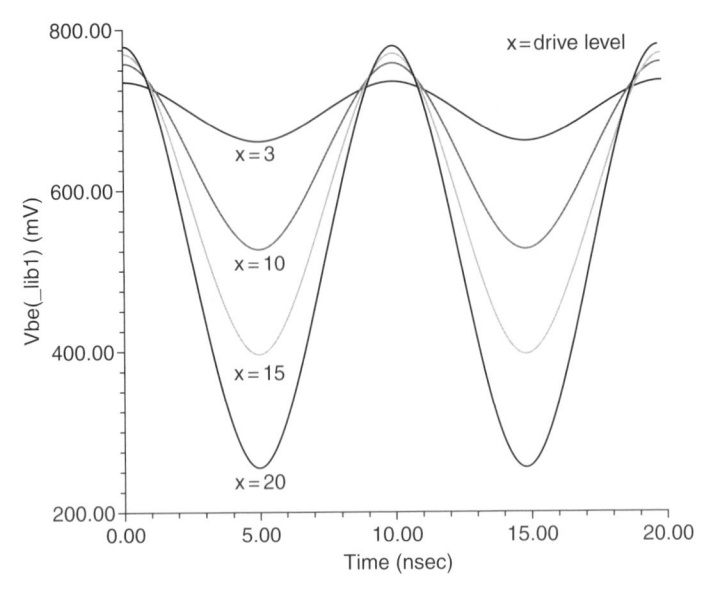

Figure 6-21 RF voltage Vbe across the base-emitter junction as a function of the normalized drive level x.

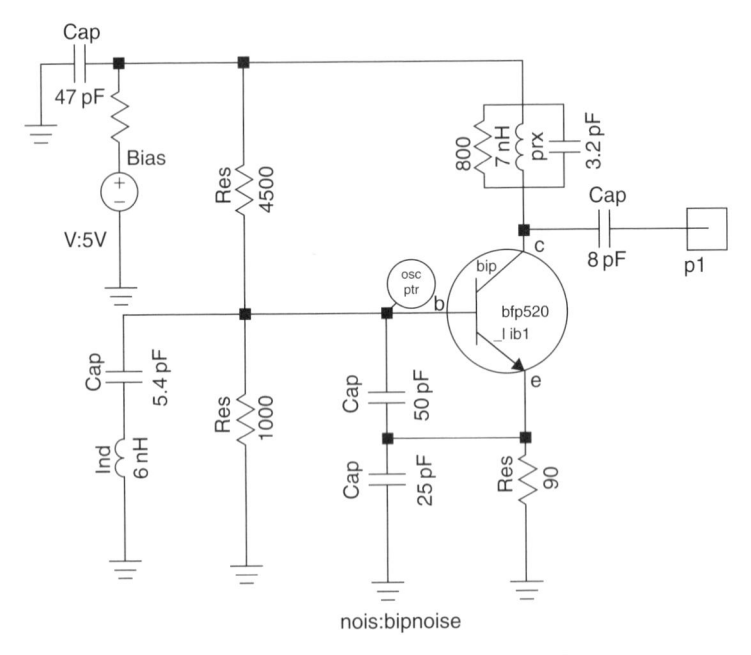

Figure 6-22 Test oscillator at 1000 MHz.

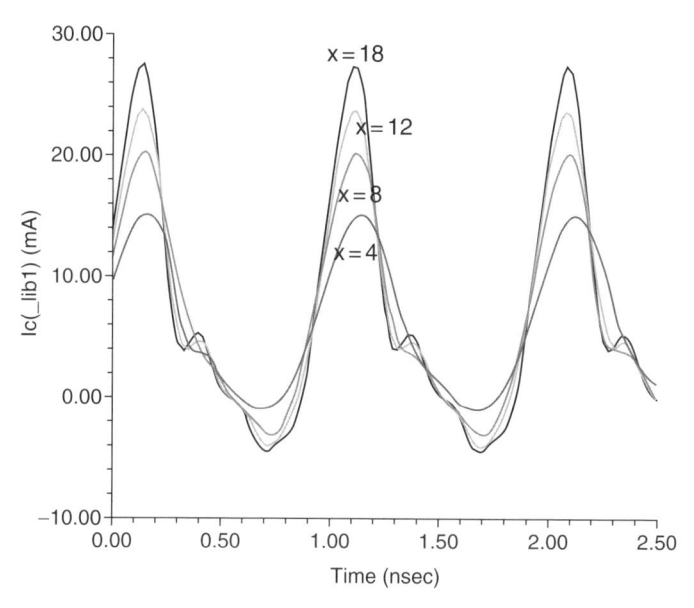

Figure 6-23 Collector current pulses as a function of the normalized drive level x.

There is already some ringing at the negative current of the collector. The base-emitter voltage, based on the tuned circuit, remains less distorted, as shown in Figure 6-24. The change of phase noise in this case for close-in phase noise is no longer so dramatic, but at frequencies above 100 kHz from the carrier, there is a big difference in the phase noise. This can be observed in Figure 6-25.

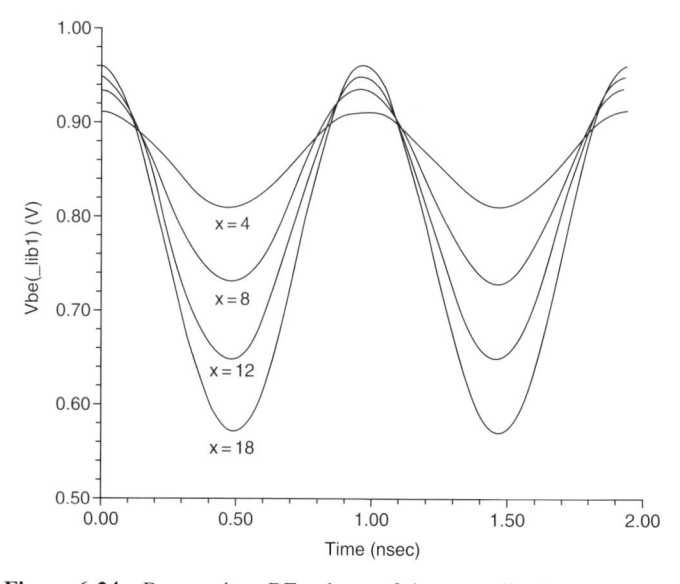

Figure 6-24 Base emitter RF voltage of the normalized drive level x.

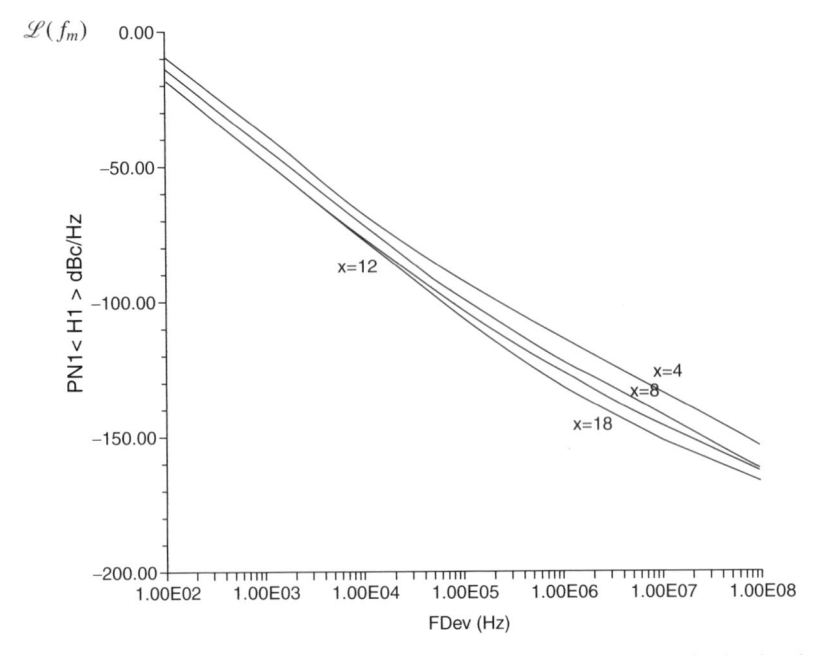

Figure 6-25 Simulated SSB phase noise as a function of the normalized drive level x.

TABLE 6-3 Drive Level for Different Values of C_2 for a 100 MHz Oscillator

Drive Level: x	C_1	C_2	L	Phase Noise at 10 kHz Offset	Frequency
3	500 pF	50 pF	80 nH	−98 dBc/Hz	100 MHz
10	500 pF	100 pF	55 nH	−113 dBc/Hz	100 MHz
15	500 pF	150 pF	47 nH	−125 dBc/Hz	100 MHz
20	500 pF	200 pF	42 nH	−125 dBc/Hz	100 MHz

TABLE 6-4 Drive Level for Different Values of C_2 for a 1000 MHz Oscillator

Drive Level: x	C_1	C_2	C_c	L	Phase Noise at 10 kHz Offset	Frequency
4	50 pF	5 pF	10 pF	6 nH	−68 dBc/Hz	1000 MHz
8	50 pF	10 pF	6.5 pF	6 nH	−72 dBc/Hz	1000 MHz
12	50 pF	15 pF	5.7 pF	6 nH	−75 dBc/Hz	1000 MHz
18	50 pF	20 pF	5.4 pF	6 nH	−77 dBc/Hz	1000 MHz

6.2 LARGE-SIGNAL OSCILLATOR DESIGN   119

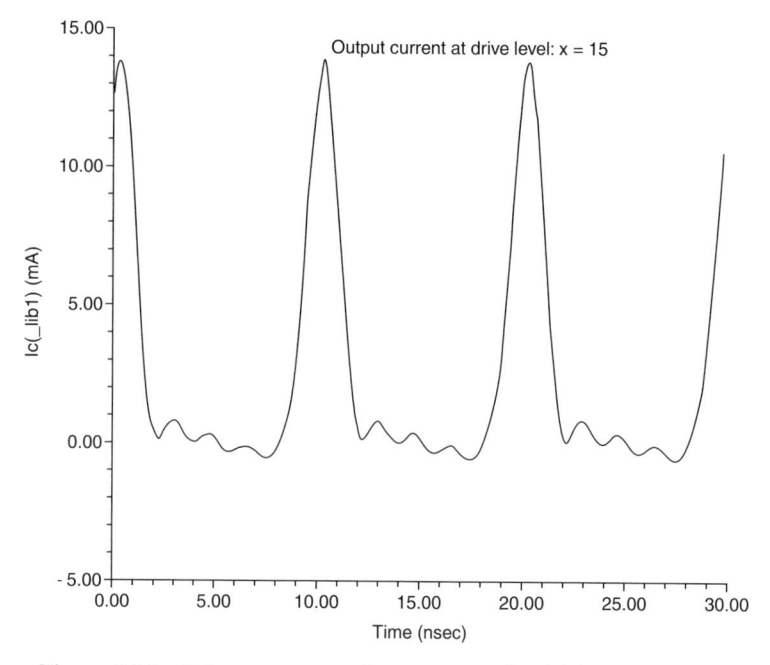

Figure 6-26 Collector current pulses at a normalized drive level $x = 15$.

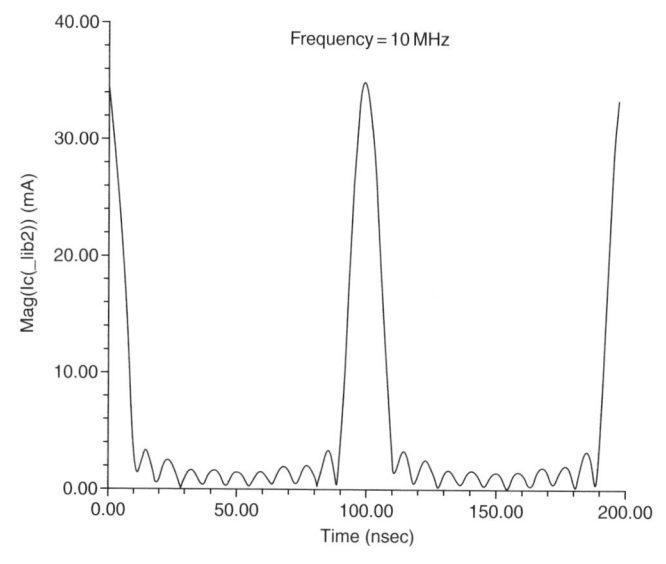

Figure 6-27 Collector current for a 10 MHz LC oscillator at a normalized drive level $x = 15$. The ringing shown is due to the harmonic content.

120   GENERAL THEORY OF OSCILLATORS

Tables 6-3 and 6-4 show the drive level for different values of C_2 for a 100 MHz and a 1000 MHz oscillator.

Now a further examination of the phase noise will be performed. Figure 6-26 shows the plot of the output current of the 100 MHz oscillator circuit for $x = 15$. The ringing of the current at the off-portion of the device is due to the device parasitic.

Figure 6-27 shows the same oscillator circuit frequency scaled to 10 MHz, and Figure 6-28 shows it scaled to 1000 MHz to verify the parasitic and packaging effect at higher frequency. At low frequencies, the device parasitics do not have much influence compared to those at higher frequencies. The noise current for the 10 MHz oscillator during off-cycle has little variation and is more or less the same throughout the off-window. The noise currents for the 100 MHz and 1000 MHz oscillators during the off-cycle vary greatly in magnitude, and the variation is more pronounced at 1000 MHz. The root cause of the phase noise lies in the noise sources of the active device used in the oscillator. Shot noise, burst noise, thermal noise, and $1/f$ noise are the major transistor noise sources, and all of them except thermal noise exist only when there is current in the device. It can be controlled to some extent by adjusting the duty cycle of the current. The basic process responsible for oscillation is due to feedback, and uses a resonant circuit in which a series of periodic current pulses charge the tuned circuit. Between all

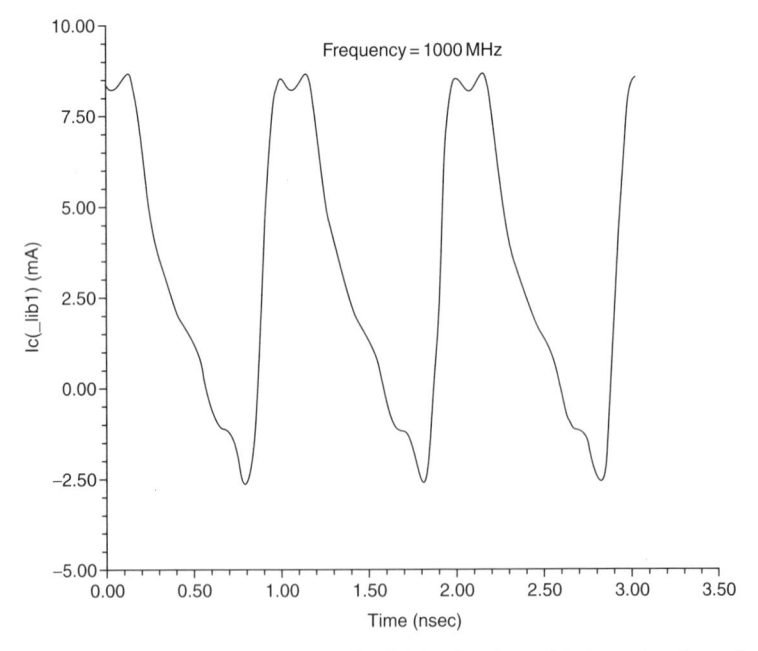

Figure 6-28 Collector current at a normalized drive level $x = 15$. At peak values, the third harmonic becomes a contributing factor (a dip in the curve), and there is a negative collector current.

charging pulses, the bipolar transistor conducts zero current and is considered off. The phase noise produced depends on the shape of the current pulse when the transistor is on. If the current is a relatively narrow pulse, existing for a very short time, less phase noise will be produced than from a wider pulse.

The simulated results support Lee and Hajimiri's theory [64–67], which states that narrowing the current pulse width will decrease phase noise. It is important to understand that the optimum drive level will generate higher harmonics and the device may go into saturation, which will degrade the phase noise performance. Lee and Hajimiri's theory does not emphasize the optimum phase noise at a given power output, which is a strong function of the ratio and the absolute values of the feedback capacitors at given drive level and resonator Q. Appendix B shows a numerical example.

7 Noise in Oscillators

Chapter 6 and Appendix A derive all the necessary equations that are needed to design power-optimized microwave oscillators. While the concept of noise has already been mentioned, oscillator noise theory will be dealt with here and the design for best phase noise will be derived. The noise in an oscillator is determined by the noise (losses) of the resonator and the noise contributions of the active device. The noise of the transistor comes from several sources; the derivation of noise theory is found in Appendices B, C, and D. In the English language, frequently there is no clear difference between the noise figure in logarithm terms and absolute terms. The noise figure is defined as

$$\text{NF} = 10 \times \log(F)$$

The equations in the appendices, therefore, correctly refer to the noise factor, F, instead of the noise figure NF typically stated.

The best signal-to-noise ratio a system can have is the available output power (in dBm) divided by the noise floor (also in dBm). The resulting noise figure is a strong function of the source impedance, which differs from the optimal noise impedance. The optimum noise source calculation is also provided in Appendix D. While the oscillator is a nonlinear circuit, noise theory is assumed to be linear. This means that the noise mechanism in an oscillator has to be treated under steady-state conditions at points of the RF current and voltage waveforms. As a result, the noise figure average. Computation of the noise in a system such as an oscillator is best done using the noise correlation matrix approach. Since noise itself consists of small currents and voltages, one can begin to describe the noise of an oscillator using the linear approach.

7.1 LINEAR APPROACH TO THE CALCULATION OF OSCILLATOR PHASE NOISE

Since an oscillator can be viewed as an amplifier with feedback, as shown in Figure 2-1, it is helpful to examine the phase noise added to an amplifier that has

---

The Design of Modern Microwave Oscillators for Wireless Applications: Theory and Optimization, by Ulrich L. Rohde, Ajay Kumar Poddar, Georg Böck
Copyright © 2005 John Wiley & Sons, Inc.

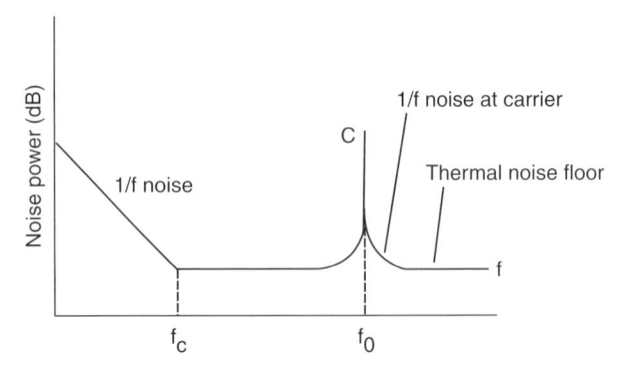

Figure 7-1 Noise power versus frequency of a transistor amplifier with an input signal applied.

a noise factor F. With F defined as

$$F = \frac{(S/N)_{in}}{(S/N)_{out}} = \frac{N_{out}}{N_{in}G} = \frac{N_{out}}{GkTB} \tag{7-1}$$

$$N_{out} = FGkTB \tag{7-2}$$

$$N_{in} = kTB \tag{7-3}$$

where N_{in} is the total input noise power to a noise-free amplifier. The input phase noise in a 1Hz bandwidth at any frequency $f_0 + f_m$ from the carrier produces a phase deviation given by Figure 7-2.

$$\Delta\theta_{1peak} = \frac{V_{nRMS1}}{V_{savRMS}} = \sqrt{\frac{FkT}{P_{sav}}} \tag{7-4}$$

$$\Delta\theta_{RMS} = \frac{1}{\sqrt{2}}\sqrt{\frac{FkT}{P_{sav}}} \tag{7-5}$$

7.1.1 Leeson's Approach

Since a correlated random phase noise relation exists at $f_0 - f_m$, the total phase deviation becomes

$$\Delta\theta_{RMStotal} = \sqrt{FkT/P_{sav}} \quad \text{(SSB)} \tag{7-6}$$

The spectral density of phase noise becomes

$$S_\theta(f_m) = \Delta\theta_{RMS}^2 = FkTB/P_{sav} \tag{7-7}$$

7.1 CALCULATION OF OSCILLATOR PHASE NOISE 125

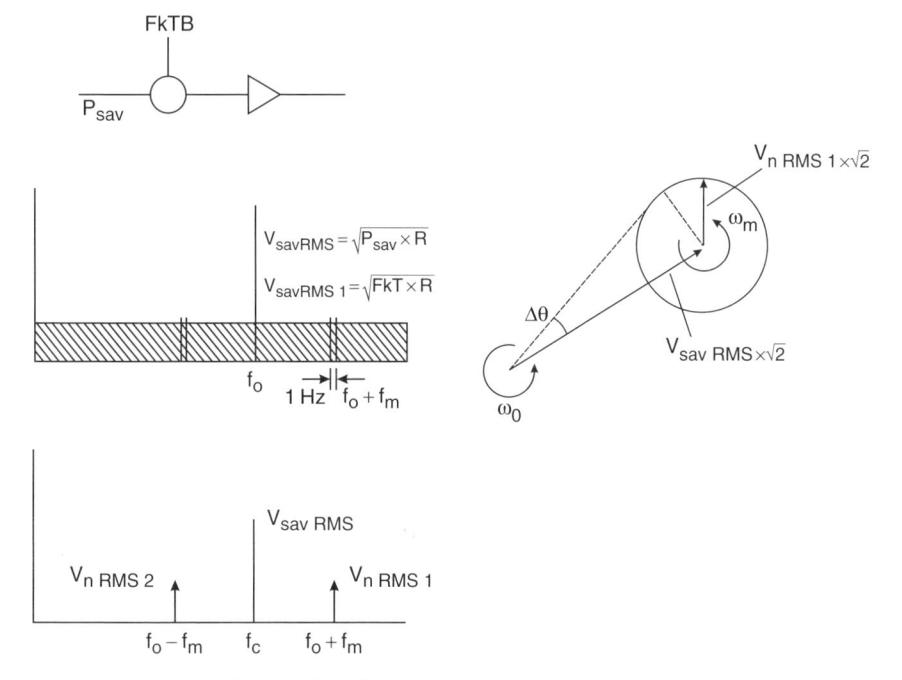

Figure 7-2 Phase noise added to the carrier.

where $B = 1$ for a 1 Hz bandwidth. Using

$$kTB = -174\,\text{dBm} \quad (B = 1\,\text{Hz},\, T = 300\,\text{K}) \tag{7-8}$$

allows a calculation of the spectral density of phase noise that is far away from the carrier (that is, at large values of f_m). This noise is the theoretical noise floor of the amplifier. For example, an amplifier with $+10$ dBm power at the input and a noise figure of 6 dB gives

$$S_\theta(f_m > f_c) = -174\,\text{dBm} + 6\,\text{dB} - 10\,\text{dBm} = -178\,\text{dBm} \tag{7-9}$$

Only if P_{out} is > 0 dBm can we expect $\mathscr{L}$ (the signal-to-noise ratio) to be greater than 174 dBc/Hz (1 Hz bandwidth). For a modulation frequency close to the carrier, $S_\theta(f_m)$ shows a flicker or $1/f$ component, which is empirically described by the corner frequency f_c. The phase noise can be modeled by a noise-free amplifier and a phase modulator at the input, as shown in Figure 7-3.

The purity of the signal is degraded by the flicker noise at frequencies close to the carrier. The phase noise can be described by

$$S_\theta(f_m) = \frac{FkTB}{P_{sav}}\left(1 + \frac{f_c}{f_m}\right) \quad (B = 1) \tag{7-10}$$

126 NOISE IN OSCILLATORS

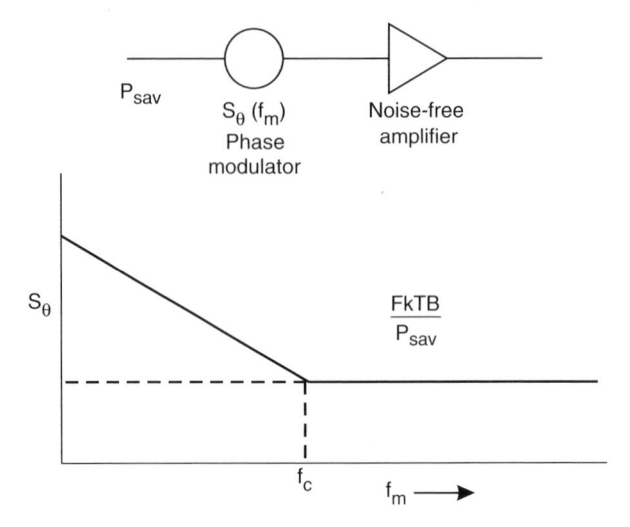

Figure 7-3 Phase noise modeled by a noise-free amplifier and a phase modulator.

No amplitude-phase (AM-to-PM) conversion is considered in this equation. The oscillator may be modeled as an amplifier with feedback, as shown in Figure 7-4.

The phase noise at the input of the amplifier is affected by the bandwidth of the resonator in the oscillator circuit in the following way. The tank circuit or bandpass

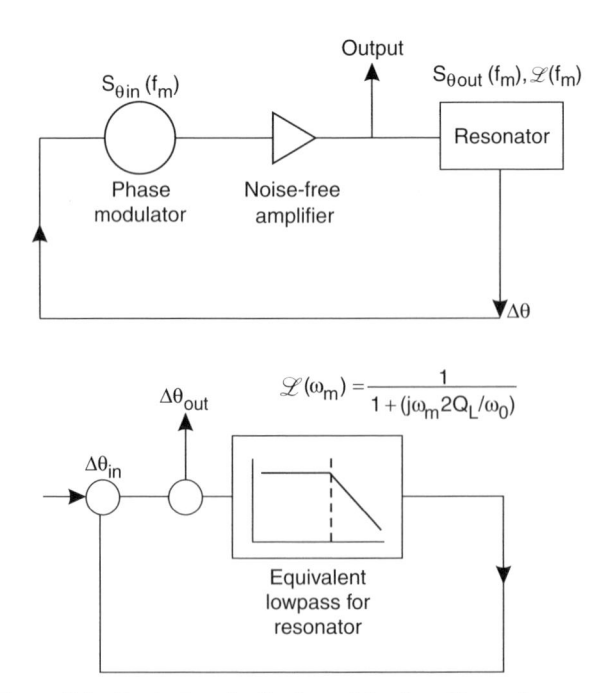

Figure 7-4 Equivalent feedback models of oscillator phase noise.

7.1 CALCULATION OF OSCILLATOR PHASE NOISE   127

resonator has a low-pass transfer function

$$\mathscr{L}(\omega_m) = \frac{1}{1 + j(2Q_L\omega_m/\omega_0)} \qquad (7\text{-}11)$$

where

$$\omega_0/2Q_L = 2\pi B/2 \qquad (7\text{-}12)$$

is the half bandwidth of the resonator. These equations describe the amplitude response of the bandpass resonator; the phase noise is transferred unattenuated through the resonator up to the half bandwidth [70–72].

The closed-loop response of the phase feedback loop is given by

$$\Delta\theta_{out}(f_m) = \left(1 + \frac{\omega_0}{j2Q_L\omega_m}\right)\Delta\theta_{in}(f_m) \qquad (7\text{-}13)$$

The power transfer becomes the phase spectral density

$$S_{\theta\ out}(f_m) = \left[1 + \frac{1}{f_m^2}\left(\frac{f_0}{2Q_L}\right)^2\right]S_{\theta\ in}(f_m) \qquad (7\text{-}14)$$

where $S_{\theta\ in}$ was given by equation (7-10). Finally, $\mathscr{L}(f_m)$, which is the single sideband phase noise (1/2), is

$$\mathscr{L}(f_m) = \frac{1}{2}\left[1 + \frac{1}{f_m^2}\left(\frac{f_0}{2Q_L}\right)^2\right]S_{\theta\ in}(f_m) \qquad (7\text{-}15)$$

This equation describes the phase noise at the output of the amplifier (flicker corner frequency, and AM-to-PM conversion are not considered). The phase perturbation $S_{\theta\ in}$ at the input of the amplifier is enhanced by the positive phase feedback within the half bandwidth of the resonator, $f_0/2Q_L$.

Depending on the relation between f_c and $f_0/2Q_L$, there are two cases of interest, as shown in Figure 7-5. For the low-Q case, the spectral phase noise is unaffected by the Q of the resonator, but the $\mathscr{L}(f_m)$ spectral density will show a $1/f^3$ and $1/f^2$ dependence close to the carrier. For the high-Q case, a region of $1/f^3$ and $1/f$ should be observed near the carrier. Substituting equation (7-10) in (7-15) gives

128 NOISE IN OSCILLATORS

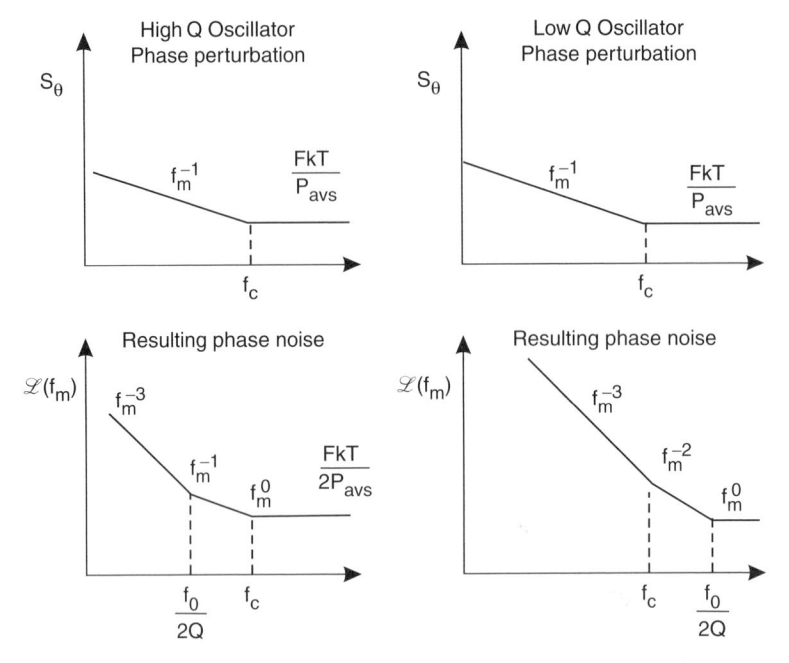

Figure 7-5 Equivalent feedback models of oscillator phase noise.

an overall noise of

$$\mathscr{L}(f_m) = \frac{1}{2}\left[1 + \frac{1}{f_m^2}\left(\frac{f}{2Q_L}\right)^2 \frac{FkT}{P_{sav}}\left(1 + \frac{f_c}{f_m}\right)\right]$$

$$= \frac{FkT}{2P_{sav}}\left[\frac{1}{f_m^3}\frac{f^2 f_c}{4Q_L^2} + \frac{1}{f_m^2}\left(\frac{f}{2Q_L}\right)^2 + \left(1 + \frac{f_c}{f_m}\right)\right] \text{dBc/Hz} \quad (7\text{-}16)$$

Examination of equation (7-16) reveals the four major causes of oscillator noise: the up-converted $1/f$ noise or flicker FM noise, the thermal FM noise, the flicker phase noise, and the thermal noise floor, respectively.

Q_L (loaded Q) can be expressed as

$$Q_L = \frac{\omega_o W_e}{P_{diss,\,total}} = \frac{\omega_o W_e}{P_{in} + P_{res} + P_{sig}} = \frac{\text{reactive power}}{\text{total dissipated power}} \quad (7\text{-}17)$$

where W_e is the reactive energy stored in L and C:

$$W_e = 1/2 CV^2 \quad (7\text{-}18)$$

$$P_{res} = \frac{\omega_o W_e}{Q_{unl}} \quad (7\text{-}19)$$

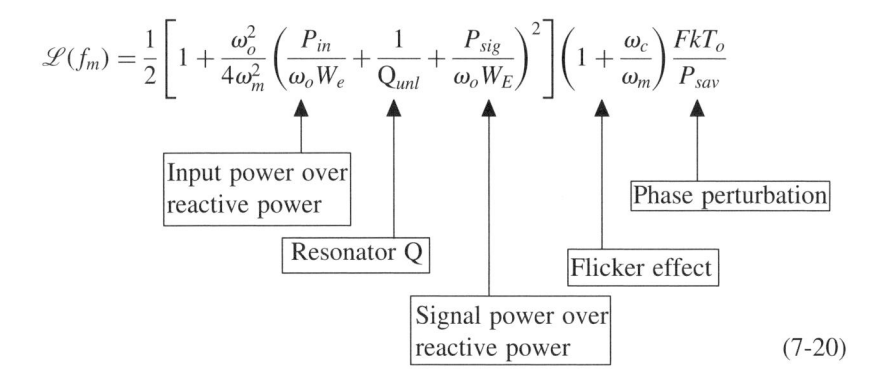

7.1 CALCULATION OF OSCILLATOR PHASE NOISE

$$\mathscr{L}(f_m) = \frac{1}{2}\left[1 + \frac{\omega_o^2}{4\omega_m^2}\left(\frac{P_{in}}{\omega_o W_e} + \frac{1}{Q_{unl}} + \frac{P_{sig}}{\omega_o W_E}\right)^2\right]\left(1 + \frac{\omega_c}{\omega_m}\right)\frac{FkT_o}{P_{sav}}$$

- $\frac{P_{in}}{\omega_o W_e}$: Input power over reactive power
- $\frac{1}{Q_{unl}}$: Resonator Q
- $\frac{P_{sig}}{\omega_o W_E}$: Signal power over reactive power
- $\frac{\omega_c}{\omega_m}$: Flicker effect
- $\frac{FkT_o}{P_{sav}}$: Phase perturbation

(7-20)

More comments on the Leeson formula are found in [70, 109, 161]. The practical oscillator will experience a frequency shift when the supply voltage is changed. This is due to the voltage and current dependent capacitances of the transistor. To calculate this effect, we can assume that the fixed tuning capacitor of the oscillator is a semiconductor junction, which is reverse biased. This capacitor becomes a tuning diode. This tuning diode itself generates a noise voltage and modulates its capacitance by a small amount, and therefore modulates the frequency of the oscillator by minute amounts. The following calculates the phase noise generated from this mechanism, which needs to be added to the phase noise calculated above.

It is possible to define an equivalent noise R_{aeq} that, inserted in Nyquist's equation,

$$V_n = \sqrt{4kT_o R_{aeq} \Delta f} \tag{7-21}$$

where $kT_o = 4.2 \times 10^{-21}$ at 300 K, R is the equivalent noise resistor, and Δf is the bandwidth, determines an open noise voltage across the tuning diode. Practical values of R_{aeq} for carefully selected tuning diodes are in the vicinity of 100 Ω or higher. If we now determine the voltage $V_n = \sqrt{4 \times 4.2 \times 10^{-21} \times 100}$, the resulting voltage value is 1.265×10^{-9} V$\sqrt{\text{Hz}}$.

This noise voltage generated from the tuning diode is now multiplied with the VCO gain, resulting in the rms frequency deviation:

$$(\Delta f_{rms}) = K_o \times (1.265 \times 10^{-9} \text{V}) \quad \text{in a 1 Hz bandwidth} \tag{7-22}$$

In order to translate this into the equivalent peak phase deviation,

$$\theta_d = \frac{K_o\sqrt{2}}{f_m}(1.265 \times 10^{-9} \text{ rad}) \quad \text{in a 1 Hz bandwidth} \tag{7-23}$$

or for a typical oscillator gain of 10 MHz/V,

$$\theta_d = \frac{0.00179}{f_m} \quad \text{rad in a 1 Hz bandwidth} \qquad (7\text{-}24)$$

For $f_m = 25$ kHz (typical spacing for adjacent channel measurements for FM mobile radios), $\theta_d = 7.17 \times 10^{-8}$. This can be converted into the SSB signal-to-noise ratio

$$\mathscr{L}(f_m) = 20 \log_{10} \frac{\theta_c}{2}$$
$$= -149 \text{ dBc/Hz} \qquad (7\text{-}25)$$

Figure 7-6 shows a plot with an oscillator sensitivity of 10 kHz/V, 10 MHz/V, and 100 MHz/V. The center frequency is 2.4 GHz. The lowest curve is the contribution of the Leeson equation. The second curve shows the beginning of the noise contribution from the diode, and the third curve shows that at this tuning sensitivity, the noise from the tuning diode by itself dominates as it modulates the VCO. This is valid regardless of the Q. This effect is called *modulation noise* (AM-to-PM conversion), while the Leeson equation deals with the conversion noise.

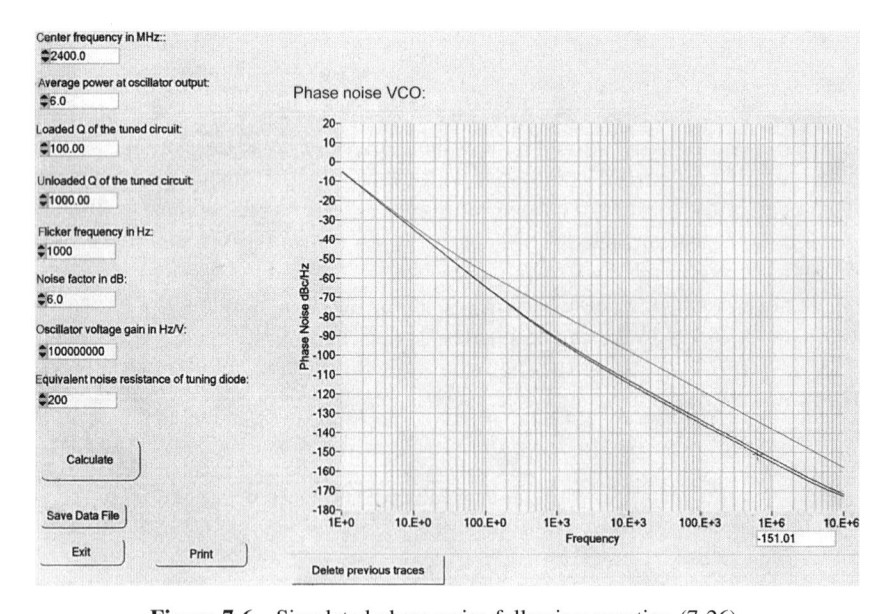

Figure 7-6 Simulated phase noise following equation (7-26).

7.1 CALCULATION OF OSCILLATOR PHASE NOISE

If we combine the Leeson formula with the tuning diode contribution, the following equation allows us to calculate the noise of the oscillator completely:

$$\mathscr{L}(f_m) = 10\log\left\{\left[1 + \frac{f_0^2}{(2f_m Q_L)^2}\right]\left(1 + \frac{f_c}{f_m}\right)\frac{FkT}{2P_{sav}} + \frac{2kTRK_0^2}{f_m^2}\right\} \quad (7\text{-}26)$$

where

$\mathscr{L}(f_m)$ = ratio of sideband power in a 1 Hz bandwidth at f_m to total power in dB
f_m = frequency offset
f_0 = center frequency
f_c = flicker frequency
Q_L = loaded Q of the tuned circuit
F = noise factor
$kT = 4.1 \times 10^{-21}$ at 300 K_0 (room temperature)
P_{sav} = average power at oscillator output
R = equivalent noise resistance of the tuning diode (typically 50 Ω–10 kΩ)
K_0 = oscillator voltage gain

The limitation of this equation is that the loaded Q in most cases has to be estimated, as does the noise factor. The microwave harmonic-balance simulator, which is based on the noise modulation theory (published by Rizzoli), automatically calculates the loaded Q and the resulting noise figure, as well as the output power [73]. The following equations, based on this equivalent circuit, are the exact values for P_{sav}, Q_L, and F which are needed for the Leeson equation. This approach shown here is novel. It calculates the output power based on equations (8-66) to (8-76). The factor of 1000 is needed since the result is expressed in dBm and is a function of n and C_1.

$$[P_o(n, C_1)]_{dBm} = 10\log\left\{\left[\frac{(V_{ce} = -0.7)^2}{4(\omega_0 L)^2}\right]\right.$$

$$\left. \times \left[\frac{Q_L^2\left[C_1^2\left(\frac{C_1}{(n-1)^2}\right)\omega_0^4 L^2\right]}{Q_L^2\left(C_1 + \left(\frac{C_1}{(n-1)^2}\right)\right) + \omega_0^4 L^2 C_1^2\left(\frac{C_1}{(n-1)^2}\right)}\right] R_L^* 1000\right\}$$

(7-27)

0.7 = high current saturation voltage;
V_{ce} collector emitter voltage $< V_{cc}$

To calculate the loaded Q_L, we have to consider the unloaded Q_0 and the loading effect of the transistor. There we have to consider the influence of Y_{21}^+. The inverse of this is responsible for the loading and reduction of the Q.

$$Q_L = \frac{Q_0 \times Q^*}{Q_0 + Q^*}; \quad Q^* = \frac{\omega_0 \times |1/Y_{21}^+|(C_1 + C_2)}{1 - \omega_0^2 C_1 L(Q = Q_0)} \tag{7-28}$$

Based on Figure 8-7, which also shows the transformation of the loading of the differential emitter impedance (resistance), we can also calculate the noise factor of the transistor under large-signal conditions. Considering Y_{21}^+, this noise calculation, while itself uses a totally new approach, is based on the general noise calculations such as the one shown by Hawkins [117] and Hsu and Snapp [118]. An equivalent procedure can be found for FETs rather than bipolar transistors.

7.1.2 Noise Factor of the Oscillator

Figure 7-7 shows the equivalent configuration of the oscillator circuit for the purpose of analyzing the noise factor with respect to the oscillator feedback components (C_1 and C_2) for better insight into the noise improvement. The objective of this chapter is to find the oscillator circuit parameters that influence the noise factor, thereby influencing the phase noise of the oscillator circuit.

From [117, 118] the noise factor F for the circuit shown in Figure 7-7 can be given by

$$F = 1 + \frac{Y_{21}^+ C_2 C_C}{(C_1 + C_2)C_1} \left[r_b + \frac{1}{2r_e \beta} \left(r_b + \frac{(C_1 + C_2)C_1}{Y_{21}^+ C_2 C_C} \right)^2 \right.$$
$$\left. + \frac{r_e}{2} + \frac{1}{2r_e} \left(r_b + \frac{(C_1 + C_2)C_1}{Y_{21}^+ C_2 C_C} \right)^2 \left(\frac{f^2}{f_T^2} \right) \right] \tag{7-29}$$

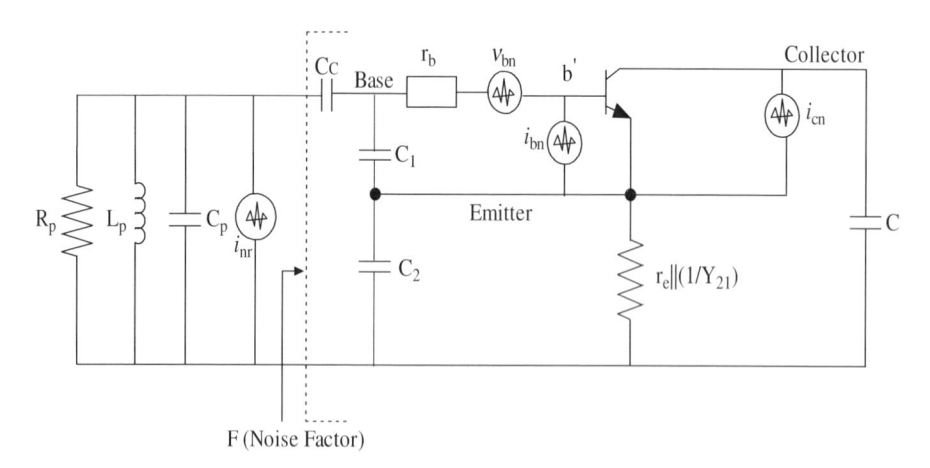

Figure 7-7 Equivalent configuration of the oscillator circuit.

7.1 CALCULATION OF OSCILLATOR PHASE NOISE

$$F = 1 + \frac{Y_{21}^+ C_2 C_C}{(C_1 + C_2)C_1}\left[r_b + \frac{1}{2r_e}\left(r_b + \frac{(C_1 + C_2)C_1}{Y_{21}^+ C_2 C_C}\right)^2\left(\frac{1}{\beta} + \frac{f^2}{f_T^2}\right) + \frac{r_e}{2}\right] \quad (7\text{-}30)$$

Figure 7-8 shows the transistor in the package parameters for the calculation of the oscillator frequency and noise factor.

Table 7-1 shows the SPICE and package parameters of NE68830 from data sheets.

The frequency of the oscillation for the oscillator circuit given in Figure 7-7 is

$$\omega_0 = \sqrt{\left\{\frac{\left(\frac{(C_1^* + C_p)C_2}{(C_1^* + C_p + C_2)} + C_C\right)}{L\left[\left(\frac{(C_1^* + C_p)C_2 C_c}{(C_1^* + C_p + C_2)}\right) + C\left(\frac{(C_1^* + C_p)C_2}{(C_1^* + C_p + C_2)} + C_C\right)\right]}\right\}} \approx 1000 \text{ MHz} \quad (7\text{-}31)$$

with

$L = 5$ nH (inductance of the parallel resonator circuit), $C_1^* = 2.2$ pF, $C_1 = C_1^* + C_p$

$C_p = 1.1$ pF (C_{BEPKG} + contribution from layout)

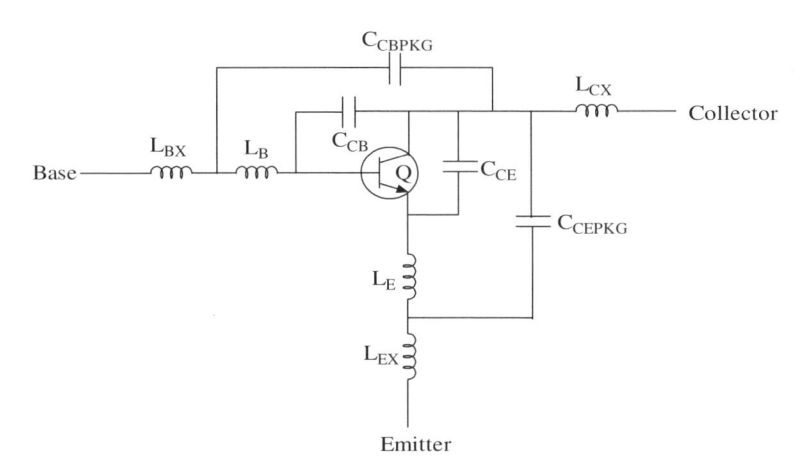

$C_1 = 2.2$ pF + 1.1 pF (*package*) = 3.3 pF

$C_2 = 2.2$ pF, $C_C = 0.4$ pF, $C = 4.7$ pF, $R_P = 18,000$

Figure 7-8 NE68830 with package parasitics. Q is the intrinsic bipolar transistor.

TABLE 7-1 SPICE Type Parameters of the NE68830

Parameters	Q	Parameters	Q	Parameters	Package
IS	3.8E−16	MJC	0.48	C_{CB}	0.24E−12
BF	135.7	XCJC	0.56	C_{CE}	0.27E−12
NF	1	CJS	0	L_B	0.5E−9
VAF	28	VJS	0.75	L_E	0.86E−9
IKF	0.6	MJS	0	C_{CBPKG}	0.08E−12
NE	1.49	TF	11E−12	C_{CEPKG}	0.04E−12
BR	12.3	XTF	0.36	C_{BEPKG}	0.04E−12
NR	1.1	VTF	0.65	L_{BX}	0.2E−9
VAR	3.5	ITF	0.61	L_{CX}	0.1E−9
IKR	0.06	PTF	50	L_{EX}	0.2E−9
ISC	3.5E−16	TR	32E−12		
NC	1.62	EG	1.11		
RE	0.4	XTB	0		
RC	4.2	KF	0		
CJE	0.79E−12	AF	1		
CJC	0.549E−12	VJE	0.71		
XTI	3	RB	6.14		
RBM	3.5	RC	4.2		
IRB	0.001	CJE	0.79E−12		
CJC	0.549E−12	MJE	0.38		
VJC	0.65				

7.1.3 Noise Factor

From equation (7-29)

$$F(NF) = 1 + (9.7E-5)\left[6 + \frac{1}{2*0.9}\left(6 + \frac{500*5.5*3.3}{2.2*0.4}\right)^2 (0.02) + 0.45\right]$$

$$= 104.47 \Longrightarrow 20.18\,\text{dB} \tag{7-32}$$

with

$$Y_{21}^+ = \frac{1}{500} = 2E-3 \text{ (large-signal } Y\text{-parameter for transistor)}$$

$$r_b = 6\,\Omega,\ r_e = 0.9\,\Omega\ @\ 28\,\text{mA}, \beta = 100$$

$$f = 1000\,\text{MHz}, f_T = 10\,\text{GHz}$$

Figures 7-9 and 7-10 illustrate the dependency of the noise factor on the feedback capacitors C_1 and C_2. From Leeson's phase noise equation (4.13), the phase noise of the oscillator circuit can be optimized by optimizing the noise factor terms as given in equation (7-29) with respect to the feedback capacitors C_1 and C_2.

7.1 CALCULATION OF OSCILLATOR PHASE NOISE

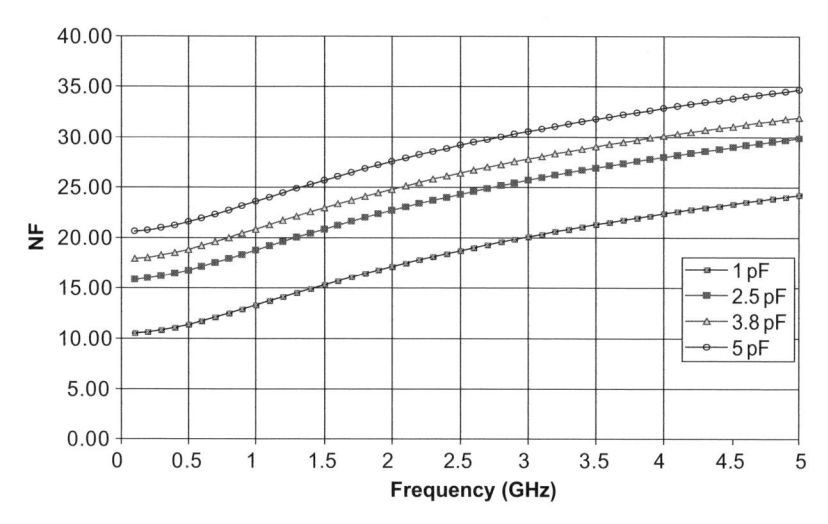

Figure 7-9 Plot of the noise figure versus frequency with respect to feedback capacitor C_1 ($C_2 = 2.2$ pF).

Figure 7-10 illustrates the strong dependency of the noise factor on C_2, thereby agreeing with the improvement in the phase noise, as shown in Tables 6-3 and 6-4. By substituting $1/r_e$ for Y_{21}^+, we obtain

$$F = 1 + \frac{C_2 C_C}{(C_1 + C_2)C_1 r_e} \left[r_b + \frac{1}{2r_e \beta}\left(r_b + \frac{(C_1 + C_2)C_1 r_e}{C_2 C_C}\right)^2 \right.$$
$$\left. + \frac{r_e}{2} + \frac{1}{2r_e}\left(r_b + \frac{(C_1 + C_2)C_1 r_e}{C_2 C_C}\right)^2 \left(\frac{f^2}{f_T^2}\right) \right] \tag{7-33}$$

When an isolating amplifier is added, the noise of an LC oscillator system is determined by

$$S_\phi(f_m) = [a_R F_0^4 + a_E(F_0/(2Q_L))^2]/f_m^3$$
$$+ [(2GFkT/P_0)(F_0/(2Q_L))^2]/f_m^2$$
$$+ (2a_R Q_L F_0^3)/f_m^2$$
$$+ a_E/f_m + 2GFkT/P_0 \tag{7-34}$$

where

$G =$ compressed power gain of the loop amplifier
$F =$ noise factor of the loop amplifier
$k =$ Boltzmann's constant

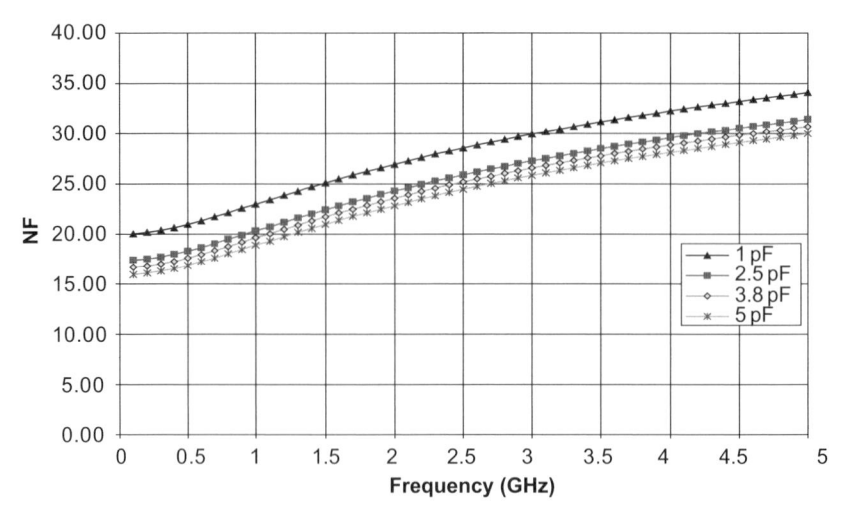

Figure 7-10 Plot of the noise figure versus frequency with respect to feedback capacitor C_2 ($C_1 = 3.3$ pF).

T = temperature in Kelvin
P_0 = carrier power level (in watts) at the output of the loop amplifier
F_0 = carrier frequency in Hz
f_m = carrier offset frequency in Hz
$Q_L = (\pi F_0 \tau_g)$ = loaded Q of the resonator in the feedback loop
a_R and a_E = flicker noise constants for the resonator and loop amplifier, respectively [74]

It is important to understand that the Leeson model is the best case since it assumes that the tuned circuit filters out of all the harmonics. In all practical cases, it is hard to predict the operating Q and the noise figure. When comparing the measured results of oscillators with the assumptions made in Leeson's equation, one frequently obtains a *de facto* noise figure in the vicinity of 20 to 30 dB and an operating Q that is different from the assumed loaded Q. In attempting to match the Leeson calculated curve, the measured curve requires totally different values from those assumed. The basic concept of the Leeson equation, however, is correct, and if each of the three unknown terms is inserted properly, the computed results will agree with the measurements. The information that is not known prior to measurement is

the output power
the noise figure under large-signal conditions
the loaded (operational) noise figure

Example: The following is a validation example for this approach.

For an output power of 13 dBm, $C_1 = 3.3$ pF, $C_2 = 2.2$ pF, $Y_{21}^+ = 2$ mS, $Q_0 = 1000$, and $Q^+ = 618$ loading from the transistor, the resulting noise factor is 104, or roughly 20 dB, and the total loaded Q is 380. Figure 7-11 shows the phase noise, including the flicker corner frequency of 10 kHz. This is one way of calculating the phase noise. The result is very close to the CAD simulations and measurements, but it is incomplete because many transistor parameters are not considered, which would increase the accuracy. The formula, however, does not allow us to enter more parameters.

7.2 THE LEE AND HAJIMIRI NOISE MODEL

The Hajimiri and Lee noise model [65] is based on the nonlinear time-varying (NLTV) properties of the oscillator current waveform. The phase noise analysis is based on the effect of the noise impulse on a periodic signal. Figure 7-12 shows the noise signal in response to the injected impulse current at two different times, peak and zero crossing.

As illustrated in Figure 7-12, if an impulse is injected into the tuned circuit at the peak of the signal, it will cause maximum amplitude modulation and no phase

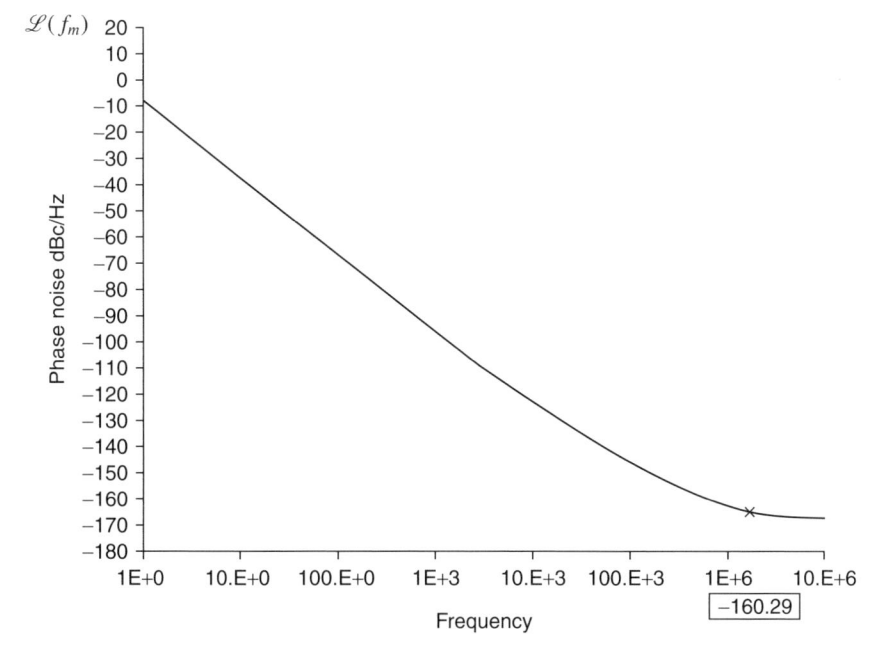

Figure 7-11 Predicted phase noise for an oscillator using the values above. It agrees well with actual measurements.

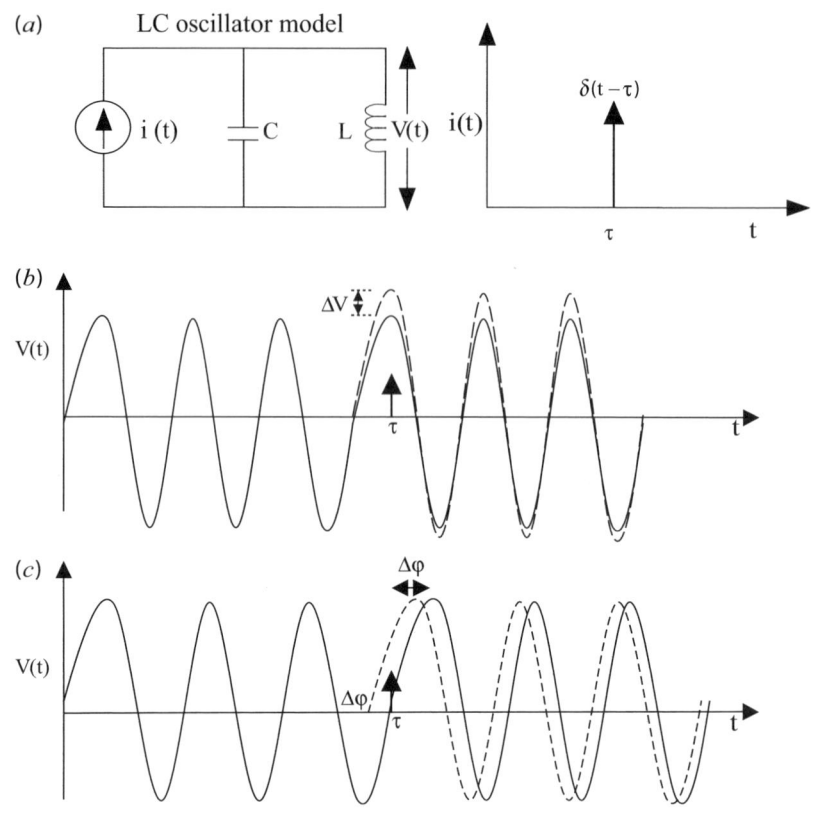

Figure 7-12 (a) LC oscillator excited by the current pulse, (b) impulse injected at the peak of the oscillation signal, and (c) impulse injected at zero crossing of the oscillation signal.

modulation, whereas if an impulse is injected at the zero crossing of the signal, there will be no amplitude modulation, only maximum phase modulation. If noise impulses are injected between the zero crossing and the peak, there will be components of both the phase and the amplitude modulation. Variations in amplitude are generally ignored because they are limited by the gain control mechanism of the oscillator. Therefore, according to this theory, to obtain the minimal phase noise, special techniques have to be adopted so that any noise impulse coincides in time with the peaks of the output voltage signal rather than at the zero crossing or between the zero crossing and the peak.

Hajimiri and Lee introduced an impulse sensitivity function (ISF) based on the injected impulse, which is different for each topology of the oscillator. It has the largest value when the maximum phase modulation occurs and the smallest value when only the amplitude modulation occurs. The calculation of the ISF is tedious and depends upon the topology of the oscillator. Based on this theory, the phase

7.3 NONLINEAR APPROACH TO THE CALCULATION OF OSCILLATOR

noise equation is expressed as [64]

$$\mathscr{L}(f_m) = \begin{cases} 10\log\left[\dfrac{C_0^2}{q_{max}^2} * \dfrac{i_n^2/\Delta f}{8f_m^2} * \dfrac{\omega_{1/f}}{f_m}\right] & \dfrac{1}{f^3} \to \text{region} \\ 10\log\left[10\log\left[\dfrac{\Gamma_{rms}^2}{q_{max}^2} * \dfrac{i_n^2/\Delta f}{4f_m^2}\right]\right] & \dfrac{1}{f^2} \to \text{region} \end{cases} \quad (7\text{-}35)$$

where

$i_n^2/\Delta f$ = noise power spectral density
Δf = noise bandwidth
$\Gamma_{rms}^2 = 1/\pi \int_0^{2\pi} |\Gamma(x)|^2 dx = \sum_{n=0}^{\infty} C_n^2$ = root mean square (RMS) value of $\Gamma(x)$
$\Gamma(x) = C_0/2 + \sum_{n=1}^{\infty} C_n \cos(nx + \theta_n)$ = ISF
C_n = Fourier series coefficient
C_0 = 0th order of the ISF (Fourier series coefficient)
θ_n = phase of the nth harmonic
f_m = offset frequency from the carrier
$\omega_{1/f}$ = flicker corner frequency of the device
q_{max} = maximum charge stored across the capacitor in the resonator.

Equation (7-35) gives good results once all the data are known, but it does not lead to exact design rules.

7.3 NONLINEAR APPROACH TO THE CALCULATION OF OSCILLATOR PHASE NOISE

The mechanism of noise generation in an oscillator combines the equivalent of frequency conversion (mixing) with the effect of AM-to-PM conversion. Therefore, to calculate oscillator phase noise, we must first be able to calculate the noise figure of a mixer [75]. This section describes the use of an algorithm for the computation of SSB carrier noise in free-running oscillators using the harmonic balance (HB) nonlinear technique.

Traditional approaches relying on frequency conversion analysis are not sufficient to describe the complex physical behavior of a noisy oscillator. The accuracy of this nonlinear approach is based on the dynamic range of the HB simulator and the quality of the parameter extraction for the active device. The algorithm described has also been verified with several examples up to millimeter wavelengths. This is the only algorithm that provides a complete and rigorous treatment of noise analysis for autonomous circuits without restrictions such as the inability to handle memory effects.

7.3.1 Noise Generation in Oscillators

As shown above, the qualitative linearized picture of noise generation in oscillators is very well known. The physical effects of random fluctuations taking place in the circuit differ, depending on their spectral allocation with respect to the carrier.

- Noise components at low-frequency deviations result in frequency modulation of the carrier through mean square frequency fluctuation proportional to the available noise power.
- Noise components at high-frequency deviations result in phase modulation of the carrier through mean square phase fluctuation proportional to the available noise power.

We will demonstrate that the same conclusions can be quantitatively derived from the HB equations for an autonomous circuit.

7.3.2 Equivalent Representation of a Noisy Nonlinear Circuit

A general noisy nonlinear network can be described by the equivalent circuit shown in Figure 7-13 and Appendix D. The circuit is divided into linear and nonlinear subnetworks as noise-free multiports. Noise generation is accounted for by connecting a set of noise voltage and noise current sources at the ports of the linear subnetwork [76–80].

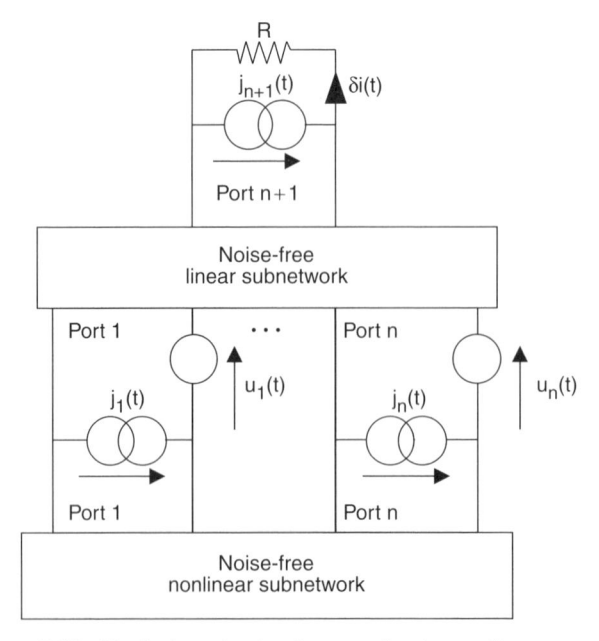

Figure 7-13 Equivalent circuit of a general noisy nonlinear network.

7.3.3 Frequency Conversion Approach

The circuit supports a large-signal, time-periodic steady state of fundamental angular frequency ω_0 (carrier). Noise signals are small perturbations superimposed on the steady state, represented by families of pseudo-sinusoids located at the sidebands of the carrier harmonics. Therefore, the noise performance of the circuit is determined by the exchange of power among the sidebands of the unperturbed steady state through frequency conversion in the nonlinear subnetwork. Due to the perturbative assumption, the nonlinear subnetwork can be replaced with a multifrequency linear multiport described by a conversion matrix. The flow of noise signals can be computed by means of conventional linear circuit techniques.

The frequency conversion approach frequently used has the following limitations:

The frequency conversion approach is not sufficient to predict the noise performance of an autonomous circuit. The spectral density of the output noise power, and consequently the PM noise computed by the conversion analysis, are proportional to the available power of the noise sources.

- In the presence of both thermal and flicker noise sources, PM noise rises as ω^{-1} for $\omega \rightarrow 0$ and tends toward a finite limit for $\omega \rightarrow \infty$.
- Frequency conversion analysis correctly predicts the far carrier noise behavior of an oscillator, and in particular the oscillator noise floor. It does not provide results consistent with the physical observations at low deviations from the carrier.

This inconsistency can be removed by adding the modulation noise analysis. In order to determine the faraway noise using the autonomous circuit perturbation analysis, the following applies:

The circuit supports a large-signal, time-periodic autonomous regime. The circuit is perturbed by a set of small sources located at the carrier harmonics and at the sidebands at a deviation ω from carrier harmonics. The perturbation of the circuit state $(\delta \mathbf{X}_B, \delta \mathbf{X}_H)$ is given by the uncoupled sets of equations

$$\frac{\partial \mathbf{E}_B}{\partial \mathbf{E}_B} \delta \mathbf{X}_B = \mathbf{J}_B(\omega) \qquad (7\text{-}36)$$

$$\frac{\partial \mathbf{E}_H}{\partial \mathbf{E}_H} \delta \mathbf{X}_H = \mathbf{J}_H(\omega) \qquad (7\text{-}37)$$

where

$\mathbf{E}_B, \mathbf{E}_H$ = vectors of HB errors

$\mathbf{X}_B, \mathbf{X}_H$ = vectors of state variable harmonics (since the circuit is autonomous, one of the entries $\mathbf{X}$ is replaced by the fundamental frequency ω_0)

$\mathbf{J}_B, \mathbf{J}_H$ = vectors of forcing terms

The subscripts B and H denote sidebands and carrier harmonics, respectively.

For a spot noise analysis at a frequency ω, the noise sources can be interpreted in either of two ways:

- Pseudo-sinusoids with random amplitude and phase located at the sidebands. Noise generation is described by equation (7-36), which is essentially a frequency conversion equation relating the sideband harmonics of the state variables and of the noise sources. This description is exactly equivalent to the one provided by the frequency conversion approach. This mechanism is referred to as *conversion noise* [70–79].
- Sinusoids located at the carrier harmonics, randomly phase- and amplitude-modulated by pseudo-sinusoidal noise at frequency ω. Noise generation is described by equation (7-37), which describes noise-induced jitter of the circuit state, represented by the vector $\delta \mathbf{X}_H$. The modulated perturbing signals are represented by replacing the entries of $\mathbf{J}_H$ with the complex modulation laws. This mechanism is referred to as *modulation noise*. One of the entries of $\delta \mathbf{X}_H$ is $\delta \omega_0$, where $\delta \omega_0(\omega)$ = phasor of the pseudo-sinusoidal components of the fundamental frequency fluctuations in a 1 Hz band at frequency ω. Equation (7-37) provides a frequency jitter with a mean square value proportional to the available noise power. In the presence of both thermal and flicker noise, PM noise rises as ω^{-3} for $\omega \to 0$ and tends to 0 for $\omega \to \infty$. Modulation noise analysis correctly describes the noise behavior of an oscillator at low deviations from the carrier and does not provide results consistent with physical observations at high deviations from the carrier.

The combination of both phenomena explains the noise in the oscillator shown in Figure 7-14, where near carrier noise dominates below ω_X and far carrier noise dominates above ω_X. Figure 7-15 (itemized form) shows the noise sources as they are applied at the IF. We have arbitrarily defined the low oscillator output as IF. This applies to the conversion matrix calculation.

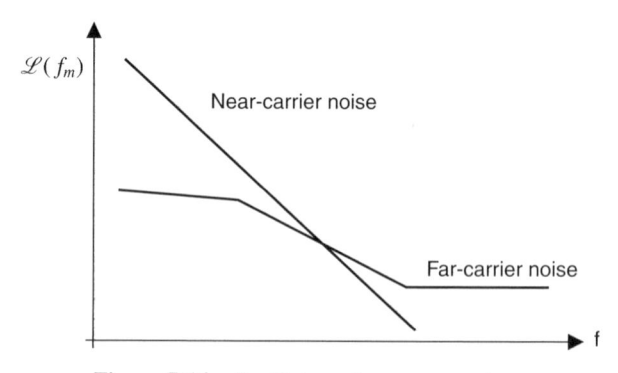

Figure 7-14 Oscillator noise components.

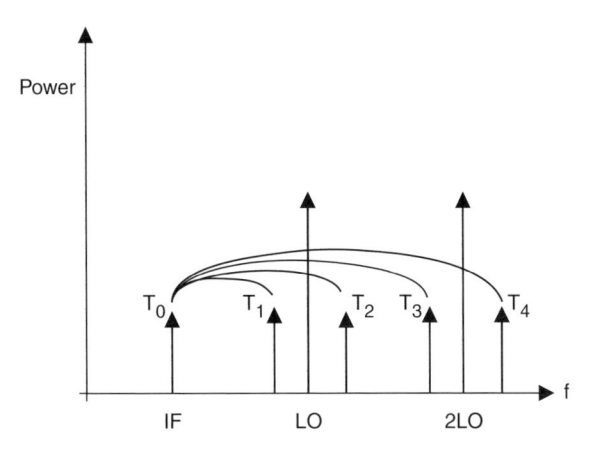

7.3 NONLINEAR APPROACH TO THE CALCULATION OF OSCILLATOR 143

Figure 7-15 Noise sources where the noise at each sideband contributes to the output noise at the IF through frequency conversion.

Figure 7-16 shows the total contribution which has to be taken into consideration for calculation of the noise at the output. The accuracy of the phase noise calculation depends strongly on the quality of the parameter extraction for the nonlinear device; in particular, high-frequency phenomena must be properly modeled. In addition, the flicker noise contribution is essential. This is also true in mixer noise analysis.

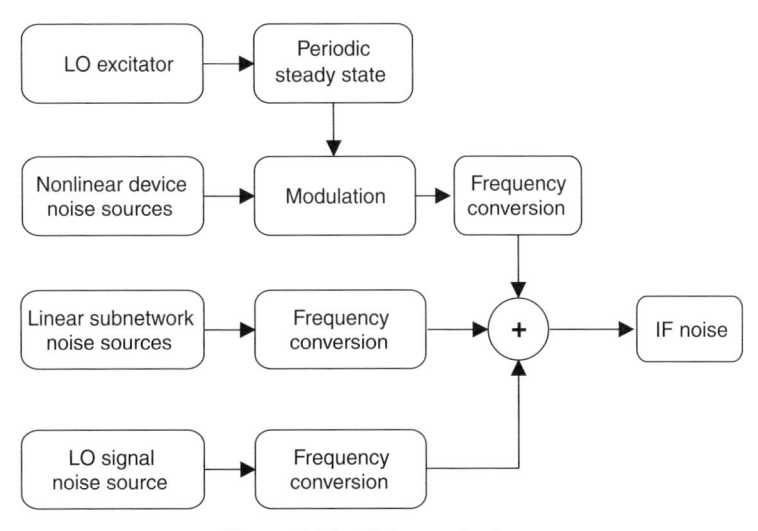

Figure 7-16 Noise mechanisms.

7.3.4 Conversion Noise Analysis

The mathematics used to calculate the noise result (Ansoft Serenade 8.x) are as follows:

kth harmonic PM noise:

$$\langle|\delta\Phi_k(\omega)|^2\rangle = \frac{N_k(\omega) - N_{-k}(\omega) - 2\operatorname{Re}[C_k(\omega)]}{R|I_k^{SS}|^2} \quad (7\text{-}38)$$

kth harmonic AM noise:

$$\langle|\delta A_k(\omega)|^2\rangle = 2\frac{N_k(\omega) - N_{-k}(\omega) + 2\operatorname{Re}[C_k(\omega)]}{R|I_k^{SS}|^2} \quad (7\text{-}39)$$

kth harmonic PM-AM correlation coefficient:

$$C_k^{PMAM}(\omega) = \langle\delta\Phi_k(\omega)\delta A_k(\omega)^*\rangle = -\sqrt{2}\frac{2\operatorname{Im}[C_k(\omega)] + j[N_k(\omega) - N_{-k}(\omega)]}{R|I_k^{SS}|^2} \quad (7\text{-}40)$$

where

$N_k(\omega), N_{-k}(\omega)$ = noise power spectral densities at the upper and lower sidebands of the kth harmonic

$C_k(\omega)$ = normalized correlation coefficient of the upper and lower sidebands of the kth carrier harmonic

R = load resistance

I_k^{SS} = kth harmonic of the steady-state current through the load

7.3.5 Modulation Noise Analysis

kth harmonic PM noise:

$$\langle|\delta\Phi_k(\omega)|^2\rangle = \frac{k^2}{\omega^2}\mathbf{T}_F\rangle\mathbf{J}_H(\omega)\mathbf{J}_H^t(\omega)\langle\mathbf{T}_F^t \quad (7\text{-}41)$$

kth harmonic AM noise:

$$\langle|\delta A_k(\omega)|^2\rangle = \frac{2}{|I_k^{SS}|^2}\mathbf{T}_{Ak}\rangle\mathbf{J}_H(\omega)\mathbf{J}_H^t(\omega)\langle\mathbf{T}_{Ak}^t \quad (7\text{-}42)$$

7.3 NONLINEAR APPROACH TO THE CALCULATION OF OSCILLATOR   145

kth harmonic PM-AM correlation coefficient:

$$C_k^{PMAM}(\omega) = \rangle \delta\Phi_k(\omega)\delta A_k(\omega)^* \left\langle = \frac{k\sqrt{2}}{j\omega|I_k^{SS}|^2} \mathbf{T}_F \right\rangle \mathbf{J}_H(\omega)\mathbf{J}_H^t(\omega) \langle \mathbf{T}_{Ak}^t \quad (7\text{-}43)$$

where

$\mathbf{J}_H(\omega)$ = vector of the Norton equivalent of the noise sources
$\mathbf{T}_F$ = frequency transfer matrix
I_k^{SS} = kth harmonic of the steady-state current through the load

7.3.6 Experimental Variations

We will look at two examples comparing predicted and measured data. Figure 7-17 shows the abbreviated circuit of a 10 MHz crystal oscillator. It uses a high-precision, high-Q crystal made by companies such as Bliley. Oscillators like this are intended for use as frequency and low-phase noise standards. In this case, the circuit under consideration is part of the Hewlett Packard (HP) 3048 phase noise measurement system.

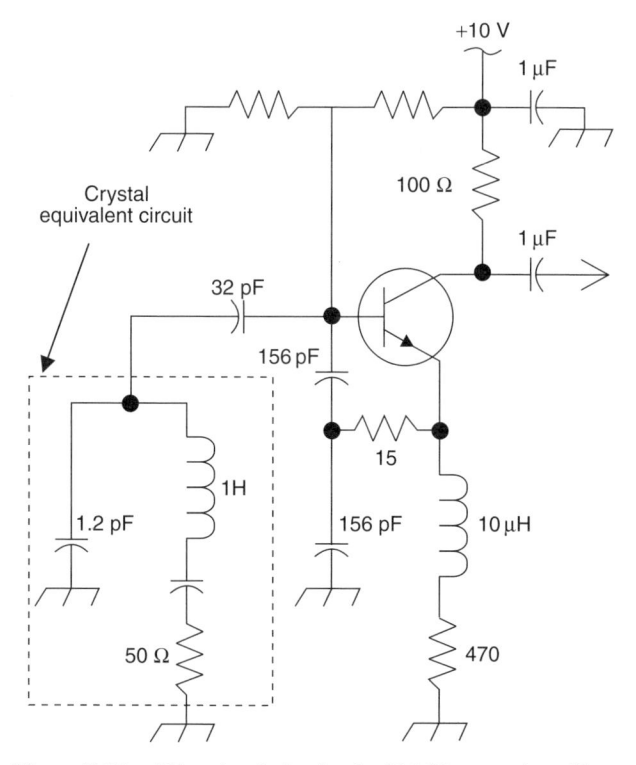

Figure 7-17 Abbreviated circuit of a 10 MHz crystal oscillator.

146   NOISE IN OSCILLATORS

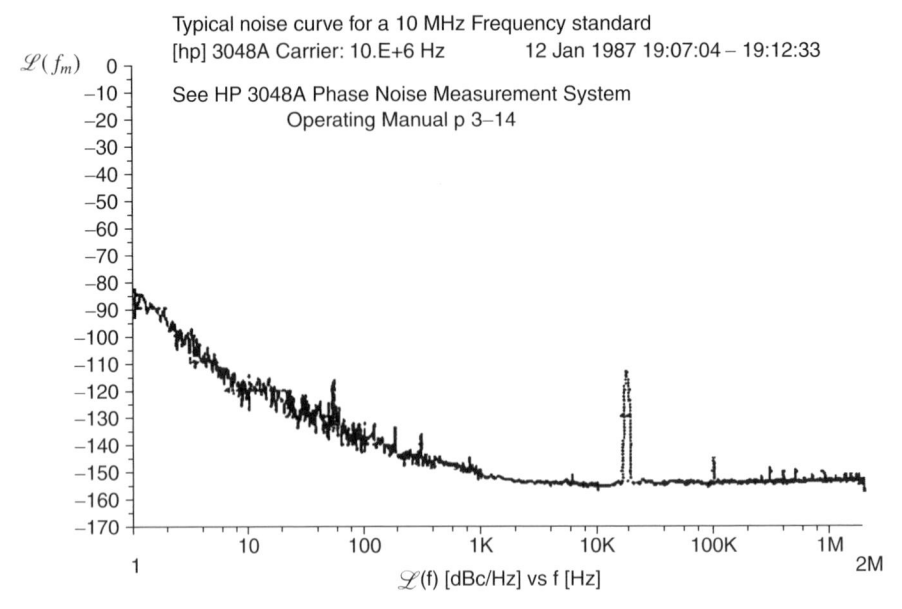

Figure 7-18 Measured phase noise for this frequency standard by HP.

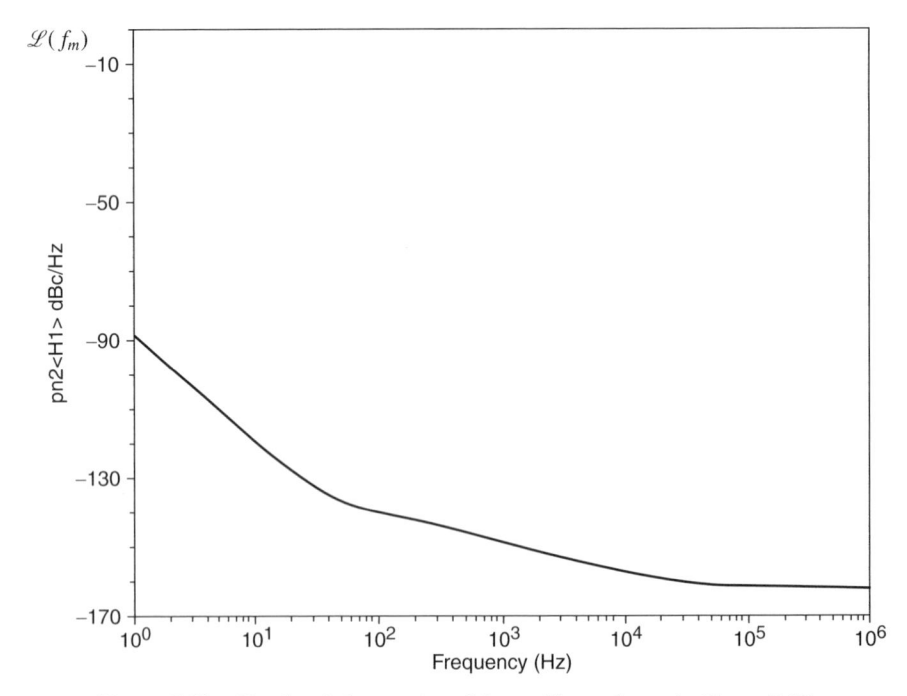

Figure 7-19 Simulated phase noise of the oscillator shown in Figure 7-17.

7.3 NONLINEAR APPROACH TO THE CALCULATION OF OSCILLATOR   147

Figure 7-18 shows the measured phase noise of this HP frequency standard, and Figure 7-19 shows the phase noise predicted using the mathematical approach outlined above.

In cooperation with Motorola, an 800 MHz VCO was analyzed. The parameter extraction for the Motorola transistor was done. Figure 7-20 shows the circuit, which is a Colpitts oscillator that uses RF feedback in the form of a 15 Ω resistor between the emitter and the capacitive voltage divider. The tuned circuit is loosely coupled to this part of the transistor circuit. Figure 7-21 shows a plot of the measured and predicted phase of this circuit.

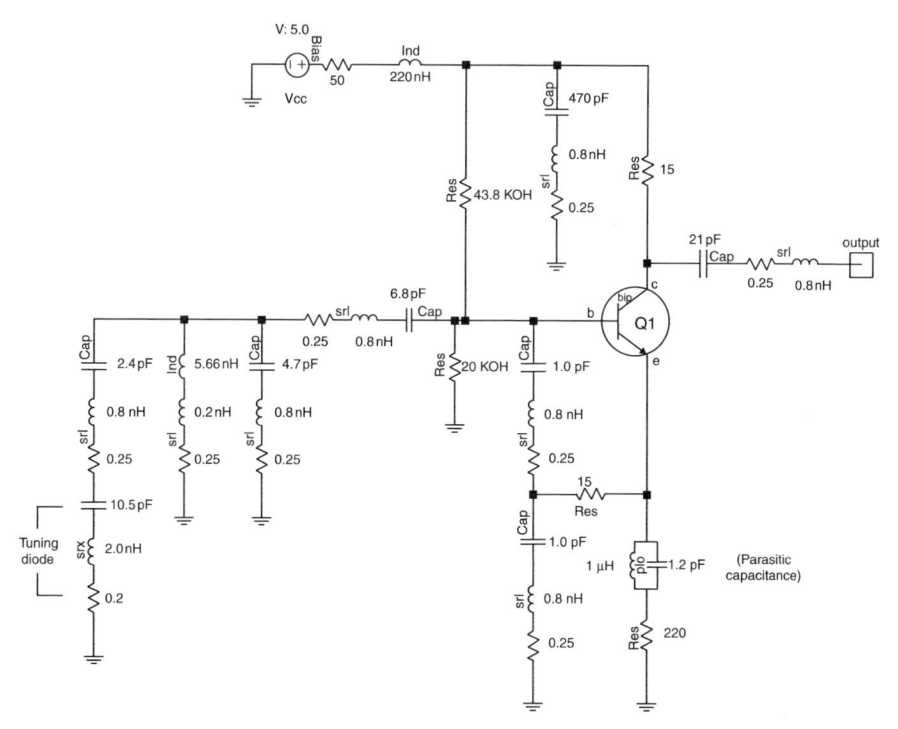

Figure 7-20 Colpitts oscillator that uses RF negative feedback between the emitter and capacitive voltage divider. To be realistic, we have also used real components rather than ideal ones. The suppliers for the capacitors and inductors provide some typical values for the parasitics. The major changes are 0.8 nH and 0.25 Ω in series with the capacitors. The same applies to the main inductance, which has a parasitic connection inductance of 0.2 nH in series with a 0.25 Ω resistance. These types of parasitics are valid for a fairly large range of components assembled in surface mount applications. Most engineers model the circuit only by assuming lossy devices, not adding these important parasitics. One of the side effects we have noticed is that the output power is more realistic and, needless to say, the simulated phase noise agrees quite well with measured data. This circuit can also serve as an example for modeling amplifiers and mixers using surface mount components.

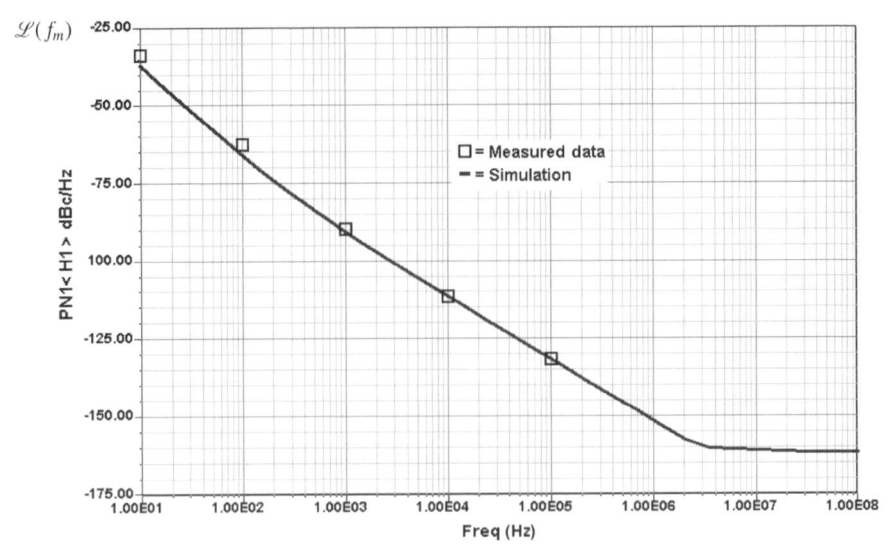

Figure 7-21 Comparison between predicted and measured phase noise for the oscillator shown in Figure 7-20.

The HB simulator Microwave Harmonica/Designer, which has been used for many cross-checks, is based on the principle shown above and is extremely accurate in making noise predictions. The HB simulator cannot be substituted for by a set of equations because the HB process works iteratively. Therefore, such a simulator is necessary in parallel to all the equations which are being derived to be used as a validation tool. An extreme example of its usefulness was the case where one of the previously shown oscillators was optimized for best output power with an assumed Q of 200 of the resonator circuit itself. In a following simulation, the value of Q was increased from 200 to 400 to examine the results. The output power remained the same, but the phase noise deteriorated. This result contradicts practical experience with optimized oscillators. The surprising result can be explained by the fact that a particular parallel resonant value was assumed, and the values of C_1 and C_2 were calculated. When the Q was increased, the feedback ratio generated too much gain, more than necessary, and therefore, the phase noise became worse. This experiment is normally not done when building oscillators, but it shows the power of the CAD tool and the insight it provides into the workings of the oscillator.

7.4 PHASE NOISE MEASUREMENTS

The single-sideband phase noise of an oscillator has been the subject of many discussions, but how can it be measured?

7.4.1 Spectrum Analyzer

The first and simplest approach is to use a spectrum analyzer which has a sufficiently low phase noise oscillator. The phase noise is measured on the screen as a function of the offset off the carrier. Modern spectrum analyzers, such as the Rohde & Schwarz FSU series, have a carrier phase noise option built in. The phase noise is measured in a normalized 1 Hz bandwidth. A bandwidth of 1 Hz is best realizable in a DSP-based IF stage, but then the measurement time would be huge. A better way to do this is to adjust the bandwidth to be approximately 10% or more off the carrier. As an example, when measuring at 100 Hz off the carrier, a 10 Hz bandwidth should be used; at 100 kHz off the carrier, a wider bandwidth, such as 10 kHz, can be used. Most analyzers have an intelligent built-in option (program) which automatically selects the proper sweep speed and bandwidth for this purpose. Since all high-performance spectrum analyzers have synthesized local oscillators with wide loop bandwidths, the phase noise of the analyzers at frequencies deviations of about 10 kHz typically exceeds the performance of the device under test. High-Q oscillators become a problem because they can have better phase noise performance than the spectrum analyzer. Crystal oscillators, ceramic resonator oscillators, SAW oscillators, and dielectric resonator oscillators fall into this category.

7.4.2 Phase Noise Test Setup

Several phase noise test setups are available. The best instruments are made by Agilent (HP) and Komstron (EuroTest). Figure 7-22 shows the popular HP phase noise setup.

This system consists of a base unit with a phase detector and an amplifier, several signal generators, and a DC output. The test setup shown in Figure 7-22 is configured to determine the minimum noise floor of the system to make sure that there

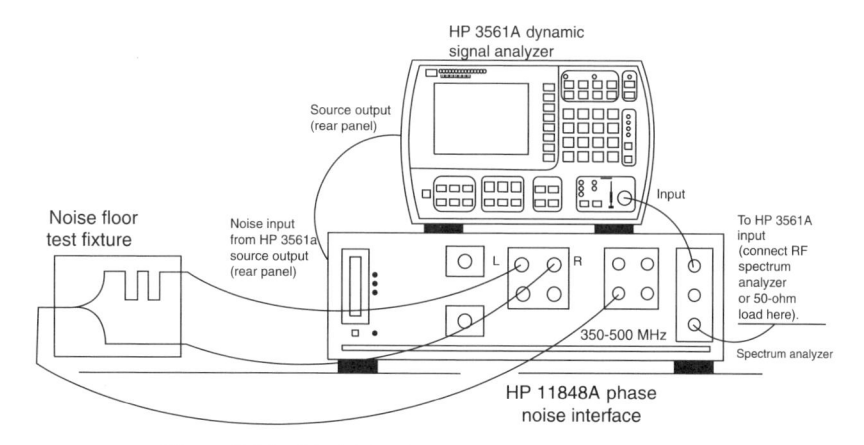

Figure 7-22 Model 3048A-based phase noise test setup.

is enough dynamic range. This is accomplished by taking the output of one of the built-in signal sources, splitting the power, and feeding the outputs into the built-in high-level, double-balanced mixer. One of the outputs is delayed in phase by the delay line shown on the left side of the figure. As a result, the double-balanced mixer receives the identical frequency into the RF and LO input, and there is a phase difference between the two signals. Because there is no difference in frequency between the two signals, the resulting output is a DC voltage with the signal generator's noise on top of it. Because of the conversion to zero IF, the analyzer connected to the output needs to cover DC (10 Hz or less) to 1–10 MHz to analyze the noise. Since there are no requirements for linearity for the mixing process, the RF levels, both of the RF and LO input, can be the same.

Using a fast Fourier transform (FFT) analyzer, Figure 7-23 shows the noise floor of the system. This is low because the signal source in question is a high-performance crystal oscillator with a much better phase noise performance than any practical device under test (DUT).

Figure 7-24 shows the block diagram of the test setup with a small modification. Instead of having one oscillator with a power splitter and a delay line, it uses two separate oscillators. This arrangement is commonly used to measure medium-quality oscillators. As long as signal generator 1 is at least 10 dB better than signal generator 2, this approach is valid. If both signal generators are identical, then there would be a 3 dB correction factor. It is good practice to synchronize oscillator 1 against oscillator 2. The phase noise setup system shown in Figure 7-24 has several frequency standards built in. The high-quality 10 MHz reference oscillator output can be used to synchronize oscillator 1 if this is a synthesized signal generator. This is valid in most cases. If oscillator 2 is a VCO, the test setup provides a DC control voltage, which can phase lock oscillator 2 in the system. Depending on the measurement offset, the system adjusts the loop bandwidth.

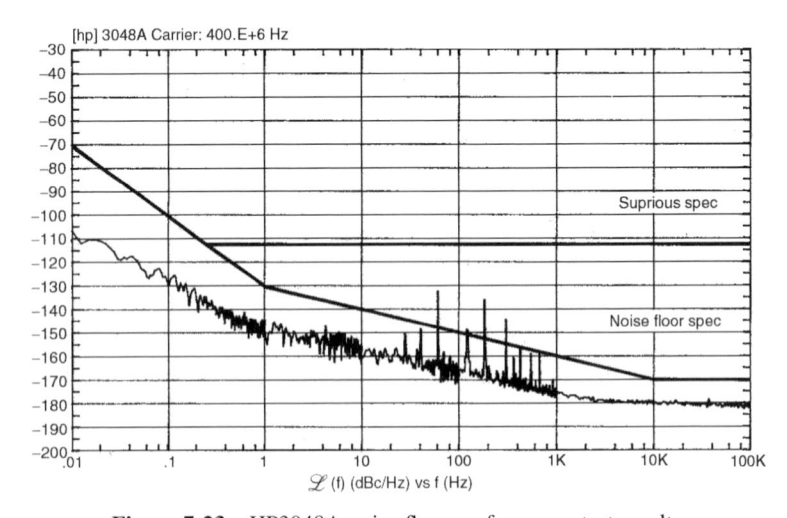

Figure 7-23 HP3048A noise floor performance test results.

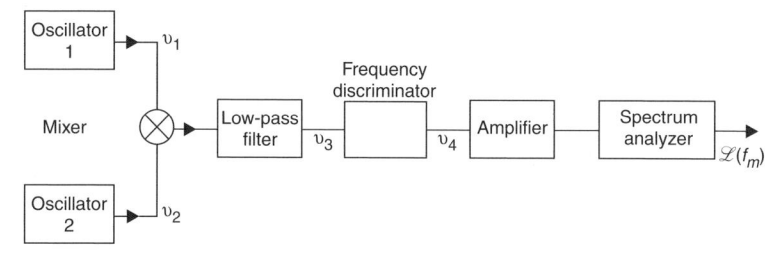

Figure 7-24 Block diagram of the principle of the HP3048A system. There is an additional DC FM feedback loop to phase lock one of the oscillators.

Both systems have advantages and disadvantages. The delay line principle is limited by the delay, which is a sine x/x function that repeats. The delay line measurement system requires several delay lines to cover a wide range of measurements. More details about the limitation of the cable measurement can be found in the system reference manual.

Figure 7-25 shows the area for which the delay line measurements are valid. This is frequency dependent and, of course, depends on the length of the delay line.

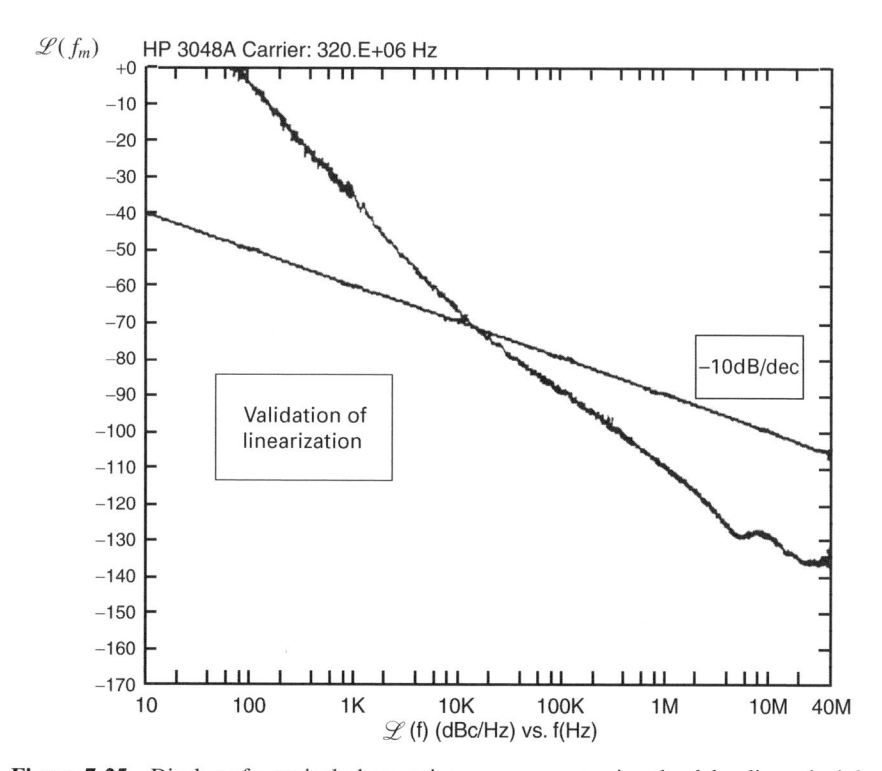

Figure 7-25 Display of a typical phase noise measurement using the delay line principle. This method is applicable only where $x \approx \sin(x)$. The measured values above the solid line violate this relationship and therefore are not valid.

152 NOISE IN OSCILLATORS

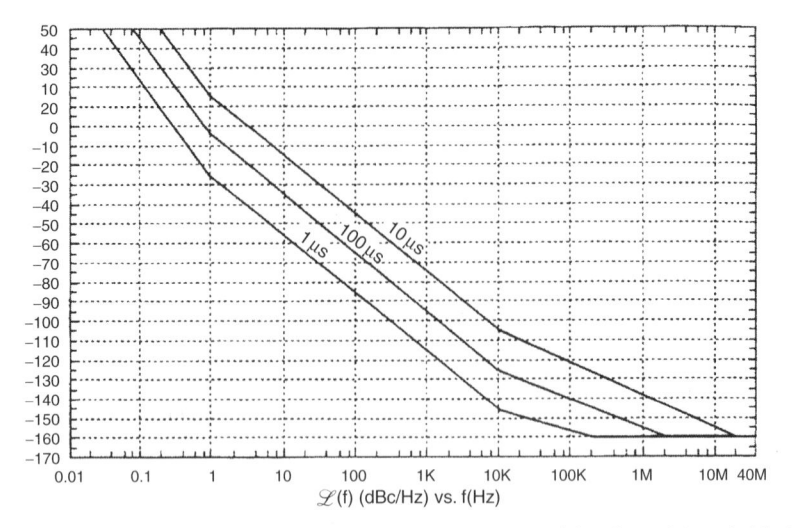

Figure 7-26 Dynamic range as a function of cable delay. A delay line of 1 μs is ideal for microwave frequencies.

By choosing the appropriate delay length, the dynamic range can be controlled. As an example, the 1 μS delay line is ideal for most microwave frequencies. The limit of −160 dBc/Hz was due to the system's performance. This can be seen in Figure 7-26. A longer delay line, which is mechanically very bulky and lossy, allows measurements closer in with better resolution. The delay line ideally is adjustable, which guarantees that small phase changes will be made since the signal fed to the mixer should be in quadrature compared to the other input. A more detailed description can be found in the system's manual [81].

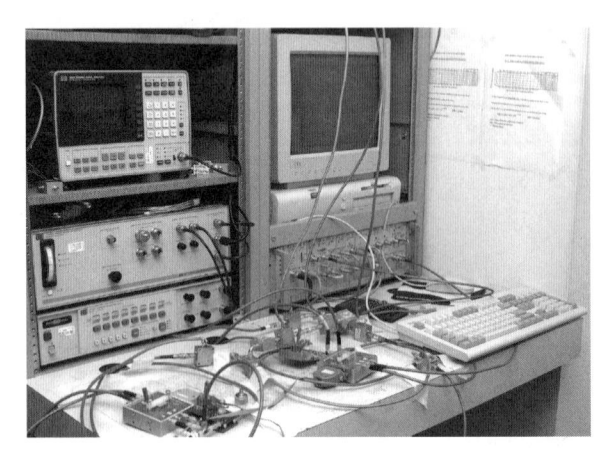

Figure 7-27 Synergy Microwave Corporation's in-house automated test system used to measure oscillator phase noise.

7.5 SUPPORT CIRCUITS

In the case of the two-oscillator (signal generator) measurement, the typical problem is that the synthesized signal generator is not always as good as the high-Q oscillators under test. While for most cases the setup with one signal generator and the DUT is sufficient, the additional delay line should be available to have a complete system.

The measurements shown in this work were taken with the HP3048A and the Euro Test system. Figure 7-27 shows the test station, which houses both systems.

7.5 SUPPORT CIRCUITS

7.5.1 Crystal Oscillators

Phase noise measurements for VCOs and synthesizers require special support circuits which are not readily available. In the case of synthesizers, a very clean reference oscillator is needed. Some of the phase noise test equipment has high-performance 10 or 100 MHz oscillators available. For those engineers who are less fortunate, we show a 20 and a 50 MHz crystal oscillator and their performance.

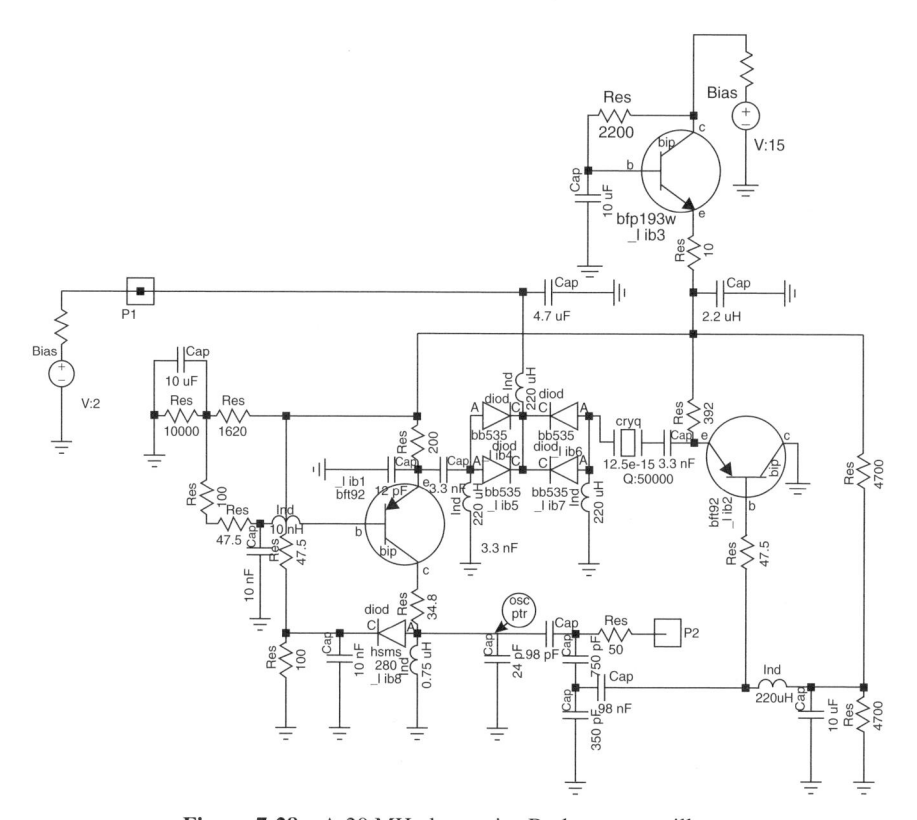

Figure 7-28 A 20 MHz low-noise Butler-type oscillator.

154 NOISE IN OSCILLATORS

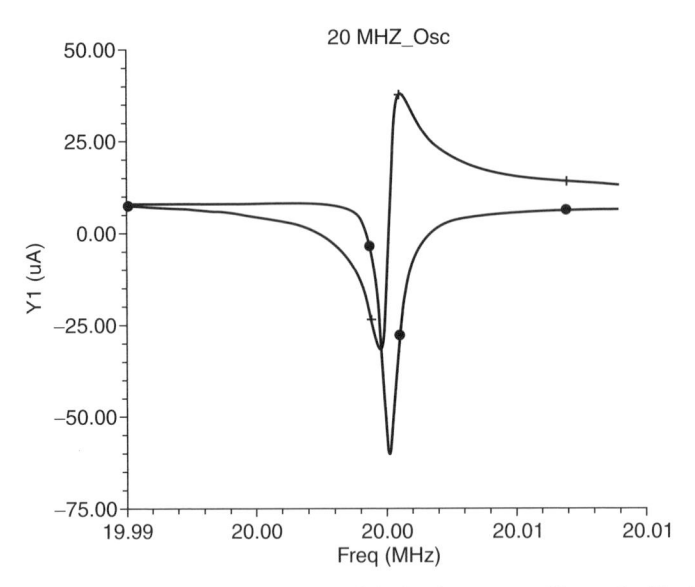

Figure 7-29 Resistive and reactive currents of the Butler-type oscillator. Oscillation occurs at the zero crossing of the reactive current.

Figure 7-28 shows a Butler-type crystal oscillator. This oscillator has the crystal and the frequency adjustment circuit between the two emitters. The hot carrier diode in the collector of the first transistor is responsible for constant output power and operating point for best phase noise. Figure 7-29 shows the resonant condition,

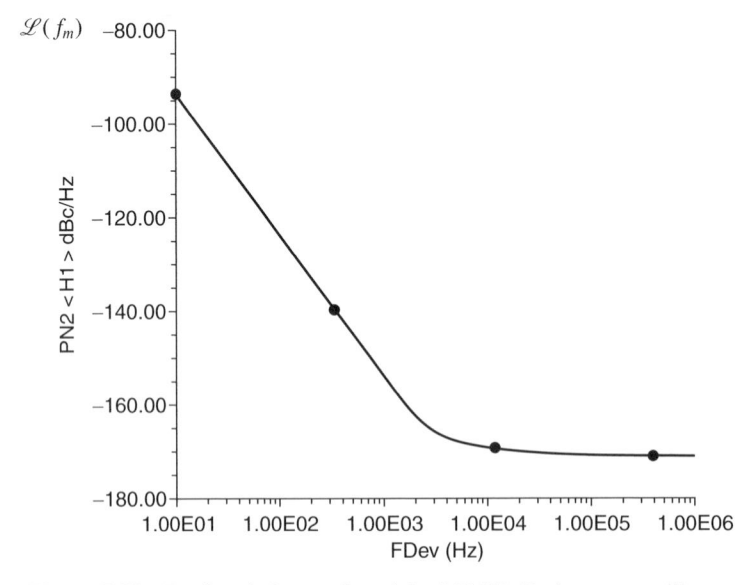

Figure 7-30 Predicted phase noise of the 20 MHz Butler-type oscillator.

7.5 SUPPORT CIRCUITS

which is identified by having the most negative resistance at the zero crossing of the reactive current. Finally, the resulting phase noise prediction is shown in Figure 7-30.

Another popular frequency is 50 MHz. Figure 7-31 shows a 50 MHz crystal oscillator. The crystal in this schematic is shown by its equivalent circuit. The tuning circuit is provided to lock this oscillator against a master standard. The output power is about +13 dBM. The amplifier shown as the output stage is one

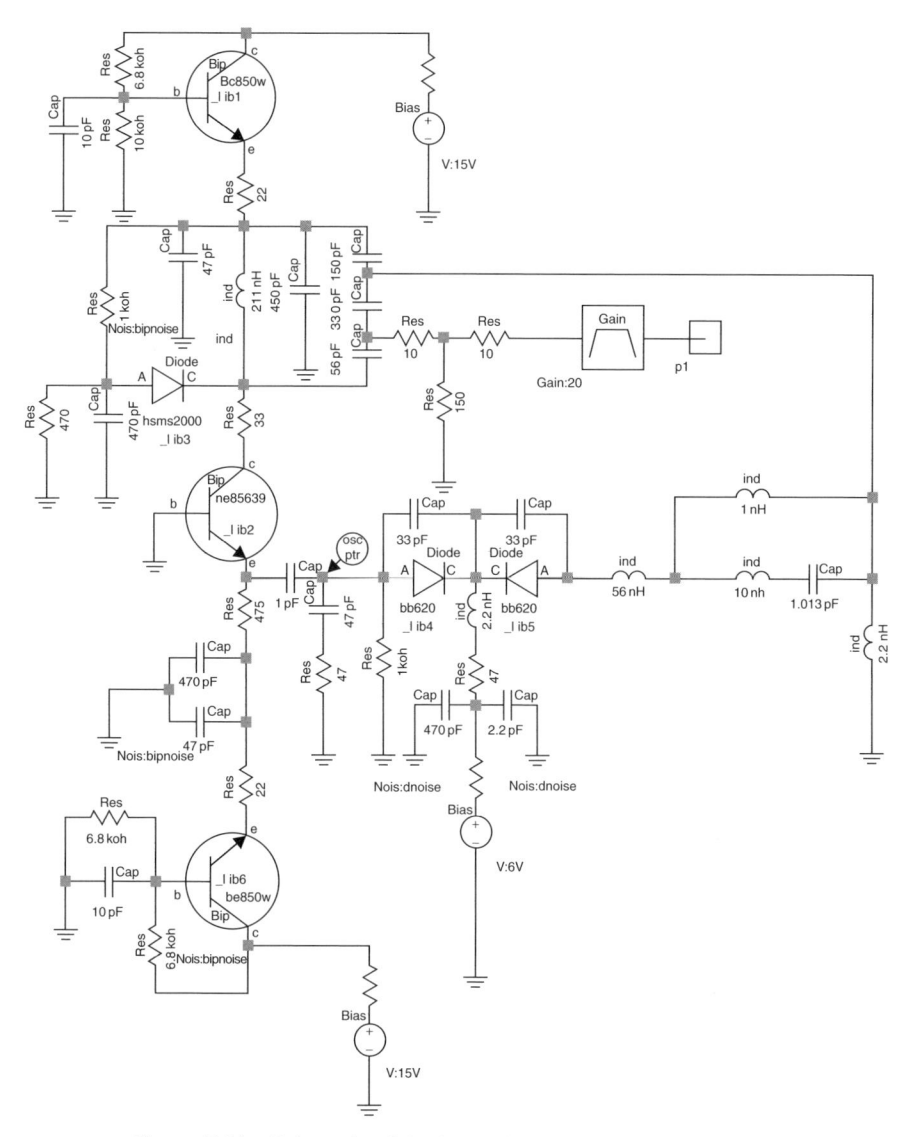

Figure 7-31 Schematic of the low-noise 50 MHz crystal oscillator.

156 NOISE IN OSCILLATORS

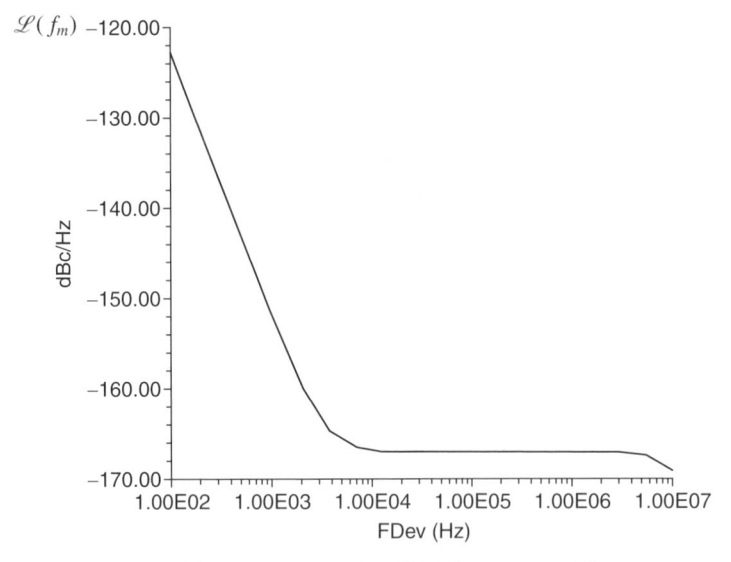

Figure 7-32 Low phase noise 50 MHz crystal oscillator.

of the popular three-terminal, dual Darlington amplifier. The predicted phase noise, which can be seen from Figure 7-32, tracks well with the measured data. These crystal oscillators can be very tricky to build, as most of their performance depends on the crystal itself. Synthetic crystals are not well settled and are much noisier than natural crystals, which are more expensive. While less temperature dependent, the name for a particular crystal cut portions doubly rotated cut for minimum temperature sensitivity center frequency (SC) cut has many spurious resonances and is not recommended.

7.5.2 Comb Generator

Another important circuit is a wideband frequency multiplier, typically in the form of a comb generator. To extend the frequency range of oscillators, it is a good idea to

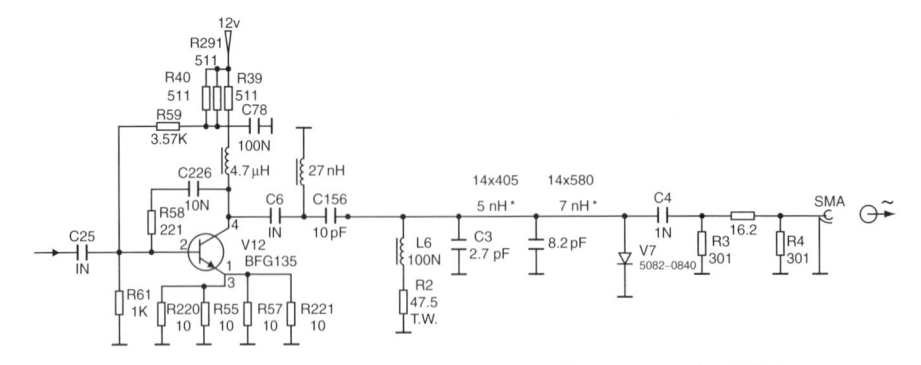

Figure 7-33 Comb generator used to multiply oscillators up to the 5 GHz range.

multiply them up and measure them at the higher frequency. As an example, if a 500 MHz oscillator has very good performance, and therefore is close to the limits of the phase noise measurement system, it is useful to multiply up 10 times and do the same measurement at 5 GHz. By multiplying by 10, the phase noise becomes 20 dB worse. Many modern spectrum analyzers, based on fractional-N synthesizers and the use of YIG oscillators, have low phase noise even at frequencies up to 26 GHz. It is, therefore, useful to take such a spectrum analyzer in the phase noise measurement mode and determine the phase noise of the oscillator that has been multiplied up.

A typical circuit diagram of such a comb generator is shown in Figure 7-33.

It is recommended that a three-stage, three-terminal device amplifier be put ahead of the multiplier. A good combination would be to start with an MSA0986, followed by two MSA1105. In this case, using a 12 V power supply, the DC current is 160 mA. The multiplier works well up to 5 GHz output and accepts input frequencies from 10 MHz to 1 GHz.

8 Calculation and Optimization of Phase Noise in Oscillators

8.1 INTRODUCTION

This chapter develops the computation for the noise properties of an oscillator, and a design guide for best performance will be given. Two methods will be used. One is the calculation of a time-domain-dependent negative resistance that is necessary to enable and sustain oscillation. A bias-dependent noise calculation is possible with this approach. The second approach is a loop gain approach in which the oscillator is considered a closed loop. This allows for the calculation of the impact of the various noise sources in the transistor, a bipolar transistor.

8.2 OSCILLATOR CONFIGURATIONS

The most relevant circuit configuration used for microwave applications is the Colpitts oscillator with a parallel tuned circuit operating in an inductive mode, as shown in Figure 8-1.

8.3 OSCILLATOR PHASE NOISE MODEL FOR THE SYNTHESIS PROCEDURE

8.3.1 Device Phase Noise

Phase noise in an active device is generated by white additive noise as well as by shot noise.

1. For white additive noise, the power spectral density is flat with the frequency. For a device having a noise factor of F, the half sided power spectral density of the phase noise is given as

$$S_{\Delta\theta}(\omega) = \frac{FkT}{P_s} \qquad (8\text{-}1)$$

The Design of Modern Microwave Oscillators for Wireless Applications: Theory and Optimization,
by Ulrich L. Rohde, Ajay Kumar Poddar, Georg Böck
Copyright © 2005 John Wiley & Sons, Inc.

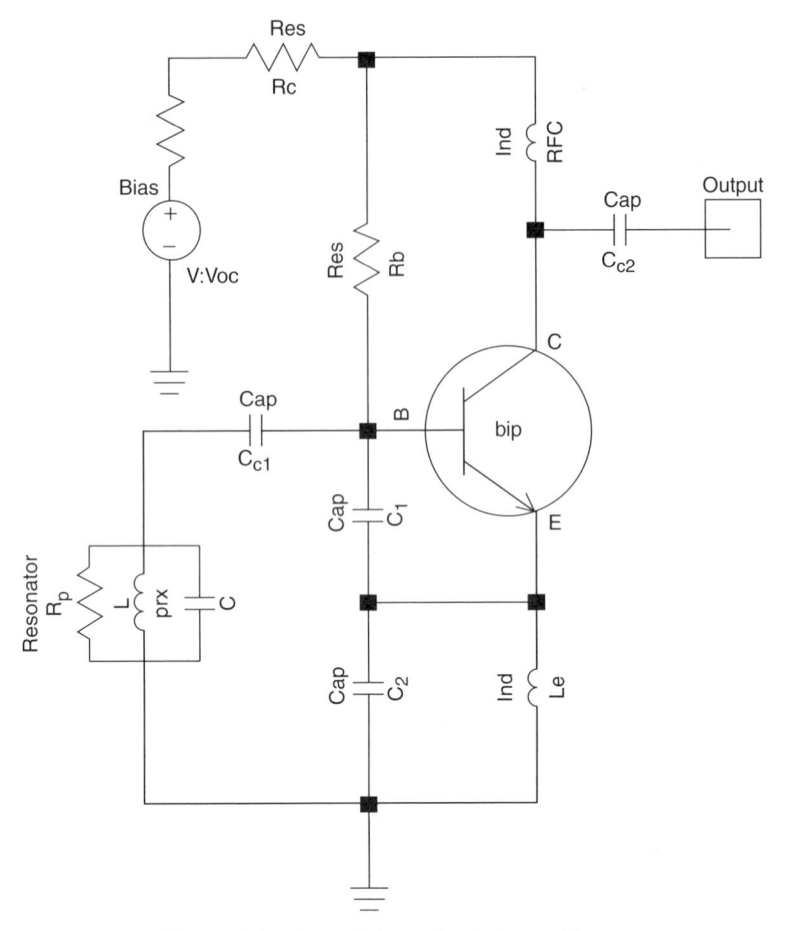

Figure 8-1 A parallel tuned Colpitts oscillator.

where k is Boltzman's constant, T is absolute temperature, and P_s is the signal level at the active device input.

2. For shot noise, the power spectral density of the phase noise representation varies inversely with the frequency and is given by

$$S_{\Delta\theta}(\omega) = \frac{K_a}{\omega} \qquad (8\text{-}2)$$

where K_a is a constant.

3. The total power spectral density of the input phase error can then be written as

$$S_{\Delta\theta}(\omega) = \frac{K_a}{\omega} + \frac{FkT}{P_s} \qquad (8\text{-}3)$$

8.3 OSCILLATOR PHASE NOISE MODEL

For small phase deviations at a frequency offset less than the resonator half bandwidth $\omega_0/2Q$, a phase error at the input to the active element of the oscillator results in a frequency error. This frequency error is determined by the phase frequency relationship of the feedback network.

$$\frac{d\varphi}{dt} = \frac{\omega_0}{2Q_L} \Delta\theta \tag{8-4}$$

Case 1 For noise modulation rates less than the half bandwidth of the feedback loop, the spectrum of the frequency error is identical to the spectrum of the oscillator input phase noise, $S_{\Delta\theta}(\omega)$. The spectrum of the phase error can be given as

$$S_\varphi(\omega) = \left[\frac{\omega_0}{2Q_L}\right]^2 \frac{S_{\Delta\theta}(\omega)}{\omega^2} \quad \text{for } f < \left[\frac{\omega_0}{2Q_L}\right] \tag{8-5}$$

Case 2 For noise modulation rates that are large compared to the half bandwidth of the feedback loop, the series feedback is not effective, and the power spectral density of the output phase, $S_\varphi(\omega)$, is identical to the spectrum of the oscillator input phase noise, $S_{\Delta\theta}(\omega)$.

$$S_\varphi(\omega) = S_{\Delta\theta}(\omega) \tag{8-6}$$

A composite expression for the power spectral density of the output phase is

$$S_\varphi(\omega) = S_{\Delta\theta}(\omega) + \left[\frac{\omega_0}{2Q_L}\right]^2 \frac{S_{\Delta\theta}(\omega)}{\omega^2} \Longrightarrow \left\{1 + \left[\frac{\omega_0}{2Q_L}\right]^2\right\} \frac{S_{\Delta\theta}(\omega)}{\omega^2} \tag{8-7}$$

$$S_{\Delta\theta}(\omega) = \frac{K_a}{\omega} + \frac{FkT}{P_s} \tag{8-8}$$

$$S_\varphi(\omega) = \left[\frac{K_a}{\omega} + \frac{FkT}{P_s}\right]\left\{1 + \left[\frac{\omega_0}{2Q_L\omega}\right]^2\right\} \tag{8-9}$$

The phase noise can be described as a short-term random frequency fluctuation of a signal which is measured in the frequency domain, and is expressed as a ratio of signal power to noise measured in a 1 Hz bandwidth at a given offset from the desired signal frequency.

$$\mathcal{L}(\Delta\omega) = 10\log\left[\frac{P_{sideband}(\omega_0 + \Delta\omega, 1\,\text{Hz})}{P_{carrier}}\right] = 10\log[S_\varphi(f)] \tag{8-10}$$

The term K_a/ω is typically omitted in the phase noise equations, specifically derivatives of the Leeson equation.

8.4 PHASE NOISE ANALYSIS BASED ON THE NEGATIVE RESISTANCE MODEL

The following noise analysis for the oscillator, while based on the approach of Kurokawa [82], is an attempt to introduce the concept of a *noisy* negative resistance, which is time dependent. Kurokawa, addressing the question of synchronized oscillators, provided insight into the general case of a series oscillator. The method introduced here is specific for a real oscillator and real noise sources.

This method starts by connecting a parallel tuned circuit to a transistor in the Colpitts configuration. Since the two capacitors C_1 and C_2 are similar in value, different by no more than a factor of 2 or 3 and connected to a parallel tuned circuit via a small coupling capacitor C_c, the output impedance of the emitter follower circuit is transformed to the base. The differential output impedance at the emitter is $1/Y_{21}^+$ (large signal), while the input impedance itself is $1/Y_{21}^+ \times \beta$. Because of this, the contribution of Y_{11} can be neglected for the basic analysis. This is valid only for this particular case. Consistent with equation (6-1), which is based on the same approximation but includes the parasitics, the transistor circuit now provides negative resistance (or negative conductance). This negative conductance cancels the losses concentrated in the loss resistor R_p, which for infinite Q would also be infinite.

Figure 8-2 is a Colpitts oscillator arrangement, simplified for the purpose of showing the circuit components. On the left side is the resonator tank circuit with the loss resistor R_p, and on the right side is the negative conductance, which is time dependent. The time dependence comes from the fact that the collector

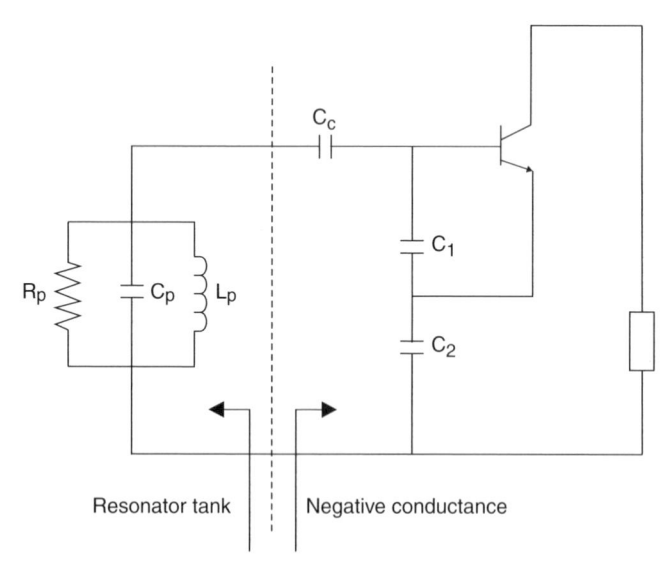

Figure 8-2 Colpitts oscillator configuration for the intrinsic case, no parasitics assumed, and an ideal transistor considered.

8.4 NEGATIVE RESISTANCE MODEL

current is a series of pulses and the negative conductance is present for only a short time. This explanation is necessary to justify the existence of a loaded Q. If a negative resistance or conductance was present all the time, the compensating circuit would reduce the bandwidth to essentially zero or an infinite Q. In reality, however, most of the time the transistor "loads" the tuned circuit; therefore, the operating Q is less than the unloaded Q. Another time-domain noise analysis was shown by Anzill et al. [80] but is not useful for HB simulators.

The following two circuits show the transition from a series tuned circuit connected with the series time-dependent negative resistance as outlined in equation (6-1) and the resulting input capacitance marked C_{IN}. Translated, the resulting configuration consists of a series circuit with inductance L and the resulting capacitance C'. The noise voltage $e_N(t)$ describes a small perturbation, which is the noise resulting from R_L and $-R_N(t)$.

Figure 8-3 shows the equivalent representation of the oscillator circuit in the presence of noise.

The circuit equation of the oscillator circuit of Figure 8-3 can be given as

$$L\frac{di(t)}{dt} + (R_L - R_N(t))i(t) + \frac{1}{C'}\int i(t)\,dt = e_N(t) \tag{8-11}$$

where $i(t)$ is the time-varying resultant current. Due to the noise voltage $e_N(t)$, equation (8-11) is a nonhomogeneous differential equation. If the noise voltage is zero, it translates into a homogeneous differential equation.

For a noiseless oscillator, the noise signal $e_N(t)$ is zero and the expression of the free-running oscillator current $i(t)$ can be assumed to be a periodic function of time and can be given as

$$i(t) = I_1 \cos(\omega t + \varphi_1) + I_2 \cos(2\omega t + \varphi_2) + I_3 \cos(3\omega t + \varphi_3)$$
$$+ \cdots I_n \cos(n\omega t + \varphi_n) \tag{8-12}$$

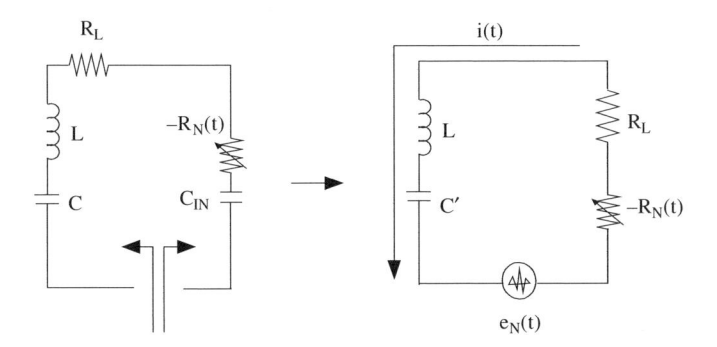

Figure 8-3 Equivalent representation of the oscillator circuit in the presence of noise.

where $I_1, I_2 \ldots I_n$ are peak harmonic amplitudes of the current and $\varphi_1, \varphi_2 \ldots \varphi_n$ are time-invariant phases.

In the presence of the noise perturbation $e_N(t)$, the current $i(t)$ may no longer be a periodic function of time and can be expressed as

$$i(t) = I_1(t)\cos[\omega t + \varphi_1(t)] + I_2(t)\cos[2\omega t + \varphi_2(t)] + I_3(t)\cos[3\omega t + \varphi_3(t)]$$
$$+ \cdots I_{n-2}(t)\cos[(n-2)\omega t + \varphi_{n-2}(t)] + I_{n-1}(t)\cos[(n-1)\omega t + \varphi_{n-1}(t)]$$
$$+ I_n(t)\cos[n\omega t + \varphi_n(t)] \qquad (8\text{-}13)$$

where $I_1(t), I_2(t) \ldots I_n(t)$ are time-varying amplitudes of the current and $\varphi_1(t), \varphi_2(t) \ldots \varphi_n(t)$ are time-varying phases.

Considering that $I_n(t)$ and $\varphi_n(t)$ do not change much over the period $2\pi/n\omega$, each corresponding harmonic over one period of oscillation cycle remains small and more or less variant. The solution of the differential equation becomes easy since the harmonics are suppressed due to a $Q > 10$, which prevents $i(t)$ to flow for the higher terms.

After substitution of the value of di/dt and $\int i(t)dt$, the complete oscillator circuit equation, as given in equation (8-11), can be rewritten as

$$L\left\{ -I_1(t)\left(\omega + \frac{d\varphi_1(t)}{dt}\right)\sin[\omega t + \varphi_1(t)] + \frac{dI_1(t)}{dt}\cos[\omega t + \varphi_1(t)]\right.$$

$$- I_2(t)\left(2\omega + \frac{d\varphi_2(t)}{dt}\right)\sin[2\omega t + \varphi_2(t)] + \frac{dI_2(t)}{dt}\cos[2\omega t + \varphi_2(t)]$$

$$- I_3(t)\left(3\omega + \frac{d\varphi_3(t)}{dt}\right)\sin[3\omega t + \varphi_3(t)] + \frac{dI_3(t)}{dt}\cos[3\omega t + \varphi_3(t)]$$

$$+ \cdots - I_n(t)\left(n\omega + \frac{d\varphi_n(t)}{dt}\right)\sin[n\omega t + \varphi_n(t)] + \left.\frac{dI_n(t)}{dt}\cos[n\omega t + \varphi_n(t)]\right\}$$

$$+ [(R_L - R_N(t))i(t)] + \frac{1}{C'}\left\{\left[\frac{I_1(t)}{\omega} - \frac{I_1(t)}{\omega^2}\left(\frac{d\varphi_1(t)}{dt}\right)\right]\sin[\omega t + \varphi_1(t)]\right.$$

$$+ \frac{1}{\omega^2}\left(\frac{dI_1(t)}{dt}\right)\cos[\omega t + \varphi_1(t)]\bigg\}$$

$$+ \frac{1}{C'}\left\{\left[\frac{I_2(t)}{2\omega} - \frac{I_2(t)}{4\omega^2}\left(\frac{d\varphi_2(t)}{dt}\right)\right]\sin(2\omega t + \varphi_2(t))\right.$$

$$+ \left.\frac{1}{4\omega^2}\left(\frac{dI_2(t)}{dt}\right)\cos(2\omega t + \varphi_2(t))\right\}$$

8.4 NEGATIVE RESISTANCE MODEL

$$+ \frac{1}{C'}\left\{\left[\frac{I_3(t)}{3\omega} - \frac{I_3(t)}{9\omega^2}\left(\frac{d\varphi_3(t)}{dt}\right)\right]\sin[3\omega t + \varphi_3(t)]\right.$$

$$+ \frac{1}{9\omega^2}\left(\frac{dI_3(t)}{dt}\right)\cos[3\omega t + \varphi_3(t)]\bigg\}$$

$$+ \cdots \frac{1}{C'}\left\{\left[\frac{I_n(t)}{n\omega} - \frac{I_n(t)}{n^2\omega^2}\left(\frac{d\varphi_n(t)}{dt}\right)\right]\sin[n\omega t + \varphi_n(t)]\right.$$

$$+ \frac{1}{n^2\omega^2}\left(\frac{dI_n(t)}{dt}\right)\cos[n\omega t + \varphi_n(t)]\bigg\} = e_N(t) \tag{8-14}$$

Because Q > 10 we approximate:

$$\frac{di(t)}{dt} = -I_1(t)\left(\omega + \frac{d\varphi_1(t)}{dt}\right)\sin[\omega t + \varphi_1(t)] + \frac{dI_1(t)}{dt}\cos[\omega t + \varphi_1(t)]$$

$$+ \text{(slowly varying function at very small amounts of}$$

$$\text{higher-order harmonics)}$$

$$\int i(t)dt = \left[\frac{I_1(t)}{\omega} - \frac{I_1(t)}{\omega^2}\left(\frac{d\varphi_1(t)}{dt}\right)\right]\sin[\omega t + \varphi_1(t)]$$

$$+ \frac{1}{\omega^2}\left(\frac{dI_1(t)}{dt}\right)\cos[\omega t + \varphi_1(t)] + \text{(slowly varying function at}$$

$$\text{very small amounts of higher-order harmonics)}$$

After substitution of the value of di/dt and $\int i(t)dt$, the oscillator circuit equation (8-14) can be rewritten as

$$L\left[-I_1(t)\left(\omega + \frac{d\varphi_1(t)}{dt}\right)\sin[\omega t + \varphi_1(t)] + \frac{dI_1(t)}{dt}\cos[\omega t + \varphi_1(t)]\right]$$

$$+ [(R_L - R_N(t))I(t)] + \frac{1}{C}\left\{\left[\frac{I_1(t)}{\omega} - \frac{I_1(t)}{\omega^2}\left(\frac{d\varphi_1(t)}{dt}\right)\right]\sin[\omega t + \varphi_1(t)]\right.$$

$$+ \frac{1}{\omega^2}\left(\frac{dI_1(t)}{dt}\right)\cos[\omega t + \varphi_1(t)]\bigg\} = e_N(t) \tag{8-15}$$

Following [82], and for simplification purposes, the equations above are multiplied by $\sin[\omega t + \varphi_1(t)]$ or $\cos[\omega t + \varphi_1(t)]$ and integrated over one period of the oscillation cycle, which will give an approximate differential equation for phase

$\varphi(t)$ and amplitude $i(t)$:

$$\left[\frac{2}{IT_0}\right]\int_{t-T_0}^{t} e_N(t)\sin[\omega t + \varphi(t)]dt = -\frac{d\varphi}{dt}\left[L + \frac{1}{\omega^2 C'}\right] + \left[-\omega L + \frac{1}{\omega C'}\right] \quad (8\text{-}16)$$

$$\left[\frac{2}{T_0}\right]\int_{t-T_0}^{t} e_N(t)\cos[\omega t + \varphi(t)]dt = \frac{dI(t)}{dt}\left[L + \frac{1}{\omega^2 C'}\right] + \left[R_L - \overline{R_N(t)}\right]I(t) \quad (8\text{-}17)$$

where $\overline{R_N(t)}$ is the average negative resistance under large-signal conditions.

$$\overline{R_N(t)} = \left[\frac{2}{T_0 I}\right]\int_{t-T_0}^{t} R_N(t)I(t)\cos^2[\omega t + \varphi]dt \quad (8\text{-}18)$$

Since the magnitude of the higher harmonics is not significant, the subscripts of $\varphi(t)$ and $I(t)$ are dropped. Based on [82], we now determine the negative resistance.

8.4.1 Calculation of the Region of Nonlinear Negative Resistance

Under steady-state free-running oscillation conditions,

$$\frac{dI(t)}{dt} \rightarrow 0$$

which implies steady current, and

$$e_N(t) \rightarrow 0$$

with $I =$ fundamental RF current. Solving the now homogeneous differential equation for $R_L - R_N(t)$ and inserting the two terms into equation (8-17), we obtain

$$\left[\frac{2}{T_0}\right]\int_{t-T_0}^{t} e_N(t)\cos[\omega t + \varphi(t)]dt = \underbrace{\frac{dI}{dt}\left[L + \frac{1}{\omega^2 C'}\right]}_{\text{term} \rightarrow 0} + \left[R_L - \overline{R_N(t)}\right]I(t) \quad (8\text{-}19)$$

Now we introduce γ; $\gamma = \Delta R/\Delta I$; for $\Delta \rightarrow 0$, $\gamma \rightarrow 0$ and

$$[R_L - \overline{R_N(t)}] = \gamma \Delta I, \quad \gamma \rightarrow 0 \implies [R_L - \overline{R_N(t)}]I(t) \rightarrow 0 \quad (8\text{-}20)$$

$$R_L - \overline{R_N(t)} = R_{Load} - \left[\frac{2}{T_0}\right]\int_{t-T_0}^{t} R_N(t)\cos^2[\omega t + \varphi(t)]dt \rightarrow 0 \quad (8\text{-}21)$$

$[R_L - \overline{R_N(t)}]I(t) \rightarrow 0$ gives the intersection of $[\overline{R_N(t)}]$ and $[R_L]$. This value is defined as I_0, which is the minimum value of the current needed for the steady-state sustained oscillation condition.

8.4 NEGATIVE RESISTANCE MODEL

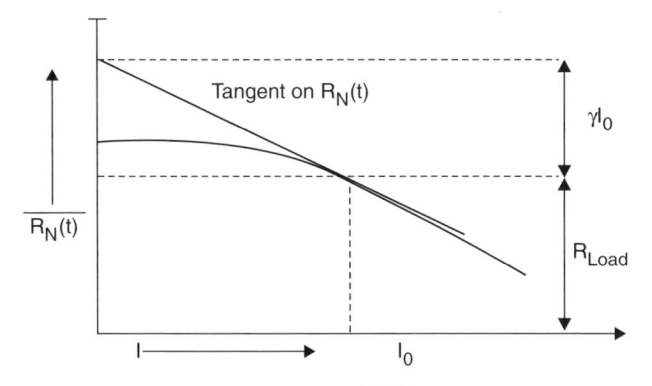

Figure 8-4 Plot of negative resistance of $[\overline{R_N(t)}]$ versus amplitude of current I.

Figure 8-4 shows the plot of the nonlinear negative resistance, which is a function of the amplitude of the RF current. As the RF amplitude gets larger, the conducting angle becomes more narrow.

For a small variation of the current ΔI from I_0, the relation above is expressed as

$$[R_L - \overline{R_N(t)}] = \gamma \, \Delta I \tag{8-22}$$

$\gamma \Delta I$ can be found from the intersection on the vertical axis by drawing the tangential line on $[\overline{R_N(t)}]$ at $I = I_0$. $|\Delta I|$ decreases exponentially with time for $\gamma > 0$.

Hence, I_0 represents the stable operating point. On the other hand, if $[\overline{R_N(t)}]$ intersects $[R_L]$ from the other side for $\gamma < 0$, then $|\Delta I|$ grows indefinitely with time. Such an operating point does not support stable operation [82].

8.4.2 Calculation of the Noise Signal in the Time Domain

To solve the two orthogonal equations, we need to obtain information about current $I(t)$ and $\varphi(t)$.

$$\left[\frac{2}{IT_0}\right] \int_{t-T_0}^{t} e_N(t) \sin[\omega t + \varphi(t)] dt = -\frac{d\varphi(t)}{dt}\left[L + \frac{1}{\omega^2 C'}\right] + \left[-\omega L + \frac{1}{\omega C'}\right] \tag{8-23}$$

$$\left[\frac{2}{T_0}\right] \int_{t-T_0}^{t} e_N(t) \cos[\omega t + \varphi(t)] dt = \frac{dI(t)}{dt}\left[L + \frac{1}{\omega^2 C'}\right] + [R_L - \overline{R_N(t)}]I(t) \tag{8-24}$$

The noise signal can be analyzed by decomposing the noise signal $e_N(t)$ to an infinite number of random noise pulses represented by

$$\varepsilon \delta(t - t_0) \tag{8-25}$$

where ε is the strength of the pulse at the time instant t_0, and both ε and t_0 are independent random variables from one pulse to the next.

The time average of the square of the current pulses over a period of time can be shown to be

$$\frac{1}{2T}\int_{-T}^{T}\left[\sum \varepsilon\delta(t-t_0)\right]^2 dt = \overline{e_N^2(t)} \qquad (8\text{-}26)$$

The mean square noise voltage $\overline{e_N^2(t)}$ is generated in the circuit in Figure 8-3. Figure 8-5 shows the noise pulse at time instant $t = t_0$.

The integral of the single noise pulse above gives the rectangular pulse with the height of $[2/T_0]\varepsilon \sin[\omega t + \varphi(t)]$ and the length of T_0, as shown in Figure 8-6.

The integration of the single elementary noise pulse, following the Dirac Δ function, results in

$$\left[\frac{2}{T_0}\right]\int_{t-T_0}^{t} e_N(t)\sin[\omega t + \varphi(t)]dt \approx \left[\frac{2}{T_0}\right]\int_{t-T_0}^{t} \varepsilon\delta(t-t_0)\sin[\omega t + \varphi(t)]dt] \qquad (8\text{-}27)$$

$$\left[\frac{2}{T_0}\right]\int_{t-T_0}^{t} \varepsilon\delta(t-t_0)\sin[\omega t + \varphi(t)]dt \approx \left[\frac{2}{T_0}\right]\varepsilon\sin[\omega t_0 + \varphi(t)] \qquad (8\text{-}28)$$

since the length of time T_0 is considered to be sufficiently small for any variation of $\varphi(t)$ and $I(t)$ during time T_0. The corresponding rectangular pulse of the magnitude $(2/T_0)\varepsilon \sin[\omega t_0 + \varphi(t)]$ is considered to be another pulse located at $t = t_0$ and can be expressed in the form of an impulse function with the amplitude $2\varepsilon \sin[\omega t_0 + \varphi(t)]$ located at $t = t_0$ for calculating the effect using equations (8-23) and (8-24).

The effect of $[2/T_0]\int_{t-T_0}^{t} e_N(t)\sin[\omega t + \varphi(t)]dt$ is given by $[n_1(t)]$, which consists of a number of rectangular pulses. The time average of the square of these pulses, following [82], can be calculated as

$$\frac{1}{2T}\int_{t=-T}^{t=T}\left[\sum 2\varepsilon \sin(\omega t_0 + \varphi(t))\delta(t-t_0)\right]^2 dt = \frac{1}{T}\int_{t=-T}^{t=T}\left[\sum \varepsilon\delta(t-t_0)\right]^2 dt \qquad (8\text{-}29)$$

$$\overline{e_N^2(t)} = \frac{1}{2T}\int_{-T}^{T}\left[\sum \varepsilon\delta(t-t_0)\right]^2 dt \qquad (8\text{-}30)$$

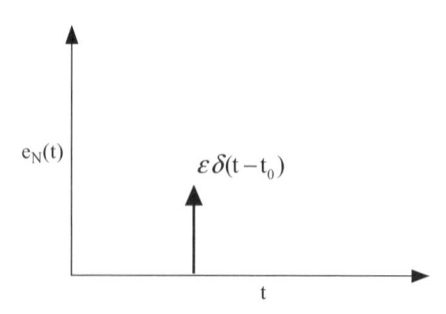

Figure 8-5 The noise pulse at $t = t_0$.

8.4 NEGATIVE RESISTANCE MODEL

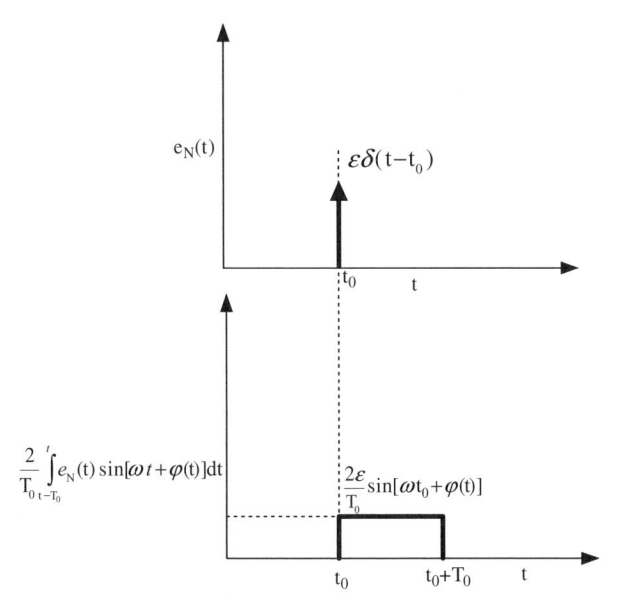

Figure 8-6 The amplitude of the rectangular pulse.

From the equation above

$$\overline{n_1^2(t)} = 2\overline{e_N^2(t)} \tag{8-31}$$

Similarly, the total response of $2/T_0 \int_{t-T_0}^{t} e_N(t) \cos[\omega t + \varphi(t)] dt$ can be expressed by $[n_2(t)]$, which consists of a large number of such pulses, and the time average of the square of these pulses is

$$\overline{n_2^2(t)} = 2\overline{e_N^2(t)} \tag{8-32}$$

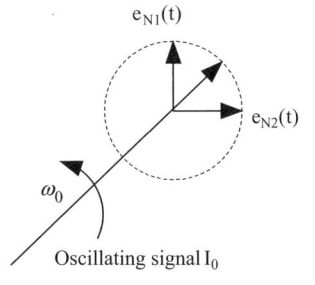

Figure 8-7 Vector presentation of the oscillator signal and its modulation by the voltages e_{N1} and e_{N2}.

since $2/T_0 \int_{t-T_0}^{t} e_N(t)\sin[\omega t + \varphi(t)]dt$ and $2/T_0 \int_{t-T_0}^{t} e_N(t)\cos[\omega t + \varphi(t)]dt$ are orthogonal functions, and in the frequency domain are the upper and lower sidebands relative to the carrier, and the correlation of $[n_1(t)]$ and $[n_2(t)]$ is

$$\overline{n_1(t)n_2(t)} = 0 \tag{8-33}$$

Now consider the narrowband noise signal, which is

$$e_N(t) = e_{N1}(t) + e_{N2}(t) \tag{8-34}$$

$$e_{N1}(t) = e_1(t)\sin[\omega_0 t + \varphi(t)] \tag{8-35}$$

$$e_{N2}(t) = -e_2(t)\cos[\omega_0 t + \varphi(t)] \tag{8-36}$$

where $e_{N1}(t)$ and $e_{N2}(t)$ are orthogonal functions, and $e_1(t)$ and $e_2(t)$ are slowly varying functions of time.

The calculation of $I_n(t)$ and $\varphi_n(t)$ for the free-running oscillator can be derived from equations (8-23) and (8-24):

$$\left[\frac{2}{IT_0}\right]\int_{t-T_0}^{t} e_N(t)\sin[\omega t + \varphi(t)]dt = -\frac{d\varphi(t)}{dt}\left[L + \frac{1}{\omega^2 C'}\right] + \left[-\omega L + \frac{1}{\omega C'}\right] \tag{8-37}$$

$$\left[\frac{2}{IT_0}\right]\int_{t-T_0}^{t} e_N(t)\sin[\omega t + \varphi(t)]dt \Longrightarrow \left[\frac{1}{I}\right]n_1(t) \tag{8-38}$$

at resonance frequency $\omega = \omega_0$,

$$\left\{-\frac{d\varphi(t)}{dt}\left[L + \frac{1}{\omega^2 C'}\right] + \left[-\omega L + \frac{1}{\omega C'}\right]\right\}_{\omega=\omega_0} = -2L\frac{d\varphi(t)}{dt} \tag{8-39}$$

and

$$\frac{1}{I}n_1(t) = -2L\frac{d\varphi(t)}{dt} \tag{8-40}$$

$$\frac{d\varphi(t)}{dt} = -\left[\frac{1}{2LI}\right]n_1(t) \tag{8-41}$$

If equation (8-41) is transformed in the frequency domain, $\varphi(t)$ can be expressed as

$$\varphi(f) = \frac{n_1(f)}{2\omega LI} \tag{8-42}$$

8.4 NEGATIVE RESISTANCE MODEL

Now the spectral density of $[\varphi(f)]$ is

$$|\varphi(f)|^2 = \frac{1}{4\omega^2 L^2 I^2}|n_1(f)|^2 \tag{8-43}$$

$$\frac{1}{4\omega^2 L^2 I^2}|n_1(f)|^2 = \frac{2|e_N(f)|^2}{4\omega^2 L^2 I^2} \Longrightarrow |\varphi(f)|^2 = \frac{2|e_N(f)|^2}{4\omega^2 L^2 I^2} \tag{8-44}$$

where f varies from $-\infty$ to $+\infty$.

The amplitude of the current can be written as $I(t) = I_0 + \Delta I(t)$, where I_0 represents the stable operating point of the free-running oscillator with a loop gain slightly greater than 1.

From equation (8-24)

$$\frac{2}{T_0}\int_{t-T_0}^{t} e_N(t)\cos[\omega t + \varphi(t)]dt = \frac{dI(t)}{dt}\left(L + \frac{1}{\omega^2 C'}\right) + \left[R_L - \overline{R_N(t)}\right]I(t)$$

we can calculate

$$\left[\frac{2}{T_0}\int_{t-T_0}^{t} e_N(t)\cos[\omega t + \varphi(t)]dt\right]_{\omega=\omega_0} = \left[2L\frac{\partial}{\partial t}[\Delta I(t)] + \Delta I(t)I_0\gamma + \Delta I^2(t)\gamma\right] \tag{8-45}$$

Since the amplitude of $\Delta I^2(t)$ is negligible, its value can be set to 0:

$$\left[2L\frac{\partial}{\partial t}[\Delta I(t)] + \Delta I(t)I_0\gamma + \Delta I^2(t)\gamma\right] = 2L\frac{\partial}{\partial t}[\Delta I(t)] + \Delta I(t)I_0\gamma \tag{8-46}$$

$$n_2(t) = \frac{2}{T_0}\int_{t-T_0}^{t} e_N(t)\cos[\omega t + \varphi(t)]dt \tag{8-47}$$

$$n_2(t) = 2L\frac{\partial}{\partial t}[\Delta I(t)] + \Delta I(t)I_0\gamma \tag{8-48}$$

$$n_2(f) = 2L\omega\Delta I(f) + \Delta I(f)I_0\gamma \tag{8-49}$$

The spectral density of $[n_2(f)]$ is

$$|n_2(f)|^2 = [4L^2\omega^2 + (I_0\gamma)^2]|\Delta I(f)|^2 \tag{8-50}$$

and the spectral density of $\Delta I(f)$ can be expressed in terms of $|n_2(f)|^2$ as

$$|\Delta I(f)|^2 = \frac{1}{[4L^2\omega^2 + (I_0\gamma)^2]}|n_2(f)|^2 \tag{8-51}$$

$$|n_2(f)|^2 = 2|e_N(f)|^2 \Longrightarrow |\Delta I(f)|^2 = \frac{2|e_N(f)|^2}{[4L^2\omega^2 + (I_0\gamma)^2]} \tag{8-52}$$

since $n_1(t)$ and $n_2(t)$ are orthogonal functions and there is no correlation between current and phase:

$$\overline{n_1(t)n_2(t)} = 0 \implies \overline{I(t)\varphi(t)} = 0 \tag{8-53}$$

The output power noise spectral density of the current is given as

$$P_{noise}(f) = 2R_L|I(f)|^2 \tag{8-54}$$

The noise spectral density of the current is given as

$$|I(f)|^2 = \int_{-\infty}^{\infty} R_I(\tau)\exp(-j\omega\tau)d\tau \tag{8-55}$$

where $R_I(\tau)$ is the autocorrelation function of the current and can be written as

$$R_I(\tau) = \left[\overline{I(t)I(t+\tau)\cos[\omega_0 t + \varphi(t)]\cos[\omega_0(t+\tau) + \varphi(t+\tau)]}\right] \tag{8-56}$$

Since $I(t)$ and $\varphi(t)$ are uncorrelated, the autocorrelation function of the current $R_I(\tau)$ can be given as

$$R_I(\tau) = \frac{1}{2}[I_0^2 + R_{\Delta I}(\tau)]\cos(\omega_0\tau)\overline{[\cos(\varphi(t+\tau) - \varphi(t))]} \tag{8-57}$$

From [82], but taking into consideration that both sidebands are correlated, we can write

$$R_I(\tau) = \frac{1}{2}\left[I_0^2 + \frac{2|e_N(\tau)|^2}{2L\gamma I_0}\exp\left(-\frac{\gamma I_0}{2L}|\tau|\right)\right]\exp\left(-\frac{|e_N(\tau)|^2}{4L^2 I_0^2}|\tau|\right)\cos(\omega_0\tau) \tag{8-58}$$

Since [82] skipped many stages of the calculation up to this point, a more complete and detailed flow is shown. These results are needed to calculate the noise performance at the component level later. Note the factor of 2, which results from the correlation.

Considering $\gamma I_0/2L \gg 2|e_N(\tau)|^2/4L^2 I_0^2$, the noise spectral density of the current is given by

$$|I(f)|^2 = \int_{-\infty}^{\infty} R_I(\tau)\exp(-j\omega\tau)d\tau \tag{8-59}$$

8.4 NEGATIVE RESISTANCE MODEL

with $I = I_0 + \Delta I(t)$; all are RF currents.

$$|I(f)|^2 = \frac{|e_N(f)|^2}{8L^2}\left[\frac{1}{(\omega-\omega_0)^2+(|e_N(f)|^2/4L^2I_0^2)^2}+\frac{1}{(\omega+\omega_0)^2+(|e_N(f)|^2/4L^2I_0^2)^2}\right]$$
$$+\frac{|e_N(f)|^2}{8L^2}\left[\frac{1}{(\omega-\omega_0)^2+(\gamma I_0/2L)^2}+\frac{1}{(\omega+\omega_0)^2+(\gamma I_0/2L)^2}\right] \quad (8\text{-}60)$$

with

$$\frac{|e_N(f)|^2}{8L^2}\left[\frac{1}{(\omega-\omega_0)^2+(|e_N(f)|^2/4L^2I_0^2)^2}+\frac{1}{(\omega+\omega_0)^2+(|e_N(f)|^2/4L^2I_0^2)^2}\right] \rightarrow \text{FM noise}$$
$$(8\text{-}61)$$

$$\frac{|e_N(f)|^2}{8L^2}\left[\frac{1}{(\omega-\omega_0)^2+(\gamma I_0/2L)^2}+\frac{1}{(\omega+\omega_0)^2+(\gamma I_0/2L)^2}\right] \rightarrow \text{AM noise} \quad (8\text{-}62)$$

Since $\gamma I_0/2L \gg 2|e_N(\tau)|^2/4L^2I_0^2$ for $\omega \rightarrow \omega_0$, FM noise predominates over AM noise. For $\omega \gg \omega_0$, both FM and AM noise terms give equal contributions.

Considering $\omega + \omega_0 \gg \omega - \omega_0$, then,

$$|I(f)|^2 = \frac{|e_N(f)|^2}{8L^2}\left[\frac{1}{(\omega-\omega_0)^2+(|e_N(f)|^2/4L^2I_0^2)^2}+\frac{1}{(\omega-\omega_0)^2+(\gamma I_0/2L)^2}\right]$$
$$(8\text{-}63)$$

$$P_{noise}(f) = 2R_L|I(f)|^2 \quad (8\text{-}64)$$

$$P_{noise}(f) = 2R_L\left(\frac{|e_N(f)|^2}{8L^2}\right)\left[\frac{1}{(\omega-\omega_0)^2+(|e_N(f)|^2/4L^2I_0^2)^2}+\frac{1}{(\omega-\omega_0)^2+(\gamma I_0/2L)^2}\right]$$
$$(8\text{-}65)$$

Since $R_{Load} = R_L + R_0$, the effective dynamic resistance of the free running oscillator is given by

$$\sum_{\text{effective}} |R_{tot}| = R_N(t) - R_{Load} = R_0 \quad (8\text{-}66)$$

where R_0 is the output resistance; $R_0 - R_{tot} = 0$.

The Q of the resonator circuit is expressed as

$$Q_L = \frac{\omega L}{R_0} \quad (8\text{-}67)$$

174 CALCULATION AND OPTIMIZATION OF PHASE NOISE

The oscillator output noise power in terms of Q is given by

$$P_{noise}(f) = \frac{\omega_0^2}{2Q_L^2} \frac{|e|^2}{2\overline{R_N(t)}} \left[\frac{1}{(\omega - \omega_0)^2 + (\omega_0^2/4Q_L^2)^2 (|e|^2/2\overline{R_N(t)}P_{out}^2)^2} \right. \\ \left. + \frac{1}{(\omega - \omega_0)^2 + (\omega_0^2/Q_L^2)(\gamma I_0/2\overline{R_N(t)})^2} \right] \quad (8\text{-}68)$$

Figure 8-8 shows the Colpitts oscillator with a series resonator and the small-signal AC equivalent circuit. From the analytical expression of the noise analysis above, the influence of the circuit components on the phase noise can be explicitly calculated as

$$|\varphi(f)|^2 = \frac{1}{4\omega^2 L^2 I_0^2(f)} |n_1(f)|^2 \quad (8\text{-}69)$$

$$\frac{1}{4\omega^2 L^2 I_0^2(f)} |n_1(f)|^2 = \frac{2|e_N(f)|^2}{4\omega^2 L^2 I_0^2(f)} \implies |\varphi(f)|^2 = \frac{2|e_N(f)|^2}{4\omega^2 L^2 I_0^2(f)} \quad (8\text{-}70)$$

where the frequency f varies from $-\infty$ to $+\infty$.

The resulting single-sideband phase noise is

$$\mathscr{L} = \frac{|e_N(f)|^2}{4\omega^2 L^2 I_0^2(f)} \quad (8\text{-}71)$$

The unknown variables are $|e_N(f)|^2$ and $I_0^2(f)$, which need to be determined next. $I_0^2(f)$ is transformed into $I_{c0}^2(f)$ by multiplying $I_0^2(t)$ by the effective current gain $Y_{21}^+/Y_{11}^+ = \beta^+$.

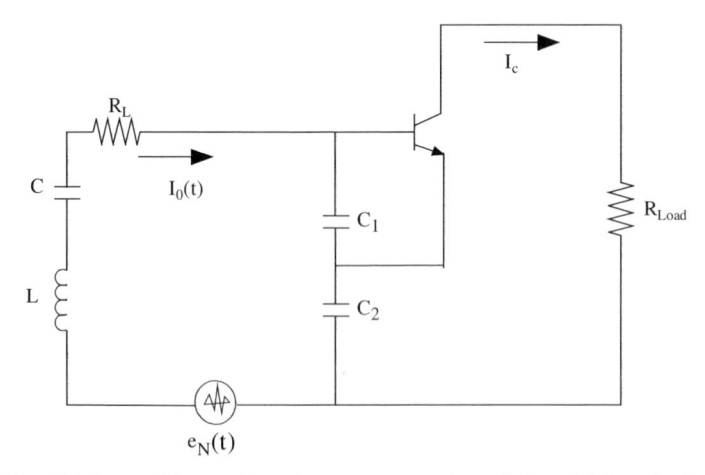

Figure 8-8 Colpitts oscillator with series resonator and small-signal AC equivalent circuit.

8.4.3 Calculation of $I_{c0}^2(f)$

From Figure 8-8, the LC-series resonant circuit is in shunt between the base and the emitter with the capacitive negative conductance portion of the transistor. We now introduce a collector load R_{Load} at the output or, better yet, an impedance Z.

The oscillator base current $i(t)$ is

$$i(t) = |I_0|\cos(\omega t) = \frac{V_{bc}(t)}{Z} \tag{8-72}$$

and the collector current is

$$|I_{c0}| = \left|\frac{[0.7 - V_{ce}]}{R_{Load} + j(\omega L - (1/\omega C_{IN}))}\right| \approx \left|\frac{V_{ce}}{R_{Load} + j(\omega L - (1/\omega C_{IN}))}\right|, \tag{8-73}$$

or

$$\overline{I_{c0}^2(f)} \approx \left\{\frac{\overline{V_{ce}^2(f)}}{[R_{Load}]^2 + (\omega L - (1/\omega C_{IN}))^2}\right\} = \left\{\frac{\overline{V_{ce}^2(f)}}{[\omega L/Q]^2 + (\omega L - (1/\omega C_{IN}))^2}\right\} \tag{8-74}$$

The voltage V_{ce} is the RF voltage across the collector-emitter terminals of the transistor. Considering the steady-state oscillation $\omega \to \omega_0$, the total loss resistance is compensated for by the negative resistance of the active device as $R_L = \overline{R_N(t)}$. The expression of $|\overline{I_{c0}^2(f)}|_{\omega=\omega_0}$ is

$$\left|\overline{I_{c0}^2(f)}\right|_{\omega=\omega_0} = \left|\frac{\overline{V_{ce}^2(f)}}{[\omega_0 L/Q]^2 + (\omega_0 L - (1/\omega_0 C_{IN}))^2}\right| = \left|\frac{\overline{V_{ce}^2(f)}}{(\omega_0 L)^2[1/Q^2 + (1 - (1/\omega_0^2 L C_{IN}))^2]}\right| \tag{8-75}$$

$$\left|\overline{I_{c0}^2(f)}\right|_{\omega=\omega_0} = \left|\frac{\overline{V_{ce}^2(f)}}{(\omega_0 L)^2[1/Q^2 + (1 - (1/\omega_0^2 L)(C_1 + C_2)/(C_1 C_2))^2]}\right| \tag{8-76}$$

where C_{IN} is the equivalent capacitance of the negative resistor portion of the oscillator circuit.

$$C' = \frac{CC_{IN}}{C + C_{IN}}, \quad C_{IN} = \frac{C_1 C_2}{C_1 + C_2} \tag{8-77}$$

$$Q = \frac{\omega L}{R_L} \tag{8-78}$$

For a reasonably high Q resonator $|\overline{I_{c0}^2(f)}|_{\omega=\omega_0} \propto [C_{IN}]_{\omega=\omega_0}$.

8.4.4 Calculation of the Noise Voltage $e_N(f)$

The equivalent noise voltage from the negative resistance portion of the oscillator circuit is given an open-circuit noise voltage electro motorical force (EMF) of the circuit, as shown in Figure 8-9 below.

The noise voltage associated with the resonator loss resistance R_s is

$$\overline{|e_R^2(f)|}_{\omega=\omega_0} = 4kTBR_s \tag{8-79}$$

R_s denotes the equivalent series loss resistor, which can be calculated from the parallel loading resistor R_{load}; see Figure 8-9.

$$\overline{|e_R^2(f)|}_{\omega=\omega_0} = 4kTR \quad \text{for } B = 1 \text{ Hz bandwidth} \tag{8-80}$$

The total noise voltage power within a 1 Hz bandwidth can be given as

$$\overline{|e_N^2(f)|}_{\omega=\omega_0} = \overline{e_R^2(f)} + \overline{e_{NR}^2(f)} = [4kTR] + \left[\frac{4qI_c g_m^2 + (K_f I_b^{AF}/\Delta\omega)\, g_m^2}{\omega_0^2 C_1^2(\omega_0^2(\beta^+)^2 C_2^2 + g_m^2(C_2^2/C_1^2))}\right] \tag{8-81}$$

Derivation of equation (8.81): The total noise voltage power within a 1 Hz bandwidth can be given as

$$\overline{|e_N^2(f)|}_{\omega=\omega_0} = \overline{e_R^2(f)} + \overline{e_{NR}^2(f)} \tag{8-82}$$

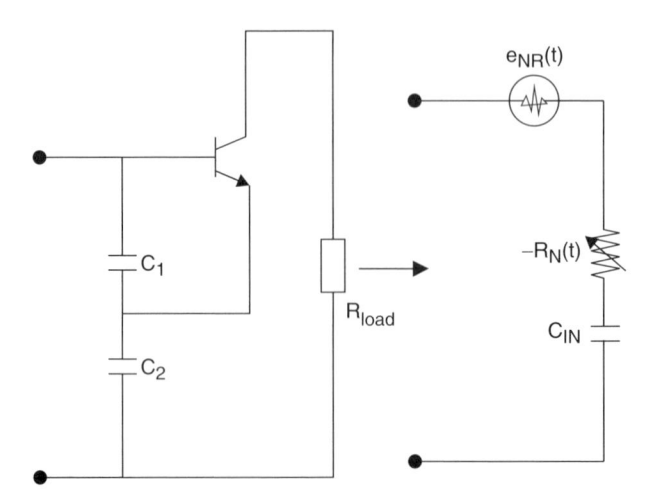

Figure 8-9 Equivalent representation of the negative resistance portion of the circuit at the input for the open-circuit noise voltage.

8.4 NEGATIVE RESISTANCE MODEL   177

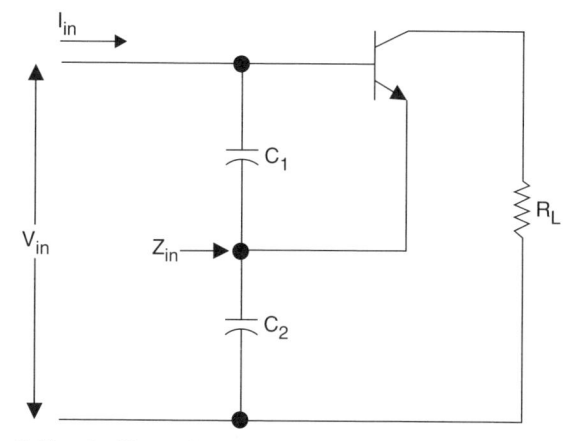

Figure 8-10 Oscillator circuit for calculation of the negative resistance.

The first term in equation (8-82) is the noise voltage power due to the loss resistance R, and the second term is associated with the negative resistance of the active device R_N.

Figures 8-10 and 8-11 illustrate the oscillator circuit for the purpose of calculating the negative resistance.

From Figure 8-11, the circuit equation is given from Kirchoff's voltage law (KVL) as

$$V_{in} = I_{in}(X_{C_1} + X_{C_2}) - I_b(X_{C_1} - \beta X_{C_2}) \tag{8-83}$$
$$0 = -I_{in}(X_{C_1}) + I_b(X_{C_1} + h_{ie}) \tag{8-84}$$

Considering, $1/Y_{11} = h_{ie}$

$$Z_{in} = \frac{V_{in}}{I_{in}} = \frac{(1+\beta)X_{C_1}X_{C_2} + h_{ie}(X_{C_1} + X_{C_2})}{X_{C_1} + h_{ie}} \tag{8-85}$$

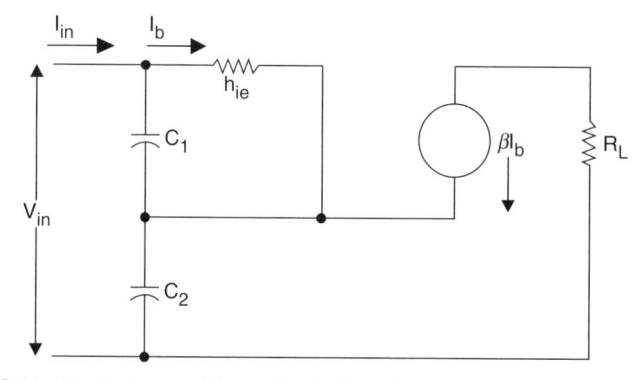

Figure 8-11 Equivalent oscillator circuit for calculation of the negative resistance.

CALCULATION AND OPTIMIZATION OF PHASE NOISE

$$Z_{in} = \frac{\left(-(1+\beta)/\omega^2 C_1 C_2 + (C_1+C_2)/j\omega C_1 C_2(1/Y_{11})\right)}{(1/Y_{11} + 1/j\omega C_1)} \tag{8-86}$$

$$Z_{in} = \frac{-jY_{11}(1+\beta) + \omega(C_1+C_2)}{\omega C_2(Y_{11}+j\omega C_1)} \tag{8-87}$$

$$Z_{in} = \frac{[\omega(C_1+C_2) - jY_{11}(1+\beta)][Y_{11} - j\omega C_1]}{\omega C_2(Y_{11}^2 + \omega^2 C_1^2)} \tag{8-88}$$

$$Z_{in} = \left[\frac{\omega Y_{11}(C_1+C_2) - (1+\beta)\omega C_1 Y_{11}}{\omega C_2(Y_{11}^2 + \omega^2 C_1^2)}\right] - j\left[\frac{Y_{11}^2(1+\beta) + \omega^2 C_1(C_1+C_2)}{\omega C_2(Y_{11}^2 + \omega^2 C_1^2)}\right] \tag{8-89}$$

$$Z_{in} = -R_n - jX \tag{8-90}$$

$$R_n = \frac{(1+\beta)\omega C_1 Y_{11} - \omega Y_{11}(C_1+C_2)}{\omega C_2(Y_{11}^2 + \omega^2 C_1^2)} = \frac{(1+\beta)C_1 Y_{11} - Y_{11}(C_1+C_2)}{C_2(Y_{11}^2 + \omega^2 C_1^2)} \tag{8-91}$$

$$R_n = \frac{\beta C_1 Y_{11} - Y_{11} C_2}{C_2(Y_{11}^2 + \omega^2 C_1^2)} = \frac{\beta Y_{11}}{(C_2/C_1)(Y_{11}^2 + \omega^2 C_1^2)} - \frac{Y_{11}}{(Y_{11}^2 + \omega^2 C_1^2)} \tag{8-92}$$

Considering $\beta = Y_{21}/Y_{11} \approx g_m/Y_{11}$

$$R_n = \frac{g_m}{(C_2/C_1)(g_m^2/\beta^2 + \omega^2 C_1^2)} - \frac{g_m/\beta}{(g_m^2/\beta^2 + \omega^2 C_1^2)} \tag{8-93}$$

$$R_n = \frac{g_m \beta^2 C_1}{(g_m^2 C_2 + \omega^2 \beta^2 C_1^2 C_2)} - \frac{g_m \beta}{(g_m^2 + \beta^2 \omega^2 C_1^2)} \tag{8-94}$$

$$R_n = \frac{g_m \beta^2 \omega^2 C_1 C_2}{\omega^2 C_1^2((C_2^2/C_1^2)g_m^2 + \omega^2 \beta^2 C_2^2)} - \frac{g_m \beta \omega^2 C_2^2}{\omega^2 C_1^2((C_2^2/C_1^2)g_m^2 + \beta^2 \omega^2 C_2^2)} \tag{8-95}$$

$$R_n = \left[\frac{g_m^2}{\omega^2 C_1^2((C_2^2/C_1^2)g_m^2 + \omega^2 \beta^2 C_2^2)}\right] \times \left[\frac{\beta^2 \omega^2 C_1 C_2}{g_m} - \frac{\beta \omega^2 C_2^2}{g_m}\right] \tag{8-96}$$

$$R_n = \left[\frac{g_m^2}{\omega^2 C_1^2((C_2^2/C_1^2)g_m^2 + \omega^2 \beta^2 C_2^2)}\right] \times \left[g_m\left[\left(\frac{\omega C_1}{Y_{11}}\right)\left(\frac{\omega C_2}{Y_{11}}\right) - \frac{\omega^2 C_2^2}{\beta Y_{11}^2}\right]\right] \tag{8-97}$$

$$R_n = \left[\frac{g_m^2}{\omega^2 C_1^2((C_2^2/C_1^2)g_m^2 + \omega^2 \beta^2 C_2^2)}\right] \times \left[g_m\left[\left(\frac{\omega C_1}{Y_{11}}\right)\left(\frac{\omega C_2}{Y_{11}}\right) - \frac{1}{\beta}\left(\frac{\omega C_2}{Y_{11}}\right)\left(\frac{\omega C_2}{Y_{11}}\right)\right]\right] \tag{8-98}$$

8.4 NEGATIVE RESISTANCE MODEL

Considering

$$\left(\frac{\omega C_1}{Y_{11}}\right)\left(\frac{\omega C_2}{Y_{11}}\right) \gg \frac{1}{\beta}\left(\frac{\omega C_2}{Y_{11}}\right)\left(\frac{\omega C_2}{Y_{11}}\right) \quad \text{and} \quad \left(\frac{\omega C_1}{Y_{11}}\right)\left(\frac{\omega C_2}{Y_{11}}\right) \approx 1 \quad (8\text{-}99)$$

$$\left[g_m\left[\left(\frac{\omega C_1}{Y_{11}}\right)\left(\frac{\omega C_2}{Y_{11}}\right) - \frac{1}{\beta}\left(\frac{\omega C_2}{Y_{11}}\right)\left(\frac{\omega C_2}{Y_{11}}\right)\right]\right] \cong \frac{I_C}{V_T} = \frac{I_C}{kT/q} \Rightarrow \frac{qI_C}{kT} \quad (8\text{-}100)$$

From equations (8-98) and (8-100),

$$R_n = \left[\frac{g_m^2}{\omega^2 C_1^2((C_2^2/C_1^2)g_m^2 + \omega^2\beta^2 C_2^2)}\right]\frac{qI_C}{kT} \quad (8\text{-}101)$$

From equation (8-82), the total noise voltage power within a 1 Hz bandwidth can be given as

$$|\overline{e_N^2(f)}|_{\omega=\omega_0} = \overline{e_R^2(f)} + \overline{e_{NR}^2(f)} \quad (8\text{-}102)$$

$$|\overline{e_N^2(f)}|_{\omega=\omega_0} = [4kTR] + \left[\frac{4qI_c g_m^2 + (K_f I_b^{AF}/\omega)g_m^2}{\omega_0^2 C_1^2(\omega_0^2(\beta^+)^2 C_2^2 + g_m^2(C_2^2/C_1^2))}\right] \quad (8\text{-}103)$$

where

$$\beta^+ = \left[\frac{Y_{21}^+}{Y_{11}^+}\right]\left[\frac{C_1}{C_2}\right]^p, \quad g_m = [Y_{21}^+]\left[\frac{C_1}{C_2}\right]^q \quad \text{redefined}$$

$$\omega = 2\pi f$$

The values of p and q depend upon the drive level.

$$|\overline{e_N^2(f)}|_{\omega=\omega_0} = [4kTR] + \left[\frac{4qI_{c0} g_m^2 + (K_f I_b^{AF}/\Delta\omega)g_m^2}{\omega_0^2 C_1^2(\omega_0^2(\beta^+)^2 C_2^2 + g_m^2(C_2^2/C_1^2))}\right] \quad (8\text{-}104)$$

where

$$\beta^+ = \left[\frac{Y_{21}^+}{Y_{11}^+}\right]\left[\frac{C_1}{C_2}\right]^p$$

$$g_m = [Y_{21}^+]\left[\frac{C_1}{C_2}\right]^q \quad \text{redefined}$$

The values of p and q depend upon the drive level.

The flicker noise contribution in equation (8–82) is introduced by adding the term $K_f I_b^{AF}/\omega$ in I_{c0}, where K_f is the flicker noise coefficient and AF is the flicker

180 CALCULATION AND OPTIMIZATION OF PHASE NOISE

noise exponent. This is valid only for the bipolar transistor. For an FET, the equivalent currents have to be used.

The first term in the expression above is related to the thermal noise due to the loss resistance of the resonator tank, and the second term is related to the shot noise and flicker noise in the transistor.

Now the phase noise of the oscillator can be expressed as

$$\left|\overline{\varphi^2(\omega)}\right| = \frac{2\left|\overline{e_N^2(\omega)}\right|}{4\omega_0^2 L^2 I_0^2(\omega)} \tag{8-105}$$

$$\left|\overline{\varphi^2(\omega)}\right|_{SSB} = \frac{1}{2}\left|\overline{\varphi^2(\omega)}\right| = \frac{\left|\overline{e_N^2(\omega)}\right|}{4\omega_0^2 L^2 I_0^2(\omega)} \tag{8-106}$$

$$\left|\overline{\varphi^2(\omega)}\right|_{SSB} = \left\{[4kTR] + \left[\frac{4qI_c g_m^2 + (K_f I_b^{AF}/\omega)g_m^2}{\omega_0^2 C_1^2(\omega_0^2(\beta^+)^2 C_2^2 + g_m^2(C_2^2/C_1^2))}\right]\right\}$$

$$\times \left[\frac{(\omega_0)^2\left[1/Q^2 + \left(1 - (1/\omega_0^2 L)(C_1 + C_2)/C_1 C_2\right)^2\right]}{4\omega^2 |V_{ce}^2(\omega)|}\right] \tag{8-107}$$

$$\left|\overline{\varphi^2(\omega)}\right|_{SSB} = \left[4kTR + \left[\frac{4qI_c g_m^2 + (K_f I_b^{AF}/\omega)g_m^2}{\omega_0^2 C_1^2(\omega_0^2(\beta^+)^2 C_2^2 + g_m^2(C_2^2/C_1^2))}\right]\right]$$

$$\times \left[\frac{\omega_0^2}{4\omega^2 V_{ce}^2}\right]\left[\frac{1}{Q^2} + \left(1 - \frac{1}{\omega_0^2 L}\frac{C_1 + C_2}{C_1 C_2}\right)^2\right] \tag{8-108}$$

Considering $[(1/\omega_0^2 L)(C_1 + C_2)/C_1 C_2] \gg 1$; for $\omega_0 = 2\pi f = 6.28\text{E}9$ Hz, $L = 1\text{E-9F}$, $C_1 = 1\text{E-12F}$, $C_2 = 1\text{E-12F}$:

$$\left(\frac{1}{\omega_0^2 L}\frac{C_1 + C_2}{C_1 C_2}\right) = 50.7$$

Since the phase noise is always expressed in dBc/Hz, we now calculate, after simplification of equation (8-86),

$$\mathscr{L}(\omega) = 10\log\left\{\left[4kTR + \left[\frac{4qI_c g_m^2 + (K_f I_b^{AF}/\omega)g_m^2}{\omega_0^2 C_1^2\left(\omega_0^2(\beta^+)^2 C_2^2 + g_m^2 \frac{C_2^2}{C_1^2}\right)}\right]\right]\right.$$

$$\left.\times \left[\frac{\omega_0^2}{4\omega^2 V_{cc}^2}\right]\left[\frac{1}{Q^2} + \frac{[C_1 + C_2]^2}{C_1^2 C_2^2 \omega_0^4 L^2}\right]\right\} \tag{8-109}$$

8.4 NEGATIVE RESISTANCE MODEL

For the bias condition (which is determined from the output power requirement), the loaded Q, and the device parameters [transconductance and [β^+]], the best phase noise can be found by differentiating $|\varphi^2(\omega)|_{SSB}$ with respect to (C_1/C_2).

Considering that all the parameters of $|\varphi^2(\omega)|_{SSB}$ are constants for a given operating condition (except the feedback capacitor), the minimum value of the phase noise can be determined for any fixed value of C_1 as

$$\overline{|\varphi^2(\omega)|} = \left[k_0 + \frac{k_1}{k_2 C_1^2 C_2^2 + k_3 C_2^2}\right]\left[\frac{C_1 + C_2}{C_1 C_2}\right]^2 \quad (8\text{-}110)$$

$$k_0 = \frac{kTR}{\omega^2 \omega_0^2 L^2 V_{ce}^2} \quad (8\text{-}111)$$

$$k_1 = \frac{qI_{c0}g_m^2 + (K_f I_b^{AF}/\omega)g_m^2}{\omega^2 \omega_0^2 L^2 V_{ce}^2} \quad (8\text{-}112)$$

$$k_2 = \omega_0^4 (\beta^+)^2 \quad (8\text{-}113)$$

$$k_3 = g_m^2 \quad (8\text{-}114)$$

where k_1, k_2, and k_3 are constant only for a particular drive level, with $y = (C_1/C_2)$. Making k_2 and k_3 also dependent on y, as the drive level changes, the final noise equation is

$$\mathscr{L}(\omega) = 10 \times \log\left[\left[k_0 + \left(\frac{k^3 k_1 [Y_{21}^+/Y_{11}^+]^2 [y]^{2p}}{[Y_{21}^+]^3 [y]^{3q}}\right)\left(\frac{1}{y^2 + k}\right)\right]\left[\frac{[1+y]^2}{y^2}\right]\right] \quad (8\text{-}115)$$

where

$$k_0 = \frac{kTR}{\omega^2 \omega_0^2 L^2 V_{cc}^2}$$

$$k_1 = \frac{qI_c g_m^2 + (K_f I_b^{AF}/\omega)g_m^2}{\omega^2 \omega_0^4 L^2 V_{cc}^2} \quad k_2 = \omega_0^2 (\beta^+)^2$$

Figure 8-12 shows the simulated phase noise and its minimum for two values of C_1, 2 pF and 5 pF. The 5 pF value provides a better phase noise and a flatter response. For larger C_1, the oscillator will cease to oscillate.

$$\frac{\partial |\phi^2(\omega, y, k)|}{\partial y} \Longrightarrow 0$$

$$\frac{\partial}{\partial y}\left\{\left[k_0 + \left(\frac{k^3 k_1 [Y_{21}^+/Y_{11}^+]^2 [y]^{2p}}{[Y_{21}^+]^3 [y]^{3q}}\right)\left(\frac{1}{y^2 + k}\right)\right]\left[\frac{[1+y]^2}{y^2}\right]\right\}_{y=m} \Longrightarrow 0 \quad (8\text{-}116)$$

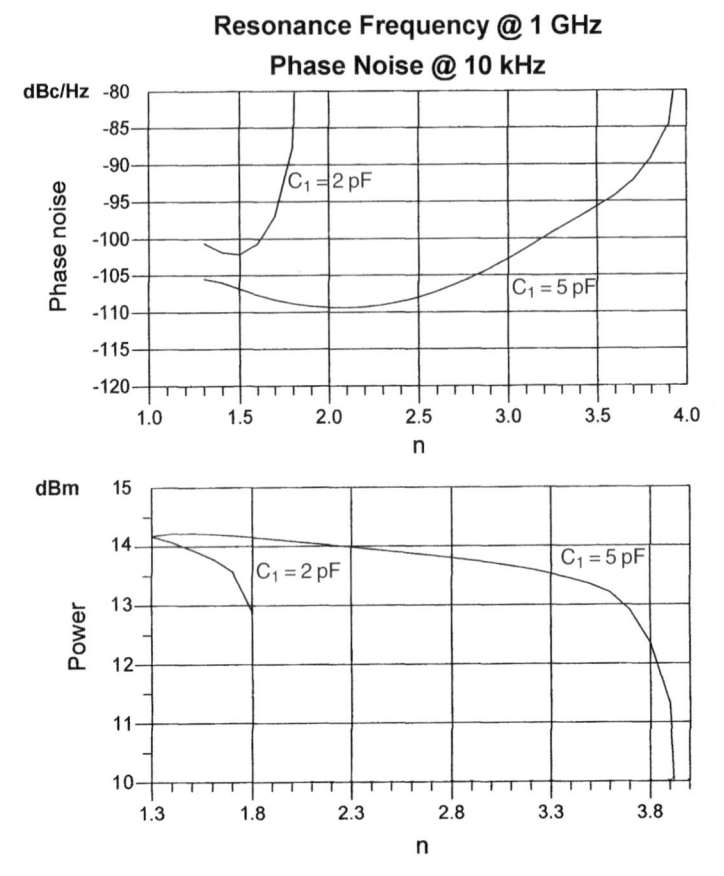

182 CALCULATION AND OPTIMIZATION OF PHASE NOISE

Figure 8-12 Phase noise versus n and output power.

From curve-fitting attempts, the following values for q and p in equation (8-116) were determined:

$$q = 1 \text{ to } 1.1; \quad p = 1.3-1.6$$

q and p are functions of the normalized drive level x and need to be determined experimentally.

The transformation factor n is defined as

$$n = 1 + \frac{C_1}{C_2} \rightarrow 1 + y \tag{8-117}$$

The plot in Figure 8-13 shows the predicted phase noise resulting from equation (8-116). For the first time, the flicker corner frequency was properly implemented and gives answers consistent with the measurements. In the following chapter, all

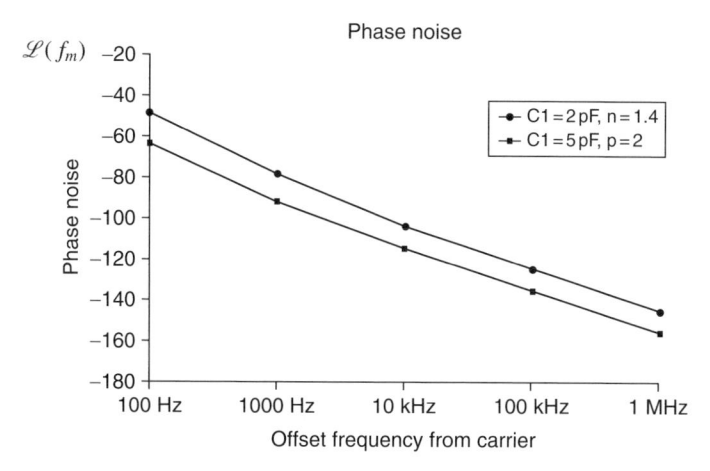

Figure 8-13 Using equation (8-116), the phase noise for different values of n for constant C_2 can be calculated.

the noise sources will be added, but the key contributors are still the resonator noise and the flicker noise. Shottky noise dominates further out. The breakpoint for the flicker noise can be clearly seen.

8.4.5 Summary Results

The analysis of the oscillator in the time domain has given us design criteria to find the optimum value of $y = (C_1/C_2)$ with values for $y + 1$ or n ranging from 1.5 to 4. For values above 3.5, the power is reduced significantly.

Consistent with the previous chapters, we note

$$C_1 = C_1^* \pm X(C_p \text{ or } L_p) \tag{8-118}$$

$$X(C_{be} \text{ or } L_b) \rightarrow C_p \text{ or } L_p \tag{8-119}$$

With a large value of C_p ($C_p > C_1$), X_1 has to be inductive to compensate for extra contributions of the device package capacitance to meet the desired value of C_1.

The following is a set of design guides to calculate the parameters of the oscillator:

$$\omega = \sqrt{\frac{1}{L[C_1C_2/(C_1+C_2)+C]}} \tag{8-120}$$

$$|R_n(L_p = 0)| = \frac{Y_{21}}{w^2 C_1 C_2} \tag{8-121}$$

$$C_1 = \frac{1}{\omega_0}\sqrt{\frac{Y_{11}}{K}} \tag{8-122}$$

C_2 is best determined graphically from the noise plot:

$$C_C > \left\{ \frac{(\omega^2 C_1 C_2)(1+\omega^2 Y_{21}^2 L_p^2)}{[(Y_{21}^2 C_2 - \omega^2 C_1 C_2)(1+\omega^2 Y_{21}^2 L_p^2)(C_1 + C_p + C_2)]} \right\} \quad (8\text{-}123)$$

$$\frac{C}{10} \geq [C_C]_{L_p=0} > \left[\frac{\omega^2 C_1 C_2}{[(Y_{21}^2 C_2 - \omega^2 C_1 C_2)(C_1 + C_p + C_2)]} \right] \quad (8\text{-}124)$$

The phase noise in dBc/Hz is shown as

$$\pounds(\omega) = 10 \times \log \left[\left[k_0 + \left(\frac{k^3 k_1 [Y_{21}^+ / Y_{11}^+]^2 [y]^{2p}}{[Y_{21}^+]^3 [y]^{3q}} \right) \left(\frac{1}{y^2 + k} \right) \right] \left[\frac{[1+y]^2}{y^2} \right] \right] \quad (8\text{-}125)$$

The phase noise improves with the square of the loaded Q_L; 10% higher $Q \rightarrow 20\%$ better phase noise.

$$\mathscr{L}(\omega) \propto \frac{1}{C_{IN}^2} \quad (8\text{-}126)$$

The loaded Q of the resonator determines the minimum possible level of the oscillator phase noise for given bias voltage and oscillator frequency.

To achieve a value close to this minimum phase noise level set by the loaded Q_L of the resonator, the optimum (rather, how large the value of the C_{IN} can be) value of C_{IN} is to be fixed.

To achieve the best possible phase noise level, the feedback capacitors C_1 and C_2 should be made as large as possible, but should still generate sufficient negative resistance to sustain steady-state oscillation.

$$[-R_N]_{negative\ resistance} \propto \frac{1}{\omega_0^2} \frac{1}{C_1 C_2} \quad \text{(no parasitics)} \quad (8\text{-}127)$$

The negative resistance of the oscillator circuit is inversely proportional to the feedback capacitors. Therefore, the limit of the feedback capacitor value is determined by the minimum negative resistance for a loop gain greater than unity.

From the phase noise equation discussed, feedback capacitor C_2 has more influence than C_1. The drive level and the conduction angle of the Colpitts oscillator circuit are strong functions of C_2.

The time-domain approach provided us with the design guide for the key components of the oscillator; however, it did not include all the noise sources of the transistor. Using the starting parameters, such as C_1 and C_2 and the bias point, as well as the information about the resonator and the transistor, a complete noise model/analysis will be shown in Section 8.5.

8.5 PHASE NOISE ANALYSIS BASED ON THE FEEDBACK MODEL

Up to now, we have calculated both the large-signal drive condition and the optimum choice of the feedback capacitance. Now we will consider the oscillator as a feedback loop with a noisy transistor, looking at all typical noise contributions. Based on a fixed set of values for C_1 and C_2, we can calculate the accurate phase noise behavior of the oscillator and analyze the various noise contributions.

First, the noisy bipolar transistor will be introduced. Figure 8-14 shows the familiar hybrid-π transistor circuit, and Figure 8-15 shows the equivalent circuit with the relevant noise sources included.

The mean square values of the noise generators in Figure 8-15, in a narrow frequency offset Δf, are given by

$$\overline{i_{bn}^2} = 2qI_b\Delta f \qquad (8\text{-}128)$$

$$\overline{i_{cn}^2} = 2qI_c\Delta f \qquad (8\text{-}129)$$

$$\overline{i_{con}^2} = 2qI_{cob}\Delta f \qquad (8\text{-}130)$$

$$\overline{v_{bn}^2} = 4kTR_b\Delta f \qquad (8\text{-}131)$$

$$\overline{v_{sn}^2} = 4kTR_S\Delta f \qquad (8\text{-}132)$$

where I_b, I_c, and I_{cob} are average DC currents over the Δf noise bandwidth.

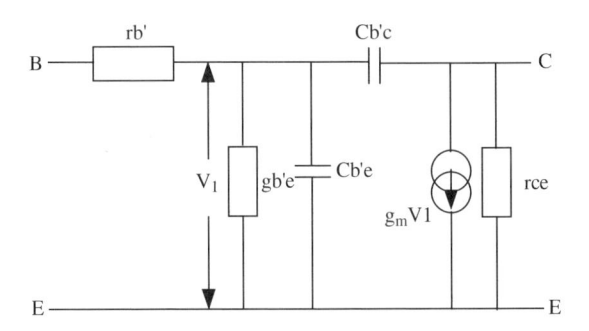

Figure 8-14 Grounded emitter bipolar transistor.

CALCULATION AND OPTIMIZATION OF PHASE NOISE

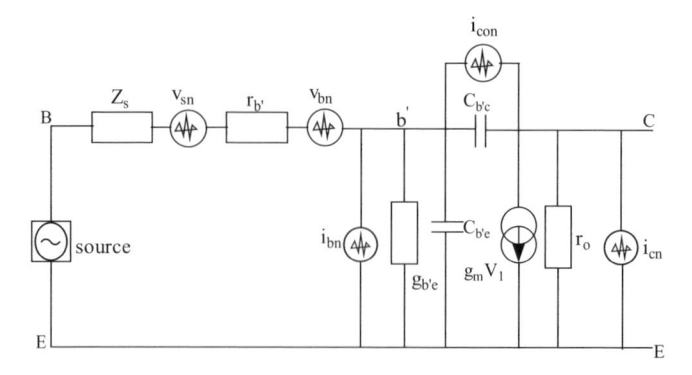

Figure 8-15 Hybrid-π configuration of the grounded bipolar transistor with noise sources.

The noise power spectral densities due to these noise sources are

$$S(i_{cn}) = \frac{\overline{i_{cn}^2}}{\Delta f} = 2qI_c = 2KTg_m \tag{8-133}$$

$$S(i_{bn}) = \frac{\overline{i_{bn}^2}}{\Delta f} = 2qI_b = \frac{2KTg_m}{\beta} \tag{8-134}$$

$$S(i_{fn}) = \frac{K_f I_b^{AF}}{f} \tag{8-135}$$

$$S(v_{bn}) = \frac{\overline{v_{bn}^2}}{\Delta f} = 4KTr_b' \tag{8-136}$$

$$S(v_{sn}) = \frac{\overline{v_{sn}^2}}{\Delta f} = 4KTR_s \tag{8-137}$$

where r_b' and R_s are base and source resistance and Z_s is the complex source impedance.

Figure 8-16 shows the feedback arrangement for the Colpitts oscillator with the noise sources.

The transistor acts like a gain block. The feedback network includes the load conductance, and a small part of the output signal goes to the input of the bipolar transistor through the resonant circuit. The [ABCD] chain matrix will be used for the analysis.

Figure 8-17 shows the linear representation of the Colpitts oscillator with the input white noise source $i_n(\omega)$.

The input noise power spectral density can be given as

$$S_{in} = \frac{|\overline{i_n^2}|}{\Delta f} \tag{8-138}$$

8.5 THE FEEDBACK MODEL   187

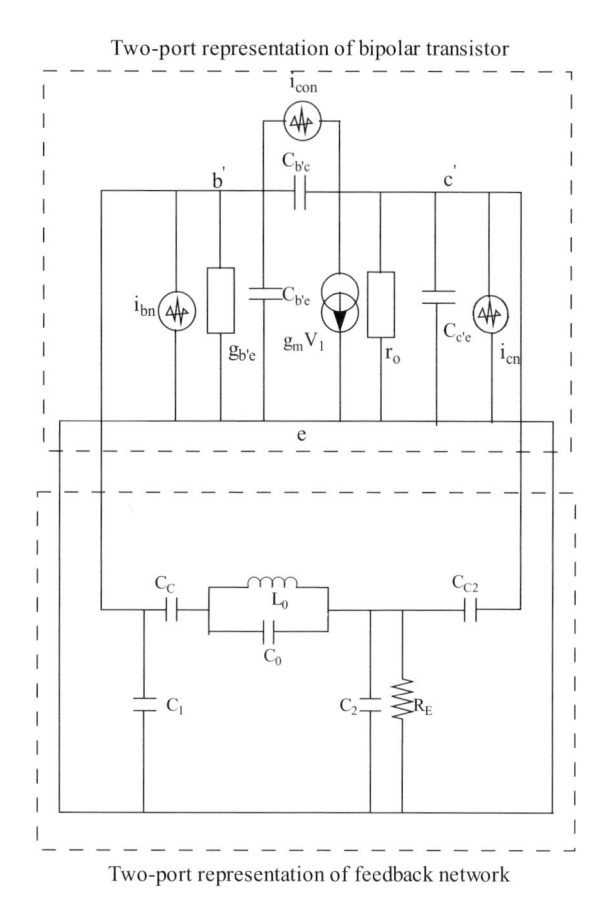

Figure 8-16 Feedback arrangement for the Colpitts oscillator with the noise sources.

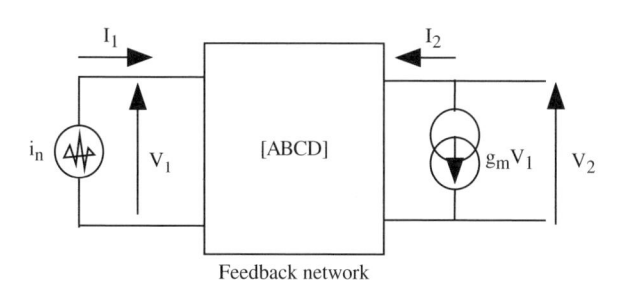

Figure 8-17 Linear representation of feedback for the Colpitts oscillator with input white noise source $i_n(\omega)$. This is not consistent with Figure 8-16, but it is useful because all nonactive components are now in the feedback network.

CALCULATION AND OPTIMIZATION OF PHASE NOISE

where

$$|\overline{i_n^2}| = \sum_{i=1}^{i=N} |\overline{i_{ni}^2}| = |\overline{i_{n1}^2}| + |\overline{i_{n2}^2}| + |\overline{i_{n3}^2}| + \cdots + 2C_{ii}[i_{ni}i^{\bullet}_{n(i+1)}] \quad (8\text{-}139)$$

and C_{ii} is the noise correlation coefficient.

The [ABCD] matrix of the above oscillator circuit can be given as

$$[A] = 1 + \left[\left(\frac{1}{j\omega C_c} + \frac{j\omega L_0}{1 - \omega^2 L_0 C_0}\right)\left(j\omega C_2 + \frac{1}{R_E}\right)\right]$$

$$[B] = \frac{1}{j\omega C_{c2}} + \left(\frac{1}{j\omega C_c} + \frac{j\omega L_0}{1 - \omega^2 L_0 C_0}\right)\left[1 + \left(j\omega C_2 + \frac{1}{R_E}\right)\left(\frac{1}{j\omega C_{c2}}\right)\right] \quad (8\text{-}140)$$

$$[C] = j\omega C_1 + \left(j\omega C_2 + \frac{1}{R_E}\right)\left[1 + j\omega C_1\left(\frac{1}{j\omega C_c} + \frac{j\omega L_0}{1 - \omega^2 L_0 C_0}\right)\right]$$

$$[D] = \frac{C_1}{C_{c2}} + \left[\left(1 + j\omega C_1\left(\frac{1}{j\omega C_c} + \frac{j\omega L_0}{1 - \omega^2 L_0 C_0}\right)\right) \times \left(1 + \left(j\omega C_2 + \frac{1}{R_E}\right)\left(\frac{1}{j\omega C_{c2}}\right)\right)\right]$$

$$\begin{bmatrix} V_1 \\ I_1 \end{bmatrix} = \begin{bmatrix} A & B \\ C & D \end{bmatrix} = \begin{bmatrix} V_2 \\ -I_2 \end{bmatrix} \quad (8\text{-}141)$$

$$V_1 = AV_2 - BI_2 \quad (8\text{-}142)$$

$$I_1 = CV_2 - DI_2 \quad (8\text{-}143)$$

$$Z_{in} = \left[\frac{V_1}{I_1}\right]_{I_2=0} = \frac{A}{C} \quad (8\text{-}144)$$

where

$$I_1 = i_n \quad (8\text{-}145)$$

$$I_2 = -g_m V_1 \quad (8\text{-}146)$$

The equivalent input noise voltage due to the input noise current, $I_1 = i_n$, is

$$v_n(\omega) = I_1 Z_{in} = I_1 \left[\frac{V_1}{I_1}\right]_{I_2=0} = I_1 \left[\frac{A(\omega)}{C(\omega)}\right] = i_n \left[\frac{A(\omega)}{C(\omega)}\right] \quad (8\text{-}147)$$

The input noise voltage $v_n(\omega)$ will produce two narrowband (1 Hz) uncorrelated components in the frequency domain located at $\omega - \omega_0$ and $\omega + \omega_0$ as $[v_n(\omega)]_{\omega=\omega_0-\Delta\omega}$ and $[v_n(\omega)]_{\omega=\omega_0+\Delta\omega}$.

8.5 THE FEEDBACK MODEL

In the presence of the two uncorrelated components of the input noise voltage, $[v_n(\omega)]_{\omega=\omega_0-\Delta\omega}$ and $[v_n(\omega)]_{\omega=\omega_0+\Delta\omega}$, the peak carrier signal of amplitude V_c at frequency $\omega = \omega_0$ is modulated with a input phase noise signal $S_{\Delta\varphi_{in}}(\omega)$.

The input phase noise spectral density at an offset of $\Delta\omega$ is

$$S_{\Delta\varphi_{in}}(\Delta\omega) = \frac{\left|[v_n(\omega)]^2_{\omega=\omega_0-\Delta\omega}\right| + \left|[v_n(\omega)]^2_{\omega=\omega_0+\Delta\omega}\right|}{\left|V_c^2(\omega)\right|} \quad (8\text{-}148)$$

$$S_{\Delta\varphi_{in}}(\Delta\omega) \cong \frac{2\left|[v_n(\omega)]^2\right|}{\left|V_c^2(\omega)\right|} \quad (8\text{-}149)$$

$$S_{\Delta\varphi_{in}}(\Delta\omega) = \frac{2\left|[v_n(\omega)]^2\right|}{\left|V_c^2(\omega)\right|} = 2\frac{\left|[i_n(\omega)]^2\right|}{\left|V_c^2(\omega)\right|}\frac{\left|A^2(\omega)\right|}{\left|C^2(\omega)\right|} \quad (8\text{-}150)$$

$$\left|\overline{i_n^2}\right| = S_{in}\Delta f \quad (8\text{-}151)$$

$$\left|\overline{i_n^2}\right|_{\Delta f=1Hz} = S_{in} \quad (8\text{-}152)$$

$$S_{\Delta\varphi_{in}}(\Delta\omega) = 2\frac{S_{in}}{\left|V_c^2(\omega)\right|}\frac{\left|A^2(\omega)\right|}{\left|C^2(\omega)\right|} \quad (8\text{-}153)$$

where S_{in} and $S_{\Delta\varphi_{in}}$ are the input noise power and phase noise spectral density.

Based on [70, 83]

$$S_{\Delta\varphi_{out}}(\omega) = S_{\Delta\varphi_{in}}(\omega)\left[1 + \frac{1}{(\omega^2)}\left(\frac{\omega_0}{2Q_L}\right)^2\right] \quad (8\text{-}154)$$

$$Q_L(\omega = \omega_0) = \frac{\omega_0}{2}\left|\frac{d\varphi}{d\omega}\right|_{\omega=\omega_0} \quad (8\text{-}155)$$

The open-loop gain is

$$G_{open}(\omega = \omega_0) = -\left[\frac{g_m}{C(\omega_0)}\right] \quad (8\text{-}156)$$

For sustained oscillation $G_{open}(\omega = \omega_0) = 1$. $-[g_m/C(\omega_0)] = 1 \Rightarrow C(\omega)_{\omega=\omega_0}$ is real and negative.

$$C(\omega_0) = C_{Real}(\omega_0) + jC_{Imag}(\omega_0) \quad (8\text{-}157)$$

$$C_{Imag}(\omega_0) = 0 \tag{8-158}$$

$$C_{Real}(\omega_0) = -g_m \tag{8-159}$$

$$\left[\frac{d\varphi}{d\omega}\right]_{\omega=\omega_0} \approx -\frac{1}{C_{Real}(\omega_0)}\left[\frac{dC_{Imag}(\omega)}{d\omega}\right]_{\omega=\omega_0} \tag{8-160}$$

$$Q_L(\omega=\omega_0) = \frac{\omega_0}{2}\left|\frac{d\varphi}{d\omega}\right|_{\omega=\omega_0} \tag{8-161}$$

$$Q_L(\omega=\omega_0) = \frac{\omega_0}{2}\left|\frac{1}{C_{Real}(\omega_0)}\right|\left[\frac{dC_{Imag}(\omega)}{d\omega}\right]_{\omega=\omega_0} \tag{8-162}$$

$$S_{\Delta\varphi_{out}}(\Delta\omega) = S_{\Delta\varphi_{in}}(\Delta\omega)\left[1 + \frac{1}{(\Delta\omega^2)}\left[\frac{C_{Real}(\omega_0)}{(dC_{Imag}(\omega)/d\omega)}\right]^2_{\omega=\omega_0}\right] \tag{8-163}$$

$$S_{\Delta\varphi_{in}}(\Delta\omega) = 2\frac{S_{in}}{|V_c^2(\omega)|}\frac{|A^2(\omega)|}{|C^2(\omega)|} \tag{8-164}$$

$$S_{\Delta\varphi_{out}}(\Delta\omega) = 2\frac{S_{in}}{|V_c^2(\omega_0)|}\frac{|A^2(\omega_0)|}{|C^2(\omega_0)|}\left[1 + \frac{1}{(\Delta\omega^2)}\left[\frac{C_{Real}(\omega_0)}{(dC_{Imag}(\omega)/d\omega)}\right]^2_{\omega=\omega_0}\right] \tag{8-165}$$

We will now perform the noise analysis of the Colpitts oscillator.

8.5.1 Individual Contribution of All Four Noise Sources

The following contribute to the noise of the oscillator

- thermal noise associated with the loss resistance of the resonator
- thermal noise associated with the base resistance of the transistor
- shot noise associated with the base bias current
- shot noise associated with the collector bias current

If we now use the oscillator circuit with a noisy resonator, we can calculate the total noise of the oscillator, as shown in Figure 8-18.

8.5.2 Noise-Shaping Function of the Resonator

For phase noise analysis, the oscillator is considered a feedback system and a noise source is present in the input, as shown in the Figure 8-19.

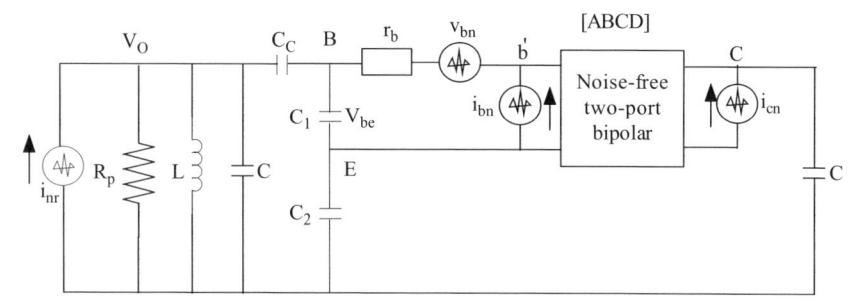

Figure 8-18 The oscillator circuit with a two-port [ABCD] matrix, consistent with the approach of Figure 8-15.

Oscillator output phase noise is a function of

- the amount of source noise present at the input of the oscillator circuit and
- how much the feedback system rejects or amplifies various noise components.

The unity-gain system closed-loop transfer function is

$$[TF(j\omega)]_{closed\ loop} = \frac{Y(j\omega)}{X(j\omega)} = \frac{H(j\omega)}{1 + H(j\omega)} \quad (8\text{-}166)$$

$$[H(j\omega)]_{\omega=\omega_0} = -1 \quad (8\text{-}167)$$

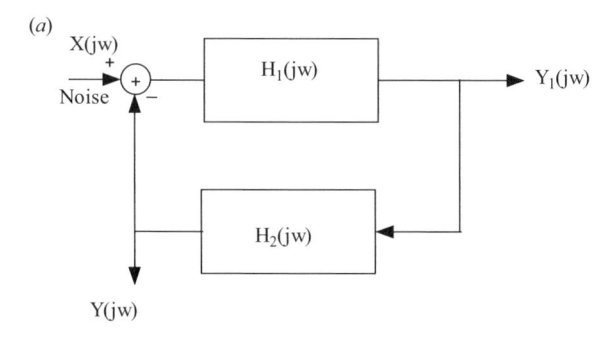

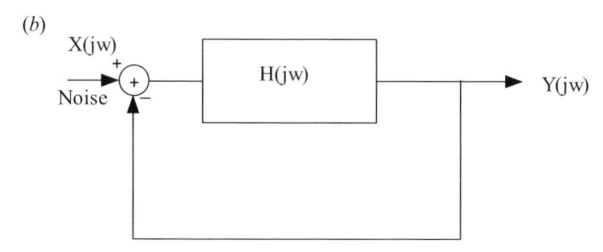

Figure 8-19 Feedback oscillator with noise source. (*a*) Nonunity gain feedback oscillator; (*b*) unity gain feedback oscillator.

192 CALCULATION AND OPTIMIZATION OF PHASE NOISE

For frequencies close to $\omega = \Delta\omega + \omega_0$, the open-loop transfer function is

$$[H(j\omega)]_{\omega=\omega_0+\Delta\omega} \approx \left[H(j\omega_0) + \Delta\omega \frac{dH(j\omega)}{d\omega}\right] \quad (8\text{-}168)$$

The noise transfer function is

$$\left[\frac{Y(j\omega+j\Delta\omega)}{X(j\omega+j\Delta\omega)}\right] = \left[\frac{H(j\omega_0) + \Delta\omega(dH(j\omega)/d\omega)}{1 + H(j\omega_0) + \Delta\omega(dH(j\omega)/d\omega)}\right] \quad (8\text{-}169)$$

Since $H(j\omega_0) = -1$ and for most practical cases $\Delta\omega(dH(j\omega))/d\omega \ll 1$, we can write

$$\left[\frac{Y(j\omega+j\Delta\omega)}{X(j\omega+j\Delta\omega)}\right] \approx \left[\frac{-1}{\Delta\omega(dH(j\omega)/d\omega)}\right] \quad (8\text{-}170)$$

From the noise transfer function, it appears that the noise component at $\omega = \Delta\omega + \omega_0$ is multiplied by the term $[-1/\Delta\omega(dH(j\omega)/d\omega)]$ relative to the output. The broadband white noise is shaped by the resonator, as seen in Figure 8-20. Therefore, the noise power spectral density can be explained as

$$\left|\frac{Y(j\omega+j\Delta\omega)}{X(j\omega+j\Delta\omega)}\right|^2 = \left|\frac{-1}{\Delta\omega(dH(j\omega)/d\omega)}\right|^2 \quad (8\text{-}171)$$

for

$$H(j\omega) = A(j\omega)\exp[j\phi(j\omega)] \quad (8\text{-}172)$$

$$\frac{dH(j\omega)}{d\omega} = \left[\frac{dA(j\omega)}{d\omega} + jA(j\omega)\frac{d\phi(j\omega)}{d\omega}\right]\exp[j\phi(j\omega)] \quad (8\text{-}173)$$

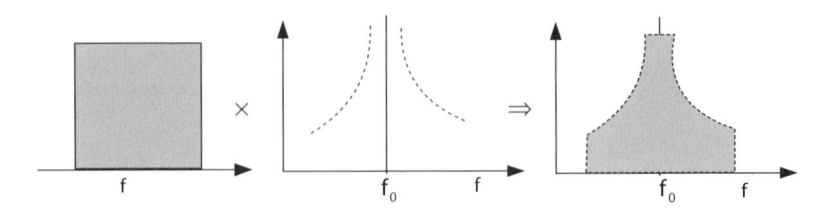

Figure 8-20 Noise shaping in the oscillator.

Assume that $\omega = \Delta\omega + \omega_0$, $\omega \to \omega_0$, and $|A(j\omega_0)| \to 1$; then the above equation is reduced to

$$\left|\frac{Y(j\omega + j\Delta\omega)}{X(j\omega + j\Delta\omega)}\right|^2 = \left[\frac{1}{(\Delta\omega)^2\{[dA(j\omega)/d\omega]^2 + [d\phi(j\omega)/d\omega]^2\}}\right]_{\omega = \Delta\omega + \omega_0} \quad (8\text{-}174)$$

The open-loop Q_L becomes

$$Q_L = \frac{\omega_0}{2}\sqrt{\left[\frac{dA(j\omega)}{d\omega}\right]^2 + \left[\frac{d\phi(j\omega)}{d\omega}\right]^2} \quad (8\text{-}175)$$

and

$$\left|\frac{Y(j\omega + j\Delta\omega)}{X(j\omega + j\Delta\omega)}\right|^2 = \left[\frac{1}{(\Delta\omega)^2\{[dA(j\omega)/d\omega]^2 + [d\phi(j\omega)/d\omega]^2\}}\right]_{\omega = \Delta\omega + \omega_0}$$

$$= \frac{1}{4Q_L^2}\left[\frac{\omega_0}{\Delta\omega}\right]^2 \quad (8\text{-}176)$$

For the LC resonator $[dA(j\omega)/d\omega]$ at resonance ($\omega \to \omega_0$) becomes zero and $Q_L = (\omega_0/2)(d\varphi/d\omega)$.

8.5.3 Nonunity Gain

For the nonunity gain feedback case where

$$H(j\omega) = H_1(j\omega)H_2(j\omega) \quad \text{(equation 8-1)}$$

it follows that

$$\left[\frac{Y(j\omega + j\Delta\omega)}{X(j\omega + j\Delta\omega)}\right]_{\omega = \Delta\omega + \omega_0} \approx \left[\frac{-1}{\Delta\omega(dH(j\omega)/d\omega)}\right] \quad (8\text{-}177)$$

and

$$\frac{Y_1(j\omega)}{X(j\omega)} = \frac{H_1(j\omega_0)}{1 + H(j\omega_0)} \quad (8\text{-}178)$$

Then the noise power is shaped by the transfer function as

$$\left|\frac{Y_1(j\omega + j\Delta\omega)}{X(j\omega + j\Delta\omega)}\right|^2 = \frac{|H_1(j\omega)|^2}{(\Delta\omega)^2|dH(j\omega)/d\omega|^2} \quad (8\text{-}179)$$

For the lossy resonator consisting of a parallel resistor, inductor, and capacitor (RLC) see Figure 8-21. Then

$$H(\omega_0 + \Delta\omega) = \left[\frac{V_{out}(\omega_0 + \Delta\omega)}{i_n(\omega_0 + \Delta\omega)}\right]_{\omega=\Delta\omega+\omega_0} = \left[\frac{1}{g_{resonator}}\right]\left[\frac{\omega_0}{\Delta\omega}\right]\left[\frac{1}{2Q_L}\right] \quad (8\text{-}180)$$

$$g_{resonator} = \frac{1}{R_P} \quad (8\text{-}181)$$

where R_P is the equivalent loss resistance of the resonator.

8.5.4 Noise Transfer Function and Spectral Densities

The noise transfer function for the relevant sources is:

$$NFT_{inr}(\omega_0) = \frac{1}{2}\left[\frac{1}{2j\omega_0 C_{eff}}\right]\left[\frac{\omega_0}{\Delta\omega}\right] \rightarrow \quad (8\text{-}182)$$

Noise transfer function of the thermal loss resistance of the resonator:

$$NFT_{V_{bn}}(\omega_0) = \frac{1}{2}\left[\frac{C_1 + C_2}{C_2}\right]\left[\frac{1}{2jQ}\right]\left[\frac{\omega_0}{\Delta\omega}\right] \rightarrow \quad (8\text{-}183)$$

Noise transfer function of the transistor's base resistance noise:

$$NFT_{i_{bn}}(\omega_0) = \frac{1}{2}\left[\frac{C_2}{C_1 + C_2}\right]\left[\frac{1}{2j\omega_0 QC_{eff}}\right]\left[\frac{\omega_0}{\Delta\omega}\right] \rightarrow \quad (8\text{-}184)$$

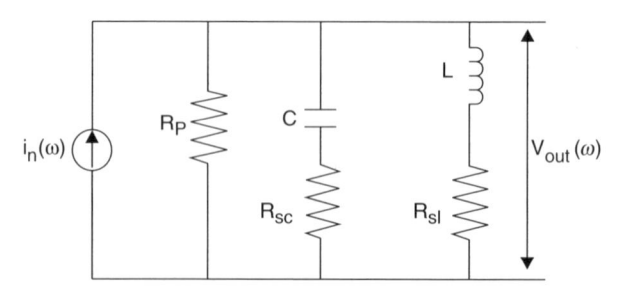

Figure 8-21 Noise response of the RLC resonator.

8.5 THE FEEDBACK MODEL   195

Noise transfer function of the transistor's base current flicker noise:

$$NFT_{ifn}(\omega_0) = \frac{1}{2}\left[\frac{C_2}{C_1+C_2}\right]\left[\frac{1}{2j\omega_0 QC_{eff}}\right]\left[\frac{\omega_0}{\Delta\omega}\right] \rightarrow \qquad (8\text{-}185)$$

Noise transfer function of the transistor's flicker noise:

$$NFT_{i_{cn}}(\omega_0) = \frac{1}{2}\left[\frac{C_1}{C_1+C_2}\right]\left[\frac{1}{2j\omega_0 QC_{eff}}\right]\left[\frac{\omega_0}{\Delta\omega}\right] \rightarrow \qquad (8\text{-}186)$$

Noise transfer function of the collector current shot noise, where

$$C_{eff} = C + \frac{C_1 C_2}{C_1 + C_2} \qquad (8\text{-}187)$$

and

$$V_0(\omega_0) = nV_{be}(\omega_0) \qquad (8\text{-}188)$$

$NFT_{in}(\omega_0)$, $NFT_{V_{bn}}(\omega_0)$, $NFT_{i_{bn}}(\omega_0)$, and $NFT_{i_{cn}}(\omega_0)$ are the noise transfer functions, as explained.

Figure 8-16 showed the four noise sources of the oscillator circuit whereby the flicker noise current is added to the base current and their noise spectral density is $K_f I_b^{AF}/f_m$.

- $NSD_{inr} = 4KT/R_P \rightarrow$ noise spectral density of the thermal noise current from the loss resistance of the resonator
- $NSD_{Vbn} = 4KTr_b \rightarrow$ noise spectral density of the thermal noise voltage from the base resistance
- $NSD_{ibn} = 2qI_b \rightarrow$ noise spectral density of the shot noise current from the base current
- $NSD_{ifn} = K_f I_b^{AF}/f_m \rightarrow$ noise spectral density due to $1/f$ flicker noise
- $NSD_{icn} = 2qI_c \rightarrow$ noise spectral density of the shot noise current from the collector current

The phase noise contribution now is

$$PN(\omega_0 + \Delta\omega) = NSD_{noise\ source}[NFT_{noise\ source}(\omega_0)]^2 \qquad (8\text{-}189)$$

$$PN_{inr}(\omega_0 + \Delta\omega) = \frac{4KT}{R_P}[NF_{inr}(\omega_0)]^2 \qquad (8\text{-}190)$$

$$PN_{Vbn}(\omega_0 + \Delta\omega) = 4KTr_b[NF_{Vbn}(\omega_0)]^2 \qquad (8\text{-}191)$$

$$PN_{ibn}(\omega_0 + \Delta\omega) = 2qI_B[NF_{ibn}(\omega_0)]^2 \quad (8\text{-}192)$$

$$PN_{ifn}(\omega_0 + \Delta\omega) = \frac{K_f I_b^{AF}}{f_m}[NF_{ibn}(\omega_0)]^2 \quad (8\text{-}193)$$

$$PN_{icn}(\omega_0 + \Delta\omega) = 2qI_c[NF_{icn}(\omega_0)]^2 \quad (8\text{-}194)$$

where $PN(\omega_0 + \Delta\omega)$ is the phase noise at the offset frequency $\Delta\omega$ from the carrier frequency ω_0 and $NSD_{noise\ source}$ is the noise spectral density of the noise sources. The phase noise contribution is

$$PN_{inr}(\omega_0 + \Delta\omega) = \frac{4KT}{R_P}[NFT_{inr}(\omega_0)]^2 = \frac{4KT}{R_P}\left\{\frac{1}{2}\left[\frac{1}{2j\omega_0 C_{eff}}\right]\left[\frac{\omega_0}{\Delta\omega}\right]\right\}^2$$

$\rightarrow$ phase noise contribution from the resonator tank

$$PN_{Vbn}(\omega_0 + \Delta\omega) = 4KTr_b[NFT_{Vbn}(\omega_0)]^2$$

$$= 4KTr_b\left\{\frac{1}{2}\left[\frac{C_1 + C_2}{C_2}\right]\left[\frac{1}{2jQ}\right]\left[\frac{\omega_0}{\Delta\omega}\right]\right\}^2$$

$\rightarrow$ phase noise contribution from the base resistance

$$PN_{ibn}(\omega_0 + \Delta\omega) = 2qI_b[NFT_{ibn}(\omega_0)]^2$$

$$= 2qI_b\left\{\frac{1}{2}\left[\frac{C_2}{C_1 + C_2}\right]\left[\frac{1}{j\omega_0 QC_{eff}}\right]\left[\frac{\omega_0}{\Delta\omega}\right]\right\}^2$$

$\rightarrow$ phase noise contribution from the base current

$$PN_{ifn}(\omega_0 + \Delta\omega) = \frac{K_f I_b^{AF}}{f_m}[NF_{ibn}(\omega_0)]^2$$

$$= \frac{K_f I_b^{AF}}{f_m}\left\{\frac{1}{2}\left[\frac{C_2}{C_1 + C_2}\right]\left[\frac{1}{j2\omega_0 QC_{eff}}\right]\left[\frac{\omega_0}{\Delta\omega}\right]\right\}^2$$

$\rightarrow$ phase noise contribution from the flicker noise of the transistor

$$PN_{icn}(\omega_0 + \Delta\omega) = 2qI_c[NFT_{icn}(\omega_0)]^2$$

$$= 2qI_c\left\{\frac{1}{2}\left[\frac{C_1}{C_1 + C_2}\right]\left[\frac{1}{2j\omega_0 QC_{eff}}\right]\left[\frac{\omega_0}{\Delta\omega}\right]\right\}^2$$

$\rightarrow$ phase noise contribution from the collector current

8.5 THE FEEDBACK MODEL

The total effect of all the four noise sources can be expressed as

$$PN(\omega_0 + \Delta\omega) = [PN_{inr}(\omega_0 + \Delta\omega)] + [PN_{Vbn}(\omega_0 + \Delta\omega)]$$
$$+ [PN_{ibn}(\omega_0 + \Delta\omega)] + [PN_{icn}(\omega_0 + \Delta\omega)] \quad (8\text{-}195)$$

$$PN(\omega_0 + \Delta\omega) = \frac{4KT}{R_P}\left\{\frac{1}{2}\left[\frac{1}{2j\omega_0 C_{eff}}\right]\left[\frac{\omega_0}{\Delta\omega}\right]\right\}^2 + 4KTr_b\left\{\frac{1}{2}\left[\frac{C_1+C_2}{C_2}\right]\left[\frac{1}{2jQ}\right]\left[\frac{\omega_0}{\Delta\omega}\right]\right\}^2$$
$$+ \left[2qI_b + \frac{2\pi K_f I_b^{AF}}{\Delta\omega}\right]\left\{\frac{1}{2}\left[\frac{C_2}{C_1+C_2}\right]\left[\frac{1}{j2Q\omega_0 C_{eff}}\right]\left[\frac{\omega_0}{\Delta\omega}\right]\right\}^2$$
$$+ 2qI_c\left\{\frac{1}{2}\left[\frac{C_1}{C_1+C_2}\right]\left[\frac{1}{2j\omega_0 QC_{eff}}\right]\left[\frac{\omega_0}{\Delta\omega}\right]\right\}^2 \quad (8\text{-}196)$$

where

K_f = flicker noise constant
AF = flicker noise exponent

$$C_{eff} = C + \frac{C_1 C_2}{C_1 + C_2} \quad (8\text{-}197)$$

Note: The effect of the Q loading of the resonator is calculated by the noise transfer function multiplied by the noise sources.

The phase noise contribution from the different noise sources for the parallel tuned Colpitts oscillator circuit at $\Delta\omega = 0$ kHz.2π from the oscillator frequency $\omega_0 = 1000$ MHz.2π will now be computed. The circuit parameters are as follows:

Base resistance of transistor $r_b = 6.14$ ohm
Parallel loss resistance of the resonator $R_P = 12{,}000$ ohm
Q of the resonator $= 380$
Resonator inductance $= 5$ nH
Resonator capacitance $= 4.7$ pF
Collector current of the transistor $I_c = 28$ mA
Base current of the transistor $I_b = 250$ μA
Flicker noise exponent $AF = 2$
Flicker noise constant $K_f = 1E\text{-}7$
Feedback factor $n = 2.5$

Phase noise at 10 kHz:

$PN_{inr}(\omega_0 + 10\,\text{kHz}) \approx -125\,\text{dBc/Hz}$
$PN_{Vbn}(\omega_0 + 10\,\text{kHz}) \approx -148\,\text{dBc/Hz}$

$PN_{(ibn+ifn)}(\omega_0 + 10\,\text{kHz}) \approx -125\,\text{dBc/Hz}$

$PN_{icn}(\omega_0 + 10\,\text{kHz}) \approx -142\,\text{dBc/Hz}$

Note: The noise contribution from the resonator at this offset is the same as the flicker noise contribution from the transistor. For low-Q cases, this can be identified as the flicker corner frequency.

Phase noise at 100 Hz:

$PN_{inr}(\omega_0 + 100\,\text{Hz}) \approx -85\,\text{dBc/Hz}$

$PN_{Vbn}(\omega_0 + 100\,\text{Hz}) \approx -108\,\text{dBc/Hz}$

$PN_{(ibn+ifn)}(\omega_0 + 100\,\text{Hz}) \approx -68\,\text{dBc/Hz}$

$PN_{icn}(\omega_0 + 100\,\text{Hz}) \approx -102\,\text{dBc/Hz}$

It appears that the flicker noise and the noise from the resonator are the limiting factors for the overall phase noise performance of the oscillator circuit.

The dependence of the phase noise performance due to different noise sources present in the oscillator circuits is

$$PN_{imr}(\omega_0 + \Delta\omega) \propto \frac{1}{R_P} \qquad (8\text{-}198)$$

$$PN_{Vbn}(\omega_0 + \Delta\omega) \propto = r_b \left\{ \frac{1}{Q} \left[1 + \frac{C_1}{C_2} \right] \right\}^2 \qquad (8\text{-}199)$$

$$PN_{ibn}(\omega_0 + \Delta\omega) \propto I_b \left\{ \frac{1}{QC_{eff}} \left[\frac{C_2}{C_1 + C_2} \right] \right\}^2 \qquad (8\text{-}200)$$

$$PN_{icn}(\omega_0 + \Delta\omega) \propto = I_c \left\{ \frac{1}{QC_{eff}} \left[\frac{C_1}{C_1 + C_2} \right] \right\}^2 \qquad (8\text{-}201)$$

Once the resonator Q is known (parallel loss resistance is fixed), the only step left is to select a device with a low flicker noise. The base resistance, current, and collector current add little to the performance. Finally, optimization of the phase noise can be done by proper selection of the feedback capacitor under the constraints of the loop gain so that it maintains oscillation.

The combined phase noise, a result of all the noise contributions, depends on the semiconductor, the resonator losses, and the feedback capacitors. Figure 8-22 shows the simulated phase noise for a given set of semiconductor parameters and various levels of n. While the values for $n = 1.5$ and 2 provide similar results and converge for frequencies more than 1 MHz off the carrier, the results for $n = 3$ also provide a much noisier condition, even at far-out frequencies. The reason is the reduced output power and a heavier loading of the resonator.

The next chapter shows a variety of test circuits which were built and measured to validate the theory shown here.

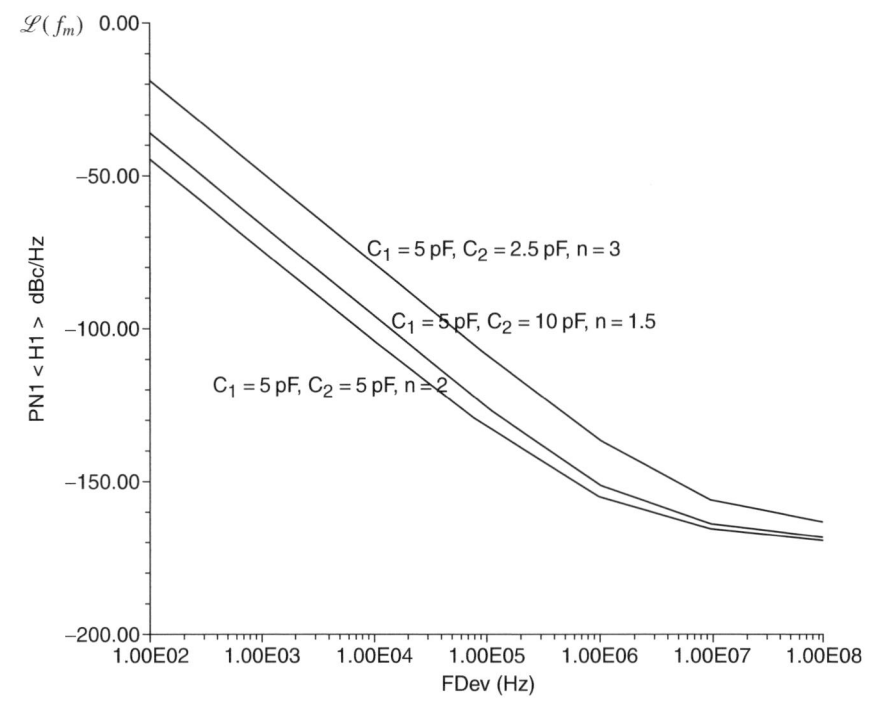

Figure 8-22 Phase noise as a function of feedback factor n.

8.6 2400 MHz MOSFET-Based Push–Pull Oscillator

Wireless applications are extremely cost sensitive, and when implemented as an RFIC, they are designed using silicon technology. Most mixers in RFICs are built on the principle of differential amplifiers (Gilbert cell) and require a phase and an out-of-phase signal (symmetrical drive). For these symmetrical requirements, this is best achieved using push–pull technology with two outputs. The design choices are SiGe transistors or BiCMOS transistors. The submicron devices in 0.35 μm and smaller technology are ideally suited for this frequency application. The 0.25 μm and smaller technology is more costly and does not provide a significant advantage. As will be seen, the critical phase noise is determined by the Q of the inductor and other elements of the resonator and by the flicker noise from the device.

Figure 8-23 shows the circuit of the 2400 MHz integrated CMOS 0.35 μm in a oscillator cross-coupled (push-pull) configuration [192–210].

The circuit above uses a cross-coupled CMOS-NMOS pair as an oscillator. The advantage compared to an all-NMOS structure is that it generates a large, symmetrical signal swing and balances out the pull-up and pull-down signals, resulting in better noise. This type of topology rejects the common mode noise and substrate noise.

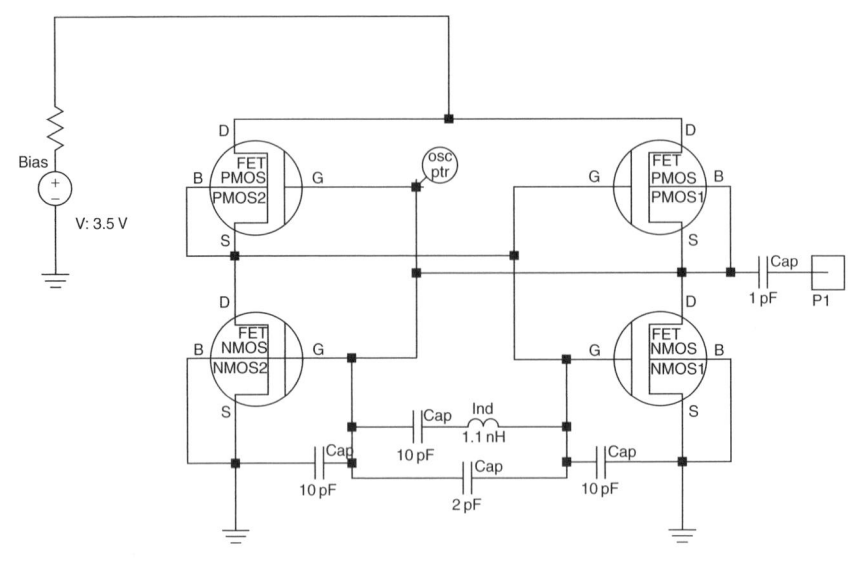

Figure 8-23 Circuit of the 2400 MHz integrated CMOS oscillator.

Figure 8-24 shows the starting condition, which requires negative resistance and cancellation of the reactances at the frequency of oscillation. The currents shown in Figure 8-24 indication that these conditions have been met.

It is important to note that the condition of zero reactance does not quite occur at the point of most negative current. Since the circuit is totally symmetrical, only the condition $C_1 = C_2$ can be met (see Section 8.2). C_1 and C_2 refer to the gate source

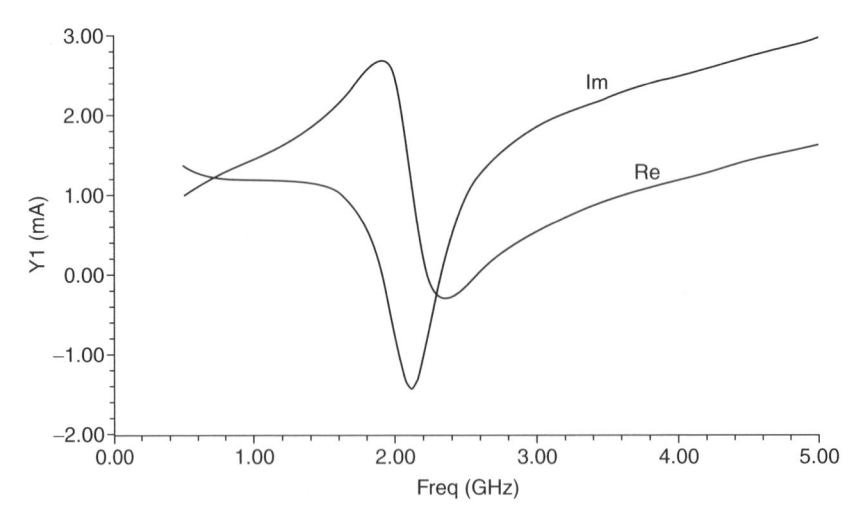

Figure 8-24 The real and imaginary currents which cause the negative resistance for oscillation.

capacitance of the FETs. As outlined previously, this is not necessarily the best condition for phase noise.

Figures 8-25 and 8-26 show the predicted phase and RF output power, including harmonic contents.

8.6.1 Design Equations

1. The transconductance

    Figure 8-27 shows a cross-coupled positive gate voltage MOS transistor (PMOS) and a cross-coupled negative gate voltage MOS transistor (NMOS) pair using CMOS devices.

    According to the literature, PMOS transistors offer lower $1/f$ and thermal noise, while NMOS transistors exhibit higher f_T and higher transconductance for the same operating point.

    The total transconductance is

$$[g_m]_{large\ signal} = -\frac{[g_m]_{PMOS} + [g_m]_{NMOS}}{2} \tag{8-202}$$

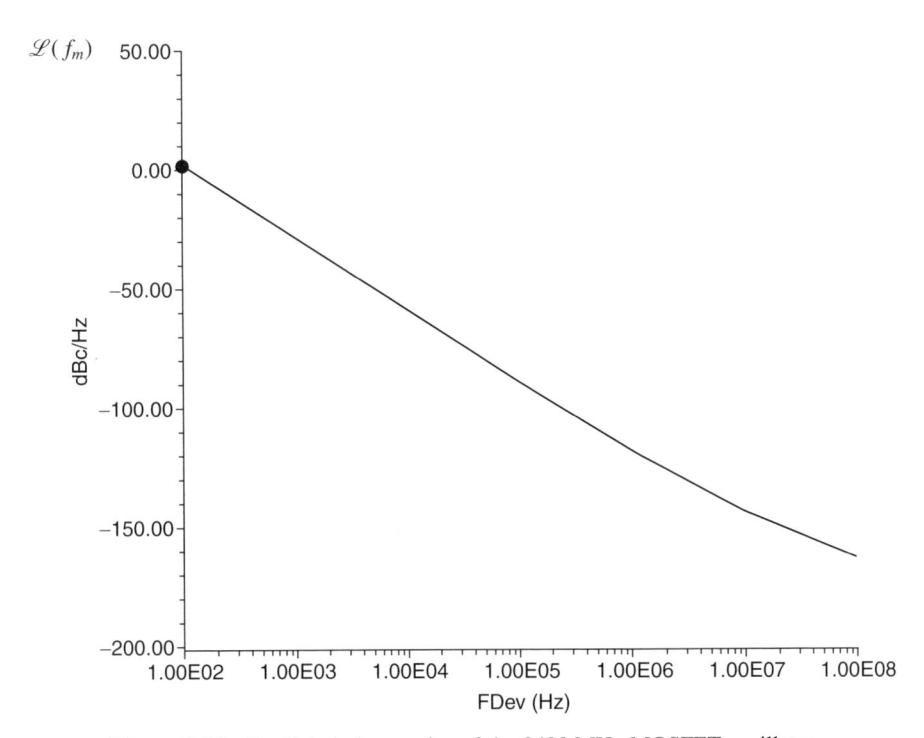

Figure 8-25 Predicted phase noise of the 2400 MHz MOSFET oscillator.

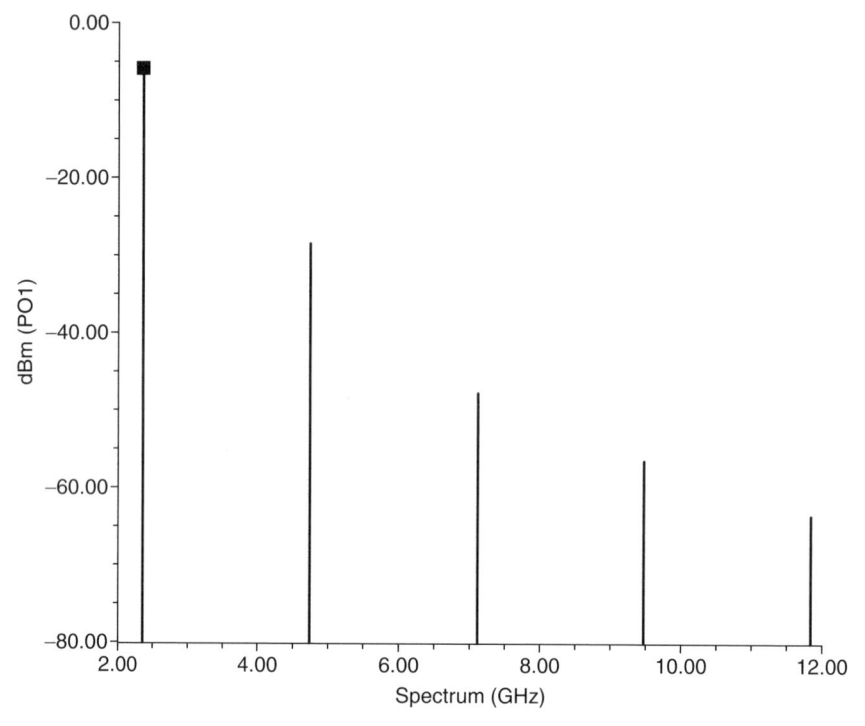

Figure 8-26 Predicted output spectrum of the 2400 MOSFET oscillator.

$$[g_m]_{large\ signal} = \frac{\partial I_{ds}}{\partial V_{gs}} \tag{8-203}$$

$$[g_m]_{NMOS} = \sqrt{2I_{ds}\mu_{nmos}C_{ox\,nmos}\left[\frac{W}{L}\right]_{nmos}} \tag{8-204}$$

$$[g_m]_{PMOS} = \sqrt{2I_{ds}\mu_{pmos}C_{ox\,nmos}\left[\frac{W}{L}\right]_{pmos}} \tag{8-205}$$

$$\frac{\partial I_{ds}}{\partial V_{gs}} = K_p \frac{w}{L}(V_{gs} - V_{th}) \tag{8-206}$$

$$[g_m]_{large\ signal} = \sqrt{\frac{2wK_pI_{ds}}{L}} \tag{8-207}$$

where

K_p = transconductance parameter
μ_{pmos} = carrier mobility of the PMOS device
μ_{nmos} = carrier mobility of the NMOS device
C_{ox} = unit capacitance of the gate oxide

8.6 2400 MHz MOSFET-Based Push–Pull Oscillator

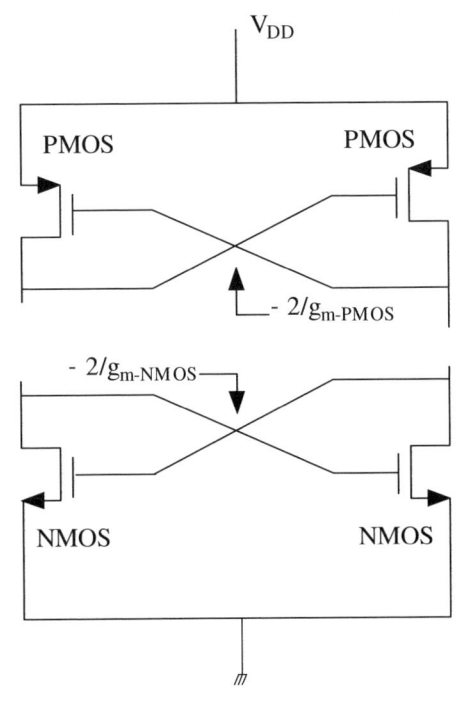

Figure 8-27 Determining the transconductance of the differential circuit of the cross-coupled PMOS and NMOS pair.

2. The transconductance parameter
   The transconductance parameter is defined as

$$K_p = \mu C_{ox} \tag{8-208}$$

   where μ is the carrier mobility and C_{ox} is the unit capacitance of the gate oxide.
3. The Gate-oxide capacitance of the device
   The unit capacitance of the gate oxide C_{ox} is given as

$$C_{ox} = \varepsilon_{ox} \left[\frac{w_i l_i}{t_{ox}} \right] \tag{8-209}$$

   where
   ε_{ox} = permitivity of the oxide
   t_{ox} = thickness of the oxide layer between the spiral and substrate
   w_i = width of the spiral line
   l_i = length of the spiral line

4. The drain current
The drain current is

$$I_{ds} = \frac{1}{2}\left[K_p \frac{w}{L}(V_{gs} - V_{th})^2\right] \qquad (8\text{-}210)$$

$$I_{ds} = \frac{g_m}{2}(V_{gs} - V_{th}) \qquad (8\text{-}211)$$

where $(V_{gs} - V_{th})$ is defined as

$$(V_{gs} - V_{th}) = \sqrt{\frac{2I_{ds}L}{K_p w}} \qquad (8\text{-}212)$$

5. The size of the device
The size of the device determines the transconductance of the transistor, and the large-signal transconductance needs to be large enough to sustain oscillation and compensate for the losses of the resonator.

The expression of the ratio of the channel width (gate) and channel length (gate) is

$$\frac{w}{L} = \frac{(2SG_P)^2}{2K_p I_{ds}} \rightarrow \frac{(g_m)^2}{2K_p I_{ds}} \qquad (8\text{-}213)$$

where w is the width of the channel (gate) and L is the length of the channel (gate) of the device.

6. The total equivalent resistance at resonant frequency
Figures 8-28a and 8-28b show the equivalent cross-coupled oscillator resonant circuit and the corresponding equivalent resistances in the resonance condition.

The total equivalent parallel resistor at resonance frequency is

$$R_T = \frac{2R_P}{2 - R_P(g_{m\,NMOS} + g_{m\,PMOS})} \qquad (8\text{-}214)$$

$$Q_L = \frac{R_P}{\omega L} \qquad (8\text{-}215)$$

where $g_{m\,NMOS}$ and $g_{m\,PMOS}$ are the corresponding large-signal transconductances of the NMOS and PMOS devices.

For a symmetrical output signal, the large-signal transconductances of the NMOS and PMOS transistors ideally should be equal, as $g_{m\,NMOS} = g_{m\,PMOS} = g_m$, and the equivalent resistance at resonance condition is

$$R_T = \frac{1}{[(1/R_P) - g_m]} \qquad (8\text{-}216)$$

8.6 2400 MHz MOSFET-Based Push–Pull Oscillator

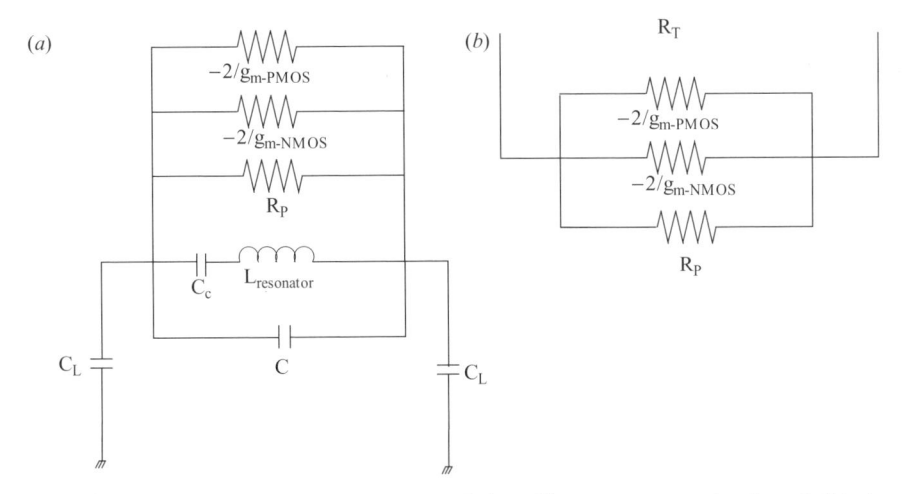

Figure 8-28 (a) The equivalent cross-coupled oscillator resonator circuit and (b) the equivalent resistances in the resonance condition.

The differential negative resistance generated by the cross-coupled NMOS and PMOS transistors-pair compensates for the parallel loss resistance R_P of the resonator circuit.

7. Startup conditions
   For the startup condition and guaranteeing sustained oscillation, the value of R_T must be negative and

$$R_T = \frac{1}{[(1/R_P) - g_m]} < 0 \implies g_m > \frac{1}{R_P} \tag{8-217}$$

From the loop gain criteria using a stability factor of 2 (loop gain = 2; the gain is adjusted to 1 by self-adjusting the conducting angle of the circuit), the startup condition is

$$\frac{g_m}{SG_P} \to 2 \tag{8-218}$$

where

$$G_P = \frac{1}{R_P} \tag{8-219}$$

where
   S = stability factor
   R_P = equivalent parallel loss resistance of the resonator

8.6.2 Design Calculations

8.6.2.1 Parallel Loss Resistance of the Resonator The equivalent parallel loss resistance of the resonator is given as

$$R_P = (1 + Q^2)R_s \Rightarrow 101 R_s \quad \text{for } Q = 10 \tag{8-220}$$

where R_s is series loss resistance.

$$[R_P]_{f=2400\,\text{MHz}} = Q\omega L_{ind} = 190\,\Omega \tag{8-221}$$

where $Q = 10$, $L_{ind} = 1.1$ nH, and $G_P = \frac{1}{R_P} \Rightarrow 6.577$ mS

8.6.2.2 Large-Signal Transconductance

$$[g_m]_{large\ signal} = \sqrt{\frac{2wK_p I_{ds}}{L}} = \sqrt{\frac{2*(250\text{E}-6)*(35.6\text{E}-6)*(14.8\text{E}-3)}{0.35\text{E}-6}}$$

$$= 27.435\,\text{mS} \tag{8-222}$$

8.6.2.3 Size of the Device The width of the CMOS is given as

$$\frac{w}{L} = \frac{(2SG_P)^2}{2K_p I_{ds}} \rightarrow \frac{(g_m)^2}{2K_p I_{ds}} \tag{8-223}$$

$$\frac{w}{L} = \frac{(2SG_P)^2}{2K_p I_{ds}} = 714.3 \tag{8-224}$$

for $L = 0.35$ μm and $w = 250$ μm, where

$K_p = 35.6\text{E} - 6$
$I_{ds} = 14.8$ mA
$G_P = 6.577$ mS
$S = 2$

8.6.2.4 Oscillation Frequency The frequency of the oscillation is given as

$$f_0 = \frac{1}{2\pi\sqrt{L_{resonator\ tank} C_{resonator\ tank}}} \tag{8-225}$$

$$C_{resonator\ tank} = \frac{1}{2}[C_{NMOS} + C_{PMOS} + C_L + C] \tag{8-226}$$

where

$$C_{NMOS} = 4C_{gd\,nmos} + C_{gs\,nmos} + C_{db\,nmos} \qquad (8\text{-}227)$$

$$C_{PMOS} = 4C_{gd\,pmos} + C_{gs\,pmos} + C_{db\,pmos} \qquad (8\text{-}228)$$

For the cross-coupled configuration, $C_{NMOS\,Pair}$ is the series combination of the two C_{NMOS} and is given as

$$C_{NMOS\,Pair} = 2C_{gd\,nmos} + \frac{1}{2}C_{gs\,nmos} + \frac{1}{2}C_{db\,nmos} \qquad (8\text{-}229)$$

Similarly, $C_{PMOS\,Pair}$ is the series combination of the two C_{PMOS} and is given as

$$C_{PMOS\,Pair} = 2C_{gd\,pmos} + \frac{1}{2}C_{gs\,pmos} + \frac{1}{2}C_{db\,pmos} \qquad (8\text{-}230)$$

The capacitance of the resonator is given as

$$C_{resonator\,tank} = \frac{1}{2}[C_{NMOS} + C_{PMOS} + C_L + C] \qquad (8\text{-}231)$$

where

$C_L = 10\,\text{pF}$ (load capacitance)
$C = 1/2$ resonator—parallel capacitance

$$f_0 = \frac{1}{2\pi\sqrt{L_{resonator\,tank}C_{resonator\,tank}}} = \frac{1}{2\pi\sqrt{1.1E-9 * 3.3E-12}} = 2400\,\text{MHz} \qquad (8\text{-}232)$$

where

$L_{resonator\,tank} = 1.1\,\text{nH}$
$C_{resonator\,tank} = 3.3\,\text{pF}$

8.6.3 Phase Noise

The phase noise of CMOS oscillators has been subject to endless discussions. The main contributors are still the resonant circuit with a low Q and the flicker frequency contribution from the device. From Section 8.5.4 we take the following equations and adapt them to the CMOS device:

$$PN_{inr}(\omega_0+\omega) = \frac{4KT}{R_P}[NFT_{inr}(\omega_0)]^2 = \frac{4KT}{R_P}\left\{\frac{1}{2}\left[\frac{1}{2j\omega_0 C_{eff}}\right]\left[\frac{\omega_0}{\omega}\right]\right\}^2$$
→ phase noise contribution from the resonator

$$PN_{Vbn}(\omega_0+\omega) = 4KTr_b[NFT_{Vbn}(\omega_0)]^2 = 4KTr_b\left\{\frac{1}{2}\left[\frac{C_1+C_2}{C_2}\right]\left[\frac{1}{2jQ_0}\right]\left[\frac{\omega_0}{\omega}\right]\right\}^2$$
→ phase noise contribution from the gate resistance

$$PN_{ibn}(\omega_0+\omega) = 2qI_b[NFT_{ibn}(\omega_0)]^2 = 2qI_b\left\{\frac{1}{2}\left[\frac{C_2}{C_1+C_2}\right]\left[\frac{1}{j\omega_0 C_{eff}}\right]\left[\frac{\omega_0}{\omega}\right]\right\}^2$$
→ phase noise contribution from the gate current

$$PN_{ibn}(\omega_0+\omega) = \frac{K_f I_b^{AF}}{f_m}[NF_{ibn}(\omega_0)]^2 = \frac{K_f I_b^{AF}}{\Delta\omega}\left\{\frac{1}{2}\left[\frac{C_2}{C_1+C_2}\right]\left[\frac{1}{j2\omega_0 Q_0 C_{eff}}\right]\left[\frac{\omega_0}{\omega}\right]\right\}^2$$
→ phase noise contribution from the flicker noise of the transistor

$$PN_{icn}(\omega_0+\omega) = 2qI_c[NFT_{icn}(\omega_0)]^2 = 2qI_c\left\{\frac{1}{2}\left[\frac{C_1}{C_1+C_2}\right]\left[\frac{1}{2j\omega_0 C_{eff}}\right]\left[\frac{\omega_0}{\omega}\right]\right\}^2$$
→ phase noise contribution from the drain current

The following values were used:

$R_P = 190\ \Omega$
$f_0 = 2.4$ GHz
$L = 1.1$ nH
$C_0 = 2$ pF
$C_1 = C_2 = 0.2$ pF
$n = 2$
$I_g = 100\ \mu A$
$I_d = 14$ mA
$AF = 2$
$KF = 5E-5$
$q = 1.6E-19$
$T = 290K$

and the following contributions were obtained at 1 MHz offset:

PN1 = -117.78 dBc/Hz
PN2 = -146.37 dBc/Hz

8.6 2400 MHz MOSFET-Based Push–Pull Oscillator

$PN3 = -123.4 \text{ dBc/Hz}$

$PN4 = -140.9 \text{ dBc/Hz}$

These calculations show that the phase noise contribution from the tuned circuit dominates and sets the value at -117.78 dBc/Hz.

The circuit was then analyzed using Microwave Harmonica/Ansoft Designer, employing a lossy circuit with a Q_0 of 10 and using the SPICE-type parameters obtained from the manufacturer.

The output power measured single-ended was -7 dBm. Figure 8-29 shows the simulated output power and harmonic contents. The accuracy of the prediction is within 1 dB.

Figure 8-30 shows the predicted phase noise from Ansoft Designer and the phase noise prediction from the set of equations given above. It should be pointed out that close in the flicker noise contribution dominates, in the medium range the resonator Q dominates, and for high currents the drain current adds significant noise.

This approach has shown very good agreement between the simulations and calculations, as demonstrated. Table 8-1 shows a list of oscillators implemented in various technologies. It is apparent from the list that this design is state of the art

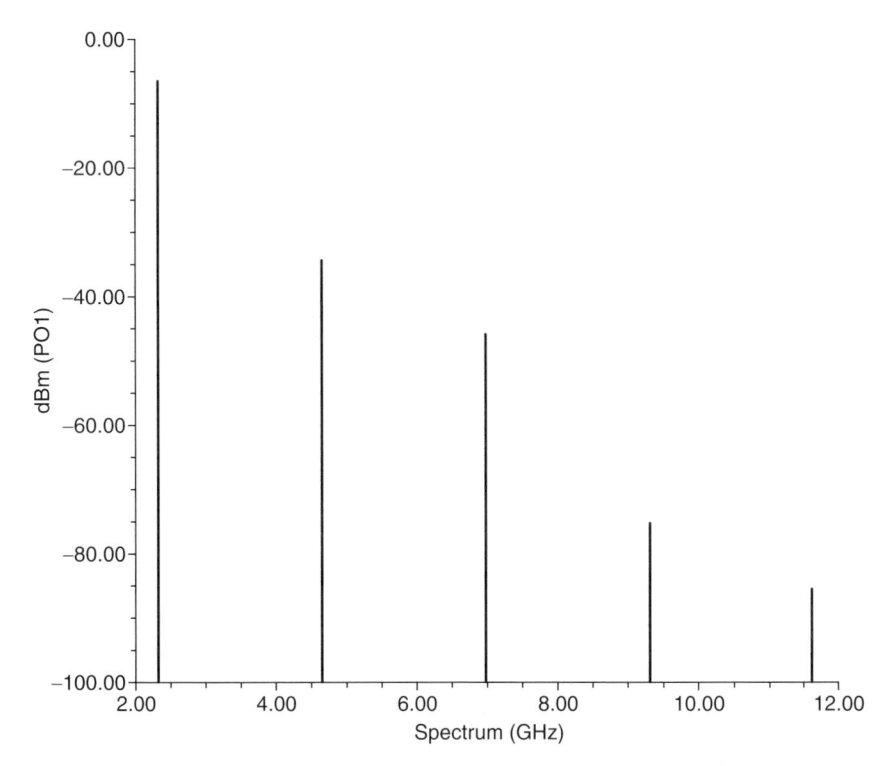

Figure 8-29 The predicted output spectrum of the CMOS oscillator.

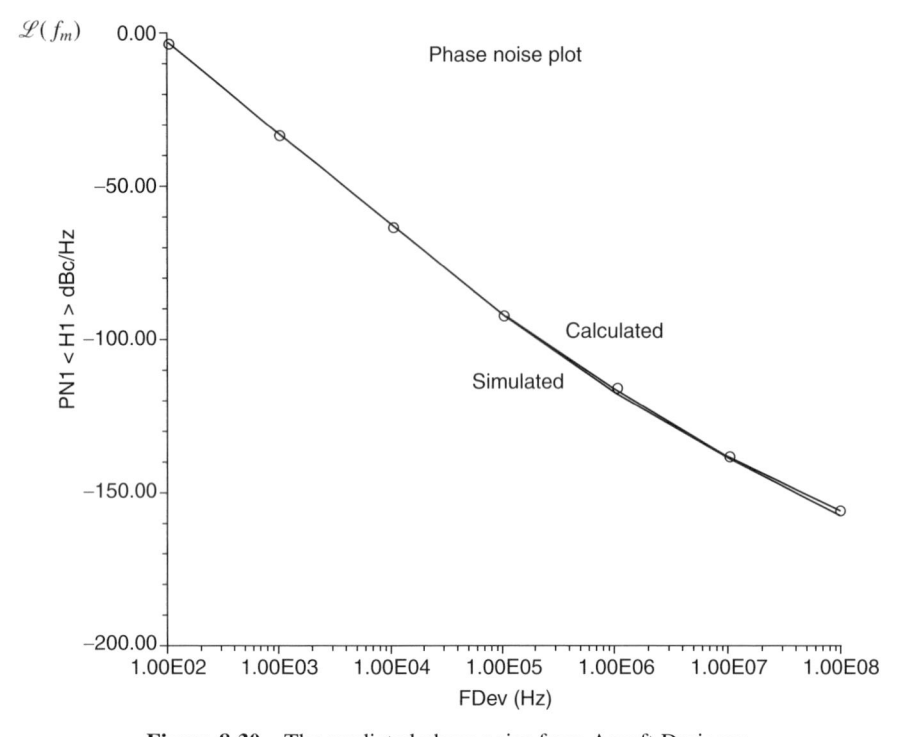

Figure 8-30 The predicted phase noise from Ansoft Designer.

[99–106]. The publications on this topic have analyzed various other contributions, both from the transistor and from the tuning mechanism. When FETs are used as varactors, the average Q is in the vicinity of 30, which means that the low-Q inductor still is responsible for the overall phase noise. The three areas of improvement are the power supply voltage, the Q, and the device selection. So far, the power supply voltage has not been addressed; however, the latest designs operating at 1.5 V show a poorer noise performance. Their distinct trade-offs and the application determine if such degradation is allowable.

8.7 PHASE NOISE, BIASING, AND TEMPERATURE EFFECTS

The biasing circuit plays an important role in the oscillator/VCO design, and in the resulting phase noise and temperature stability. As the frequency band for the wireless communication increases, thermal stable DC sources with ultra-low-noise, a wide tuning range, and low cost become more and more challenging, as the active device parameters (bipolar/FET) vary over the temperature range.

With a poorly designed biasing network, the variation of the collector current can have the same ratio as the β variation from lot to lot. The collector current could double when β is doubled and a change in temperature occurs. Therefore, the goal of the biasing network is to maximize the circuit's tolerance to β variation

TABLE 8-1

Technology	Type	Frequency	Bias	Phase Noise
0.8 μm BiCMOS	LC	200 MHz	40 μA@2 V	−104 dBc/Hz@24 kHz
0.35 μm CMOS	Ring	450 MHz	?@3,3 V	−99 dBc/Hz@100 kHz
1.2 μm CMOS	Ring	900 MHz	1 mA@2 V	−83 dBc/Hz@100 kHz
0.8 μm CMOS	Ring	800 MHz	?@5 V	−80 dBc/Hz@100 kHz
0.6 μm CMOS	Ring	900 MHz	10 mA@3 V	−101 dBc/Hz@100 kHz
0.6 μm CMOS	LC	900 MHz	1.2 mA@2.5 V	−102 dBc/Hz100 kHz
0.5 μm CMOS	Ring	900 MHz	6.2 mA@2.5 V	−105 dBc/Hz@600 kHz
25 GHz bipolar	LC	800 MHz	1 mA@2.7 V	−106 dBc/Hz@100 kHz
0.8 μm CMOS	LC	1.1 GHz	4.7 mA@2.7 V	−112 dBc/Hz@100 kHz
0.8 μm CMOS	Ring	1.27 GHz	2.64 mA@5 V	?
0.35 μm CMOS	LC	1.2 GHz	3.7 mA@2.5 V	−153 dBc/Hz@3 MHz
0.35 μm CMOS	LC	1.3 GHz	6 mA@2 V	−119 dBc/Hz@600 kHz
0.6 μm CMOS	LC	1.6 GHz	12 mA@3 V	−99 dBc/Hz@100 kHz
0.7 μm CMOS	LC	1.8 GHz	8 mA@3 V	−115 dBc/Hz@200 kHz
0.7 μm CMOS	LC	1.8 GHz	4 mA@1.5 V	−116 dBc/Hz@600 kHz
0.6 μm CMOS	LC	1.8 GHz	3.8 mA@2.7 V	−100 dBc/Hz@100 kHz
0.6 μm CMOS	LC	1.8 GHz	2.7 mA@2.7 V	−101 dBc/Hz@100 kHz
0.4 μm CMOS	LC	1.8 GHz	3.7 mA@3 V	−118 dBc/Hz@600 kHz
0.35 μm CMOS	LC	2 GHz	4 mA@2.5 V	−117 dBc/Hz@600 kHz
0.35 μm CMOS	LC	1.9 GHz	9.2 mA@3 V	−122 dBc/Hz@600 kHz
0.35 μm CMOS	LC	2 GHz	7.5 mA@3 V	−87 dBc/Hz@100 kHz
0.35 μm CMOS	LC	2 GHz	7.5 mA@3 V	−87 dBc/Hz@100 kHz
0.25 μm CMOS	LC	1.9 GHz	5 mA@2.5 V	−100 dBc/Hz@100 kHz
0.25 μm CMOS	LC	1.8 GHz	7.8 mA@2.5 V	−143 dBc/Hz@3 MHz
0.2 μm SIMOX	LC	2 GHz	6 mA@0.5 V	−110 dBc/Hz@1 MHz
0.65 μm BiCMOS	LC	1.8 GHz	19 mA@1.8 V	−125 dBc/Hz@600 kHz
18 GHz BiCMOS	Ring	2 GHz	0.3 mA@3 V	−75 dBc/Hz@1 MHz
25 GHz 0,5 μm bipolar	LC	1.9 GHz	6.4 mA@3.3 V	−100 dBc/Hz@100 kHz
20 GHz bipolar	LC	2.5 GHz	7 mA@2 V	−104 dBc/Hz@100 kHz
35 GHz BiCMOS	LC	4.4 GHz	4 mA@2.5 V	−102 dBc/Hz@600 kHz
0.25 μm CMOS	LC	5.3 GHz	4.7 mA@1.5 V	−93 dBc/Hz@100 kHz

and temperature variation. Also, as the bias changes, sometimes the phase noise changes dramatically. It can become better or worse.

8.7.1 Active Bias Circuits

Figure 8-31 shows a simplified schematic of the oscillator using the half-butterfly resonator.

Biasing is accomplished via a traditional scheme using a voltage divider between the power supply and the ground and an emitter resistor. The temperature dependency is caused by the changes in the transistor parameters as well as the diodes. The most significant change in this scheme occurs because of the tuning diodes. Figure 8-32 shows the predicted phase noise as a function of temperature.

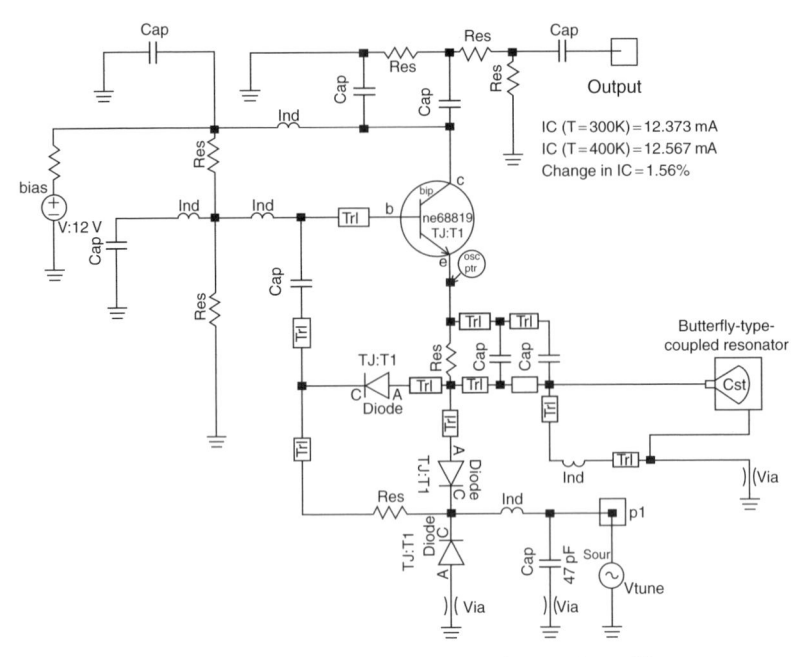

212   CALCULATION AND OPTIMIZATION OF PHASE NOISE

Figure 8-31   A 900–1800 MHz half-butterfly oscillator.

Phase noise deteriorates at higher temperatures, specifically in the close-in phase noise. In order to get better resolution, the exact values, in increments of 10 degrees Kelvin, are shown in Table 8-2.

The most attractive solution to the biasing dilemma is the use of an active biasing network that effectively sets the DC operating condition practically, regardless of the variations in device β and temperature, as discussed in Section 9.1 [87, 88]. Although active biasing would be the best option for controlling the variation, the associated costs and space requirements are usually higher than those of the passive bias network. Infineon sells a ready-to-go integrated circuit, model BCR400W, which is easy to use. Figure 8-33 shows a typical application for this circuit. More details can be taken from the data sheet. The same data sheet also shows an application for stabilizing an FET.

A somewhat simpler implementation of active stabilization is shown in Figure 8-34. It uses either the same transistor, configured as a diode, or a suitable diode. As both semiconductors have a negative temperature coefficient, the change of V_{BE} is mostly compensated for. A temperature change from 300 to 400 degrees Kelvin results in a 0.9% change in the collector current. A 10% change of the supply voltage, however, changes the DC current by 10.8%.

The resulting phase noise, which is fairly low is shown in Figure 8-35. Only at lower offsets from the carrier does the phase noise change. As a point of reference, at 100 Hz off the carrier, the phase noise is approximately −63 dBc/Hz. We will use this value for comparison to other circuits.

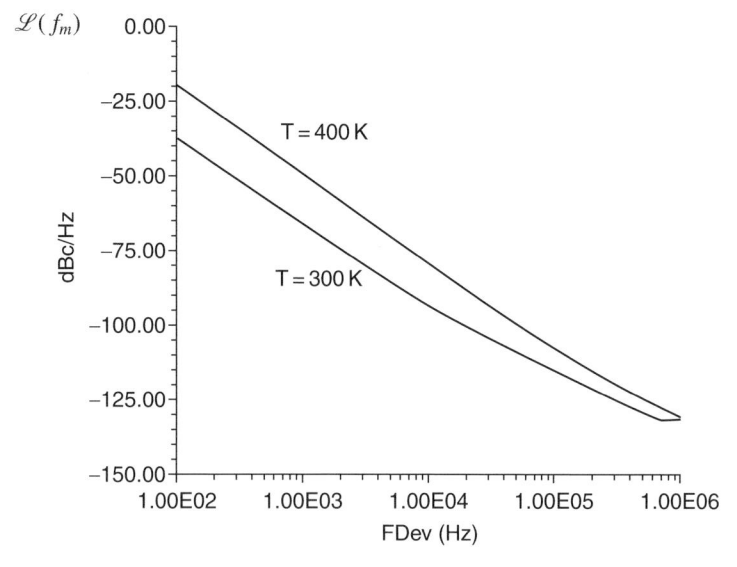

8.7 PHASE NOISE, BIASING, AND TEMPERATURE EFFECTS 213

Figure 8-32 Simulated phase noise of the VCO in Figure 8-31.

TABLE 8-2

Temperature	Phase noise at 10 kHz offset
200 K	−82.7 dBc/Hz
210 K	−82.9 dBc/Hz
220 K	−83.1 dBc/Hz
230 K	−83.4 dBc/Hz
240 K	−83.6 dBc/Hz
250 K	−83.8 dBc/Hz
260 K	−95.6 dBc/Hz
270 K	−95.4 dBc/Hz
280 K	−95.1 dBc/Hz
290 K	−94.5 dBc/Hz
300 K	−93.5 dBc/Hz
310 K	−92.0 dBc/Hz
320 K	−89.9 dBc/Hz
330 K	−86.7 dBc/Hz
340 K	−83.7 dBc/Hz
350 K	−82.4 dBc/Hz
360 K	−81.0 dBc/Hz
370 K	−79.3 dBc/Hz
380 K	−77.9 dBc/Hz
390 K	−82.6 dBc/Hz
400 K	−78.9 dBc/Hz

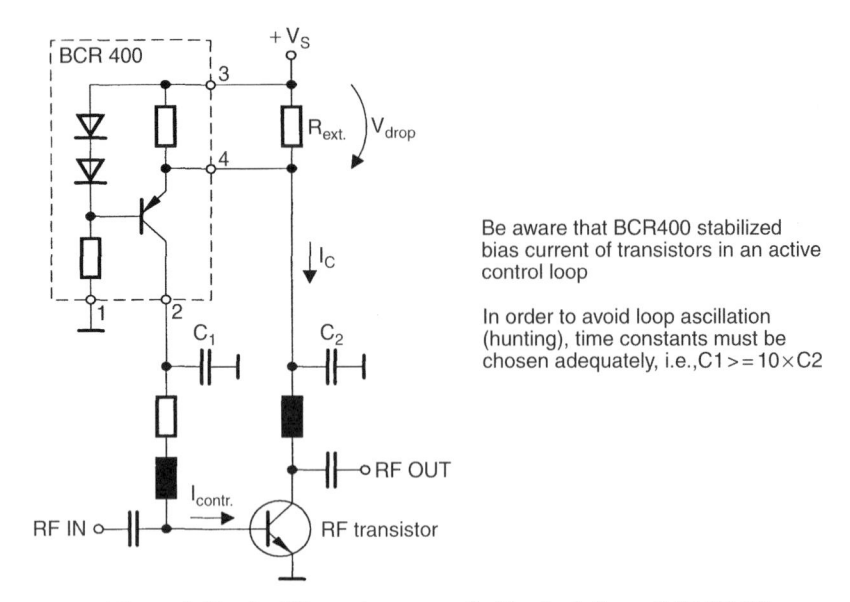

214 CALCULATION AND OPTIMIZATION OF PHASE NOISE

Be aware that BCR400 stabilized bias current of transistors in an active control loop

In order to avoid loop ascillation (hunting), time constants must be chosen adequately, i.e., C1 >= 10×C2

Figure 8-33 An RF transistor controlled by the Infineon BCR400 IC.

8.7.2 Passive Bias Circuits

In order to design a temperature-stable oscillator circuit, different types of passive bias networks [189–191] are discussed, including their stability factor. This provides greater insight and understanding. This section describes the circuit

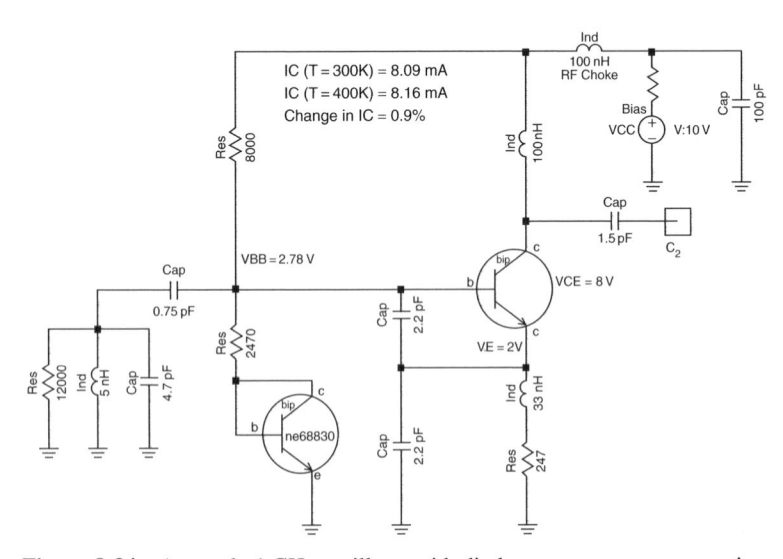

Figure 8-34 A sample 1 GHz oscillator with diode temperature compensation.

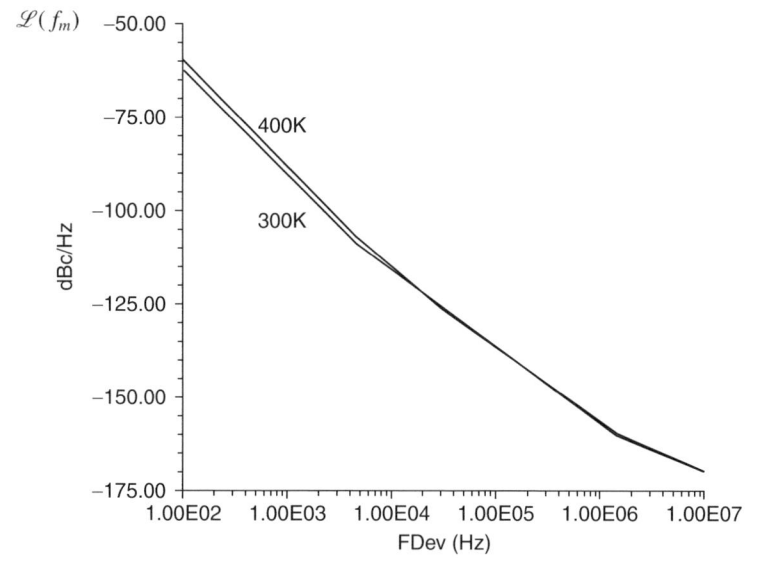

8.7 PHASE NOISE, BIASING, AND TEMPERATURE EFFECTS 215

Figure 8-35 Simulated phase noise for the oscillator shown in Figure 8-34.

analysis of four commonly used stabilized bias circuits and one nonstabilized bias circuit (a simple circuit, which should not be used under any circumstances) for the bipolar transistor. The effects of the networks on the phase noise will also be shown.

Figures 8-36a shows the passive form of the bias network; this is a nonstabilized version (not recommended, but included for completeness). Figure 8-36b shows a voltage feedback bias network. As the transistor draws more current, the collector voltage increases, the collector base voltage increases, and therefore, the circuit is stabilized. Figure 8-36c shows a voltage feedback with a current source bias network. This arrangement further increases the temperature stability. Figure 8-36d shows a voltage feedback with a voltage source bias circuit. This is a subset of Figure 8-36c. The final configuration, Figure 8-36e, shows the most frequently used bias scheme, whereby the base voltage depends on the voltage division between R_{B1} and R_{B2} with an additional emitter resistor. Each of these circuits has plusses and minuses, as the following discussion shows.

Figure 8-36a shows the simplest form of a (nonstabilized) bias circuit, which is determined by the internal parameters β_{dc} (DC current gain), I_{CBO} (reverse collector current), h_{ie} (hybrid input impedance), and V'_{BE} (base emitter voltage) and can be modeled with two current sources, such as $\beta_{dc} I_B$ and $I_{CBO}(1 + \beta_{dc})$.

The device parameters β_{dc}, V'_{BE}, and I_{CBO} are temperature-dependent variables. β_{dc} (current gain) typically increases with temperature at the rate of 0.5%/°C, V'_{BE} has a negative temperature coefficient of -2 mV/°C, and I_{CBO} typically doubles for every 10°C rise in temperature.

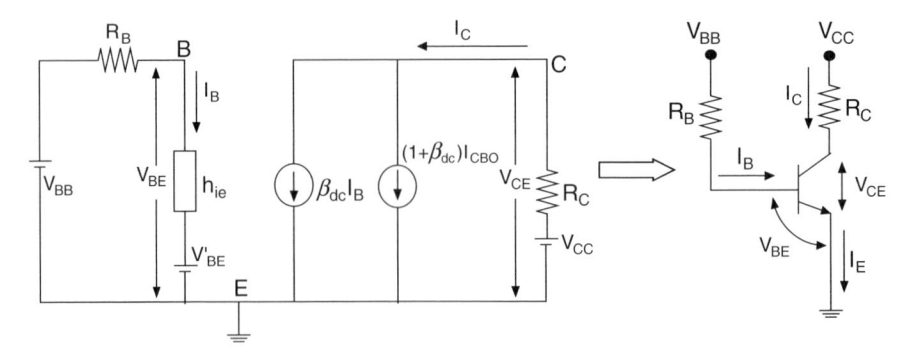

Figure 8-36a Nonstabilized bias circuit (bias circuit shows only DC components).

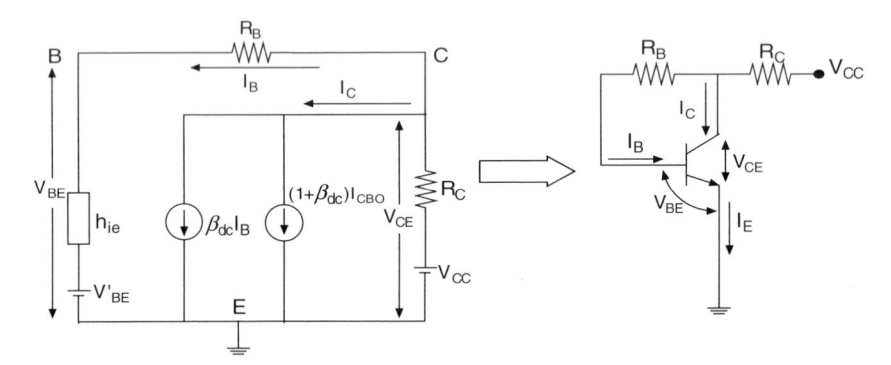

Figure 8-36b Voltage feedback bias circuit (bias circuit shows only DC components).

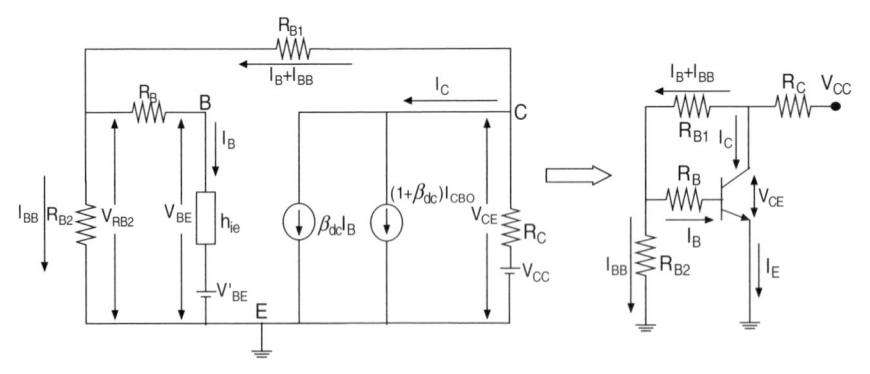

Figure 8-36c Voltage feedback with constant base current source bias circuit (bias circuit shows only DC components).

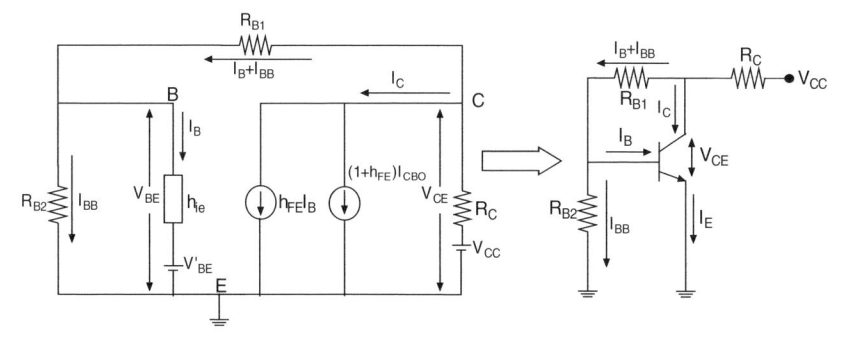

8.7 PHASE NOISE, BIASING, AND TEMPERATURE EFFECTS 217

Figure 8-36d Voltage feedback with voltage source-bias circuit (bias circuit shows only DC components).

The DC value of the current gain is defined as the ratio of the collector current to the base current:

$$h_{FE} = \beta_{dc} = \frac{I_C}{I_B}\bigg|_{V_{CE}\ \text{has const}} \tag{8-233}$$

From the circuit shown in Figure 8-36a,

$$I_B = \frac{V_{BB} - V_{BE}}{R_B} = \frac{V_{BB} - V'_{BE}}{h_{ie} + R_B} \tag{8-234}$$

$$R_B = \frac{V_{BB} - V_{BE}}{I_B} \tag{8-235}$$

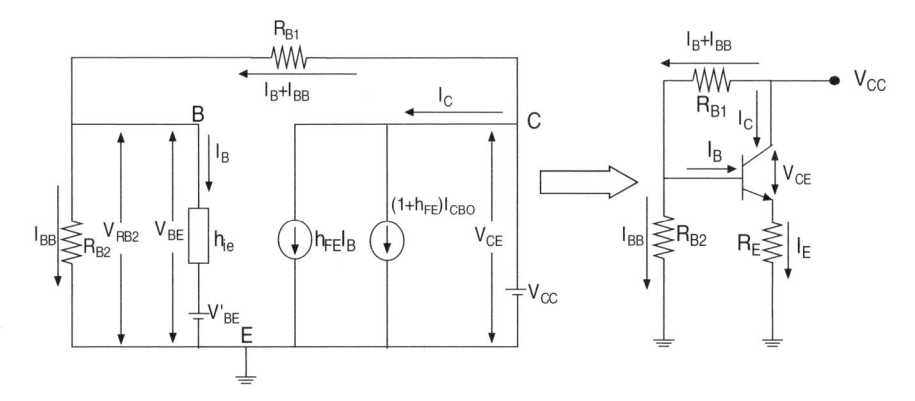

Figure 8-36e Emitter feedback-bias circuit (bias circuit shows only DC components).

$$R_C = \frac{V_{CC} - V_{CE}}{I_C} \tag{8-236}$$

$$V_{CE} = V_{CC} - I_C * R_C \tag{8-237}$$

$$I_C = \frac{\beta_{dc}(V_{BB} - V'_{BE})}{h_{ie} + R_B} + I_{CBO}(1 + \beta_{dc}) \tag{8-238}$$

In equation (8-238), all three temperature-dependent variables, β_{dc}, I_{CBO}, and V'_{BE}, influence the collector current due to the change in temperature. Therefore, a partial derivative of the collector current I_C, with respect to each of these parameters (I_{CBO}, β_{dc}, and V'_{BE}), will give a stability factor of the biasing circuit.

8.7.3 Calculation of the Stability Factors and Their Combined Effect on I_C

The stability factor for I_{CBO} is given by

$$S_{I_{CBO}} = \left.\frac{\partial I_C}{\partial I_{CBO}}\right|_{\beta_{dc}, V'_{BE} = const} = (1 + \beta_{dc}) \tag{8-239}$$

The stability factor for V'_{BE} is given by

$$S_{V'_{BE}} = \left.\frac{\partial I_C}{\partial V'_{BE}}\right|_{I_{CBO}, \beta_{dc} = const} = \left[\frac{-\beta_{dc}}{h_{ie} + R_B}\right] \tag{8-240}$$

The stability factor for β_{dc} is given by

$$S_{\beta_{dc}} = \left.\frac{\partial I_C}{\partial \beta_{dc}}\right|_{I_{CBO}, V'_{BE} = const} = \left[\frac{V_{BB} - V'_{BE}}{h_{ie} + R_B}\right] + I_{CBO} \tag{8-241}$$

The change in the collector current is given by

$$\Delta I_C = S_{I_{CBO}} * \Delta I_{CBO} + S_{V'_{BE}} * \Delta V'_{BE} + S_{\beta_{dc}} * \Delta \beta_{dc} \tag{8-242}$$

From equations (8-237) and (8-238), for constant supply voltage V_{CC} and V_{BB}, the collector current I_C will vary in direct proportion to β_{dc}. Therefore, the circuit shown in Figure 8-36a does not compensate for variation in β_{dc} due to the change in temperature.

Figure 8-36b shows the voltage feedback form of the stabilized biasing network, which provides voltage feedback to the bias current source resistor R_B. In this case, the base current source is fed from the voltage across the collector-emitter of the bipolar transistor V_{CE}, as opposed to the supply voltage V_{CC}. The value of R_B is calculated based upon the nominal device base-emitter voltage V_{BE} and the desired collector-emitter V_{CE} as per the design requirement.

8.7 PHASE NOISE, BIASING, AND TEMPERATURE EFFECTS

As shown in Figure 8-36b, collector resistor R_C has both I_C and I_B flowing through it. The variation in β_{dc}, due to the change in the temperature, will tend to cause changes in the collector current I_C. Any increase in voltage across R_C causes V_{CE} to decrease, and the decrease in V_{CE} causes I_B to decrease because the potential difference across the base bias resistor R_B is decreased. This configuration of the biasing circuit provides negative feedback that tends to reduce the amount by which the collector current increases as β_{dc} is increased due to the rise in temperature.

From the circuit shown in Figure 8-36b,

$$I_B = \frac{V_{CE} - V_{BE}}{R_B} = \frac{V_{CE} - V'_{BE}}{h_{ie} + R_B} \tag{8-243}$$

$$R_B = \frac{V_{CE} - V_{BE}}{I_B} \tag{8-244}$$

$$R_C = \frac{V_{CC} - V_{CE}}{I_C + I_B} \tag{8-245}$$

$$V_{CE} = V_{CC} - (I_C + R_B) * R_C \tag{8-246}$$

$$I_C = \frac{\beta_{dc}(V_{CC} - V'_{BE}) + I_{CBO}(1 + \beta_{dc})(h_{ie} + R_B + R_C)}{h_{ie} + R_B + R_C(1 + \beta_{dc})} \tag{8-247}$$

From equation (8-247), all three temperature-dependent variables, β_{dc}, I_{CBO}, and V'_{BE}, influence the collector current due to the change in temperature. Therefore, a partial derivative of the collector current I_C with respect to each of these parameters will give the stability factor of the biasing circuit.

The stability factor for I_{CBO} is given by

$$S_{I_{CBO}} = \left.\frac{\partial I_C}{\partial I_{CBO}}\right|_{\beta_{dc}, V'_{BE}=const} = \frac{(1 + \beta_{dc})(h_{ie} + R_B + R_C)}{h_{ie} + R_B + R_C(1 + \beta_{dc})} \tag{8-248}$$

The stability factor for V'_{BE} is given by

$$S_{V'_{BE}} = \left.\frac{\partial I_C}{\partial V'_{BE}}\right|_{I_{CBO}, \beta_{dc}=const} = \left[\frac{-\beta_{dc}}{h_{ie} + R_B + R_C(1 + \beta_{dc})}\right] \tag{8-249}$$

220   CALCULATION AND OPTIMIZATION OF PHASE NOISE

The stability factor for β_{dc} is given by

$$\begin{aligned}S_{h_{FE}} &= \left.\frac{\partial I_C}{\partial \beta_{dc}}\right|_{I_{CBO}, V'_{BE}=const} \\ &= \left[\frac{[V_{CC} - V'_{BE} + (h_{ie} + R_B + R_C)I_{CBO}][\beta_{dc}R_C + h_{ie} + R_B + R_C]}{[\beta_{dc}R_C + h_{ie} + R_B + R_C]^2}\right] \\ &\quad - R_C\left[\frac{\beta_{dc}[V_{CC} - V'_{BE} + (h_{ie} + R_B + R_C)I_{CBO}] + [h_{ie} + R_B + R_C]I_{CBO}}{[\beta_{dc}R_C + h_{ie} + R_B + R_C]^2}\right]\end{aligned}$$

(8-250)

The change in the collector current is given by

$$\Delta I_C = S_{I_{CBO}} * \Delta I_{CBO} + S_{V'_{BE}} * \Delta V'_{BE} + S_{\beta_{dc}} * \Delta \beta_{dc} \qquad (8\text{-}251)$$

Figure 8-36c shows the voltage feedback with a constant base current source biasing network that provides voltage feedback similar to that given in Figure 8-36b. The voltage divider network consisting of R_{B1} and R_{B2} provides a potential divider, from which resistor R_B is connected that determines the base current I_B. The potential drop across the resistor R_C is determined by the collector current I_C, base current I_B, and current through the voltage divider consisting of R_{B1} and R_{B2}.

From circuit shown in Figure 8-36c,

$$R_B = \frac{V_{RB2} - V_{BE}}{I_B} \qquad (8\text{-}252)$$

$$R_{B1} = \frac{V_{CE} - V_{RB2}}{I_B + I_{BB}} \qquad (8\text{-}253)$$

$$R_{B2} = \frac{V_{RB2}}{I_{BB}} \qquad (8\text{-}254)$$

$$R_C = \frac{V_{CC} - V_{CE}}{I_C + I_B + I_{BB}} \qquad (8\text{-}255)$$

For $V_{CE} > V_{RB2} > V_{BE}$

$$I_C = \beta_{dc}\left[\frac{-V'_{BE}(R_{B1} + R_{B2} + R_C) - R_{B2}[R_C I_{CBO}(1 + \beta_{dc}) - V_{CC}]}{(h_{ie} + R_B)(R_{B1} + R_{B2} + R_C) + R_{B2}(\beta_{dc}R_C + R_{B1} + R_C)}\right]$$
$$+ I_{CBO}(1 + \beta_{dc}) \qquad (8\text{-}256)$$

From equation (8-256), all three temperature-dependent variables, β_{dc}, I_{CBO}, and V'_{BE}, influence the collector current due to the change in temperature. Therefore, a partial derivative of the collector current I_C, with respect to each of these parameters, will give a stability factor of the biasing circuit.

8.7 PHASE NOISE, BIASING, AND TEMPERATURE EFFECTS

The stability factor for I_{CBO} is given by

$$S_{I_{CBO}} = \frac{\partial I_C}{\partial I_{CBO}}\bigg|_{\beta_{dc}, V'_{BE}=const}$$

$$= (1 + \beta_{dc}) - \left[\frac{R_{B2}\beta_{dc}R_C(1 + \beta_{dc})}{(R_{B1} + R_{B2} + R_C)(h_{ie} + R_B) + R_{B2}(\beta_{dc}R_C + R_C + R_{B1})}\right] \quad (8\text{-}257)$$

The stability factor for V'_{BE} is given by

$$S_{V'_{BE}} = \frac{\partial I_C}{\partial V'_{BE}}\bigg|_{I_{CBO}, \beta_{dc}=const}$$

$$= \left[\frac{-\beta_{dc}(R_{B1} + R_{B2} + R_C)}{(h_{ie} + R_B)(R_{B1} + R_{B2} + R_C) + R_{B2}(R_C\beta_{dc} + R_C + R_{B1})}\right] \quad (8\text{-}258)$$

The stability factor for β_{dc} is given by

$$S_{\beta_{dc}} = \frac{\partial I_C}{\partial \beta_{dc}}\bigg|_{I_{CBO}, V'_{BE}=const}$$

$$= \beta_{dc}\left[\frac{R_{B2}R_C[(R_{B2}V_{CC} + V'_{BE}(R_{B1} + R_{B2} + R_C)) - R_{B2}R_C I_{CBO}(1 + \beta_{dc})]}{[(h_{ie} + R_B)(R_{B1} + R_{B2} + R_C) + R_{B2}(\beta_{dc}R_C + R_{B1} + R_C)]^2}\right]$$

$$+ \left[\frac{-V'_{BE}(R_{B1} + R_{B2} + R_C) - R_{B2}[R_C I_{CBO}(1 + \beta_{dc}) + V_{CC} + \beta_{dc}R_C I_{CBO}]}{[(h_{ie} + R_B)(R_{B1} + R_{B2} + R_C) + R_{B2}(\beta_{dc}R_C + R_{B1} + R_C)]}\right] + I_{CBO}$$

$$(8\text{-}259)$$

The change in the collector current is given by

$$\Delta I_C = S_{I_{CBO}} * \Delta I_{CBO} + S_{V'_{BE}} * \Delta V'_{BE} + S_{\beta_{dc}} * \Delta \beta_{dc} \quad (8\text{-}260)$$

Figure 8-36d shows the voltage feedback with the voltage source stabilized biasing circuit. This configuration is similar to that in Figure 8-36c, except that the series current source resistor R_B is omitted. In this topology, the current passes through the resistor R_{B1} and is shared by resistor R_{B2} and the voltage across the base emitter junction of the transistor V_{BE}. The regulation depends on the amount of current flowing through R_{B2}. Therefore, the more current passes through resistor R_{B2}, the greater the regulation of the base emitter voltage V_{BE}.

From the circuit shown in Figure 8-36d,

$$R_{B2} = \frac{V_{BE}}{I_{BB}} \tag{8-261}$$

$$R_{B1} = \frac{V_{CE} - I_{BB}R_{B2}}{I_B + I_{BB}} \tag{8-262}$$

$$R_C = \frac{V_{CC} - V_{CE}}{I_C + I_B + I_{BB}} \tag{8-263}$$

$$I_C = \frac{I_{CBO}(-A) + I_{CBO}h_{ie}(-B) + D - V_{CC}}{C} \tag{8-264}$$

From equation (8-264), all three temperature-dependent variables, β_{dc}, I_{CBO}, and V_{BE}', influence the collector current due to the change in temperature. Therefore, a partial derivative of the collector current I_C, with respect to each of these parameters, will give a stability factor of the biasing circuit.

The stability factor for I_{CBO} is given by

$$S_{I_{CBO}} = \left.\frac{\partial I_C}{\partial I_{CBO}}\right|_{\beta_{dc}, V_{BE}'=const} = \frac{(A + h_{ie}B)}{C} \tag{8-265}$$

The stability factor for V_{BE}' is given by

$$S_{V_{BE}'} = \left.\frac{\partial I_C}{\partial V_{BE}'}\right|_{I_{CBO},\beta_{dc}=const} = \frac{-(R_C + R_{B1} + R_{B2})}{CR_{B2}} \tag{8-266}$$

The stability factor for β_{dc} is given by

$$S_{\beta_{dc}} = \left.\frac{\partial I_C}{\partial \beta_{dc}}\right|_{I_{CBO}, V_{BE}'=const} = \left[\frac{-(R_C + R_{B1})I_{CBO} + I_{CBO}\beta_{dc}^2 h_{ie}E}{\beta_{dc}^2 C}\right]$$

$$- \left[\frac{\beta_{dc}^2 h_{ie}E - R_C - R_{B1}}{\beta_{dc}^2}\right]\left[\frac{I_{CBO}A + I_{CBO}h_{ie}B - D + V_{CC}}{C^2}\right] \tag{8-267}$$

where

$$A = \left[\frac{R_C}{\beta_{dc}} + R_C + \frac{R_{B1}}{\beta_{dc}} + R_{B1}\right]$$

$$B = \left[\frac{R_C}{R_{B2}\beta_{dc}} + \frac{R_C}{R_{B2}} + \frac{R_{B1}}{R_{B2}\beta_{dc}} + \frac{R_{B1}}{R_{B2}} + \frac{1}{\beta_{dc}} + 1\right]$$

$$C = \left[R_C + \frac{R_C}{\beta_{dc}} + \frac{R_{B1}}{\beta_{dc}}\right] + h_{ie}\left[\frac{R_C}{R_{B2}\beta_{dc}} + \frac{R_{B1}}{R_{B2}\beta_{dc}} + \frac{1}{\beta_{dc}}\right]$$

8.7 PHASE NOISE, BIASING, AND TEMPERATURE EFFECTS 223

$$D = \left[\frac{R_C}{R_{B2}} V'_{BE} + \frac{R_{B1}}{R_{B2}} V'_{BE} + V'_{BE}\right]$$

$$E = -\left[\frac{R_C}{R_{B2}\beta_{dc}^2} + \frac{R_{B1}}{R_{B2}\beta_{dc}^2} + \frac{1}{\beta_{dc}^2}\right]$$

The change in the collector current is given by

$$\Delta I_C = S_{I_{CBO}} * \Delta I_{CBO} + S_{V'_{BE}} * \Delta V'_{BE} + S_{\beta_{dc}} * \Delta \beta_{dc} \tag{8-268}$$

Figure 8-36e is the emitter feedback bias network, where a resistor is connected in series with the device emitter lead to provide the voltage feedback. This configuration provides optimum control of variations of β_{dc} over the variations in temperature and also from device to device (due to process variations from lot to lot).

The emitter resistor R_E must be properly bypassed for RF to avoid a regenerative effect. The typical bypass capacitor often has an internal lead inductance that can create unwanted regenerative feedback.

From the circuit shown in Figure 8-36e,

$$R_E = \frac{V_{CC} - V_{CE}}{I_C\left(1 + \frac{1}{\beta_{dc}}\right)} \tag{8-269}$$

$$R_{B1} = \frac{V_{CC} - I_{BB}R_{B2}}{I_B + I_{BB}} \tag{8-270}$$

$$R_{B2} = \frac{V_{RB2}}{I_{BB}} \tag{8-271}$$

$$I_C = -\left[\frac{Ah_{ie}I_{CBO} + BI_{CBO} - D}{C}\right] \tag{8-272}$$

The stability factor for I_{CBO} is given by

$$S_{I_{CBO}} = \left.\frac{\partial I_C}{\partial I_{CBO}}\right|_{\beta_{dc}, V'_{BE} = \text{const}} = \left[\frac{Ah_{ie} + B}{C}\right] \tag{8-273}$$

The stability factor for V'_{BE} is given by

$$S_{V'_{BE}} = \left.\frac{\partial I_C}{\partial V'_{BE}}\right|_{I_{CBO}, \beta_{dc} = \text{const}} = -\left[\frac{R_{B1} + R_{B2}}{CR_{B2}}\right] \tag{8-274}$$

224   CALCULATION AND OPTIMIZATION OF PHASE NOISE

The stability factor for V'_{BE} is given by

$$S_{h_{FE}} = \frac{\partial I_C}{\partial \beta_{dc}}\bigg|_{I_{CBO},V'_{BE}=const} = \left\{\frac{I_{CBO}E + I_{CBO}h_{ie}\left[-\frac{1}{\beta_{dc}^2} - \frac{R_{B1}}{R_{B2}\beta_{dc}^2}\right]}{C}\right.$$
$$\left. - \left\{\left[\frac{I_{CBO}B + I_{CBO}h_{ie}A - D}{C^2}\right]\left[-h_{ie}\left(\frac{1}{\beta_{dc}^2} + \frac{R_{B1}}{R_{B2}\beta_{dc}^2}\right) + E\right]\right\}\right\}$$

(8-275)

where

$$A = \frac{R_{B1}}{R_{B2}\beta_{dc}} + \frac{R_{B1}}{R_{B2}} + \frac{1}{\beta_{dc}} + 1$$

$$B = \left(\frac{R_{B1}}{R_{B2}}\right)\left(\frac{R_E}{\beta_{dc}}\right) + R_E\left(\frac{R_{B1}}{R_{B2}}\right) + \left(\frac{R_E}{\beta_{dc}}\right) + \left(\frac{R_{B1}}{\beta_{dc}}\right) + R_E + R_{B1}$$

$$C = h_{ie}\left[\frac{1}{\beta_{dc}} + \frac{R_{B1}}{R_{B2}\beta_{dc}}\right] + \left[\frac{R_E}{\beta_{dc}}\right]$$
$$+ \left[\frac{R_{B1}}{R_{B2}}\right]\left[\frac{R_E}{\beta_{dc}}\right] + R_E\left[\frac{R_{B1}}{R_{B2}}\right] + \left[\frac{R_{B1}}{\beta_{dc}}\right] + R_E$$

$$D = V'_{BE} + \left[\frac{R_{B1}}{R_{B2}}\right]V'_{BE} - V_{CC}$$

$$E = -\left[\frac{R_E}{\beta_{dc}^2} + \frac{R_{B1}R_E}{R_{B2}\beta_{dc}^2} + \frac{R_{B1}}{\beta_{dc}^2}\right]$$

The change in the collector current is given by

$$\Delta I_C = S_{I_{CBO}} * \Delta I_{CBO} + S_{V'_{BE}} * \Delta V'_{BE} + S_{\beta_{dc}} * \Delta \beta_{dc} \qquad (8\text{-}276)$$

8.7.4 A Design Example

The degree of influence that each biasing circuit has on controlling I_C due to the variation in β_{dc} and the intrinsic temperature-dependent parameter is defined by the biasing topologies. A comparative analysis of the different biasing circuits is tabulated for better understanding of which circuits compensate best for each parameter over the variation of process and temperature. The Agilent HBFP-0405 transistor is used as a test example for each biasing topology based on the application notes described in [191]. The Agilent HBFP-0405 transistor is biased at $V_{CC} = 2.7$ V, $V_{CE} = 2$ V, $I_C = 5$ mA. The value of β_{dc} for minimum collector current is 50, for nominal/typical is 80, and for maximum is 150 (this data is taken from the manufacturer).

8.7 PHASE NOISE, BIASING, AND TEMPERATURE EFFECTS

TABLE 8-3 Biasing Resistor Values for Agilent HBFP-0405 Transistor Biased at $V_{CC} = 2.7$ V, $V_{CE} = 2$ V, $V_{BB} = 2.7$ V, $I_C = 5$ mA, $\beta_{dc} = 80$ for the Different Biasing Topologies [191]

Biasing Resistor	Nonstabilized Biasing Circuit (Option 1)	Voltage Feedback Biasing Circuit (Option 2)	Voltage Feedback With Current Source Biasing Circuit (Option 3)	Voltage Feedback With Voltage Source Biasing Circuit (Option 4)	Emitter Feedback Biasing Circuit (Option 5)
R_C	140 Ω	138 Ω	126 Ω	126 Ω	
R_B	30,770 Ω	19,552 Ω	11,539 Ω		
R_{B1}			889 Ω	2,169 Ω	2,169
R_{B2}			3,000 Ω	1,560 Ω	2,960
R_E					138

Table 8-3 shows the calculated bias resistors for values for an Agilent HBFP-0405 transistor.

Table 8-4 shows the variation of the collector current with respect to β_{dc} (minimum, typical, and maximum).

Each option described in Table 8-2 has positive and negative percentage change in the collector current I_C due to variation in β_{dc} with respect to β_{dc} (maximum = 150) and β_{dc} (minimum = 50) from β_{dc} (nominal = 80) at +25°C.

TABLE 8-4 Variation of Collector Current I_C Due to the variation in β_{dc} (Minimum = 50, Typical = 80 and Maximum = 150) for Different Biasing Topologies for Agilent HBFP-0405 Transistor Biased at $V_{CC} = 2.7$ V, $V_{CE} = 2$ V, $V_{BB} = 2.7$ V, $I_C = 5$ mA, $T_j = +25°C$ [191]

Biasing Current I_C (mA)	Nonstabilized Biasing Circuit (Option 1)	Voltage Feedback Biasing Circuit (Option 2)	Voltage Feedback With Current Source Biasing Circuit (Option 3)	Voltage Feedback With Voltage Source Biasing Circuit (Option 4)	Emitter Feedback Biasing Circuit (Option 5)
I_C (mA) for $[\beta_{dc}]_{minimum}$	3.14	3.63	3.66	4.53	4.7
I_C (mA) for $[\beta_{dc}]_{typical}$	5.0	5.0	5.0	5.0	5.0
I_C (mA) for $[\beta_{dc}]_{maximum}$	9.27	7.09	6.98	5.44	5.27
% change in I_C from nominal I_C	+85.4 −37.2	+41.8 −27.4	+39.6 −26.8	+8.8 −9.4	+5.4% −6%

226 CALCULATION AND OPTIMIZATION OF PHASE NOISE

The biasing network for option 1 does not support the compensation due to the variation of β_{dc}, allowing the collector current I_C to increase 85.4% for β_{dc} (maximum) = 150 and -37.2% for β_{dc} (minimum) = 50.

$$[\text{percentage change in } I_C \text{ from nominal } I_C]_{option\ 1}$$
$$= \frac{[I_C]_{\beta_{dc}(maximum)} - [I_C]_{nominal}}{[I_C]_{nominal}} = \frac{4.27}{5} = 85.4\%$$

$$[\text{percentage change in } I_C \text{ from nominal } I_C]_{option\ 1}$$
$$= \frac{[I_C]_{\beta_{dc}(minimum)} - [I_C]_{nominal}}{[I_C]_{nominal}} = \frac{-1.86}{5} = -37.2\%$$

The biasing network for option 2 offers compensation due to the variation of β_{dc}, allowing the collector current I_C to increase only 41.8% for β_{dc} (maximum) = 150 and -27.4% for β_{dc} (minimum) = 50.

$$[\text{percentage change in } I_C \text{ from nominal } I_C]_{option\ 2}$$
$$= \frac{[I_C]_{\beta_{dc}(maximum)} - [I_C]_{nominal}}{[I_C]_{nominal}} = \frac{2.029}{5} = 41.8\%$$

$$[\text{percentage change in } I_C \text{ from nominal } I_C]_{option\ 2}$$
$$= \frac{[I_C]_{\beta_{dc}(minimum)} - [I_C]_{nominal}}{[I_C]_{nominal}} = \frac{-1.37}{5} = -27.4\%$$

The biasing network for option 3 offers little improvement over option 2 due to the variation of β_{dc}, allowing the collector current I_C to increase 39.6% for β_{dc} (maximum) = 150 and -26.8% for β_{dc} (minimum) = 50.

$$[\text{percentage change in } I_C \text{ from nominal } I_C]_{option\ 3}$$
$$= \frac{[I_C]_{\beta_{dc}(maximum)} - [I_C]_{nominal}}{[I_C]_{nominal}} = \frac{1.98}{5} = 39.6\%$$

$$[\text{percentage change in } I_C \text{ from nominal } I_C]_{option\ 3}$$
$$= \frac{[I_C]_{\beta_{dc}(minimum)} - [I_C]_{nominal}}{[I_C]_{nominal}} = \frac{-1.34}{5} = -26.8\%$$

The biasing network for option 4 offers an improvement over options 1, 2, and 3 due to the variation of β_{dc}, allowing the collector current I_C to increase only 8.8% for

8.7 PHASE NOISE, BIASING, AND TEMPERATURE EFFECTS

β_{dc} (maximum) = 150 and -9.4% for β_{dc} (minimum) = 50.

$$[\text{percentage change in } I_C \text{ from nominal } I_C]_{option\,4}$$
$$= \frac{[I_C]_{\beta_{dc}(maximum)} - [I_C]_{nominal}}{[I_C]_{nominal}} = \frac{0.44}{5} = 8.8\%$$

$$[\text{percentage change in } I_C \text{ from nominal } I_C]_{option\,4}$$
$$= \frac{[I_C]_{\beta_{dc}(minimum)} - [I_C]_{nominal}}{[I_C]_{nominal}} = \frac{-0.47}{5} = -9.4\%$$

The biasing network for option 5 offers the best performance due to the variation of β_{dc}, allowing the collector current I_C to increase only 5.4% for β_{dc} (maximum) = 150 and -6% for β_{dc} (minimum) = 50.

$$[\text{percentage change in } I_C \text{ from nominal } I_C]_{option\,5}$$
$$= \frac{[I_C]_{\beta_{dc}(maximum)} - [I_C]_{nominal}}{[I_C]_{nominal}} = \frac{0.27}{5} = 5.4\%$$

$$[\text{percentage change in } I_C \text{ from nominal } I_C]_{option\,4}$$
$$= \frac{[I_C]_{\beta_{dc}(minimum)} - [I_C]_{nominal}}{[I_C]_{nominal}} = \frac{-0.3}{5} = -6\%$$

Table 8-5 shows the stability factors for the different biasing topologies for the Agilent HBFP-0405 transistor.

Each option described in Table 8-5 has three different stability factors ($S_{I_{CBO}}$, $S_{V'_{BE}}$, and $S_{\beta_{dc}}$) that are multiplied with the corresponding changes (ΔI_{CBO}, $\Delta V'_{BE}$, and $\Delta \beta_{dc}$) and summed for the total change in the collector current as $\Delta I_C = S_{I_{CBO}} * \Delta I_{CBO} + S_{V'_{BE}} * \Delta V'_{BE} + S_{\beta_{dc}} * \Delta \beta_{dc}$.

These changes (ΔI_{CBO}, $\Delta V'_{BE}$, and $\Delta \beta_{dc}$) are calculated based on the variations in these parameters and the manufacturing process [191]. As a test example, the variation in the collector current I_C is calculated over the change in temperature from $+25°C$ to $+65°C$.

For the Agilent HBFP-0405 transistor, the change in reverse leakage current flowing through a reverse biased PN junction over the temperature is given as

$[I_{CBO}]_{T=25°C} = 100\,\text{nA}$ (doubles for every 10°C rise in temperature)

$[I_{CBO}]_{T=65°C} = 1600\,\text{nA}$

$\Delta I_{CBO} = [I_{CBO}]_{T=65°C} - [I_{CBO}]_{T=25°C} = 1500\,\text{nA}$

$[\Delta I_C]_{I_{CBO}} = S_{I_{CBO}} * \Delta I_{CBO}$

TABLE 8-5 Bias Stability Analysis for Different Biasing Topologies for the Agilent HBFP-0405 Transistor Biased at $V_{CE} = 2$ V, $V_{CC} = 2.7$ V, $I_C = 5$ mA, $T_j = +65°C$ [191]

Biasing Networks	Nonstabilized Biasing Circuit (Option 1)	Voltage Feedback Biasing Circuit (Option 2)	Voltage Feedback With Current Source Biasing Circuit (Option 3)	Voltage Feedback With Voltage Source Biasing Circuit (Option 4)	Emitter Feedback Biasing Circuit (Option 5)
$S_{I_{CBO}} = \left.\dfrac{\partial I_C}{\partial I_{CBO}}\right|_{\beta_{dc}, V'_{EE}=constant}$	81	52.238	50.865	19.929	11.286
$S_{V'_{EE}} = \left.\dfrac{\partial I_C}{\partial V'_{EE}}\right|_{I_{CBO}, \beta_{dc}=constant}$	$-2.56653\text{E}-3$	$-2.56801\text{E}-3$	$-3.956\text{E}-3$	-0.015	$-6.22437\text{E}-3$
$S_{\beta_{dc}} = \left.\dfrac{\partial I_C}{\partial \beta_{dc}}\right|_{I_{CBO}, V'_{EE}=constant}$	$6.249877\text{E}-5$	$4.031\text{E}-5$	$3.924702\text{E}-5$	$1.537669\text{E}-5$	$8.707988\text{E}-6$
ΔI_C due to I_{CBO} (mA)	0.120	0.078	0.076	0.030	0.017
ΔI_C due to V'_{EE} (mA)	0.210	0.205	0.316	1.200	0.497
ΔI_C due to β_{dc} (mA)	0.999	0.645	0.628	0.246	0.140
Total ΔI_C (mA)	1.329	0.928	1.020	1.476	0.654
% change in I_C from nominal I_C	26.58	18.56%	20.4%	29.52%	13.032%

8.7 PHASE NOISE, BIASING, AND TEMPERATURE EFFECTS

For the Agilent HBFP-0405 transistor, the change in the base-emitter potential over the temperature is given as

$$\lfloor \Delta V'_{BE} \rfloor_{T=25°C} = 0.755 \text{ V} \quad \text{(negative temperature coefficient of} -2 \text{ mV per } °C\text{)}$$
$$\lfloor \Delta V'_{BE} \rfloor_{T=65°C} = 0.675 \text{ V}$$
$$\Delta V'_{BE} = \lfloor V'_{BE} \rfloor_{T=65°C} - \lfloor V'_{BE} \rfloor_{T=25°C} = -0.08 \text{ V}$$
$$[\Delta I_C]_{V'_{BE}} = S_{V'_{BE}} * \Delta V'_{BE}$$

For the Agilent HBFP-0405 transistor, the change in β_{dc} over the temperature is given as

$$[\Delta \beta_{dc}]_{T=25°C} = 80 \quad \text{(typically increases at a rate of 0.5\% per } °C\text{)}$$
$$[\Delta \beta_{dc}]_{T=65°C} = 96$$
$$\Delta \beta_{dc} = [\beta_{dc}]_{T=65°C} - [\beta_{dc}]_{T=25°C} = 16$$
$$[\Delta I_C]_{\beta_{dc}} = S_{\beta_{dc}} * \Delta \beta_{dc}$$

After all three stability terms are evaluated, they can be summed to give a net change in the collector current at $+65°C$ from the nominal value at $+25°C$.

$$[\Delta I_C]_{total} = S_{I_{CBO}} * \Delta I_{CBO} + S_{V'_{BE}} * \Delta V'_{BE} + S_{\beta_{dc}} * \Delta \beta_{dc}$$
$$[\Delta I_C]_{total\ option\ 1} = [S_{I_{CBO}} * \Delta I_{CBO} + S_{V'_{BE}} * \Delta V'_{BE} + S_{\beta_{dc}} * \Delta \beta_{dc}]_{option\ 1} = 1.329$$
$$[\Delta I_C]_{total\ option\ 2} = [S_{I_{CBO}} * \Delta I_{CBO} + S_{V'_{BE}} * \Delta V'_{BE} + S_{\beta_{dc}} * \Delta \beta_{dc}]_{option\ 2} = 0.928$$
$$[\Delta I_C]_{total\ option\ 3} = [S_{I_{CBO}} * \Delta I_{CBO} + S_{V'_{BE}} * \Delta V'_{BE} + S_{\beta_{dc}} * \Delta \beta_{dc}]_{option\ 3} = 1.020$$
$$[\Delta I_C]_{total\ option\ 4} = [S_{I_{CBO}} * \Delta I_{CBO} + S_{V'_{BE}} * \Delta V'_{BE} + S_{\beta_{dc}} * \Delta \beta_{dc}]_{option\ 4} = 1.476$$
$$[\Delta I_C]_{total\ option\ 5} = [S_{I_{CBO}} * \Delta I_{CBO} + S_{V'_{BE}} * \Delta V'_{BE} + S_{\beta_{dc}} * \Delta \beta_{dc}]_{option\ 5} = 0.654$$

$$[\text{percentage change in } I_C \text{ from } nominal\ I_C]_{option\ 1} = \frac{[\Delta I_C]_{total\ option\ 1}}{[I_C]_{nominal}} = \frac{1.329}{5} = 26.58\%$$

$$[\text{percentage change in } I_C \text{ from } nominal\ I_C]_{option\ 2} = \frac{[\Delta I_C]_{total\ option\ 2}}{[I_C]_{nominal}} = \frac{0.928}{5} = 18.56\%$$

$$[\text{percentage change in } I_C \text{ from } nominal\ I_C]_{option\ 3} = \frac{[\Delta I_C]_{total\ option\ 3}}{[I_C]_{nominal}} = \frac{1.020}{5} = 20.4\%$$

$$[\text{percentage change in } I_C \text{ from } nominal\ I_C]_{option\ 4} = \frac{[\Delta I_C]_{total\ option\ 4}}{[I_C]_{nominal}} = \frac{1.476}{5} = 29.52\%$$

$$[\text{percentage change in } I_C \text{ from } nominal\ I_C]_{option\ 5} = \frac{[\Delta I_C]_{total\ option\ 5}}{[I_C]_{nominal}} = \frac{0.654}{5} = 13.08\%$$

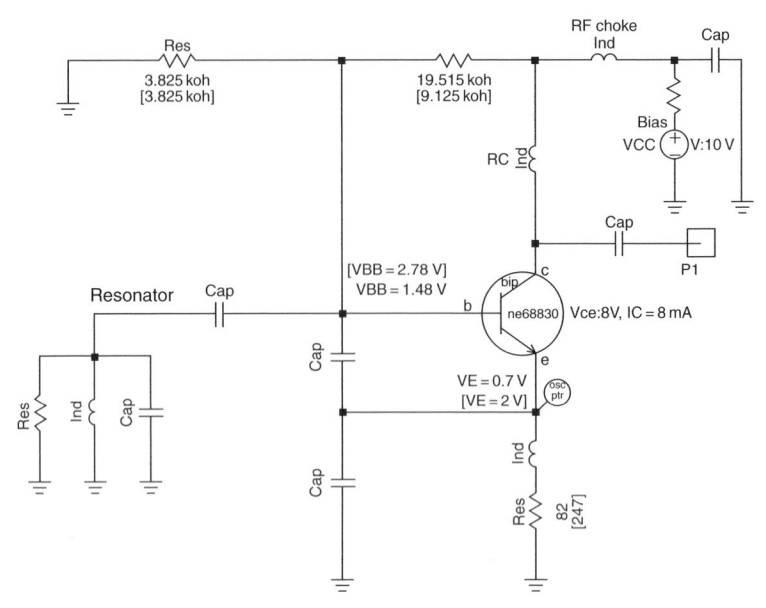

Figure 8-37 The 1 GHz oscillator with 0.7 V voltage drop between the emitter and ground.

As described in Table 8-5, option 1 allows the collector current I_C to increase about 27%, while options 2 and 3 offer a 19–20% increase. Option 4 allow a nearly 30% increase in I_C with temperature (from 25°C to 65°C) due to the major contribution from $S_{V'_{BE}}$. This may be due to the impedance of the divider network R_{B1} and R_{B2} acting against the base-emitter potential V'_{BE}. Options 2 and 3 have similar performance over the temperature range and offer a significant improvement over options 1 and 4. Option 5 offers the best performance over the temperature range due to the emitter feedback.

8.7.5 Choosing the Bias Values

The absolute values of the bias circuit can have a major consequence. The 1 GHz VCO shown in Figure 8-37 adjusts the voltage drop at 0.7 V between the emitter and ground. The base voltage divider is calculated to fit this voltage. The bias point is 8 V V_{CE} and 8 mA. This circuit is functionally the same as the circuit in Figure 8-34; however, it is a purely passive biasing circuit.

The predicted phase noise shown in Figure 8-38, which agrees well with the measured data, indicates a value of -40 dBc/Hz compared to the reference example with active biasing, which shows a phase noise of -63 dBc/Hz. It is interesting to see that there is essentially no difference in phase noise as a function of temperature.

By increasing the emitter resistor from 82 Ω to 247 Ω and adjusting the base voltage divider to maintain the same 8 mA, the noise improves. Why is this so?

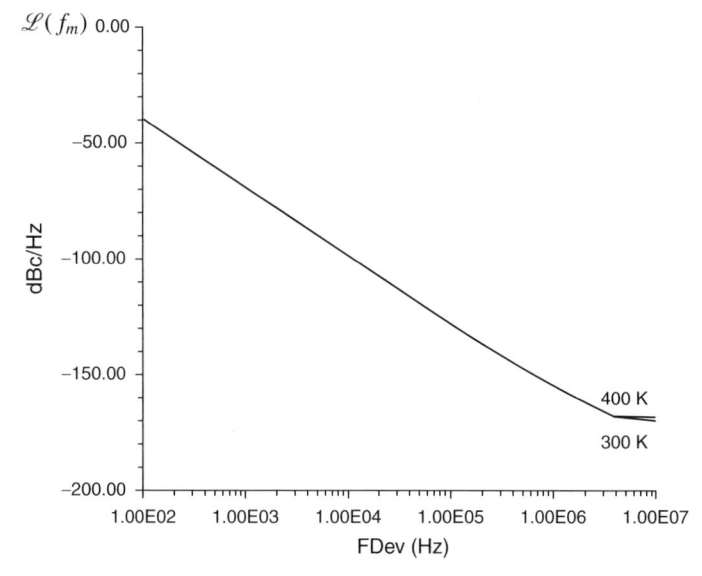

8.7 PHASE NOISE, BIASING, AND TEMPERATURE EFFECTS

Figure 8-38 Predicted phase noise of the oscillator shown in Figure 8-37.

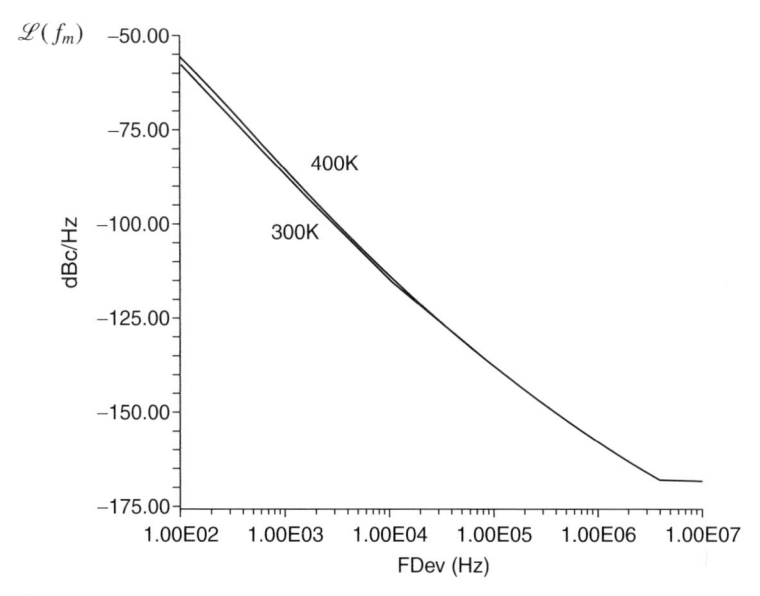

Figure 8-39 Simulated phase noise of the oscillator shown in Figure 8-37 with the emitter resistor increased from 82 Ω to 247 Ω at the same collector current of 8 mA.

The answer is that the emitter now experiences a more constant current generator and therefore counteracts the AM-to-PM conversion, resulting in better phase noise. Figure 8-39 shows the resulting phase noise with a larger emitter resistor. The phase noise at 100 Hz off the carrier is now -57 dBc/Hz, compared to -63 dBc/Hz with active biasing and -40 dBc/Hz with the smaller emitter resistor. Since many of these tests cannot easily be done by experiment, the use of a good HB simulator is highly recommended. Both Ansoft's Designer and Agilent's ADS system gave the same answer and agreed with the measurements. For some applications, however, the Ansoft product gives a more accurate result. ADS 2003 can deviate from the measured data. As new versions of ADS become available, this problem may be corrected.

9 Validation Circuits

Chapter 7 developed the mathematical background for optimizing microwave oscillators. The next step is to validate the synthesis of the circuits. The following circuits have been chosen for validation:

- 1000 MHz bipolar transistor–based oscillator with ceramic resonator
- 4100 MHz bipolar transistor–based oscillator with transmission line resonators
- 2000 MHz GaAs FET-based oscillator with transmission line resonators

9.1 1000 MHz CRO

Many applications require a very-low-noise microwave oscillator in the 1000 MHz region; this is best accomplished with a ceramic resonator [85, 86]. An operating Q in the vicinity of 500 is available in this material. An oscillator using an NEC NE68830 transistor has been selected because of its superior flicker noise. The Colpitts oscillator uses an 8.2 Ω resistor between the emitter and the capacitive feedback. Rather than being taken at the collector, the RF signal is taken from a tap of the emitter inductor. The collector circuit, using PNP transistors, has been designed to set the DC current. The necessary equations for this DC bias are found in [88].

Class A common-emitter amplifiers are usually very sensitive to stray impedance in the emitter circuit. Any small inductance in series with the emitter will cause instability; for this reason, the emitter needs to be grounded as directly as possible, and bias components in the emitter are generally undesirable. In the schematic in Figure 9-1, Q_1 is the RF amplifier, and Q_2 provides its base current required for the constant voltage difference across R_c. This constant voltage difference ensures constant collector current.

Diode D_1 provides some measure of temperature compensation. R_b should be high in order not to affect base impedance but not high enough to cause Q_2 to saturate over temperature and β_1 variations. Neglecting the base current of Q_2,

---

The Design of Modern Microwave Oscillators for Wireless Applications: Theory and Optimization, by Ulrich L. Rohde, Ajay Kumar Poddar, Georg Böck
Copyright © 2005 John Wiley & Sons, Inc.

234 VALIDATION CIRCUITS

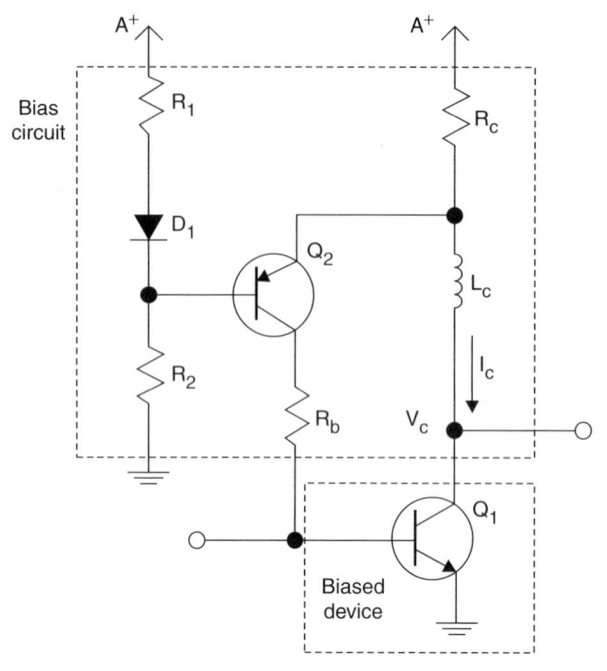

Figure 9-1 Active bias network for a common-emitter RF amplifier stage.

the design equations are

$$I_c = \frac{R_1(A^+ - V_d)}{R_c(R_1 + R_2)} \tag{9-1}$$

$$V_c = A^+ - I_c R_c \tag{9-2}$$

Assuming that we are designing the bias circuit to provide a certain device bias current I_c and collector voltage V_c, select a convenient supply voltage $A^+ > V_c$. The component values are then supplied by the following equations:

$$R_c = \frac{A^+ - V_c}{I_c} \tag{9-3}$$

$$R_1 = \frac{A^+ - V_c}{I_d} \tag{9-4}$$

$$R_2 = \frac{V_c - V_d}{I_d} \tag{9-5}$$

$$R_b < \beta_{\min} \frac{V_c - V_d - 0.2}{I_c} \tag{9-6}$$

where

I_c = desired collector current of Q_1 (A)
V_c = desired collector voltage of Q_1 (V)
V_d = diode, or base-emitter voltage drop, nominally 0.7 (V)
A^+ = chosen supply voltage (V)
R_i = resistor values as shown in Figure 9-1 (Ω)
I_d = bias current through R_1, R_2, and D_1 (A)
β_{min} = minimum beta of Q_1

The bias circuit shown has to be carefully bypassed at both high and low frequencies. There is one inversion from base to collector of Q_1, and another inversion may be introduced by L_c matching components and stray capacitances, resulting in positive feedback around the loop at low frequencies. Low equivalent series resistance (ESR) electrolytic or tantalum capacitors from the collector of Q_2 to ground are usually adequate to ensure stability.

The ceramic resonator is coupled loosely to the transistor with a capacitor of 0.9 pF. The resonator has a parallel capacitor of 0.6 pF, which reduces the manufacturing tolerances of the resonator. The tuning diode assembly, two diodes in parallel, is coupled to the resonator with 0.8 pF. The reason for using two diodes is that there is no single diode available with the necessary capacitance and Q. Figure 9-2 shows the schematic of the oscillator.

It has been pointed out that the best operating condition exists where the most negative resistance occurs at the point of resonance to achieve the best phase noise. This is shown in Figure 9-3. The Im curve starting below zero shows the imaginary current which resonates at 1000 MHz, while the Re curve shows the negative resistance. Its maximum negative peak occurs at exactly 1000 MHz, as it should.

Figure 9-4 shows the measured phase noise of this oscillator. The measurements were performed with the Aeroflex Euro Test system. At 1 kHz the phase noise

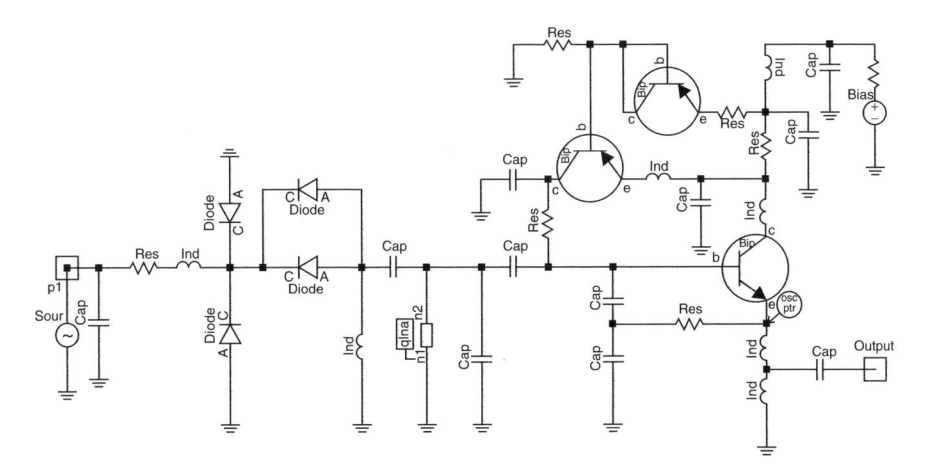

Figure 9-2 The 1000 MHz ceramic resonator oscillator.

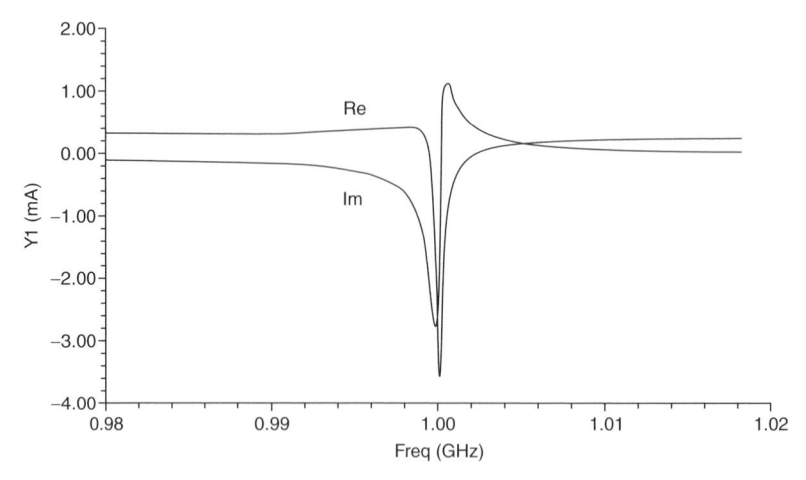

Figure 9-3 Plot of real and imaginary oscillator currents as a function of frequency.

is ~95 dBc/Hz, and at 10 kHz it is ~124 dBc/Hz. This is a 30 dB/decade slope, which is triggered by the flicker corner frequency of the transistor. From 10 to 100 kHz the slope is 20 dB/decade, with a phase noise of −145.2 dBc/Hz at 100 kHz. At 1 MHz off the carrier, it is −160 dBc/Hz.

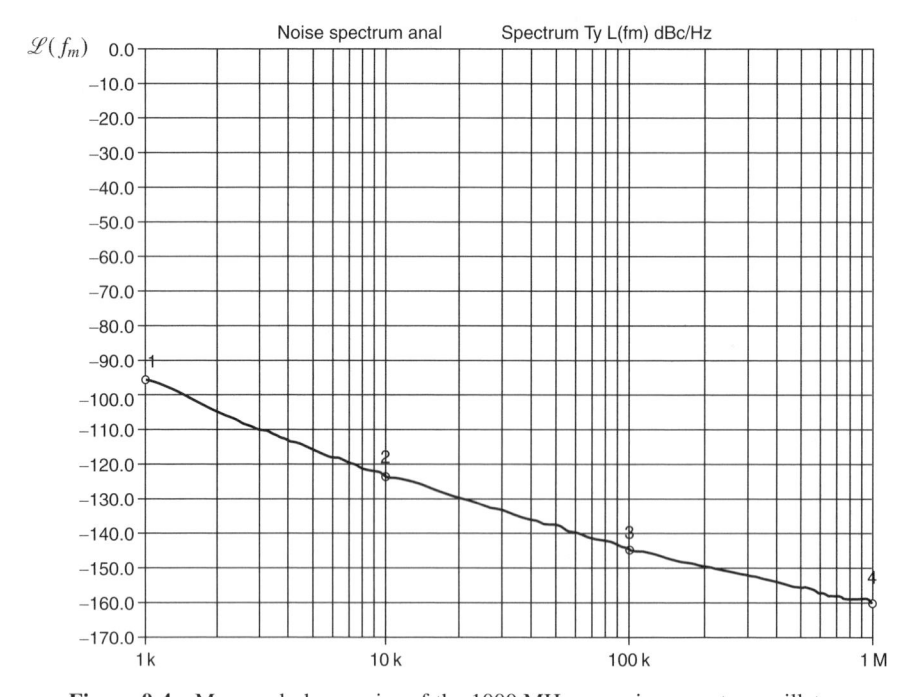

Figure 9-4 Measured phase noise of the 1000 MHz ceramic resonator oscillator.

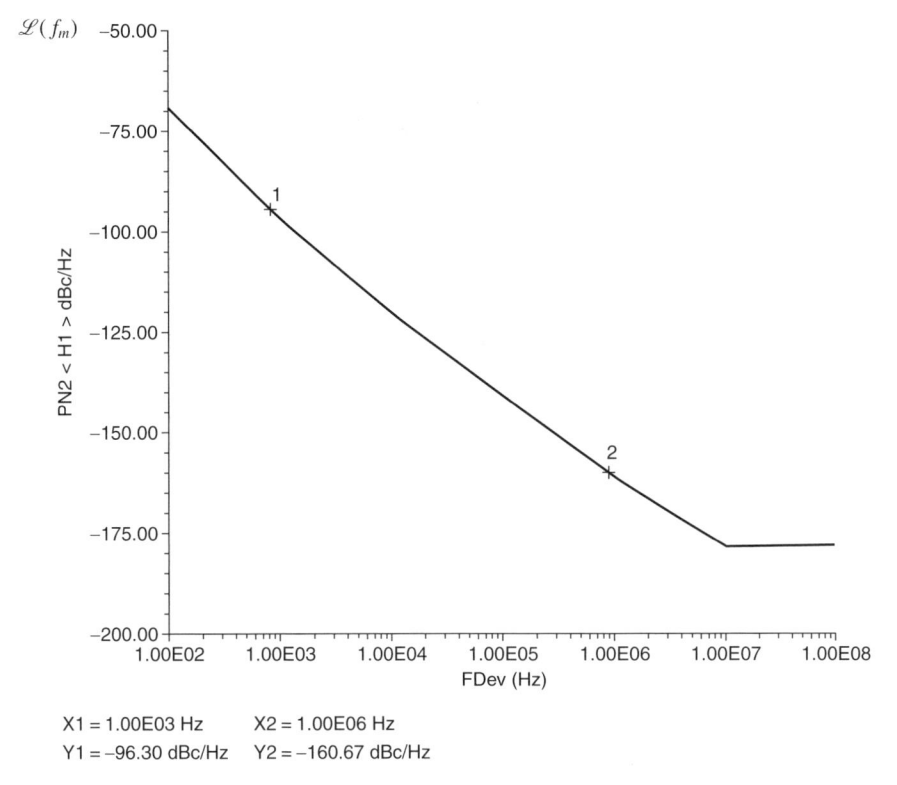

Figure 9-5 Predicted phase noise of the CRO at 1 GHz shown in Figure 9-2.

X1 = 1.00E03 Hz   X2 = 1.00E06 Hz
Y1 = −96.30 dBc/Hz   Y2 = −160.67 dBc/Hz

Because of the narrow tuning range and the loose coupling of the tuning diode, the noise contribution of the diode is negligible.

This circuit has been designed using the synthesis procedure and also has been analyzed with the HB simulator Microwave Harmonica from Ansoft. Figure 9-5 shows the predicted performance of the phase noise. The actual circuit arrangement is shown in Figure 9-6. The ceramic resonator can be spotted easily.

9.2 4100 MHz OSCILLATOR WITH TRANSMISSION LINE RESONATORS

For less demanding applications, it is possible to design oscillators using transmission line resonators. These resonators were discussed in Chapter 5. Their Q depends on the material and implementation of the resonator. Figure 9-7 shows the circuit of the oscillator. While the previous example was a Colpitts parallel resonant circuit, this circuit operates in series resonant mode. The NPN transistor

238 VALIDATION CIRCUITS

Figure 9-6 Photograph of the 1 GHz CRO of the schematic shown in Figure 9-2.

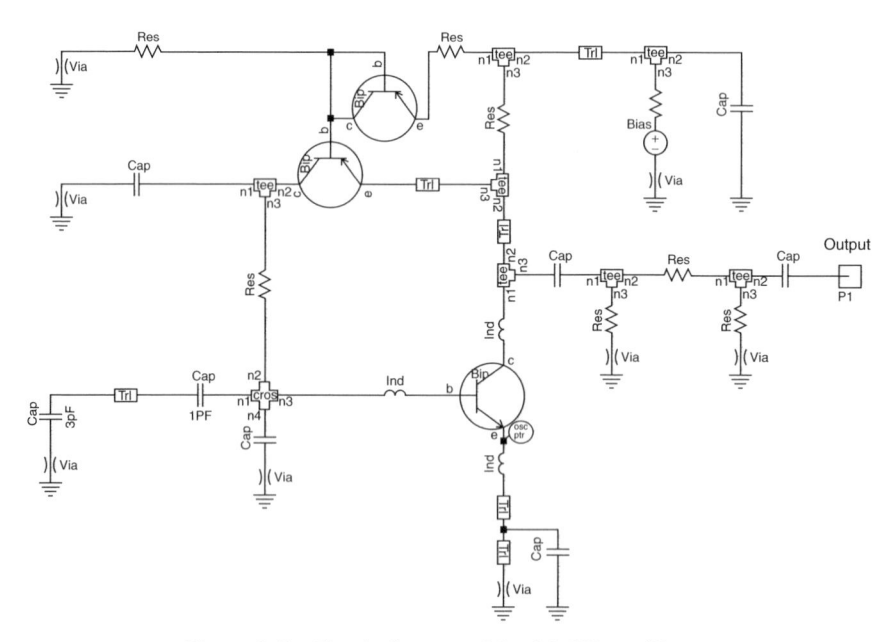

Figure 9-7 Circuit diagram of the 4.1 GHz oscillator.

9.2 4100 MHz OSCILLATOR WITH TRANSMISSION LINE RESONATORS   239

NE68830 has parasitic inductance in the emitter, base, and collector lines. For accurate modeling, T-junction and cross-junction models were used, as well as transmission lines where applicable. The DC stabilization circuit uses the same technique shown in Figure 9-2. This time the RF power is taken from the collector and uses a 10 dB attenuator to minimize frequency pulling. The ground connections for the capacitors are done using via holes. A via hole is the electrical equivalent of a small inductor.

The phase noise of this oscillator was simulated using the values of the synthesis program. Figure 9-8 shows the predicted phase noise.

The output power of this oscillator is 6.8 dBm. This oscillator was built and measured. Figure 9-9 shows the printed circuit board of the oscillator.

Because of the pad-like microstrips, the simulation needs to be done very carefully; the soldering of the component is also critical. This frequency range makes the assembly very difficult because it is not high enough for an RFIC and still needs to be done on a printed circuit board. The measured phase noise is shown in Figure 9-10. It agrees well with the predicted phase noise. At 100 kHz the difference is about 3 dB. The same is true at 10 kHz. At 1 kHz there is a larger difference.

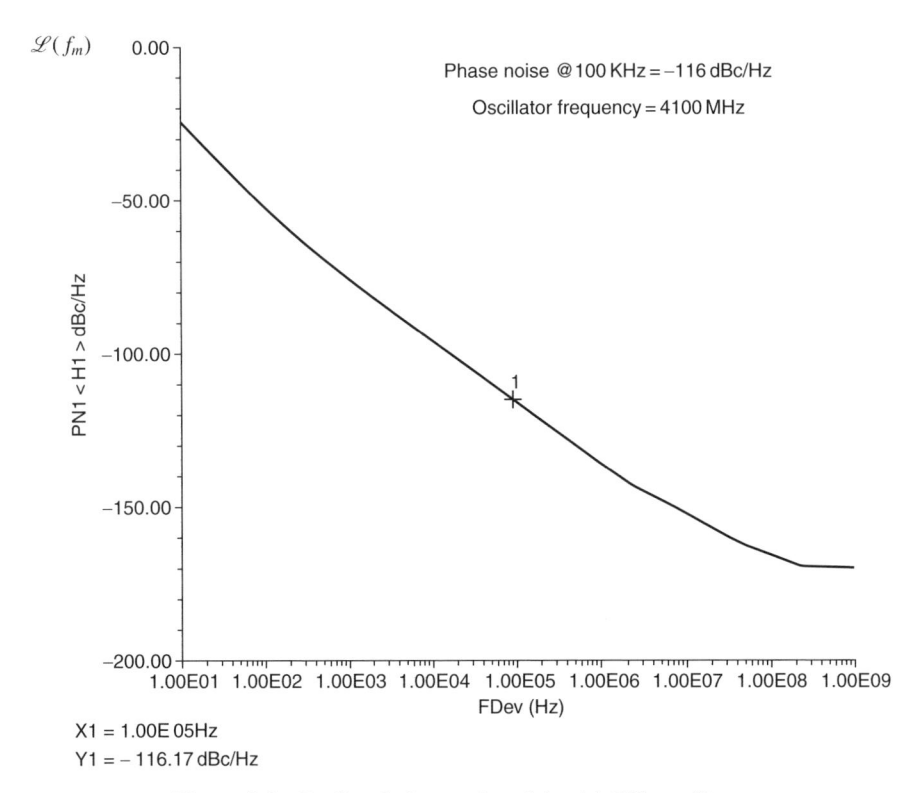

Figure 9-8 Predicted phase noise of the 4.1 GHz oscillator.

240 VALIDATION CIRCUITS

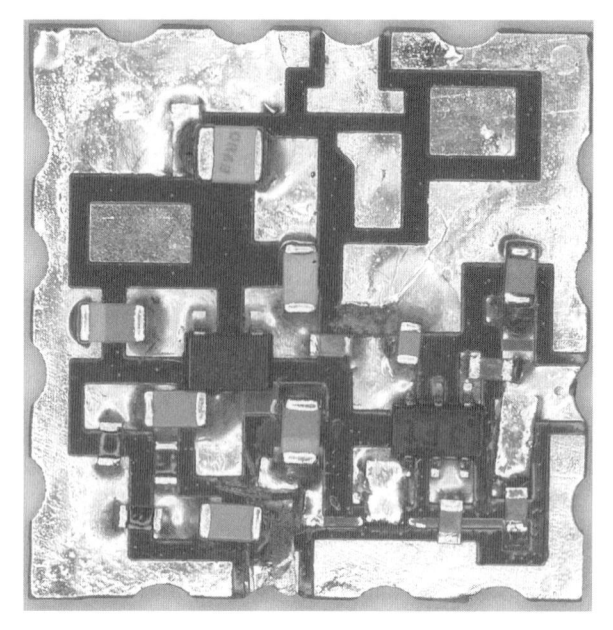

Figure 9-9 Printed circuit board of the 4.1 GHz oscillator shown in Figure 9-8.

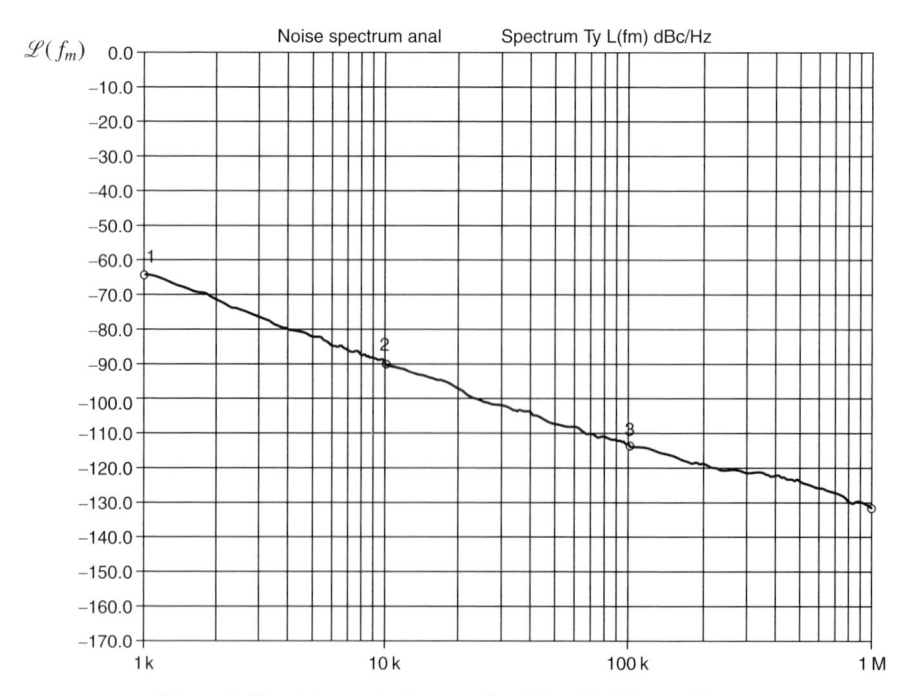

Figure 9-10 Measured phase noise of the 4.1 GHz oscillator.

9.3 2000 MHz GaAs FET-BASED OSCILLATOR

Low-cost applications are frequently implemented as an RFIC. For further validation, a GaAs FET-based 2000 MHz Colpitts oscillator was designed and built. Figure 9-11 shows the circuit diagram of the oscillator. It uses a combination of transmission lines and rectangular inductors as resonators. The inductor in the middle of the schematic in Figure 9-11, connected to a via hole, is needed as a DC return. If a tuning diode is connected to the capacitor on the left side of the schematic in Figure 9-11, then a DC control voltage can be applied and the center inductor becomes an RF choke. The output is taken from the source. An additional external DC decoupling capacitor will be needed because of the DC coupling. The transistor and the circuit were constructed using the TriQuint GaAs Foundry, and the transistor was optimized for the DC current. Figure 9-12 shows the predicted phase noise of this oscillator.

It is interesting to examine the load line of this oscillator, which is shown in Figure 9-13. This circuit is operated in a fairly linear range.

Figure 9-14 shows the layout of the 2 GHz GaAs FET oscillator. Its output power is 1.8 dBm.

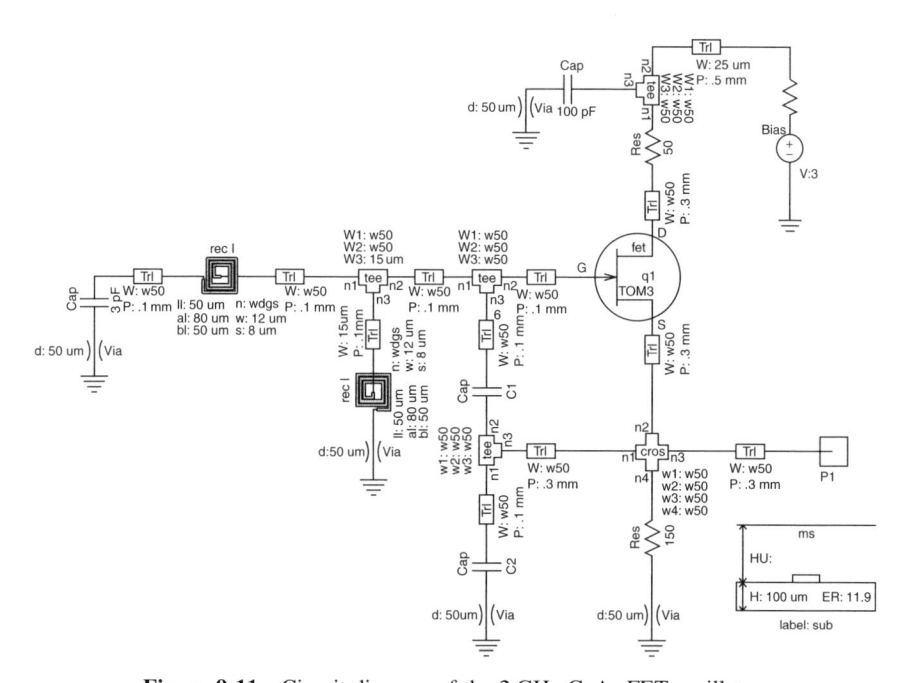

Figure 9-11 Circuit diagram of the 2 GHz GaAs FET oscillator.

242 VALIDATION CIRCUITS

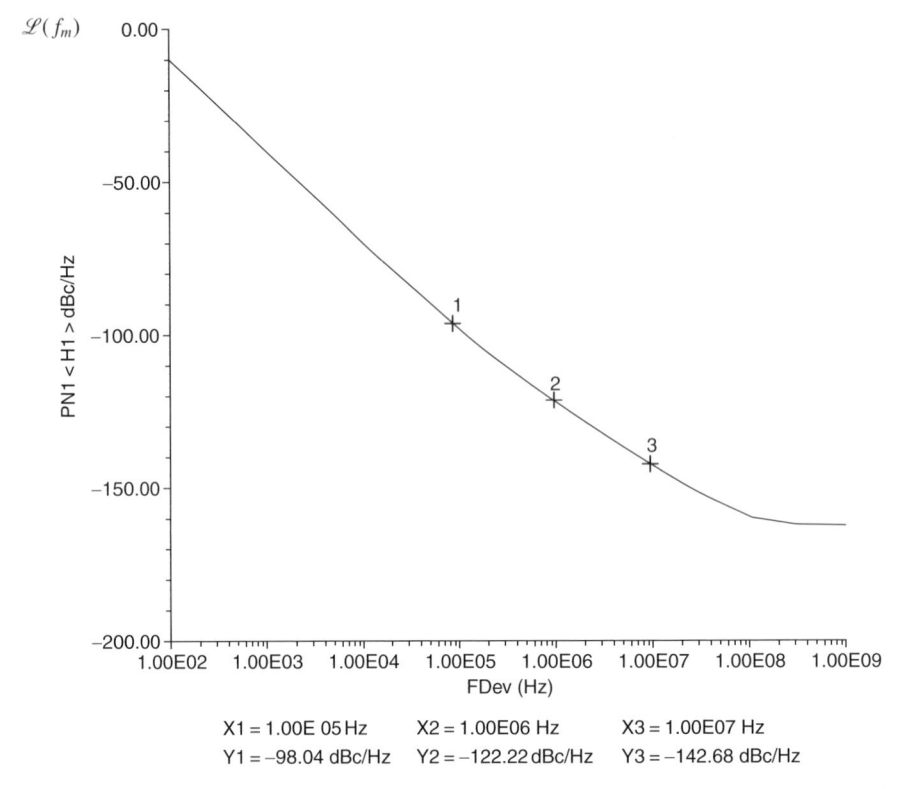

Figure 9-12 Predicted phase noise of the oscillator shown in Figure 9-11. The measured values were 100 dBc/Hz at 100 Hz and 120 dBc/Hz at 1 MHz. There is a deviation of about 2 dB compared to the simulation.

X1 = 1.00E 05 Hz   X2 = 1.00E06 Hz   X3 = 1.00E07 Hz
Y1 = −98.04 dBc/Hz   Y2 = −122.22 dBc/Hz   Y3 = −142.68 dBc/Hz

9.4 77 GHz SiGe OSCILLATOR

Millimeterwave oscillators have been published using SiGe bipolar transistors. A lot of data on the output power and phase noise of these oscillators is found in the literature. Therefore, it is interesting to synthesize a 77 GHz oscillator using lossy, lumped elements, which later can be translated into distributed elements, specifically coplanar waveguides.

Figure 9-15 shows a Colpitts oscillator designed around an advanced product of the BFP620 family. It is the typical Colpitts arrangement with a capacitive divider. The resonant circuit consists of a 0.07 pF capacitor and a 100 pH inductor with a Q of 70. Figure 9-16 shows the predicted phase noise at 77 GHz, which agrees well with published data. Here $n = (1 + 1(C_1/C_2)) = 3$ (low-Q case, as described earlier). The literature shows that such values are obtainable [89–98].

9.4 77 GHz SiGe OSCILLATOR   **243**

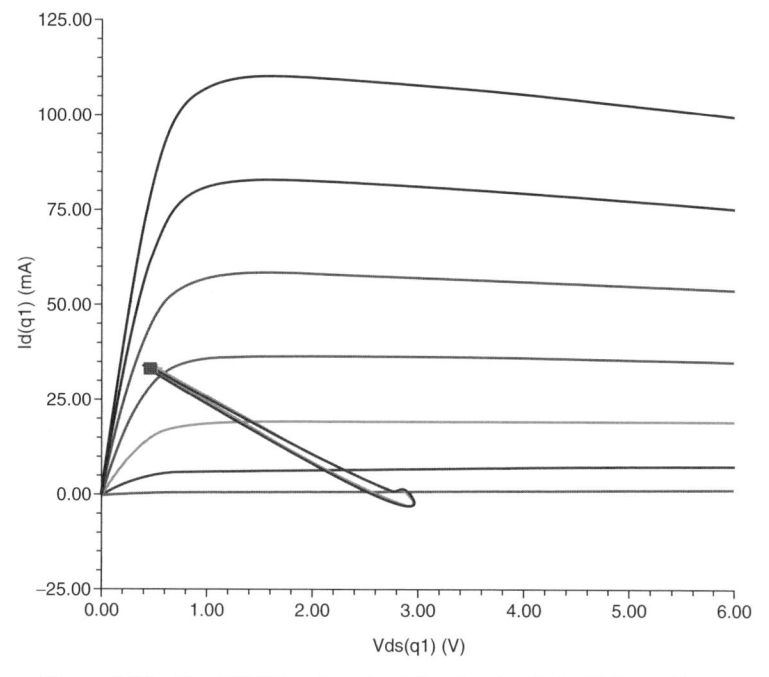

Figure 9-13 The DC-IV and the load line for the GaAs FET oscillator.

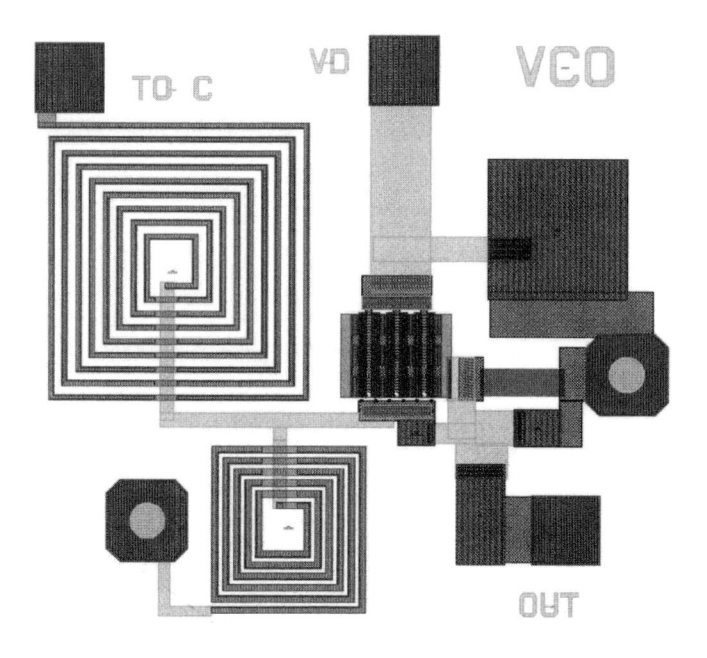

Figure 9-14 Layout of the 2 GHz GaAs FET oscillator.

244   VALIDATION CIRCUITS

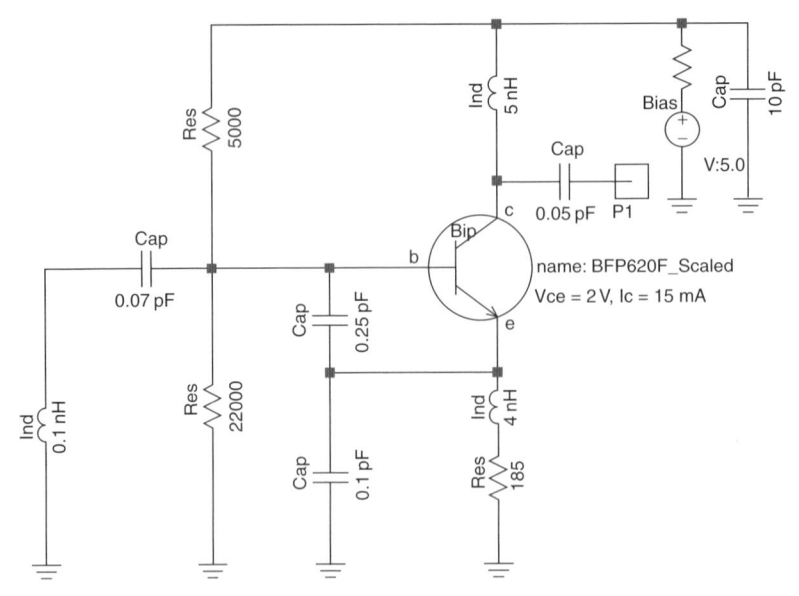

Figure 9-15   The 77 GHz modified Colpitts oscillator.

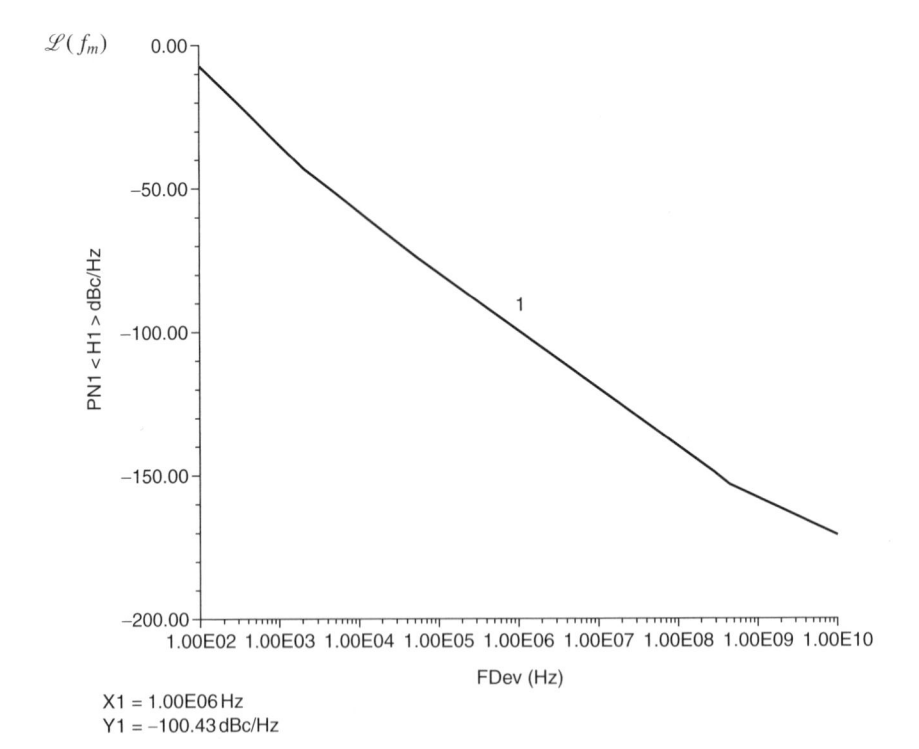

X1 = 1.00E06 Hz
Y1 = −100.43 dBc/Hz

Figure 9-16   Predicted phase noise of the oscillator shown in Figure 9-15.

9.5 900–1800 MHz HALF-BUTTERFLY RESONATOR-BASED OSCILLATOR

This oscillator can only be analyzed, built, and optimized using electromagnetic (EM) tools. The resonator here is a quarter-wave-length resonator at 1800 MHz, which is pulled down by the tuning diodes. Its schematic is shown in Figure 9-17.

For a better understanding of the circuit, depends highly on the layout information; the breakdown of components is given in Figure 9-18. The resonator is shown at the top right of the layout, a half-butterfly arrangement. The hole on the lower right side is the marking for the ground via.

This was the first attempt to build an EM-based oscillator from which the coupled resonator activity evolved. Figure 9-18 shows the actual built oscillator, and Figure 9-19 shows the phase noise achieved. Given the fact that this is a one- to two-range oscillator (900–1800 MHz), the phase noise compares favorably with other efforts in this frequency range.

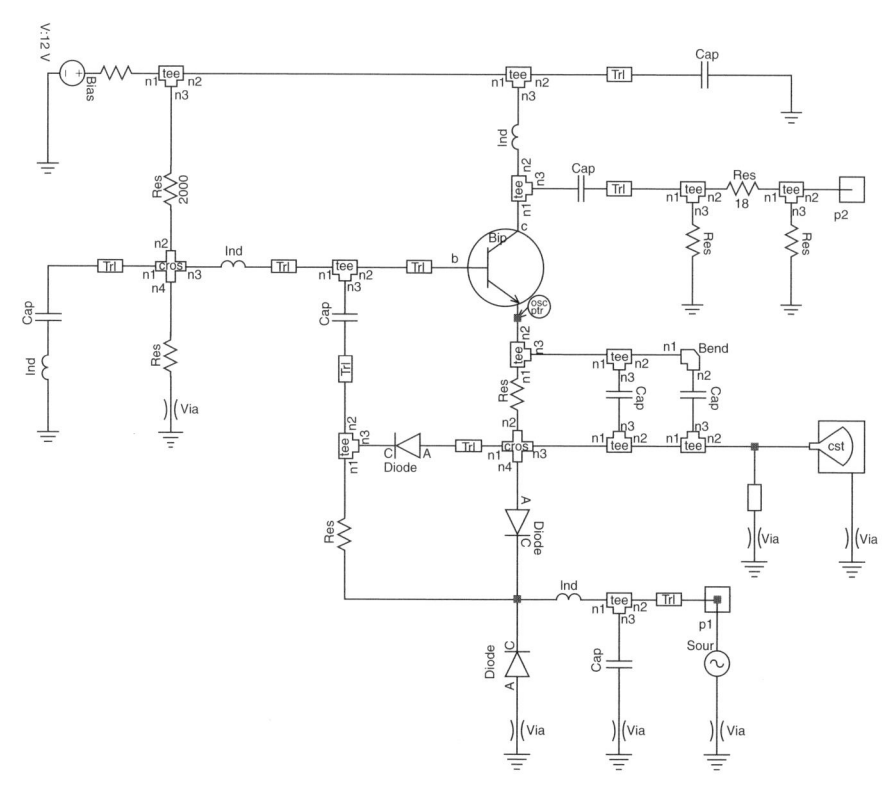

Figure 9-17 Schematic of the oscillator used to demonstrate the multiple-coupled resonator. Its technique is described in Chapter 10.

Figure 9-18 Photograph of the 900–1800 MHz VCO.

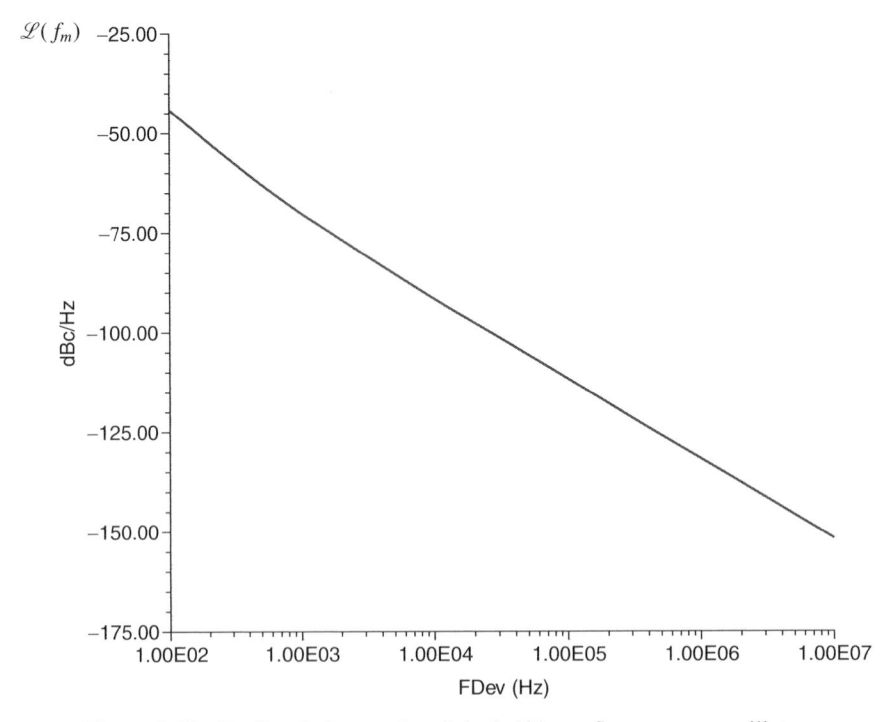

Figure 9-19 Predicted phase noise of the half-butterfly resonator oscillator.

10 Systems of Coupled Oscillators

10.1 MUTUALLY COUPLED OSCILLATORS USING THE CLASSICAL PENDULUM ANALOGY

A system of coupled oscillators/VCOs possesses a wealth of interesting and useful nonlinear dynamic phenomena, and the design of such systems requires a detailed understanding of the behavior of the coupled oscillator systems. The purpose of this analysis is to provide a basic understanding of a class of coupled oscillator systems based on the classical analogy of the coupled pendulum and to apply these techniques to practical ultra-low-phase-noise wideband oscillators/VCOs. In this chapter we develop a general method of analysis for finding the synchronized frequency of the coupled oscillator systems in the mutually locked state conditions. Figure 10-1 shows the classic example of a pendulum connected with a spring.

In the presence of the initial excitation, one pendulum starts swinging with small amplitude and the other slowly increases its amplitude as the spring feeds energy from the first pendulum into the second. Then energy flows back into the first pendulum, and the cycle repeats due to the transfer of energy back and forth.

There can be two modes of motion: swinging together in the same direction or swinging in opposite directions. If the pendulums swing in the same direction, they swing in unison at their natural frequency. If they swing in opposite directions, they will swing at a higher frequency than if they are uncoupled. When the pendulums are not identical, there are still two normal modes, but the motions are more

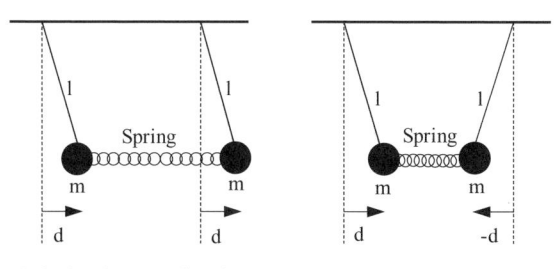

Figure 10-1 Pendulum connected with a spring.

The Design of Modern Microwave Oscillators for Wireless Applications: Theory and Optimization, by Ulrich L. Rohde, Ajay Kumar Poddar, Georg Böck
Copyright © 2005 John Wiley & Sons, Inc.

complicated and neither mode is at the uncoupled frequency. A similar analogy is given for a coupled oscillator. Figures 10-2a and 10-2b show the simplified block diagram and the schematic of the series tuned coupled oscillator circuit for the purpose of the analysis. Oscillators 1 and 2 are the series tuned oscillators; L_1, C_1 and L_2, C_2 are the components of the LC resonant circuit of the series tuned oscillators 1 and 2.

$R_{1\text{-}Loss}$ and $R_{2\text{-}Loss}$ are the loss resistances of the resonator, and R_{n1} and R_{n2} are the negative resistances of the active device, respectively.

The simplest coupling network can be realized by capacitor C_c, which couples the two oscillator circuits. The coupling strength depends upon the value of C_c. The value of $C_c \rightarrow 0$ indicates strong coupling, and the circuits behave like a single LC oscillator circuit. The equivalent values of L and C are given as

$$L = L_1 + L_2 \tag{10-1}$$

$$C = \frac{C_1 C_2}{C_1 + C_2} \tag{10-2}$$

The value of $C_c \rightarrow \infty$ indicates zero coupling; the two oscillator circuits are uncoupled, and capacitor C_c behaves like a short-circuit for RF signals. Since oscillators 1 and 2 are oscillating at their free-running frequencies and the negative resistance of the active devices compensates for the corresponding loss resistance of the resonators, Figure 10-2b is reduced to Figure 10-3, showing the coupled oscillator circuit without the series loss resistance, as the negative resistance of the active device compensates for it.

In a real application, intermediate coupling strength ($\infty > C_c > 0$) is more meaningful for the analysis of mutually coupled oscillator systems.

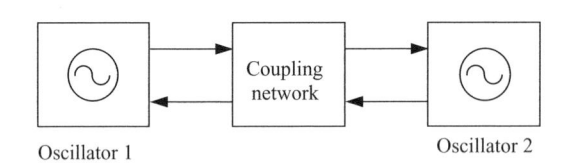

Figure 10-2a Block diagram of the coupled oscillator.

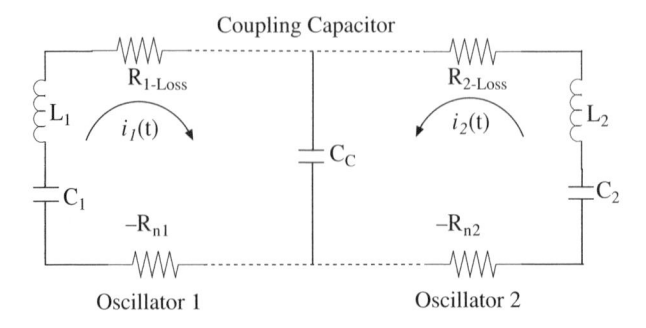

Figure 10-2b Schematic of the coupled oscillator.

10.1 MUTUALLY COUPLED OSCILLATORS

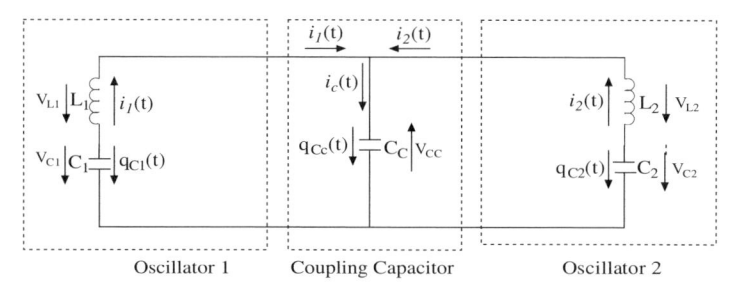

Figure 10-3 Coupled oscillator circuit without series loss resistance.

The circuit equation of Figure 10-3 is given from the Kirchoff's voltage law (KVL):

$$V_{C_1}(t) + V_{L_1}(t) + V_{C_c}(t) = 0 \tag{10-3}$$

$$\frac{q_{C_1}(t)}{C_1} - L_1 \frac{\partial i_1(t)}{\partial t} - \frac{q_{C_c}(t)}{C_c} \Rightarrow \frac{q_{C_1}(t)}{C_1} + L_1 \frac{\partial^2 [q_{C_1}(t)]}{\partial t^2} - \frac{q_{C_c}(t)}{C_c} = 0 \tag{10-4}$$

$$q_{C_c}(t) = -[q_{C_1}(t) + q_{C_2}(t)] \tag{10-5}$$

$$\frac{q_{C_1}(t)}{C_1} + L_1 \frac{\partial^2 [q_{C_1}(t)]}{\partial t^2} + \frac{1}{C_c}[q_{C_1}(t) + q_{C_2}(t)] = 0 \tag{10-6}$$

$$\frac{\partial^2 [q_{C_1}(t)]}{\partial t^2} + \left[\frac{1}{L_1 C_1} + \frac{1}{L_1 C_c}\right][q_{C_1}(t)] + \frac{1}{L_1 C_c}[q_{C_2}(t)] = 0 \tag{10-7}$$

$$\frac{\partial^2 [q_{C_1}(t)]}{\partial t^2} + \omega_1^2 [q_{C_1}(t)] + \omega_{1c}^2 [q_{C_2}(t)] = 0 \tag{10-8}$$

where

$$V_{C_1}(t) = \frac{q_{C_1}(t)}{C_1}, \quad V_{L_1}(t) = -L_1 \frac{\partial i_1(t)}{\partial t}, \quad V_{C_c}(t) = -\frac{q_{C_c}(t)}{C_c}$$

$$i_1(t) = -\frac{\partial q_{C_1}(t)}{\partial t}, \quad i_2(t) = -\frac{\partial q_{C_2}(t)}{\partial t}, \quad i_{C_c}(t) = i_1(t) + i_2(t) = -\frac{\partial q_{C_c}(t)}{\partial t}$$

$$q_{C_c}(t) = -[q_{C_1}(t) + q_{C_2}(t)] = -\int \left[\frac{\partial q_{C_c}(t)}{\partial t}\right] dt$$

$$\omega_1 = \sqrt{\frac{1}{L_1}\left[\frac{1}{C_1} + \frac{1}{C_c}\right]}, \quad \omega_2 = \sqrt{\frac{1}{L_2}\left[\frac{1}{C_2} + \frac{1}{C_c}\right]}, \quad \omega_{1c} = \sqrt{\frac{1}{L_1 C_c}}, \quad \omega_{2c} = \sqrt{\frac{1}{L_2 C_c}}$$

SYSTEMS OF COUPLED OSCILLATORS

Similarly,

$$V_{C_2}(t) + V_{L_2}(t) + V_{C_c}(t) = 0 \tag{10-9}$$

$$\frac{q_{C_2}(t)}{C_2} - L_2 \frac{\partial i_2(t)}{\partial t} - \frac{q_{C_c}(t)}{C_c} \Rightarrow \frac{q_{C_2}(t)}{C_2} + L_2 \frac{\partial^2 [q_{C_2}(t)]}{\partial t^2} - \frac{q_{C_c}(t)}{C_c} = 0 \tag{10-10}$$

$$q_{C_c}(t) = -[q_{C_1}(t) + q_{C_2}(t)] \tag{10-11}$$

$$\frac{q_{C_2}(t)}{C_2} + L_2 \frac{\partial^2 [q_{C_2}(t)]}{\partial t^2} + \frac{1}{C_c}[q_{C_1}(t) + q_{C_2}(t)] = 0 \tag{10-12}$$

$$\frac{\partial^2 [q_{C_2}(t)]}{\partial t^2} + \left[\frac{1}{L_2 C_2} + \frac{1}{L_2 C_c}\right][q_{C_2}(t)] + \frac{1}{L_2 C_c}[q_{C_1}(t)] = 0 \tag{10-13}$$

$$\frac{\partial^2 [q_{C_2}(t)]}{\partial t^2} + \omega_2^2 [q_{C_2}(t)] + \omega_{2c}^2 [q_{C_1}(t)] = 0 \tag{10-14}$$

Equations (10-8) and (10-14) are the second-order homogeneous differential equations of the mutually coupled oscillator circuit shown in Figure 10-3.

The natural frequencies ω_{r1} and ω_{r2} of the coupled oscillator can be derived by solving the second-order homogeneous equations (10-8) and (10-14). The solutions are

$$q_{C_1}(t) = K_1 \exp[j\omega_{r1} t] + K_2 \exp[j\omega_{r2} t] \tag{10-15}$$

$$q_{C_2}(t) = K_3 \exp[j\omega_{r1} t] + K_4 \exp[j\omega_{r2} t] \tag{10-16}$$

where K_1, K_2, K_3, and K_4 are the coefficients of the $q_{C_1}(t)$ and $q_{C_2}(t)$.

After substituting the value of $q_{C_1}(t)$ and $q_{C_2}(t)$ in equations (10-8) and (10-14) and equating the coefficient of $\exp[j\omega_{r1} t]$ and $\exp[j\omega_{r2} t]$, we obtain four linear equations:

$$\omega_1^2 K_1 + \omega_{1c}^2 K_3 - \omega_{r1}^2 K_1 = 0 \tag{10-17}$$

$$\omega_1^2 K_2 + \omega_{1c}^2 K_4 - \omega_{r2}^2 K_2 = 0 \tag{10-18}$$

$$\omega_2^2 K_3 + \omega_{2c}^2 K_1 - \omega_{r1}^2 K_3 = 0 \tag{10-19}$$

$$\omega_2^2 K_4 + \omega_{2c}^2 K_2 - \omega_{r2}^2 K_4 = 0 \tag{10-20}$$

After solving equations (10-17), (10-18), (10-19), and (10-20), the expression of the coefficients K_1, K_2, K_3, and K_4 are

$$K_1 = \left[\frac{\omega_{r1}^2 - \omega_2^2}{\omega_{2c}^2}\right] K_3 \tag{10-21}$$

$$K_2 = \left[\frac{\omega_{r2}^2 - \omega_2^2}{\omega_{2c}^2}\right] K_4 \qquad (10\text{-}22)$$

$$K_3 = \left[\frac{\omega_{r1}^2 - \omega_1^2}{\omega_{1c}^2}\right] K_1 \qquad (10\text{-}23)$$

$$K_4 = \left[\frac{\omega_{r2}^2 - \omega_1^2}{\omega_{1c}^2}\right] K_2 \qquad (10\text{-}24)$$

From equations (10-21), (10-22), (10-23), and (10-24), the coefficients K_1, K_2, K_3, and K_4 can be eliminated, and ω_{r1} and ω_{r2} are

$$(\omega_{r1}^2 - \omega_1^2)(\omega_{r1}^2 - \omega_2^2) - \omega_{1c}^2 \omega_{2c}^2 = 0 \qquad (10\text{-}25)$$

$$(\omega_{r2}^2 - \omega_1^2)(\omega_{r2}^2 - \omega_2^2) - \omega_{2c}^2 \omega_{1c}^2 = 0 \qquad (10\text{-}26)$$

$$\omega_{r1}^2 = \frac{1}{2}(\omega_1^2 + \omega_2^2) + \frac{1}{2}\sqrt{(\omega_1^2 - \omega_2^2) + 4\omega_{1c}^2 \omega_{2c}^2} \qquad (10\text{-}27)$$

$$\omega_{r2}^2 = \frac{1}{2}(\omega_1^2 + \omega_2^2) - \frac{1}{2}\sqrt{(\omega_1^2 - \omega_2^2) + 4\omega_{1c}^2 \omega_{2c}^2} \qquad (10\text{-}28)$$

where ω_{r1} and ω_{r2} are the natural frequencies of oscillation corresponding to the two normal modes of the oscillation of the coupled oscillator.

From equations (10-15) and (10-16), $q_{C_1}(t)$ and $q_{C_2}(t)$ can be expressed as

$$q_{C_1}(t) = K_1 \exp[j\omega_{r1} t] + K_2 \exp[j\omega_{r2} t] \qquad (10\text{-}29)$$

$$q_{C_2}(t) = K_1 \left[\frac{\omega_{r1}^2 - \omega_1^2}{\omega_{1c}^2}\right] \exp[j\omega_{r1} t] + K_2 \left[\frac{\omega_{r2}^2 - \omega_1^2}{\omega_{1c}^2}\right] \exp[j\omega_{r2} t] \qquad (10\text{-}30)$$

Defining ω_{av} and ω_{diff} as

$$\omega_{av} = \frac{\omega_{r1} + \omega_{r2}}{2} \quad \text{and} \quad \omega_{diff} = \frac{\omega_{r1} - \omega_{r2}}{2}$$

Equations (10-29) and (10-30) can be further simplified to

$$q_{C_1}(t) = \exp[j\omega_{av} t]\left(K_1 \exp[j\omega_{diff} t] + K_2 \exp[-j\omega_{diff} t]\right) \qquad (10\text{-}31)$$

$$q_{C_2}(t) = \exp[j\omega_{av} t]\left(K_1 \frac{\omega_{r1}^2 - \omega_1^2}{\omega_{1c}^2} \exp[j\omega_{diff} t] + K_2 \left[\frac{\omega_{r2}^2 - \omega_1^2}{\omega_{1c}^2}\right] \exp[-j\omega_{diff} t]\right)$$

$$(10\text{-}32)$$

For the case of the identical oscillator ($L = L_1 = L_2$ and $C = C_1 = C_2$), the coupled oscillator parameters are reduced to

$$\omega_1^2 = \frac{1}{L}\left[\frac{1}{C} + \frac{1}{C_c}\right] \tag{10-33}$$

$$\omega_2^2 = \frac{1}{L}\left[\frac{1}{C} + \frac{1}{C_c}\right] \tag{10-34}$$

$$\omega_{1c}^2 = \frac{1}{LC_c} \tag{10-35}$$

$$\omega_{2c}^2 = \frac{1}{LC_c} \tag{10-36}$$

$$\omega_{r1}^2 = \frac{1}{2}(\omega_1^2 + \omega_2^2) + \frac{1}{2}\sqrt{(\omega_1^2 - \omega_2^2) + 4\omega_{1c}^2\omega_{2c}^2} \tag{10-37}$$

$$\omega_{r1}^2 = \frac{1}{2}\left\{\frac{1}{L}\left[\frac{1}{C} + \frac{1}{C_c}\right] + \frac{1}{L}\left[\frac{1}{C} + \frac{1}{C_c}\right]\right\} + \frac{1}{2}\sqrt{4\left(\frac{1}{LC_c}\right)^2} \tag{10-38}$$

$$\omega_{r1}^2 = \frac{1}{L}\left[\frac{1}{C} + \frac{1}{C_c}\right] + \left(\frac{1}{LC_c}\right) = \left[\frac{1}{LC} + \frac{2}{LC_c}\right] \tag{10-39}$$

$$\omega_{r2} = \frac{1}{2}(\omega_1^2 + \omega_2^2) - \frac{1}{2}\sqrt{(\omega_1^2 - \omega_2^2) + 4\omega_{1c}^2\omega_{2c}^2} \tag{10-40}$$

$$\omega_{r2}^2 = \frac{1}{2}\left\{\frac{1}{L}\left[\frac{1}{C} + \frac{1}{C_c}\right] + \frac{1}{L}\left[\frac{1}{C} + \frac{1}{C_c}\right]\right\} - \frac{1}{2}\sqrt{4\left(\frac{1}{LC_c}\right)^2} = \frac{1}{LC} \tag{10-41}$$

For ($L = L_1 = L_2$ and $C = C_1 = C_2$), equations (10-31) and (10-32) can be given by

$$q_{C_1}(t) = \exp[j\omega_{av}t](K_1 \exp[j\omega_{diff}t] + K_2 \exp[-j\omega_{diff}t]) \tag{10-42}$$

$$q_{C_2}(t) = \exp[j\omega_{av}t](K_1 \exp[j\omega_{diff}t] - K_2 \exp[-j\omega_{diff}t]) \tag{10-43}$$

10.1.1 Coupled Oscillator Frequency and Modulation Frequency

Let us assume that oscillator 1 starts oscillating first, and oscillator 2 starts oscillating due to the injection mechanism from oscillator 1 in the coupled oscillator circuit shown in Figure 10-3. This condition forces zero current at $t = 0$ in the circuit of oscillator 2. The corresponding classical analogy of this mechanism, shown in Figure 10-1, can be given by pulling one mass to the side of the pendulum and releasing it from the resting point.

10.1 MUTUALLY COUPLED OSCILLATORS

Under these constraints, the initial conditions of the charge distribution on capacitors C_1 and C_2 are given as

$$q_{C_1}(t)\big|_{t=0} \neq 0 \tag{10-44}$$

$$q_{C_2}(t)\big|_{t=0} = 0 \tag{10-45}$$

Assuming $K_1 = K_2 = K$, equations (10-42) and (10-43) can be rewritten as

$$q_{C_1}(t) = K\exp[j\omega_{av}t]\big(\exp[j\omega_{diff}t] + \exp[-j\omega_{diff}t]\big) \Rightarrow q_{C_1}(t)\big|_{t=0} = 2K \tag{10-46}$$

$$q_{C_2}(t) = K\exp[j\omega_{av}t]\big(\exp[j\omega_{diff}t] - \exp[-j\omega_{diff}t]\big) \Rightarrow q_{C_2}(t)\big|_{t=0} \Rightarrow 0 \tag{10-47}$$

Equations (10-46) and (10-47) can be expressed in the form of cosine and sine functions as

$$q_{C_1}(t) = 2K\{\cos(\omega_{diff}t)(\cos(\omega_{av}t) + j\sin(\omega_{av}t))\} \tag{10-48}$$

$$q_{C_2}(t) = -2jK\{\sin(\omega_{diff}t)(\cos(\omega_{av}t) + j\sin(\omega_{av}t))\} \tag{10-49}$$

The real and imaginary part of equations (10-48) and (10-49) are

$$\text{Re}[q_{C_1}(t)] = 2K[\cos(\omega_{av}t)\cos(\omega_{diff}t)] \tag{10-50}$$

$$\text{Im}[q_{C_1}(t)] = 2K[\sin(\omega_{av}t)\cos(\omega_{diff}t)] \tag{10-51}$$

$$\text{Re}[q_{C_2}(t)] = 2K[\sin(\omega_{av}t)\sin(\omega_{diff}t)] \tag{10-52}$$

$$\text{Im}[q_{C_2}(t)] = -2K[\cos(\omega_{av}t)\sin(\omega_{diff}t)] \tag{10-53}$$

The real part of equations (10-48) and (10-49) gives the equation of the oscillation for $q_{C_1}(t)$ and $q_{C_2}(t)$ of the identical mutually (synchronized) coupled oscillator. From equations (10-50) and (10-52), $q_{C_1}(t)$ and $q_{C_2}(t)$ oscillate at the frequency ω_{av} and are modulated by the lower frequency ω_{diff}. From equation (10-50) and (10-52), we can see that the modulation associated with $q_{C_1}(t)$ and $q_{C_2}(t)$ is 90° out of phase, which indicates that the total energy is flowing back and forth between the two coupled oscillator circuits, oscillator 1 and oscillator 2.

Defining the coupled oscillator frequency as ω_c, we obtain

$$\omega_c = \omega_{av} = \frac{\omega_{r1} + \omega_{r2}}{2} \tag{10-54}$$

Defining the modulation frequency ω_m, we obtain

$$\omega_m = \omega_{diff} = \frac{\omega_{r1} - \omega_{r2}}{2} \tag{10-55}$$

Further, from equations (10-54) and (10-55),

$$\frac{\omega_m}{\omega_c} = \frac{\omega_{r1} - \omega_{r2}}{\omega_{r1} + \omega_{r2}} \tag{10-56}$$

$$\omega_m = \left[\frac{\omega_{r1} - \omega_{r2}}{\omega_{r1} + \omega_{r2}}\right]\omega_c \tag{10-57}$$

From equations (10-38) and (10-41),

$$\omega_m = \left[\frac{(\omega_{r1} + \omega_{r2})(\omega_{r1} - \omega_{r2})}{(\omega_{r1} + \omega_{r2})(\omega_{r1} + \omega_{r2})}\right]\omega_c = \left[\frac{(\omega_{r1}^2 - \omega_{r2}^2)}{(\omega_{r1} + \omega_{r2})(\omega_{r1} + \omega_{r2})}\right]\omega_c \tag{10-58}$$

$$\omega_m = \left[\frac{(\omega_{r1}^2 - \omega_{r2}^2)}{(\omega_{r1} + \omega_{r2})(\omega_{r1} + \omega_{r2})}\right]\omega_{av} = \left[\frac{1}{\omega_{r1}^2 + \omega_{r2}^2 + 2\omega_{r1}\omega_{r2}}\right]\left[\frac{2}{LC_c}\right]\omega_c \tag{10-59}$$

The modulation frequency, therefore, is

$$\omega_m \approx \frac{C}{2C_c}\sqrt{\frac{1}{LC}} = \left[\frac{C}{2C_c}\right]\omega_c \implies \omega_m = \omega_c\left[\frac{C}{2C_c}\right] \tag{10-60}$$

10.2 PHASE CONDITION FOR MUTUALLY LOCKED (SYNCHRONIZED) COUPLED OSCILLATORS

Figure 10-4 shows the equivalent circuit of the mutually coupled oscillators (parallel-tuned oscillators) coupled through the transmission line as a coupling network. The equivalent model of the oscillator circuit can be given by either a

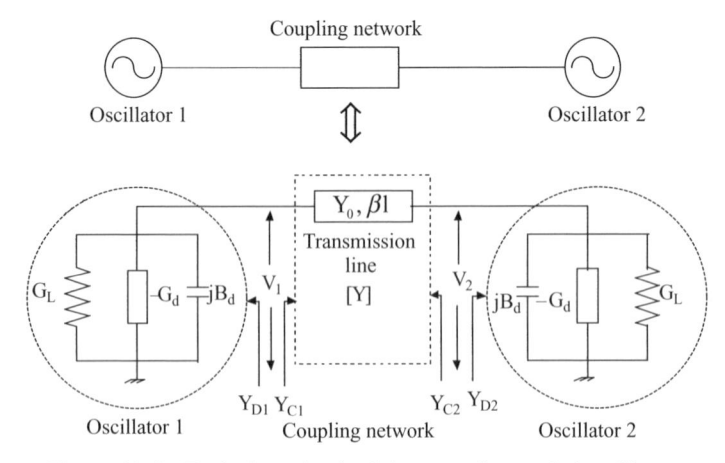

Figure 10-4 Equivalent circuit of the mutually coupled oscillators.

series-tuned or parallel-tuned configuration, and the condition for mutual locking is valid for both cases. Since the transmission line acts as a resonant circuit in addition to the coupling network, and to simplify the analysis, the loss associated with the transmission line is assumed to be part of the load admittances G_L.

In Figure 10-4, Y_{D1} and Y_{D2} represent the equivalent admittance of the two active devices and their loads, while Y_{C1} and Y_{C2} represent the admittance seen by the devices of the mutually coupled oscillator circuit. Based on the extended resonance technique [133], the length of the transmission line is selected such that the two devices resonate with each other at the common resonance frequency. This can be done by choosing the length of the transmission line of the coupling network so that each device's susceptance is transformed into a susceptance with the same magnitude, but opposite in sign, thereby creating a virtual short-circuit at the midpoint of the transmission line, ensuring that the devices are injection locked with respect to out of phase.

For analysis purposes, the transmission line is characterized as a two-port network with its terminal voltages represented by a voltage phasor as $V_1 = |V_1|e^{j\varphi_1}$ and $V_2 = |V_2|e^{j\varphi_2}$, where $|V_1|$, $|V_2|$, φ_1 and φ_2 are the magnitudes and phases of the voltage phasor.

The circuit equation for Figure 10-4 at the transmission line terminals can be expressed in matrix form as

$$\begin{bmatrix} -Y_{D1}\,V_1 \\ -Y_{D1}\,V_2 \end{bmatrix} = \begin{bmatrix} Y_{11} & Y_{12} \\ Y_{21} & Y_{22} \end{bmatrix} \begin{bmatrix} V_1 \\ V_2 \end{bmatrix} \qquad (10\text{-}61)$$

$$[Y]_{transmission\ line} = \begin{bmatrix} Y_{11} & Y_{12} \\ Y_{21} & Y_{22} \end{bmatrix} = \begin{bmatrix} -jY_0 \cot(\beta l) & jY_0 \operatorname{cosec}(\beta l) \\ jY_0 \operatorname{cosec}(\beta l) & -jY_0 \cot(\beta l) \end{bmatrix} \qquad (10\text{-}62)$$

where β, Y_0, and l are the phase constant, characteristic admittance, and length of the transmission line.

For identical coupled oscillators:

$$V_1 = V_2 = V, \quad Y_{D1} = Y_{D2} = Y_D, \quad \text{and} \quad Y_{C1} = Y_{C2} = Y_C$$

From equation (10-61), the phase difference, we obtain

$$e^{\pm j(\varphi_2 - \varphi_1)} = e^{\pm j(\Delta\varphi)} = \frac{[Y_D + Y_{11}]}{-[Y_{12}]} \qquad (10\text{-}63)$$

The admittance Y_D is composed of the device admittance and load conductance as

$$Y_D = G_L - G_D + jB_D; \quad G_D > 0 \qquad (10\text{-}64)$$

The admittance Y_C is given from the transmission line equation as

$$Y_C = Y_0 \left[\frac{Y_D + jY_0 \tan(\beta l)}{Y_0 + jY_D \tan(\beta l)} \right] \quad (10\text{-}65)$$

Applying the extended resonance technique [133], the length of the transmission line is selected such that the real and imaginary parts of the admittance Y_C are given by

$$\text{Re}[Y_C] = G_L - G_D \quad (10\text{-}66)$$

$$\text{Im}[Y_C] = -jB_D \quad (10\text{-}67)$$

From equations (10-64) and (10-65),

$$\tan(\beta l) = \frac{2B_D Y_0}{B_D^2 - Y_0^2 + (G_L - G_D)^2} \quad (10\text{-}68)$$

During the startup of oscillation, the real parts of the admittances Y_C and Y_D are negative, and as the signal level increases, the device gain drops until the losses are compensated for. Under steady-state oscillation conditions, $G_L - G_D = 0$ and the electrical length θ of the transmission line is given from equation (10-68) as

$$\theta = (\beta l) = \tan^{-1}\left[\frac{2B_D Y_0}{B_D^2 - Y_0^2}\right] \quad (10\text{-}69)$$

The $[Y]$ parameter of the transmission line can be rewritten from equation (10-63) as

$$[Y]_{transmission\ line} = \begin{bmatrix} -jY_0 \cot(\beta l) & jY_0 \csc(\beta l) \\ jY_0 \csc(\beta l) & -jY_0 \cot(\beta l) \end{bmatrix} = \begin{bmatrix} -j\dfrac{B_D^2 - Y_0^2}{2B_D} & j\dfrac{B_D^2 + Y_0^2}{2B_D} \\ j\dfrac{B_D^2 + Y_0^2}{2B_D} & -j\dfrac{B_D^2 - Y_0^2}{2B_D} \end{bmatrix}$$

$$(10\text{-}70)$$

From (10-63) and (10-70), the phase difference is given by

$$e^{\pm j(\varphi_2 - \varphi_1)} = \frac{[Y_D + Y_{11}]}{-[Y_{12}]} = \left[\frac{Y_D - j(B_D^2 - Y_0^2)/2B_D}{j(B_D^2 + Y_0^2)/2B_D}\right] = \left[\frac{jB_D - j(B_D^2 - Y_0^2)/2B_D}{j(B_D^2 + Y_0^2)/2B_D}\right]$$

$$= -1 \Rightarrow \Delta\varphi = \varphi_2 - \varphi_1 = \pm 180° \quad (10\text{-}71)$$

Equation (10-71) gives the necessary phase condition for the mutually locked conditions of the coupled oscillator system. Thus, the outputs of the mutually synchronized coupled oscillator are in anti-phase ($\Delta\varphi = 180°$) [136].

Figures 10-5a and 10-5b show an example of a mutually synchronized coupled oscillator at a frequency of 2000 MHz ($2f_0$) in which two oscillators oscillate at 1000 MHz (f_0). This figure provides some insight into the phase relationship between the two identical mutually coupled oscillator circuits, oscillators 1 and 2. Figure 10-5a shows the circuit diagram of the mutually coupled oscillator. The circuit is fabricated on 32-mil thickness Roger substrate of dielectric constant 3.38 and loss tangent 2.7E−4. Figure 10-5b shows the simulated (Ansoft Designer) plot of the base current I_{b1} and I_{b2}, which is phase shifted by 180° in a mutually synchronized condition. The transmission line MSL2 shown in Figure 10-5a provides the phase shift for the mutually synchronized coupled oscillators; see equation (10-71).

10.3 DYNAMICS OF COUPLED OSCILLATORS

The purpose of this analysis is to develop the time-domain dynamics of the coupled oscillator systems based on the Van der Pol (VDP) model [134]. The time-dependent characteristic of this model is the basis for using this model in our problem for analyzing the system dynamics of coupled oscillators.

Figure 10-6 shows the simple model of the two identical coupled oscillators where β_{12} and β_{21} are coupling coefficients.

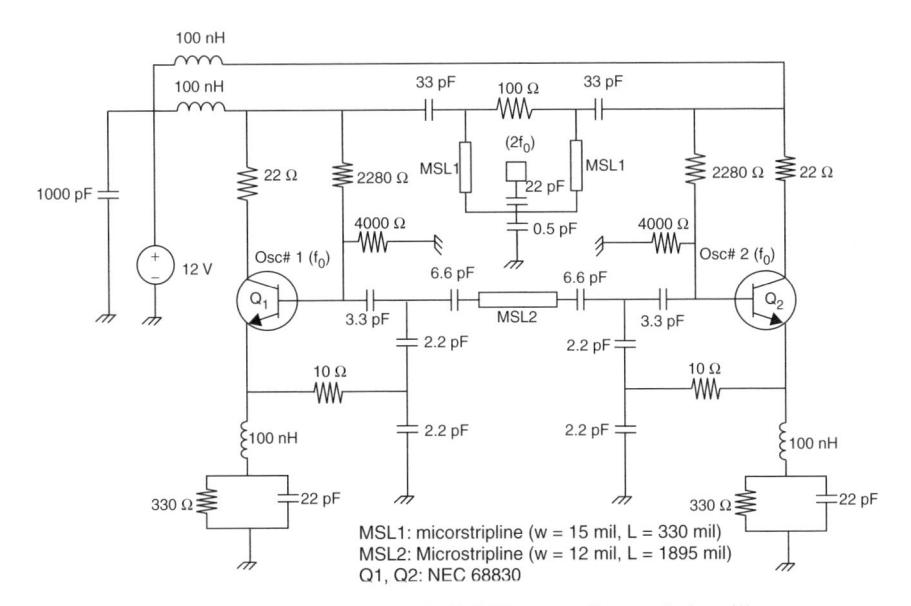

Figure 10-5a Schematic of a 2000 MHz mutually coupled oscillator.

258 SYSTEMS OF COUPLED OSCILLATORS

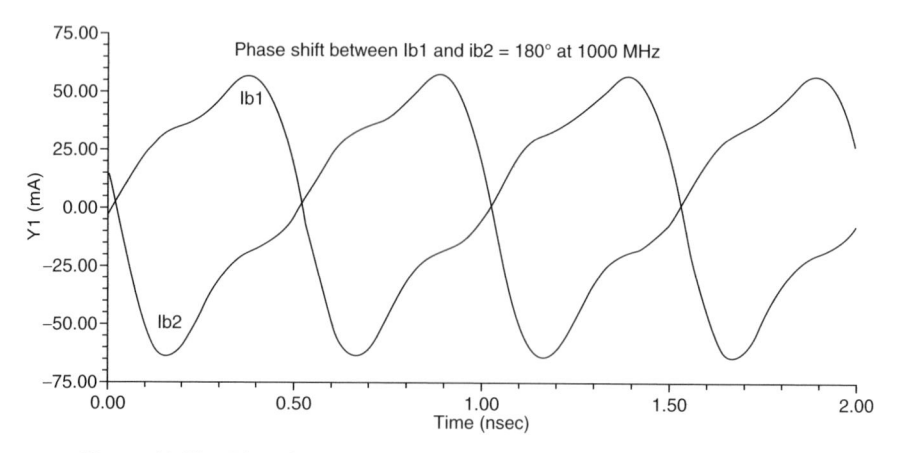

Figure 10-5b Plot of the RF base current of the mutually coupled oscillator.

Figure 10-7 shows the equivalent circuit model of Figure 10-6, where the series-tuned resonator circuit models are part of the coupled oscillator system. The equivalent model of the oscillator circuit can be given by either a series-tuned or parallel-tuned configuration, and the condition for the system dynamics is valid for both cases. Regardless of the topology, the system of the coupled oscillators must synchronize to a common frequency and maintain a desired phase relationship in the steady-state oscillating condition.

For analysis of the dynamics of the coupled oscillator system, Figure 10-7 is reduced to an equivalent injection-locked, coupled oscillator system, as shown in Figure 10-8, by replacing oscillator 1 with the additional source $V_{inj}(t)$, which accounts for the interaction and coupling with the adjacent oscillator 2.

The negative resistance of the device can be described as a time-averaged value and is given by (8-18).

$$R_n(t) = \frac{1}{T}\int_0^T \frac{V_{Rn}(t)}{I_{Rn}(t)} dt = \frac{1}{T}\int_0^T \frac{V_{Rn}(t)}{i(t)} dt \qquad (10\text{-}72)$$

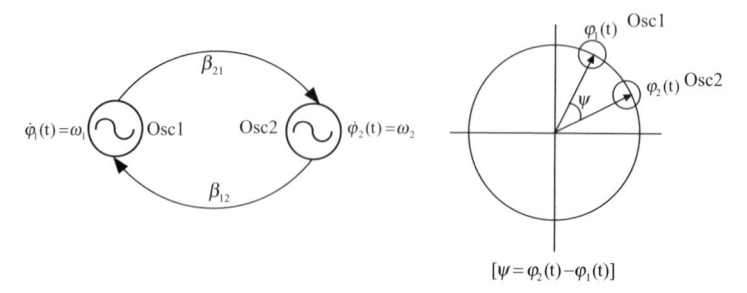

Figure 10-6 Simple model of the system of two-coupled oscillators.

10.3 DYNAMICS OF COUPLED OSCILLATORS

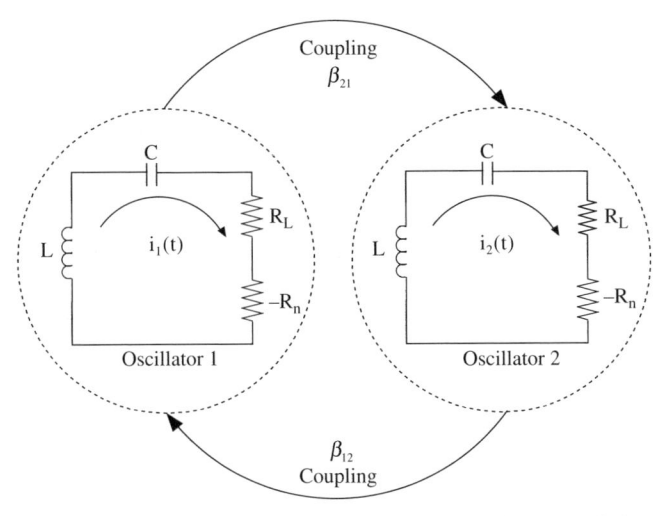

Figure 10-7 Equivalent circuit representation of Figure 10-4.

For a series-tuned resonator circuit, defining

$$\omega_0 = \sqrt{\frac{1}{LC}}, \quad Q = \frac{\omega_0 L}{R_L}, \quad Q = \frac{1}{\omega_0 C R_L}, \quad R_n \approx R_n[|V_{out}(t)|]$$

meaning that R_n is a strong function of $V_{out}(t)$, ω_0 is the resonance frequency and Q is the quality factor of the embedded resonator circuit. The quality factor of the series-tuned resonator is assumed to be greater than 10, so that the output signal $V_{out}(t)$ is close to the natural resonance frequency of the oscillator circuit.

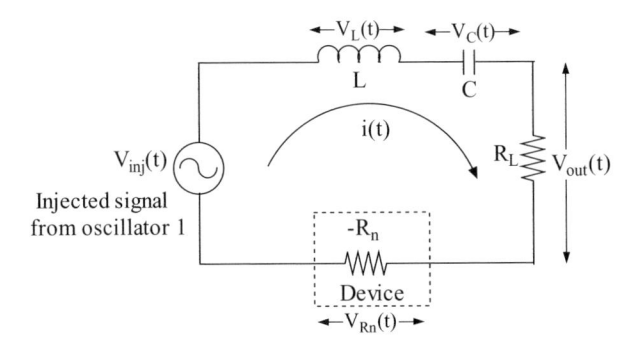

Figure 10-8 Equivalent representation of Figure 10-7, where $V_{inj}(t)$ accounts for the interaction and coupling.

SYSTEMS OF COUPLED OSCILLATORS

The circuit equations of Figure 10-8 are given from the KVL as

$$V_{inj}(t) = V_L(t) + V_C(t) + V_{Rn}(t) + V_{out}(t)$$

$$= L\frac{\partial i(t)}{\partial t} + \frac{1}{C}\int i(t)dt - R_n(|V_{out}(t)|)i(t) + R_L i(t) \quad (10\text{-}73)$$

$$i(t) = \frac{V_{out}(t)}{R_L} \quad (10\text{-}74)$$

The output signal $V_{out}(t)$ can be expressed as

$$V_{out}(t) = A(t)\exp\{j[\omega_0 t + \varphi(t)]\} = A(t)\exp[j\theta(t)] \quad (10\text{-}75)$$

where $\theta(t)$ is the instantaneous phase and $A(t)$ and $\varphi(t)$ are the amplitude and phase terms of the output signal, which vary slowly with respect to time in comparison to the output periodic oscillation.

To solve equation (10-73), $\int V_{out}(t)dt$ and $\partial V_{out}(t)/\partial t$ are evaluated first and can be given by

$$\int V_{out}(t)dt = -\frac{j2A(t)\exp[j\theta(t)]}{\omega_0} + \frac{1}{\omega_0^2}\frac{\partial[A(t)\exp[j\theta(t)]]}{\partial t} + \ldots, \text{ (higher-order terms)}$$

$$(10\text{-}76)$$

$$\frac{\partial}{\partial t}[V_{out}(t)] = \left[\frac{\partial A(t)}{\partial t}\right][\exp j[\theta(t)]] + A(t)\left[\exp\{j[\theta(t)]\}\cdot j\left\{\omega_0 + \frac{\partial\varphi(t)}{\partial t}\right\}\right] \quad (10\text{-}77)$$

The higher-order term in equation (10-76) can be neglected in comparison to the output periodic oscillation and

$$\int V_{out}(t)dt = -\frac{j2V_{out}(t)}{\omega_0} + \frac{1}{\omega_0^2}\frac{\partial V_{out}(t)}{\partial t} \quad (10\text{-}78)$$

From equations (10-73) and (10-74),

$$V_{inj}(t) = \left[\frac{L}{R_L}\right]\frac{\partial V_{out}(t)}{\partial t} + \left[\frac{1}{CR_L}\right]\int V_{out}(t)dt + \left[1 - \frac{R_n(|V_{out}(t)|)}{R_L}\right]V_{out}(t) \quad (10\text{-}79)$$

By multiplying equation (10-79) by ω_0/Q

$$\left[\frac{\omega_0}{Q}\right]V_{inj}(t) = \frac{\partial V_{out}(t)}{\partial t} + \omega_0^2\int V_{out}(t)dt + \left[1 - \frac{R_n(|V_{out}(t)|)}{R_L}\right]\left[\frac{\omega_0}{Q}\right]V_{out}(t) \quad (10\text{-}80)$$

where

$$V_{inj}(t) = A_{inj}(t)\exp\{j[\omega_{inj}(t) + \psi_{inj}(t)]\} \Rightarrow A_{inj}(t)\exp\{j\theta_{inj}(t)\}$$

Following Van der Pol [134], the device saturation and amplitude dependence of the negative resistance are modeled by a quadratic function such that $[1 - (R_n(|V_{out}(t)|)/R_L)] \Rightarrow -\mu(\alpha_0^2 - |V_{out}(t)|^2)$, where α_0 is the free-running amplitude of the oscillation and μ is an empirical nonlinear parameter describing the oscillator. For the uncoupled free-running oscillator $[V_{inj}(t) = 0]$ and the value of $[1 - (R_n(|V_{out}(t)|)/R_L)]$ is zero $(R_n \to R_L)$, whereas for the coupled oscillator $[V_{inj}(t) \neq 0]$ and the value of $[1 - (R_n(|V_{out}(t)|)/R_L)]$ is nonzero to compensate for the additional energy being supplied due to the injection mechanism from the other oscillator.

From equations (10-78) and (10-80),

$$\frac{\partial V_{out}(t)}{\partial t} = V_{out}(t)\left[\frac{\mu\omega_0}{2Q}(\alpha_0^2 - |V_{out}(t)|^2) + j\omega_0\right] + \frac{\omega_0}{2Q}V_{inj}(t) \quad (10\text{-}81)$$

From equations (10-77) and (10-81),

$$V_{out}(t)\left[\frac{\mu\omega_0}{2Q}(\alpha_0^2 - |V_{out}(t)|^2) + j\omega_0\right] + \frac{\omega_0}{2Q}V_{inj}(t)$$
$$= \left\{\left[\frac{\partial A(t)}{\partial t}\right][\exp j[\theta(t)]] + A(t)\left[\exp\{j[\theta(t)]\}\cdot j\left\{\omega_0 + \frac{\partial\varphi(t)}{\partial t}\right\}\right]\right\} \quad (10\text{-}82)$$

10.3.1 Amplitude and Phase Dynamics of the Coupled Oscillator

By calculating the real and imaginary parts of equation (10-82), the amplitude and phase dynamics of the coupled oscillator system can be given as [151]

$$\frac{\partial A(t)}{\partial t} = \mu\left[\frac{\omega_0}{2Q}\right]A(t)(\alpha_0^2 - |A(t)|^2) + \left[\frac{\omega_0}{2Q}\right]A(t)\,\text{Re}\left[\frac{V_{inj}(t)}{V_{out}(t)}\right] \to \text{amplitude dynamics}$$

(10-83)

$$\frac{\partial \theta(t)}{\partial t} = \omega_0 + \left[\frac{\omega_0}{2Q}\right]\text{Im}\left[\frac{V_{inj}(t)}{V_{out}(t)}\right] \to \text{phase dynamics} \quad (10\text{-}84)$$

For $|V_{out}(t)| \gg |V_{inj}(t)|$, the oscillator amplitude remains close to its free-running value and the system dynamics of the coupled oscillator are predominantly determined by the phase dynamics given in equation (10-84).

From equation (10-84),

$$\frac{\partial \theta(t)}{\partial t} = \omega_0 + \left[\frac{\omega_0}{2Q}\right] \text{Im}\left[\frac{V_{inj}(t)}{V_{out}(t)}\right] = \omega_0 + \left[\frac{\omega_0}{2Q}\right] \text{Im}\left[\frac{A_{inj}(t)\exp[j\theta_{inj}(t)]}{A(t)\exp[j\theta(t)]}\right] \quad (10\text{-}85)$$

$$\frac{\partial \theta(t)}{\partial t} = \omega_0 + \left[\frac{\omega_0}{2Q}\right]\left[\frac{A_{inj}(t)}{A(t)}\right]\sin[\theta_{inj}(t) - \theta(t)] \quad (10\text{-}86)$$

$$\frac{\partial \theta(t)}{\partial t} = \omega_0 + \left[\frac{\omega_0}{2Q}\right]\left[\frac{A_{inj}(t)}{A(t)}\right]\sin\{[\omega_{inj}t + \psi_{inj}(t)] - [\omega_0 t + \varphi(t)]\} \quad (10\text{-}87)$$

$$\frac{\partial \theta(t)}{\partial t} = \omega_0 + \left[\frac{\omega_0}{2Q}\right]\left[\frac{A_{inj}(t)}{A(t)}\right]\sin[(\omega_{inj} - \omega_0)t + (\psi_{inj}(t) - \varphi(t))] \quad (10\text{-}88)$$

where

$$\theta_{inj}(t) = [\omega_{inj}t + \psi_{inj}(t)]; \quad \theta(t) = [\omega_0 t + \varphi(t)] \quad \text{and} \quad V_{out}(t) = A(t)\exp[j\theta(t)]$$

10.3.2 Locking Bandwidth of the Coupled Oscillator

In the presence of the injection signal from the neighboring oscillator, the oscillator locks onto the injected signal as $\partial \theta(t)/\partial t \to \omega_{inj}$, and at steady state the equation of the phase dynamics is given from equation (10-88) as

$$\frac{\partial \theta(t)}{\partial t} \to \omega_{inj} \Rightarrow \omega_{inj} = \omega_0 + \left[\frac{\omega_0}{2Q}\right]\left[\frac{A_{inj}(t)}{A(t)}\right]\sin[\theta_{inj}(t) - \theta(t)] \quad (10\text{-}89)$$

From equation (10-89),

$$\omega_{inj} - \omega_0 = \left[\frac{\omega_0}{2Q}\right]\left[\frac{A_{inj}(t)}{A(t)}\right]\sin[\theta_{inj}(t) - \theta(t)] \quad (10\text{-}90)$$

$$\omega_{inj} - \omega_0 = (\Delta\omega_{lock})\sin(\Delta\theta) \quad (10\text{-}91)$$

where

$$\Delta\omega_{lock} = \left[\frac{\omega_0}{2Q}\right]\left[\frac{A_{inj}}{A}\right] \to \text{locking bandwidth}$$

$\Delta\theta = \theta_{inj}(t) - \theta(t) \to$ steady-state phase difference between the oscillator and the injected signal.
From equations (10-90) and (10-91),

$$[\Delta\omega_{lock}]_{locking\ bandwidth} \propto \frac{1}{Q} \quad (10\text{-}92)$$

From equation (10-92), low-Q factors are required for a wide locking range, but they will degrade the noise performance of the coupled oscillator system. Therefore, there is a trade-off between the phase noise and locking bandwidth of the VCOs.

From equation (10-91),

$$\sin(\Delta\theta) = \left[\frac{\omega_{inj} - \omega_0}{\Delta\omega_{lock}}\right] \Rightarrow \Delta\theta = \sin^{-1}\left[\frac{\omega_{inj} - \omega_0}{\Delta\omega_{lock}}\right], \quad -\frac{\pi}{2} \leq \Delta\theta \leq \frac{\pi}{2} \quad (10\text{-}93)$$

$$\omega_{inj} = \omega_0 \pm \Delta\omega_{lock} \quad \text{for } \Delta\theta = \pm\frac{\pi}{2} \quad (10\text{-}94)$$

$$\omega_{inj} - \omega_0 = \pm\Delta\omega_{lock} \Rightarrow |\omega_{inj} - \omega_0| = \Delta\omega_{lock} \quad \text{for } \Delta\theta = \pm\frac{\pi}{2} \quad (10\text{-}95)$$

From equation (10-94), the injected signal frequency ω_{inj} is tuned over the locking range of the oscillator ($\omega_0 \pm \Delta\omega_{lock}$) and the associated phase difference $\Delta\theta$ varies from $-\frac{\pi}{2}$ to $+\frac{\pi}{2}$. To determine the locking mechanism [135], equation (10-95) can be expressed as

$$\Delta\omega_{lock} > |\omega_{inj} - \omega_o| \quad \text{for } -\frac{\pi}{2} < \Delta\theta < \frac{\pi}{2} \quad (10\text{-}96)$$

The oscillator can be synchronized with an injected signal as long as $\Delta\omega_{lock} > |\omega_{inj} - \omega_0|$, where $\Delta\omega_{lock}$ represents half of the entire locking range, whereas if the frequency of the injected signal ω_{inj} is such that $\Delta\omega_{lock} \leq \omega_{inj} - \omega_0$, the oscillator cannot lock on to the injected signal and the nonlinearity of the oscillator will then generate mixing products in the coupled oscillator system.

10.4 DYNAMICS OF N-COUPLED (SYNCHRONIZED) OSCILLATORS

As discussed in Section 10.3, the amplitude and phase dynamics of the mutually coupled oscillator are given by equations (10-83) and (10-84), based on the coupled set of differential equations, and are derived by first describing the behavior of the individual oscillator with injection locking and then allowing the injection signals to be provided by the neighboring oscillator. The purpose of this analysis is to develop the general system dynamics of the N-coupled oscillators for bilateral coupling through the N port arbitrary coupling network [150]. Figure 10-9 shows the chain of the N-coupled oscillator system with bilateral coupling between the neighboring oscillators.

For an N-coupled oscillator system, the coupling between ith and jth oscillators can be described by a coupling coefficient β_{ij} as

$$\beta_{ij} = \lambda_{ij} \exp[-j\varphi_{ij}] \quad (10\text{-}97)$$

where λ_{ij} and φ_{ij} are the magnitude and phase of the coupling coefficient for the coupling between the ith and jth oscillators in the N bilateral coupled oscillator

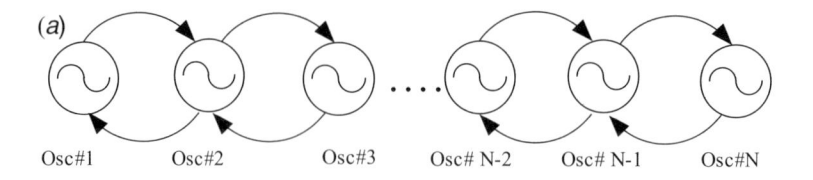

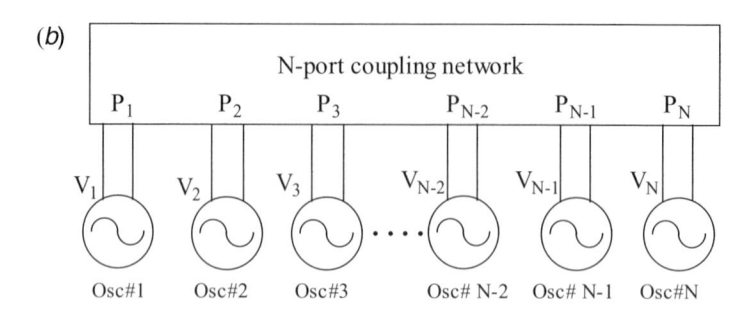

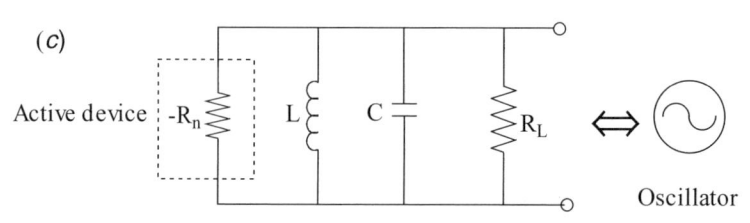

Figure 10-9 (a) N-coupled oscillator with bilateral coupling; (b) N-coupled oscillator coupled through the N port coupling network; (c) equivalent parallel model of the free-running oscillator.

systems. For a reciprocal system, the coupling coefficient is defined as $\beta_{ij} = \beta_{ji}$ and it is unitless.

The injected signal $V_{inj}(t)$ seen by the ith oscillator for the N-coupled oscillator system is given by

$$V_{inj}(t) = \sum_{\substack{j=1 \\ j \neq i}}^{N} \beta_{ij} V_j(t) \qquad (10\text{-}98)$$

where $V_j(t)$ is the output voltage of the jth oscillator.

Assuming that all the N oscillators have approximately the same Q and μ factors, the coupled oscillator system dynamics can be given from equation (10-81) as

$$\frac{\partial V_i(t)}{\partial t} = V_i(t) \left[\frac{\mu \omega_i}{2Q}(\alpha_i^2 - |V_i(t)|^2) + j\omega_i \right] + \frac{\omega_i}{2Q} V_{inj}(t) \qquad (10\text{-}99)$$

10.4 DYNAMICS OF N-COUPLED OSCILLATORS

After substituting the expression of $V_{inj}(t)$, equation (10-99) can be rewritten as

$$\frac{\partial V_i(t)}{\partial t} = V_i(t)\left[\frac{\mu\omega_i}{2Q}(\alpha_i^2 - |V_i(t)|^2) + j\omega_i\right] + \frac{\omega_i}{2Q}\sum_{\substack{j=1\\j\neq i}}^{N} B_{ij}V_j(t) \quad (10\text{-}100)$$

where $V_i(t)$ is the output voltage, α_i is the free-running amplitude, and ω_i is the free-running frequency of the ith oscillator.

The output voltage $V_i(t)$ of the ith oscillator can be expressed as

$$V_i(t) = A_i(t)\exp\{j[\omega_i t + \varphi_i(t)]\} \Rightarrow A_i(t)\exp[j\theta_i(t)] \quad (10\text{-}101)$$

where $A_i(t)$ is the amplitude and $\varphi_i(t)$ is the instantaneous phase of the ith oscillator. From equation (10-101),

$$\frac{\partial V_i(t)}{\partial t} = j\left[\omega_i + \frac{\partial \varphi_i(t)}{\partial t} - j\frac{1}{A_i(t)}\frac{\partial A_i(t)}{\partial t}\right]V_i(t) \quad (10\text{-}102)$$

From equations (10-100) and (10-102), the system dynamics of the N-coupled oscillator can be described in terms of amplitude and phase dynamics as

$$\frac{\partial A_i(t)}{\partial t} = A_i(t)\left[\frac{\mu\omega_i}{2Q}(\alpha_i^2 - |A_i(t)|^2)\right] + \frac{\omega_i}{2Q}\sum_{\substack{j=1\\j\neq i}}^{N}\lambda_{ij}A_j(t)\cos[\theta_i(t) - \theta_j(t) + \varphi_{ij}];$$

$$i = 1, 2, 3 \ldots N \quad (10\text{-}103)$$

$$\frac{\partial \theta_i(t)}{\partial t} = \omega_i - \left[\frac{\omega_i}{2Q}\right]\left\{\sum_{\substack{j=1\\j\neq i}}^{N}\lambda_{ij}\left[\frac{A_j(t)}{A_i(t)}\right]\sin[\theta_i(t) - \theta_j(t) + \varphi_{ij}]\right\}; \quad i = 1, 2, 3 \ldots N$$

$$(10\text{-}104)$$

10.4.1 Coupling Parameters

For $\beta_{ij} \to 0$ (zero coupling):

The system dynamics of the N-coupled oscillator are reduced to the dynamics of uncoupled free-running oscillators and can be given from equations (10-103) and (10-104) as

$$\left[\frac{\partial A_i(t)}{\partial t}\right]_{\beta_{ij}=0} = A_i(t)\left[\frac{\mu\omega_i}{2Q}(\alpha_i^2 - |A_i(t)|^2)\right]; \quad i = 1, 2, 3 \ldots N \quad (10\text{-}105)$$

$$\left[\frac{\partial \theta_i(t)}{\partial t}\right]_{\beta_{ij}=0} = \omega_i; \quad i = 1, 2, 3 \ldots N \quad (10\text{-}106)$$

Equations (10-105) and (10-106) are the time dynamics of the set of individual oscillators with amplitudes $A_i(t)$ and frequencies ω_i. They can be considered a generalized version of Adler's equation [135] and the basis for formulation of the system dynamics of the N-coupled oscillators.

For $1 \gg \beta_{ij} > 0$ (weak coupling):

The amplitude of the oscillators in the N-coupled oscillator system remains close to its free-running values ($A_i = \alpha_i$, $A_j = \alpha_j$), and the system dynamics of the N-coupled oscillator essentially are governed and influenced by the phase dynamics as given in equation (10-104).

For $1 \gg \beta_{ij} > 0$, equation (10-104) can be rewritten as

$$\frac{\partial \theta_i(t)}{\partial t} = \omega_i - \left[\frac{\omega_i}{2Q}\right]\left\{\sum_{\substack{j=1 \\ j \neq i}}^{N} \lambda_{ij}\left[\frac{\alpha_j}{\alpha_i}\right]\sin[\theta_i(t) - \theta_j(t) + \varphi_{ij}]\right\}; \quad i = 1, 2, 3 \ldots N$$

(10-107)

10.4.2 Synchronized Coupled Oscillator Frequency

In this case ($1 \gg \beta_{ij} > 0$), the oscillators in the N-coupled oscillator system may lock to a single frequency ω_s.

For ($1 \gg \beta_{ij} > 0$), under synchronization to a common frequency $\rightarrow \omega_s$, the value of $\partial \theta_i(t)/\partial t$ can be given by

$$\left[\frac{\partial \theta_i(t)}{\partial t}\right]_{i=1,2,3\ldots N} \rightarrow \omega_s \qquad (10\text{-}108)$$

From equations (10-107) and (10-108), the steady-state synchronized frequency ω_s is given by

$$\omega_s = \omega_i - \left[\frac{\omega_i}{2Q}\right]\left\{\sum_{\substack{j=1 \\ j \neq i}}^{N} \lambda_{ij}\left[\frac{\alpha_j}{\alpha_i}\right]\sin[\theta_i(t) - \theta_j(t) + \varphi_{ij}]\right\}; \quad i = 1, 2, 3 \ldots N$$

(10-109)

10.5 OSCILLATOR NOISE

Noise is associated with all the components of the oscillator circuit. However, the major noise contribution in an oscillator is from the active device, which introduces amplitude modulation (AM) noise and phase modulation (PM) noise. The AM component of the noise is generally ignored because of the gain-limiting effects of the active device operating under saturation, allowing little variation in output amplitude due to the noise in comparison to the PM noise component, which directly affects the frequency stability of the oscillator and creates noise sidebands.

10.5 OSCILLATOR NOISE

For greater insight into the noise effects in the oscillator design, it is necessary to understand how the noise arises in a transistor. The designer has very limited control over the noise sources in a transistor, controlling only the device selection and the operating bias point. However, knowing how noise affects oscillator waveforms, the designer can substantially improve phase noise performance by optimizing the conduction angle and drive level [132]; see Chapter 6.

10.5.1 Sources of Noise

There are mainly two noise sources in the oscillator circuit: broadband noise sources due to thermal and shot noise effects and low-frequency noise due to $1/f$ (flicker noise effects) characteristics. The current flow in a transistor is not a continuous process, but is made up of the diffusive flow of large numbers of discrete carriers. The motions of these carriers are random and account for the noise. The thermal fluctuation in the minority carrier flow and generation-recombination processes in the semiconductor device generate thermal noise, shot noise, partition noise, burst noise, and $1/f$ noise.

Figure 10-10a shows the equivalent schematic of the bipolar transistor in a grounded-emitter (GE) configuration. The high-frequency noise of a silicon bipolar transistor in a common-emitter (CE) configuration can be modeled using the three

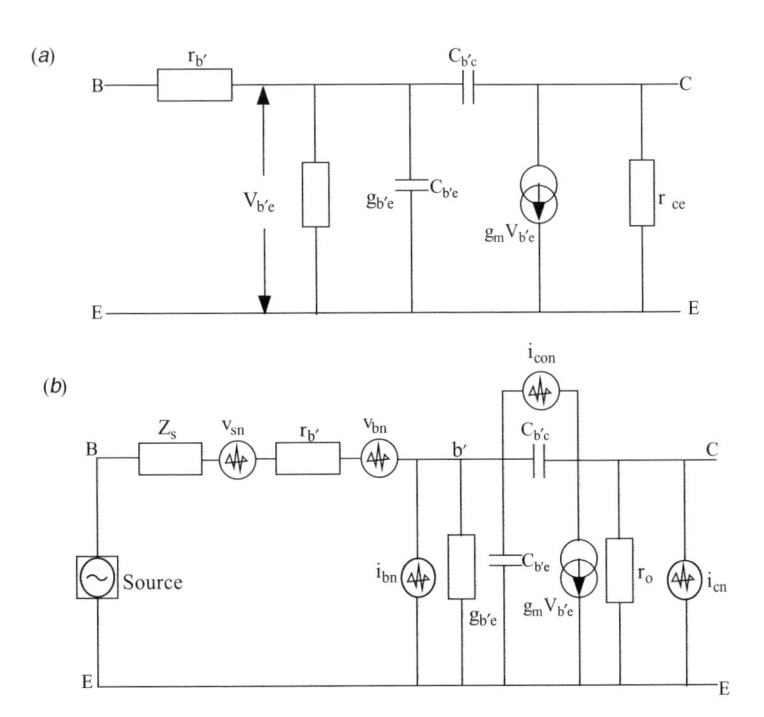

Figure 10-10 (a) π-Configuration of the GE-bipolar transistor and (b) π-configuration of the CE bipolar transistor with noise sources.

noise sources shown in the equivalent schematic (hybrid-π) in Figure 10-10b. The emitter junction in this case is conductive, generating shot noise in the emitter. The emitter current is divided into a base (I_b) and a collector current (I_c), both of which generate shot noise. Collector reverse current (I_{cob}) also generates shot noise. The emitter, base, and collector are made of semiconductor material and have a finite value of resistance associated with them, which generates thermal noise.

The value of the base resistor is relatively high in comparison to the resistance associated with the emitter and the collector, so the noise contribution of this resistor can be neglected [121]. For noise analysis, three sources are introduced in a noiseless transistor: noise generators due to fluctuation in the DC bias current (i_{bn}), the DC collector current (i_{cn}), and the thermal noise of the base resistance (v_{bn}). In a silicon transistor, the collector reverse current (I_{cob}) is very small, and the noise (i_{con}) it generates can be neglected.

For evaluation of noise performance, the signal-driving source should also be taken into consideration. Its internal conductance generates noise, and its susceptance affects the noise level through noise tuning.

The mean square value of a noise generator in a narrow frequency interval Δf is given by

$$\overline{i_{bn}^2} = 2qI_b\Delta f \tag{10-110}$$

$$\overline{i_{cn}^2} = 2qI_c\Delta f \tag{10-111}$$

$$\overline{i_{con}^2} = 2qI_{cob}\Delta f \tag{10-112}$$

$$\overline{v_{bn}^2} = 4kTr_b'\Delta f \tag{10-113}$$

$$\overline{v_{sn}^2} = 4kTR_s\Delta f \tag{10-114}$$

I_b, I_c, and I_{cob} are average DC current over the Δf noise bandwidth. The noise power spectral densities due to noise sources are given as

$$S(i_{cn}) = \frac{\overline{i_{cn}^2}}{\Delta f} = 2qI_c = 2KTg_m \tag{10-115}$$

$$S(i_{bn}) = \frac{\overline{i_{bn}^2}}{\Delta f} = 2qI_b = \frac{2KTg_m}{\beta} \tag{10-116}$$

$$S(v_{bn}) = \frac{\overline{v_{bn}^2}}{\Delta f} = 4KTr_b' \tag{10-117}$$

$$S(v_{sn}) = \frac{\overline{v_{sn}^2}}{\Delta f} = 4KTR_s \tag{10-118}$$

r_b' and R_s are base and source resistance, and Z_s is the complex source impedance.

10.5.2 Oscillator Noise Model Comments

Phase noise generation in oscillators/VCOs has been a focus of important research efforts. It is still an issue despite significant gains in practical experience and modern CAD tools for design. In the design of VCOs, minimizing phase noise is the prime task. This has been accomplished using empirical rules or numerical optimizations, which are often held as trade secrets by many manufacturers. The ability to achieve optimum phase noise performance is paramount in most RF designs, and the continued improvement of phase noise in oscillators is required for efficient use of the frequency spectrum. The degree to which an oscillator generates constant frequency throughout a specified period of time is defined as the frequency stability of the oscillator. Frequency instability is due to the presence of noise in the oscillator circuit that effectively modulates the signal, causing a change in the frequency spectrum commonly known as *phase noise*. Phase noise and timing jitter are measures of uncertainty in the output of an oscillator. Phase noise defines the frequency domain uncertainty of an oscillator, and timing jitter is a measure of oscillator uncertainty in the time domain.

The equation for an ideal sinusoidal oscillator in the time domain is given by

$$V_{out}(t) = A \cos(2\pi f_0 t + \varphi) \tag{10-119}$$

where A, f_0, and φ are the amplitude, frequency, and fixed phase of the oscillator.

The equation of the real oscillator in the time domain is given by

$$V_{out}(t) = A(t) \cos[2\pi f_0 t + \varphi(t)] \tag{10-120}$$

where $A(t)$, f_0, and $\varphi(t)$ are the time-variable amplitude, frequency, and time-variable phase of the oscillator.

Figures 10-11a and 10-11b illustrate the frequency spectrum of ideal and real oscillators, as well as the frequency fluctuation corresponding to jitter in the time domain, which is random perturbation of the zero crossing of a periodic signal.

From equations (10-119) and (10-120), the fluctuations introduced by $A(t)$ and $\varphi(t)$ are functions of time and lead to the sideband around the center frequency f_0. In the frequency domain, the spectrum of the oscillator consists of Dirac impulses at $\pm f_0$.

At present, three separate but closely related models of the oscillator phase noise exist. The first, proposed by Leeson [70], is referred to as *Leeson's model*, and the noise prediction using this model is based on LTIV (linear time invariant) properties of the oscillator, such as resonator Q, feedback gain, output power, and noise figure. This model has been enhanced by us, see equations (7-27) to (7-29). The second model was proposed by Lee and Hajimiri, based on time-varying properties of the oscillator RF current waveform. The third model was proposed by Rohde, based on the signal drive level and the conduction angle of the time-varying properties of the oscillator current waveform.

270 SYSTEMS OF COUPLED OSCILLATORS

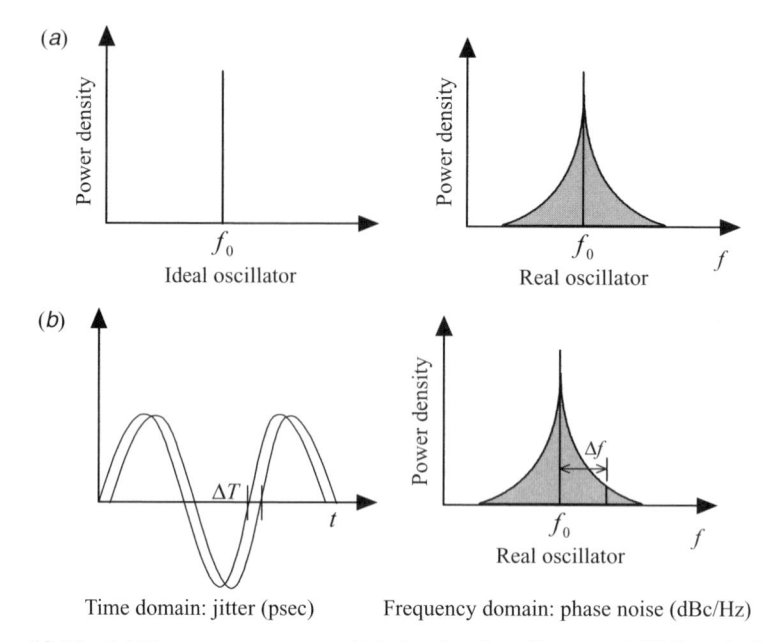

(a) Power density vs frequency for Ideal oscillator (delta at f_0) and Real oscillator (peak at f_0).

(b) Time domain: jitter (psec) showing ΔT; Frequency domain: phase noise (dBc/Hz) showing Δf around f_0 for Real oscillator.

Figure 10-11 (a) Frequency spectrum of ideal and real oscillators and (b) Jitter in the time domain related to phase noise in the frequency domain.

10.5.3 Leeson's Noise Model

Leeson's phase noise equation is given by

$$\mathscr{L}(f_m) = 10\log\left\{\left[1 + \frac{f_0^2}{(2f_m Q_L)^2\left(1 - \frac{Q_L}{Q_0}\right)^2}\right]\left(1 + \frac{f_c}{f_m}\right)\frac{FkT}{2P_0} + \frac{2kTRK_0^2}{f_m^2}\right\}$$

(10-121)

where

$\mathscr{L}(f_m)$ = ratio of sideband power in a 1 Hz bandwidth at f_m to total power in dB
f_m = frequency offset from the carrier
f_0 = center frequency
f_c = flicker frequency
Q_L = loaded Q of the tuned circuit
Q_0 = unloaded Q of the tuned circuit
F = noise factor
$kT = 4.1 \times 10^{-21}$ at 300 K (room temperature)
P_0 = average power at oscillator output
R = equivalent noise resistance of the tuning diode
K_0 = oscillator voltage gain

It is important to understand that the Leeson model is based on LTIV characteristics and is the best model since it assumes that the tuned circuit filters out all of the harmonics. In all practical cases, it is hard to predict what the operating Q and noise figure will be. The prediction power of the Leeson model is limited because the following are not known prior to measurements: the output power, the noise figure under large-signal conditions, and the loaded Q. Leeson's classic paper [70] is still an extraordinarily good design guide. The advantage of this approach is that it is easy to understand and leads to a good approximation of the phase noise. The drawback is that the values for the flicker noise contribution (a necessary input to the equation), the RF output power, the loaded Q, and the noise factor of the amplifier under large-signal conditions are not known.

10.6 NOISE ANALYSIS OF THE UNCOUPLED OSCILLATOR

The following noise analysis for the oscillator is based on the approach proposed by Rohde [132] (see Chapter 6) and is an attempt to introduce the concept of reduction of phase noise in mutually coupled oscillator systems. This chapter provides the basic noise equations for uncoupled free-running oscillator/VCOs. These equations can be extended to mutually coupled oscillator and N-coupled oscillator systems.

Figure 10-12 shows the negative resistance series-tuned model of an oscillator, Z_d is the active device impedance and the loss resistance associated with the oscillator circuit, Z_r is the resonator impedance, and $e_n(t)$ is the noise perturbation.

From the KVL, the circuit equation of Figure 10-10 is given by nonhomogeneous differential equation as

$$L\frac{\partial i(t)}{\partial t} + \frac{1}{C}\int i(t)\,dt + [R_L - R_n]i(t) = e_n(t) \tag{10-124}$$

Defining $e(t)$ as

$$e(t) = e_n(t)e^{-j(\omega_0 t + \Delta\varphi)} \tag{10-125}$$

$$V_d(t) + V_r(t) = e_n(t) \Rightarrow e_n(t)e^{-j\omega_0 t} = e^{-j\omega_0 t}[V_d(t) + V_r(t)] \tag{10-126}$$

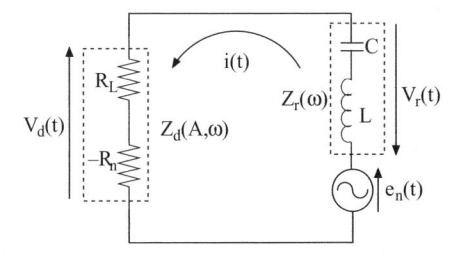

Figure 10-12 Negative resistance series tuned model of oscillator.

272   SYSTEMS OF COUPLED OSCILLATORS

Equation (10-126) can be expressed in the frequency domain as

$$V_r(\omega) + V_d(\omega) = e(\omega) = Z(\omega, A)I(\omega) \tag{10-127}$$

The circuit impedance of Figure 10-12 can be given by

$$Z(\omega, A) = Z_d(\omega, A) + Z_r(\omega) \tag{10-128}$$

Assuming that the device impedance is a function of the amplitude of the RF current, but has less variation with frequency and is assumed to be frequency independent to the first order, equation (10-128) can be rewritten as

$$Z(\omega, A) = Z_d(A) + Z_r(\omega) \tag{10-129}$$

Defining

$$\left.\frac{\partial Z(\omega, A)}{\partial \omega}\right|_{\omega_0} = \left.\frac{\partial Z_r(\omega)}{\partial \omega}\right|_{\omega_0} = \dot{Z}_{\omega_0} = |\dot{Z}_{\omega_0}|e^{j\alpha} \quad \text{and}$$

$$\left.\frac{\partial Z(\omega, A)}{\partial A}\right|_{A_0} = \left.\frac{\partial Z_d(A)}{\partial A}\right|_{A_0} = \dot{Z}_{A_0} = |\dot{Z}_{A_0}|e^{j\beta}$$

where α and β are the phases associated with resonator and device circuit parameters.

In the presence of noise perturbation $e_n(t)$, the oscillator current $i(t)$ in Figure 10-10 can be expressed as

$$i(t) = A(t)e^{j\omega_0 t} \Rightarrow i(t)|_{e_n(t)=0} = A_0 e^{j\omega_0 t} \tag{10-130}$$

$A(t)$ is defined as

$$A(t) = [A_0 + \Delta A(t)]e^{j\Delta\varphi(t)} = A_0\left[1 + \frac{\Delta A(t)}{A_0}\right]e^{j\Delta\varphi(t)} \tag{10-131}$$

where ω_0 is the free-running frequency and $\Delta A(t)$ and $\Delta \varphi(t)$ are the fluctuations in amplitude and phase of the oscillator.

The Fourier transform of $i(t)$ is derived from the time-domain equation (10-130) as

$$I(\omega) = \int_{-\infty}^{+\infty} i(t)e^{-j\omega t}dt = \int_{-\infty}^{+\infty} A(t)e^{j\omega_0 t}e^{-j\omega t}dt \tag{10-132}$$

10.6 NOISE ANALYSIS OF THE UNCOUPLED OSCILLATOR

The Fourier transform of $A(t)$ is given by

$$A(\omega) = \int_{-\infty}^{+\infty} A(t) e^{-j\omega t} dt \tag{10-133}$$

The voltage across Z_r is given by

$$V_r(t) = \int_{-\infty}^{+\infty} V_r(\omega) e^{j\omega t} d\omega = \int_{-\infty}^{+\infty} A(\omega) e^{j(\omega+\omega_0)t} Z_r(\omega+\omega_0) d\omega \tag{10-134}$$

Expanding equations (10-134) around ω_0 [145],

$$V_r(t) = e^{j\omega_0 t} \left[Z_r(\omega_0) \int_{-\infty}^{+\infty} A(\omega) e^{j\omega t} d\omega + \left.\frac{\partial Z_r}{\partial \omega}\right|_{\omega_0} \right.$$
$$\left. \times \int_{-\infty}^{+\infty} \omega A(\omega) e^{j\omega t} d\omega + \frac{1}{2} \left.\frac{\partial^2 Z_r}{\partial \omega^2}\right|_{\omega_0} \int_{-\infty}^{+\infty} \omega^2 A(\omega) e^{j\omega t} d\omega + \cdots \right] \tag{10-135}$$

From equations (10-133) and (10-135),

$$V_r(t) = e^{j\omega_0 t} \left[Z_r(\omega_0) A(t) - j \left.\frac{\partial Z_r}{\partial \omega}\right|_{\omega_0} \frac{dA(t)}{dt} - \frac{1}{2} \left.\frac{\partial^2 Z_r}{\partial \omega^2}\right|_{\omega_0} \frac{d^2 A(t)}{dt^2} + \cdots \right] \tag{10-136}$$

Similarly,

$$V_d(t) = e^{j\omega_0 t} \left[Z_d(A_0) A(t) + \left.\frac{\partial Z_d}{\partial A}\right|_{A_0} \Delta A(t) + \cdots \right] \tag{10-137}$$

From equation (10-131),

$$\frac{dA(t)}{dt} = \left[\frac{d[\Delta A(t)]}{dt} + j[A_0 + \Delta A(t)] \frac{d[\Delta \varphi(t)]}{dt} \right] e^{j\Delta \varphi(t)} \tag{10-138}$$

From equations (10-124), (10-125), (10-136), (10-137), and (10-138),

$$e_n(t) e^{-j\omega_0 t} = e^{j\Delta\varphi(t)} \left[A_0 Z(A_0, \omega_0) - j \dot{Z}_{\omega_0} \left(\frac{d[\Delta A(t)]}{dt} + jA_0 \frac{d[\Delta\varphi(t)]}{dt} \right) + \dot{Z}_{A_0} A_0 [\Delta A(t)] \right]$$

$$\Rightarrow e(t) = e_n(t) e^{-j(\omega_0 t + \Delta\varphi)} = \left[A_0 Z(A_0, \omega_0) - j\dot{Z}_{\omega_0} \right.$$
$$\left. \times \left(\frac{d[\Delta A(t)]}{dt} + jA_0 \frac{d[\Delta\varphi(t)]}{dt} \right) + \dot{Z}_{A_0} A_0 [\Delta A(t)] \right] \tag{10-139}$$

274   SYSTEMS OF COUPLED OSCILLATORS

For the case of a free-running noiseless oscillator $e_n(t) \to 0$, equation (10-139) can be rewritten as

$$A_0 Z(A_0, \omega_0) - j\dot{Z}_{\omega_0}\left(\frac{d[\Delta A(t)]}{dt} + jA_0 \frac{d[\Delta \varphi(t)]}{dt}\right) + \dot{Z}_{A_0} A_0[\Delta A(t)] = 0 \qquad (10\text{-}140)$$

At a stable point of operation, the amplitude of the oscillator is constant and the phase difference does not change with time, that is, $d[\Delta \varphi(t)]/dt = 0$ and $\Delta A(t) = 0$. From equation (10-128), the condition for oscillation, loss resistance is compensated for by the negative resistance of the active device at resonance frequency $\omega = \omega_0$, $Z(\omega_0, A_0) \to 0$, and equation (10-140) is reduced to

$$\dot{Z}_{A_0} A_0 [\Delta A(t)] - j\dot{Z}_{\omega_0}\left(\frac{d[\Delta A(t)]}{dt} + jA_0 \frac{d[\Delta \varphi(t)]}{dt}\right) = 0 \qquad (10\text{-}141)$$

From equations (10-125) and (10-139),

$$e(t) = e_n(t) e^{-j[\omega_0 + \Delta \varphi(t)]}$$

$$= A_0 Z(A_0, \omega_0) - j\dot{Z}_{\omega_0}\left(\frac{d[\Delta A(t)]}{dt} + jA_0 \frac{d[\Delta \varphi(t)]}{dt}\right) + \dot{Z}_{A_0} A_0[\Delta A(t)] \qquad (10\text{-}142)$$

At the oscillator resonance frequency, equation (10-142) can be expressed as

$$e(t) = A_0 \frac{\partial Z(\omega, A)}{\partial A}[\Delta A(t)] - j\frac{\partial Z(\omega, A)}{\partial \omega}\left(\frac{d[\Delta A(t)]}{dt} + jA_0 \frac{d[\Delta \varphi(t)]}{dt}\right) \qquad (10\text{-}143)$$

Now let us consider the noise perturbation $e(t)$ are a narrowband noise signal that can be decomposed into quadrature components $e_{n1}(t)$ and $e_{n2}(t)$. We assume they are uncorrelated, as shown in Figure 10-13.

$$e(t) = e_{n1}(t) + je_{n2}(t) \qquad (10\text{-}144)$$

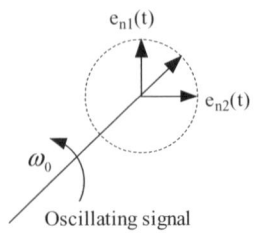

Figure 10-13 Vector presentation of the oscillator signal and its modulation by the noise signals e_{n1} and e_{n2}.

10.6 NOISE ANALYSIS OF THE UNCOUPLED OSCILLATOR 275

We define $e_{n1}(t) = e_1(t) \sin[\omega_0 t + \varphi(t)]$ and $e_{n2}(t) = -e_2(t) \cos[\omega_0 t + \varphi(t)]$, where $e_{n1}(t)$ and $e_{n2}(t)$ are orthogonal functions and $e_1(t)$ and $e_2(t)$ are slowly varying functions of time.

From equation (10-36),

$$\overline{e^2(t)} = \overline{e_{n1}^2(t)} + \overline{e_{n2}^2(t)} + \overline{2(e_{n1}(t) * e_{n2}(t))} \tag{10-145}$$

$$\overline{e_{n1}(t) * e_{n2}(t)} = 0 \Rightarrow \overline{e^2(t)} = \overline{e_{n1}^2(t)} + \overline{e_{n2}^2(t)} \tag{10-146}$$

Considering $\left|\overline{e_{n1}^2(t)}\right| = \left|\overline{e_{n2}^2(t)}\right|$, then,

$$\overline{e_{n1}^2(t)} = \frac{\overline{e^2(t)}}{2} \tag{10-147}$$

$$\overline{e_{n2}^2(t)} = \frac{\overline{e^2(t)}}{2} \tag{10-148}$$

From equation (10-144), the orthogonal components of the noise perturbation can be given by

$$e_{n1}(t) = A_0 |\dot{Z}_{A_0}| [\Delta A(t)] \cos \beta + |\dot{Z}_{\omega_0}| \left(\frac{d[\Delta A(t)]}{dt} \sin \alpha + A_0 \frac{d[\Delta \varphi(t)]}{dt} \cos \alpha \right)$$

$$\tag{10-149}$$

$$e_{n2}(t) = A_0 |\dot{Z}_{A_0}| [\Delta A(t)] \sin \beta - |\dot{Z}_{\omega_0}| \left(\frac{d[\Delta A(t)]}{dt} \cos \alpha - A_0 \frac{d[\Delta \varphi(t)]}{dt} \sin \alpha \right)$$

$$\tag{10-150}$$

From equations (10-149) and (10-150),

$$|\dot{Z}_{\omega_0}| \frac{d[\Delta A(t)]}{dt} + A_0 |\dot{Z}_{A_0}| [\Delta A(t)] \sin(\alpha - \beta) = e_{n1}(t) \sin \alpha - e_{n2}(t) \cos \alpha$$

$$\tag{10-151}$$

$$A_0 |\dot{Z}_{\omega_0}| \frac{d[\Delta \varphi(t)]}{dt} + |\dot{Z}_{A_0}| A_0 [\Delta A(t)] \cos s(\alpha - \beta) = e_{n1}(t) \cos \alpha + e_{n2}(t) \sin \alpha$$

$$\tag{10-152}$$

In the frequency domain, the operator $d/dt \to j\omega$ and noise spectral density due to amplitude and phase is obtained from equations (10-151) and (10-152) as

$$|\Delta A(\omega)|^2 = \frac{|e|^2}{2|\dot{Z}_{\omega_0}|^2 \omega^2 + 2|\dot{Z}_{A_0}|^2 A_0^2 \sin^2(\alpha - \beta)} \tag{10-153}$$

$$|\Delta\varphi(\omega)|^2 = \left[\frac{|e|^2}{2A_0^2\omega^2}\right]\left[\frac{|\dot{Z}_{\omega_0}|^2\omega^2 + |\dot{Z}_{A_0}|^2 A_0^2}{|\dot{Z}_{\omega_0}|^4\omega^2 + |\dot{Z}_{\omega_0}|^2|\dot{Z}_{A_0}|^2 A_0^2 \sin^2(\alpha-\beta)}\right] \quad (10\text{-}154)$$

where ω is the offset frequency from the carrier.

Equation (10-154) at near carrier frequency ($\omega \to 0$) is reduced to

$$|\Delta\varphi(\omega)|^2 = \frac{|e|^2}{2\omega^2 A_0^2 |\dot{Z}_{\omega_0}|^2 \sin^2(\alpha-\beta)} \quad (10\text{-}155)$$

The correlation coefficient for amplitude and phase is given by [137]

$$C = \left[\frac{\overline{\Delta A * \Delta\varphi}}{\sqrt{\overline{\Delta A^2 * \Delta\varphi^2}}}\right] = -\cos(\alpha-\beta) \quad (10\text{-}156)$$

10.7 NOISE ANALYSIS OF MUTUALLY COUPLED (SYNCHRONIZED) OSCILLATORS

Now we want to show how a simple oscillator model can be used to give a good explanation for the relative noise reduction in mutually coupled oscillator systems and to compare this with experimental results (Figure 11-8). Figure 10-14 shows

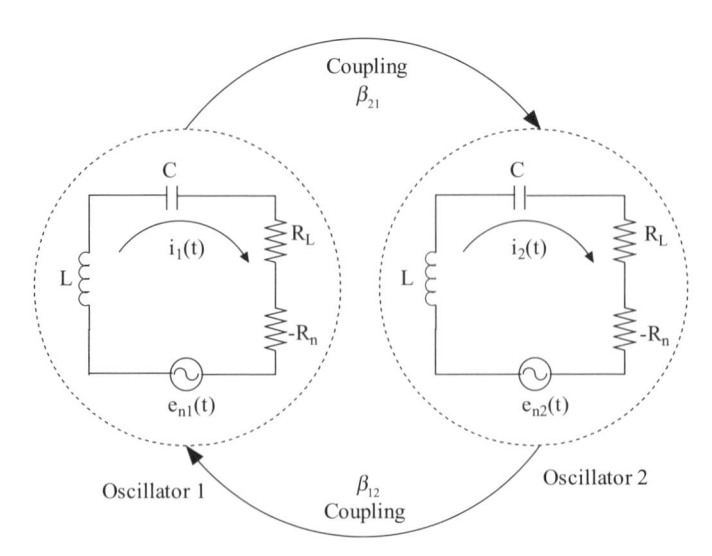

Figure 10-14 Two identical coupled oscillators coupled through an arbitrary coupling network.

10.7 NOISE ANALYSIS

the two identical mutually coupled oscillators where β_{12} and β_{21} are the coupling coefficients.

In an oscillator synchronized by a stable source, a number of theoretical and experimental treatments are available. However, relatively little is known about the mutual injection lock of two or more oscillators in coupled oscillator systems. Although the effect of noise in coupled oscillator systems has been discussed [138–148], the description was not in a convenient form for explaining the improvement in the phase noise performance of the coupled oscillator system. Figure 10-15 shows the equivalent representation of Figure 10-14, where $e_{inj}(t)$ accounts for the mutual interaction and coupling with the adjacent oscillator.

The circuit equation of Figure 10-13 is given by

$$L\frac{di(t)}{dt} + [R_L - R_n(t)]i(t) + \frac{1}{C}\int i(t)\,dt = e_{inj}(t) + e_n(t) \tag{10-157}$$

where

$$e_{n1}(t) = e_{n2}(t) = e_n(t)$$

The RF current $i(t)$ is

$$i(t) = \sum_n A_n(t)\cos[n\omega t + \varphi_n(t)] = A_1(t)\cos[\omega t + \varphi_1(t)]$$
$$+ A_2(t)\cos[2\omega t + \varphi_2(t)] + \cdots A_n(t)\cos[n\omega t + \varphi_n(t)] \tag{10-158}$$

Equation (10-157) is a nonhomogeneous differential equation. After substituting the values of $di(t)/dt$ and $\int i(t)\,dt$ and neglecting the higher-order harmonic terms, it can be rewritten as

$$L\left[-A_1(t)\left(\omega + \frac{d\varphi_1(t)}{dt}\right)\sin[\omega t + \varphi_1(t)] + \frac{dA_1(t)}{dt}\cos[\omega t + \varphi_1(t)]\right] + [(R_L - R_n(t))A_1(t)]$$
$$+ \frac{1}{C}\left\{\left[\frac{A_1(t)}{\omega} - \frac{A_1(t)}{\omega^2}\left(\frac{d\varphi_1(t)}{dt}\right)\right]\sin[\omega t + \varphi_1(t)] + \frac{1}{\omega^2}\left(\frac{dA_1(t)}{dt}\right)\cos[\omega t + \varphi_1(t)]\right\}$$
$$= e_{inj}(t) + e_n(t) \tag{10-159}$$

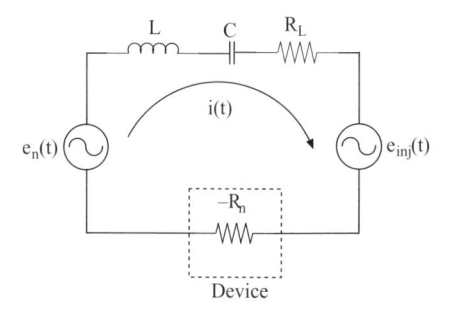

Figure 10-15 Equivalent representation of Figure 10-14.

Defining,

$$e(t) = e_{inj}(t) + e_n(t) \tag{10-160}$$

Following [11], the equations above are multiplied by $\sin[\omega t + \varphi_1(t)]$ or $\cos[\omega t + \varphi_1(t)]$ and integrated over one period of the oscillation cycle (from $t - T_0$ to t), which will give an approximate differential equation for phase $\varphi(t)$ and amplitude $A(t)$ as

$$-\frac{d\varphi(t)}{dt}\left[L + \frac{1}{\omega^2 C}\right] + \left[-\omega L + \frac{1}{\omega C}\right] = \left[\frac{2}{AT_0}\right]\left[\int_{t-T_0}^{t} [e_{inj}(t) + e_n(t)]\sin[\omega t + \varphi(t)]dt\right] \tag{10-161}$$

$$\frac{dA(t)}{dt}\left[L + \frac{1}{\omega^2 C}\right] + [R_L - \overline{R_n(t)}]A(t) = \left[\frac{2}{T_0}\right]\left[\int_{t-T_0}^{t} [e_{inj}(t) + e_n(t)]\cos[\omega t + \varphi(t)]dt\right] \tag{10-162}$$

From equation (10-53),

$$\frac{dA(t)}{dt}\left[L + \frac{1}{\omega^2 C}\right] + \gamma \Delta A A(t) = \left[\frac{2}{T_0}\right]\left[\int_{t-T_0}^{t} [e_{inj}(t) + e_n(t)]\cos[\omega t + \varphi(t)]dt\right] \tag{10-163}$$

Where $\gamma = (\Delta R / \Delta I) = [R_L - \overline{R_n(t)}]/\Delta A$, $\overline{R_n(t)}$ is the average negative resistance over one cycle and is defined as

$$\overline{R_n(t)} = \left[\frac{2}{T_0 A}\right]\int_{t-T_0}^{t} R_n(t)A(t)\cos^2[\omega t + \varphi(t)]dt \tag{10-164}$$

Since the magnitude of the higher harmonics is not significant, the subscripts of $\varphi(t)$ and $A(t)$ are dropped in equations (10-161), (10-162), and (10-163).

In equation (10-162), $[R_L - \overline{R_n(t)}]I(t) \to 0$ gives the intersection of $[\overline{R_n(t)}]$ and $[R_L]$. This value is defined as A_0, which is the minimum value of the current needed for the steady-state sustained oscillation condition. Figure 10-16 shows the plot of the nonlinear negative resistance as a function of the amplitude of the RF current.

From Figure 10-16, $\gamma \Delta I$ can be found from the intersection on the vertical axis by drawing the tangential line on $[\overline{R_n(t)}]$ at $I = A_0$, and $|\Delta I|$ decreases exponentially with time for $\gamma > 0$. Hence, A_0 represents the stable operating point. On the other hand, if $[\overline{R_n(t)}]$ intersects $[R_L]$ from the other side for $\gamma < 0$, then $|\Delta I|$ grows indefinitely with time. Such an operating point does not support stable operation [143].

$n_1(t)$ and $n_2(t)$ are defined for an uncoupled oscillator as

$$n_1(t) \to \frac{2}{T_0}\int_{t-T_0}^{t} e_n(t)\sin[\omega t + \varphi(t)]dt \tag{10-165}$$

$$n_2(t) \to \frac{2}{T_0}\int_{t-T_0}^{t} e_n(t)\cos[\omega t + \varphi(t)]dt \tag{10-166}$$

10.7 NOISE ANALYSIS

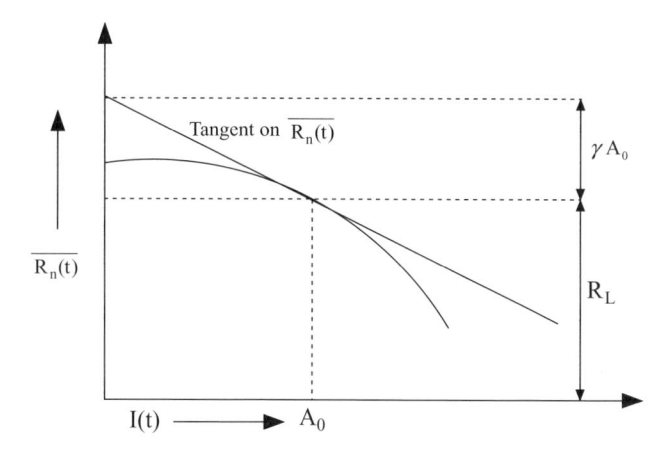

Figure 10-16 Plot of negative resistance of $[\overline{R_n(t)}]$ versus amplitude of current $I(t)$.

The time average of the square of $[n_1(t)]$ and $[n_2(t)]$ is given from equations (10-165) and (10-166) as

$$\overline{n_1^2(t)} = 2\overline{e_n^2(t)} \tag{10-167}$$

$$\overline{n_2^2(t)} = 2\overline{e_n^2(t)} \tag{10-168}$$

From equations (10-16) and (10-166), $n_1(t)$ and $n_2(t)$ are orthogonal functions, and the correlation between $n_1(t)$ and $n_2(t)$ is defined as $\rightarrow \overline{n_1(t) * n_2(t)} = 0$.

Defining, $\varphi(t) = \varphi_0 + \Delta\varphi(t)$ and $A(t) = A_0 + \Delta A(t)$, where $\Delta\varphi(t)$ and $\Delta A(t)$ are phase and amplitude fluctuations of the RF output current.

Assuming the phase and current fluctuations, $\Delta\varphi(t)$ and $\Delta A(t)$ are orthogonal function. Therefore, the correlation is given by

$$\overline{\Delta\varphi(t) * \Delta A(t)} = 0 \tag{10-169}$$

For an uncoupled free-running oscillator ($\omega \rightarrow \omega_0$), equations (10-161) and (10-163) can be given as

$$-\frac{d[\Delta\varphi(t)]}{dt}\left[L + \frac{1}{\omega^2 C}\right]_{\omega=\omega_0} + \left[-\omega L + \frac{1}{\omega C}\right]_{\omega=\omega_0}$$

$$= \left[\frac{2}{AT_0}\right]\left[\int_{t-T_0}^{t} e(t)\sin[\omega t + \varphi(t)]dt\right]_{e(t)\rightarrow e_n(t)} \tag{10-170}$$

$$\frac{d[\Delta A(t)]}{dt}\left[L+\frac{1}{\omega^2 C}\right]_{\omega=\omega_0} + \gamma\Delta AA(t)$$
$$= \left[\frac{2}{T_0}\right]\left[\int_{t-T_0}^{t} e(t)\cos[\omega t + \varphi(t)]dt\right]_{e(t)\to e_n(t)} \quad (10\text{-}171)$$

In the steady-state condition ($\omega \to \omega_0$), the phase dynamics are given from equation (10-169) as

$$-2L\frac{d\varphi(t)}{dt} = \left[\frac{1}{A}\right]n_1(t) \quad (10\text{-}172)$$

The PM noise spectral density in the frequency domain can be expressed from equation (10-169) as

$$|\Delta\varphi(\omega)|^2 = \frac{1}{4\omega^2 L^2 A^2}|n_1(\omega)|^2 \approx \frac{|e_n(\omega)|^2}{2\omega^2 L^2 A_0^2} \quad (10\text{-}173)$$

Equation (10-173) is in the same form as equation (10-155) for the case of uncorrelated amplitude and phase ($\alpha - \beta = \pi/2$).

In the steady-state condition ($\omega \to \omega_0$), the amplitude dynamics are given from equation (10-170) as

$$\left[2L\frac{dA(t)}{dt} + \gamma\Delta AA(t)\right] = n_2(t) \quad (10\text{-}174)$$

The amplitude of the RF current can be written as $i(t) = A_0 + \Delta A(t)$, where A_0 represents the stable operating point of the free-running oscillator; equation (10-174) can be given by

$$\left[2L\frac{d[\Delta A(t)]}{dt} + \Delta A(t)A_0\gamma + \Delta A^2(t)\gamma\right] = n_2(t) \quad (10\text{-}175)$$

$\Delta A^2(t)$ is negligible in comparison with A_0. Its value can be assumed to be zero for simplification in the analysis, and equation (10-175) can be rewritten as

$$2L\frac{d}{dt}[\Delta A(t)] + \Delta A(t)A_0\gamma = n_2(t) \quad (10\text{-}176)$$

The AM noise spectral density in the frequency domain can be expressed from equation (10-176) as

$$|\Delta A(\omega)|^2 = \frac{1}{[4L^2\omega^2 + (A_0\gamma)^2]}|n_2(\omega)|^2 \implies |\Delta A(\omega)|^2 = \frac{2|e_n(\omega)|^2}{[4L^2\omega^2 + (A_0\gamma)^2]}$$
$$(10\text{-}177)$$

10.7.1 Mutually Coupled (Synchronized) Oscillators

In the free-running oscillator phase, φ can take almost any value. In contrast, in the mutually coupled condition, the oscillator is synchronized with the injected signal from the adjacent oscillator, and phase φ should stay in the vicinity of φ_0 because of the restoring force due to the injected signal $e_{inj}(t)$ from the adjacent neighboring oscillator.

Defining

$$e_{inj}(t) = B\cos(\omega_{inj}t), \quad \varphi(t) = \varphi_0 + \Delta\varphi(t) \quad \text{and} \quad A(t) = A_0 + \Delta A(t)$$

For $\omega \to \omega_0$, equation (10-160) is reduced to

$$-2L\frac{d[\Delta\varphi(t)]}{dt} = \left[\frac{2}{AT_0}\right]\left[\int_{t-T_0}^{t} e_n(t)\sin[\omega t + \varphi(t)]dt + \int_{t-T_0}^{t} B\cos(\omega_{inj}t)\sin[\omega t + \varphi(t)]dt\right] \tag{10-178}$$

$$-2L\frac{d[\Delta\varphi(t)]}{dt} = \frac{1}{A_0}n_1(t) + \frac{B}{A_0}[\Delta\varphi(t)]\cos(\varphi_0) \tag{10-179}$$

where $A(t) \approx A_0$.

For $\omega \to \omega_0$, equation (10-163) is reduced to

$$2L\frac{d[\Delta A(t)]}{dt} + \gamma\Delta A A_0 = \left[\frac{2}{T_0}\right]$$
$$\left[\int_{t-T_0}^{t} e_n(t)\cos[\omega t + \varphi(t)]dt + \int_{t-T_0}^{t} B\cos(\omega_{inj}t)\cos[\omega t + \varphi(t)]dt\right] \tag{10-180}$$

$$2L\frac{d[\Delta A(t)]}{dt} + \gamma[\Delta A(t)]A_0 = n_2(t) - B[\Delta\varphi(t)]\sin(\varphi_0) \tag{10-181}$$

From equations (10-179) and (10-181), phase and amplitude noise spectral density are given by

$$|\Delta\varphi(\omega)|^2 = \frac{2|n_1(\omega)|^2}{4\omega^2 L^2 A_0^2 + B^2\cos^2(\varphi_0)} \implies |\Delta\varphi(\omega)|^2 = \frac{2|e_n(\omega)|^2}{4\omega^2 L^2 A_0^2 + B^2\cos^2(\varphi_0)} \tag{10-182}$$

$$|\Delta A(\omega)|^2 = \left[\frac{|n_2(\omega)|^2}{4L^2\omega^2 + (A_0\gamma)^2}\right] + \left[\frac{B^2\sin^2(\varphi_0)}{4L^2\omega^2 + (A_0\gamma)^2}\right]\left[\frac{|n_1(\omega)|^2}{4L^2\omega^2 A_0^2 + (B\cos\varphi_0)^2}\right] \tag{10-183}$$

$$|\Delta A(\omega)|^2 = \left[\frac{2|e_n(\omega)|^2}{4L^2\omega^2 + (A_0\gamma)^2}\right]\left[\frac{4L^2\omega^2 A_0^2 + B^2}{4L^2\omega^2 A_0^2 + (B\cos\varphi_0)^2}\right] \tag{10-184}$$

An attempt has been made to calculate the noise of two identical mutually coupled oscillators (without considering the $1/f$ noise in the oscillator).

Equations (10-173) and (10-182) show the relative improvement in phase noise when the oscillator is mutually coupled. Equations (10-177) and (10-184) show that the AM noise of the coupled oscillator is considerably increased. However, in practical applications, the increase in envelope fluctuation associated with the amplitude is controlled with the gain-limiting active device. Thus, the effect of AM noise is small in comparison to the effect of phase noise.

10.8 NOISE ANALYSIS OF N-COUPLED (SYNCHRONIZED) OSCILLATORS

Equation (10-182) shows the relative phase noise improvement for two identical mutually coupled oscillators. The purpose of this analysis is to show the relative noise improvement in the N-coupled oscillator system coupled through the arbitrary N port-coupling network. The analytical expression shows that the total phase noise of the coupled oscillator system is reduced compared to that of a single free-running oscillator in direct proportion to the number of arrays in the oscillators, provided that the coupling network is designed and optimized properly. The coupling configuration of the N-coupled oscillator system is shown in Figure 10-17.

Figure 10-18 shows the parallel negative conductance oscillator model with noise equivalent admittance Y_{noise} corresponding to the noise associated with the oscillator circuit. Noisy oscillators can be described either through the addition of equivalent noise admittance or by an equivalent noise-current generator [138, 139].

The normalized noise admittance Y_n with respect to load admittance G_L is given by

$$Y_n = \frac{Y_{noise}}{G_L} = \frac{G_{noise}}{G_L} + j\frac{B_{noise}}{G_L} = G_n + jB_n \qquad (10\text{-}185)$$

where

G_L = oscillator load admittance in the free-running state
G_n = in-phase component of the noise source
B_n = quadrature component of the noise source.

For an uncoupled free-running oscillator, G_n corresponds to the oscillator amplitude fluctuations and B_n corresponds to the phase fluctuations.

The amplitude and phase dynamics of each oscillator in the N-coupled oscillator system are given from equations (10.103) and (10.104) as

$$\frac{\partial A_i(t)}{\partial t} = A_i(t)\left[\frac{\mu_i \omega_i}{2Q_i}(\alpha_i^2 - |A_i(t)|^2)\right] + \frac{\omega_i}{2Q_i}\left[\sum_{\substack{j=1 \\ j \neq i}}^{N} \beta_{ij}A_j(t)\cos[\theta_i(t) - \theta_j(t) + \varphi_{ij}]\right]$$

$$- \left[\frac{\omega_i}{2Q_i}A_i(t)G_{ni}(t)\right]; \quad i = 1, 2, 3\ldots N \qquad (10\text{-}186)$$

10.8 N-COUPLED OSCILLATORS

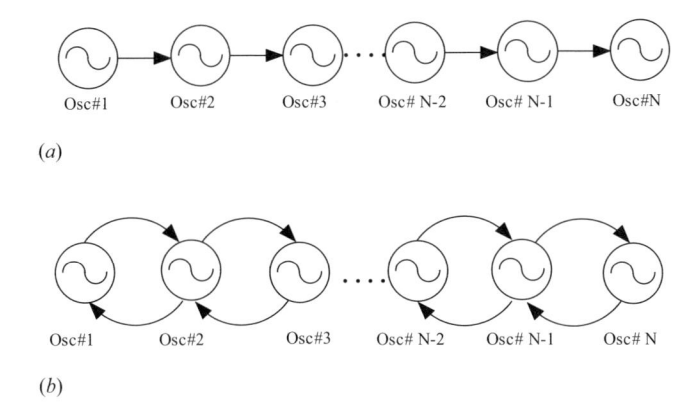

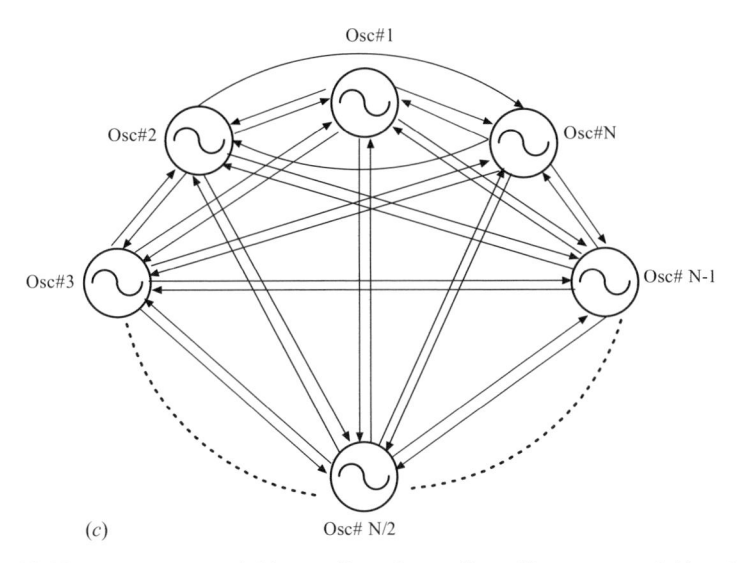

Figure 10-17 (*a*) Nearest neighbor unilateral coupling, (*b*) nearest neighbor bilateral coupling, and (*c*) global coupling.

$$\frac{\partial \theta_i(t)}{\partial t} = \omega_i - \left[\frac{\omega_i}{2Q_i}\right]\left[\sum_{\substack{j=1 \\ j \neq i}}^{N} \beta_{ij}\left(\frac{A_j(t)}{A_i(t)}\right) \sin[\theta_i(t) - \theta_j(t) + \varphi_{ij}]\right] - \left[\frac{\omega_i}{2Q_i} B_{ni}(t)\right];$$

$$i = 1, 2, 3 \ldots N \qquad (10\text{-}187)$$

where $A_i(t)$, $\theta_i(t)$, ω_i, and Q_i are the amplitude, phase, free-running frequency, and Q factor of the *i*th oscillator, respectively, and β_{ij} and φ_{ij} are the coupling parameters between the *i*th and *j*th oscillators.

Defining

$$A_i(t) = \dot{A}_i + \Delta A_i(t) \quad \text{and} \quad \theta_i(t) = \dot{\theta}_i + \Delta \theta_i(t) \qquad (10\text{-}188)$$

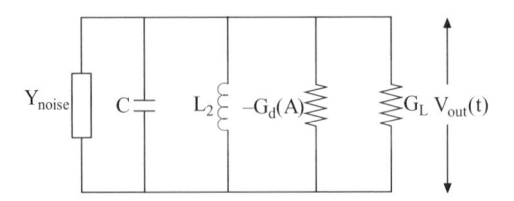

Figure 10-18 Oscillator model with noise admittance.

where $[\dot{A}, \dot{\theta}]$ are the steady-state solutions of the N-coupled oscillator coupling and $[\Delta A_i(t), \Delta\theta_i(t)]$ are the amplitude and phase fluctuations of the ith oscillator. Assuming fluctuations $\rightarrow [\Delta A_i(t), \Delta\theta_i(t)]$ are small in comparison to the steady-state solutions $\rightarrow [\dot{A}, \dot{\theta}]$, for simplification of the noise analysis, equations (10-186) and (10-187) are linearized around $(\dot{A}, \dot{\theta})$ and can be expressed as

$$\frac{\partial[\Delta A_i(t)]}{\partial t} = \mu_i \frac{\omega_i}{2Q_i}\left[[\alpha_i^2 - 3\dot{A}_i^2]\Delta A_i(t)\right] + \frac{\omega_i}{2Q_i}\left[\sum_{\substack{j=1\\j\neq i}}^{N} \beta_{ij}[\Delta A_j(t)]\cos[\dot{\theta}_i - \dot{\theta}_j + \varphi_{ij}]\right]$$

$$- \frac{\omega_i}{2Q_i}\left[\sum_{\substack{j=1\\j\neq i}}^{N} \beta_{ij}[\Delta\theta_i(t) - \Delta\theta_j(t)]\dot{A}_j \sin[\dot{\theta}_i - \dot{\theta}_j + \varphi_{ij}]\right] - \frac{\omega_i}{2Q_i}[\dot{A}_i G_{ni}(t)];$$

$$i = 1, 2, 3\ldots N \quad (10\text{-}189)$$

$$\frac{\partial[\Delta\theta_i(t)]}{\partial t} = -\frac{\omega_i}{2Q_i}\sum_{\substack{j=1\\j\neq i}}^{N}\left[\beta_{ij}\left(\frac{\dot{A}_j}{\dot{A}_i}\right)\left(\frac{\Delta A_j(t) - \Delta A_i(t)}{\dot{A}_i}\right)\sin[\dot{\theta}_i - \dot{\theta}_j + \varphi_{ij}]\right]$$

$$- \frac{\omega_i}{2Q_i}\sum_{\substack{j=1\\j\neq i}}^{N}\beta_{ij}\left([\Delta\theta_i(t) - \Delta\theta_j(t)]\left(\frac{\dot{A}_j}{\dot{A}_i}\right)\cos[\dot{\theta}_i - \dot{\theta}_j + \varphi_{ij}]\right) - \frac{\omega_i}{2Q_i}[B_{ni}(t)];$$

$$i = 1, 2, 3\ldots N \quad (10\text{-}190)$$

In the case of N identical oscillators coupled through the N port-coupling network of $\varphi_{ij} = 2\pi$, and transforming time dynamics to the frequency domain analysis by using Fourier transformation, equations (10-189) and (10-190) and can be given by

$$[\Delta A_i(\omega)] = \left[\frac{\mu\omega_i}{2Q\omega}\right][\alpha_i^2 - 3\dot{A}_i^2][\Delta A_i(\omega)] + \left[\frac{\omega_i}{2Q\omega}\right]\sum_{\substack{j=1\\j\neq i}}^{N} \beta_{ij}[\Delta A_j(\omega)]\cos[\dot{\theta}_i - \dot{\theta}_j]$$

$$- \left[\frac{\omega_i}{2Q\omega}\right]\sum_{\substack{j=1\\j\neq i}}^{N} \beta_{ij}[[\Delta\theta_i(\omega) - \Delta\theta_j(\omega)]\dot{A}_j \sin[\dot{\theta}_i - \dot{\theta}_j]] - \left[\frac{\omega_i}{2Q\omega}\right]\dot{A}_i G_{ni}(\omega);$$

$$i = 1, 2, 3\ldots N \quad (10\text{-}191)$$

$$[\Delta\dot{\theta}_i(\omega)] = -\left[\frac{\omega_i}{2Q\omega}\right]\sum_{\substack{j=1\\j\neq i}}^{N}\left[\beta_{ij}\left(\frac{\dot{A}_j}{\dot{A}_i}\right)\left(\frac{\Delta A_j(\omega)-\Delta A_i(\omega)}{\dot{A}_i}\right)\sin[\dot{\theta}_i-\dot{\theta}_j]\right]$$

$$-\left[\frac{\omega_i}{2Q\omega}\right]\sum_{\substack{j=1\\j\neq i}}^{N}\beta_{ij}([\Delta\theta_i(\omega)]-[\Delta\theta_j(\omega)]\frac{\dot{A}_j}{A_i}\cos[\dot{\theta}_i-\dot{\theta}_j])-\left[\frac{\omega_i}{2Q\omega}\right]B_{ni}(\omega);$$

$$i=1,2,3\ldots N \quad (10\text{-}192)$$

where ω is the noise offset frequency from the carrier, $\mu_i=\mu$, and $Q_i=Q$ for an N identical coupled oscillator.

From equation (10-191), the second term of the RHS represents AM noise transformed from all the $j\neq i$ oscillators to the AM noise of the ith oscillator as [138, 139]

$$\left[\frac{\omega_i}{2Q\omega}\right]\sum_{\substack{j=1\\j\neq i}}^{N}\beta_{ij}[\Delta A_j(\omega)]\cos[\dot{\theta}_i-\dot{\theta}_j] \Rightarrow \text{AM} \longrightarrow \text{AM}$$

and the third term is the conversion of PM noise to AM noise as

$$\left[\frac{\omega_i}{2Q\omega}\right]\sum_{\substack{j=1\\j\neq i}}^{N}\beta_{ij}[[\Delta\theta_i(\omega)-\Delta\theta_j(\omega)]\dot{A}_j\sin[\dot{\theta}_i-\dot{\theta}_j]] \Rightarrow \text{PM} \longrightarrow \text{AM}$$

Similarly, from equation (10-192), the first term of the RHS represents AM noise transformed to PM noise as

$$-\left[\frac{\omega_i}{2Q\omega}\right]\sum_{\substack{j=1\\j\neq i}}^{N}\left[\beta_{ij}\left(\frac{\dot{A}_j}{\dot{A}_i}\right)\left(\frac{\Delta A_j(\omega)-\Delta A_i(\omega)}{\dot{A}_i}\right)\sin[\dot{\theta}_i-\dot{\theta}_j]\right] \Rightarrow \text{AM} \longrightarrow \text{PM}$$

and the second term is conversion of PM to PM noise as

$$\left[\frac{\omega_i}{2Q\omega}\right]\sum_{\substack{j=1\\j\neq i}}^{N}\beta_{ij}\left([\Delta\theta_i(\omega)]-[\Delta\theta_j(\omega)]\frac{\dot{A}_j}{A_i}\cos[\dot{\theta}_i-\dot{\theta}_j]\right) \Rightarrow \text{PM} \longrightarrow \text{PM}$$

Equations (10-191) and (10-192) can be expressed in the following concise matrix format [18]:

$$\begin{bmatrix} \text{AM}\to\text{AM} & \text{PM}\to\text{AM} \\ \text{AM}\to\text{PM} & \text{PM}\to\text{PM} \end{bmatrix}\begin{bmatrix} \overline{\Delta A(\omega)} \\ \overline{\Delta\theta(\omega)} \end{bmatrix} = \begin{bmatrix} \overline{\dot{A}G_n(\omega)} \\ \overline{B_n(\omega)} \end{bmatrix} \quad (10\text{-}193)$$

where $\overline{\dot{A}G_n(\omega)}$ and $\overline{B_n(\omega)}$ are the in-phase or AM and quadrature or PM noise source vector of the order $N \times 1$.

The influence of PM-to-PM conversion is greater over the other terms. To simplify the analysis, contributions from the conversion of AM to AM, PM to AM and AM to PM noise conversion are considered negligible.

Assuming that all the steady-state amplitudes are identical ($\dot{A}_i = \dot{A}_j$), PM-to-PM noise conversion is considered for deriving the phase noise equation of the N-coupled oscillators. Therefore, equation (10-191) is rewritten as

$$\left[\frac{\omega}{\omega_{3dB}}\right][\Delta\theta_i(\omega)] = -\left[\sum_{\substack{j=1 \\ j \neq i}}^{N} \beta_{ij}\big([\Delta\theta_i(\omega)] - [\Delta\theta_j(\omega)]\big)\big(\cos[\dot{\theta}_i - \dot{\theta}_j]\big) - B_{ni}(\omega)\right];$$

$$i = 1, 2, 3 \ldots N \quad (10\text{-}194)$$

where

$$\omega_{3dB} = \frac{\omega_i}{2Q}$$

Equation (10-194) can be expressed in matrix form as [138]

$$[\overline{C}]\begin{bmatrix} [\Delta\theta_1(\omega)] \\ [\Delta\theta_2(\omega)] \\ \vdots \\ [\Delta\theta_{N-1}(\omega)] \\ [\Delta\theta_N(\omega)] \end{bmatrix} = \begin{bmatrix} B_{n1}(\omega) \\ B_{n2}(\omega) \\ \vdots \\ B_{N-1}(\omega) \\ B_{nN}(\omega) \end{bmatrix} \Rightarrow [\overline{C}][\overline{\Delta\theta(\omega)}] = [\overline{B_n(\omega)}] \quad (10\text{-}195)$$

where

$$[\overline{\Delta\theta(\omega)}] = \begin{bmatrix} [\Delta\theta_1(\omega)] \\ [\Delta\theta_2(\omega)] \\ \vdots \\ [\Delta\theta_{N-1}(\omega)] \\ [\Delta\theta_N(\omega)] \end{bmatrix}, \quad [\overline{B_n(\omega)}] = \begin{bmatrix} B_{n1}(\omega) \\ B_{n2}(\omega) \\ \vdots \\ B_{N-1}(\omega) \\ B_{nN}(\omega) \end{bmatrix}$$

The matrix $[\overline{C}]$ represents the arbitrary coupling topology of the N-coupled oscillator system.

The phase fluctuations of the individual oscillator can be given from equation (10-195) as

$$[\overline{\Delta\theta(\omega)}] = [\overline{C}]^{-1}[\overline{B_n(\omega)}] = [\overline{P}][\overline{B_n(\omega)}] \quad (10\text{-}196)$$

where $[\overline{P}] = [\overline{C}]^{-1}$.

Defining,

$$[\overline{P}] = \sum_{j=1}^{N} p_{ij}$$

where p_{ij} is the element of the matrix $[\overline{P}]$.
From equation (10-196),

$$[\Delta\theta_i(\omega)] = \sum_{j=1}^{N} p_{ij} B_{nj}(\omega) \qquad (10\text{-}197)$$

Noise power spectral density of the ith oscillator is calculated by taking the ensemble average of the phase fluctuation of the ith oscillator as

$$\left|[\Delta\theta_i(\omega)][\Delta\theta_i(\omega)]^*\right| = \left|\left[\sum_{j=1}^{N} p_{ij} B_{nj}(\omega)\right]\left[\sum_{j=1}^{N} p_{ij} B_{nj}(\omega)\right]^*\right| \qquad (10\text{-}198)$$

$$\left|[\Delta\theta_i(\omega)][\Delta\theta_i(\omega)]^*\right| = \sum_{j=1}^{N} |p_{ij}|^2 |B_{nj}(\omega)|^2 \qquad (10\text{-}199)$$

$$|\Delta\theta_i(\omega)|^2 = |B_n(\omega)|^2 \sum_{j=1}^{N} |p_{ij}|^2 \qquad (10\text{-}200)$$

where $|B_{ni}(\omega)|^2 = |B_{nj}(\omega)|^2 = |B_n(\omega)|^2$, assuming identical noise power spectral density from the N identical oscillator noise sources.

From equation (10-200), the phase noise of the ith oscillator is expressed by the sum of the square magnitude of the elements in the ith row of the matrix $[\overline{P}]$.

The output of the N identical coupled oscillator system synchronized to a common frequency ω_0 can be expressed as

$$V(t) = A \sum_{j=1}^{N} \cos[\omega_0 t + \Delta\theta_j(t)] = A[\cos(\omega_0 + \Delta\theta_1(t)]$$
$$+ \cdots + A[\cos(\omega_0 + \Delta\theta_{N-1}(t)] + A[\cos(\omega_0 + \Delta\theta_N(t)] \qquad (10\text{-}201)$$

Assuming small phase fluctuations, equation (10-201) can be rewritten as

$$V(t) = NA[\cos(\omega_0 t + \Delta\theta_{total}(t)] \qquad (10\text{-}202)$$

where

$$[\Delta \theta_{total}(t)] = \frac{1}{N} \sum_{j=1}^{N} [\Delta \theta_j(t)] \qquad (10\text{-}203)$$

From equations (10-197) and (10-203),

$$[\Delta \theta_{total}(\omega)] = \frac{1}{N} \sum_{j=1}^{N} \left(\sum_{i=1}^{N} p_{ij} \right) B_{nj}(\omega) \qquad (10\text{-}204)$$

From equations (10-201) and (10-203), the total phase noise of the N-coupled oscillator is given by

$$|\Delta \theta_{total}(\omega)|^2 = \frac{|B_n(\omega)|^2}{N^2} \sum_{j=1}^{N} \left| \sum_{i=1}^{N} p_{ij} \right|^2 \qquad (10\text{-}205)$$

where $\sum_{j=1}^{N} \left| \sum_{i=1}^{N} p_{ij} \right|^2 \to$ is the arbitrary coupling parameter and can be determined based on the coupling topology, which is discussed in Appendix H.

From equations (10-200) and (10-205), the total phase noise of the N-coupled oscillator in terms of the single individual oscillator is given by

$$|\Delta \theta_{total}(\omega)|^2 = \frac{1}{N} |\Delta \theta_i(\omega)|^2 \qquad (10\text{-}206)$$

Equation (10-206) shows the relative phase noise improvement for N-coupled oscillator systems, which becomes 1/N of that of the individual single oscillator. The relative noise of the N identical coupled oscillators with respect to the free-running uncoupled oscillator has been calculated (without considering the $1/f$ noise in the oscillator). For better insight into the noise analysis of the N-coupled oscillator, refer to the coupling topology discussed in Appendix H.

10.8.1 A Simple Phase Noise Analysis of the Fundamental Oscillator, Frequency Doubler/Multiplier, and Push–Push Oscillator

The following noise analysis is based on the pushing factor [183]. It is an attempt to introduce the concept of relative improvement in phase noise with respect to circuit configurations such as the fundamental oscillator, the frequency multiplier/frequency doubler, and the push–push topologies. The pushing-factor analysis is based on the assumption that the active device, low-frequency equivalent circuit is approximately constant in the LF (low-frequency) noise frequency range. In this analysis, the effect of the active device LF noise on the carrier frequency is

10.8 N-COUPLED OSCILLATORS

calculated by considering the frequency sensitivity as a small-signal noise perturbation near the nonlinear operating point of the oscillator circuit.

The pushing factor is defined as [183]

$$K_{PF} = \frac{\Delta f_n}{\Delta v_n} \quad \text{(Hz/V)} \tag{10-207}$$

where Δv_n is the input noise voltage perturbation (across the base-emitter junction of the bipolar transistor or gate-source junction of the FETs) and Δf_n is the oscillator noise frequency (square root of the frequency noise spectral density: $\Delta f_n = \sqrt{|\overline{\Delta f_n^2}|}$).

The single sideband (SSB) phase noise can be given in terms of the pushing factor [183] as

$$\mathscr{L}(f_m) = 20\log\left[\frac{\Delta f_n}{\sqrt{2}f_m}\right] = 20\log\left[\frac{K_{PF}\Delta v_n}{\sqrt{2}f_m}\right] \quad \text{(dBc/Hz)} \tag{10-208}$$

Figure 10-19 shows the Colpitts configuration of the oscillator circuit for the noise analysis based on the pushing factor.

The expression of input impedance is given as

$$Z_{IN}|_{pacakage} = -\left[\frac{Y_{21}}{\omega^2(C_1+C_p)C_2}\frac{1}{(1+\omega^2 Y_{21}^2 L_p^2)}\right] \\ -j\left[\frac{(C_1+C_p+C_2)}{\omega(C_1+C_p)C_2} - \frac{\omega Y_{21}L_p}{(1+\omega^2 Y_{21}^2 L_p^2)}\frac{Y_{21}}{\omega(C_1+C_p)C_2}\right] \tag{10-209}$$

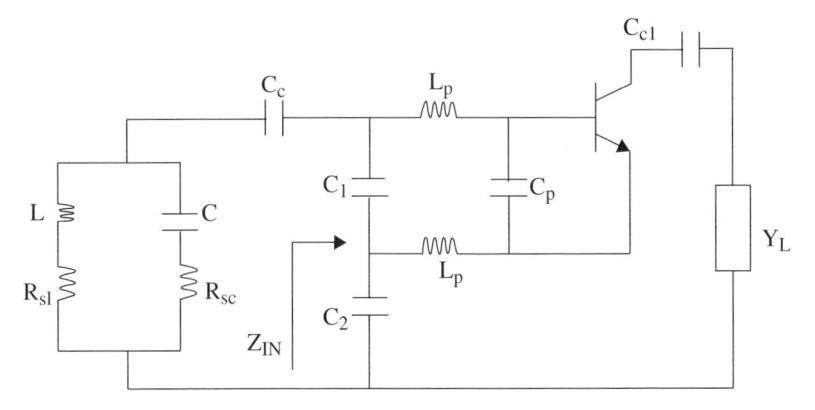

Figure 10-19 Colpitts oscillator with base-lead inductances and package capacitance.

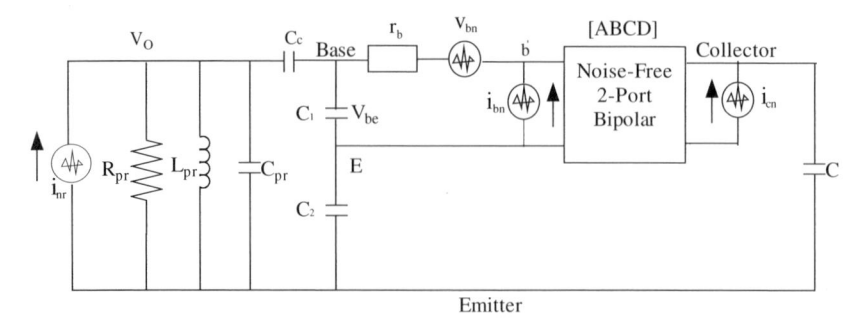

Figure 10-20 Noise equivalent model of the oscillator circuit as shown in Figure 10-18.

where L_p is the base-lead inductance and C_p is the package capacitance; Y_{21} is the large signal Y-parameter of the device and ω is the oscillator frequency.

Figure 10-20 illustrates the noise equivalent model of the oscillator circuit as shown in Figure 10-19, where the following contribute to the noise of the oscillator circuit:

- thermal noise associated with the loss resistance of the resonator (i_{nr})
- thermal noise associated with the base resistance of the transistor (i_{bn})
- shot noise associated with the base bias current (i_{bn})
- shot noise associated with the collector bias current (i_{cn})

The purpose of the present approach is to analyze the relative noise with respect to the oscillator topologies (fundamental, doubler, and push–push). Figure 10-20 can be equivalently represented in terms of a negative conductance and capacitance, which is a nonlinear function of the oscillator RF signal amplitude V_0, as shown in Figure 10-21, to simplify the noise analysis based on the pushing factor [183].

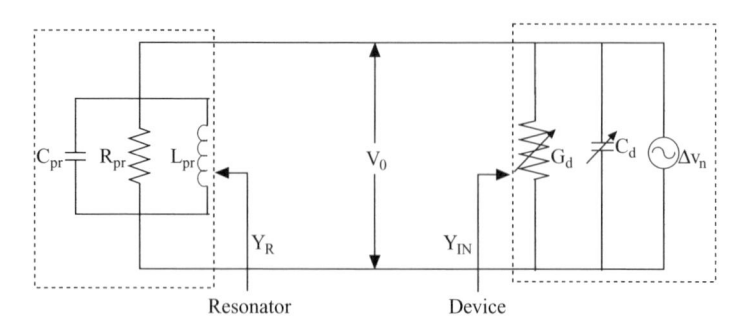

Figure 10-21 Equivalent model of the oscillator shown in Figure 10-19.

10.8 N-COUPLED OSCILLATORS

From equation (10-209), the input admittance of the circuit shown in Figure 10-21 is defined by

$$Y_{IN} = -G_d + j\omega C_d \tag{10-210}$$

where

$$G_d = \frac{\dfrac{Y_{21}}{\omega^2(C_1+C_p)C_2}\dfrac{1}{(1+\omega^2 Y_{21}^2 L_p^2)}}{\left(\dfrac{Y_{21}}{\omega^2(C_1+C_p)C_2}\dfrac{1}{(1+\omega^2 Y_{21}^2 L_p^2)}\right)^2 + \left(\dfrac{C_1+C_p+C_2}{\omega(C_1+C_p)C_2} - \dfrac{\omega Y_{21} L_p}{(1+\omega^2 Y_{21}^2 L_p^2)}\dfrac{Y_{21}}{\omega(C_1+C_p)C_2}\right)^2} \tag{10-211}$$

$$C_d = \frac{\dfrac{C_1+C_p+C_2}{\omega^2(C_1+C_p)C_2} - \dfrac{\omega Y_{21} L_p}{(1+\omega^2 Y_{21}^2 L_p^2)}\dfrac{Y_{21}}{\omega^2(C_1+C_p)C_2}}{\left(\dfrac{Y_{21}}{\omega^2(C_1+C_p)C_2}\dfrac{1}{(1+\omega^2 Y_{21}^2 L_p^2)}\right)^2 + \left(\dfrac{C_1+C_p+C_2}{\omega(C_1+C_p)C_2} - \dfrac{\omega Y_{21} L_p}{(1+\omega^2 Y_{21}^2 L_p^2)}\dfrac{Y_{21}}{\omega(C_1+C_p)C_2}\right)^2} \tag{10-212}$$

From Figure 10-21, the variation of the capacitance C_d, due to the LF noise perturbation, Δv_n, gives rise to a frequency and phase deviation of the oscillator frequency.

From equation (10-201), the pushing factor of the oscillator can be described as

$$K_{PF} = \frac{\Delta f_0}{\Delta v_n} = \frac{1}{2\pi}\left[\frac{\Delta \omega_0}{\Delta V_n}\right] = \frac{1}{2\pi}\left[\frac{\partial \omega_0}{\partial C_d}\right]\left[\frac{\partial C_d}{\partial V_n}\right] \tag{10-213}$$

where

$$\frac{\partial \omega_0}{\partial C_d} = \frac{R_p \omega_0 \omega_{reso}}{2(\omega_{reso} R_p C_p + \omega_{reso} R_p C_d)} = \frac{R_p \omega_0 \omega_{reso}}{2(Q_r + \omega_{reso} R_p C_d)}$$

$$Q_r = \omega_{reso} R_p C_p, \quad \omega_{reso} = \frac{1}{\sqrt{L_p C_p}}, \quad \omega_0 = \frac{1}{\sqrt{L_{reso}(C_d + C_r)}}$$

From equation (10-213), the pushing factor at the fundamental frequency f_0 can be given by

$$[K_{PF}]_{f=f_0} = \left[\frac{1}{2\pi}\right]\left[\frac{R_p \omega_0 \omega_{reso}}{2(Q_r + \omega_{reso} R_p C_d)}\right]\left[\frac{\partial C_d}{\partial v_n}\right] \tag{10-214}$$

Considering the same quality factor of the resonator at half of the operating frequency, $f_0/2$, the pushing factor of the oscillator circuit at $f_0/2$ can be given by

$$[K_{PF}]_{f=\frac{f_0}{2}} = \left[\frac{1}{2\pi}\right]\left[\frac{R_p \omega_0 \omega_{reso}}{4(2Q_r + \omega_{reso}R_p C_d)}\right]\left[\frac{\partial C_d}{\partial v_n}\right] \quad (10\text{-}215)$$

Assuming that the factor $\partial C_d/\partial v_n$ is independent of the operating frequency of the oscillator, the ratio of the pushing factor at f_0 and $f_0/2$ is given from equations (10-214) and (10-215) as

$$\frac{[K_{PF}]_{f=f_0}}{[K_{PF}]_{f=\frac{f_0}{2}}} = \frac{2(2Q_r + \omega_{reso}R_p C_d)}{Q_r + \omega_{reso}R_p C_d} \approx 4 \quad \text{for } Q_r > \omega_{reso}R_p C_d \quad (10\text{-}216)$$

Equation (10-216) is valid under the assumption that the resonator has a reasonably high Q factor, and the imaginary parts of the active device admittance are negligible with respect to the resonator susceptance. Therefore, $C_p > C_d$ and $Q_{reso} > \omega_{reso}R_p C_d$.

From equation (10-208),

$$\frac{[\mathscr{L}(f_m)]_{f=f_0}}{[\mathscr{L}(f_m)]_{f=f_0/2}} = \frac{20\log\{([K_{PF}]_{f=f_0}\Delta V_n)/\sqrt{2}f_m\}}{20\log\{([K_{PF}]_{f=f_0/2}\Delta V_n)/\sqrt{2}f_m\}} \Rightarrow [\mathscr{L}(f_m)]_{f=f_0}$$

$$= 12\,\text{dB} + [\mathscr{L}(f_m)]_{f=f_0/2} \quad (10\text{-}217)$$

From equation (10-217), phase noise is worsened by 12 dB/octave. This figure could be even larger because the value is calculated based on the simplified oscillator model. For real applications, the resonator and the LF noise characteristics of the active device degrade and become worse when they operate at twice the frequency. From equation (10-121), the phase noise becomes 6 dB worse at two-times multiplication (frequency doubler) with respect to the fundamental frequency of oscillation. This is the best performance of the frequency multiplier topology.

10.8.2 Push–Push Configuration

Figure 10-21 shows two identical oscillator circuits coupled through the arbitrary coupling network in the push–push configuration.

The evaluation of the pushing factor in the push–push configuration is carried out by considering uncorrelated noise voltage perturbation, Δv_{n1} and Δv_{n2}, associated with the two identical oscillator circuits, as shown in Figure 10-22. From [158–160], due to the symmetry of the push–push oscillator topology, two modes (common and differential) exist, and the corresponding pushing factor is calculated in terms of the common mode (CM) and differential mode (DM) pushing factors.

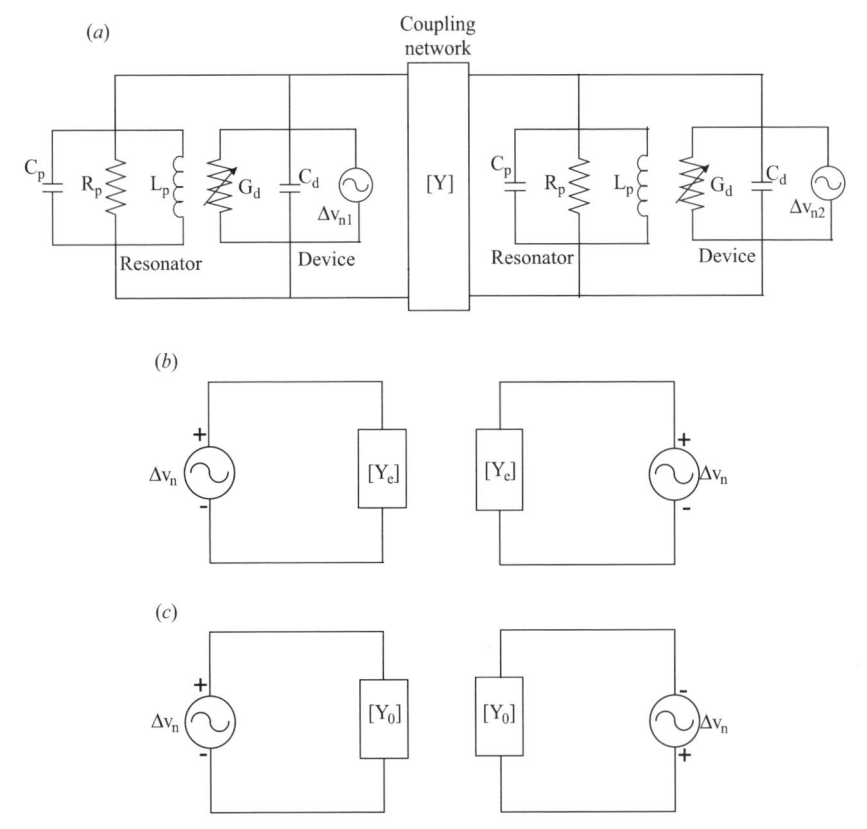

Figure 10-22 Two identical oscillator circuits coupled through the arbitrary coupling network in the push–push configuration. (a) Push–push configuration; (b) common mode (CM): $\Delta v_{n1} = \Delta v_{n2} = \Delta v_n$; (c) differential mode (DM): $\Delta v_{n1} = -\Delta v_{n2} = \Delta v_n$.

The frequency noise spectral density for the push–push topology can be given by

$$\left[\overline{\Delta f_n^2}\right]_{push-push} = \left([\Delta f_n]_{CM} + [\Delta f_n]_{DM}\right)^2 \tag{10-218}$$

$$\left[\overline{\Delta f_n^2}\right]_{push-push} = [K_{PF}]_{CM}^2 \left[\overline{\Delta v_n^2}\right]_{DM} + [K_{PF}]_{DM}^2 \left[\overline{\Delta v_n^2}\right]_{DM}$$
$$+ 2[K_{PF}]_{CM}[K_{PF}]_{DM}\left([\Delta v_n]_{CM} * [\Delta v_n]_{DM}\right) \tag{10-219}$$

where $[K_{PF}]_{CM}$ and $[K_{PF}]_{DM}$ are the CM and DM pushing factors and $[\Delta v_n]_{CM}$ and $[\Delta v_n]_{DM}$ are the CM and DM noise perturbations.

The effect of differential noise perturbation, due to the symmetry of the push–push topology, produces insignificant variation in the oscillating frequency, so $[K_{PF}]_{DM} \to 0$.

294   SYSTEMS OF COUPLED OSCILLATORS

The CM input noise perturbation of the circuit shown in Figure 10-20 can be given as

$$[\overline{\Delta v_n^2}]_{CM} = \left[\frac{\overline{\Delta v_{n1} + \Delta v_{n2}}}{2}\right]^2 = \frac{1}{4}\left([\overline{\Delta v_{n1}^2}] + [\overline{\Delta v_{n2}^2}] + [\overline{\Delta v_{n1} * \Delta v_{n2}}]\right) \quad (10\text{-}220)$$

Since the input noise voltage perturbations Δv_{n1} and Δv_{n2} associated with the two identical active devices are uncorrelated with each other, $[\overline{\Delta v_{n1} * \Delta v_{n2}}] = 0$. Because the active devices (transistors) for the two identical oscillator circuits in push–push topology operate under the same working conditions, their input noise voltage perturbation can be described by the same statistic and is given as $[\overline{\Delta v_{n1}^2}] = [\overline{\Delta v_{n2}^2}]$.

Equation (10-220) can be rewritten as

$$[\overline{\Delta V_n^2}]_{CM} = \left[\frac{\overline{\Delta v_{n1} + \Delta v_{n2}}}{2}\right]^2 = \left[\frac{[\overline{\Delta v_n^2}]}{2}\right] \quad (10\text{-}221)$$

From equation (10-219),

$$\left[\overline{\Delta f_n^2}\right]_{push-push} = [K_{PF}]_{CM}^2 [\overline{\Delta v_n^2}]_{CM} = [K_{PF}]_{CM}^2 \left[\frac{[\overline{\Delta v_n^2}]}{2}\right] \quad (10\text{-}222)$$

From equations (10-207), (10-208), and (10-222),

$$[\mathscr{L}(f_m)]_{push-push(f=f_0)} = 20\log\left\{\frac{[\Delta f_n]_{push-push}}{\sqrt{2}f_m}\right\}$$

$$= -3\,\text{dB} + 20\log\left\{\frac{[K_{PF}]_{CM}\Delta v_n}{\sqrt{2}f_m}\right\} \quad (10-223)$$

Equation (10-223) shows a 3 dB improvement in phase noise with respect to the individual oscillator oscillating at half the frequency of the push–push frequency. The analysis agrees with the general equation of the N-coupled oscillator described by Equation (10-206).

The improvement in the phase noise of the push–push topology, referring to one individual oscillator which oscillates at fundamental frequency f_0, can be expressed as

$$\frac{[\mathscr{L}(f_m)]_{push-push(f=f_0)}}{[\mathscr{L}(f_m)]_{fundamental(f=2\frac{f_0}{2})}} = \frac{20\log\left\{\frac{[\Delta f_n]_{push-push(f=f_0)}}{\sqrt{2}f_m}\right\}}{20\log\left\{\frac{[\Delta f_n]_{fundamental(f=2\frac{f_0}{2})}}{\sqrt{2}f_m}\right\}} \quad (10\text{-}224)$$

10.8 N-COUPLED OSCILLATORS

From equations (10-208) and (10-217),

$$[\mathscr{L}(f_m)]_{push-push(f=f_0)} = -9 \text{ dB} + [\mathscr{L}(f_m)]_{fundamental(f=2\frac{f_0}{2})} \quad (10\text{-}225)$$

where $f_0/2$ is the fundamental frequency of the subcircuit of the oscillator in the push–push topology.

From equation (10-225), the push–push topology gives a 9 dB improvement in the phase noise compared to the fundamental frequency of the individual oscillator oscillating at f_0, twice the designed oscillating frequency of $f_0/2$.

Example:

Figure 10-23a shows the schematic of the push–push oscillator with two identical oscillators. It consists of the two individual oscillator circuits, osc #1 and osc #2, which oscillate at half of the push–push frequency ($f_0/2$). The individual oscillator circuit corresponding to the half-resonator oscillates at $f_0/2$ (1000 MHz) and is used as a starting point to verify the above noise analysis with respect to the push–push configuration.

The circuit shown in Figure 10-23a, is fabricated on 32 mil thickness Roger substrate of dielectric constant 3.38 and loss tangent 2.7E-4. As shown in Figure 10-23a, the resonator is equivalently represented by the parallel RLC model (measured data at a self-resonance frequence of 1000 MHz as $R_p = 12000 \, \Omega$, $L_p = 5.2$ nH, $C_p = 4.7$ pF), which acts as a resonator and also provides the required phase shift of $180°$ for the synchronization.

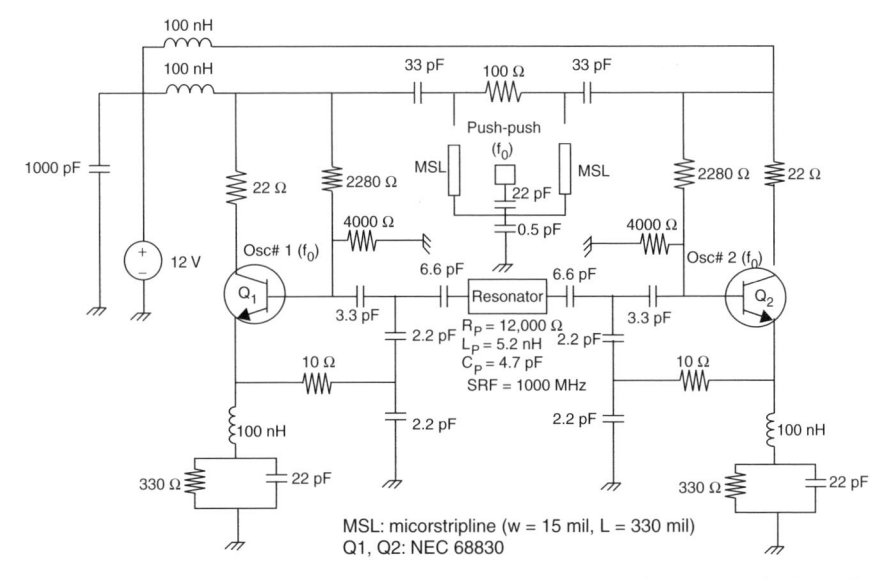

Figure 10-23a Two identical oscillator circuits coupled through the resonator in the push–push configuration.

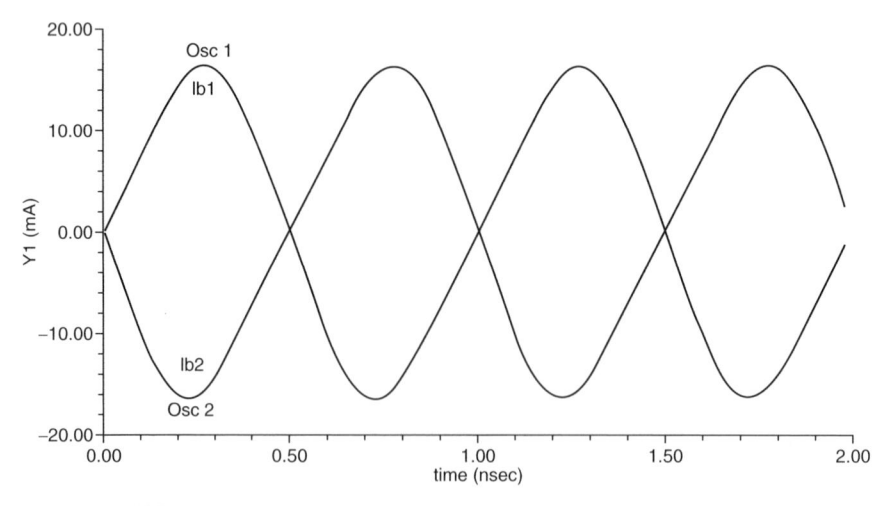

Figure 10-23b Simulated plot of the RF base current for the push–push configuration.

Figures 10-23b–d show the simulated (Ansoft Designer) plot of the RF base current, phase noise, and output power for the push–push configuration (which is phase shifted by 180° in a mutually synchronized condition ($f_0 = 2000$ MHz).

Figure 10-23e shows the schematic of the uncoupled individual oscillator circuit osc #1 for a comparative analysis of performance. The circuit is fabricated on 32 mil thickness Roger substrate of dielectric constant 3.38 and loss tangent 2.7E-4.

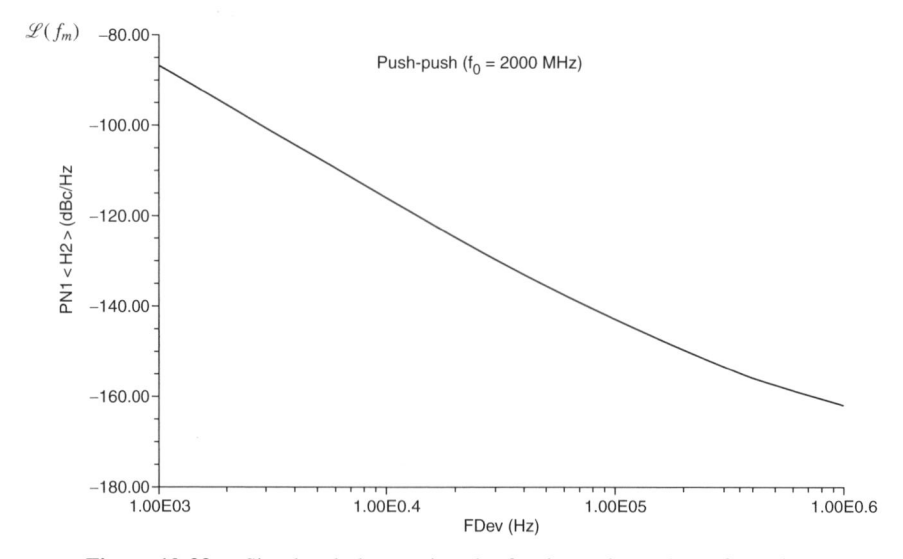

Figure 10-23c Simulated phase noise plot for the push–push configuration.

10.8 N-COUPLED OSCILLATORS   297

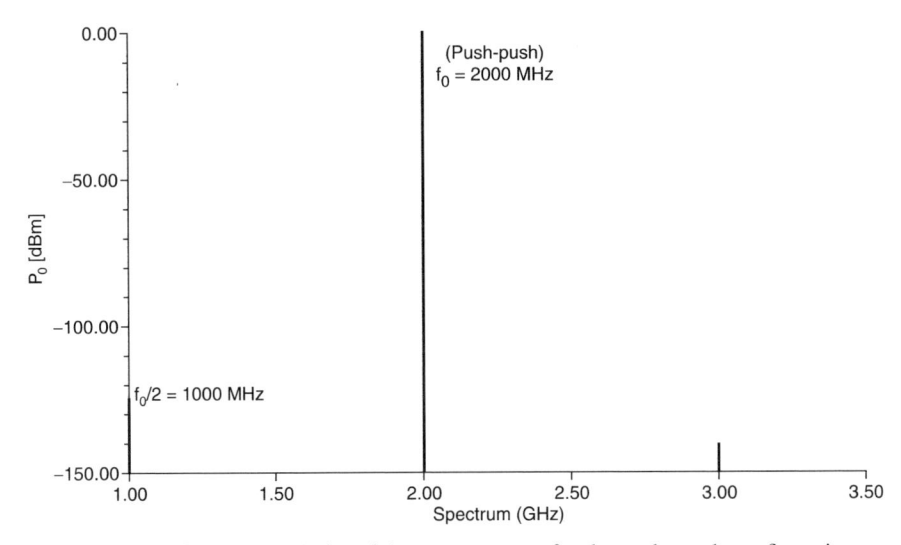

Figure 10-23d Simulated plot of the output power for the push–push configuration.

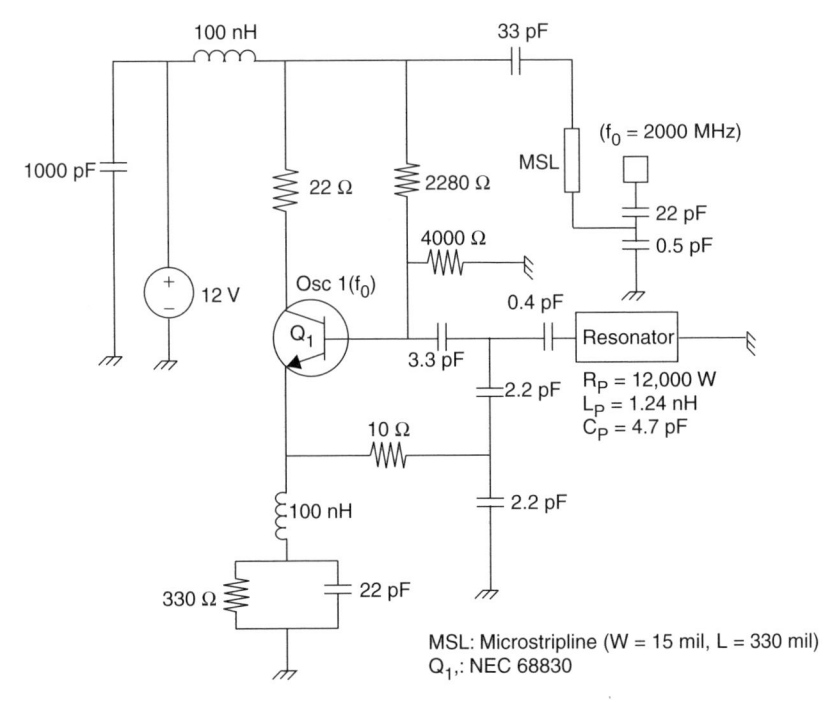

Figure 10-23e Schematic of the uncoupled individual oscillator circuit osc #1 ($f_0 = 2000$ MHz).

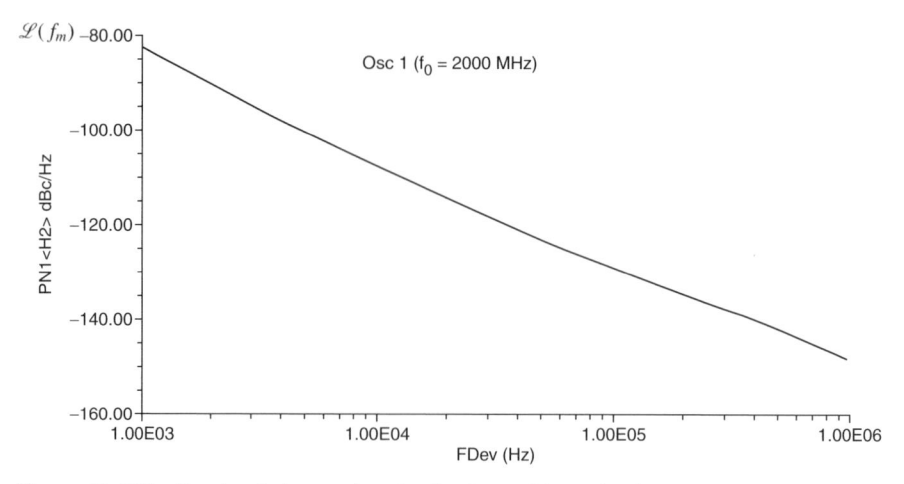

Figure 10-23f Simulated phase noise plot for the oscillator circuit shown in Figure 10-23e.

As shown in Figure 10-23e, the resonator is equivalently represented by a parallel RLC model (measured data at a self-resonance frequency of 2000 MHz as $R_p = 12000\ \Omega$, $L_p = 1.24$ nH, $C_p = 4.7$ pF).

The oscillator circuit osc #1, as shown in Figure 10-23e, oscillates at f_0 (2000 MHz) and is used as a reference for the comparative noise analysis of the push–push configuration.

Figures 10-23f and 10-23g show the simulated (Ansoft Designer) plot of the phase noise and the output power for the individual uncoupled oscillator circuit shown in Figure 10-23e.

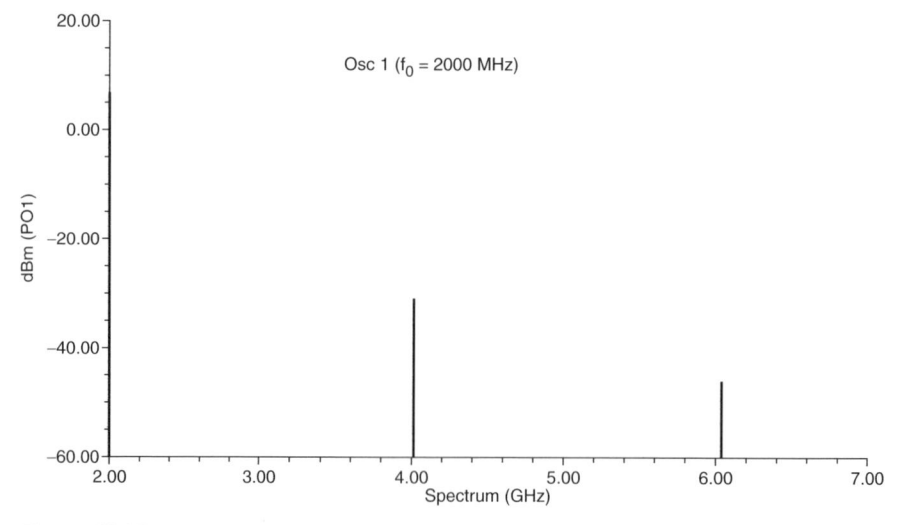

Figure 10-23g Simulated plot of the output power for the oscillator circuit shown in Figure 10-23e.

10.8 N-COUPLED OSCILLATORS 299

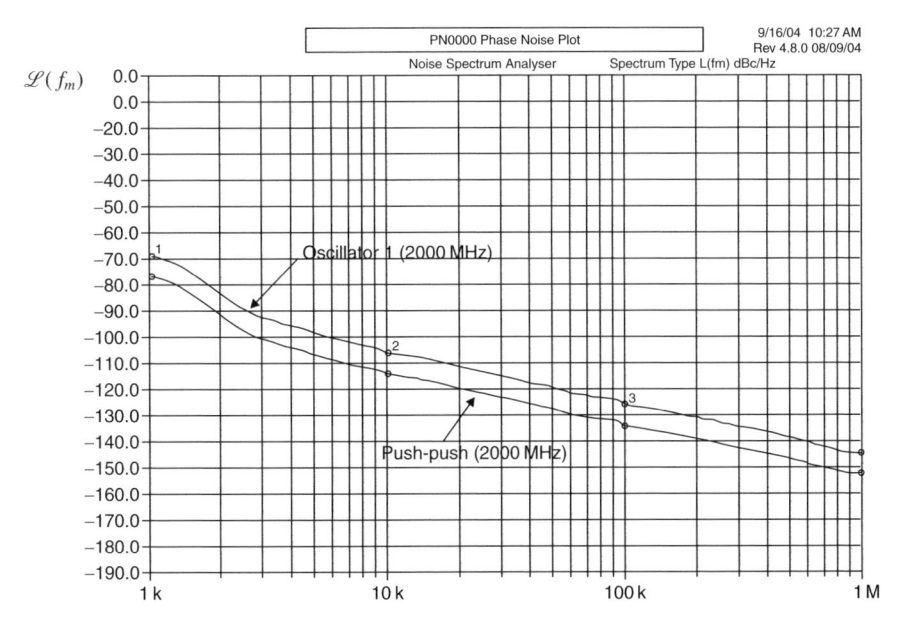

Figure 10-23h Measured phase noise plot for the push–push configuration ($f_0 = 2000$ MHz) and the uncoupled individual oscillator circuit ($f_0 = 2000$ MHz) shown in Figures 10-23a and 10-23e. Source: Eroflex Co.

From Figures 10-23c and 10-23f, the improvement of the phase noise for the push–push topology at 10 kHz offset from the carrier is 8.46 dB, which agrees with the analytical expression given within 1 dB in equation (10-225).

Figure 10-23h shows the measured phase noise plot for the push–push configuration ($f_0 = 2000$ MHz) and the uncoupled individual oscillator circuit ($f_0 = 2000$ MHz) shown in Figures 10-23a and 10-23c. The measured phase noise for the push–push configuration and the uncoupled individual oscillator at 10 kHz offset from the carrier is -113 dBc/Hz and -105 dBc/Hz, respectively.

The relative improvement in the measured phase noise of the push–push configuration, with respect to the fundamental frequency of the oscillator in the push–push topology, is 8 dB, and agrees with the simulated (Ansoft Designer) and analytical values within 1 dB.

The discrepancy of 1 dB of the measured phase noise from the analytical value, equation (10-225), can be attributed to the package parasitics, dynamic loaded Q, and tolerances of the component values of the two uncoupled individual oscillator circuits.

The above relative noise analysis gives a theoretical basis of the noise prediction as follows [187]:

- Fundamental oscillator: 12 dB degradation of the phase noise with respect to the fundamental oscillator oscillating at f_0, twice the designed oscillating frequency $f_0/2$.

300 SYSTEMS OF COUPLED OSCILLATORS

- Frequency multiplier/doubler: 6 dB degradation of the phase noise with respect to the fundamental oscillator oscillating at f_0, twice the designed oscillating frequency of $f_0/2$.
- Push–push topology ($f = 2f_0/2$): 9 dB improvement of the phase noise with respect to the fundamental oscillator oscillating at f_0, twice the designed oscillating frequency of $f_0/2$.

10.8.3 Validation

Figure 10-24 shows the schematic of the push–push oscillator with two identical individual oscillators that are oscillating at half of the push–push frequency ($f_0/2$). The individual oscillator, corresponding to the half-resonator, oscillates at $f_0/2$ (1000 MHz) and is used as a starting point to verify the above noise analysis with respect to the frequency multiplier and push–push configuration.

Figure 10-24 shows the phase noise plot of the individual oscillator operating at a fundamental frequency of $f_0/2$ (1000 MHz) and f_0 (2000 MHz), working as a frequency doubler (frequency multiplier) at 2000 MHz and a push–push configuration at 2000 MHz.

As discussed above, the phase noise of the fundamental oscillator operating at double the oscillating frequency, 2000 MHz, is worsened by 12 dB/octave with respect to the fundamental oscillator oscillating at a frequency of 1000 MHz.

The simulated graph is based on the unchanged parameters of the active device and the passive components with respect to the two-frequency $f_0/2$ and f_0. It is not easy to design the same oscillator to operate at $f_0/2$ and f_0 and maintain the same operating parameters of the active device, coupling coefficient, drive level, quality factor, and so on.

For the frequency doubler, the phase noise is degraded by 6 dB with respect to the fundamental frequency, as shown in Figure 10-25.

The relative improvement in the phase noise of the push–push configuration, with respect to the fundamental frequency of the oscillator in the push–push topology, is 9 dB. This is shown in Figure 10-25, and agrees with the theoretically predicted result [187].

10.9 N-PUSH COUPLED MODE (SYNCHRONIZED) OSCILLATORS

The increasing demand for more bandwidth to support mobile communication applications has resulted in a demand for higher operating frequencies. A high-frequency signal can be generated based either on an oscillator operating at a fundamental frequency or a harmonic oscillator. A typical oscillator operating at the fundamental frequency suffers from a low Q factor, insufficient device gain, and higher phase noise at a high frequency of operation. The cascade structure and the parallel structure are the two configurations used for harmonic oscillators. The cascade structure supports the second-harmonic oscillation based on frequency doubling. On the other

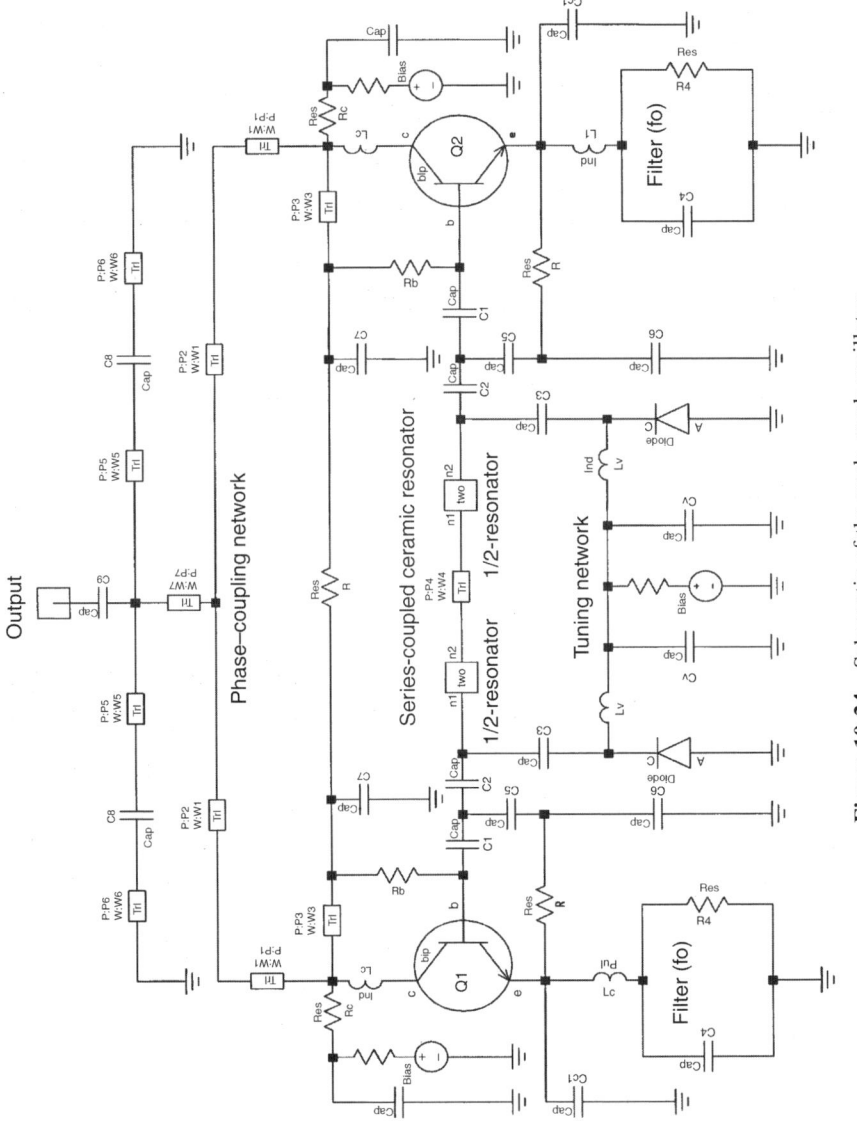

Figure 10-24 Schematic of the push–push oscillator.

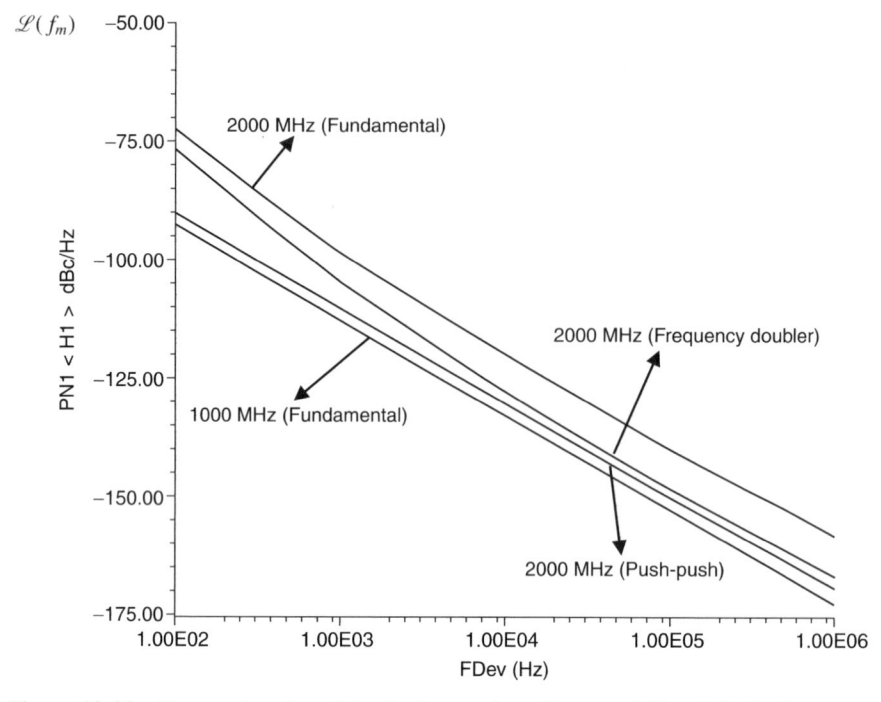

Figure 10-25 Phase noise plot of the fundamental oscillator at $f_0/2$ and f_0, the frequency doubler at f_0, and the push–push oscillator at f_0.

hand, the parallel structure supports the Nth harmonic frequency oscillation (N-push oscillator topology: double push/push–push, triple push, quadruple push ... N-push) [151] based on the N-coupled oscillator approach. The frequency doubler and other means of up-conversion may provide a practical and quick method to generate high-frequency signals from the oscillators operating at a lower frequency, but it may also introduce distortions and have poor phase noise performance. The purpose of this chapter is to provide the knowledge and working principles of the oscillators/VCOs based on the N-push coupled mode topology.

The advantages of the N-push coupled mode topology are the extended frequency generation capabilities of the transistor and the reduction in phase noise in comparison with the single oscillator by a factor of N, where N is the number of oscillator subcircuits, as given in equation (10-205). A further advantage of N-push design is that load pulling is suppressed effectively due to the separation of internal and external frequency. The drawback of the push–push oscillator is a complicated design that requires large-signal analysis to verify the odd-mode operation of the subcircuits, and the bias network has to be properly designed with respect to two critical frequencies associated with the even and odd modes of operation.

10.9.1 Push/Push–Push Oscillator

The push–push configuration [153–156] is generally used for implementing a second-harmonic oscillator in which two identical oscillators are arranged antisymmetrically. When the two out-of-phase oscillation signals are combined the fundamental frequency components are canceled out, and the second-harmonic components are enhanced and added constructively. As a push–push oscillator operates only at half the output frequency [122–131], a higher resonator Q level can be reached and low phase noise can be achieved. Furthermore, the noise sources of the two individual oscillators are uncorrelated, while the carrier powers add in phase noise. Therefore, there is a reduction of the phase noise in a push–push oscillator of 9 dB compared with the fundamental frequency of the individual oscillators in the push–push topology. Figure 10-26 shows the block diagram of the push–push topology.

The time-varying signals of the two individual oscillators are given by

$$V_1(t) = \sum_n A_n e^{jn(\omega_0 t)} = A_1 e^{j\omega_0 t} + A_2 e^{j2\omega_0 t} + A_3 e^{j3\omega_0 t} + A_4 e^{j4\omega_0 t} + \cdots + A_n e^{jn\omega_0 t}$$
(10-226)

$$V_2(t) = \sum_n A_n e^{jn[\omega_0(t-\Delta t)]} = A_1 e^{j\omega_0(t-\Delta t)} + A_2 e^{j2\omega_0(t-\Delta t)} + A_3 e^{j3\omega_0(t-\Delta t)}$$

$$+ A_4 e^{j4\omega_0(t-\Delta t)} + \cdots + A_n e^{jn\omega_0(t-\Delta t)}$$
(10-227)

$$V_{out}(t) = [V_1(t) + V_2(t)] = \left[\sum_n A_n e^{jn(\omega_0 t)} + \sum_n A_n e^{jn[\omega_0(t-\Delta t)]}\right]$$
(10-228)

$$V_{out}(t) = A_1 e^{j\omega_0 t}[1 + e^{-j\omega_0 \Delta t}] + A_2 e^{j2\omega_0 t}[1 + e^{-j2\omega_0 \Delta t}] + A_3 e^{j3\omega_0 t}[1 + e^{-j3\omega_0 \Delta t}] + \cdots$$
(10 − 229)

From equation (10-71), the phase condition for a mutually locked coupled oscillator in push–push topology is $\omega_0 \Delta t = \pi$. Therefore, equation (10-229) can

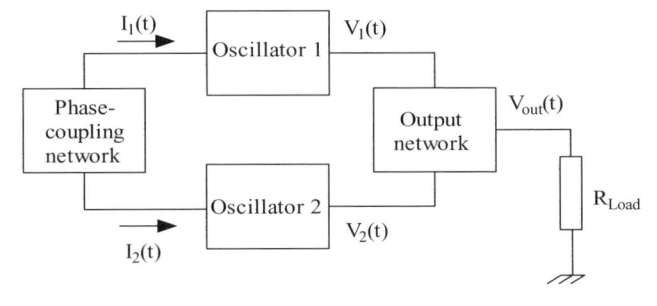

Figure 10-26 Block diagram of the push–push topology.

be rewritten as

$$[V_{out}(t)]_{push-push} = \sum_{n=1}^{2} V_n(t) = 2A_2 e^{j2\omega_0 t} + 2A_4 e^{j4\omega_0} + 2A_6 e^{j6\omega_0 t} + \cdots \quad (10\text{-}230)$$

Equation (10-230) shows the cancellation of all the odd harmonics, especially the fundamental signal, where the even harmonics are added constructively. The higher-order harmonics ($4\omega_0$, $6\omega_0$ ...) are filtered out.

10.9.2 Triple-Push Oscillator

A triple-push oscillator consists of three identical oscillators having a phase shift of 120° among the three fundamental signals oscillating at f_0 simultaneously in a mutual injection mode. The second harmonic $2f_0$ signal has a phase shift of 240°, while the third harmonic $3f_0$ signal has a phase shift of 360°, so that the fundamental signal f_0 and the second harmonic $2f_0$ signals will be canceled and the third harmonic signal $3f_0$ (the desired output signal) will combine in phase. Figure 10-27 shows the block diagram of the triple push–push topology.

The time-varying signals of each oscillator module for Figure 10-27 can be given by

$$V_1(t) = \sum_n A_n e^{jn(\omega_0 t)} = A_1 e^{j\omega_0 t} + A_2 e^{j2\omega_0 t} + A_3 e^{j3\omega_0 t} + A_4 e^{j4\omega_0 t} + \cdots + A_n e^{jn\omega_0 t}$$

$$(10\text{-}231)$$

$$V_2(t) = \sum_n A_n e^{jn(\omega_0 t - \frac{2\pi}{3})} = A_1 e^{j(\omega_0 t - \frac{2\pi}{3})} + A_2 e^{j2(\omega_0 t - \frac{2\pi}{3})} + A_3 e^{j3(\omega_0 t - \frac{2\pi}{3})} + \cdots + A_n e^{jn(\omega_0 t - \frac{2\pi}{3})}$$

$$(10\text{-}232)$$

$$V_3(t) = \sum_n A_n e^{jn(\omega_0 t - \frac{4\pi}{3})} = A_1 e^{j(\omega_0 t - \frac{4\pi}{3})} + A_2 e^{j2(\omega_0 t - \frac{4\pi}{3})} + A_3 e^{j3(\omega_0 t - \frac{4\pi}{3})} + \cdots + A_n e^{jn(\omega_0 t - \frac{4\pi}{3})}$$

$$(10\text{-}233)$$

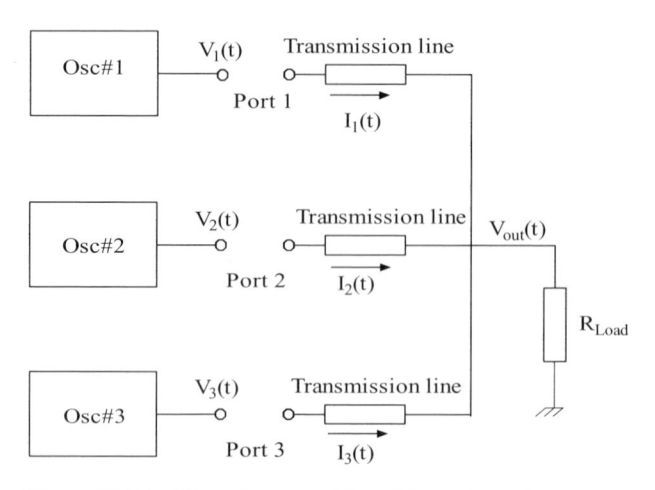

Figure 10-27 Block diagram of the triple push–push topology.

The output of the triple-push oscillator is given by

$$[V_{out}(t)]_{Triple-Push} = \sum_{n=1}^{3} V_n(t) = K_3 e^{j3\omega_0 t} + K_6 e^{j6\omega_0 t} + K_9 e^{j9\omega_0 t} + \cdots \quad (10\text{-}234)$$

Equation (10-234) shows the constructive addition of the third harmonic and its multiple higher-order harmonics. The higher-order harmonics ($6\omega_0$, $9\omega_0$ ...) are filtered out.

10.9.3 Quadruple-Push Oscillator

A quadruple-push oscillator consists of four identical oscillators having a phase shift of 90° among the four fundamental signals oscillating at f_0, simultaneously in a mutual injection mode [157]. The oscillating signal from a neighboring oscillator module is injected into another oscillator module, and is again injected into other modules, and so on, so that all the oscillator modules can oscillate in the same fundamental frequency f_0. Figure 10-28 shows a quadruple-push oscillator consisting of four identical oscillator modules using a ring resonator as a common resonator. The four identical modules oscillate in the same fundamental frequency f_0, and the fundamental oscillating signal of each suboscillator circuit has a phase difference of 90°, 180°, and 270°. Thus, the undesired fundamental signals f_0, the second harmonic signals $2f_0$, the third harmonic signals $3f_0$, the fifth harmonic signals $5f_0$, and so on can be suppressed in principle and the desired fourth harmonic signal $4f_0$ is obtained when these signals are all combined in phase.

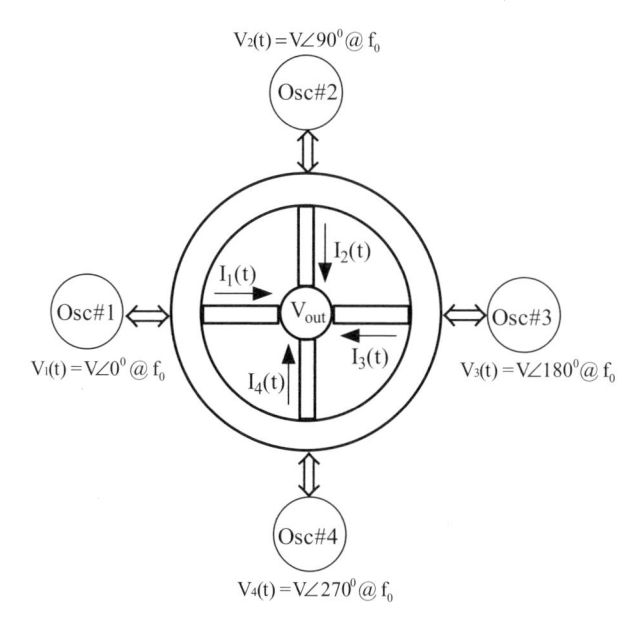

Figure 10-28 Configuration of the quadruple-push oscillator.

The time-varying oscillating signals of each oscillator module for a quadruple-push oscillator can be given as

$$V_1(t) = \sum_n A_n e^{jn(\omega_0 t)} = A_1 e^{j\omega_0 t} + A_2 e^{j2\omega_0 t}$$
$$+ A_3 e^{j3\omega_0 t} + A_4 e^{j4\omega_0 t} + \cdots + A_n e^{jn(\omega_0 t)} \quad (10\text{-}235)$$

$$V_2(t) = \sum_n A_n e^{jn(\omega_0 t - \frac{\pi}{2})} = A_1 e^{j(\omega_0 t - \frac{\pi}{2})} + A_2 e^{j2(\omega_0 t - \frac{\pi}{2})}$$
$$+ A_3 e^{j3(\omega_0 t - \frac{\pi}{2})} + A_4 e^{j4(\omega_0 t - \frac{\pi}{2})} + \cdots + A_n e^{jn(\omega_0 t - \frac{\pi}{2})} \quad (10\text{-}236)$$

$$V_3(t) = \sum_n A_n e^{jn(\omega_0 t - \pi)} = A_1 e^{j(\omega_0 t - \pi)} + A_2 e^{j2(\omega_0 t - \pi)}$$
$$+ A_3 e^{j3(\omega_0 t - \pi)} + A_4 e^{j4(\omega_0 t - \pi))} + \cdots + A_n e^{jn(\omega_0 t - \pi)} \quad (10\text{-}237)$$

$$V_4(t) = \sum_n A_n e^{jn(\omega_0 t - \frac{3\pi}{2})} = A_1 e^{j(\omega_0 t - \frac{3\pi}{2})} + A_2 e^{j2(\omega_0 t - \frac{3\pi}{2})}$$
$$+ A_3 e^{j3(\omega_0 t - \frac{3\pi}{2})} + A_4 e^{j4(\omega_0 t - \frac{3\pi}{2})} + \cdots + A_n e^{jn(\omega_0 t - \frac{3\pi}{2})} \quad (10\text{-}238)$$

The output of the quadruple-push oscillator is given by

$$[V_{out}(t)]_{quadrupule\ push} = \sum_{n=1}^{4} V_n(t) = K_4 e^{j4\omega_0 t} + K_8 e^{j8\omega_0 t} + K_{12} e^{j12\omega_0 t} + \cdots \quad (10\text{-}239)$$

The undesired fundamental signal f_0, the second harmonic signals $2f_0$, the third harmonic signals $3f_0$, and the fifth harmonic signals $5f_0$ are suppressed due to the phase relations above, while the desired fourth harmonic signals $4f_0$ are combined because of their in-phase relations. The higher-order harmonics ($8\omega_0$, $12\omega_0$ ...) are filtered out.

10.9.4 N-Push Oscillator

Figure 10-29 shows the N-push oscillator consisting of an array of N-coupled oscillator modules that share a common resonator and produce N duplicates of the oscillation signals with a 360°/N phase difference between the adjacent oscillator modules [152].

The oscillating signal from a neighboring oscillator module is injected into another oscillator module, and injected into other modules, and so on, so that all oscillator modules oscillate simultaneously in the same fundamental frequency f_0 in a mutual injection mode. After the output signals from N signal paths are combined, the desired harmonic components are added constructively and the lower-order harmonic components are canceled out due to the symmetry of the signal phases.

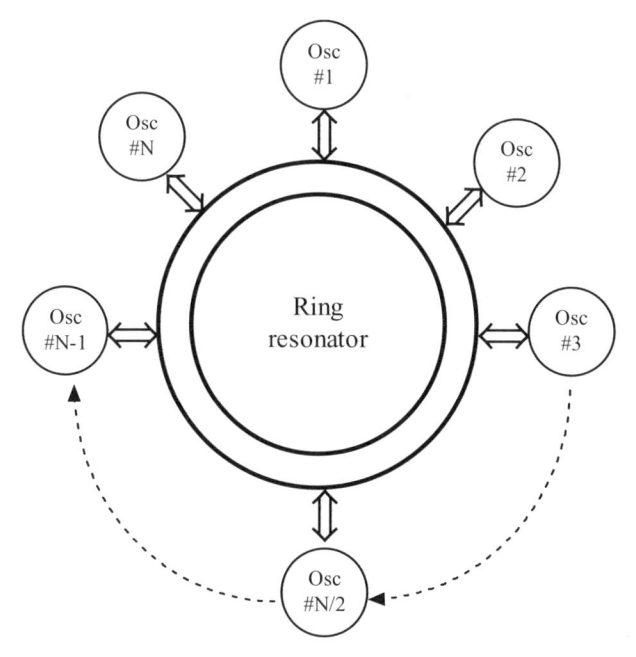

Figure 10-29 Configuration of the coupling of the N-oscillator subcircuits.

10.9.5 Mode Analysis

A system of N nonlinear oscillators can, in principle, operate in any one of N frequency modes [158–167]. However, in steady-state oscillation conditions, typically only one of these modes meets the phased criteria of the N-coupled oscillator system.

Push–push: Figure 10-30 shows the two identical oscillator circuits coupled through the arbitrary coupling network for the purpose of mode analysis.

The admittance matrix of the coupling network of Figure 10-30a is given by

$$[Y]_{coupling\ network} = \begin{bmatrix} Y_{11} & Y_{12} \\ Y_{21} & Y_{22} \end{bmatrix} \quad (10\text{-}240)$$

The circuit equation for Figure 10-28 is given in matrix form as

$$[I] = [Y]_{coupling\ network}[V] \Rightarrow \begin{bmatrix} I_1 \\ I_2 \end{bmatrix} = \begin{bmatrix} Y_{11} & Y_{12} \\ Y_{21} & Y_{22} \end{bmatrix} \begin{bmatrix} V_1 \\ V_2 \end{bmatrix} \quad (10\text{-}241)$$

assuming that the coupling network shown in Figure 10-28a is symmetrical, and from the property of symmetry $\rightarrow Y_{11} = Y_{22}$ and from the reciprocity $\rightarrow Y_{12} = Y_{21}$.

From equation (10-241),

$$I_1 = Y_{11}V_1 + Y_{12}V_2 \quad \text{and} \quad I_2 = Y_{21}V_1 + Y_{22}V_2 \quad (10\text{-}242)$$

308   SYSTEMS OF COUPLED OSCILLATORS

(a) Coupled oscillator

(b) Even mode: $V_1 = V_2$

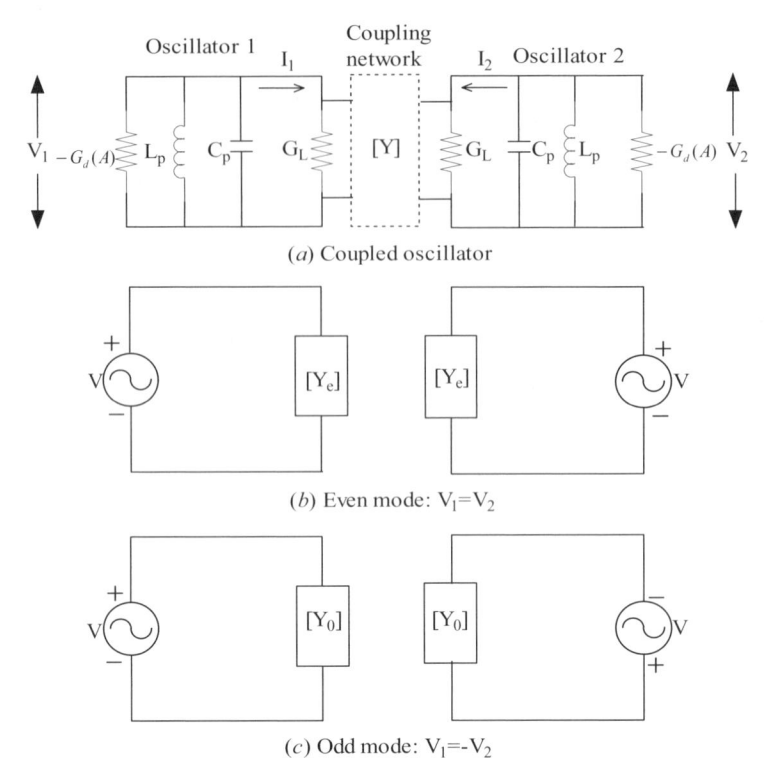

(c) Odd mode: $V_1 = -V_2$

Figure 10-30   *(a)* Two identical oscillator circuits coupled through an arbitrary coupling network; *(b)* even-mode operation; *(c)* odd-mode operation.

For even mode $\Rightarrow V_1 = V_2 = V$

$$I_1 = [Y_{11} + Y_{12}]V = [Y_e]V \quad \text{and} \quad I_2 = [Y_{21} + Y_{22}]V = [Y_e]V \qquad (10\text{-}243)$$

For odd mode $\Rightarrow V_1 = V_2 = -V$

$$I_1 = [Y_{11} - Y_{12}]V = [Y_0]V \quad \text{and} \quad I_2 = [Y_{21} - Y_{22}]V = [Y_0]V \qquad (10\text{-}244)$$

The circuit analysis for the push–push coupled oscillator is now reduced to the simpler consideration of the equivalent loads (even or odd mode) in the independent oscillators. For the calculation of eigenvalues and eigenvectors associated with the mode analysis, Figure 10-30 is transformed into a series-equivalent configuration, as shown in Figure 10-31.

10.1 N-PUSH COUPLED MODE OSCILLATORS

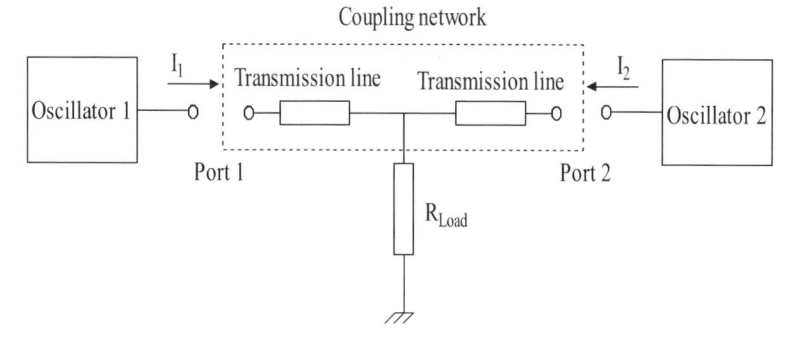

Figure 10-31 Equivalent configuration of the push–push oscillator.

The associated Z-parameter matrix equation for Figure 10-31 is given by

$$\begin{bmatrix} Z_{11} & Z_{12} \\ Z_{21} & Z_{22} \end{bmatrix} \begin{bmatrix} I_1 \\ I_2 \end{bmatrix} = \begin{bmatrix} V_1 \\ V_2 \end{bmatrix} \quad (10\text{-}245)$$

From reciprocity $Z_{12} = Z_{21}$ and from symmetry $Z_{11} = Z_{22}$.

For the even mode, the eigenvalue and the associated eigenvector are given from properties of the matrix as

$$\lambda_e = Z_{11} + Z_{12} \longrightarrow \text{eigenvalue} \quad (10\text{-}246)$$

$$\begin{bmatrix} I_1 \\ I_2 \end{bmatrix} = \begin{bmatrix} 1 \\ 1 \end{bmatrix} \longrightarrow \text{eigenvector} \quad (10\text{-}247)$$

For the odd mode, the eigenvalue and the associated eigenvector are given by

$$\lambda_o = Z_{11} - Z_{12} \longrightarrow \text{eigenvalue} \quad (10\text{-}248)$$

$$\begin{bmatrix} I_1 \\ I_2 \end{bmatrix} = \begin{bmatrix} 1 \\ -1 \end{bmatrix} \longrightarrow \text{eigenvector} \quad (10\text{-}249)$$

The eigenvectors are orthogonal [162] and linearly independent. They represent the two modes (even and odd) for the circuit configuration shown in Figures 10-30b and 10-30c.

The even mode represents the normal circuit operation where the currents and voltages at each port are in phase, whereas the odd mode represents the circuit operation in which the port currents and voltages are 180° out of phase. For a twofold symmetric structure (N = 2), one in-phase (even mode) and one out-of-phase (odd mode) exists. For push–push operation, the even mode is quenched and the odd mode is excited, so that the fundamental odd-mode currents cancel out and produce an in-phase constructive combining of the second harmonic signals.

310 SYSTEMS OF COUPLED OSCILLATORS

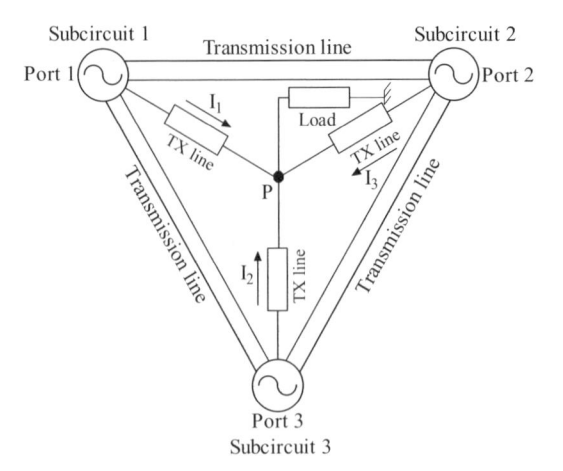

Figure 10-32 Circuit topology of the triple-push oscillator/VCOs.

Triple-push oscillator/VCOs: Figure 10-32 shows the circuit topology of the triple-push oscillator. The associated Z-parameter matrix equation of the triple-push circuit is given as

$$\begin{bmatrix} Z_{11} & Z_{12} & Z_{13} \\ Z_{21} & Z_{22} & Z_{23} \\ Z_{31} & Z_{32} & Z_{33} \end{bmatrix} \begin{bmatrix} I_1 \\ I_2 \\ I_3 \end{bmatrix} = \begin{bmatrix} V_1 \\ V_2 \\ V_3 \end{bmatrix} \qquad (10\text{-}250)$$

From the properties of symmetry and reciprocity, the three-port network is $Z_{ij} = Z_{ji}$, $i, j = 1, 2, 3$, and equation (10-250) can be rewritten as

$$\begin{bmatrix} Z_{11} & Z_{12} & Z_{13} \\ Z_{12} & Z_{11} & Z_{12} \\ Z_{12} & Z_{12} & Z_{11} \end{bmatrix} \begin{bmatrix} I_1 \\ I_2 \\ I_3 \end{bmatrix} = \begin{bmatrix} V_1 \\ V_2 \\ V_3 \end{bmatrix} \qquad (10\text{-}251)$$

For the Z-matrix in equation (10-251), the three eigenvalues and associated eigenvectors are given by

$$\lambda_e = Z_{11} + 2Z_{12}, \quad \begin{bmatrix} I_1 \\ I_2 \\ I_3 \end{bmatrix} = \begin{bmatrix} 1 \\ 1 \\ 1 \end{bmatrix} \quad \text{for the even mode} \qquad (10\text{-}252)$$

$\lambda_o = Z_{11} - Z_{12}$ (double root), $\quad I_1 + I_2 + I_3 = 0 \quad$ for the odd mode

$$(10\text{-}253)$$

From equation (10-253), two independent modes can be selected. Assuming that the magnitudes of the odd-mode currents (I_1, I_2, and I_3) are equal due to the

10.1 N-PUSH COUPLED MODE OSCILLATORS 311

symmetry of each subcircuit, each odd-mode current must have a 120° phase shift to every other to satisfy equation (10-253).

Based on this assumption, the two odd modes can be uniquely determined as

$$\begin{bmatrix} 1 \\ e^{\frac{j2\pi}{3}} \\ e^{\frac{j4\pi}{3}} \end{bmatrix} \text{ and } \begin{bmatrix} 1 \\ e^{\frac{-j2\pi}{3}} \\ e^{\frac{-j4\pi}{3}} \end{bmatrix} \quad (10\text{-}254)$$

The vector expressions of one even mode and two odd modes are shown in Figure 10-33.

10.9.6 Design Analysis

10.9.6.1 Push–Push/Two-Push Oscillator Figure 10-34 shows the equivalent schematic of the push–push voltage-controlled oscillator. For stable operation under push–push conditions, the transistors must oscillate in odd mode (+ −), and this condition can be assured by the proper choice of the terminating impedance and the gain of the transistor at the fundamental frequency. Under push–push operation, the load resistance absorbs the oscillation power for even mode; hence, even-mode oscillation is prevented. In odd mode, the common port becomes a virtual ground, whereas in even mode (unwanted), the impedance becomes $2Z_L$. In odd-mode operation the center point P becomes a virtual ground, and the circuit can be redrawn as in Figure 10-35. In this circuit, each transistor is represented as a series combination of negative resistance and its input capacitive

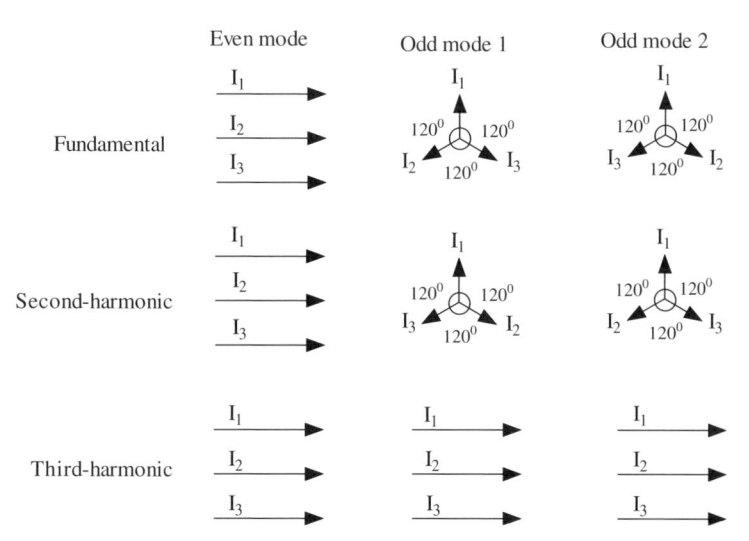

Figure 10-33 Vector expressions of the mode currents (I_1, I_2, and I_3) of a triple-push oscillator.

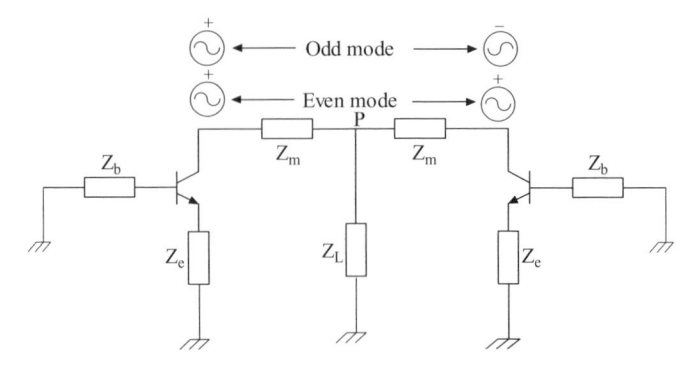

Figure 10-34 Equivalent schematic of the push–push oscillator.

reactance. Series resistance R_s equivalently represents losses in the microstrip and the varactor.

Therefore, the design of the push–push oscillator is as follows: First, one suboscillator will oscillate in the odd mode, while it is stable in the even mode. Then the combined suboscillators will also operate in the odd mode and provide a proper push–push oscillation. The steady-state oscillation condition is described in terms of the impedance of the oscillator Z_{osc} and its load Z_L. For steady-state oscillation, the sum of these impedances has to be zero at the frequency of oscillation and is given by

$$Z_{osc} + Z_L \leq 0 \qquad (10\text{-}255)$$

For the startup, the sum of the oscillator impedance Z_{osc} and the load impedance Z_L have to be less than zero, and due to the nonlinearity associated with the active device, the oscillator runs into saturation and finally reaches the steady-state condition.

The load resistance R_L can be redrawn as the parallel combination of the two $2R_L$ resistors at the center point P. Since both of the subcircuits are in phase at the

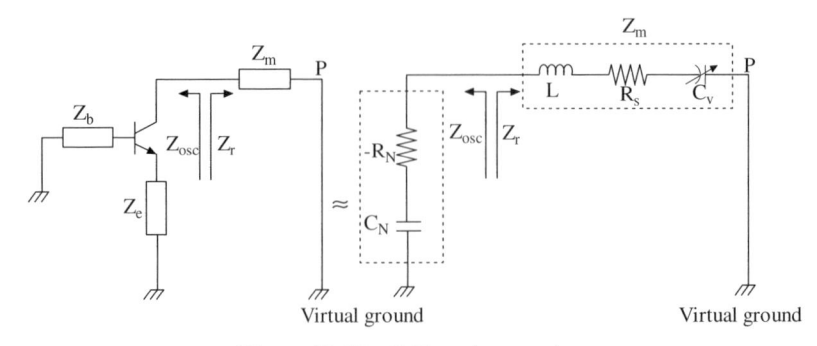

Figure 10-35 Odd-mode operation.

10.1 N-PUSH COUPLED MODE OSCILLATORS 313

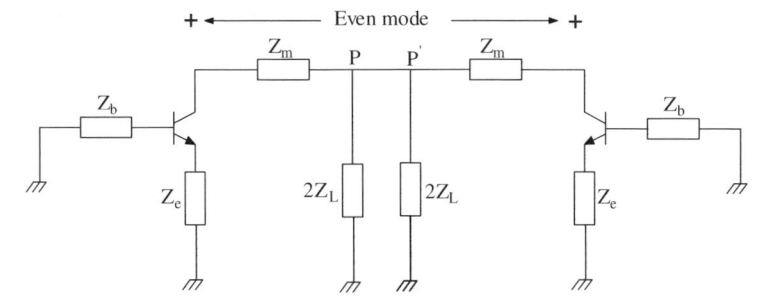

Figure 10-36 Even-mode push–push oscillator to a common load.

fundamental, the current between the center point P and P' is zero, and the circuit may be split into two independent halves, as shown in Figure 10-37.

For the push–push oscillator, equation (10-255) can be given as

$$\sum Z_{odd\ mode} = 0 \qquad (10\text{-}256)$$

For a guaranteed stable odd-mode operation, unwanted even-mode oscillation must be suppressed. This condition is given by

$$R[Z_{even\ mode} + 2Z_L] > 0 \qquad (10\text{-}257)$$

For the circuit to start oscillating, the negative resistance of the transistor must be greater than the positive losses in the tuning elements and can be given by

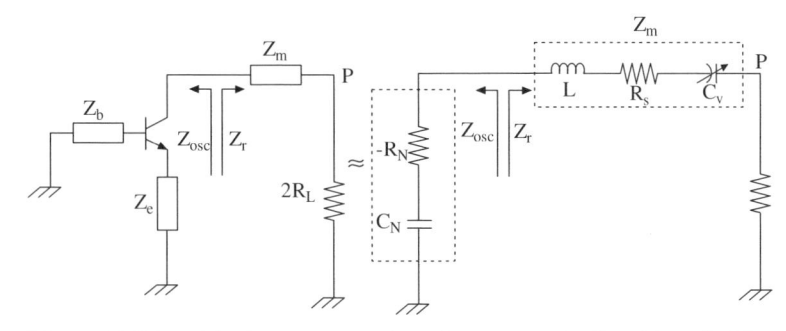

$$|-R_N| > R_s \qquad (10\text{-}258)$$

Figure 10-37 Equivalent representation of even-mode half-oscillator/VCOs.

The frequency of the oscillation for the circuit shown in Figure 10-35 is given by

$$\frac{1}{j\omega C_N} + \frac{1}{j\omega C_v} + j\omega L = 0 \qquad (10\text{-}259)$$

In order to eliminate the possibility of oscillating in the even mode (+ +), a further restriction must be placed on the negative resistance of the transistor.

Figure 10-36 shows the even mode operation of the push–push topology. The condition for this circuit corresponding to nonoscillation in the even mode is that the negative resistance of the active device must be less than the positive losses in the circuit and can be given by

$$|-R_N| < R_s + 2R_L \qquad (10\text{-}260)$$

Therefore, for stable oscillation in the push–push topology, Equations (10-258) and (10-260) yield the total condition of the active device negative resistance to maintain odd-mode oscillation.

$$R_s < |-R_N| < R_s + 2R_L \qquad (10\text{-}261)$$

The losses, R_s, are on the order of 2–5 ohms, and twice the load resistance is 100 ohms (assuming a 50 ohm load). If the device has more than $(R_s + 2R_L) \approx 110\ \Omega$ of negative resistance, matching can be used to increase the load resistance at the fundamental or the emitter capacitance can be increased to reduce the transistor gain.

10.9.6.2 Triple-Push Oscillator Triple-push oscillator/VCOs need to excite the odd mode, while the even mode is suppressed. Thus, the fundamental odd-mode currents cancel out at the output-combining node due to a 120° phase shift among I_1, I_2, and I_3, whereas the second harmonic signal currents, which have a 240° phase shift among I_1, I_2, and I_3, also cancel out, as shown in Figures 10-32 and 10-33. Consequently, the third harmonic signals combine in-phase at the output load. The even mode represents the normal circuit operation where the currents and the voltages are in-phase at the output node P, as shown in Figure 10-32, whereas the odd mode represents circuit operation in which the port currents and voltages are 180° out of phase.

For the even mode, the load resistance R_L appears to be three times 50 ohms ($R_{L\ Even} = 150$ ohms), whereas for the odd mode, the voltages that have a 120° phase shift for one another make output node P behave as a virtual ground ($R_{L\ Odd} = 0$ ohm). Figure 10-38 shows the even- and odd-mode equivalent circuits with 150 and 0 ohm load impedance, respectively.

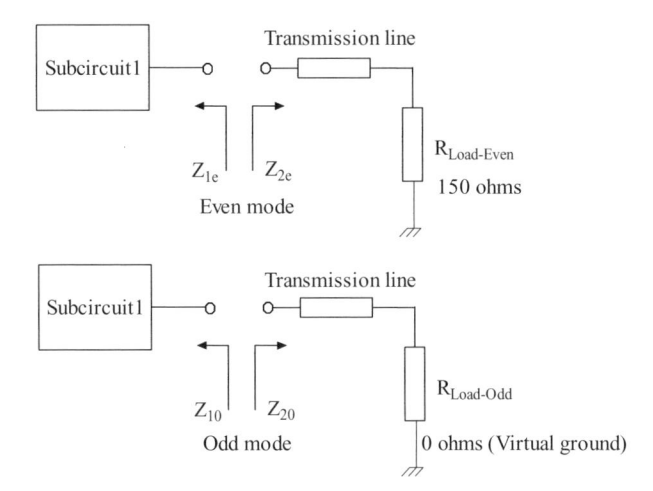

Figure 10-38 Even- and odd-mode equivalent circuits.

With a different equivalent load impedance, the following conditions must be satisfied for triple-push operation [167]:

$$\text{Re}[Z_{1e} + Z_{2e}] > 0 \quad \text{for the even mode} \tag{10-262}$$

$$\text{Re}[Z_{10} + Z_{20}] < 0 \tag{10-263}$$

$$\text{Im}[Z_{10} + Z_{20}] = 0 \quad \text{for the odd mode} \tag{10-264}$$

where Z_{1e} and Z_{2e} are the equivalent impedances looking into port 1 and port 2 for the even-mode and Z_{10} and Z_{20} are the equivalent impedances looking into port 1 and port 2 for the odd-mode equivalent circuit in Figure 10-38.

10.10 ULTRA-LOW-NOISE WIDEBAND OSCILLATORS

10.10.1 Wideband VCO Design Approach

VCOs with a wide tuning range (more than octave band) and low phase noise are essential building blocks for next-generation wireless communication systems. The VCO is one of the most important blocks of the phase-locked loop (PLL)-based frequency synthesizer because its performance is determined inside the loop bandwidth by the loop and outside the loop bandwidth by the phase noise of the VCO. It is, therefore, of major importance to build a low-phase-noise integrated wideband oscillator that operates with low power consumption.

Typical oscillator designs for wideband VCOs use a grounded base or grounded collector circuit for generating a negative resistance at one port, which is usually terminated with a parallel or series LC-resonant circuit. The main challenge in this design is to generate negative resistance over the wide tuning range, which cannot be easily extended to more than an octave band. For octave band tunability, the required negative resistance over the band is generated by the feedback base inductance (in the grounded base topology), but the polarity of the reactance may change over the frequency band and can lead to the disappearance of the negative resistance as the operating frequency exceeds its self-resonant frequency (SRF). Furthermore, the low Q of commercially available surface-mounted device (SMD) inductors and tuning diodes degrades the phase noise performance over the band. Figure 10-39 shows the series and parallel configurations of the oscillator.

Figure 10-40 shows the series feedback grounded base topology for analysis of wideband VCOs. The negative resistance is created by an inductor in the base of the transistor instead of the capacitive feedback network used in parallel feedback topology. The series feedback grounded base topology is best suited for wide tuning range applications because the loaded Q of the resonator is approximately the same over the band and is not greatly affected by the large-signal nonlinear negative resistance-generating device.

Figure 10-41 shows the general topology of the series feedback oscillator.

The steady-state oscillation condition for the series feedback configuration shown in Figure 10-41 is

$$Z_{osc}(I_L, \omega) + Z_L(\omega) = 0 \qquad (10\text{-}265)$$

where I_L is the load current amplitude and ω is the resonance frequency. Z_{osc} is the current and frequency-dependent output impedance, and Z_L is the function of frequency.

$$Z_{osc}(I_L, \omega) = R_{osc}(I_L, \omega) + jX_{osc}(I_L, \omega) \qquad (10\text{-}266)$$

$R_{osc}(I_L, \omega) \to$ negative resistance components generated by the device.

$$Z_L(\omega) = R_L(\omega) + jX_L(\omega) \qquad (10\text{-}267)$$

$$Z_{osc} = [Z_{22} + Z_2] - \frac{[Z_{12} + Z_2][Z_{21} + Z_2]}{[Z_{11} + Z_1 + Z_2]} \qquad (10\text{-}268)$$

where Z_{11}, Z_{22}, Z_{12}, and Z_{21} are [Z] parameters of the transistor.

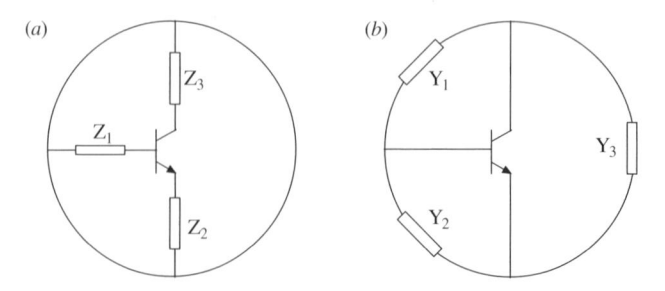

Figure 10-39 (*a*) Series and (*b*) parallel configurations of the oscillator.

10.10 ULTRA-LOW-NOISE WIDEBAND OSCILLATORS

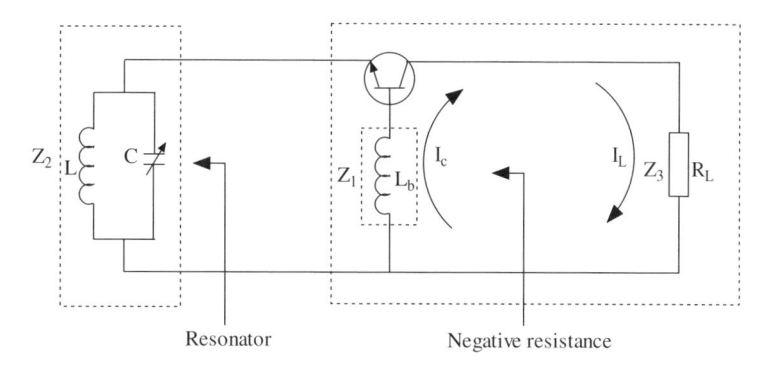

Figure 10-40 Series feedback grounded base topology.

The [Z] parameter of the transistor can be calculated in terms of the [Y] parameter as

$$Z_{11} = \frac{Y_{22}}{Y_{11}Y_{22} - Y_{12}Y_{21}}; \quad Z_{12} = \frac{-Y_{12}}{Y_{11}Y_{22} - Y_{12}Y_{21}}; \quad Z_{21} = \frac{-Y_{21}}{Y_{11}Y_{22} - Y_{12}Y_{21}};$$

$$Z_{22} = \frac{Y_{11}}{Y_{11}Y_{22} - Y_{12}Y_{21}}$$

$$[Y] = \begin{bmatrix} (g_{b'c} + g_{b'e}) + j\omega(C_{b'c} + C_{b'e}) & -(g_{b'c} + j\omega C_{b'c}) \\ g_m - (g_{b'c} + j\omega C_{b'c}) & (g_{b'c} + g_{ce}) + j\omega(C_{b'c} + C_{ce}) \end{bmatrix} \quad (10-269)$$

$$Z_{11} = \frac{Y_{22}}{Y_{11}Y_{22} - Y_{12}Y_{21}} = \frac{[(g_{b'c} + g_{ce}) + j\omega(C_{b'c} + C_{ce})]}{[(g_{b'c} + g_{b'e}) + j\omega(C_{b'c} + C_{b'e})][(g_{b'c} + g_{ce}) + j\omega(C_{b'c} + C_{ce})]}$$
$$+ [(g_{b'c} + j\omega C_{b'c})][g_m - (g_{b'c} + j\omega C_{b'c})]$$

(10-270)

$$Z_{12} = \frac{-Y_{12}}{Y_{11}Y_{22} - Y_{12}Y_{21}} = \frac{[(g_{b'c} + j\omega C_{b'c})]}{[(g_{b'c} + g_{b'e}) + j\omega(C_{b'c} + C_{b'e})][(g_{b'c} + g_{ce}) + j\omega(C_{b'c} + C_{ce})]}$$
$$+ [(g_{b'c} + j\omega C_{b'c})][g_m - (g_{b'c} + j\omega C_{b'c})]$$

(10-271)

$$Z_{21} = \frac{-Y_{21}}{Y_{11}Y_{22} - Y_{12}Y_{21}} = \frac{-[g_m - (g_{b'c} + j\omega C_{b'c})]}{[(g_{b'c} + g_{b'e}) + j\omega(C_{b'c} + C_{b'e})][(g_{b'c} + g_{ce}) + j\omega(C_{b'c} + C_{ce})]}$$
$$+ [(g_{b'c} + j\omega C_{b'c})][g_m - (g_{b'c} + j\omega C_{b'c})]$$

(10-272)

$$Z_{22} = \frac{Y_{11}}{Y_{11}Y_{22} - Y_{12}Y_{21}} = \frac{[(g_{b'c} + g_{b'e}) + j\omega(C_{b'c} + C_{b'e})]}{[(g_{b'c} + g_{b'e}) + j\omega(C_{b'c} + C_{b'e})][(g_{b'c} + g_{ce}) + j\omega(C_{b'c} + C_{ce})]}$$
$$+ [(g_{b'c} + j\omega C_{b'c})][g_m - (g_{b'c} + j\omega C_{b'c})]$$

(10-273)

318   SYSTEMS OF COUPLED OSCILLATORS

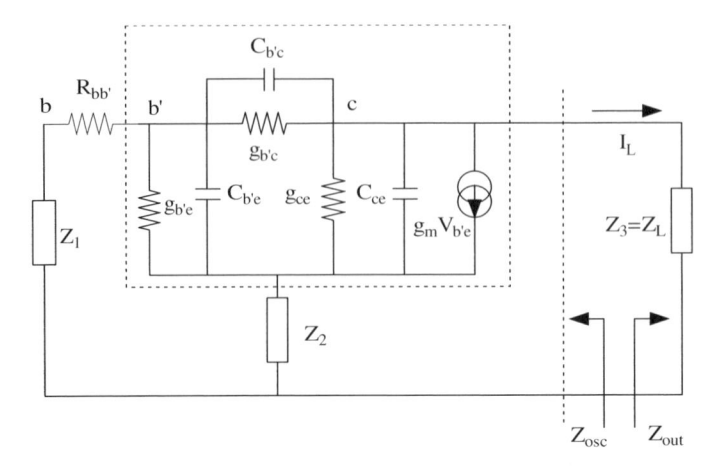

Figure 10-41 Series feedback topology of the oscillator using a bipolar transistor.

where

$$Z_1 = j\omega L_1, \quad Z_2 = \frac{1}{j\omega C_2}$$

Assuming that $g_{b'c}$, $g_{b'e}$, g_{ce}, and $C_{b'c}$ have very little effect, the [Z] parameters of the transistor are

$$[Z] = \begin{bmatrix} \dfrac{1}{j\omega C_{b'e}} & 0 \\ \dfrac{g_m}{\omega^2 C_{b'e} C_{ce}} & \dfrac{1}{j\omega C_{ce}} \end{bmatrix} \qquad (10\text{-}274)$$

where

$$Z_{11} = \frac{1}{j\omega C_{b'e}}, \quad Z_{12} = 0, \quad Z_{21} = \frac{g_m}{\omega^2 C_{b'e} C_{ce}}, \quad Z_{22} = \frac{1}{j\omega C_{ce}}$$

From equation (10-268), Z_{osc} is given by

$$Z_{osc} = \left[\frac{1}{j\omega C_{ce}} + \frac{1}{j\omega C_2}\right] - \frac{\left[\frac{1}{j\omega C_2}\right]\left[\frac{g_m}{\omega^2 C_{b'e} C_{ce}} + \frac{1}{j\omega C_2}\right]}{\left[j\omega L_1 + \frac{1}{j\omega C_{b'e}} + \frac{1}{j\omega C_2}\right]} \qquad (10\text{-}275)$$

$$Z_{osc} = -\frac{g_m}{\omega^2 C_{ce}[C_{b'e} + C_2 - L_1 C_2 C'_{be}\omega^2]}$$

$$-j\left[\left(\frac{C_{ce} + C_2}{\omega C_{ce} C_2}\right) - \left(\frac{1}{\omega\left(\frac{C_2^2}{C_{b'e}}\right) + \omega C_2 - L_1 C_2^2 \omega^3}\right)\right] \qquad (10\text{-}276)$$

10.10 ULTRA-LOW-NOISE WIDEBAND OSCILLATORS

$$Z_{osc} = R_{osc} + jX_{osc} \tag{10-277}$$

$$R_{osc} = R_n = -\left[\frac{g_m}{\omega^2 C_{ce}[C_{b'e} + C_2 - L_1 C_2 C_{b'e} \omega^2]}\right] \tag{10-278}$$

For sustained oscillation, $R_n < 0 \Rightarrow [L_1 C_2 C_{b'e} \omega^2] < [C_{b'e} + C_2] \rightarrow \omega < \sqrt{\frac{1}{L_1}[(C_{b'e} + C_2)/C_{b'e} C_2]}$.

$$X_{osc} = -\left[\left(\frac{C_{ce} + C_2}{\omega C_{ce} C_2}\right) - \left(\frac{1}{\omega\left(\frac{C_2^2}{C_{b'e}}\right) + \omega C_2 - L_1 C_2^2 \omega^3}\right)\right] \tag{10-279}$$

where R_n is the negative resistance of the series feedback oscillator.
From equation (10-279), the frequency of the oscillation is given as

$$X_{osc} = 0 \Rightarrow \omega_0 = \sqrt{\frac{1}{L_1} \frac{[C_{b'e} + C_2 + C_{ce}]}{C_{b'e}[C_{ce} + C_2]}} \tag{10-280}$$

From equation (10-278), for $C_{b'e} < C_2$

$$R_{osc} = -\left[\frac{g_m}{\omega^2 C_{ce} C_2 [1 - L_1 C_{b'e} \omega^2]}\right] \tag{10-281}$$

If $L_1 C_{b'e} \omega^2 < 1$, then

$$R_{osc} \rightarrow -\left[\frac{g_m}{\omega^2 C_{ce} C_2}\right] \tag{10-282}$$

If we now change from the small-signal transconductance g_m to a large-signal time average transconductance, then equation (10.282) can be expressed as

$$R_{osc} = R_N(t) = -\left[\frac{g_m(t)}{\omega^2 C_{ce} C_2 [1 - L_1 C_{b'e} \omega^2]}\right] \tag{10-283}$$

where $g_m(t)$ is the large-signal transconductance and is defined as

$$g_m(t) = \sum_{n=-\infty}^{n=\infty} g_m^{(n)} \exp(jn\omega t) \tag{10-284}$$

where n is the number of the harmonic considered.

320   SYSTEMS OF COUPLED OSCILLATORS

The total noise voltage power within 1 Hz bandwidth can be given as equation (8–82): $\overline{|e_N^2(f)|}\,|_{\omega=\omega_0} = \overline{e_R^2(f)} + \overline{e_{NR}^2(f)}$.

The first term in the expression above is related to the thermal noise due to the loss resistance of the resonator tank. The second term is related to the shot noise and flicker noise in the transistor.

After lengthy calculations and the addition of shot noise, flicker noise, and the loss resistor, we obtain from equations (8–103) and (8–108)

$$\mathscr{L}(\omega) = \left[4kTR + \frac{|g_m^2(t)|\left[4qI_c + \frac{K_f I_b^{AF}}{\omega}\right]}{\omega_0^4 \beta^2 C_{ce}^2 (C_2 + C_{b'e} - L_1 C_2 C_{be}' \omega_0^2)^2 + |g_m^2(t)|\omega_0^2 (C_2 + C_{b'e} - L_1 C_2 C_{b'e} \omega_0^2)^2} \right] *$$
$$\left[\frac{\omega_0^2}{4(\omega)^2 V_{cc}^2}\right]\left[\frac{1}{Q_L^2} + \left(1 - \left(\frac{1}{\omega_0^2 L_1}\right)\left(\frac{[(C_2 + C_{b'e} - L_1 C_2 C_{b'e} \omega_0^2) + C_{ce}]}{C_{ce}[(C_2 + C_{b'e} - L_1 C_2 C_{b'e} \omega_0^2)]}\right)\right)^2\right]$$

(10-285)

The flicker noise contribution in equation (10-285) is introduced by adding the term $(K_f I_b^{AF})/\omega$ in the RF collector current I_c, where K_f is the flicker noise coefficient and AF is the flicker noise exponent. This is valid only for the bipolar transistor. For an FET, the equivalent currents have to be used.

10.10.2 Design Criteria for Ultra-Low-Noise Wideband VCOs

Phase noise is the noise which results from the modulations of the oscillation frequency or carrier frequency of an oscillator.

Oscillators intended for fixed-frequency operation are relatively easy to optimize if only a parameter of particular concern is involved, but serious problems occur when they are tuned to operate over wideband frequencies. In contrast to single-frequency oscillators, wideband VCOs need to cover a range of up to 1 octave or more. For a varactor-tuned oscillator to tune continuously over a wide range, the tuning diode must exhibit a large change in capacitance in response to a small change in tuning voltage. However, the tuning diode's own capacitance is modulated by the random electronic noise signals generated internally by various oscillator circuit elements, including the tuning diode itself. The tuning range of the oscillator directly influences the phase noise, and there is a trade-off between the continuous tuning range of the VCO and the amount of phase noise generated by the varactor capacitance modulation. However, there are demanding requirements for low-noise performance over the complete frequency range. A wide tuning range with a small tuning voltage and good phase noise performance has always been needed, but it causes the problem of controlling the loop parameters and the dynamic loaded Q of the resonator over the wideband operation.

Thus, there is a need for a method which improves the phase noise performance over the wideband. This can be accomplished by using the novel coupled oscillator topology integrated with a tuning diode or switched capacitor system. The N-push/push–push configuration is another attractive option to extend the frequency domain of operation of the transistor while keeping phase noise low.

Wideband ultra-low-noise VCO design criteria are based on the choice of the active device, the tuning network, the coupled resonator circuit, and the passive component. The active device, the tuning network, the coupled resonator, and the passive device determine the noise.

10.10.3 Active Device

The designer has no control over the noise sources in a transistor, being able to select only the device selection and the operating bias point. For example, the bulk resistance of a transistor, upon which thermal noise depends, is an unchangeable, intrinsic property of the device. However, knowing how the oscillator waveforms affect the noise, the designer can substantially improve phase noise performance by selecting the optimal bias point and signal drive level.

The transistor is used as a negative resistance-generating active device. Its choice should be based on the operating frequency, output power requirement, low $1/f$ noise at the desired frequency, and bias condition. Low $1/f$ noise of the transistor is a critical parameter because it appears as the sideband noise around the carrier frequency and is directly related to the current density in the transistor. A bipolar transistor biased at a low collector current will keep the flicker corner frequency to a minimum, typically around 2 to 10 kHz. Large transistors with a high I_{cmax} used at low currents have the best $1/f$ performance. Most active devices exhibit a broad U-shaped noise curve versus the bias current curve. Based on this, the optimal operating bias current corresponding to the minimum noise factor for the device can be selected.

The overall loading of the oscillator resonator is important, and optimization of the loaded Q with respect to the insertion loss (G_{21}) is done dynamically for ultra-low-noise performance over the tuning range. Figure 10-42 shows the equivalent model of the feedback oscillator for analysis of the insertion loss (G_{21}) and needed gain (S_{21}) with respect to the Q factor of the resonator.

The transfer function for Figure 10-42 is given by

$$[TF(j\omega)]_{closed\ loop} = \frac{[Y(j\omega)]_{output}}{[X(j\omega)]_{input}} = \frac{V_o(\omega)}{V_{in}(\omega)} = \frac{H_1(j\omega)}{[1 + H_1(j\omega)H_2(j\omega)]} \quad (10\text{-}286)$$

The feedback coefficient between output and input can be expressed as

$$H_2(j\omega) = \frac{R_{in}}{(R_{Loss} + R_{in}) + j\left[\omega L - \dfrac{1}{\omega C}\right]} \quad (10\text{-}287)$$

322   SYSTEMS OF COUPLED OSCILLATORS

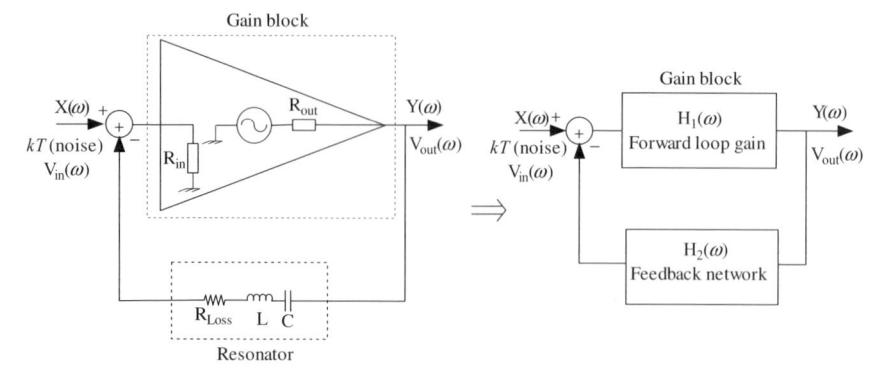

Figure 10-42   Equivalent model of the feedback oscillator.

The unloaded and loaded Q of the resonator are defined as

$$Q_0 = \frac{\omega_0 L}{R_{Loss}} \tag{10-288}$$

$$Q_L = \frac{\omega_0 L}{R_{in} + R_{Loss} + R_{out}} \tag{10-289}$$

where ω_0 is the resonance frequency and Q_0 and Q_L are the unloaded and loaded Q of the oscillator.

From equations (10-287) and (10-289),

$$H_2(j\omega) = \frac{R_{in}}{(R_{in} + R_{Loss}) + j\left[\omega L - \dfrac{1}{\omega C}\right]}$$

$$\approx \frac{R_{in}}{(R_{in} + R_{Loss} + R_{out})\left[1 \pm j2Q_L \dfrac{\Delta\omega}{\omega_0}\right]} \tag{10-290}$$

Defining

$$\omega L - \frac{1}{\omega C} \text{ is equal to } \pm 2\Delta\omega L$$

where $\Delta\omega$ is the offset frequency from the carrier frequency ω_0.

From equations (10-288) and (10-289),

$$1 - \frac{Q_L}{Q_0} = \frac{R_{in} + R_{out}}{R_{in} + R_{Loss} + R_{out}} \tag{10-291}$$

From equation (10-290), the feedback factor is given by

$$H_2(j\omega) = \frac{R_{in}}{(R_{in} + R_{Loss} + R_{out})\left[1 \pm j2Q_L \frac{\Delta\omega}{\omega_0}\right]}$$

$$= \left[\frac{R_{in}}{(R_{in} + R_{out})}\right]\left[1 - \frac{Q_L}{Q_0}\right]\left[\frac{1}{1 \pm j2Q_L \frac{\Delta\omega}{\omega_0}}\right] \quad (10\text{-}292)$$

$$G_{21} \approx H_2(j\omega) = \frac{1}{2}\left[1 - \frac{Q_L}{Q_0}\right]\left[\frac{1}{1 \pm j2Q_L \frac{\Delta\omega}{\omega_0}}\right] \quad \text{for } R_{in} = R_{out} \quad (10\text{-}293)$$

Equation (10-293) describes the general equation for the variation of the insertion loss (G_{21}) of the resonator in the oscillator circuit in terms of loaded and unloaded Q. Figure 10-43 shows the plot of the loaded Q versus the insertion loss.

For low phase noise performance, the insertion loss parameter (G_{21}) must be optimized dynamically over the tuning range for the application of the ultra-low-noise wideband VCOs, so that $m_{opt} = 0.5$, as given in Figure 10-48.

10.10.4 Tuning Network

In certain areas of microwave applications such as frequency modulated continuous wave (FMCW) radars, FM modulators, and PLL circuits, system performance is enhanced when the frequency of a VCO is linear with respect to its control voltage. YIG-tuned oscillators have good linearity and a wide tuning range, but their low tuning speed restricts their usefulness. Varactor-tuned VCOs support high tuning speed, but their linearity is inferior to that of YIG-tuned oscillators. External linearizing networks with a predetermined topology can improve the tuning linearity, but they introduce new problems, such as additional power consumption, lower input impedance, and a narrower tuning bandwidth. Since tuning linearity is mainly

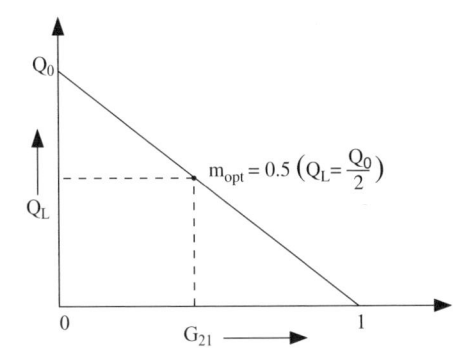

Figure 10-43 Loaded Q versus insertion loss.

determined by the voltage-to-capacitance relationship of a varactor, as well as by its coupling network, the choice of a varactor and the design of a coupling network are critical in achieving the desired tuning linearity.

10.10.5 Tuning Characteristics and Loaded Q

For a fairly wide tuning range (more than octave band), hyperabrupt tuning diodes are the best choice, but at the cost of a lower Q. Therefore, the phase noise performance is inferior to that provided by abrupt tuning diodes. In varactor-tuned VCOs, the tuning range is limited due to the resistive loading effect of the tuning diode. Figure 10-44 presents the equivalent circuit model of the varactor-tuned VCO, showing the varactor loading effect of the oscillator circuits.

The parasitics associated with the tuning diode are considered to be included in the equivalent RC of the oscillator/VCOs. The varactor capacitance C_v is a function of the tuning voltage, decreasing monotonically from C_{V0} at zero bias to C_{VB} at breakdown.

The capacitance ratio r is defined as

$$r = \frac{C_{V0}}{C_{VB}} > 1 \tag{10-294}$$

To simplify the analysis, the series resistance R_s given in Figure 10-44a is assumed to be constant at all bias levels, and the equivalent parallel resistance R_p is

$$R_p = R_s \left[1 + Q_v^2 \left(\frac{C_{v0}}{C_v}\right)^2 \right] \cong R_s Q_v^2 \left(\frac{C_{v0}}{C_v}\right)^2 \quad \text{for } Q_v \gg 1 \tag{10-295}$$

where Q_v of the varactor is defined as

$$Q_v = \frac{1}{\omega R_s C_V} \tag{10-296}$$

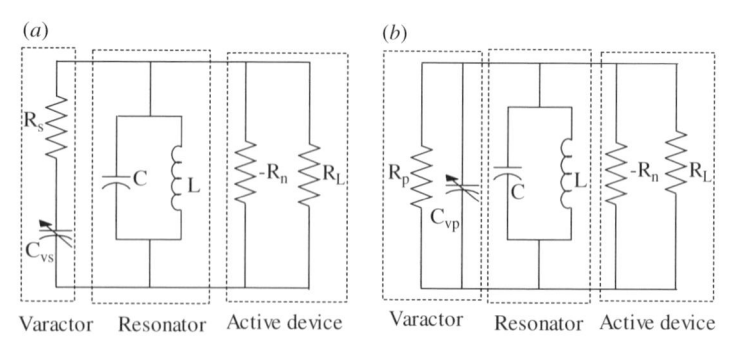

Figure 10-44 Equivalent circuit of the varactor-tuned negative resistance oscillator with the tuning diode represented by (a) a series RC and (b) a parallel RC.

10.10 ULTRA-LOW-NOISE WIDEBAND OSCILLATORS

The parallel equivalent capacitance of the tuning diode C_{vp} is given by

$$C_{vp} = \frac{C_{vs}}{\left[1 + \frac{1}{Q_v^2}\left(\frac{C_v}{C_{v0}}\right)^2\right]} \cong C_{vs} \quad \text{for } Q_v^2 \gg 1 \qquad (10\text{-}297)$$

Without a tuning diode, the RF and unloaded Q of the oscillator are given by

$$\omega_0 = \sqrt{\frac{1}{LC}} \qquad (10\text{-}298)$$

$$Q_0 = \omega_0 CR \qquad (10\text{-}299)$$

where $R = |-R_n| = R_L$

After the tuning diode is incorporated, the frequency of the oscillation can be tuned from ω_1 to ω_2:

$$\omega_1 = \sqrt{\frac{1}{L(C + C_{v0})}} \qquad (10\text{-}300)$$

$$\omega_2 = \sqrt{\frac{1}{L(C + C_{VB})}} \qquad (10\text{-}301)$$

Practically, the difference between C_{v0} and C_{VB} is a small fraction of the total capacitance $(C + C_{V0})$. The fractional tuning range with respect to ω_1 can be expressed as a function of the capacitance ratio as

$$\frac{\Delta f}{f_1} = \frac{\omega_2 - \omega_1}{\omega_1} \cong \frac{1}{2}\left[\frac{C_{V0} - C_{VB}}{C + C_{V0}}\right] = \frac{1}{2}\left[\frac{1 - \frac{1}{r}}{1 + \frac{C}{C_{V0}}}\right] \qquad (10\text{-}302)$$

The loaded Q of the oscillator at zero bias is defined as Q_L and can be given as

$$Q_L = \omega(C + C_{V0})\left[\frac{RR_s Q_v^2}{R + R_s Q_v^2}\right] = \left[\frac{Q_0 Q_v^2}{\left(\frac{R}{R_s}\right) + Q_v^2} + \frac{Q_v}{1 + \left(\frac{R_s}{R}\right)Q_v^2}\right] \qquad (10\text{-}303)$$

$$\frac{Q_L}{Q_0} = \left[1 + \left(2r\frac{\Delta f}{f_1}\right)\frac{\left(\frac{Q_0}{Q_v} - 1\right)}{(r-1)}\right]^{-1} = \left[1 + (2rp)\frac{(q-1)}{(r-1)}\right]^{-1} \qquad (10\text{-}304)$$

where

$$p = \frac{\Delta f}{f_1}, \quad q = \frac{Q_0}{Q_v}, \quad Q_v\big|_{v=0} = \frac{1}{\omega R_s C_{V0}}, \quad Q_v\big|_{v=VB} = \frac{1}{\omega R_s C_{VB}}$$

The loaded Q of the varactor-tuned VCOs reaches its lowest value at zero tuning voltage because varactor at zero bias introduces the greatest perturbation and its own Q is lowest. For octave-band tunability, the ratio of the loaded Q to the unloaded Q is to be optimized for low noise performance as given in equation (10-304).

10.10.6 Coupled Resonator

The Q factor of the resonator can be increased by introducing the coupling factor β, which is defined as the ratio of the series coupling capacitor to the resonator capacitor. Figure 10-45 shows two identical resonators with series coupling where Z_r and Z_c are the resonators and the coupling network impedance, respectively.

The effective coupled impedance of Figure 10-45 is given by

$$Z_{\mathit{eff}}(\omega) = \left[\frac{V_0}{I_{in}}\right] = \frac{Z_r(\omega)}{2 + \dfrac{Z_c(\omega)}{Z_r(\omega)}} = \frac{Z_r^2(\omega)}{Z_c(\omega) + 2Z_r(\omega)} \tag{10-305}$$

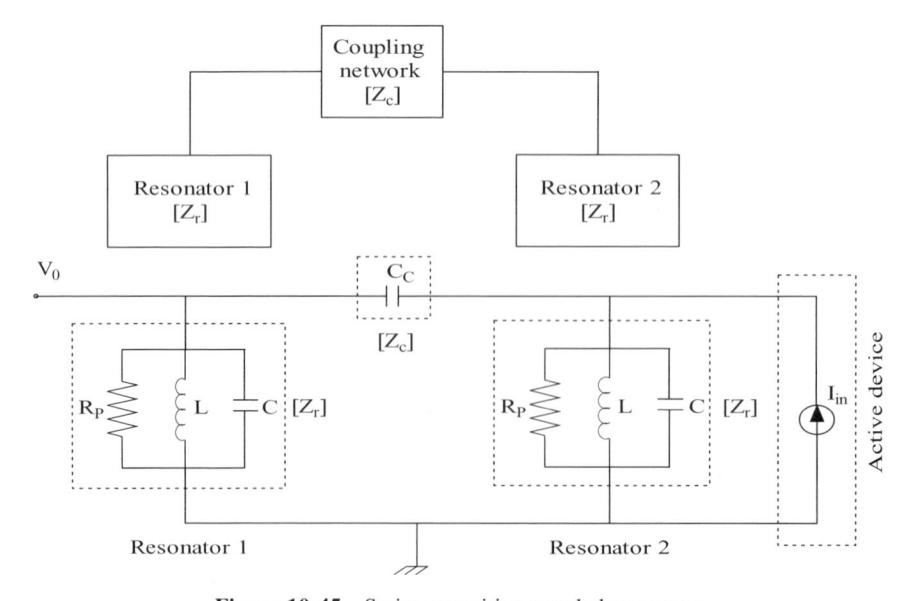

Figure 10-45 Series capacitive coupled resonator.

10.10 ULTRA-LOW-NOISE WIDEBAND OSCILLATORS

where I_{in} is the large-signal current from the active device.

$$Y_{eff}(\omega) = \frac{1}{Z_{eff}(\omega)} = \left[\frac{Z_c(\omega)}{Z_r^2(\omega)} + \frac{2}{Z_r(\omega)}\right] = \left[\frac{Y_r^2(\omega)}{Y_c(\omega)} + 2Y_r(\omega)\right] \quad (10\text{-}306)$$

For $Z_c(\omega) \gg Z_r(\omega)$, and assuming that the Q factor of $Z_r(\omega)$ is sufficiently large, the denominator of equation (10-305) may be considered constant over the frequencies within the bandwidth of $Z_r(\omega)$. The coupling admittance is defined by $Y_c(\omega) = j\omega C_c$.

The resonator admittance is given by

$$Y_r(\omega) = \left[\frac{1}{R_p} + \frac{1}{j\omega L} + j\omega C\right] = \left[\frac{j\omega L R_p}{R_p(1 - \omega^2 LC) + j\omega L}\right]^{-1} \quad (10\text{-}307)$$

From equations (10-306) and (10-307), $Y_{eff}(\omega)$ can be rewritten as

$$Y_{eff}(\omega) = \left[\frac{2}{R_p} - \frac{2R_p(1 - \omega^2 LC)}{\omega^2 L R_p^2 \beta C}\right]$$

$$+ j\left[\frac{[R_p^2(1 - \omega^2 LC)^2 - \omega^2 L^2]}{\omega^3 R_p^2 L^2 \beta C} - \frac{2R_p(1 - \omega^2 LC)}{R_p \omega L}\right] \quad (10\text{-}308)$$

From equation (10-307), the phase shift of the coupled resonator is given as

$$\varphi = \tan^{-1}\left[\frac{\left(\frac{[R_p^2(1 - \omega^2 LC)^2 - \omega^2 L^2]}{\omega^3 R_p^2 L^2 \beta C} - \frac{2R_p(1 - \omega^2 LC)}{R_p \omega L}\right)}{\left(\frac{2}{R_p} - \frac{2R_p(1 - \omega^2 LC)}{\omega^2 L R_p^2 \beta C}\right)}\right] \quad (10\text{-}309)$$

At resonance, the real part of $Y_{eff}(\omega)$ is reduced to zero, and the resonance frequency can be derived as

$$\text{Re}[Y_{eff}(\omega)]_{\omega=\omega_0} = \left[\frac{2}{R_p} - \frac{2R_p(1 - \omega^2 LC)}{\omega^2 L R_p^2 \beta C}\right]_{\omega=\omega_0} = 0 \Rightarrow \omega_0^2 LC(1 + \beta) = 1$$

$$(10\text{-}310)$$

$$[\omega_0]_{\phi=90°} = \frac{1}{\sqrt{LC(1+\beta)}} \quad (10\text{-}311)$$

$$[Y_{eff}(\omega)]_{\omega=\omega_0} = -j\left[\frac{R_p^2 \beta^2 C + (1+\beta)L}{\beta(1+\beta)\omega L R_p^2 C}\right] \quad (10\text{-}312)$$

$$[Z_{eff}(\omega)]_{\omega=\omega_0} = j\left[\frac{\beta(1+\beta)\omega L R_p^2 C}{R_p^2\beta^2 C + (1+\beta)L}\right] = j\left[\frac{\beta R_p^2 \omega C}{\frac{R_p^2\beta^2 C}{(1+\beta)L}+1}\right] \Rightarrow j\left[\frac{Q_0\beta R_p}{1+Q^2\beta^2}\right]$$

(10-313)

where

$$Q_0 = \omega C R_p = \frac{R_p}{\omega L}; \quad \beta = \frac{C_c}{C}$$

From equation (10-311), the effective quality factor of the coupled resonator is given by [174–176]

$$[Q_{eff\ coupled}(\omega)]_{\omega=\omega_0} = \frac{\omega_0}{2}\left[\frac{\partial\phi}{\partial\omega}\right] \Rightarrow \frac{2Q_0(1+\beta)}{1+Q_0^2\beta^2} \quad (10\text{-}314)$$

$$[Q_{eff\ coupled}(\omega_0)]_{\beta\ll 1} = \left[\frac{2Q_0(1+\beta)}{1+Q_0^2\beta^2}\right]_{\beta\ll 1} \approx 2Q_0 \quad (10\text{-}315)$$

Weakly coupled resonators ($\beta \ll 1$) will produce high attenuation due the large value of Z_c, so a trade-off between doubling the Q factor and doubling the permissible attenuation is required for the best phase noise performance. For octave-band tunability, the coupling factor β is dynamically adjusted over the tuning range for low noise performance.

Example: Figure 10-46a shows the schematic of a 1000 MHz parallel coupled resonator-based oscillator. This provides some insight into the improvement of the phase noise of the coupled resonator-based oscillator.

Figure 10-46b shows the NE68830 transistor with the package parameters for the calculation of the oscillator frequency. Table 10-1 shows the SPICE and package parameters of NE68830 from data sheets.

The circuit shown in Figure 10-46a is fabricated on 32 mil thick Roger substrate of dielectric constant 3.38 and loss tangent 2.7E-4.

The frequency of the oscillation for the oscillator circuit shown in Figure 10-46a is given in equation (6-33).

$$\omega_0 = \sqrt{\frac{\left(\frac{(C_1^* + C_p)C_2}{(C_1^* + C_p + C_2)} + C_{c1}\right)}{L_{PR}\left[\left(\frac{(C_1^* + C_p)C_2 C_{c1}}{(C_1^* + C_p + C_2)}\right) + C_{PR}\left(\frac{(C_1^* + C_p)C_2}{(C_1^* + C_p + C_2)} + C_{c1}\right)\right]}}$$

$$\approx 1000\ \text{MHz} \quad (10\text{-}316)$$

10.10 ULTRA-LOW-NOISE WIDEBAND OSCILLATORS 329

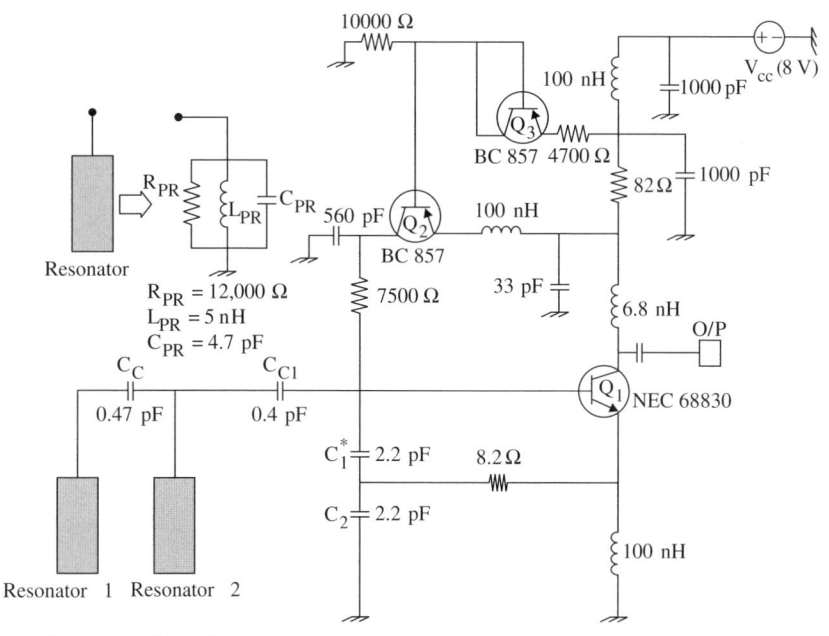

Figure 10-46a Schematic of the parallel coupled resonator-based oscillator.

with

$C_1^* = 2.2$ pF, $C_1 = C_1^* + C_p$
$C_p = 1.1$ pF (C_{BEPKG} + contribution from layout)
$C_1 = 2.2$ pF + 1.1 pF package = 3.3 pF; $C_2 = 2.2$ pF, $C_{c1} = 0.4$ pF
Resonator: $R_{PR} = 12{,}000\ \Omega$, $C_{PR} = 4.7$ pF, $L_{PR} = 5$ nH

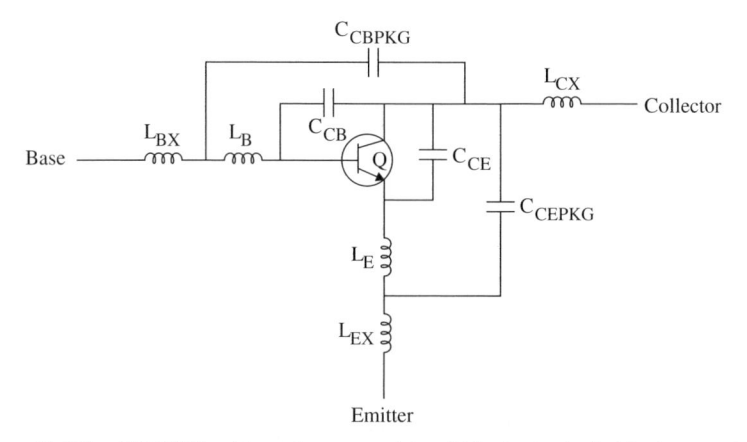

Figure 10-46b NE68830 with package parasitics (Q is the intrinsic bipolar transistor).

TABLE 10-1 SPICE Parameters and Package Parameters of NEC Transistor NE68830

Parameters	Q	Parameters	Q	Parameters	Package
IS	3.8E-16	MJC	0.48	C_{CB}	0.24E-12
BF	135.7	XCJC	0.56	C_{CE}	0.27E-12
NF	1	CJS	0	L_B	0.5E-9
VAF	28	VJS	0.75	L_E	0.86E-9
IKF	0.6	MJS	0	C_{CBPKG}	0.08E-12
NE	1.49	TF	11E-12	C_{CEPKG}	0.04E-12
BR	12.3	XTF	0.36	C_{BEPKG}	0.04E-12
NR	1.1	VTF	0.65	L_{BX}	0.2E-9
VAR	3.5	ITF	0.61	L_{CX}	0.1E-9
IKR	0.06	PTF	50	L_{EX}	0.2E-9
ISC	3.5E-16	TR	32E-12		
NC	1.62	EG	1.11		
RE	0.4	XTB	0		
RC	4.2	KF	0		
CJE	0.79E-12	AF	1		
CJC	0.549E-12	VJE	0.71		
XTI	3	RB	6.14		
RBM	3.5	RC	4.2		
IRB	0.001	CJE	0.79E-12		
CJC	0.549E-12	MJE	0.38		
VJC	0.65				

Note: SPICE parameters (Gummel-Poon Model, Berkley-Spice).

Figure 10-46c shows the simulated (Ansoft Designer) response of the single resonator (1 resonator) oscillator circuit having resonance at 999.8 MHz, or a less than 1% error. The small variation in resonant frequency may be due to the frequency-dependent packaged parameters, but it is a good starting value for the analysis of the coupled resonator-based oscillator.

Figures 10-46d and 10-46e show the simulated and measured phase noise plots for the single resonator (1-resonator) and the coupled resonator (2-resonator).

10.10.7 Optimum Phase Noise with Respect to the Loaded Q

The amount of loading on a resonator is critical for optimum phase noise in VCOs. A very lightly loaded resonator will have a higher Q factor but will pass less power through it, whereas a heavily loaded resonator will have a very low Q factor but will pass more power through it. From Figure 10-47, the equivalent loading is R_{reso} in parallel with the series combination of $1/g_m$ and R_L, and this represents the loading factor in the oscillator circuit.

From equation (7-26), the phase noise is given as

$$\mathscr{L}(f_m) = 10\log\left\{\left[1 + \frac{f_0^2}{(2f_m Q_L)^2(1-m)^2}\right]\left(1 + \frac{f_c}{f_m}\right)\frac{FkT}{2P_0} + \frac{2kTRK_0^2}{f_m^2}\right\} \quad (10\text{-}317)$$

10.10 ULTRA-LOW-NOISE WIDEBAND OSCILLATORS

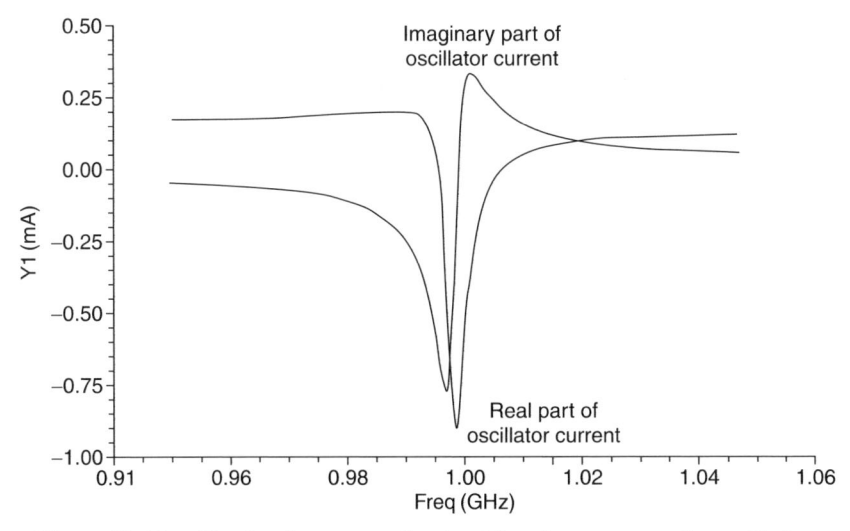

Figure 10-46c Simulated response of real and imaginary currents for oscillation.

$$\mathscr{L}(f_m) = 10\log\left\{\left[1 + \frac{f_0^2}{(2f_m Q_0)^2 m^2 (1-m)^2}\right]\left(1 + \frac{f_c}{f_m}\right)\frac{FkT}{2P_0} + \frac{2kTRK_0^2}{f_m^2}\right\} \quad (10\text{-}318)$$

where $m = (Q_L/Q_0)$.

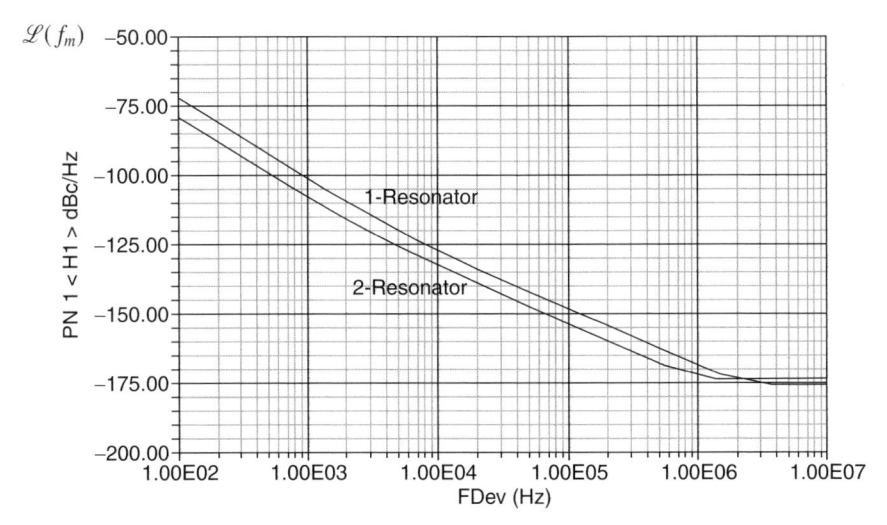

Figure 10-46d Simulated phase noise plot for the single resonator (1-resonator) and the coupled resonator (2-resonator).

332 SYSTEMS OF COUPLED OSCILLATORS

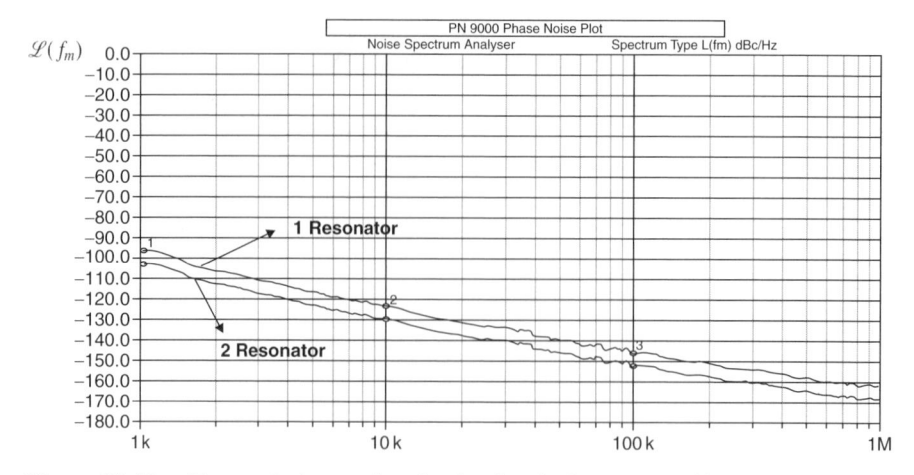

Figure 10-46e Measured phase noise plot for the single resonator (1-resonator) and the coupled resonator (2-resonator). *Source*: Synergy Microwave Corp., using the Aeroflex phase noise system.

From equation (10-318), the minimum phase noise can be found by differentiating the equation and equating to zero as $\frac{\partial}{\partial m}[\mathscr{L}(f_m)]_{m=m_{opt}} = 0$

$$\frac{d}{dm}\left[10\log\left\{\left[1 + \frac{f_0^2}{(2f_m Q_0)^2 m^2 (1-m)^2}\right]\left(1 + \frac{f_c}{f_m}\right)\frac{FkT}{2P_0} + \frac{2kTRK_0^2}{f_m^2}\right\}\right]$$
$$= 0 \Rightarrow m_{opt} \approx 0.5 \qquad (10\text{-}319)$$

Figure 10-48 shows the plot of relative phase noise versus the ratio of the loaded to the unloaded Q factor of the resonator (for an example, see Appendix C).

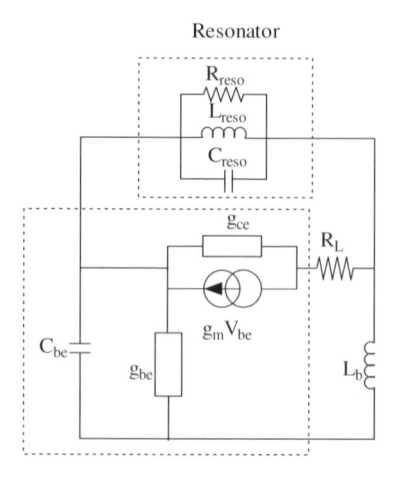

Figure 10-47 Small-signal model of the grounded base oscillator.

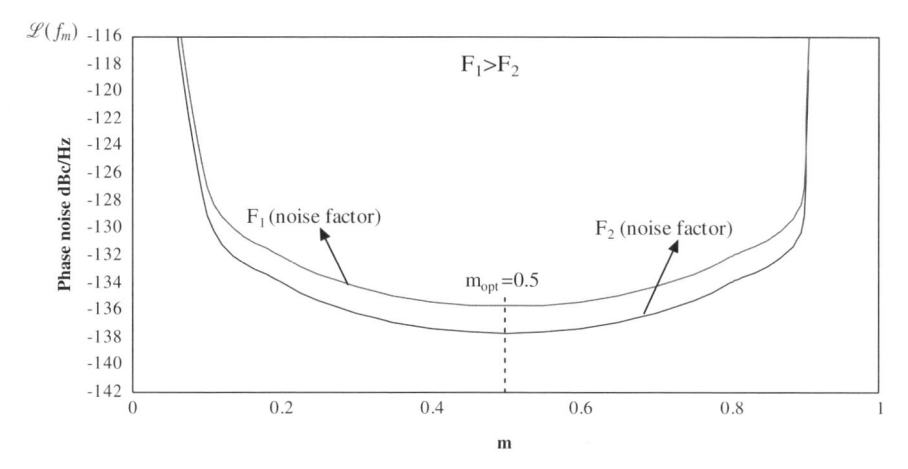

Figure 10-48 Relative phase noise versus the ratio of loaded to unloaded Q of the resonator for noise factors F_1 and F_2 ($F_1 > F_2$).

This implies that for low-noise wideband application, the value of m should be dynamically controlled over the tuning range, and it should lie in the vicinity of m_{opt} for ultra-low-phase noise performance over the frequency band [184].

10.10.8 Passive Device

Despite an ongoing trend toward higher integration levels, discrete RF components still have a place in contemporary wireless designs. Furthermore, discrete RF devices offer advantages such as superior performance, great design flexibility and versatility, faster time to market, low cost, and reduce risk. For cost reasons, surface-mounted components are used in VCO circuits. In reality, these components possess parasitics resulting in a self-resonant frequency (SRF) comparable to the operating frequency and limit the operating frequency band.

An understanding of the parasitic and packaging effects of passive SMDs [160], including characterization of the pertinent interconnect, is required for large-signal analysis of the oscillator design. RF components packaging, pad footprint, PCB layout, and substrate interaction cause local parasitic effects such as resonant coupling, cross-talk, signal loss, signal distortion, and so on. Extrinsic modeling of the devices is crucial for large-signal analysis, which takes the above-mentioned parasitic effects into account. Incomplete specification of the passive components inevitably leads to inaccurate results and considerable tuning and testing after board fabrication. If the design analysis is based on the extrinsic circuit model, the increased accuracy attained by incorporating board parasitics can potentially reduce the number of design cycles. Furthermore, by using an accurate library base of equivalent-circuit models, it is feasible to accomplish most of the design with a circuit simulator instead of the trial and error at the bench.

Active devices are shrinking and are occupying less and less chip area, whereas passive devices remained large. While it is possible to fabricate small values of capacitance on chip, small inductors are virtually impossible due to the large physical area required to obtain sufficient inductance at a given frequency. This is compounded by the losses of the substrate, which makes it virtually impossible to fabricate high-quality devices. At a low frequency, this low loss design is possible as long as package parasitics are negligible in comparison with the external electrical characteristics of the device.

For instance, at 100 MHz, a typical resonator inductance value is on the order of 100 nH, whereas at 10 GHz this value is around 1 nH. To build a 1 nH inductor is impossible in standard low-cost packaging since the bond wire inductance can exceed 1 nH. Typical examples of passive devices include linear time-invariant resistors of finite resistance $R > 0$, which dissipate energy, and ideal linear time-invariant capacitors and inductors, which only store energy.

10.10.9 Quality of the Passive Devices

The complex power delivered to a one-port network at frequency ω is given by [173]

$$P = \frac{1}{2} \oint_S E \times H^*.dS = P_n + j2\omega(W_m - W_e) \qquad (10\text{-}320)$$

where P_n represents the average power dissipated by the network, and W_m and W_e represent the time average of the stored magnetic and electrical energy. For $W_m > W_e$ the device acts inductively, and for $W_e > W_m$ it acts capacitively.

The input impedance can be defined as

$$P = R + jX = \frac{V}{I} = \frac{VI^*}{|I|^2} = \frac{P}{\frac{1}{2}|I|^2} = \frac{P_n + j2\omega(W_m - W_e)}{\frac{1}{2}|I|^2} \qquad (10\text{-}321)$$

The Q of the passive device is defined as

$$Q = 2\pi \left[\frac{E_{stored}}{E_{dissipated}}\right] = 2\pi \left[\frac{W_m + W_e}{P_n \times T}\right] = \frac{\omega(W_m + W_e)}{P_n} \qquad (10\text{-}322)$$

where E_{stored} and $E_{dissipated}$ are the energy stored and dissipated per cycle.

$$Q_{inductor} = 2\pi \left[\frac{E_{stored}}{E_{dissipated}}\right]_{W_e=0} = 2\pi \left[\frac{W_m}{P_n \times T}\right] = \frac{\omega W_m}{P_n} = \omega \left[\frac{\frac{1}{2}(LI^2)}{\frac{1}{2}(I^2R)}\right] = \frac{\omega L}{R}$$

$$(10\text{-}323)$$

$$Q_{capacitor} = 2\pi \left[\frac{E_{stored}}{E_{dissipated}}\right]_{W_m=0} = 2\pi \left[\frac{W_e}{P_n \times T}\right] = \frac{\omega W_e}{P_n} = \omega \left[\frac{\frac{1}{2}(CV^2)}{\frac{1}{2}\left(\frac{V^2}{R}\right)}\right] = \frac{1}{\omega CR}$$

$$(10\text{-}324)$$

where R is the series loss resistance in the inductor and capacitor.

10.10.10 Inductor Model

Figure 10-49 shows the equivalent circuit model for the inductor. The series resistance R accounts for the package losses, resistive effects of the winding, and pad or trace losses.

The series capacitance C represents the self-capacitance of the inductor package and the capacitance between the pads. The pad is simply a step-in-width inductive transition and augments the series inductance of the intrinsic model. The total inductance is denoted as L. The dielectric losses in the substrate introduce a shunt conductance $G_p = 1/R_p$, and the capacitance to ground of each pad introduces a shunt C_p, which is influenced by the substrate.

The direct RF measurement of the model parameters L, R, and C of an SMD inductor is difficult because of their strong interaction with the board parasitic. Typically, any of theses parameters can be obtained indirectly from measurable parameters such as impedance, quality factor, and resonant frequency. Figure 10-50 shows the closed form of the inductor model in Figure 10-49.

From Figure 10-49,

$$Z = \frac{(R+j\omega L)\frac{1}{j\omega C}}{(R+j\omega L + \frac{1}{j\omega C})} = \frac{R+j\omega L}{(1-\omega^2 LC)+j\omega RC} \tag{10-325}$$

$$Y_p = G_p + j\omega C_p \tag{10-326}$$

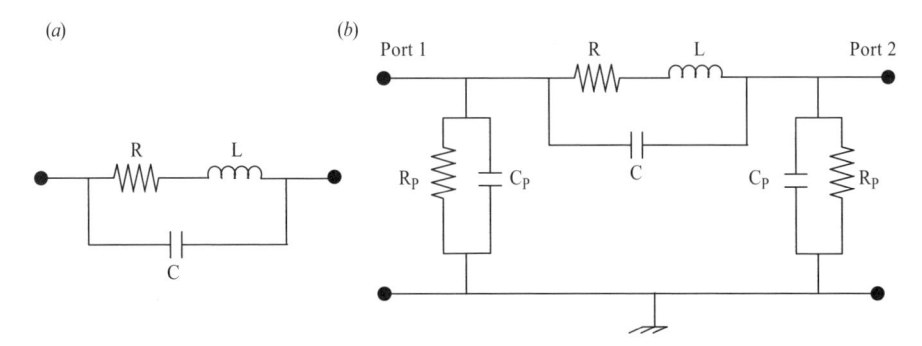

Figure 10-49 Model for an SMD inductor. (*a*) Intrinsic model; (*b*) extrinsic model.

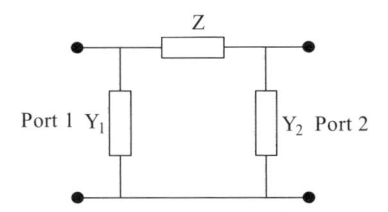

Figure 10-50 π-Model for an SMD inductor.

SYSTEMS OF COUPLED OSCILLATORS

The measured S-parameters on a vector network analyzer are converted into the two-port $[Y]$ parameters using

$$[Y] = \begin{bmatrix} Y_{11} & Y_{12} \\ Y_{21} & Y_{22} \end{bmatrix} \tag{10-327}$$

where

$$Y_{11} = Y_0 \left[\frac{(1 - S_{11})(1 + S_{22}) + S_{12}S_{21}}{\Delta} \right], \quad Y_{12} = -Y_0 \left[\frac{2S_{12}}{\Delta} \right]$$

$$Y_{22} = Y_0 \left[\frac{(1 + S_{11})(1 - S_{22}) + S_{12}S_{21}}{\Delta} \right], \quad Y_{21} = -Y_0 \left[\frac{2S_{21}}{\Delta} \right]$$

$$\Delta = (1 + S_{11})(1 + S_{22}) - S_{12}S_{21}, \quad Y_0 = \frac{1}{Z_0}$$

Z_0 is the characteristic impedance of the port. From reciprocity, $S_{12} = S_{21}$ and $Y_{12} = Y_{21}$.

From Figure 10-50,

$$Z = -\frac{1}{Y_{12}} \tag{10-328}$$

$$Y_1 = Y_{11} + Y_{12} \tag{10-329}$$

$$Y_2 = Y_{22} + Y_{21} \tag{10-330}$$

From symmetry,

$$Y_1 = Y_2 = Y_p \tag{10-331}$$

$$Y_p = G_P + j\omega C_p \tag{10-332}$$

$$G_p = \text{Re}[Y_p] \tag{10-333}$$

$$C_p = \frac{\text{Im}[Y_p]}{\omega} \tag{10-334}$$

The circuit variables for the SMD inductor are frequency dependent and can be expressed in terms of the resonant frequency, quality factor, and impedance at resonance.

From equation (10-325)

$$Z(\omega) = \frac{(R + j\omega L)(1 - \omega^2 LC - j\omega RC)}{(1 - \omega^2 LC)^2 + (\omega RC)^2} = \frac{R + j\omega L^2 C(\omega_0^2 - \omega^2)}{(1 - \omega^2 LC)^2 + (\omega RC)^2} \tag{10-335}$$

10.10 ULTRA-LOW-NOISE WIDEBAND OSCILLATORS

Defining

$$\omega_0 = \frac{1}{\sqrt{L_0 C_0}} \left[\frac{Q^2}{1+Q^2} \right]^{1/2} \quad (10\text{-}336)$$

$$Q = \left[\frac{\omega_0 L_0}{R_0} \right] = \left[\frac{\omega_0}{\Delta \omega} \right]_{3\,\text{dB}} \quad (10\text{-}337)$$

where L_0 and R_0 denote inductance and resistance at the resonant frequency. Equation (10-335) can be expressed as

$$Z(\omega) = Z_{real}(\omega) + jZ_{imag}(\omega) \quad (10\text{-}338)$$

At the resonance frequency, $Z_{imag}(\omega)\big|_{\omega=\omega_0} = 0$ and $Z(\omega)$ is a real number at $\omega = \omega_0$.

$$Z(\omega)\big|_{\omega=\omega_0} = Z_0 = \frac{R}{(1-\omega_0^2 LC)^2 + (\omega_0 RC)^2} \quad (10\text{-}339)$$

$$R_0 = \frac{Z_0}{1+Q^2} \quad (10\text{-}340)$$

$$L_0 = Q\left[\frac{R_0}{\omega_0} \right] \quad (10\text{-}341)$$

From equation (10-335), admittance $Y(\omega)$ is given as

$$Y(\omega) = \frac{1}{Z(\omega)} = Y_{real}(\omega) + jY_{imag}(\omega) \quad (10\text{-}342)$$

$$Y(\omega) = \left[\frac{R}{R^2 + \omega^2 L^2} \right] + j\omega \left[C_0 - \frac{L}{R^2 + \omega^2 L^2} \right] \quad (10\text{-}343)$$

$$R = \frac{1 + \sqrt{1 - (2Y_{real}\omega L)^2}}{2Y_{real}} \quad (10\text{-}344)$$

$$Y_{imag} = \omega C_0 - Y_{real}\left[\frac{\omega L}{R} \right] \Rightarrow \frac{\omega C_0 - Y_{imag}}{Y_{real}} = \frac{\omega L}{R} \quad (10\text{-}345)$$

From equations (10-344) and (10-345), L can be expressed as a function of frequency $L(\omega)$ as

$$L = L(\omega) = \frac{R}{\omega}\left[\frac{\omega C_0 - Y_{imag}}{Y_{real}} \right] = \frac{1}{\omega\left[\omega C_0 + \frac{|Y|^2 - \omega C_0 Y_{imag}}{\omega C_0 - Y_{imag}} \right]} \quad (10\text{-}346)$$

$$\omega L(\omega) = \left[\frac{Z_{imag}(\omega) + \omega C|Z(\omega)|^2}{[1 + \omega C Z_{imag}(\omega)]^2 + [\omega C Z_{real}(\omega)]^2} \right] \quad (10\text{-}347)$$

where

$$|Y| = \frac{1}{|Z|}; \quad Y_{real} = \frac{Z_{real}}{|Z|^2}; \quad Y_{imag} = -\frac{Z_{imag}}{|Z|^2}$$

From equation (10-345)

$$R(\omega) = \frac{Y_{real}}{(\omega C_0)^2 - 2\omega C_0 Y_{imag} + |Y|^2}$$

$$= \frac{Z_{real}(\omega)}{[1 + \omega C Z_{imag}(\omega)]^2 + [\omega C Z_{real}(\omega)]^2} \quad (10\text{-}348)$$

$$C(\omega_0) = C_0 = \left[\frac{1}{\omega_0^2 L_0}\right]\left[\frac{Q^2}{1+Q^2}\right] \quad (10\text{-}349)$$

$$\left[\frac{C_0}{C(\omega)}\right]\left[\frac{\omega_0}{\omega}\right]^2 = \left[1 - \left|\frac{Z_i(\omega)\omega_0}{\omega Q R_0}\right|\right] \quad (10\text{-}350)$$

From equations (10-346) and (10-347), L and R are functions of the frequency, whereas self-capacitance C, as given in equation (10-349), is practically independent of the frequency. Thus, a priori estimation of the resonant frequency, quality factor, and impedance at resonance frequency helps to determine the inductor SMD component variables.

10.10.11 Capacitance Model

Figure 10-51 shows the equivalent circuit model for the SMD capacitor. The measured S-parameters are transferred into the two-port [Y] parameters using equation (10-327). The series impedance and shunt admittance of the π-equivalent representation of the capacitor, as shown in Figure 10-50, are given from equations (10-328), (10-329) and (10-330). For the capacitor, the equivalent circuit consists of a series connection of R, L, and C, as shown in Figure 10-51, and by duality from the inductor model, the stray inductance L is invariant with frequency, just as the self-capacitance is invariant with frequency in the inductor.

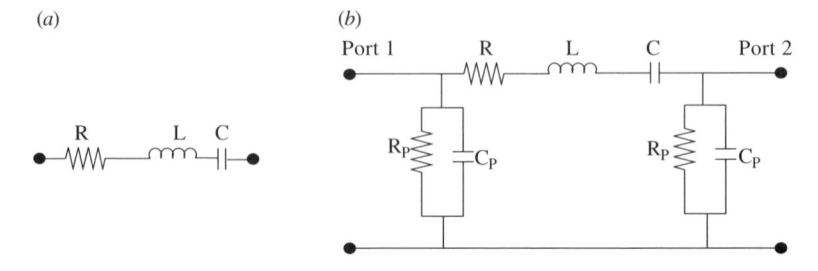

Figure 10-51 Model for an SMD capacitor. (*a*) Intrinsic model; (*b*) extrinsic model.

10.10 ULTRA-LOW-NOISE WIDEBAND OSCILLATORS

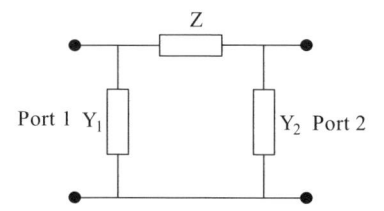

Figure 10-52 π-Model for an SMD capacitor.

From Figures 10-51 and 10-52, impedance is given as

$$Z(\omega) = Z_{real}(\omega) + jZ_{imag}(\omega)$$
$$= R(\omega) + jQ\left[\frac{\omega R_0}{\omega_0}\right]\left[1 - \left(\frac{\omega_0}{\omega}\right)^2 \frac{C_0}{C(\omega)}\right] \quad (10\text{-}351)$$

$$Z(\omega) = -\frac{1}{Y_{12}(\omega)} \quad (10\text{-}352)$$

$$R(\omega) = Z_{real}(\omega) \quad (10\text{-}353)$$

$$C(\omega) = \frac{1}{\omega(\omega L_0 - Z_{imag}(\omega))} \quad (10\text{-}354)$$

At resonance,

$$R_0 = R(\omega_0) = Z_{real}(\omega_0) \quad (10\text{-}355)$$

$$L_0 = Q\left[\frac{R_0}{\omega_0}\right] \quad (10\text{-}356)$$

$$C(\omega)|_{\omega=\omega_0} = C_0 = \frac{1}{\omega_0^2 L_0} \quad (10\text{-}357)$$

10.10.12 Resistor Model

The resistor can be implemented in several ways. Figure 10-53 shows the two-port extrinsic equivalent circuit model for a resistor. The measured S-parameters are transformed into the two-port [Y] parameters using equation (10-327). The series impedance and shunt admittance of the π-equivalent representation of the capacitor shown in Figure 10-54 are from equations (10-328), (10-329), and (10-330).

$$Z(\omega) = -\frac{1}{Y_{12}(\omega)} \quad (10\text{-}358)$$

$$Z(\omega) = Z_{real}(\omega) + jZ_{imag}(\omega) \quad (10\text{-}359)$$

$$R(\omega) = Z_{real}(\omega)$$

SYSTEMS OF COUPLED OSCILLATORS

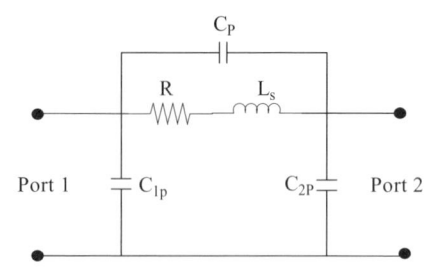

Figure 10-53 Two port extrinsic equivalent circuit model for a resistor.

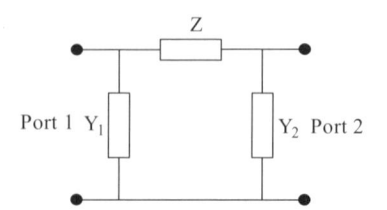

Figure 10-54 π-Model for an SMD resistor.

At resonance,

$$R_0 = R(\omega_0) = Z_{real}(\omega_0) \tag{10-360}$$

The two-port [Y] parameter of the resistor in Figure 10-54 is given by

$$[Y] = \begin{bmatrix} Y_{11} & Y_{12} \\ Y_{21} & Y_{22} \end{bmatrix}$$

$$= \begin{bmatrix} \left(j\omega[C_{1p}+C_p] + \frac{1}{[R+j\omega L_s]}\right) & -\left(j\omega C_p + \frac{1}{[R+j\omega L_s]}\right) \\ -\left(j\omega C_p + \frac{1}{[R+j\omega L_s]}\right) & \left(j\omega[C_{1p}+C_p] + \frac{1}{[R+j\omega L_s]}\right) \end{bmatrix}$$

$$\tag{10-361}$$

11 Validation Circuits for Wideband Coupled Resonator VCOs

This chapter demonstrates the state of the art in designing ultra-low-noise wideband VCOs having a tuning bandwidth more than octave band and amenable for integration in integrated chip form. This work deals with the design, fabrication, and testing of wideband oscillators/VCOs which can satisfy the demand for low noise, low cost, wide tuning range, power efficiency, manufacturing tolerance, and miniaturization.

The following circuits have been chosen for validation:

- wideband VCO (300–1100 MHz) based on a coupled resonator
- wideband VCO (1000–2000/2000–4000 MHz) based on the push–push approach
- wideband VCO (1500–3000/3000–6000 MHz) based on a dual coupled resonator
- hybrid-tuned ultra-low-noise wideband VCO (1000–2000/2000–4000 MHz)

11.1 300–1100 MHz COUPLED RESONATOR OSCILLATOR

A number of operational parameters may be considered, depending on the oscillator's intended applications, but phase noise is the important figure for measurement and instrument applications. In view of the limitations of wideband tunability, the circuit topology and layout of the resonator are selected in such a way that they support uniform negative resistance more than octave band [114]. Figure 11-1 illustrates the block diagram of the wideband VCO (300–1100 MHz) [8].

The circuit topology, as shown in Figure 11-2a, is based on coupled resonators. The VCO can be dynamically tuned to operate over a fairly wide range of frequencies while maintaining low phase noise. Replacing a simple microstripline resonator with a coupled resonator optimizes the dynamic loaded Q of the resonator and

---

The Design of Modern Microwave Oscillators for Wireless Applications: Theory and Optimization, by Ulrich L. Rohde, Ajay Kumar Poddar, Georg Böck
ISBN 0-471-72342-8 Copyright © 2005 John Wiley & Sons, Inc.

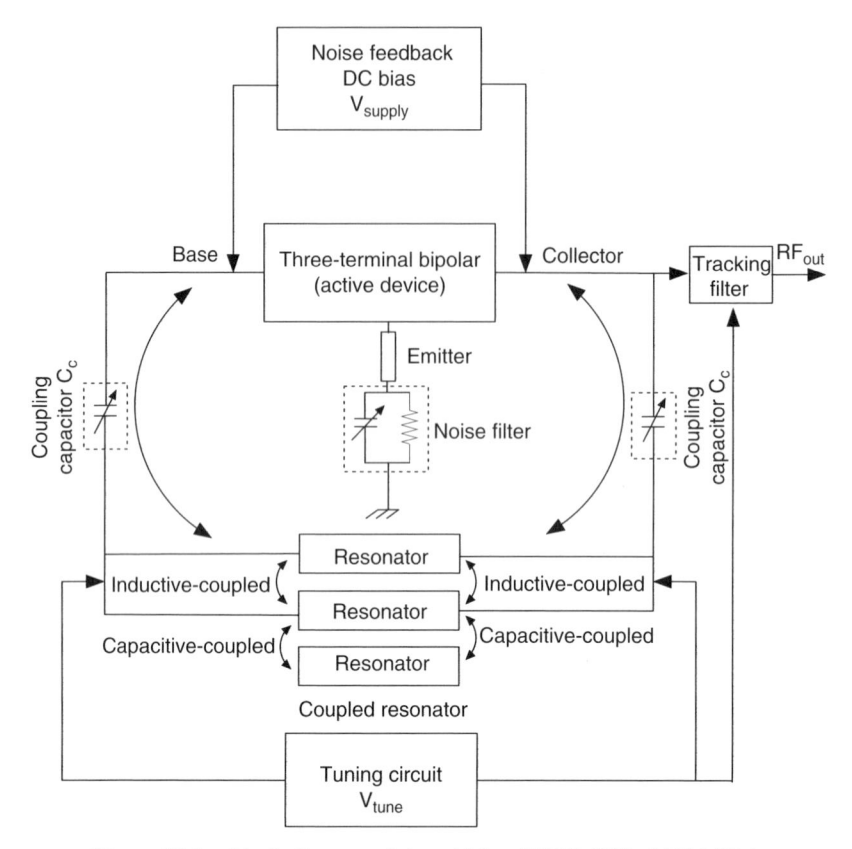

Figure 11-1 Block diagram of the wideband VCO (300–1100 MHz).

thereby increases the average loaded Q [equation (10-315)], which depends upon the coupling factor. In particular, by choosing appropriate spacing between the coupled resonators, the coupling factor can be optimized, thus improving the phase noise. Further improvement is accomplished by incorporating noise filtering at the emitter of the bipolar transistor [119].

To support a uniform negative resistance over the tuning range, the varactor-tuned coupled resonator shown in Figure 11-2a is connected across the base and collector of the active device, and the loss resistance is compensated for by the negative resistance. It is dynamically adjusted as a change of oscillator frequency occurs dynamically, tuning the phase shift of the negative resistance-generating network to meet the phase shift criteria for the resonance over the operating frequency band. Furthermore, by incorporating a tracking filter at the output, the VCOs are controlled.

The variable coupling capacitor C_c is designed for optimum loading of the resonator network across the active device (the base and the collector of the transistor). As shown in Figure 11-1, the coupled resonator is connected across the base and the

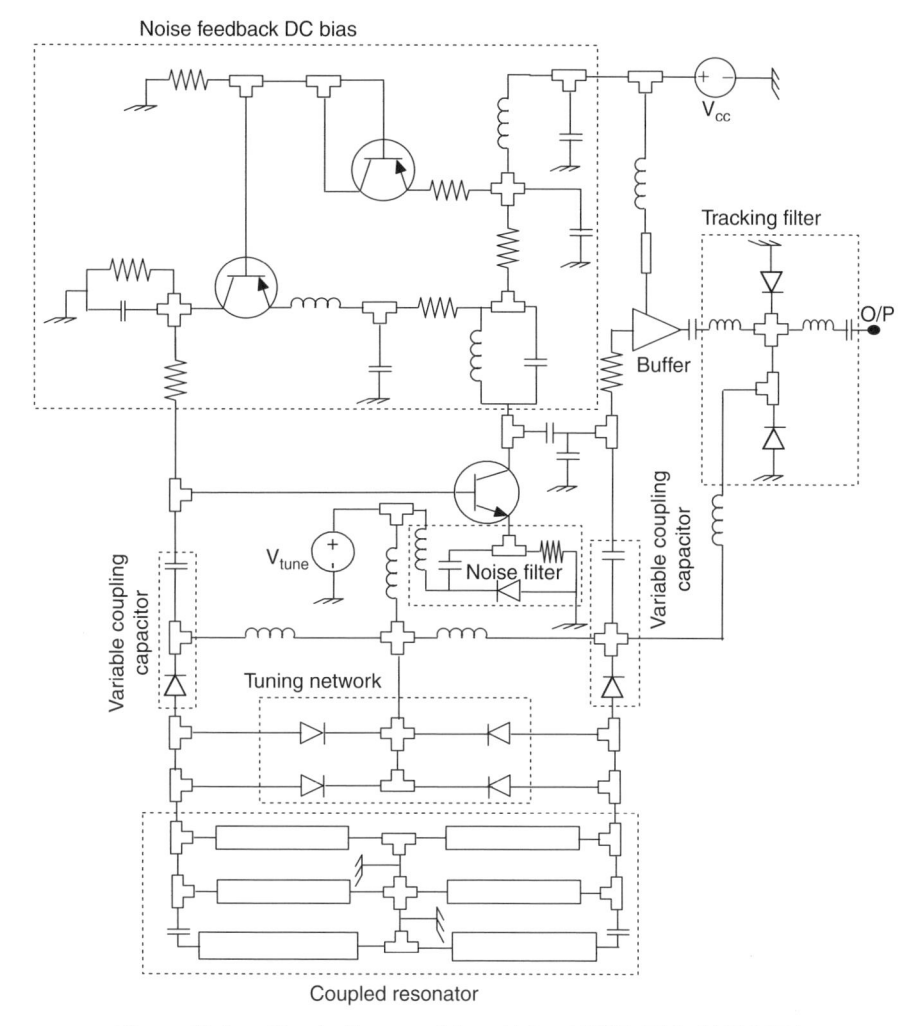

11.1 300–1100 MHz Coupled Resonator Oscillator   **343**

Figure 11-2a   Circuit diagram of the wideband VCO (300–1100 MHz).

collector of the three-terminal device through the coupling capacitor, which is electronically tuned by applying the tuning voltage to the tuning network integrated with the coupled resonator. The values of the coupling capacitor C_c are derived from the input stability circle. It should be within the input stability circle so that the circuit will oscillate at a particular frequency for the given lowest possible value of C_c. The additional feature of this topology is that the user-defined frequency band is obtained by adjusting the length and spacing of the coupled resonator. Figures 11-2a and 11-2b depict the corresponding circuit and layout diagram, which allows for a miniaturization and is amenable to integrated chip form. This structure and application is covered by U.S. patent application no. 60/564173.

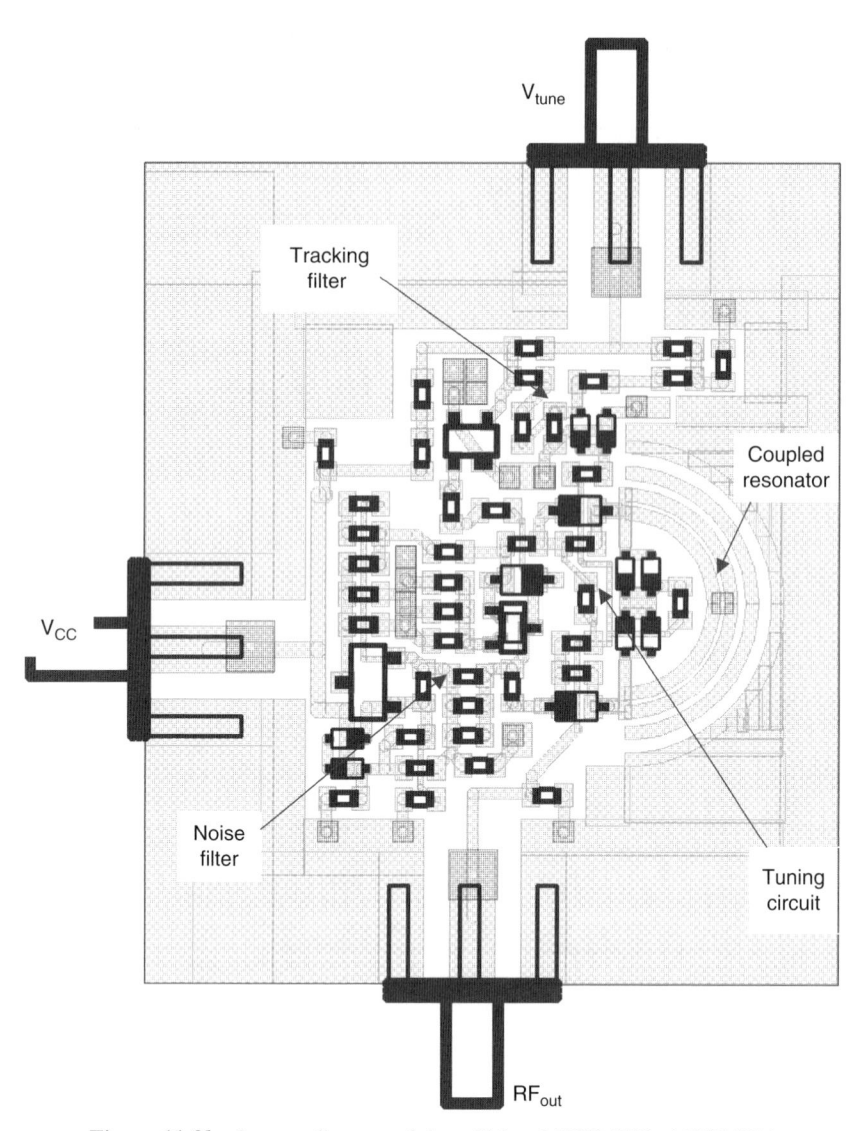

Figure 11-2b Layout diagram of the wideband VCO (300–1100 MHz).

Figures 11-3a and 11-3b show the simulated and measured phase noise plot for the wideband VCO (300–1100 MHz).

The simulated plot of the phase noise agrees with the measured phase noise plot. The measured phase noise is better than -110 dBc/Hz at 10 kHz for the tuning range of 300–1100 MHz. The difference of 3 dB over the band is due to the change in the component characteristics over the tuning range.

Figures 11-4 to 11-6 show the measured tuning sensitivity, output power, and harmonic suppression plot for the 300–1100 MHz VCO.

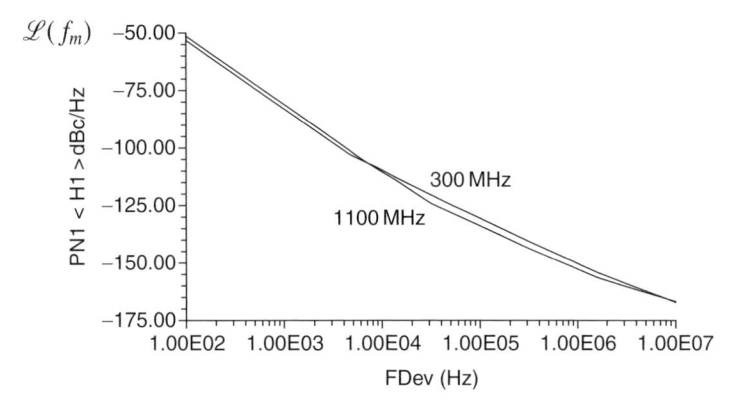

11.1 300–1100 MHz Coupled Resonator Oscillator   345

Figure 11-3a Simulated phase noise plot for the wideband VCO (300–1100 MHz).

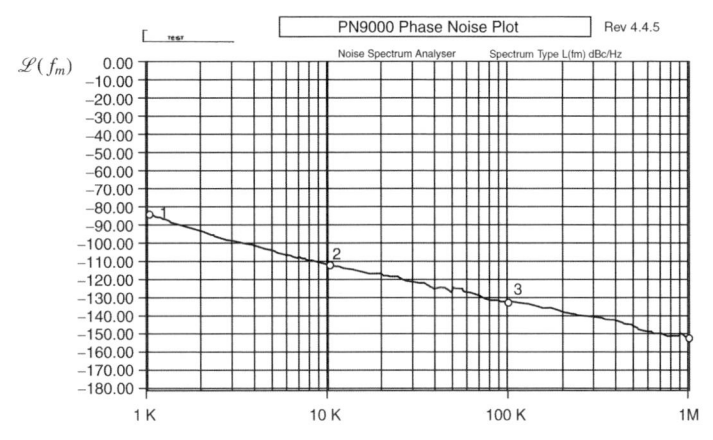

Figure 11-3b Measured phase noise for the wideband VCO (300–1100 MHz) for a frequency of 1100 MHz. *Source*: Synergy Microwave Corp. using the Aeroflex phase noise system.

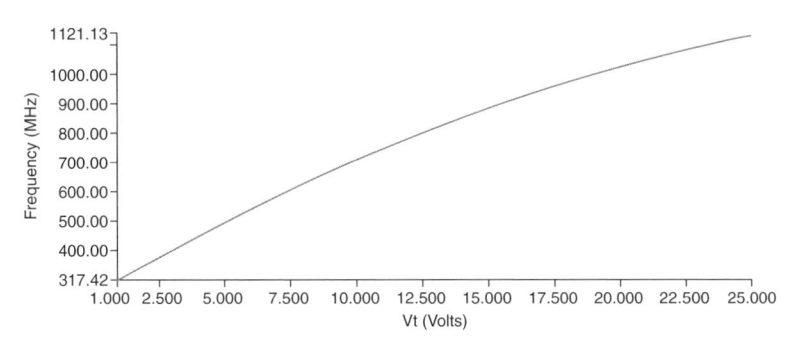

Figure 11-4 Frequency versus tuning voltage.

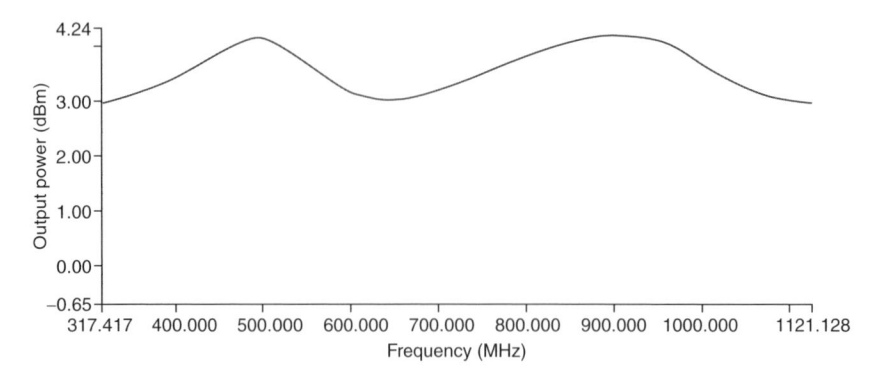

Figure 11-5 Output power versus frequency.

This work offers a cost-effective and power-efficient solution (5 V, 28 mA), and efficiency can be improved further by replacing the buffer amplifier with a matching network. The oscillator will operate at 5 V and 18 mA and is best suited for an application where power efficiency and phase noise are criteria for a selection of wideband VCOs.

Wideband VCO (300–1100 MHz) Features

Oscillator frequency:	300–1100 MHz
Tuning voltage:	0–25 VDC (nom)
Tuning sensitivity:	24–48 MHz/V
Bias voltage:	+5 VDC at 30 mA (nom)
Output power:	+2 dBm (min)
Harmonic suppression:	25 dBc (typ)
Voltage standing wave ratio (VSWR):	1.5:1
Phase noise:	−112 dBc/Hz at 10 kHz offset (typ)
Phase noise:	−132 dBc/Hz at 100 kHz offset (typ)
Frequency pulling:	4 MHz (max) at 1.75:VSWR
Frequency pushing:	2 MHz/V (max)
Output impedance:	50 ohms
Operating temp:	−40°C to 85°C
Size:	0.9 × 0.9 in.

11.2 1000–2000/2000–4000 MHz PUSH–PUSH OSCILLATOR

This work demonstrates the state of the art of ultra-low-phase noise wideband VCOs using push–push topology [187]. Figure 11-7 depicts the block diagram of the wideband VCOs (1000–2000/2000–4000 MHz), in which all the modules are

11.2 1000–2000/2000–4000 MHz Push–Push Oscillator   347

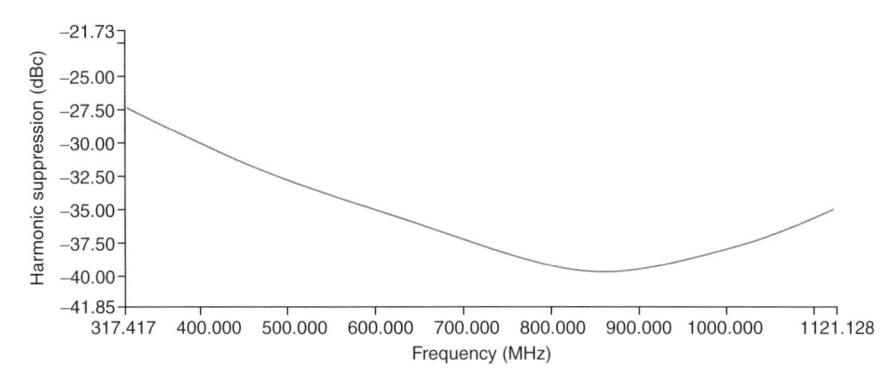

Figure 11-6 Second-order harmonic versus frequency.

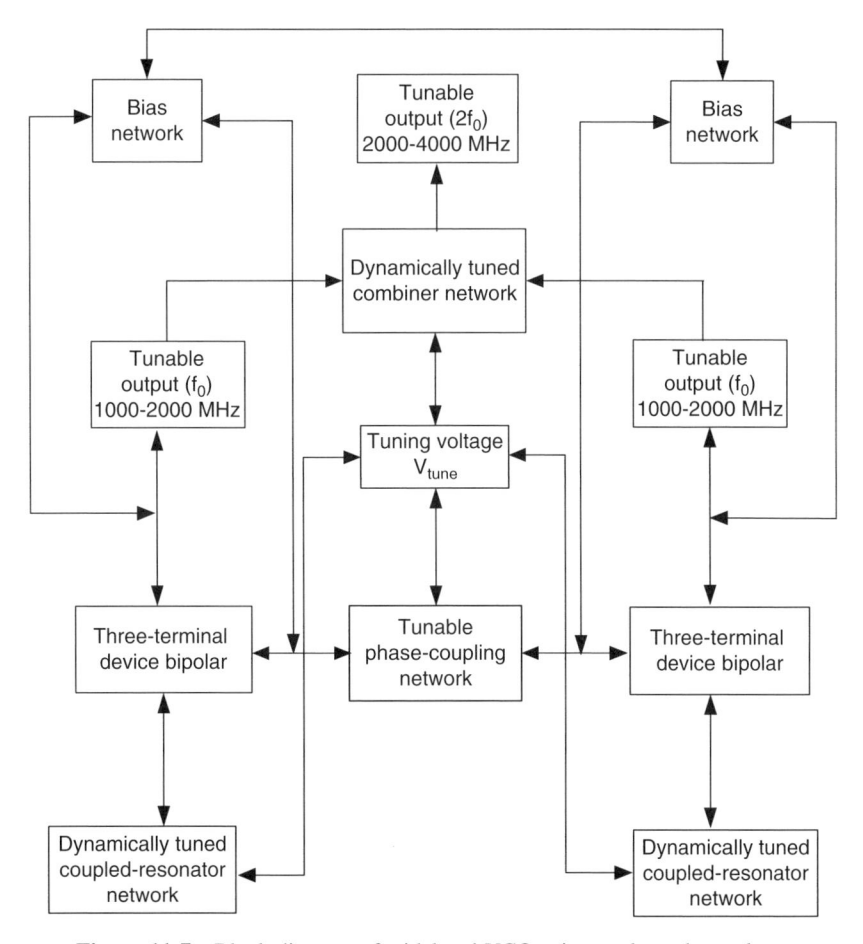

Figure 11-7 Block diagram of wideband VCO using push–push topology.

348   VALIDATION CIRCUITS FOR WIDEBAND COUPLED RESONATOR VCOs

self-explanatory. In push–push topology, two subcircuits of a symmetrical topology operate in opposite phase at the fundamental frequency, and the outputs of the two signals are combined through the dynamically tuned combiner network so that the fundamental cancels out while the first harmonic is available over the tuning range.

As shown in Figure 11-8, each subcircuit runs at one-half of the desired output frequency (f_0). Thus, the second harmonic ($2f_0$) is combined with the help of the dynamically tuned combiner network. Wideband tunability is achieved by incorporating a dynamically tuned phase coupling network so that 180° phase difference, mutually locked condition [equation (10.71)] is maintained over the tuning range for push–push operation.

The push–push topology has several advantages over single-ended versions other than the improvement in phase noise. The usable frequency range of the transistors can be extended, and this can be exploited, for instance, using transistors that are larger than usual and have lower $1/f$ noise due to reduced current density.

Figure 11-8 shows the schematic of the wideband VCO (1000–2000/2000–4000 MHz) in the push-push configuration. The various modules depicted in

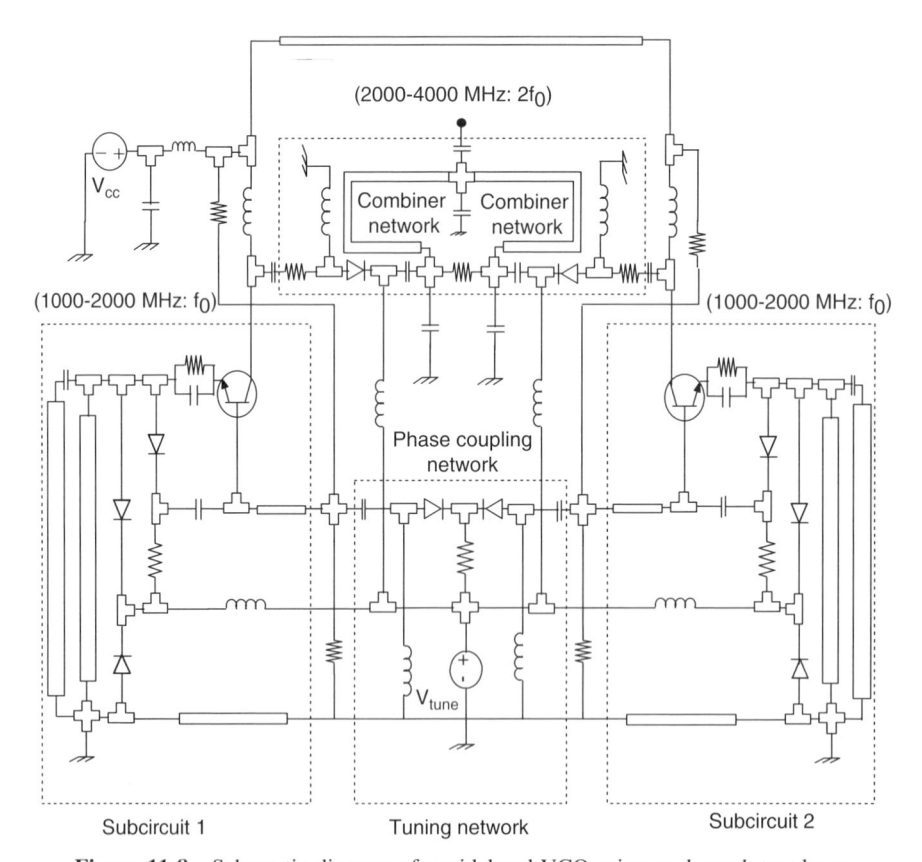

Figure 11-8   Schematic diagram of a wideband VCO using push–push topology.

11.2   1000–2000/2000–4000 MHz Push–Push Oscillator   349

Figure 11-7 are implemented in a way that allows miniaturization and is amenable for integrated chip design. This structure and application is covered by U.S. patent application nos. 60/501371 and 60/501790.

Figure 11-9 shows the layout of Figure 11-8 [187]. Experimental results have shown that a poor mismatch at the fundamental results in discontinuous tuning due to the nonuniform phase shift over the tuning range. This mismatch in phase shift between the two subcircuits is due to possible component tolerances, package parameters, and the phase associated with the path difference over the tuning range. Therefore, the system goes out of the locking range.

Figure 11-10 shows the compact layout of Figure 11-9 so that the phase shift between the two subcircuits can be minimized. The effective phase shift is reduced in comparison to the layout in Figure 11-9.

The layout shown in Figure 11-9 minimizes the phase shift due to the path difference between the two subcircuits over the tuning range, but still shows discontinuous tuning at some point over the band due to the package parasitics and component tolerances associated with the discrete components of the circuit. The problem of discontinuous tuning can be overcome by incorporating a phase detector (patent application no. 60/563481). Then the tuning range can be extended to a multi-octave band. Figure 11-11 shows the plot of the RF current of subcircuits 1 and 2, which are out of phase in a mutually synchronized condition, as given by equation (10-71).

Figures 11-12a and 11-12b show the simulated phase noise plot for a wideband push–push VCO at a frequency of 2 GHz and 4 GHz.

Figures 11-12c–f show the measured phase noise plot for a wideband push–push VCO for the oscillation frequency of 2 GHz and 4 GHz.

Figures 11-12a and 11-12b show a simulated phase noise with an 8–9 dB improvement over the frequency band (2000–4000 MHz). The measured phase noise show 5–7 dB improvement with respect to the fundamental oscillator subcircuit and is better than -112 dBc/Hz at 100 kHz for the tuning range of 2000–4000 MHz in the push–push configuration.

From equation (10-206), the phase noise relative to the carrier decreases by a factor of N, where N is the number of coupled oscillator circuits. For a bilateral coupled oscillator (push–push), the improvement of the simulated phase noise is 9 dB with respect to the fundamental oscillator, as given in equation (10-226). Measured results show a 5–7 dB improvement of the phase noise over the band. The discrepancy can be attributed to the package parasitics, dynamic loaded Q, and tolerances of the component values of the two subcircuits over the tuning range.

Further improvement in phase noise can be achieved by implementing N-push approach with an integrated phase detector for minimizing and correcting the phase shift over the tuning range; this is a further extension of this research work for the purpose of covering the millimeter frequency range. Figure 11-13 shows the layout of a multioctave integrated VCO (1–2 GHz, 2–4 GHz, 4–8 GHz, 8–16 GHz) and is amenable for integration with wideband synthesizer systems. This structure and application are covered by U.S. patent application no. 60/563481.

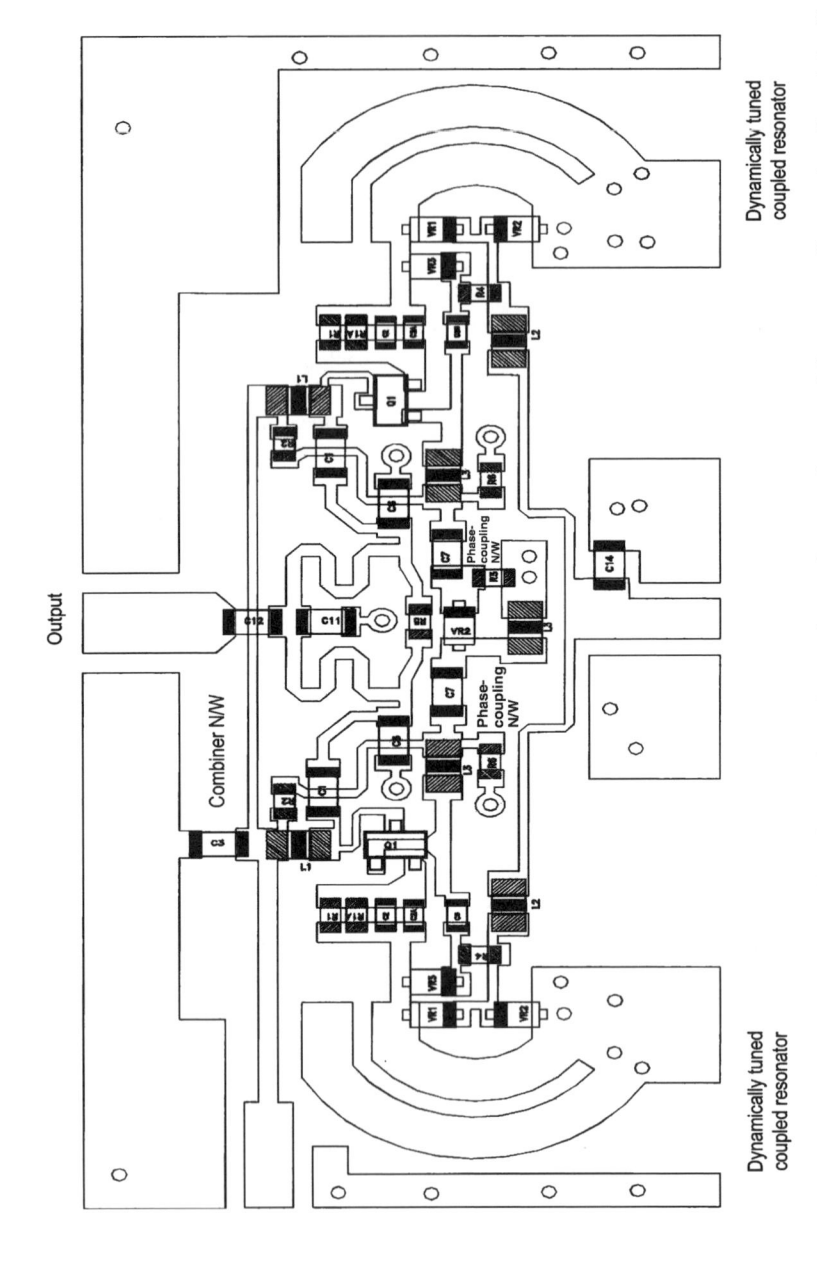

Figure 11-9 Layout of a wideband VCO using push–push topology. *Source*: Synergy Microwave Corporation-Copyright Registration No. Vau-603-982.

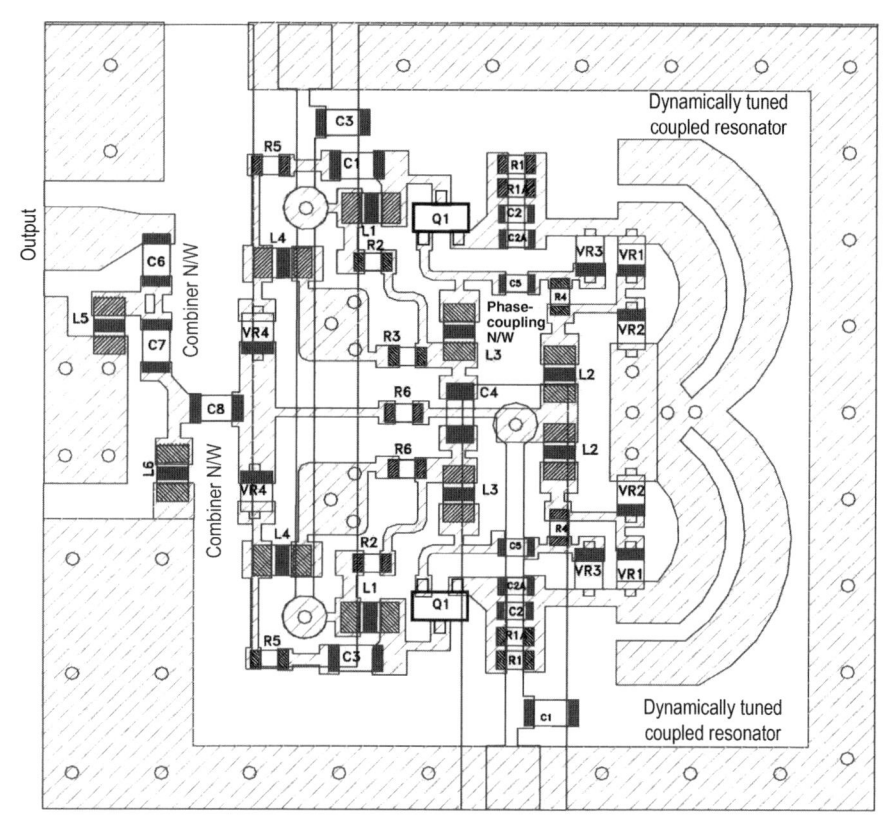

11.2 1000–2000/2000–4000 MHz Push–Push Oscillator   **351**

Figure 11-10 Compact layout of a wideband VCO using push–push topology. *Source*: Synergy Microwave Corporation-Copyright Registration No. Vau-603-982.

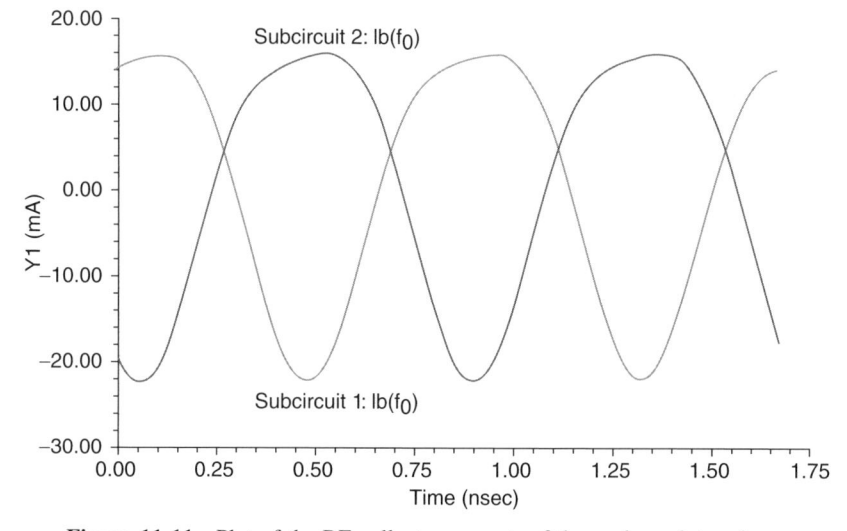

Figure 11-11 Plot of the RF-collector currents of the push–push topology.

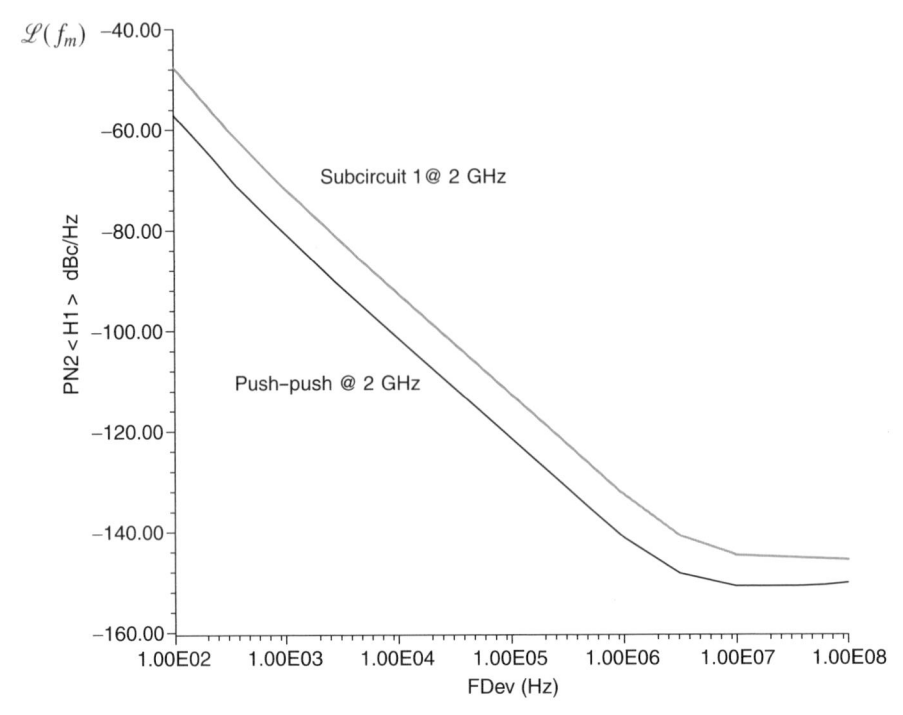

352 VALIDATION CIRCUITS FOR WIDEBAND COUPLED RESONATOR VCOs

Figure 11-12a Simulated phase noise plot for the wideband push–push VCO (2000–4000 MHz) for the oscillation frequency of 2 GHz.

Wideband VCO (1000–2000/2000–4000 MHz) Features

Oscillator frequency:	1000–2000/2000–4000 MHz
Tuning voltage:	0–25 VDC (nom)
Tuning sensitivity:	40–100 MHz/V
Bias voltage:	+12 VDC at 30 mA (nom)
Output power:	+5 dBm (typ)
Harmonic suppression:	20 dBc (typ)
VSWR:	1.5:1
Phase noise:	−95 dBc/Hz at 10 kHz offset (typ)
Phase noise:	−115 dBc/Hz at 100 kHz offset (typ)
Frequency pulling:	14 MHz (max) at 1.75:VSWR
Frequency pushing:	7 MHz/V (max)
Output impedance:	50 ohms
Operating temp:	−40°C to 85°C
Size:	0.9 × 0.9 in.

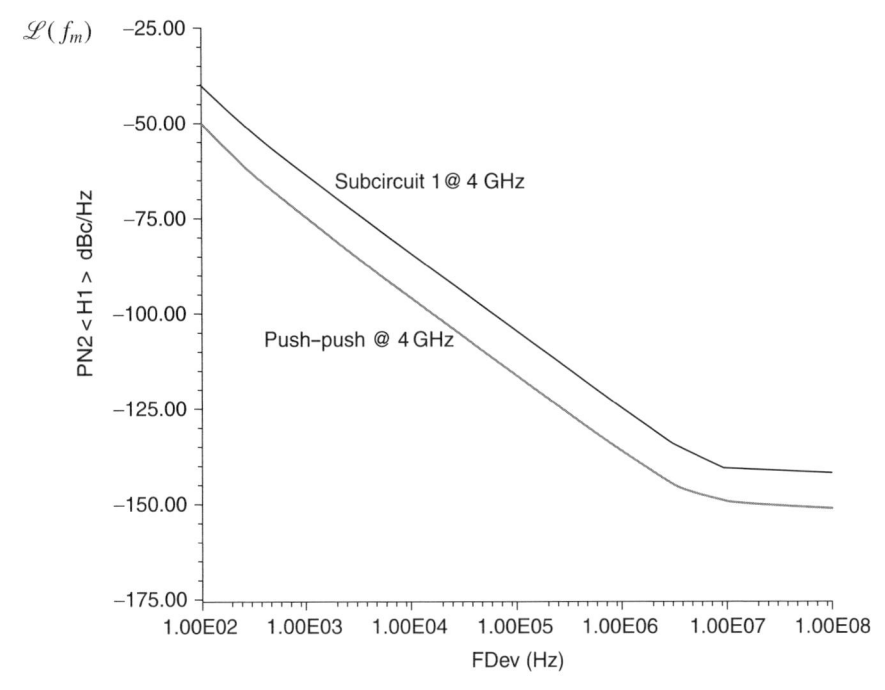

Figure 11-12b Simulated phase noise plot for the wideband push–push VCO (2000–4000 MHz) for the oscillation frequency of 4 GHz.

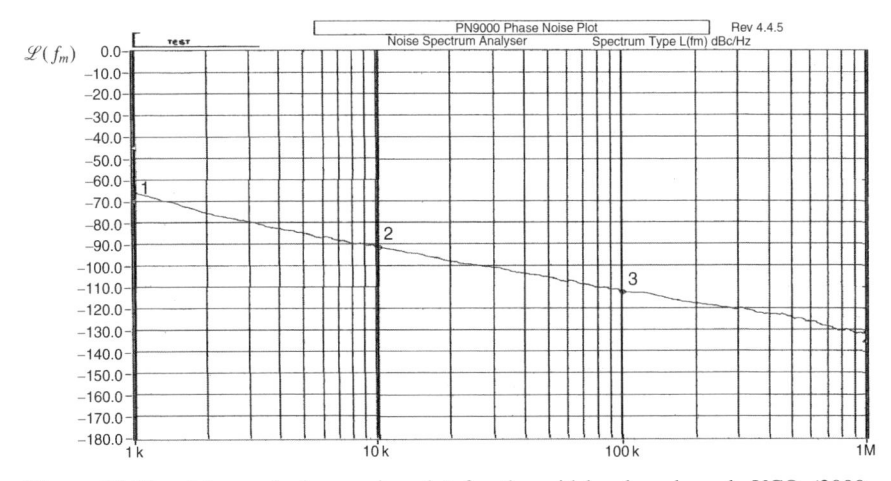

Figure 11-12c Measured phase noise plot for the wideband push–push VCO (2000–4000 MHz) of subcircuit 1 for the oscillation frequency of 2000 MHz. *Source*: Synergy Microwave Corp. using the Aeroflex phase noise system.

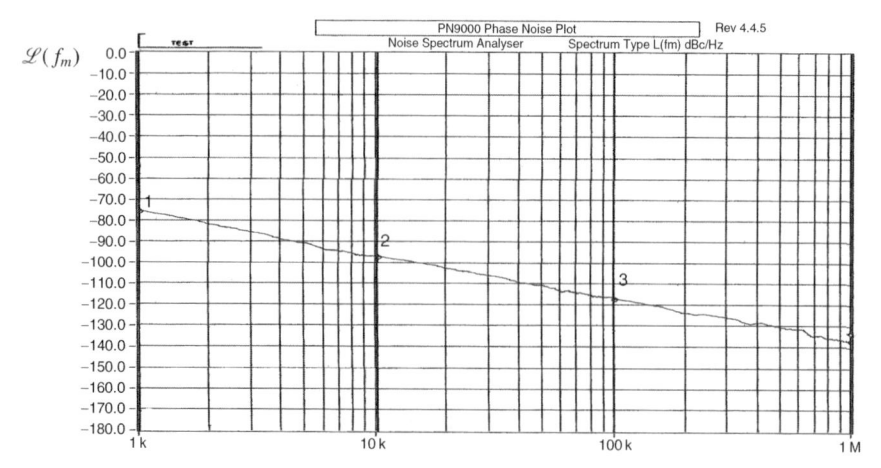

Figure 11-12d Measured phase noise plot for the wideband push–push VCO (2000–4000 MHz) of combiner output for the oscillation frequency of 2000 MHz. *Source*: Synergy Microwave Corp. using the Aeroflex phase noise system.

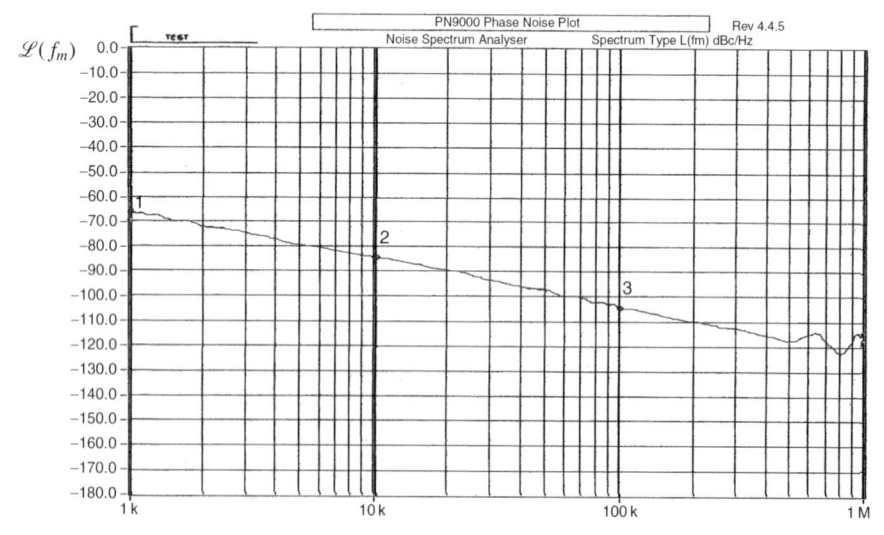

Figure 11-12e Measured phase noise plot for the wideband push–push VCO (2000–4000 MHz) of subcircuit 1 for the oscillation frequency of 4000 MHz. *Source*: Synergy Microwave Corp. using the Aeroflex phase noise system.

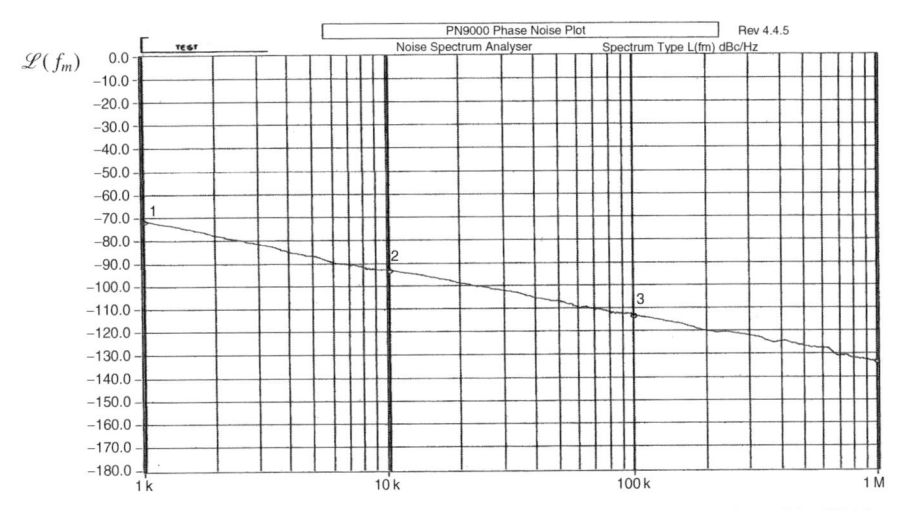

Figure 11-12f Measured phase noise plot for the wideband push–push VCO (2000–4000 MHz) of combiner output for the oscillation frequency of 4000 MHz. *Source*: Synergy Microwave Corp. using the Aeroflex phase noise system.

11.3 1500–3000/3000–6000 MHz DUAL COUPLED RESONATOR OSCILLATOR

Figure 11-14 shows the schematic diagram of the wideband VCO (1500–3000/3000–6000 MHz) based on a dual-coupled resonator.

For wideband applications such as the octave band, the grounded base topology is preferred due to the broadband tunability factor.

For wideband tunability, the required negative resistance over the band is generated by the feedback base inductance, but the polarity of the reactance may change over the frequency band and can lead to the disappearance of the negative resistance as the operating frequency exceeds its SRF. This problem is overcome by incorporating a coupled line topology instead of the lumped inductance used as a feedback element for generating negative resistance over the tuning range, as shown in Figure 11-14.

Designing wideband VCOs at high frequency is challenging due to the change in the characteristics of the RF components over the tuning range. Biasing is critical and can have a negative effect on the overall performance if implemented poorly because the oscillator/VCO circuits are highly sensitive to voltage fluctuations and up-convert any noise on the bias points.

Biasing with a resistor provides a resonant-free solution for wideband applications, but it may unnecessarily load the output of the oscillator, thereby lowering the gain and degrading the phase noise performance. A normal biasing scheme (see Section 8.7), such as a voltage divider or a current divider technique using resistors,

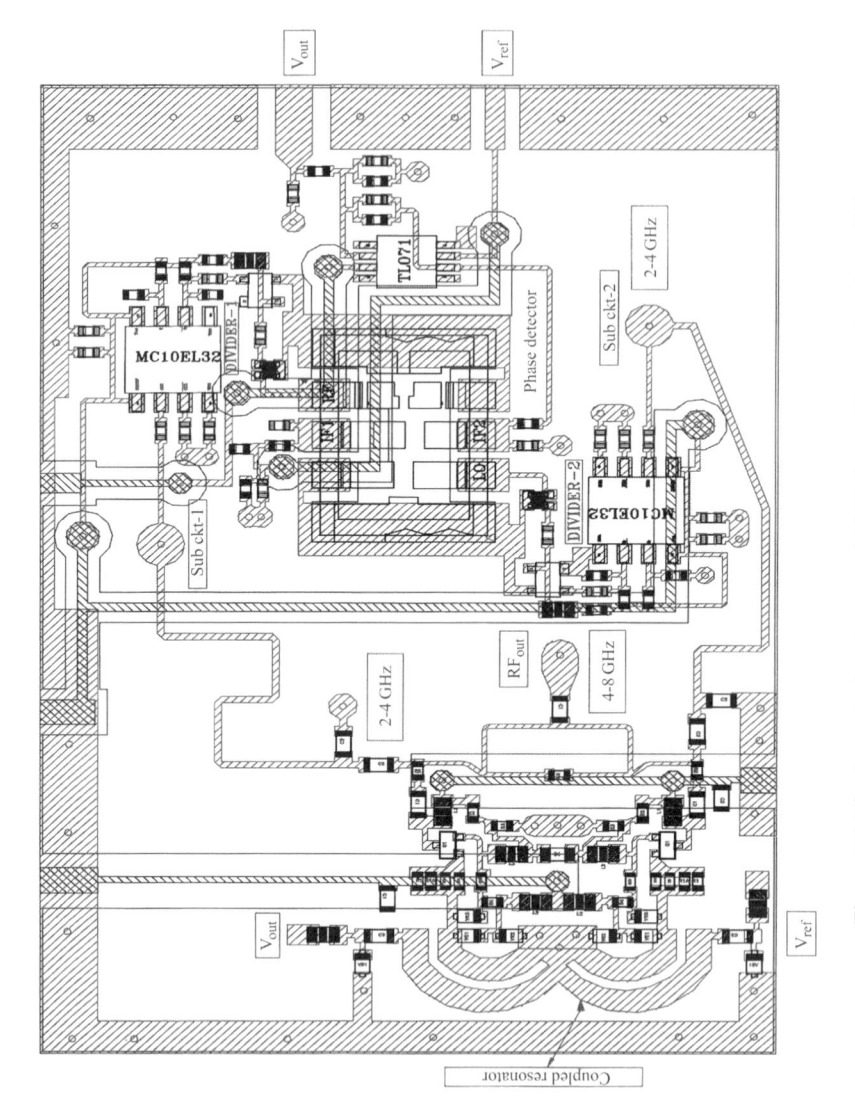

Figure 11-13 Layout of an integrated VCO for multioctave band capability.

11.3 1500–3000/3000–6000 MHz Dual Coupled Resonator Oscillator

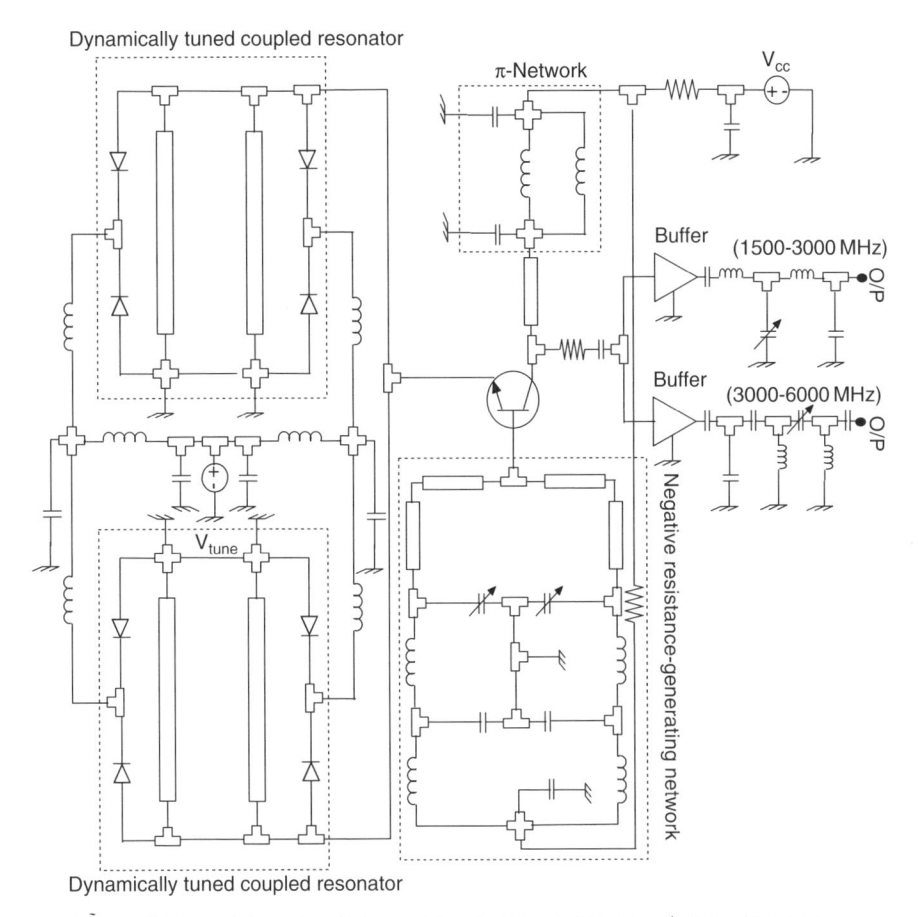

Figure 11-14 Schematic of the wideband VCO (1500–3000/3000–6000 MHz).

is good for amplifier design but may contribute significantly to noise in the oscillator circuit. Biasing with the potential divider with the resistors effectively increases the noise, which may be up-converted to the fundamental frequency of oscillation as an additional phase noise.

For VCO circuits, the collector is generally biased using an inductor with an SRF as high as possible above the oscillation frequency so that it will not affect the fundamental frequency of oscillation. The value of the inductor must be high enough to resemble a broadband RF open circuit for a desired tuning range. This can be achieved by incorporating two parallel inductor series with a transmission line to create the broadband open circuit shown in Figure 11-14. Similarly, multiple capacitors are used in parallel to create a broadband RF short-circuit with respect to the SRF for the lower and higher ends of the tuning range.

Biasing for the tuning network is also critical for overall phase noise performance. The tuning diode exhibits a change in capacitor as a function of voltage. A resistor provides resonant-free biasing, but it introduces noise to the oscillator due to the voltage gain (integration gain) K_0 (MHz/V) of the oscillator. Therefore, a π-network is incorporated to reduce the noise in the tuning network, as shown in Figure 11-14, which minimizes the noise at the tuning port. Figure 11-15 shows the layout diagram of the wideband VCO (1500–3000/3000–6000 MHz) frequency band. This structure and application are covered by U.S. patent application nos. 60/501371 and 60/501790.

As shown in Figures 11-14 and 11-15, the symmetrical coupled resonator is connected across the two emitters of the transistor (Infineon BFP 520) to provide a uniform loaded Q over the tuning range, thereby providing a uniform phase noise over the tuning range. Figure 11-16 shows the plot of the Q factor (loaded and unloaded) versus the tuning voltage.

Figures 11-17a and 11-17b show the simulated and measured phase noise plot of the wideband VCO (1500–3000 MHz).

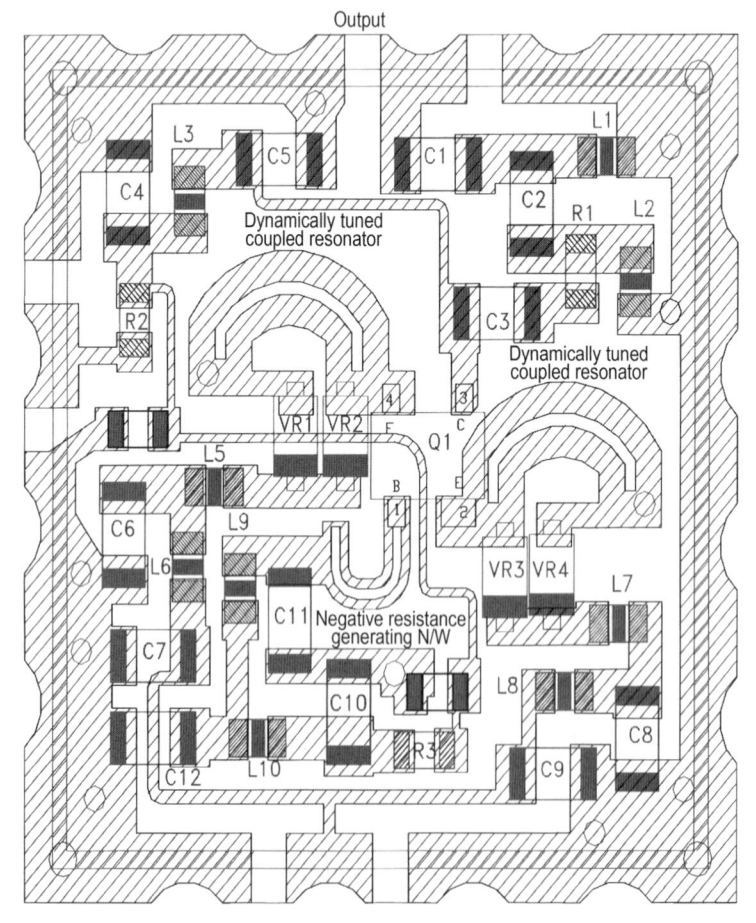

Figure 11-15 Layout of the wideband VCO (1500–3000/3000–6000 MHz). *Source*: Synergy Microwave Corporation-Copyright Registration No. Vau-603-984.

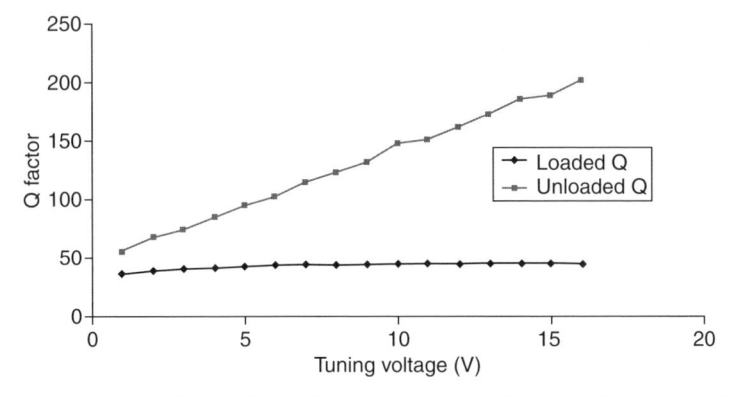

Figure 11-16 Plot of the Q factor (loaded and unloaded) versus the tuning voltage.

The circuit of Figure 11-14 is rich of the second harmonic because of the large drive signal level, and the second harmonic (3000–6000 MHz) can be filtered out as an extension of the fundamental frequency band without an external multiplier module. Figure 11-18 shows the simulated phase noise plot of the wideband VCO (3000–6000 MHz).

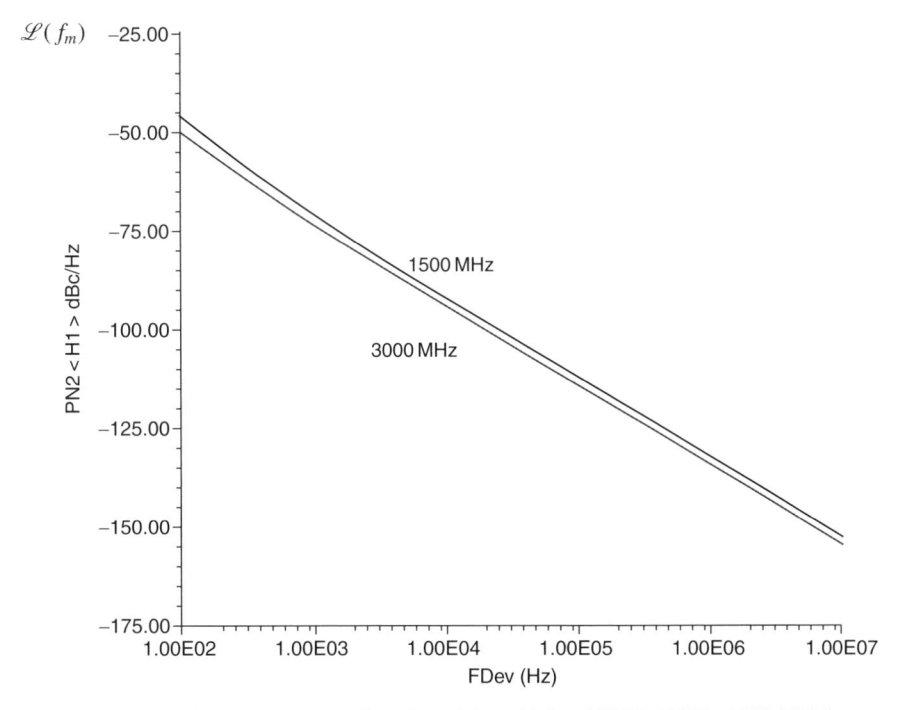

Figure 11-17a Simulated phase noise plot of the wideband VCO (1500–3000 MHz).

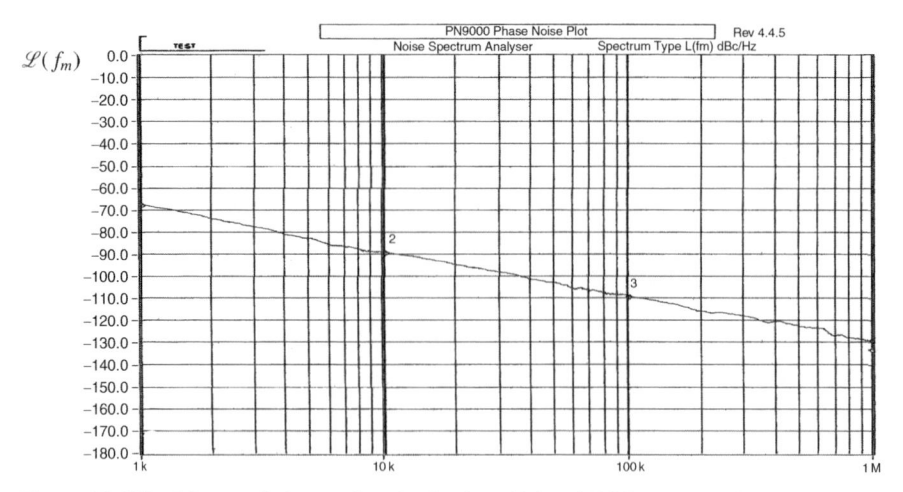

360 VALIDATION CIRCUITS FOR WIDEBAND COUPLED RESONATOR VCOs

Figure 11-17b Measured phase noise plot for the wideband VCO (1500–3000 MHz) for the oscillation frequency 2000 MHz. *Source*: Synergy Microwave Corp. using the Aeroflex phase noise system.

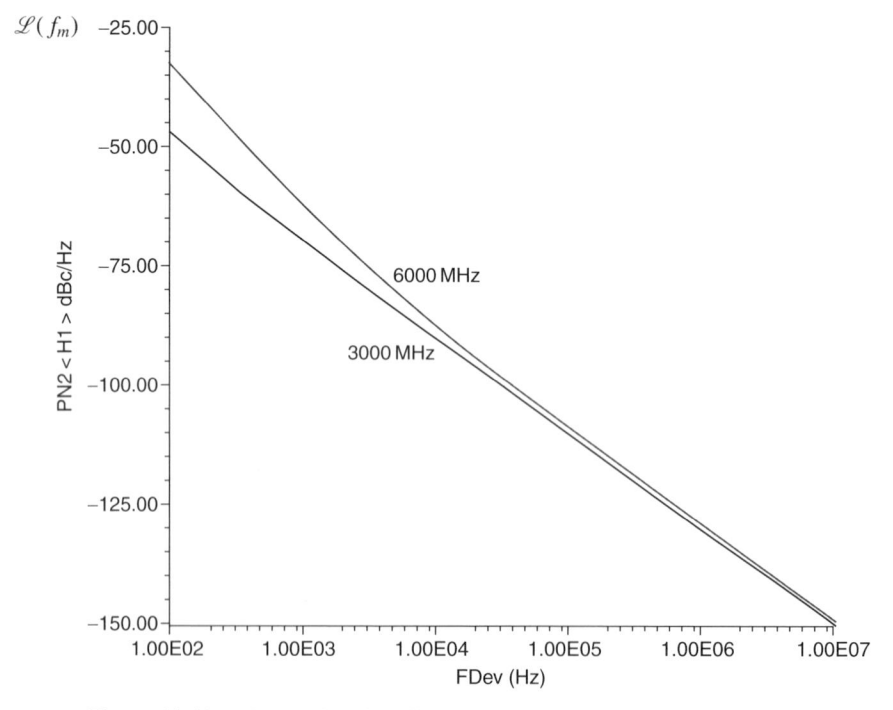

Figure 11-18 Phase noise plot of the wideband VCO (3000–6000 MHz).

Wideband VCO (1500–3000/3000–6000 MHz) Features

Oscillator frequency:	1500–3000/3000–6000 MHz
Tuning voltage:	0–20 VDC (nom)
Tuning sensitivity:	75–150 MHz/V
Bias voltage:	+5 VDC at 30 mA(nom)
Output power:	+3 dBm (typ)
Harmonic suppression:	15 dBc (typ)
VSWR:	1.75:1
Phase noise:	−95 dBc/Hz at 10 kHz offset (typ)
Phase noise:	−115 dBc/Hz at 100 kHz offset (typ)
Frequency pulling:	10 MHz (max) at 1.75:VSWR
Frequency pushing:	6 MHz/V (max)
Output impedance:	50 ohms
Operating temp:	−40°C to 85°C
Size:	0.5 × 0.5 in.

11.4 1000–2000/2000–4000 MHz HYBRID TUNED VCO

The design is based on an innovative topology which supports fast convergence by using both coarse-tuning and fine-tuning networks. The oscillator with the proposed topology, as shown in Figure 11-19, has remarkable low phase noise performance over the whole tuning range. This structure and application are covered by U.S. patent application no. 60/601823.

As a PLL requires a continuous tuning range from the VCO, the fine-tuning network must be implemented in parallel with a coarse-tuning network. In this tuning approach, large changes in capacitance are provided by the coarse-tuning network, and the fine-tuning network only needs to provide a small tuning range. To ensure that there are no "dead zones" of tuning (i.e., intermediate frequencies that cannot be tuned) the fine-tuning network provides the change in capacitance to meet the requirement. The advantage of a hybrid-tuning network is that the VCO is less susceptible to large jumps in frequency due to noise. Thus, the hybrid (coarse/fine) tuning approach minimizes the trade-off between a wide tuning range and voltage control–induced phase noise. The overall performance of the PLL is related to the frequency sensitivity of the tuning control voltage. A VCO with voltage gain has less phase jitter but also a narrower tuning range. In applications where coarse and fine tuning is a prime requirement, the phase noise becomes drastically degraded in the fine-tuning network if incorporated with the coarse-tuning network.

Specifically, due to the fairly large bandwidth (octave band) tuning range for PLL applications, a fine-tuning network adds extra noise and loading to the resonator circuit. This problem can be solved by introducing an innovative topology for a dual tuning network, which incorporates coarse and fine tuning without

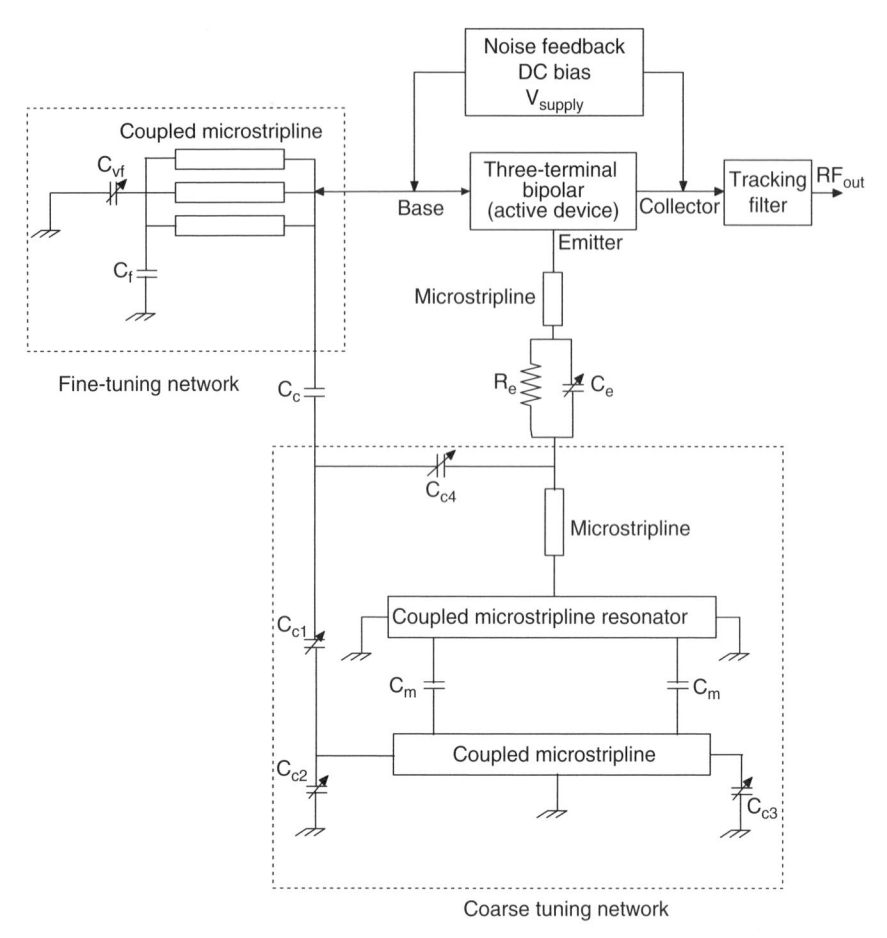

Figure 11-19 This block diagram shows the discrete-device approach with a coupled resonator used in octave band (1000–2000/2000–4000 MHz) hybrid-tuned VCOs.

compromising the phase noise performance over the tuning range. This novel proprietary design approach facilitates coarse and fine tuning and maintains ultra-low-noise performance over the tuning range (1000–2000/2000–4000 MHz). The typically phase noise is −105 dBc/Hz at 10 kHz offset for the frequency 1000–2000 MHz and −98 dBc/Hz for the frequency 2000–4000 MHz. Figure 11-20 shows the measured phase noise plot for the frequency 2500 MHz. The circuit operates at 8 V and 25 mA and gives a power output of more than 10 dBm over the tuning range. Coarse tuning for 1000–2000 MHz is 1.5 to 22 V, and fine-tuning is 1–5 V (10–20 MHz/V). The profile of the tuning sensitivity (coarse and fine) is user-defined, and the step size is selectable. Figure 11-21 shows the plot of the coarse-tuning sensitivity for the tuning range 1–25 V. Figure 11-22 shows the plot of harmonic suppression, which is better than

11.4 1000–2000/2000–4000 MHz Hybrid Tuned VCO

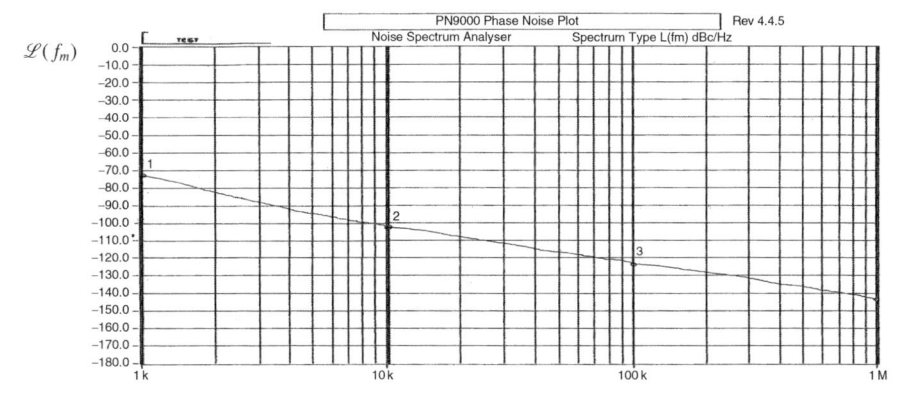

Figure 11-20 Measured phase noise plot of broadband hybrid-tuned VCOs for the frequency 2500 MHz (1000–2000/2000–4000 MHz). *Source*: Synergy Microwave Corp. using the Aeroflex phase noise system

−16 dB over the tuning range. Figure 11-23 shows the plot of the output power, with less than 3 dB for the octave band (1000–2000 MHz).

Furthermore, to achieve a tuning range of more than 100% and to cover process and temperature variations, VCO coarse tuning (with high gain) would make the circuit more sensitive to coupling from nearby circuits and power supply noise. To overcome this problem, the fine-tuning network is incorporated at the proper node of the oscillator circuit, which needs less gain to cover temperature and supply variations. The DC bias selection for oscillators and temperature effects on oscillator performance is discussed in Section 8.7.

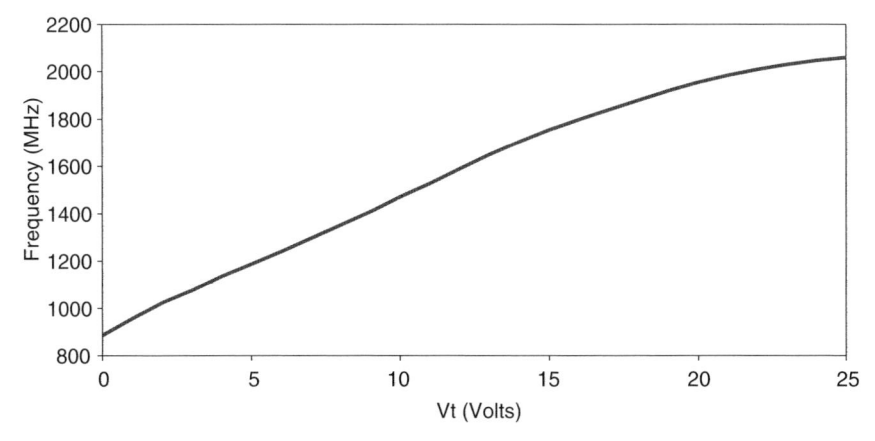

Figure 11-21 Plot showing the extremely linear tuning response of broadband hybrid-tuned VCOs with a tuning voltage from 1 to 25 V.

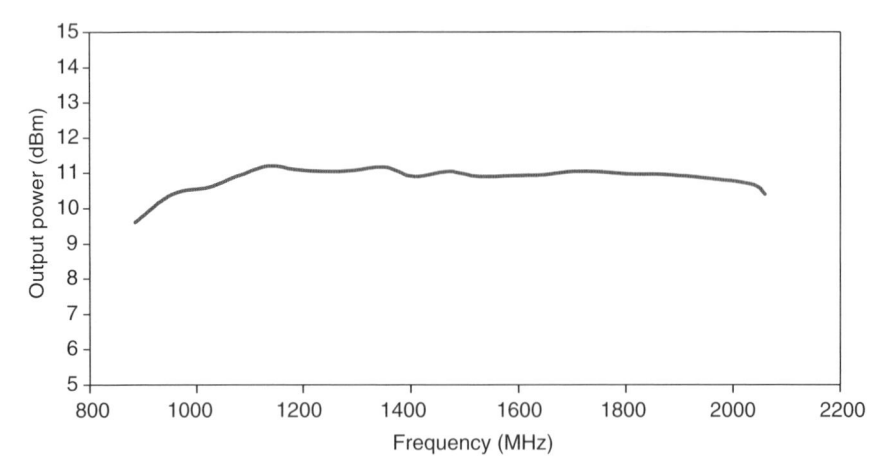

Figure 11-22 Plot showing the output power of broadband hybrid-tuned VCOs.

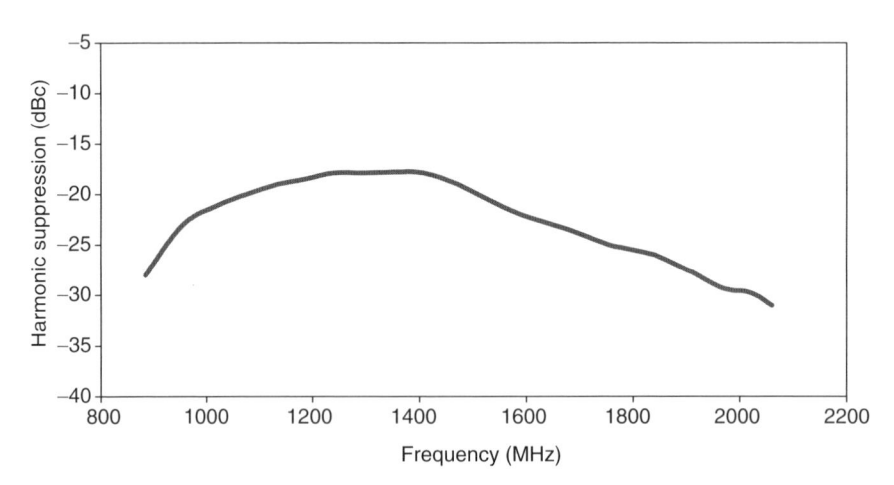

Figure 11-23 Plot showing the harmonic suppression of broadband hybrid-tuned VCOs.

Dual Tuned Wideband VCO (1000–2000 MHz) Features

Oscillator frequency:	1000–2000/2000–4000 MHz
Coarse tuning voltage:	0–25 VDC (nom)
Coarse tuning sensitivity:	40–100 MHz/V
Fine tuning voltage:	1–5 VDC (nom)
Fine tuning sensitivity:	10–20 MHZ/V
Modulation bandwidth:	>10 MHz
Bias voltage:	+8 VDC at 30 mA (nom)
Output power:	+5 dBm (min)

Harmonic suppression: 15 dBc (typ)
VSWR: 1.5:1
Phase noise: −105 dBc/Hz at 10 kHz offset (typ)
Phase noise: −125 dBc/Hz at 100 kHz offset (typ)
Frequency pulling: 10 MHz (max) at 1.75:VSWR
Frequency pushing: 5 MHz/V (max)
Output impedance: 50 ohms
Operating temp: −40°C to 85°C
Size: 0.5 × 0.5 in.

References

1. The references used in Chapter 1 in the discussion of the early oscillators are in H.E. Hollmann, Physik und Technik der ultrakurzen Wellen, Erster Band, Erzeugung ultrakurzwelliger (Schwingungen; Verlag von Julius Springer, 1936) and by themselves are probably no longer available.
2. G.R. Basawapatna, R.B. Stancliff, "A Unified Approach to the Design of Wide-Band Microwave Solid-State Oscillators," IEEE Transactions on Microwave Theory and Techniques, Vol. 27, pp. 379–385, May 1979.
3. M. Sebastein, N. Jean-Christophe, Q. Raymond, P. Savary, J. Obregon, "A Unified Approach for the Linear and Nonlinear Stability Analysis of Microwave Circuits Using Commercially Available Tolls," IEEE Transactions on Microwave Theory and Techniques, Vol. 47, pp. 2403–2409, December 1999.
4. M. Prigent, M. Camiade, J.C. Nallatamby, J. Guittard, J. Obregon, "An Efficient Design Method of Microwave Oscillator Circuits for Minimum Phase Noise," IEEE Transactions on Microwave Theory and Techniques, Vol. 47, pp. 1122–1125, July 1999.
5. P. Bolcato, J.C. Nallatamby, R. Larcheveque, M. Prigent, J. Obregon, "A Unified Approach of PM Noise Calculation in Large RF Multitude Autonomous Circuits," IEEE Microwave Theory and Techniques, Symposium Digest, 2000.
6. S. Mons, M.A. Perez, R. Quere, J. Obregon, "A Unified Approach for the Linear and Nonlinear Stability Analysis of Microwave Circuits Using Commercially Available Tools," IEEE Microwave Theory and Techniques, Symposium Digest, Vol. 3, pp. 993–996, 1999.
7. A. Demir, A. Mehotra, J. Roychowdhury, "Phase Noise in Oscillators: A Unifying Theory and Numerical Methods for Characterization," IEEE Transactions on Circuits and Systems, Vol. 47, pp. 655–674, May 2000.
8. U.L. Rohde, K. Juergen Schoepf, A.K. Poddar, "Low Noise VCOs Conquer Widebands," Microwaves and RF, pp. 98–106, June 2004.
9. T. Hajder, "Higher Order Loops Improve Phase Noise of Feedback Oscillators," Applied Microwave and Wireless, pp. 24–31, October 2002.
10. R.J. Gilmore, M.B. Steer, "Nonlinear Circuit Analysis Using the Method of Harmonic Balance—A Review of the Art. Part I. Introductory Concepts," International Journal of Microwave and Millimeter-Wave Computer-Aided Engineering, Vol. 1, No. 1, pp. 22–37, January 1991.
11. R.J. Gilmore, M.B. Steer, "Nonlinear Circuit Analysis Using the Method of Harmonic Balance—A Review of the Art. Part II. Advanced Concepts," International Journal of

Microwave and Millimeter-Wave Computer-Aided Engineering, Vol. 1, No. 2, pp. 159–180, April 1991.

12. U.L. Rohde, "Frequency Synthesizers," *The Wiley Encyclopedia of Telecommunications*, ed. J.G. Proakis, John Wiley & Sons, 2002.

13. U.L. Rohde, *Microwave and Wireless Synthesizers: Theory and Design*, John Wiley & Sons, 1997.

14. U.L. Rohde, D.P. Newkirk, *RF/Microwave Circuit Design for Wireless Applications*, John Wiley & Sons, 2000.

15. U.L. Rohde, J. Whitaker, *Communications Receivers*, third edition, McGraw-Hill, 2000.

16. G. Vendelin, A.M. Pavio, U.L. Rohde, *Microwave Circuit Design Using Linear and Nonlinear Techniques*, John Wiley & Sons, 1990.

17. K. Kurokawa, "Some Basic Characteristics of Broadband Negative Resistance Oscillator Circuits," Bell Systems Technology Journal, University of Berkeley, pp. 1937–1955, July–August 1969.

18. User's manual for Berkley-SPICE.

19. User's manual for H-SPICE.

20. User's manual for P-SPICE

21. User's manual for RF Spectre.

22. H.C. de Graaff, W.J. Kloosterman, IEEE Transactions on Electron Devices, Vol. ED-32, 2415, 1986.

23. H.C. de Graaff, W.J. Kloosterman, T.N. Jansen, Ext. Abstr. Solid State Dev. Mat. Conference, Tokyo, 1986.

24. H.C. de Graaff, W.J. Kloosterman, J.A.M. Geelen, M.C.A.M. Koolen, "Experience with the New Compact MEXTRAM Model for Bipolar Transistors," Phillips Research Laboratories, 1989.

25. U.L. Rohde, "New Nonlinear Noise Model for MESFETS Including MM-Wave Application," First International Workshop of the West German IEEE MTT/AP Joint Chapter on Integrated Nonlinear Microwave and Millimeterwave Circuits (INMMC'90) Digest, October 3–5, 1990, Duisburg University, Duisburg, Germany.

26. U.L. Rohde, "Improved Noise Modeling of GaAs FETS: Using an Enhanced Equivalent Circuit Technique," Microwave Journal, pp. 87–101, November 1991; pp. 87–95, December 1991.

27. R.A. Pucel, W. Struble, R Hallgren, U.L. Rohde, "A General Noise De-embedding Procedure for Packaged Two-Port Linear Active Devices," IEEE Transactions on Microwave Theory and Techniques, Vol. 40, No. 11, pp. 2013–2024, November 1992.

28. U.L. Rohde, "Parameter Extraction for Large Signal Noise Models and Simulation of Noise in Large Signal Circuits Like Mixers and Oscillators," 23rd European Microwave Conference, Madrid, Spain, September 6–9, 1993.

29. P.R. Gray, R.G. Meyer, *Analysis and Design of Analog Integrated Circuits*, third edition, John Wiley and Sons, 1993.

30. R.A. Pucel, U.L. Rohde, "An Exact Expression for the Noise Resistance R_n of a Bipolar Transistor for Use with the Hawkins Noise Model," IEEE Microwave and Guided Wave Letters, Vol. 3, No. 2, pp. 35–37, February 1993.

31. H.K. Gummel, H.C. Poon, "An Integral Charge Control Model of Bipolar Transistors," Bell Systems Technology Journal, pp. 827–852, May–June 1970.

32. G.M. Kull, L.W. Nagel, S. Lee, P. Lloyd, E.J. Prendergast, H. Dirks, "A Unified Circuit Model for Bipolar Transistors Including Quasi-Saturation Effects," IEEE Transactions on Electron Devices, Vol. 32, No. 6, June 1985.
33. H.C. de Graff, W.J. Koostermann, "Modeling of the Collector Epilayer of a Bipolar Transistor in the Mextram Model," IEEE Transactions on Electron Devices, Vol. 36, No. 7, pp. 274–282, February 1995.
34. M. Schroeter, H.M. Rein, "Investigation of Very Fast and High-Current Transients in Digital Bipolar IC's Using Both a New Model and a Device Simulator," IEEE Journal of Solid-State Circuits, Vol. 30, pp. 551–562, May 1995.
35. M. Rudolph, R. Doerner, K. Beilenhoff, P. Heymann, "Scalable GaIn/GaAs HBT Large-Signal Model," IEEE Transactions on Microwave Theory and Techniques, Vol. 48, pp. 2370–2376, December 1990.
36. C.C. McAndrew, J.A. Seitchik, D.F. Bowers, M. Dunn, M. Foisy, I. Getreu, M. McSwain, S. Moinian, J. Parker, D.J. Roulston, M. Schroeter, P. van Wijnen, L.F. Wagner, "VBIC95, The Vertical Bipolar Inter-Company Model," IEEE Journal of Solid-State Circuits, Vol. 31, No. 10, pp. 1476–1483, October 1996.
37. X. Cao, J. McMacken, K. Stiles, P. Layman, J.J. Liou, A. Ortiz-Conde, S. Moinian, "Comparison of the New VBIC and Conventional Gummel-Poon Bipolar Transistor Models," IEEE Transactions on Electron Devices, Vol. 47, No. 2, pp. 427–433, February 2000.
38. R.A. Pucel, H.A. Haus, H. Statz, "Signal and Noise Properties of Gallium Arsenide Microwave Field-Effect Transistors," in *Advances in Electronics and Electron Physics*, Vol. 38, Academic Press, 1975.
39. A. Cappy, "Noise Modeling and Measurement Techniques," IEEE Transactions and Microwave Theory and Techniques, Vol. 36, pp. 1–10, January 1988.
40. M.W. Pospieszalski, "Modeling of Noise Parameters of MESFETs and MODFETs and Their Frequency and Temperature Dependence," IEEE Transactions on Microwave Theory and Techniques, Vol. 37, pp. 1340–1350, September 1989.
41. V. Rizzoli, F. Mastri, C. Cecchetti, "Computer-Aided Noise Analysis of MESFET and HEMT Mixers," IEEE Transactions on Microwave Theory and Techniques, Vol. 37, pp. 1401–1411, September 1989.
42. A. Riddle, "Extraction of FET Model Noise Parameters from Measurement," IEEE Microwave Theory and Techniques, International Microwave Symposium Digest, pp. 1113–1116, 1991.
43. R.A. Pucel, W. Struble, R. Hallgren, U.L. Rohde, "A General Noise De-Embedding Procedure for Packaged Two-Port Linear Active Devices," IEEE Transactions on Microwave Theory and Techniques, Vol. 40, pp. 2013–2024, November 1992.
44. J. Portilla, R. Quere, J. Obregon, "An Improved CAD Oriented FET Model for Large-Signal and Noise Applications," IEEE Microwave Theory and Techniques, International Microwave Symposium Digest, pp. 849–852, 1994.
45. T. Felgentreff, G. Olbrich, P. Russer, "Noise Parameter Modeling of HEMTs with Resistor Temperature Noise Sources," IEEE Microwave Theory and Techniques, International Microwave Symposium Digest, pp. 853–856, 1994.
46. J. Stenarson, M. Garcia, I. Angelov, H. Zirath, "A General Parameter-Extraction Method for Transistor Noise Models," IEEE Transactions on Microwave Theory and Techniques, Vol. 47, pp. 2358–2363, December 1999.

47. D. Lee, Y. Kwon, H.-S. Min, "Physical-Based FET Noise Model Applicable to Millimeter-Wave Frequencies," IEEE Microwave theory and Techniques, International Microwave Symposium Digest, pp. 01–04, 1999.
48. M.R. Murti, J. Laskar, S. Nuttinck, S. Yoo, A. Raghavan, J.I. Bergman, J. Bautista, R. Lai, R. Grundbacher, M. Barsky, P. Chin, P. H. Liu, "Temperature-Dependent Small-Signal and Noise Parameters Measurement and Modeling on InP HEMTs," IEEE Transactions on Microwave Theory and Techniques, Vol. 48, pp. 2579–2586, December 2000.
49. A. Pascht, G. Markus, D. Wiegner, M. Berroth, "Small-Signal and Temperature Noise Model for MOSFETs," IEEE Transactions on Microwave Theory and Techniques, Vol. 50, pp. 927–1934, August 2002.
50. Y.P. Tsividis, *Operation and Modeling of the MOS Transistor*, McGraw-Hill, 1987.
51. BISIM3v3 Manual, Department of Electrical Engineering and Computer Sciences, University of California, Berkeley.
52. Z. Liu, C. Hu, J. Huang, T. Chan, M. Jeng, P.K. Ko, Y.C. Cheng, "Threshold Voltage Model for Deep-Submicronmeter MOSFET's," IEEE Transactions Electron Devices, Vol. 40, No. 1, pp. 86–95, January 1993.
53. Y. Cheng, M. Jeng, Z. Liu, J. Juang, M. Chan, K. Chen, P.K. Ko, C. Ho, "A Physical and Scalable *I-V* Model in BISIM3v3 for Analog/Digital Circuit Simulation," IEEE Transactions on Electron Devices, Vol. 44, No. 2, pp. 277–287, February 1997.
54. U.L. Rohde, Schwarz, "Introduction Manual for Z-G Diagraph, Models ZGU and ZDU," Munich, 1963.
55. "*S*-Parameters, Circuit Analysis and Design," Hewlett-Packard Application Note 95, September 1968.
56. P. Bodharamik, L. Besser, R.W. Newcomb, "Two Scattering Matrix Programs for Active Circuit Analysis," IEEEE Transactions on Circuit Theory, Vol. CT-18, pp. 610–619, November 1971.
57. Integrated Circuits (RFIC) Symposium, "*S*-Parameter Design," Hewlett-Packard Application Note 154, April 1972.
58. C. Arnaud, D. Basataud, J. Nebus, J. Teyssier, J. Villotte, D. Floriot, "An Active Pulsed RF and Pulsed DC Load-Pull System for the Characterization of HBT Power Amplifiers Used in Coherent Radar and Communication Systems," IEEE Transactions on Microwave Theory and Techniques, Vol. 48, No. 12, pp. 2625–2629, December 2000.
59. F.M. Ghannouchi, R. Larose, R.G. Bosisio, "A New Multiharmonic Loading Method for Large-Signal Microwave and Millimeter-Wave Transistor Characterization," IEEE Transactions on Microwave Theory and Techniques, Vol. 39, No. 6, pp. 986–992, June 1991.
60. H. Abe, Y. Aono, "11 GHz GaAs Power MESFET Load-Pull Measurements Utilizing a New Method of Determining Tuner *Y*-Parameters," IEEE Transactions on Microwave Theory and Techniques, Vol. 27, No. 5, pp. 394–399, May 1979.
61. Q. Cai, J. Gerber, S. Peng, "A Systematic Scheme for Power Amplifier Design Using a Multi-Harmonic Load-Pull Simulation Technique," Microwave Symposium Digest, 1998 IEEE Microwave Theory and Techniques, Symposium International, Vol. 1, pp. 161–165, June 7–12, 1998.
62. P. Berini, M. Desgagne, F.M. Ghannouchi, R.G. Bosisio, "An Experimental Study of the Effects of Harmonic Loading on Microwave MESFET Oscillators and Amplifiers,"

IEEE Transactions on Microwave Theory and Techniques, Vol. 42, No. 6, pp. 943–950, June 1994.
63. K.K. Clarke, D.T. Hess, *Communication Circuits: Analysis and Design*, Chapter 4, Addison-Wesley, 1971.
64. T. Lee, A. Hajimiri, "Linearity, Time Variation, and Oscillator Phase Noise", in *RF and Microwave Oscillator Design*, ed. M. Odyniec, Artech House, 2002.
65. A. Hajimiri, T. Lee, "A General Theory of Phase Noise in Electrical Oscillators," IEEE Journal of Solid-State Circuits, Vol. 33, No. 2, pp. 179–194, February 1998.
66. D. Ham, A. Hajimiri, "Concepts and Methods in Optimization of Integrated LC VCOs," IEEE Journal of Solid-State Circuits, June 2001.
67. A. Hajimiri, S. Limotyrakis, T. Lee, "Jitter and Phase Noise in Ring Oscillators," IEEE Journal of Solid-State Circuits, Vol. 34, No. 6, pp. 790–804, June 1999.
68. W.H. Haywood, *Introduction to Radio Frequency Design*, Prentice Hall, 1982.
69. Chris O' Connor, "Develop a Trimless Voltage-Controlled Oscillator," Microwaves and RF, January 2000.
70. D.B. Leeson, "A Simple Model of Feedback Oscillator Noise Spectrum," Proceedings of the IEEE, Vol. 54, pp. 329–330, 1966.
71. T. Hajder, "Higher Order Loops Improve Phase Noise of Feedback Oscillators," Applied Microwave and Wireless, October 2002.
72. M. Vidmar, "A Wideband, Varactor-Tuned Microstrip VCO," Microwave Journal, June 1999.
73. V. Rizzoli, A. Neri, A. Costanzo, F. Mastri, "Modern Harmonic-Balance Techniques for Oscillator Analysis and Optimization," in *RF and Microwave Oscillator Design*, ed. M. Odyniec, Artech House, 2002.
74. U.L. Rohde, G. Klage, "Analyze VCOs and Fractional-N Synthesizers," Microwaves and RF, pp. 57–78, August 2000.
75. U.L. Rohde, C.R Chang, J. Gerber, "Design and Optimization of Low-Noise Oscillators Using Nonlinear CAD Tools," IEEE Frequency Control Symposium Proceedings, pp. 548–554, 1994.
76. U.L. Rohde, C.R. Chang, J. Gerber, "Parameter Extraction for Large Signal Noise Models and Simulation of Noise in Large Signal Circuits Like Mixers and Oscillators," Proceedings of the 23rd European Microwave Conference, Madrid, Spain, September 6–9, 1993.
77. V. Rizzoli, F. Mastri, C. Cecchefti, "Computer-Aided Noise Analysis of MESFET and HEMT Mixers," IEEE Transactions on Microwave Theory and Techniques, Vol. MTT-37, pp. 1401–1410, September 1989.
78. V. Rizzoli, A. Lippadni, "Computer-Aided Noise Analysis of Linear Multiport Networks of Arbitrary Topology," IEEE Transactions on Microwave Theory and Techniques, Vol. MTT-33, pp. 1507–1512, December 1985.
79. V. Rizzoli, F. Mastri, D. Masotti, "General-Purpose Noise Analysis of Forced Nonlinear Microwave Circuits," Military Microwave, 1992.
80. W. Anzill, F.X. Kärtner, P. Russer, "Simulation of the Single-Sideband Phase Noise of Oscillators," Second International Workshop of Integrated Nonlinear Microwave and Millimeterwave Circuits, 1992.
81. Hewlett-Packard, 3048A Operating Manual.

82. K. Kurokawa, "Noise in Synchronized Oscillators," IEEE Transactions on Microwave Theory and Techniques, Vol. 16, pp. 234–240, April 1968.
83. J.C. Nallatamby, M. Prigent, M. Camiade, J. Obregon, "Phase Noise in Oscillators—Leeson Formula Revisted," IEEE Transactions on Microwave Theory and Techniques, Vol. 51, No. 4, pp. 1386–1394, April 2003.
84. U.L. Rohde, "A Novel RFIC for UHF Oscillators (Invited)," 2000 IEEE Radio Frequency, Boston, June 11–13, 2000.
85. D.J. Healy III, "Flicker of Frequency and Phase and White Frequency and Phase Fluctuations in Frequency Sources," Proceedings of the 26th Annual Symposium on Frequency Control, Fort Monmouth, NJ, pp. 43–49, June 1972.
86. M.M. Driscoll, "Low Noise Oscillator Design Using Acoustic and Other High Q Resonators," 44th Annual Symposium on Frequency Control, Baltimore, May 1990.
87. U.L. Rohde, K. Danzeisen, Feedback Oscillators, Patent No. DE10033741A1 (Germany, United States, Japan).
88. P. Vizmuller, *RF Design Guide: Systems, Circuits, and Equations*, Artech House, p. 76, 1995.
89. C. Rheinfelder, Ein Gross-signal-Modell des SiGe-Hetero-Bipolar-Transistors fuer den Entwurf monolithisch intergrierter Mikrowellen-Schaltungen, Fortschritt-Berichte VDI, 1998. Also, personal communication with the author.
90. C. Rheinfelder, M. Rudolph, F. Beißwanger, W. Heinrich, "Nonlinear Modeling of SiGe HBTs Up to 50 GHz," IEEE International Microwave Symposium Digest, Vol. 2, pp. 877–880, 1997.
91. C.N. Rheinfelder, K.M. Strohm, L. Metzger, J.-F. Luy, W. Heinrich, "A SiGe-MMIC Oscillator at 47 GHz," International Microwave Symposium Digest, Vol. 1, pp. 5–8, 1999.
92. F. Lenk, M. Schott, J. Hilsenbeck, J. Würfl, W. Heinrich, "Low Phase-Noise Monolithic GaInP/GaAs-HBT VCO for 77 GHz," 2003 IMS.
93. F. Lenk, J. Hilsenbeck, H. Kuhnert, M. Schott, J. Würfl, W. Heinrich, "GaAs-HBT MMIC-Oscillators for Frequencies up to 40 GHz," Second Symposium on Opto- and Microelectronic Devices and Circuits (SODC 2002), Stuttgart, Germany, March 17–23, 2002.
94. H. Kuhnert, W. Heinrich, W. Schwerzel, A. Schüppen, "25 GHz MMIC Oscillator on Commercial SiGe-HBT Process," IEE Electronics Letters, Vol. 36, No. 3, pp. 218–220, February 2000.
95. H. Kuhnert, F. Lenk, J. Hilsenbeck, J. Würfl, W. Heinrich, "Low Phase-Noise GaInP/GaAs-HBT MMIC Oscillators up to 36 GHz," International Microwave Symposium Digest, Vol. 3, pp. 1551–1554, 2001.
96. H. Kuhnert, W. Heinrich, "5 to 25 GHz SiGe MMIC Oscillators on a Commercial Process," Proceedings of the Second Topical Meeting on Silicon Monolithic Integrated Circuits in RF Systems, pp. 60–63, April 2000.
97. H. Kuhnert, W. Heinrich, "Coplanar SiGe VCO MMICs Beyond 20 GHz," IEEE Topical Meeting on Silicon Monolithic Integrated Circuits in RF Systems, Ann Arbor, Digest, pp. 231–233, September 12–14, 2001.
98. F. Lenk, M. Schott, W. Heinrich, "Modeling and Measurement of Phase Noise in GaAs HBT Ka-Band Oscillators," European Microwave Conference Digest, Vol. 1, pp. 181–184, 2001.

99. M. Nemes, "Entwicklung eins UHF-Transceivers fuer das ISM-Frequenzband 868–870 MHz in einer 0,8 μm-CMOS-Technologie," Ph.D. dissertation, University of Cottbus, 2003.
100. A. Roufougaran, J. Rael, M. Rofougaran, A. Abidi, "A 900 MHz CMOS LC-Oscillator with Quadrature Outputs," International Solid-State Current Conference Digest of Technology Papers, pp. 392–393, February 1996.
101. B. Razvi, "Study of Phase Noise in CMOS Oscillators," IEEE Journal of Solid-State Circuits, Vol. 31, pp. 331–343, March 1996.
102. J. Craninckx, M. Steyaert, "A Fully Integrated Spiral-LC CMOS VCO Set with Prescaler for GSM and DCS-1800 Systems," Proceedings of the Custom Integrated Circuits Conference, pp. 403–406, May 1997.
103. M. Thanmsirianunt, T.A. Kwasniewski, "CMOS VCO's for PLL Frequency Synthesis in GHz Digital Mobile Radio Communications," IEEE Journal of Solid-State Circuits, Vol. 32, pp. 1511–1518, October 1997.
104. F. Herzel, M. Pierschel, P. Weger, M. Tiebout, "Phase Noise in Differential CMOS Voltage-Controlled Oscillator for RF Applications," IEEE Transactions on Circuits and Systems, Vol. 47, No. 1, pp. 11–15, January 2000.
105. F. Herzel, H. Erzgraber, P. Weger, "Integrated CMOS Wideband Oscillator for RF Applications," Electronics Letters, Vol. 37, March 2001.
106. A. Hajimiri, T. Lee, "Design Issues in CMOS Differential LC Oscillators," IEEE Journal of Solid-State Circuits, May 1999.
107. T.H. Lee, *The Design of CMOS Radio-Frequency Integrated Circuits*, Cambridge University Press, 1998.
108. K. Cheng, K. Chan, "Power Optimization of High-Efficiency Microwave MESFET Oscillators," IEEE Transactions on Microwave Theory and Techniques, Vol. 48, No. 5, pp. 787–790, May 2000.
109. G. Sauvage, "Phase Noise in Oscillators: A Mathematical Analysis of Leeson's Model," IEEE Transactions on Instrumentation and Measurement, Vol. IM-26, No. 4, December 1977.
110. G. Böck, P.W. v. Basse, J.E. Müller, W. Kellner, "Accurate SPICE Modeling of GaAs-MESFETs for Use in GaAs-IC Design," Proceedings of the 11th Workshop of Compound Semiconductor Devices and Integrated Circuits, S. 66–69, Grainau (Garmisch-Partenkirchen), West Germany, May 4–6, 1987.
111. W. Stiebler, M. Matthes, G. Böck, T. Koppel, A. Schäfer, "Bias Dependent Cold-(H)FET Modeling," IEEE International Microwave Symposium Digest, San Francisco, pp. 1313–1316, June 1996.
112. L. Klapproth, G. Böck, A. Schäfer, W. Stiebler, "On the Determination of HFET Noise Parameters from 50 Ohm Noise Figure Measurements," Sindelfingen, Kongreßunterlagen, S., pp. 348–352, 1997.
113. P. Heymann, M. Rudolph, H. Prinzler, R. Doerner, L. Klapproth, G. Böck, "Experimental Evaluation of Microwave Field-Effect-Transistor Noise Models," IEEE Transactions on Microwave Theory and Techniques, February 1999.
114. N.H.W. Fong, J.O. Plouchart, N. Zamdmer, D. Liu, L.F. Wagner, C. Plett, N.G. Tarr, "A 1-V 3.8–5.7-GHz Wide-Band VCO with Differentially Tuned Accumulation MOS Varactors for Common-Mode Noise Rejection in CMOS SOI Technology," IEEE Transactions on Microwave Theory and Techniques, February 2003.

115. Application Notes, Temperature Stable Microwave Ceramics, Products for RF/Microwave Application, Transactions on Techniques, pp. 6–85.
116. D. Kajfez, "Q Factor Measurement with a Scalar Network Analyzer," IEE Proceedings on Microwave Antennas Propagation, Vol. 142, pp. 389–372, October 1995.
117. R.J. Hawkins, "Limitations of Nielsen's and Related Noise Equations Applied to Microwave Bipolar Transistors and a New Expression for the Frequency and Current Dependent Noise Figure," Solid-State Electronics, Vol. 20 pp. 191–196, March 1977.
118. T.H. Hsu, C.P. Snapp, "Low-Noise Microwave Bipolar Transistor with Sub-Half-Micrometer Emitter Width," IEEE Transactions on Electron Devices, Vol. ED-25, pp. 723–730, June 1978.
119. R. Aparicio, A. Hajimiri, "A CMOS Noise Shifting Colpitts VCO," Digest of Technical Papers International Solid-State Circuit Conference, Session 17.2, pp. 288–299, 2002.
120. R. Rhea, "A New Class of Oscillators," IEEE Microwave Magazine, pp. 72–83, June 2004.
121. S. Maas, "Why I Hate Base Resistance," IEEE Microwave Magazine, pp. 54–60, June 2004.
122. F.X. Sinnesbichler, "Hybrid Millimeter-Wave Push-Push Oscillators using Silicon-Germanium HBTs," IEEE Transactions on Microwave Theory and Techniques, Vol. 51, February 2003.
123. S. Kudszus, W.H. Haydl, A. Tessmann, W. Bronner, M. Schlechtwez, "Push-Push Oscillators for 94 and 140 GHz Applications Using Standard Pseudomorphic GaAs HEMTs," IEEE Microwave Theory and Techniques, Microwave Symposium Digest, pp. 1571–1574, 2001.
124. Y. Baeyens et al., "Compact InP-Based HBT VCOs with a Wide Tuning Range at W-Band," IEEE Transactions on Microwave Theory and Techniques, Vol. 48, pp. 2403–2408, December 2001.
125. Y. Sun, T. Tieman, H. Pflung, W. Velthius, "A Fully Integrated Dual Frequency Push-Push VCO for 5.2 and 5.8 GHz Wireless Applications," Microwave Journal, pp. 64–74, April 2001.
126. M. Schott, H. Kuhnert, J. Hilsenbeck, J. Wurlf, H. Heinrich, "38 GHz Push-Push GaAs-HBT MMIC Oscillator," IEEE Microwave Theory and Techniques, Symposium Digest, pp. 839–842, 2002.
127. F.X. Sinnesbichler, G.R. Olbrich, "SiGe HBT Push-Push Oscillators for V-Band Operation," IEEE Microwave Theory and Techniques Symposium, Silicon Monolithic Integrated Circuits in RF Systems Symposium, Garmisch, Germany, pp. 55–59, April 26–28, 2000.
128. F.X. Sinnesbichler, H. Geltinger, G.R. Olbrich, "A 38 GHz Push-Push Oscillator Based on 25 GHz-f_T BJTs," IEEE Microwave Guided Wave Letters, Vol. 9, pp. 151–153, April 1999.
129. K.W. Kobayashi et al., "A 108-GHz InP-based HBT Monolithic Push-Push VCO with Low Phase Noise and Wide Tuning Bandwidth," IEEE Journal of Solid-State Circuits, Vol. 34, pp. 1225–1232, September 1999.
130. L. Dussopt, D. Guillois, G. Rebeiz, "A Low Phase Noise Silicon 9 GHz VCO and an 18 GHz Push-Push Oscillator," IEEE Microwave Theory and Techniques, Symposium Digest, pp. 695–698, 2002.

131. F.X. Sinnesbichler, B. Hauntz, G.R. Olbrich, "A Si/SiGe HBT Dielectric Resonator Push-Push Oscillators at 58 GHz," IEEE Microwave Guided Wave Letters, Vol. 10, pp. 145–147, April 2000.
132. U.L. Rohde, "A New and Efficient Method of Designing Low Noise Oscillators," Ph.D. dissertation, Technical University of Berlin, February 2004.
133. A. Mortazawi, B.C. De Loach, Jr., "Multiple Element Oscillators Utilizing a New Power Combining Technique," IEEE Microwave Theory and Techniques, International Microwave Symposium Digest, pp. 1093–1096, 1992.
134. B. Van der Pol, "The Nonlinear Theory of Electrical Oscillators," Proceedings of the IRE, Vol. 22 No. 9, pp. 1051–1086, September 1934.
135. R. Adler, "A Study of Locking Phenomena in Oscillators," Proceedings of the IEEE, Vol. 61, pp. 180–1385, October 1973.
136. K.S. Ang, M.J. Underhill, I.D. Robertson, "Balanced Monolithic Oscillators at K- and Ka-Band," IEEE Transactions on Microwave Theory and Techniques, Vol. 48, No. 2, pp. 187–194, February 2000.
137. M. Ahdjoudj, "Conception de VCO a Faible Bruit de Phase en Technologie Monolithique PHEMT Dans les Bandes K et Ka," Ph.D. thesis, Doctorat de l' Universite Pierre Et Marie Curie Paris VI, December 1997.
138. H.-C. Chang, X. Cao, U.K. Mishra, R. York, "Phase Noise in Coupled Oscillators: Theory and Experiment," IEEE Transactions on Microwave Theory and Techniques, International Microwave Symposium Digest, Vol. 45, pp. 604–615, May 1997.
139. H.-C. Chang, X. Cao, M.J. Vaughan, U.K. Mishra, R. York, "Phase Noise in Externally Injection-Locked Oscillator Arrays," IEEE Transactions on Microwave Theory and Techniques, Vol. 45, pp. 2035–2042, November 1997
140. A. Borgioli, P. Yeh, R.A. York, "Analysis of Oscillators with External Feedback Loop for Improved Locking Range and Noise Reduction," IEEE Transactions on Microwave Theory and Techniques, Vol. 47, pp. 1535–1543, August 1999.
141. J.J. Lynch, R.A. York, "Synchronization of Oscillators Coupled Through Narrow-Band Networks," IEEE Transactions on Microwave Theory and Techniques, Vol. 49, pp. 238–249, February 2001.
142. K.F. Schunemann, K. Behm, "Nonlinear Noise Theory for Synchronized Oscillators," IEEE Transactions on Microwave Theory and Techniques, Vol. 27, pp. 452–458, May 1979.
143. K. Kurokawa, "The Single Cavity Multiple Device Oscillator," IEEE Transactions on Microwave Theory and Techniques, Vol. 19, pp. 793–801, October 1971.
144. K. Kurokawa, "Noise in Synchronized Oscillators," IEEE Transactions on Microwave Theory and Techniques, Vol. 16, pp. 234–240, April 1968.
145. H.-C. Chang, "Analysis of Coupled Phase-Locked Loops with Independent Oscillators for Beam Control Active Phased," IEEE Transactions on Microwave Theory and Techniques, Vol. 52, pp. 1059–1065, March 2004.
146. W.O. Schlosser, "Noise in Mutually Synchronized Oscillators," IEEE Transactions on Microwave Theory and Techniques, Vol. MTT-16, pp. 732–737, September 1968.
147. U.L. Rohde and A.K. Poddar, "A Unifying Theory and Characterization of Microwave Oscillators/VCOs," 18th IEEE CCECE05, May 1–4, 2005, Canada.

148. R.L. Kuvas, "Noise in Single Frequency Oscillators and Amplifiers," IEEE Transactions on Microwave Theory and Techniques, Vol. MTT-21, pp. 127–134, March 1973.
149. H. Stark, J.W. Woods, *Probability, Random Processes, and Estimation Theory for Engineers*, Prentice Hall, 1986.
150. R.A. York, P. Liao, J.J. Lynch, "Oscillator Array Dynamics with Broad-Band N-Port Coupling Networks," IEEE Transactions on Microwave Theory and Techniques, Vol. 42, pp. 2040–2045, November 1994.
151. R.A. York, "Nonlinear Analysis of Phase Relationship in Quasi-Optical Oscillator Arrays," IEEE Transactions on Microwave Theory and Techniques, Vol. 41, pp. 1799–1809, October 1993.
152. S.-C. Yen, T.-H. Chu, "An Nth-Harmonic Oscillator Using an N-Push Coupled Oscillator Array with Voltage-Clamping Circuits," IEEE Microwave Theory and Techniques, Symposium Digest, pp. 545–548, 1992.
153. J.R. Bender, C. Wong, "Push-Push Design Extends Bipolar Frequency Range," Microwave and RF, pp. 91–98, October 1983.
154. F. Ramirez, J.L. Garcia H., T. Fernandez, A. Suarez, "Nonlinear Simulation Techniques for the optimized Design of Push-Push Oscillators," IEEE Microwave Theory and Techniques, Symposium Digest, pp. 2157–2160.
155. J.-G. Kim, D.-H. Baek, S.-H. Jeon, J.-W. Park, S. Hong, "A 60 GHz InGaP/GaAs HBT Push-Push MMIC VCO," IEEE Microwave Theory and Techniques, Symposium Digest, pp. 885–888, 2003.
156. F.X. Sinnesbichler, B. Hautz, G.R. Olbrich, "A Low Phase Noise 58 GHz SiGe HBT Push-Push, Oscillator with Simultaneous 29 GHz Output," IEEE Microwave Theory and Techniques, Symposium Digest, pp. 35–38, 2000.
157. H. Xiao, T. Tanka, M. Aikawa, "A Ka-Band Quadruple-Push Oscillator," IEEE Microwave Theory and Techniques, Symposium Digest, pp. 889–892, 2003.
158. R.G. Freitag, S.H. Lee, D.M. Krafcsik, D.E. Dawson, J.E. Degenford, "Stability and Improved Circuit Modeling Considerations for High Power MMIC Amplifiers," IEEE Microwave and Millimeter-Wave Monolithic Circuits Symposium, pp. 2169–2172, 2003.
159. J. Heinbockel, A. Mortazawi, "A Periodic Spatial Power Combining MESFET Oscillator," IEEE Microwave Theory and Techniques, Symposium Digest, pp. 545–548, 1992.
160. K. Naishadham, T. Durak, "Measurement-Based Closed-Form Modeling of Surface Mounted RF Components," IEEE Transactions on Microwave Theory and Techniques, Vol. 50, No. 10, pp. 2276–2286, October 2002.
161. M. Kuramitsu, F. Takasi, "Analytical Method for Multimode Oscillators Using the Averaged Potential," Electrical Communication, Japan, Vol. 66-A, pp. 10–19, 1983.
162. R.G. Freitag, "A Unified Analysis of MMIC Power Amplifier Stability," IEEE Microwave Theory and Techniques, Symposium Digest, pp. 297–300, 1992.
163. S. Nogi, J. Lin, T. Itoh, "Mode Analysis and Stabilization of a Spatial Power Combining Array with Strongly Coupled Oscillators," IEEE Microwave Theory and Techniques, Vol. 41, pp. 1827–1837, October 1993.
164. A. Mortazawi, H.D. Foltz, T. Itoh, "A Periodic Second Harmonic Spatial Power Combining Oscillator," IEEE Microwave Theory and Techniques, Vol. 40, pp. 851–856, May 1992.

165. J.J. Lynch, R.A. York, "An Analysis of Mode-Locked Arrays of Automatic Level Control Oscillators," IEEE Transactions on Circuits and Systems-I, Vol. 41, pp. 859–865, December 1994.

166. R.A. York, R.C. Compton, "Mode-Locked Oscillator Arrays," IEEE Microwave and Guided Letter, Vol. 1, pp. 215–218, August 1991.

167. Y.-L. Tang, H. Wang, "Triple-Push Oscillator Approach: Theory and Experiments," IEEE Journal of Solid-State Circuits, Vol. 36, pp. 1472–1479, October 2001.

168. C. Arnaud, D. Basataud, J. Nebus, J. Teyssier, J. Villotte, D. Floriot, "An Active Pulsed RF and Pulsed DC Load-Pull System for the Characterization of HBT Power Amplifiers Used in Coherent Radar and Communication Systems," IEEE Transactions on Microwave Theory and Techniques, Vol. 48, No. 12, pp. 2625–2629, December 2000.

169. F.M. Ghannouchi, R. Larose, R.G. Bosisio, "A New Multi-Harmonic Loading Method for Large-Signal Microwave and Millimeterwave Transistor Characterization," IEEE Transactions on Microwave Theory and Techniques, Vol. 39, No. 6, pp. 986–992, June 1991.

170. H. Abe, Y. Aono, "11 GHz GaAs Power MESFET Load-Pull Measurements Utilizing a New Method of Determining Tuner Y-Parameters," IEEE Transactions on Microwave Theory and Techniques, Vol. 27, No. 5, pp. 394–399, May 1979.

171. Q. Cai, J. Gerber, S. Peng, "A Systematic Scheme for Power Amplifier Design Using a Multi-Harmonic Load-Pull Simulation Technique," 1998 IEEE, Microwave Theory and Techniques, Symposium Digest, pp. 161–165, June 7–12, 1998.

172. P. Berini, M. Desgagne, F.M. Ghannouchi, R.G. Bosisio, "An Experimental Study of the Effects of Harmonic Loading on Microwave MESFET Oscillators and Amplifiers," IEEE Transactions on Microwave Theory and Techniques, Vol. 42, No. 6, pp. 943–950, June 1994.

173. D.M. Pozar, *Microwave Engineering*, second edition, John Wiley & Sons, 1998.

174. A.M. Elsayed, M.I. Elmasry, "Low-Phase-Noise LC Quadrature VCO Using Coupled Tank Resonators in Ring," IEEE Journal of Solid-State Circuits, Vol. 36, pp. 701–705, April 2001.

175. M. Ticbout, "Low Power, Low Phase Noise, Differentially Tuned Quadrature VCO Design in Standard CMOS," IEEE Journal of Solid-State Circuit, Vol. 36, pp. 1018–1024, July 2001.

176. K.O., "Estimation Methods for Quality Factors of Inductors Fabricated in Silicon Integrated Circuit Process Technologies," IEEE Journal of Solid-State Circuits, Vol. 44, pp. 1565–1567, September 1997.

177. A.V. Grebennikov, "Microwave Transistor Oscillators: An Analytic Approach to Simplify Computer-Aided Design," Microwave Journal, pp. 292–299, May 1999.

178. A.V. Grebennikov, "Microwave FET Oscillators: An Analytic Approach to Simplify Computer-Aided Design," Microwave Journal, pp. 102–111, January 2000.

179. J.-S. Sun, "Design Analysis of Microwave Varactor-Tuned Oscillators," Microwave Journal, pp. 302–308, May 1999.

180. C. Enz, F. Krummenacher, E. Vittoz, "An Analytical MOS Transistor Model Valid in All Regions of Operation and Dedicated to Low-Voltage and Low-Current Applications," Journal on Analog Integrated Circuits and Signal Processing, pp. 83–114, July 1995.

181. M. Bucher, D. Kazizis, F. Krummenacher, "Geometry- and Bias-Dependence of Normalized Tranconductances in Deep Submicron CMOS," Seventh International Conference on Modeling and Simulation of Microsystems, March 7–11, 2004.
182. 30+ papers by team members: http://legwww.epfl.ch/ekv/people.html#references.
183. M. Regis, O. Llopis, J. Graffeuil, "Nonlinear Modeling and Design of Bipolar Transistor Ultra Low Phase Noise Dielectric Resonator Oscillators," IEEE Transactions on Microwave Theory and Techniques, Vol. 46, No. 10, pp. 1589–1593, October 1998.
184. J. Everard, *Fundamentals of RF Circuit Design with Low Noise Oscillators*, John Wiley & Sons, 2001.
185. U.L. Rohde, "Oscillators and Frequency Synthesizers," in *Handbook of RF/Microwave Components and Engineering*, ed. K. Chang, John Wiley & Sons, 2003. See other detailed references in this book.
186. S.J. Heinen, "CAD-Verfahren fuer die Signal-und Rauschanalyse Analoger Linearer und Nichtlinearer Schaltungen im Eingeschwungenen Zustand," Fachbereich Elektrotechnik der Universitaet Gesamthochschule, Duisburg, May 26, 1992.
187. U.L Rohde, A.K. Poddar, "Noise Analysis of Systems of Coupled Oscillators," Integrated Nonlinear Microwave and Millimeterwave Circuits (INMMIC) Workshop, Monte Porzo Catone, Italy, November 15–16, 2004.
188. H. Von Barkhausen, *Lehrbuch der Elecktronen-Roehren, Band 3*, Rueckkopplung, 1935.
189. "Microwave Transistor Bias Considerations," Hewlett-Packard Application Note 944–1, August 1980.
190. "1800 to 1900 MHz Amplifier Using the HBFP-0405 and HBFP-0420 Low Noise Silicon Bipolar Transistors", Hewlett-Packard Application Note 1160, Publication No. 5968–2387E, November 1998.
191. A. Ward, B. Ward, "A Comparison of Various Bipolar Transistor Biasing Circuits," Applied Microwave and Wireless, Vol. 13, pp. 30–52, 2001.
192. A.S. Porret, T. Melly, C.C. Enz, E.A Vittoz, "Design of High Q Varactors for Low-Power Wireless Applications Using a Standard CMOS Process," IEEE Journal of Solid-State Circuits, Vol. 35, pp. 337–345, March 2000.
193. R.B. Merrill, T.W. Lee, H. You, R. Rasmussen, L.A. Moberly, "Optimization of High Q Integrated Inductors for Multi-Level Metal CMOS," IEEE International Electron Devices Meeting, pp. 983–986, 1995.
194. D. Ham, A. Hajimiri, "Concepts and Methods in Optimization of Integrated LC VCOs," IEEE Journal of Solid-State Circuits, Vol. 36, No. 6, pp. 896–909, June 2001.
195. L.F. Tiemeijer, D. Leenaerts, N. Pavlovic, R.J. Havens, "Record Q Spiral Inductors in Standard CMOS," IEEE Electron Devices Meeting Technical Digest, pp. 40.7–40.7.3, December 2001.
196. S.S. Song, H. Shin, "An RF Model of the Accumulation-Mode MOS Varactor Valid in Both Accumulation and Depletion Regions," IEEE Transactions on Electron Devices, Vol. 50, No. 9, pp. 1997–1999, September 2003.
197. K. Holnar, G. Rappitsch, Z. Huszka, E. Seebacher, "MOS Varactor Modeling with a Subcircuit Utilizing the BSIM3v3 Model," IEEE Transactions on Electron Devices, Vol. 49, No. 7, pp. 1206–1211, July 2002.

198. N.H.W. Fong, J.O. Plouchart, N. Zamdamer, D. Liu, L.F. Wagner, C. Plett, N.G. Tarr, "Design of Wide-Band CMOS VCO for Multiband Wireless LAN Applications," IEEE Journal of Solid-State Circuits, Vol. 38, No. 8, pp. 1333–1342.

199. C.M. Hung, B.A. Floyd, N. Park, K.K.O, "Fully Integrated 5.35 GHz CMOS VCOs and Prescalers," IEEE Transactions on Microwave Theory and Techniques, Vol. 49, No. 1, pp. 17–22, January 2001.

200. J. Maget, M. Tiebout, R. Kraus, "Influence of Novel MOS Varactors on the Performance of a Fully Integrated UMTS VCO in Standard 0.25 μm CMOS Technology," IEEE Journal of Solid-State Circuits, Vol. 37, No. 7, pp. 953–958, July 2002.

201. P. Andreani, S. Mattisson, "On the Use of MOS Varactors in RF VCOs," IEEE Journal of Solid-State Circuits, Vol. 35, No. 6, pp. 905–910, June 2000.

202. X. Zhang, K. Ding, "Capacitance-Time Transient Characteristics of Pulsed MOS Capacitor Application in Measurement of Semiconductor Parameters," IEEE Proceedings-G, Vol. 140, No. 6, pp. 449–452, December 1993.

203. R.L. Bunch, S. Raman, "Large-Signal Analysis of MOS Varactors in CMOS—$Gm\ LC$ VCOs," IEEE Journal of Solid-State Circuits, Vol. 38, No. 8, pp. 1325–1332, August 2003.

204. A. Kral, F. Behbahani, A.A. Abidi, "RF-CMOS Oscillators with Switched Tuning," IEEE Custom Integrated Circuits Conference, pp. 555–558, 1998.

205. J. Maget, M. Tiebout, R. Kraus, "MOS Varactors with n- and p-Type Gates and Their Influence on an LC-VCO in Digital CMOS," IEEE Journal of Solid-State Circuits, Vol. 38, No. 7, pp. 1139–1147, July 2003.

206. S.K. Magierowski, K. Iniewski, S. Zukotynski, "Differentially Tunable Varactor with Built-In Common-Mode Rejection," IEEE 45th Midwest Symposium on Circuits and Systems, Vol. 1, pp. 559–562, August 4–7, 2002.

207. Y. Tang, A. Aktas, M. Ismail, S. Bibyk, "A Fully Integrated Dual-Mode Frequency Synthesizer for GSM and Wideband CDMA in 0.5 μm CMOS," 44th IEEE Symposium on Circuits and Systems, Vol. 2, pp. 866–869, August 2001.

208. S. Levantino, C. Samori, A. Bonfanti, S.L.J. Gierkink, A.L. Lacaita, V. Boccuzzi, "Frequency Dependence on Bias Current in 5-GHz CMOS VCOs: Impact on Tuning Range and Flicker Noise Upconversion," IEEE Journal of Solid-State Circuits, Vol. 37, No. 8, pp. 1003–1011, August 2002.

209. N.H.W. Fong, J.O. Plouchart, N. Zamdamer, D. Liu, L.F. Wagner, C. Plett, N.G. Tarr, "A 1-V 3.8–5.7-GHz Wide-Band VCO with Differentially Tuned Accumulation MOS Varactors for Common-Mode Noise Rejection in CMOS SOI Technology," IEEE Transactions on Microwave Theory and Techniques, Vol. 51, No. 8, pp. 1952–1959, August 2003.

210. J. Cheah, E. Kwek, E. Low, C. Quek, C. Yong, R. Enright, J. Hirbawi, A. Lee, H. Xie, L. Wei, L. Luong, J. Pan, S. Yang, W. Lau, W. Ngai, "Design of a Low-Cost Integrated 0.25 μm CMOS Bluetooth SOC in 16.5 mm^2 Silicon Area," IEEE International Solid-State Circuits Conference, pp. 90–91, February 2002.

211. U.L. Rohde and A.K. Poddar, "Ultra Low Noise, Low Cost Octave-Band Hybrid-Tuned Microwave VCOs," 18th IEEE CCECE05, May 1–4, 2005, Canada.

APPENDIX A
Design of an Oscillator Using Large-Signal S-Parameters

A.1 PARALLEL RESONATOR OSCILLATOR

Figure A-1 is a numerical calculation of a 3000 MHz oscillator based on parallel feedback using large-signal S-parameters. This example is of particular interest

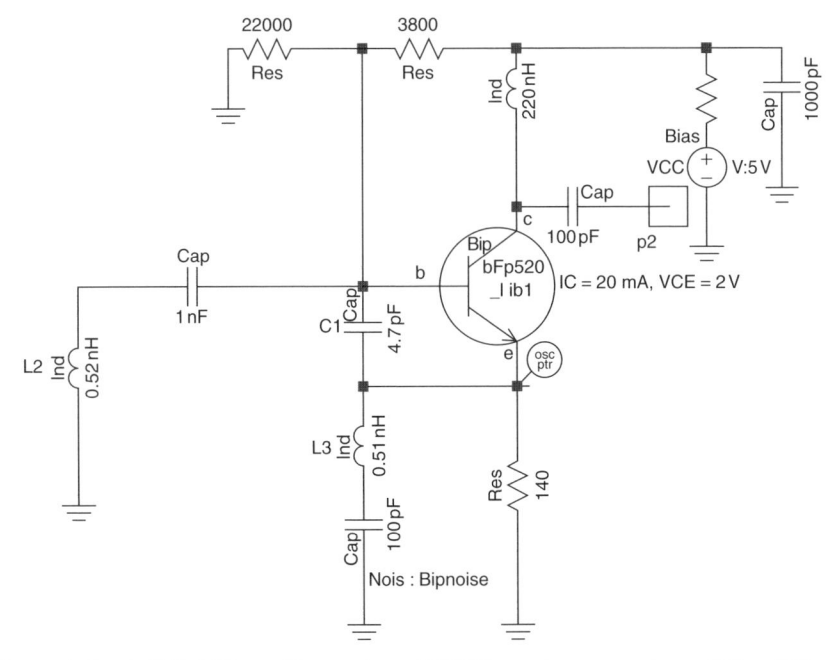

Figure A-1 A 3000 MHz oscillator using a BFP520 transistor operating at 2 V and 20 mA. In this case, capacitor C_2 needs to be replaced by an inductor L_3 which tunes out the collector emitter capacitance to achieve the optimum value. The 1 nF on the left is a DC separation capacitor. This design is optimized for output power.

The Design of Modern Microwave Oscillators for Wireless Applications: Theory and Optimization, by Ulrich L. Rohde, Ajay Kumar Poddar, Georg Böck
Copyright © 2005 John Wiley & Sons, Inc.

because it requires an inductor instead of the familiar capacitor, C_2, between base and emitter. The circuit as such is a Colpitts oscillator.

The measured large-signal Y-parameter data ($I_c = 20$ mA, $V_{ce} = 2$ V) at 3000 MHz are:

$$Y_{11} = G_{11} + jB_{11} = (11.42 + j8.96) \text{ mS} \tag{A-1}$$

$$Y_{21} = G_{21} + jB_{21} = (4.35 - j196.64) \text{ mS} \tag{A-2}$$

$$Y_{12} = G_{12} + jB_{12} = (-433.09 - j1.5643) \text{ mS} \tag{A-3}$$

$$Y_{22} = G_{22} + jB_{22} = (4.41 + j9.10) \text{ mS} \tag{A-4}$$

The optimum values of feedback elements calculated from the given expression of B_1^* and B_2^* are

$$B_1^* = -\left\{ B_{11} + \left[\frac{B_{12} + B_{21}}{2}\right] + \left[\frac{G_{21} - G_{12}}{B_{21} - B_{12}}\right]\left[\frac{G_{12} + G_{21}}{2} + G_{11}\right] \right\} \tag{A-5}$$

$$jB_1^* = 89.8\text{E}{-3} \tag{A-6}$$

$$jB_1^* = j\omega C_1 \tag{A-7}$$

$$C_1 = \frac{89.8\text{E}{-3}}{2\pi f} = 4.77 \text{ pF} \tag{A-8}$$

$$B_2^* = \left[\frac{B_{12} + B_{21}}{2}\right] + \left[\frac{(G_{12} + G_{21})(G_{21} - G_{12})}{2(B_{21} - B_{12})}\right] \tag{A-9}$$

$$jB_2^* = -103.5\text{E}{-3} \tag{A-10}$$

$$jB_2^* = \frac{1}{j\omega L_2} \tag{A-11}$$

$$L_2 = \frac{1}{(2\pi f) \times 103.5\text{E}{-3}} 0.515 \text{ nH} \tag{A-12}$$

The optimum values of the real and imaginary part of the output admittance are

$$Y_{out}^* = [G_{out}^* + jB_{out}^*] \tag{A-13}$$

A.1 PARALLEL RESONATOR OSCILLATOR

where G_{out}^* and B_{out}^* are given as

$$G_{out}^* = G_{22} - \left[\frac{(G_{12} + G_{21})^2(B_{21} - B_{12})^2}{4G_{11}}\right] \quad \text{(A-14)}$$

$$G_{out}^* = -823.53\text{E}-3 \quad \text{(A-15)}$$

$$B_{out}^* = B_{22} + \left[\frac{G_{21} - G_{12}}{B_{21} - B_{12}}\right] - \left[\frac{(G_{12} + G_{21})}{2} + G_{22} - G_{out}^*\right] + \left[\frac{B_{21} + B_{12}}{2}\right] \quad \text{(A-16)}$$

$$B_{out}^* = -105.63\text{E}-3 \quad \text{(A-17)}$$

$$jB_{out}^* = \frac{1}{j\omega L_3} \quad \text{(A-18)}$$

$$L_3 = 0.502 \text{ nH} \quad \text{(A-19)}$$

Figure A-2 shows the simulated response of the oscillator circuit with resonance at 3120 MHz or a 5% error. The small variation in resonant frequency may be due to the frequency-dependent packaged parameters, but it is a good starting value for tuning and optimization for the best phase noise and output power. The best phase noise at a given power output is basically dependent upon the ratio and absolute value of the feedback capacitors, which in turn depend upon the optimum drive level.

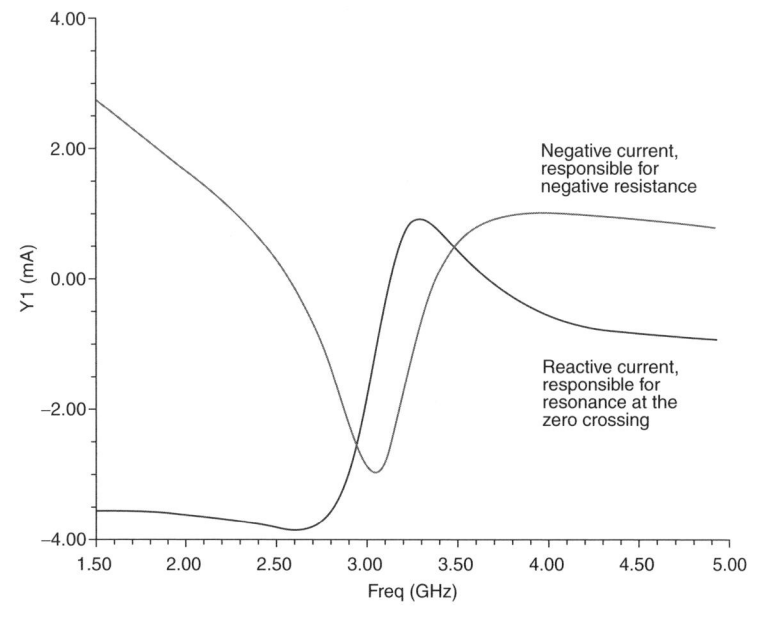

Figure A-2 The real and imaginary currents for oscillation. The reactive current crosses the zero line at 3120 MHz. This is close but not exactly at the point of most negative resistance current. The shift of 120 MHz is due to the use of small-signal rather than large-signal analysis.

A.2 SERIES FEEDBACK OSCILLATOR

The steady-state oscillation condition for the series feedback configuration can be expressed as

$$Z_{out}(I,\omega) + Z_L(\omega) = 0 \tag{A-20}$$

$$Z_L(\omega) \rightarrow Z_3(\omega) \tag{A-21}$$

where I is the load current amplitude and w is the resonance frequency. Z_{out} is current, and frequency-dependent output impedance, whereas Z_L is only a function of frequency.

$$Z_{out}(I,\omega) = R_{out}(I,\omega) + jX_{out}(I,\omega) \tag{A-22}$$

$$Z_L(\omega) = R_L(\omega) + jX_L(\omega) \tag{A-23}$$

The expression of output impedance Z_{out} can be written as

$$Z_{out} = -Z_3 \Rightarrow [Z_{22} + Z_2] - \frac{[Z_{12} + Z_2][Z_{21} + Z_2]}{[Z_{11} + Z_1 + Z_2]} \tag{A-24}$$

where Z_{ij} ($i,j = 1,2$) are the Z-parameters of the hybrid transistor model and can be written as

$$Z_{i,j} = [R_{ij} + jX_{ij}]_{i,j=1,2} \tag{A-25}$$

According to optimum criteria, the negative real part of the output impedance Z_{out} has to be maximized, and the possible optimal values of feedback reactance under which the negative value R_{out} are maximized by setting

$$\frac{\partial \text{Re}[Z_{out}]}{\partial X_1} = 0 \quad \text{and} \quad \frac{\partial \text{Re}[Z_{out}]}{\partial X_2} = 0 \tag{A-26}$$

$$\Rightarrow \frac{\partial [R_{out}]}{\partial X_1} = 0 \quad \text{and} \quad \frac{\partial [R_{out}]}{\partial X_2} = 0 \tag{A-27}$$

The optimal values X_1^* and X_2^*, based on the above condition, can be expressed in terms of a two-port parameter of the active device (BJT/FET) as [177, 178]

$$X_1^* = -X_{11} + \left[\frac{X_{12} + X_{21}}{2}\right] + \left[\frac{R_{21} - R_{12}}{X_{21} - X_{12}}\right]\left[\frac{R_{12} + R_{21}}{2} - R_{11} - R_1\right] \tag{A-28}$$

$$X_2^* = -\left[\frac{X_{12} + X_{21}}{2}\right] - \left[\frac{(R_{21} - R_{12})(2R_2 + R_{12} + R_{21})}{2(X_{21} - X_{12})}\right] \tag{A-29}$$

A.2 SERIES FEEDBACK OSCILLATOR

By substituting values of X_1^* and X_2^* into the above equation, the optimal real and imaginary parts of the output impedance Z_{out}^* can be expressed as

$$Z_{out}^* = R_{out}^* + X_{out}^* \tag{A-30}$$

$$R_{out}^* = R_2 + R_{22} - \left[\frac{(2R_2 + R_{21} + R_{12})^2 + (X_{21} - X_{12})^2}{4(R_{11} + R_2 + R_1)}\right] \tag{A-31}$$

$$X_{out}^* = X_2^* + X_{22} - \left[\frac{R_{21} - R_{12}}{X_{21} - X_{12}}\right]\left[R_{out}^* - R_2 - R_{22}\right] \tag{A-32}$$

where

$$X_2^* = -\left[\frac{X_{12} + X_{21}}{2}\right] - \left[\frac{(R_{21} - R_{12})(2R_2 + R_{12} + R_{21})}{2(X_{21} - X_{12})}\right] \tag{A-33}$$

Thus, in the steady-state operation mode of the oscillator, amplitude and phase balance conditions can be written as

$$R_{out}^* + R_L = 0 \tag{A-34}$$

$$X_{out}^* + X_L^* = 0 \tag{A-35}$$

The output power of the oscillator can be expressed in terms of load current and load impedance as

$$P_{out} = \frac{1}{2}I^2 \, \text{Re}[Z_L] \tag{A-36}$$

where I and V are the corresponding load current and voltage across the output.

$$I = \left[\frac{Z_{11} + Z_1 + Z_2}{Z_{22}(Z_{11} + Z_1 + Z_2) - Z_{21}(Z_{12} + Z_2)}\right]V \tag{A-37}$$

The expression of phase noise for the series feedback oscillator, following the approach for the Colpitts oscillator, is

$$|\mathscr{L}|_{SSB} = \left[4KTR + \frac{4qI_c g_m^2(t)}{\omega_0^4 \beta^2 C_{ce}^2 (C_2 + C_{be} - L_1 C_2 C_{be} \omega_0^2)^2 + g_m^2 \omega_0^2 (C_2 + C_{be} - L_1 C_2 C_{be} \omega_0^2)^2}\right]$$
$$\times \left[\frac{\omega_0^2}{4(\Delta\omega)^2 V_{cc}^2}\right]\left[\frac{1}{Q_L^2} + \left(1 - \left(\frac{1}{\omega_0^2 L_1}\right)\right)\right]$$
$$\times \left(\frac{(C_2 + C_{be} - L_1 C_2 C_{be} \omega_0^2) + C_{ce}}{C_{ce}[(C_2 + C_{be} - L_1 C_2 C_{be} \omega_0^2)]}\right)^2\right] \tag{A-38}$$

For large values of Q_1,

$$|\overline{\mathscr{L}}|_{SSB} = \left[4KTR + \frac{4qI_c g_m^2(t)}{\omega_0^4 \beta^2 C_{ce}^2 (C_2 + C_{be} - L_1 C_2 C_{be}\omega_0^2)^2 + g_m^2 \omega_0^2 (C_2 + C_{be} - L_1 C_2 C_{be}\omega_0^2)^2}\right]$$
$$\times \left[\frac{\omega_0^2}{4\omega^2 V_{cc}^2}\right] \frac{1}{\omega_0^4 L_1^2} \left(\frac{(C_2 + C_{be} - L_1 C_2 C_{be}\omega_0^2) + C_{ce}}{C_{ce}(C_2 + C_{be} - L_1 C_2 C_{be}\omega_0^2)}\right) \quad \text{(A-39)}$$

The important information that can be derived from this calculation is that the parasitics now dominate the design. The negative resistance which was proportional to $1/\omega^2$ is now $1/\omega^4$. The rule of thumb is to use a large device for lower frequencies and operate it at medium DC currents. In the millimeterwave area, this would be fatal. The large device would have excessive parasitic elements such as inductors and capacitors, and the optimum design would no longer be possible since the parasitics would be larger than the values required for optimum performance. These parasitics are the major reason why at millimeterwave and wide tuning ranges the phase noise is not as good as that provided by a narrowband Colpitts oscillator.

Example: A 3000 MHz oscillator is designed based on the above analytical series feedback approach and is validated with the simulated results. Figure A-3 shows the series feedback oscillator.

Large-signal Z-parameter measured data ($I_c = 20$ mA, $V_{ce} = 2$ V) at 3000 MHz are given as

$$Z_{11} = R_{11} + jX_{11} = (22.96 + j27.30)\Omega \quad \text{(A-40)}$$
$$Z_{21} = R_{21} + jX_{21} = (140 + j670)\Omega \quad \text{(A-41)}$$
$$Z_{12} = R_{12} + jX_{12} = (2.72 + j4.99)\Omega \quad \text{(A-42)}$$
$$Z_{22} = R_{22} + jX_{22} = (46.04 + j21.45)\Omega \quad \text{(A-43)}$$

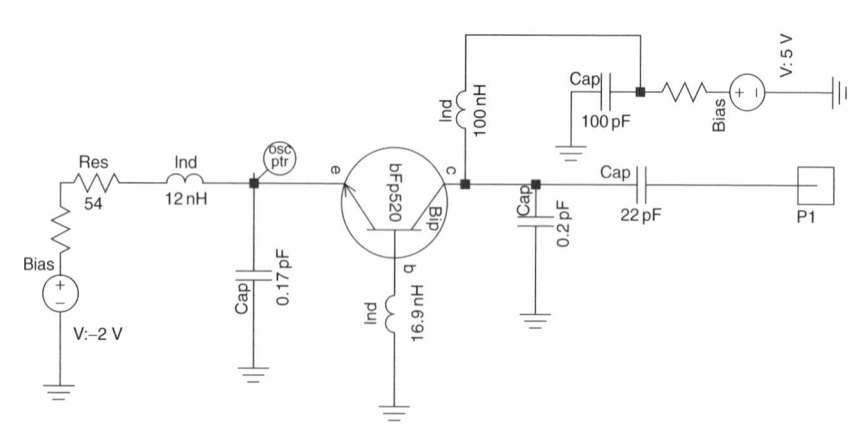

Figure A-3 A series feedback oscillator. For the oscillation condition, base-to-ground inductance and emitter-to-ground capacitance are required. The 12 nH inductor acts as a choke. The output is tuned to and terminated at 50 Ω.

A.2 SERIES FEEDBACK OSCILLATOR

$$X_1^* = -X_{11} + \left[\frac{X_{12} + X_{21}}{2}\right] + \left[\frac{R_{21} - R_{12}}{X_{21} - X_{12}}\right]\left[\frac{R_{12} + R_{21}}{2} - R_{11} - R_1\right] \quad \text{(A-44)}$$

$$X_1^* = 319.9654\,\Omega \Rightarrow L_1 = 16.9\,\text{nH} \quad \text{(A-45)}$$

$$X_2^* = -\left[\frac{X_{12} + X_{21}}{2}\right] - \left[\frac{(R_{21} - R_{12})(2R_2 + R_{12} + R_{21})}{2(X_{21} - X_{12})}\right] \quad \text{(A-46)}$$

$$X_2^* = -311.67084 \Rightarrow C_2 = 0.17\,\text{pF} \quad \text{(A-47)}$$

$$X_{out}^* = X_2^* + X_{22} - \left[\frac{R_{21} - R_{12}}{X_{21} - X_{12}}\right][R_{out}^* - R_2 - R_{22}] \quad \text{(A-48)}$$

$$X_{out}^* = -259.31176 \Rightarrow C_3 = 0.2\,\text{pF} \quad \text{(A-49)}$$

The simulated response of the oscillator circuit, having resonance at 2980 MHz or 1% error, is a good starting value for tuning and optimization for optimum phase noise and output power. The best phase noise at a given power output is basically dependent upon the ratio and absolute value of the feedback capacitor, which in turn depend upon the optimum drive level. Detailed analysis for designing the best phase noise, based on a unified approach, is presented in Appendix B. Figure A-4 shows the real and imaginary currents under oscillating conditions for

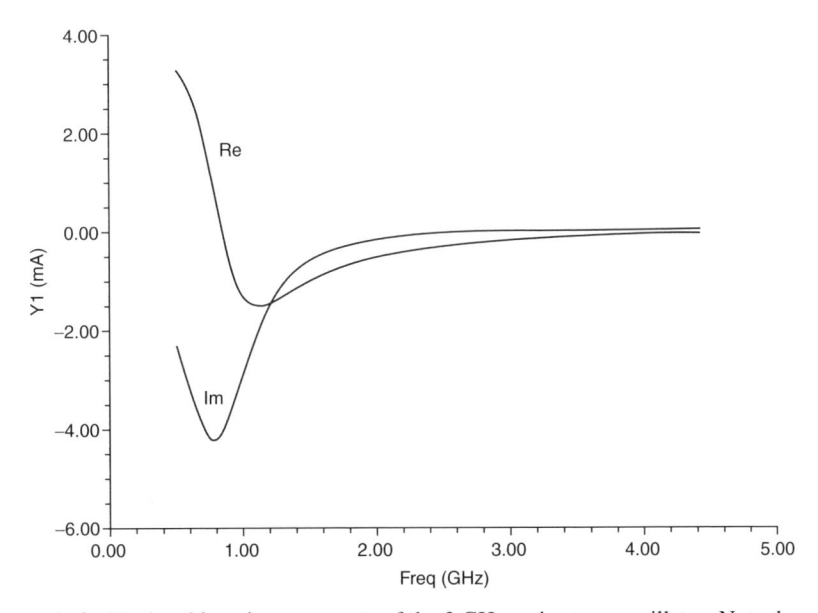

Figure A-4 Real and imaginary currents of the 3 GHz series-type oscillator. Note the very shallow curve.

optimum output power. In this case, the operating Q is very low, as can be seen from the shallow curve at which the imaginary current crosses the zero line, while the real current is still negative. To optimize this circuit for phase noise, the imaginary curve should go through the zero line at the point of steepest ascent while maintaining a negative real current. The low-Q resonator guarantees that the maximum output power is available and the resonator is heavily loaded.

APPENDIX B
Example of a Large-Signal Design Based on Bessel Functions

Frequency = 1000 MHz
Power output = 5 mW
Load = 500 Ω

Figure B-1 shows the schematic of the 1 GHz oscillator described in Chapter 9. The output termination is 500 Ω.

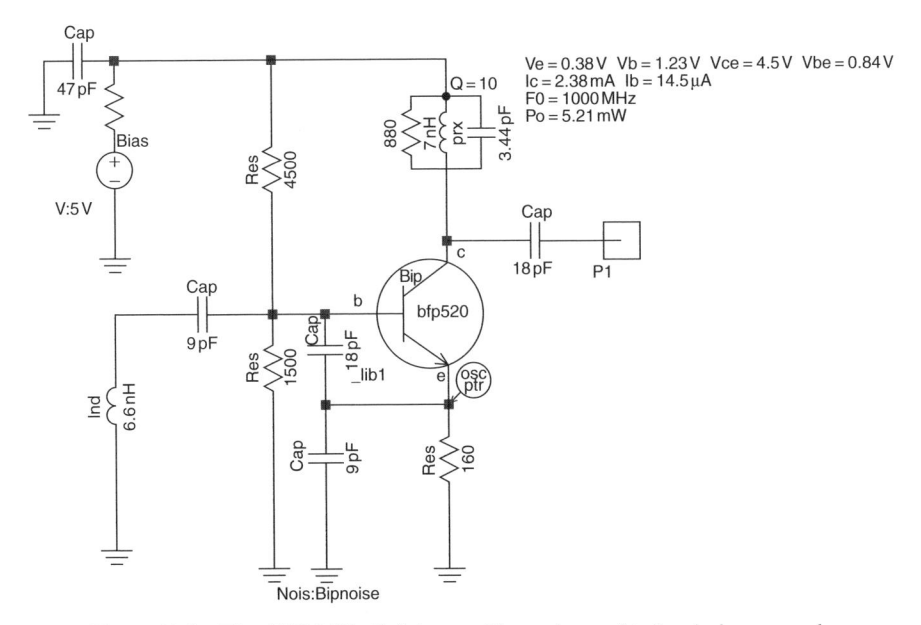

Figure B-1 The 1000 MHz Colpitts oscillator chosen for the design example.

The Design of Modern Microwave Oscillators for Wireless Applications: Theory and Optimization,
by Ulrich L. Rohde, Ajay Kumar Poddar, Georg Böck
Copyright © 2005 John Wiley & Sons, Inc.

B.1 DESIGN STEPS

Step 1

The 1000 MHz oscillator, using the bipolar transistor BFP520 (Infineon), is designed based on analytical equations and is later verified by actual results. Based on the output power requirement and harmonics at a given load, the drive level is fixed.

The normalized drive level of $x = 15$ is chosen to allow an adequate drive level to sustain oscillation but not to produce excessive harmonic content. For drive level $x = 15$, the fundamental peak current is given from a graph/table as

$$I_1(fundamental) = 1.932 I_{dc} \tag{B-1}$$

I_1 is the fundamental current specified by the output power needed for the designated load.

$$R_L = 500 \, \Omega \tag{B-2}$$

$$V_{out} = \sqrt{P_{out}(\text{mW}) \times 2R_L} = \sqrt{5\text{E}{-}3 \times 2 \times 500} = 2.236 \text{ V}$$

$$\text{(no saturation voltage assumed)} \tag{B-3}$$

$$I_1 = \frac{V_{out}}{500} = \frac{2.236}{500} = 4.472 \text{ mA} \tag{B-4}$$

$$I_e = I_{dc} = \frac{I_1}{1.932} = \frac{4.472}{1.932} = 2.314 \text{ mA} \tag{B-5}$$

Step 2

To avoid saturation in the transistor, select an emitter resistor R_e to maintain a sufficiently small emitter signal voltage of approximately half the base-emitter drop. The DC emitter voltage also provides a reasonable offset to the variations in the base-emitter bias voltage. R_e is set to 160 Ω. Using the common equation for biasing, the expression for the voltage at the base is given as

$$V_b = I_e \left[R_e + \frac{R_e}{\beta + 1} \right] + V_{be} = 1.23 \text{ V} \tag{B-6}$$

β is assumed to be around 100, and V_{be} is approximately 0.8 V. Bias resistors R_1 and R_2 are given as

$$V_b = \frac{R_2}{R_1 + R_2} V_{cc} = 1.23 \text{ V} \Rightarrow \frac{R_1}{R_2} \approx 3 \tag{B-7}$$

$$R_1 = 1500 \, \Omega \tag{B-8}$$

$$R_2 = 4500 \, \Omega \tag{B-9}$$

$$V_{cc} = 5 \text{ V} \tag{B-10}$$

Step 3

The large-signal transconductance is determined as

$$Y_{21} = \frac{I_1}{V_1}\bigg|_{fundamental\ freq} = \frac{1.932 I_{dc}}{260\ \text{mV}} = \frac{4.472\ \text{mA}}{260\ \text{mV}} = 17.2\ \text{mS} \quad \text{(B-11)}$$

Step 4

The value of the n factor is calculated from the equation above as

$$n^2(G_2 + G_3) - n(2G_3 + Y_{21}\alpha) + (G_1 + G_3 + Y_{21}) = 0 \quad \text{(B-12)}$$

$$G_1 = 0 \quad \text{(B-13)}$$

$$G_2 = \frac{1}{R_1 \| R_2} = 0.88\ \text{mS} \quad \text{(B-14)}$$

$$G_3 = \frac{1}{R_e} = \frac{1}{160} = 6.25\ \text{mS} \quad \text{(B-15)}$$

$$Y_{21} = 17.2\ \text{mS} \quad \text{(B-16)}$$

where

$$\alpha = 0.99 \quad \text{(B-17)}$$

The quadratic equation above is reduced to

$$n^2(G_2 + G_3) - n(2G_3 + Y_{21}\alpha) + (G_1 + G_3 + Y_{21}) = 0 \quad \text{(B-18)}$$

$$n^2(0.88 + 6.25) - n(2 \times 6.25 + 17.2 \times 0.99) + (0 + 6.25 + 17.2) = 0 \quad \text{(B-19)}$$

$$7.13n^2 - 29.528n + 23.45 = 0 \quad \text{(B-20)}$$

$$n = \frac{29.528 \pm \sqrt{(29.528)^2 - 4 \times 7.13 \times 23.45}}{2 \times 7.13}$$

$$= \frac{29.528 \pm \sqrt{871.9 - 668.794}}{2 \times 7.13} \quad \text{(B-21)}$$

$$n = \frac{29.528 \pm 14.25}{14.26} \Rightarrow n_1 = 3.06 \quad \text{and} \quad n_2 = 1.071 \quad \text{(B-22)}$$

The higher value of the transformation factor, n, is selected as $n = 3$.

EXAMPLE OF A LARGE-SIGNAL DESIGN BASED ON BESSEL FUNCTIONS

The values of C_1 and C_2 are calculated as

$$\frac{C_2}{C_1+C_2} = \frac{1}{n} \Rightarrow C_2 = \frac{C_1}{n-1} \tag{B-23}$$

$$C_2 = \frac{C_1}{n-1} = \frac{C_1}{2} \Rightarrow \frac{C_1}{C_2} = 2 \tag{B-24}$$

The ratio of the capacitor C_1 to C_2 is 2. The absolute values of the capacitors are determined from the loop-gain condition of the oscillator as

$$Y_{21}|_{large\ signal} = G_m(x) = \frac{qI_{dc}}{kTx}\left[\frac{2I_1(x)}{I_0(x)}\right]_{n=1} = \frac{g_m}{x}\left[\frac{2I_1(x)}{I_0(x)}\right]_{n=1} \tag{B-25}$$

$$G_m(x) = \frac{1}{R_P}\frac{[C_1+C_2]^2}{C_1 C_2} \tag{B-26}$$

$$\frac{g_m}{x}\left[\frac{2I_1(x)}{I_0(x)}\right]_{n=1} = \frac{1}{R_P}\frac{[C_1+C_2]^2}{C_1 C_2} = \frac{1}{R_P}\frac{C_1}{C_2}\left[1+\frac{C_2}{C_1}\right]^2 \tag{B-27}$$

$$\frac{88\ \text{mS}}{10} \times 1.932 = \frac{1}{R_P}\frac{C_1}{C_2}\left[1+\frac{C_2}{C_1}\right]^2 \tag{B-28}$$

$$17.01\text{E}-3 = \frac{4.50}{R_P} \tag{B-29}$$

$$R_P = \frac{4.50}{17.01\text{E}-3} = 264.5\ \Omega \tag{B-30}$$

The quality factor of the inductor is assumed to be 10 at 1000 MHz, a low-Q case. The value of the inductor is obtained as

$$Q_T = \frac{R_P}{\omega_0 L} \Rightarrow L = \frac{264.5}{10 \times \omega_0} \tag{B-31}$$

$$L = \frac{264.5}{10 \times 2\pi \times 1000\text{E}6} = 4.2\ \text{nH} \tag{B-32}$$

$$\omega = \sqrt{\frac{1}{L}\left[\frac{1}{C_1}+\frac{1}{C_2}\right]} \tag{B-33}$$

$$\omega^2 = \frac{1}{L}\left[\frac{1}{C_1}+\frac{1}{C_2}\right] = \frac{C_1+C_2}{LC_1 C_2} \tag{B-34}$$

The value of the capacitor is determined as

$$C_2 = \frac{3}{\omega^2 \times 8.4\text{E}{-}9} \Rightarrow C_2 = \frac{3}{331.5\text{E}{-}9} = 9 \text{ pF} \quad \text{(B-35)}$$

$$C_1 = 2C_2 = 18 \text{ pF} \quad \text{(B-36)}$$

Step 5

The value of the coupling capacitor, C_s, is assumed to be 10 pF, and the effect of C_s on the series reactance of the inductor L must be considered. Therefore, the inductor value is adjusted to

$$j\omega_0 \times L = j\omega_0 \left(L' - \frac{1}{\omega_0^2 C_s} \right) \quad \text{(B-37)}$$

$$j\omega_0 \times 4.2 \text{ nH} = j\omega_0 \left(L' - \frac{1}{\omega_0^2 \times 10\text{E}{-}12} \right) \quad \text{(B-38)}$$

$$L' = 6.6 \text{ nH} \quad \text{(B-39)}$$

The base-lead inductance of the BFP520 is approximately 0.4 nH. After this is corrected, the effective value of the inductor is 6.2 nH.

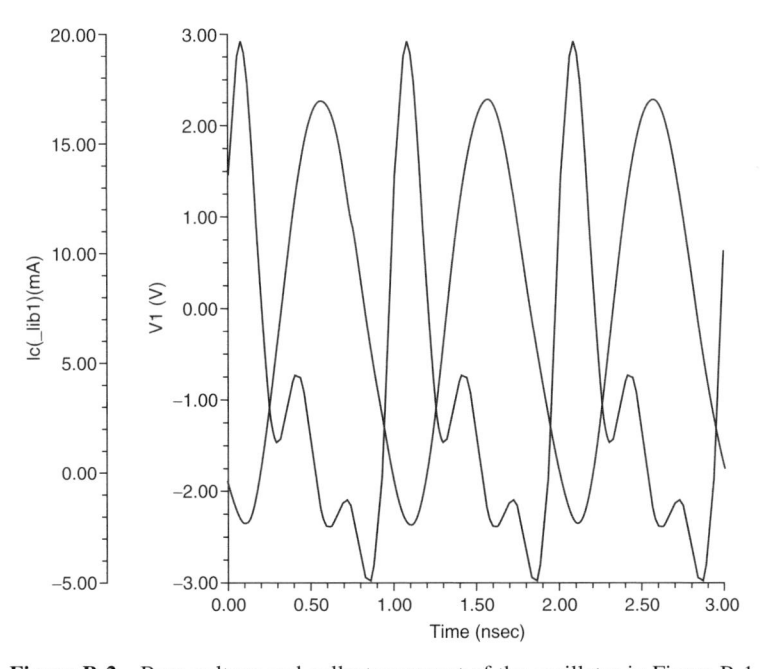

Figure B-2 Base voltage and collector current of the oscillator in Figure B-1.

Step 6

The harmonic content can be calculated from the table of Bessel functions as

$$x = 15 \Rightarrow I_1 = 1.932 I_{dc}; \quad I_2 = 1.742 I_{dc}; \quad I_3 = 1.272 I_{dc}; \quad I_4 = 0.887 I_{dc} \quad \text{(B-40)}$$

The parallel tank circuit at the output of the oscillator is designed to filter out higher harmonics.

$$Q = \frac{R}{wL} \quad \text{(B-41)}$$

$$Q = 20 \quad \text{(B-42)}$$

$$R = 880 \quad \text{(B-43)}$$

$$L = 7 \text{ nH} \quad \text{(B-44)}$$

$$C = 3.60 \text{ pF} \quad \text{(B-45)}$$

The analytically calculated values show good agreement with the simulated and published results. Figure B-2 shows both the base-emitter voltage, which looks

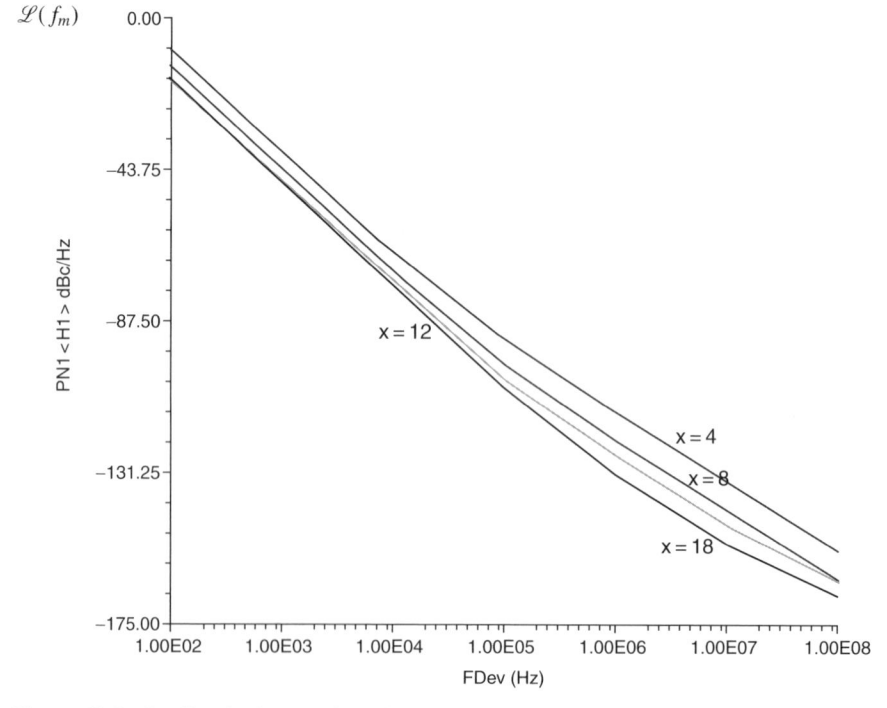

Figure B-3 Predicted phase noise of the circuit shown in Figure B-1 with different normalized drive levels.

sinusoidal, and the collector current under the given operating condition. As previously shown, due to the harmonic contents, there is a certain amount of ringing, as well as negative collector current. This is due to the tuned collector circuit.

Figure B-3 shows the predicted phase noise as a function of the normalized drive level using values of x between 4 and 18. The phase noise is not the optimized phase noise for this configuration. The best phase noise can be achieved by adjusting the proper ratio and absolute values of the feedback capacitors at a given drive level and required output power.

APPENDIX C
Design Example of Best Phase Noise and Good Output Power

Figure C-1 shows the parallel-tuned Colpitts oscillator circuit, which must be designed with the following specifications. The unit was also built and measured. It uses a ceramic resonator, and its equivalent circuit is shown.

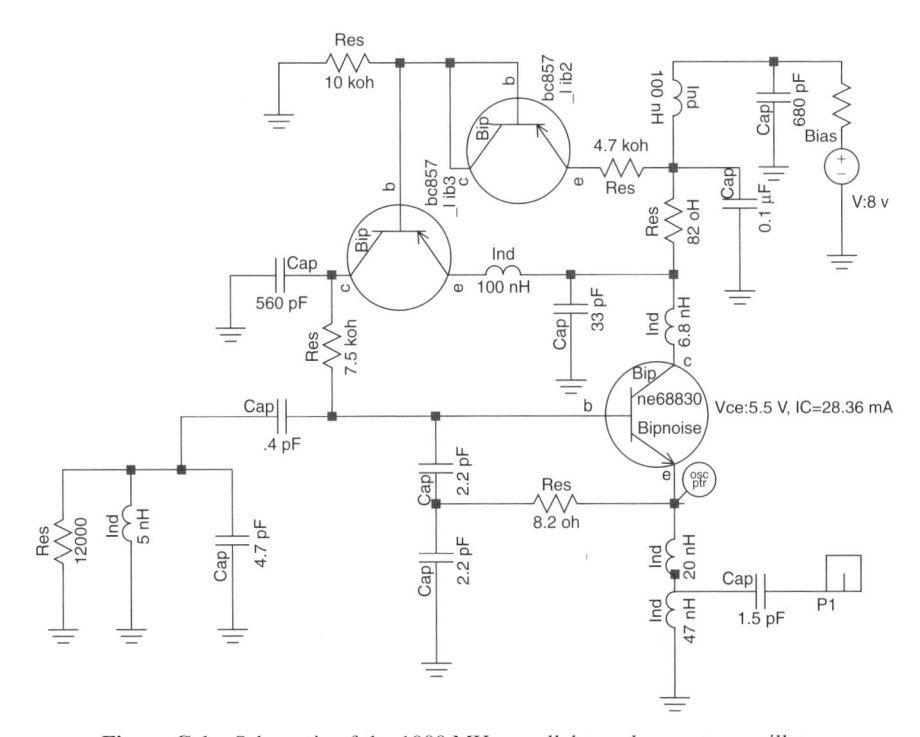

Figure C-1 Schematic of the 1000 MHz parallel-tuned resonator oscillator.

The Design of Modern Microwave Oscillators for Wireless Applications: Theory and Optimization, by Ulrich L. Rohde, Ajay Kumar Poddar, Georg Böck
Copyright © 2005 John Wiley & Sons, Inc.

DESIGN EXAMPLE OF BEST PHASE NOISE AND GOOD OUTPUT POWER

C.1 REQUIREMENTS

- output power requirement: 13 dBm
- operating frequency: 1000 MHz
- load: 50 Ω
- phase noise: −124 dBc/Hz at 10 kHz

C.2 DESIGN STEPS

Step 1

For calculation of the operating point for a fixed, normalized drive of $x = 20$ (high output power), see Table 6-1.

Based on the output power requirement, the following are calculated:

1. The oscillator output voltage at the fundamental frequency is

$$V_{out}(\omega_0) = \sqrt{P_{out}(\omega_0) * 2R_L} = \sqrt{20\text{E}-3 * 2 * 50} \approx 1.414 \text{ V} \quad \text{(C-1)}$$

2. The fundamental current is

$$I_{out}(\omega_0) = \frac{V_{out}(\omega_0)}{50} = \frac{1.414}{50} = 28.3 \text{ mA} \quad \text{(C-2)}$$

3. The DC operating point is calculated based on the normalized drive level $x = 20$. The expression for the emitter DC current given in terms of the Bessel function with respect to the drive level is

$$[I_E(\omega_0)] = 2I_{DC}\left[\frac{I_1(x)}{I_0(x)}\right]_{x=normalized\ drive\ level} \quad \text{(C-3)}$$

For the normalized drive level $x = 20$, the output emitter current at the fundamental frequency can be given as

$$[I_E(\omega_0)]_{x=20} = [I_{E1}(\omega_0)]_{x=20} + [I_{E2}(\omega_0)]_{x=20} = 2I_{DC}\left[\frac{I_1(x)}{I_0(x)}\right]_{x=20} \approx 56 \text{ mA} \quad \text{(C-4)}$$

$$[I_{E1}(\omega_0)]_{x=20} = I_{out}(\omega_0) = 28.3 \text{ mA} \quad \text{(output current to the load)} \quad \text{(C-5)}$$

Figure C-2 shows the predicted output of the oscillator and Figure C-3 shows the predicted resulting phase noise.

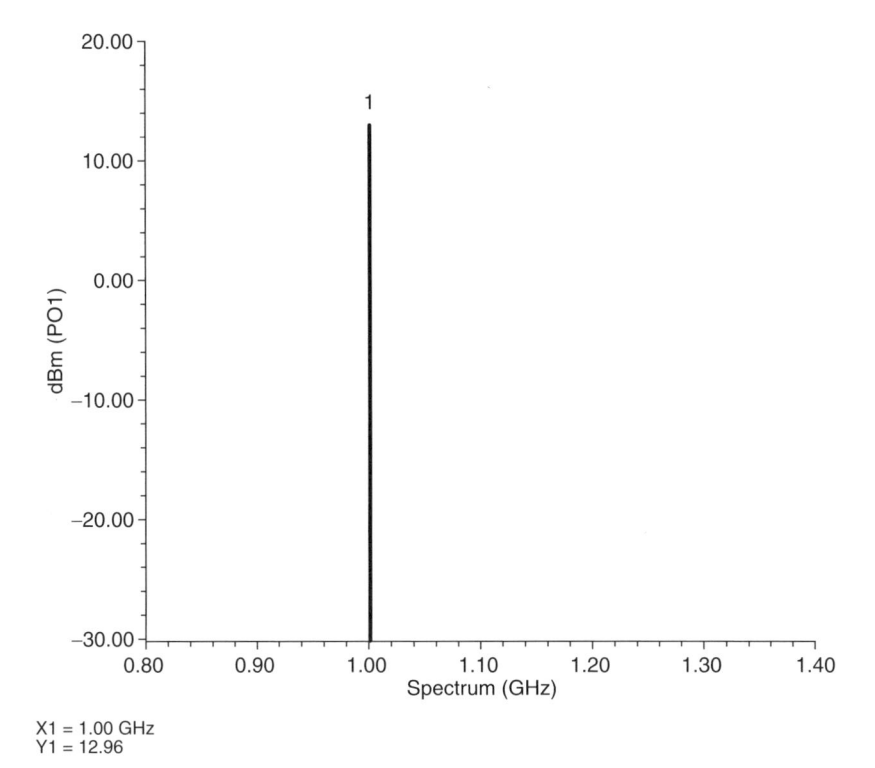

Figure C-2 Predicted output power of the oscillator.

Figure C-4 shows the oscillator circuit configuration in which DC and RF current distribution is shown, divided into its components.

$$[I_{E2}(\omega_0)]_{x=20} = [I_E(\omega_0)]_{x=20} - [I_{E1}(\omega_0)]_{x=20} = 27.3 \text{ mA} \tag{C-6}$$

$$I_{E-DC} = \frac{[I_E(\omega_0)]_{x=20}}{2[I_1(x)/I_0(x)]_{x=20}} = 28.3 \text{ mA} \tag{C-7}$$

For this application, the NE68830 was selected.

Step 2: Biasing Circuit

For the best phase noise close-in, a DC/AC feedback circuit is incorporated, which provides the desired operating DC condition [84]:

$I_E = 28.3$ mA
$V_{CE} = 5.5$ V, supply voltage $V_{cc} = 8$ V
$\beta = 120$
$I_B \approx 0.23$ mA

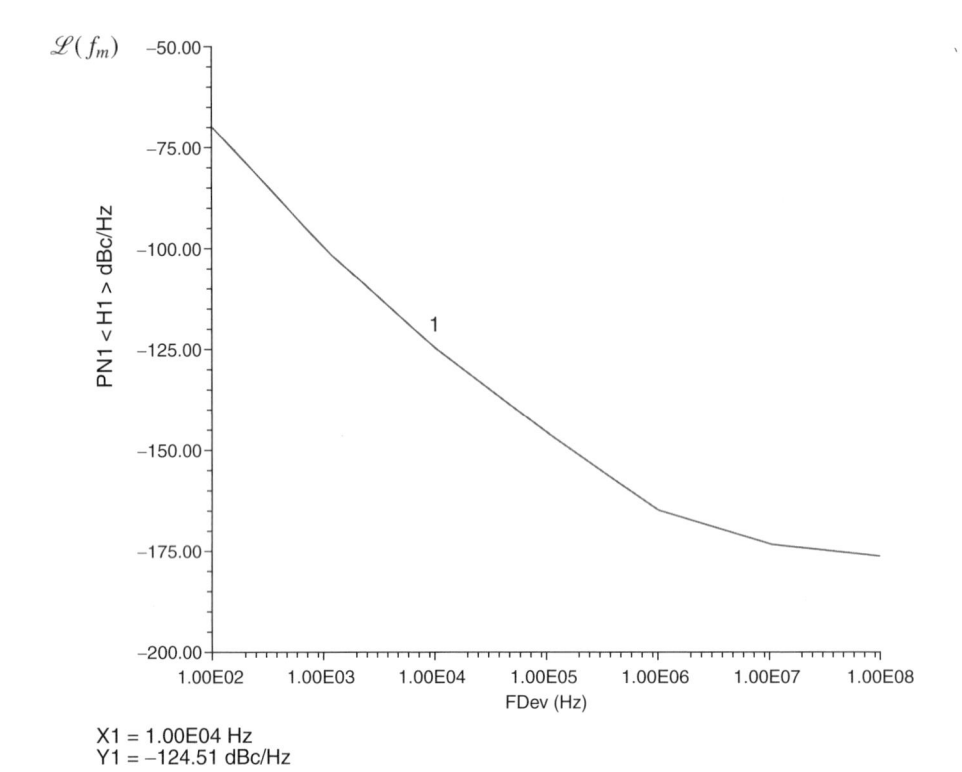

400 DESIGN EXAMPLE OF BEST PHASE NOISE AND GOOD OUTPUT POWER

X1 = 1.00E04 Hz
Y1 = −124.51 dBc/Hz

Figure C-3 Predicted phase noise of the oscillator.

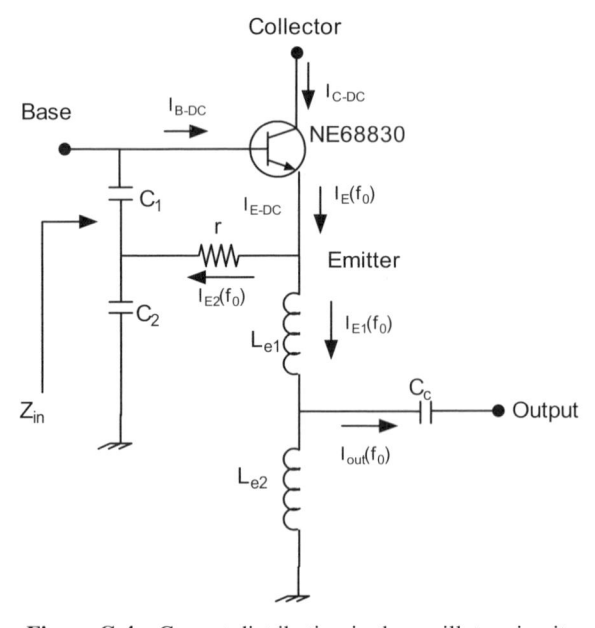

Figure C-4 Current distribution in the oscillator circuit.

Step 3: Calculation of the Large-Signal Transconductance

$$Y_{21}|_{large\ signal} = G_m(x) = \frac{qI_{dc}}{kTx}\left[\frac{2I_1(x)}{I_0(x)}\right]_{fundamental} \tag{C-8}$$

$$[Y_{21}]_{\omega=\omega_0} = \left[\frac{1.949 I_{E-DC}}{520\,\text{mV}}\right] = 0.107 \tag{C-9}$$

Step 4: Loop Gain

The loop gain is

$$\text{Loop Gain} = [LG]_{sustained\ condition} = \left[\frac{R_P Y_{21}(x)}{n}\right]$$

$$= \left[\frac{R_P g_m}{x}\right]\left[\frac{2I_1(x)}{I_0(x)}\right]\left[\frac{1}{n}\right] > 1 \tag{C-10}$$

$$R_{PEQ}(f_0) = R_P \parallel \text{Bias circuit} \Rightarrow 50.73\,\Omega \tag{C-11}$$

As derived earlier, the loop gain should be 2.1 to have good starting conditions.

$$n = \left[\frac{R_{PEQ} Y_{21}(x)}{2.1}\right] = \frac{0.107 * 50.73}{2.1} \approx 2.523 \tag{C-12}$$

Step 5: Calculation of the Feedback Capacitor Ratio

$$n = 1 + \left[\frac{C_1}{C_2}\right] = 2.523 \Rightarrow \left[\frac{C_1}{C_2}\right]_{x=20} = 1.523\ (2.523 - 1) \tag{C-13}$$

Step 6: Calculation of Absolute Values of the Feedback Capacitor

The expression of Z_{in} (looking into the base of the transistor) can be given as

$$Z_{in} \cong -\left[\left(\frac{Y_{21}}{\omega^2(C_1^* + C_p)C_2}\right)\left(\frac{1}{1 + \omega^2 Y_{21}^2 L_p^2}\right)\right]$$

$$-j\left[\left(\frac{C_1^* + C_p + C_2}{\omega(C_1^* + C_p)C_2}\right) - \left(\frac{\omega Y_{21} L_p}{1 + \omega^2 Y_{21}^2 L_p^2}\right)\left(\frac{Y_{21}}{\omega(C_1^* + C_p)C_2}\right)\right] \tag{C-14}$$

where

$C_p = (C_{BEPKG} + \text{contribution from layout}) = 1.1\,\text{pF}$
$L_p = (L_B + L_{BX} + \text{contribution from layout}) = 2.2\,\text{nH}$

The expression for negative resistance R_n is

$$R_{neq} = \frac{R_n}{1 + \omega^2 Y_{21}^2 L_p^2} = \frac{R_n}{[1 + (2\pi * 1E9)^2 * (0.107)^2 * (2.2\,\text{nH})^2]} \quad \text{(C-15)}$$

$$R_{neq} \approx \frac{R_n}{3.65} \quad \text{(C-16)}$$

$$R_n = -\left[\frac{Y_{21}^+}{\omega^2 C_1 C_2}\right]_{x=20} = \frac{0.107}{(2\pi * 1E9)^2 C_1 C_2} \quad \text{(C-17)}$$

R_n is the negative resistance without parasitics (C_p, L_p).
For sustained oscillation → $R_{neq} \geq 2R_{PEQ} \cong 101.4$ ohms.

$$R_n = 3.65 * 101.4 \approx 371 \text{ ohms} \quad \text{(C-18)}$$

$$C_1 C_2 = \left[\frac{1}{\omega^2}\right]\left[\frac{0.107}{371}\right] \approx 7.26 \quad \text{(C-19)}$$

$$\left[\frac{C_1}{C_2}\right]_{x=20} \approx 1.52 \quad \text{(C-20)}$$

$$C_1 = 3.3 \text{ pF} \quad \text{(C-21)}$$

$$C_2 = 2.2 \text{ pF} \quad \text{(C-22)}$$

Step 7: Calculation of the Coupling Capacitor r_e

The expression for the coupling capacitor is

$$\frac{C}{10} > C_c > \left\{\frac{(\omega^2 C_1 C_2)(1 + \omega^2 Y_{21}^2 L_p^2)}{[(Y_{21}^2 C_2 - \omega^2 C_1 C_2)(1 + \omega^2 Y_{21}^2 L_p^2)(C_1 + C_p + C_2)]}\right\} \quad \text{(C-23)}$$

$$C_c \rightarrow 0.4 \text{ pF} \quad \text{(C-24)}$$

Figure C-5 shows the transistor in the package parameters for the calculation of the oscillator frequency and loop gain.

Tables C-1 and C-2 show NE68830 nonlinear parameters and package parameters taken from the NEC data sheets.

C.3 DESIGN CALCULATIONS

C.3.1 Frequency of Oscillation

Frequency of oscillation is

$$\omega_0 = \sqrt{\frac{1}{L\left[\frac{(C_1^* + C_p)C_2 C_c/(C_1^* + C_p + C_2)}{[(C_1^* + C_p)C_2/(C_1^* + C_p + C_2) + C_c]} + C\right]}} \approx 1000 \text{ MHz} \quad \text{(C-25)}$$

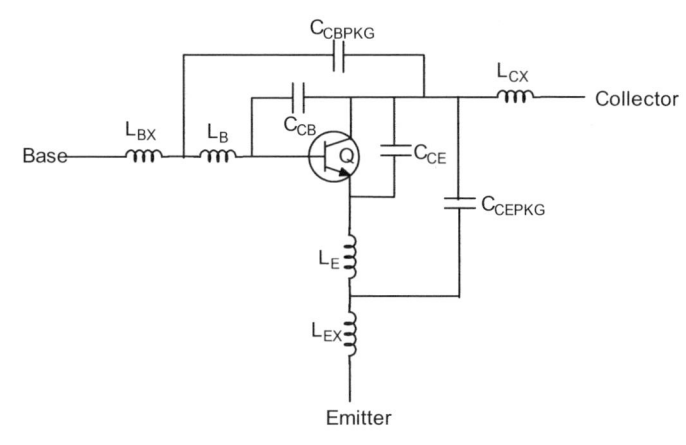

Figure C-5 NE68830 with package parasitics. Q is the intrinsic bipolar transistor.

with

$L = 5$ nH (inductance of the parallel resonator circuit)
$C_1^* = 2.2$ pF
$C_1 = C_1^* + C_p$
$C_p = 1.1$ pF (C_{BEPKG} + contribution from layout)

TABLE C-1 Nonlinear Parameters of NE68830

Parameters	Q	Parameters	Q
IS	3.8E−16	MJC	0.48
BF	135.7	XCJC	0.56
NF	1	CJS	0
VAF	28	VJS	0.75
IKF	0.6	MJS	0
NE	1.49	TF	11E−12
BR	12.3	XTF	0.36
NR	1.1	VTF	0.65
VAR	3.5	ITF	0.61
IKR	0.06	PTF	50
ISC	3.5E−16	TR	32E−12
NC	1.62	EG	1.11
RE	0.4	XTB	0
RB	6.14	XTI	3
RBM	3.5	KF	0
IRB	0.001	AF	1
RC	4.2	VJE	0.71
CJE	0.79E−12	MJE	0.38
CJC	0.549E−12	VJC	0.65

TABLE C-2 Package Parameters of NE68830

Parameters	NE68830
C_{CB}	0.24E−12
C_{CE}	0.27E−12
L_B	0.5E−9
L_E	0.86E−9
C_{CBPKG}	0.08E−12
C_{CEPKG}	0.04E−12
C_{BEPKG}	0.04E−12
L_{BX}	0.2E−9
L_{CX}	0.1E−9
L_{EX}	0.2E−9

$C_2 = 2.2 \, \text{pF}$

$C_c = 0.4 \, \text{pF}$

$C = 4.7 \, \text{pF}$

$R_P = 12{,}000$ (measured)

$Q_{unloaded} = [R_P/\omega L] = 380$

C.3.2 Calculation of the Phase Noise

The noise equations determined in Section 8.4, equations (8-109), (8-115), and (8-117), and which contain resonator noise, shot noise, and flicker noise, can now be used to graphically determine the best phase noise as a function of n. Figure C-6 shows a plot of this curve. It gives 2.5 as the best number of n, which

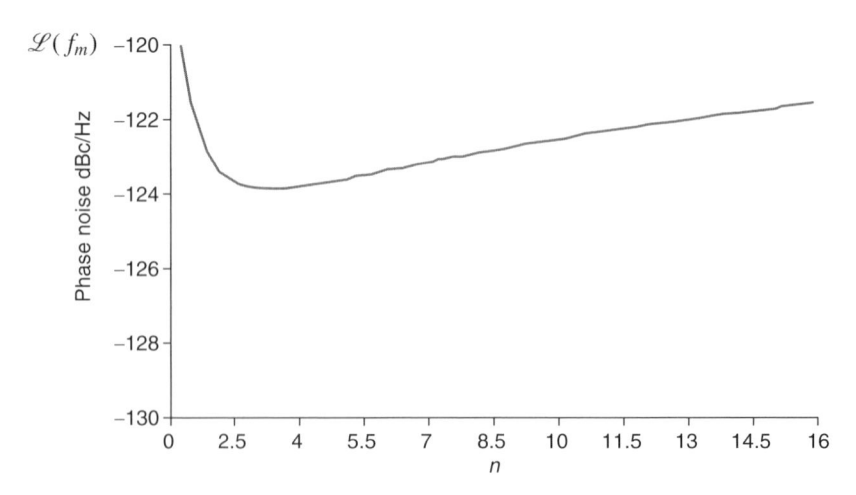

Figure C-6 The phase noise contribution of the lossy resonator at 10 kHz offset.

is consistent with the calculation done for the large-signal condition. Equation (C-12) gives the same result.

The calculated phase noise at 10 kHz off the carrier is $-124\,\text{dBc/Hz}$, which agrees with the measurements within 1 dB. The other values are $-140\,\text{dBc/Hz}$ at 100 kHz offset and $-160\,\text{dBc/Hz}$ at 1 MHz offset.

This circuit is shown in Figure 9-2 in Section 9.10. The actual measured phase noise is shown in Figure 9-4, and the simulation is shown in Figure 9-5. Considering that equation (8-109) contains only shot and flicker noise as well as resonator noise, it has been proven that this calculation by itself is a very accurate formula for practical use. Figure 9-5 has been generated by Ansoft Designer, which includes all noise sources and is based on the HB principle.

The important conclusion in Chapter 8 is that for the first time we have a complete mathematical synthesis procedure for best phase noise that covers both flicker noise and white noise for the oscillator. In the past, most publications referenced an oscillator built with many shortcuts, and then the author found that the measured results agreed with the expectations. A complete synthesis approach had not appeared previously.

APPENDIX D
A Complete Analytical Approach for Designing Efficient Microwave FET and Bipolar Oscillators

For large-signal operation, it is necessary to obtain the exact nonlinear device parameters of the active two-port network and calculate the external feedback elements of the oscillator circuits. Initially, the feedback element values are unknown. There is no good or efficient experimental solution for this task, only a small-signal approach, which does not handle power and noise.

D.1 SERIES FEEDBACK (MESFET)

The best way to calculate series or parallel feedback oscillators with external elements is to use an analytical approach, designing a microwave oscillator that determines the explicit expression for the optimum feedback elements and the load impedance in terms of the transistor equivalent circuit parameters. These equations also provide a better understanding of the fundamental limitations involved in obtaining high output power for a given microwave oscillator topology.

Maximizing oscillator output power and efficiency is also the focus of many ongoing applications, such as the active phase-array antenna and others.

Figure D-1 shows the series feedback topology of the oscillator using a metal semiconductor field-effect transistor (MESFET). External feedback elements Z_1, Z_2, and Z_3 are shown outside the dotted line.

The optimum values of Z_1, Z_2, and Z_3 are given as

$$Z_1^{opt} = R_1^* + jX_1^* \tag{D-1}$$

$$Z_2^{opt} = R_2^* + jX_2^* \tag{D-2}$$

$$Z_3^{opt} = Z_L^{opt} = R_3^* + jX_3^* \tag{D-3}$$

The Design of Modern Microwave Oscillators for Wireless Applications: Theory and Optimization, by Ulrich L. Rohde, Ajay Kumar Poddar, Georg Böck
Copyright © 2005 John Wiley & Sons, Inc.

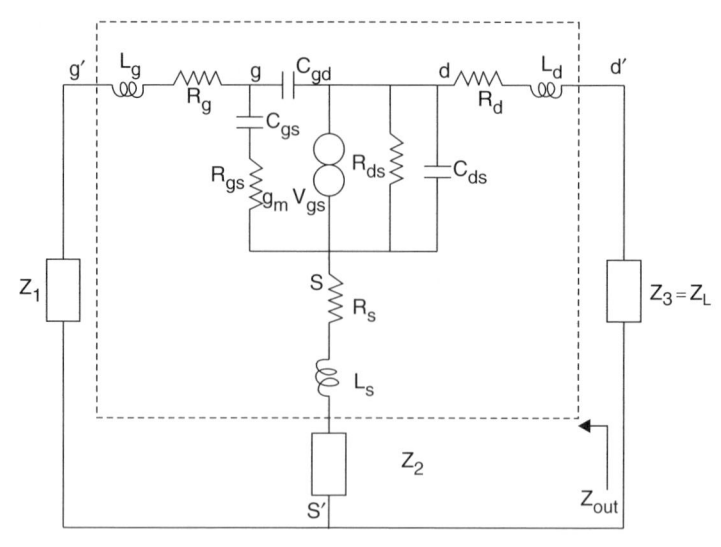

Figure D-1 Series feedback topology.

$$Z_{out}^{opt} + Z_L^{opt} \Rightarrow 0 \tag{D-4}$$

$$Z_{out}^{opt} = R_{out}^* + jX_{out}^* \tag{D-5}$$

The general approach for designing an oscillator providing maximum output power at a given frequency is based on the optimum values of the feedback element and the load under steady-state large-signal operation. The steady-state oscillation condition for a series feedback configuration can be expressed as

$$[Z_{out}(I_0, \omega_0) + Z_L(\omega_0)]_{w=w_0} = 0 \tag{D-6}$$

I_0 is the amplitude of the load current, and w_0 is the oscillator frequency. Assuming that the steady-state current entering the active circuit is near sinusoidal, a medium- to high-Q case, the output impedance $Z_{out}(I_0, \omega_0)$ and the load impedance $Z_L(\omega_0)$ can be expressed in terms of real and imaginary parts as

$$Z_{out}(I_0, \omega_0) = R_{out}(I_0, \omega_0) + jX_{out}(I_0, \omega_0) \tag{D-7}$$

$$Z_L(\omega) = R_L(\omega) + jX_L(\omega) \tag{D-8}$$

$Z_{out}(I_0,\omega_0)$ is the current amplitude and the frequency-dependent function, and $Z_L(w)$ is a function of the frequency.

The common source [Z] parameter of the MESFET is given as

$$[Z]_{cs} = \begin{bmatrix} Z_{11} & Z_{12} \\ Z_{21} & Z_{22} \end{bmatrix}_{cs} \tag{D-9}$$

with

$$Z_{11} = R_{11} + jX_{11} \tag{D-10}$$

$$R_{11} = R_{gs}\left[\frac{a}{(a^2+b^2)} + \frac{b\omega R_{ds}C_{ds}(1+C_{gd}/C_{ds})}{(a^2+b^2)}\right]$$
$$+ \left[\frac{a\omega R_{ds}C_{ds}(1+C_{gd}/C_{ds})}{\omega C_{gs}(a^2+b^2)} - \frac{b}{\omega C_{gs}(a^2+b^2)}\right] \tag{D-11}$$

$$X_{11} = R_{gs}\left[\frac{a\omega R_{ds}C_{ds}(1+C_{gd}/C_{ds})}{(a^2+b^2)} - \frac{b}{(a^2+b^2)}\right]$$
$$- \left[\frac{a}{\omega C_{gs}(a^2+b^2)} + \frac{b\omega R_{ds}C_{ds}(1+C_{gd}/C_{ds})}{\omega C_{gs}(a^2+b^2)}\right] \tag{D-12}$$

$$Z_{12} = R_{12} + jX_{12} \tag{D-13}$$

$$R_{12} = \frac{aR_{ds}C_{gd}}{C_{gs}(a^2+b^2)} + \frac{b\omega R_{ds}C_{gd}R_{gs}}{(a^2+b^2)} \tag{D-14}$$

$$X_{12} = \frac{a\omega R_{ds}C_{gd}R_{gs}}{(a^2+b^2)} - \frac{bR_{ds}C_{gd}}{C_{gs}(a^2+b^2)} \tag{D-15}$$

$$Z_{21} = R_{21} + jX_{21} \tag{D-16}$$

$$R_{21} = R_{ds}\left[\frac{C_{gd}}{C_{gs}}\frac{a}{(a^2+b^2)} + \frac{b\omega R_{gs}C_{gd}}{(a^2+b^2)} + \frac{g_m(b\cos\omega\tau + a\sin\omega\tau)}{\omega C_{gs}(a^2+b^2)}\right] \tag{D-17}$$

$$X_{21} = R_{ds}\left[\frac{a\omega R_{gs}C_{gd}}{(a^2+b^2)} - \frac{C_{gd}}{C_{gs}}\frac{b}{(a^2+b^2)} + \frac{g_m(a\cos\omega\tau - b\sin\omega\tau)}{\omega C_{gs}(a^2+b^2)}\right] \tag{D-18}$$

$$Z_{22} = R_{22} + jX_{22} \tag{D-19}$$

$$R_{22} = R_{ds}\left[\frac{a}{(a^2+b^2)} + \frac{C_{gd}}{C_{gs}}\frac{a}{(a^2+b^2)} + \frac{b}{(a^2+b^2)}\omega R_{gs}C_{gd}\right] \tag{D-20}$$

$$X_{22} = R_{ds}\left[\frac{a\omega R_{gs}C_{gd}}{(a^2+b^2)} - \frac{C_{gd}}{C_{gs}}\frac{b}{(a^2+b^2)} - \frac{b}{(a^2+b^2)}\right] \tag{D-21}$$

with

$$a = 1 + \frac{C_{gd}}{C_{gs}}(1 - \omega^2 R_{gs}C_{gs}R_{ds}C_{ds}) + \frac{g_m R_{ds}C_{gd}}{C_{gs}}\cos(\omega\tau) \tag{D-22}$$

$$b = \omega(R_{ds}C_{ds} + R_{ds}C_{gd}) + \omega\frac{C_{gd}}{C_{gs}}(R_{gs}C_{gs} + R_{ds}C_{ds}) - \frac{g_m R_{ds}C_{gd}}{C_{gs}}\sin(\omega\tau) \tag{D-23}$$

The expression of the output impedance Z_{out} can be written as

$$Z_{out} = \left[[Z_{22} + Z_2] - \frac{[Z_{12} + Z_2][Z_{21} + Z_2]}{[Z_{11} + Z_1 + Z_2]}\right] \tag{D-24}$$

$$Z_{out} + Z_3 \Rightarrow Z_{out} + Z_L = 0 \tag{D-25}$$

where $Z_{ij}(i,j = 1, 2)$ are the Z-parameters of the transistor model and can be expressed as

$$\left[Z_{i,j}\right]_{i,j=1,2} = [R_{ij} + jX_{ij}]_{i,j=1,2} \tag{D-26}$$

According to the criterion for maximum output power at a given oscillator frequency, the negative real part of the output impedance Z_{out} has to be maximized. The optimal values of the feedback reactance under which the negative value of R_{out} is maximized are given by the following [178, 179]:

$$\frac{\partial \mathrm{Re}}{\partial X_1}[Z_{out}(I,\omega)] = 0 \Rightarrow \frac{\partial}{\partial X_1}[R_{out}] = 0 \tag{D-27}$$

$$\frac{\partial \mathrm{Re}}{\partial X_2}[Z_{out}(I,\omega)] = 0 \Rightarrow \frac{\partial}{\partial X_2}[R_{out}] = 0 \tag{D-28}$$

The values of X_1 and X_2, which will satisfy the differential equations above, are given as X_1^* and X_2^*, which can be expressed in terms of a two-port parameter of the active device (MESFET) as

$$X_1^* = -X_{11} + \left[\frac{X_{12} + X_{21}}{2}\right] + \left[\frac{R_{21} - R_{12}}{X_{21} - X_{12}}\right]\left[\frac{R_{12} + R_{21}}{2} - R_{11} - R_1\right] \tag{D-29}$$

$$X_1^* = \frac{(1 - \omega\tau_g \tan \omega\tau)}{\omega C_{gs}(a - b\tan \omega\tau)} - \frac{(b + a\tan \omega\tau)(R_1 + R_g)}{(a - b\tan \omega\tau)}$$

$$- \left[\frac{R_{ds}C_{ds}(\omega\tau_g + \tan \omega\tau)}{C_{gs}(a - b\tan \omega\tau)} - \frac{g_m R_{ds}}{2\omega C_{gs} \cos \omega\tau(a - b\tan \omega\tau)}\right] \tag{D-30}$$

where τ is the transit time in the MESFET channel and $\tau_g = R_{gs}$ and $\tau_d = R_{ds}$.

$$X_2^* = -\left[\frac{X_{12} + X_{21}}{2}\right] - \left[\frac{(R_{21} - R_{12})(2R_2 + R_{12} + R_{21})}{2(X_{21} - X_{12})}\right] \tag{D-31}$$

$$X_2^* = \frac{R_{ds}C_{gd}(\omega\tau_g + \tan \omega\tau)}{C_{gs}(a - b\tan \omega\tau)} - \frac{(b + a\tan \omega\tau)(R_2 + R_s)}{(a - b\tan \omega\tau)}$$

$$- \frac{g_m R_{ds}}{(a - b\tan \omega\tau)2\omega C_{gs} \cos \omega\tau} \tag{D-32}$$

D.1 SERIES FEEDBACK (MESFET)

$\lfloor R_{i,j} \rfloor_{i,j=1,2}$ and $= [X_{ij}]_{i,j=1,2}$ are the real and imaginary parts of $[Z_{i,j}]_{i,j=1,2}$ of the transistor.

The output impedance can be given as $Z_{out}(I,w) = R_{out}(I,w) + X_{out}(I,w)$, and the corresponding optimum output impedance for the given oscillator frequency can be derived analytically by substituting values of the optimum values of susceptance under which the negative value of R_{out} is maximized as

$$\lfloor Z^*_{out}(I,\omega) \rfloor_{\omega - \omega_0} = \lfloor R^*_{out}(I,\omega) + X^*_{out}(I,\omega) \rfloor_{\omega = \omega_0} \tag{D-33}$$

$$[R^*_{out}(I,\omega_0)]_{X^*_1, X^*_2} = \left\{ R_2 + R_{22} - \left[\frac{(2R_2 + R_{21} + R_{12})^2 + (X_{21} - X_{12})^2}{4(R_{11} + R_2 + R_1)} \right] \right\} \tag{D-34}$$

$$[X^*_{out}(I,\omega)] = \left\{ \frac{X_{22} - X_{12} - X_{21}}{2} - \frac{(R_{21} - R_{12})(2R_2 + R_{12} + R_{21})}{2(X_{21} - X_{12})} \right.$$

$$\left. - \frac{(R_{21} - R_{12})(R^*_{out} - R_2 - R_{22})}{X_{21} - X_{12}} \right\} \tag{D-35}$$

$$[X^*_{out}(I,\omega)]_{X^*_1, X^*_2} = X^*_2 + X_{22} - \left[\frac{R_{21} - R_{12}}{X_{21} - X_{12}} \right] [R^*_{out} - R_2 - R_{22}] \tag{D-36}$$

$$X^*_{out} = \frac{R_{ds}}{(a - b\tan w\tau)} \left[\tan w\tau - \frac{g_m}{2wC_{gs}\cos w\tau} \right] - \frac{(b + a\tan w\tau)R^*_{out}}{(a - b\tan w\tau)} \tag{D-37}$$

X^*_1 and X^*_2 in the equations above are the optimal values of the external feedback susceptance.

For easier analysis, the effects of transit time and gate drain capacitance are neglected for preliminary calculation of an optimum value of the feedback element, and the simplified expressions are given as

$$X^*_1 = \frac{1}{\omega C_{gs}} + R_{ds} \left[-\omega C_{ds}(R_1 + R_g + R_{gs}) + \frac{g_m}{2\omega C_{gs}} \right] \tag{D-38}$$

$$X^*_2 = -R_{ds} \left[\omega C_{ds}(R_2 + R_s) + \frac{g_m}{2\omega C_{gs}} \right] \tag{D-39}$$

$$X^*_{out} = -R_{ds} \left[\omega C_{ds} R^*_{out} + \frac{g_m}{2\omega C_{gs}} \right] \tag{D-40}$$

$$R^*_{out} = (R_2 + R_s) + \frac{R_{ds}}{1 + (\omega C_{ds} R_{ds})^2} \left[1 - \frac{R_{ds}}{R_g + R_s + R_1 + R_2 + R_{gs}} \left(\frac{g_m}{2\omega C_{gs}} \right)^2 \right] \tag{D-41}$$

The simplified expressions above show accuracy with the HB-based simulated results for a gate length less than 1 μm at an operating frequency range up to 20 GHz.

A COMPLETE ANALYTICAL APPROACH

The differential drain resistance R_{ds} can be expressed in terms of the optimum output resistance as

$$R_{ds} = \frac{\left(1 + \sqrt{1 - 4(R_{out}^* - R_2 - R_s)G_{dso}}\right)}{2G_{dso}} \quad \text{(D-42)}$$

where

$$G_{dso} = \frac{1}{R_g + R_s + R_2 + R_{gs}} \left(\frac{g_m}{2\omega C_{gs}}\right)^2 + (R_{out}^* - R_2 - R_s)(\omega C_{ds})^2 \quad \text{(D-43)}$$

Alternatively, a differential drain resistance can be obtained from a quasi-linear analysis. Under large-signal operation, the transistor parameters vary with the drive level. If we restrict our interest to the fundamental signal frequency component, then V_{gs} and V_{ds} can be expressed as

$$V_{gs}(t) = V_{gso} + V_{gs}\sin(\omega t + \varphi) \quad \text{(D-44)}$$

$$V_{ds}(t) = V_{dso} + V_{ds}\sin(\omega t) \quad \text{(D-45)}$$

V_{gso} and V_{dso} are the DC operating bias voltages, V_{gs} and V_{ds} are the amplitude of the signal frequency components, and φ is the phase difference between the gate and drain voltages.

The drain current I_d can be expressed as

$$I_{ds} = I_{ds}(V_{gs}, V_{dso}) \quad \text{(D-46)}$$

Under the assumption of linear superposition of the DC and RF currents, an instantaneous drain current can be expressed as

$$I_{ds}(t) = I_{dso} + g_m v_{gs}\cos(\omega t + \varphi) + G_d v_{ds}\cos(\omega t) \quad \text{(D-47)}$$

where I_{dso} is the DC bias drain current.

The transconductance g_m and the drain conductance G_d are defined as

$$g_m = \left[\frac{I_{ds}}{V_{gs}}\right]_{V_{ds}=0} \quad \text{(D-48)}$$

$$G_D = \left[\frac{I_{ds}}{V_{gs}}\right]_{V_{gs}=0} \quad \text{(D-49)}$$

Under large-signal conditions, transconductance and drain conductance are given as

$$g_m = \frac{\omega}{\pi V_{gs} \sin \varphi} \int_0^{2\pi/w} I_{ds} \sin(\omega t) dt \qquad \text{(D-50)}$$

$$G_d = \frac{\omega}{\pi V_{ds} \sin \varphi} \int_0^{2\pi/w} I_{ds} \sin(\omega t + \varphi) dt \qquad \text{(D-51)}$$

The drain current can be expressed in terms of V_{gs}, V_p, and V_{ds} as

$$I_d = I_{dss} \left[1 - \frac{V_g}{V_p} \right]^2 \tanh \left[\frac{\alpha V_d}{V_g - V_p} \right] \qquad \text{(D-52)}$$

$$V_p = V_{p0} + \gamma V_d \qquad \text{(D-53)}$$

I_{dss} is the saturation current, and V_p is the gate pinch-off voltage. α, γ, and V_{p0} are the model parameters of the MESFET.

Applying a Taylor-series expansion of the equation on the operating DC point and considering the fundamental frequency component terms, the large-signal drain resistance, as a function of the small-signal drain voltage amplitude, can be given as

$$R_{DS}|_{large\ signal} = \frac{R_{ds}}{(1 + AV_d^2)} \qquad \text{(D-54)}$$

where R_{DS} and R_{ds} are the large- and small-signal differential resistances.
A is defined as

$$A = \left\{ \frac{3 \tanh^2[\alpha V_{d0}/(V_{g0} - V_p)] - 1}{4[(V_{g0} - V_p)/\alpha]^2} \right\} \qquad \text{(D-55)}$$

$$R_{ds} = \left\{ \frac{\cosh^2[\alpha V_{d0}/(V_{g0} - V_p)]}{I_{dss}[1 - (V_{g0}/V_p)]^2} \left[\frac{V_{g0} - V_p}{\alpha} \right] \right\} \qquad \text{(D-56)}$$

From the expression above, $R_{DS}|_{large\ signal}$ has a maximum value in the absence of the RF drive signal and decreases as the amplitude of the RF signal increases. Consequently, oscillator output impedance and oscillator output power are functions of the change in drain resistance under large-signal operation. To support the steady-state operation mode, the amplitude and phase balance conditions can be written as

$$\left[R_{out}^*(I,\omega) + R_L(\omega) \right]_{\omega=\omega_0} = 0 \qquad \text{(D-57)}$$

$$\left[X_{out}^*(I,\omega) + X_L^*(w) = 0 \right]_{\omega=\omega_0} = 0 \qquad \text{(D-58)}$$

The output power of the oscillator can be expressed in terms of load current and load impedance as

$$P_{out} = \frac{1}{2} I_{out}^2 \operatorname{Re}[Z_L] \tag{D-59}$$

where I_{out} and V_{out} are the corresponding load current and drain voltage across the output.

$$I_{out} = \left[\frac{Z_{11} + Z_1 + Z_2}{Z_{22}(Z_{11} + Z_1 + Z_2) - Z_{21}(Z_{12} + Z_2)} \right] V_{out} \tag{D-60}$$

$$P_{out} = \frac{1}{2} I_{out}^2 \operatorname{Re}[Z_L] \Rightarrow \frac{1 + (R_{21} - R_{12}/X_{21} - X_{12})^2}{(R_{22} + R)^2 + (X_{22} + X)^2} (R_{out} + R_d) \frac{V_d^2}{2} \tag{D-61}$$

where

$$R = \frac{X_{21}(X_{12} + X_2^*) - R_{21}(R_{12} + R_2 + R_s) - X_{22}(X_{11} + X_1^* + X_2^*)}{R_{11} + R_1 + R_2 + R_g + R_s} \tag{D-62}$$

$$X = \frac{R_{21}(X_{11} + X_1^* + X_2^*) - R_{21}(X_{12} + X_2^*) - X_{21}(R_{12} + R_2 + R_s)}{R_{11} + R_1 + R_2 + R_g + R_s} \tag{D-63}$$

D.2 PARALLEL FEEDBACK (MESFET)

Figure D-2 shows the parallel feedback topology of the oscillator using the MESFET, in which the external feedback elements Y_1, Y_2, and Y_3 are shown outside the dotted line.

The optimum values of the feedback elements Y_1, Y_2, and Y_3 are given as

$$Y_1^{opt} = R_1^* + jX_1^* \tag{D-64}$$

$$Y_2^{opt} = R_2^* + jX_2^* \tag{D-65}$$

$$Y_3^{opt} = Z_L^{opt} = R_3^* + jX_3^* \tag{D-66}$$

$$Y_{out}^{opt} + Z_L^{opt} \Rightarrow 0 \tag{D-67}$$

$$Y_{out}^{opt} = R_{out}^* + jX_{out}^* \tag{D-68}$$

D.2 PARALLEL FEEDBACK (MESFET)

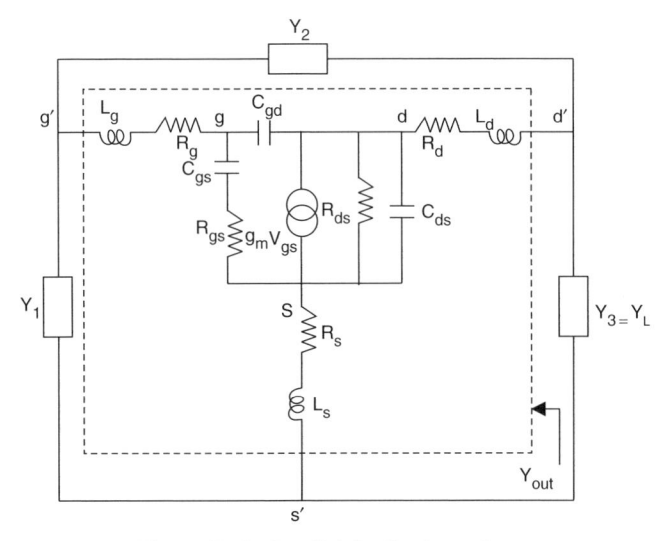

Figure D-2 Parallel feedback topology.

The common source [Y] parameter of the MESFET is given as

$$[Y]_{cs} = \begin{bmatrix} Y_{11} & Y_{12} \\ Y_{21} & Y_{22} \end{bmatrix} \tag{D-69}$$

$$Y_{11} = \frac{j\omega C_{gs}}{1 + j\omega C_{gs} R_{gs}} + j\omega C_{gd} \Rightarrow G_{11} + jB_{11} \tag{D-70}$$

$$Y_{21} = \frac{g_m \exp(-j\omega\tau)}{1 + j\omega C_{gs} R_{gs}} - j\omega C_{gd} \Rightarrow G_{21} + jB_{21} \tag{D-71}$$

$$Y_{12} = -j\omega C_{gd} \Rightarrow G_{12} + jB_{12} \tag{D-72}$$

$$Y_{22} = \frac{1}{R_{ds}} + j\omega(C_{ds} + C_{gd}) \Rightarrow G_{22} + jB_{22} \tag{D-73}$$

The optimum values of the output admittance Y^*_{out} and the feedback susceptances B^*_1 and B^*_2, which can be expressed in terms of the two-port Y-parameter of the active device, are given as

$$B^*_1 = -\left\{ B_{11} + \left[\frac{B_{12} + B_{21}}{2}\right] + \left[\frac{G_{21} - G_{12}}{B_{21} - B_{12}}\right]\left[\frac{G_{12} + G_{21}}{2} + G_{11}\right] \right\} \tag{D-74}$$

$$B^*_1 = \frac{g_m}{2\omega C_{gs} R_{gs}} \tag{D-75}$$

$$B_2^* = \left[\frac{B_{12} + B_{21}}{2}\right] + \left[\frac{(G_{12} + G_{21})(G_{21} - G_{12})}{2(B_{21} - B_{12})}\right] \quad \text{(D-76)}$$

$$B_2^* = -\omega C_{dg} - \frac{g_m}{2\omega C_{gs} R_{gs}} \quad \text{(D-77)}$$

The optimum values of the real and imaginary parts of the output admittance are given as

$$Y_{out}^* = [G_{out}^* + jB_{out}^*] \quad \text{(D-78)}$$

where G_{out}^* and B_{out}^* are given as

$$G_{out}^* = G_{22} - \left[\frac{(G_{12} + G_{21})^2 (B_{21} - B_{12})^2}{4G_{11}}\right] \quad \text{(D-79)}$$

$$G_{out}^* = \frac{1}{R_{ds}} - \frac{1}{R_{gs}}\left[\frac{g_m}{2\omega C_{gs}}\right]^2 \quad \text{(D-80)}$$

$$B_{out}^* = B_{22} + \left[\frac{G_{21} - G_{12}}{B_{21} - B_{12}}\right] - \left[\frac{(G_{12} + G_{21})}{2} + G_{22} - G_{out}^*\right] + \left[\frac{B_{21} + B_{12}}{2}\right] \quad \text{(D-81)}$$

$$B_{out}^* = \omega C_{gd} - \frac{1}{R_{gs}}\left[\frac{g_m}{2\omega C_{gs}}\right]\left[1 - \frac{1}{R_{gs}\omega C_{gs}}\frac{g_m}{2\omega C_{gs}}\right] \quad \text{(D-82)}$$

The value of the output susceptance B_{out}^* may be positive or negative, depending on the values of the transistor transconductance and $\tau_{gs} = R_{gs}C_{gs}$.

The voltage feedback factor n and phase φ_n can be expressed in terms of transistor Y-parameters as

$$n(V_{ds}/V_{gs}) = \frac{\sqrt{(G_{12} + G_{21} - 2G_2)^2 + (B_{21} - B_{12})^2}}{2(G_{12} + G_{21} - G_2)} \Rightarrow \frac{1}{2}\sqrt{1 + (\omega R_s C_{gs})^2} \quad \text{(D-83)}$$

$$\Phi_n(\text{phase}) = \tan^{-1}\frac{B_{21} - B_{12}}{G_{12} + G_{21} - 2G_2} \Rightarrow -\tan^{-1}(\omega R_s C_{gs}) \quad \text{(D-84)}$$

The output power of the oscillator can be expressed in terms of the load current and the load impedance as

$$P_{out} = \frac{1}{2}I_{out}^2 \text{Re}[Z_L] \quad \text{(D-85)}$$

where I_{out} and V_{out} are the corresponding load current and drain voltage across the output.

$$I_{out} = \left[\frac{Z_{11} + Z_1 + Z_2}{Z_{22}(Z_{11} + Z_1 + Z_2) - Z_{21}(Z_{12} + Z_2)}\right] V_{out} \qquad \text{(D-86)}$$

D.3 SERIES FEEDBACK (BIPOLAR)

Figure D-3 shows the series feedback oscillator topology for deriving explicit analytical expressions for the optimum values of the external feedback elements and the load impedance for maximum power output at a given oscillator frequency through [Z]-parameters of a bipolar transistor. The external feedback elements Z_1, Z_2, and Z_3 are shown outside the dotted line.

The optimum values of the feedback element Z_1, Z_2, and Z_3 are given as

$$Z_1^{opt} = R_1^* + jX_1^* \qquad \text{(D-87)}$$

$$Z_2^{opt} = R_2^* + jX_2^* \qquad \text{(D-88)}$$

$$Z_3^{opt} = Z_L^{opt} = R_3^* + jX_3^* \qquad \text{(D-89)}$$

$$Z_{out}^{opt} + Z_L^{opt} \Rightarrow 0 \qquad \text{(D-90)}$$

$$Z_{out}^{opt} = R_{out}^* + jX_{out}^* \qquad \text{(D-91)}$$

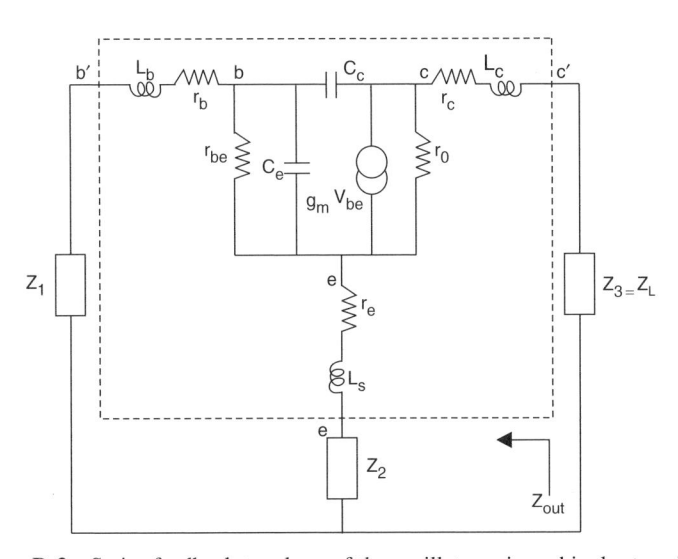

Figure D-3 Series feedback topology of the oscillator using a bipolar transistor.

The [Z]-parameters of the internal bipolar transistor in a common-emitter, small-signal condition are given as [177]

$$[Z]_{ce} = \begin{bmatrix} Z_{11} & Z_{12} \\ Z_{21} & Z_{22} \end{bmatrix} \tag{D-92}$$

$$Z_{11} = R_{11} + jX_{11} \Rightarrow a\left[\frac{1}{g_m} + r_b\left(\frac{\omega}{\omega_T}\right)^2\right] - ja\frac{\omega}{\omega_T}\left[\frac{1}{g_m} - r_b\right] \tag{D-93}$$

$$Z_{12} = R_{12} + jX_{12} \Rightarrow a\left[\frac{1}{g_m} + r_b\left(\frac{\omega}{\omega_T}\right)^2\right] - ja\frac{\omega}{\omega_T}\left[\frac{1}{g_m} - r_b\right] \tag{D-94}$$

$$Z_{21} = R_{21} + jX_{21} \Rightarrow a\left[\frac{1}{\omega_T C_c} + \frac{1}{g_m} + r_b\left(\frac{\omega}{\omega_T}\right)^2\right] - ja\frac{\omega}{\omega_T}\left[\frac{1}{g_m} - \frac{1}{\omega_T C_c} - r_b\right] \tag{D-95}$$

$$Z_{22} = R_{22} + jX_{22} \Rightarrow a\left[\frac{1}{\omega_T C_c} + \frac{1}{g_m} + r_b\left(\frac{\omega}{\omega_T}\right)^2\right] - ja\frac{\omega}{\omega_T}\left[\frac{1}{g_m} + \frac{1}{\omega_T C_c} - r_b\right] \tag{D-96}$$

where

$$a = \left\{\frac{1}{1 + \left[\frac{\omega}{\omega_T}\right]^2}\right\} \tag{D-97}$$

$$\omega_T = 2\pi f_T \tag{D-98}$$

$$f_T = \frac{g_m}{2\pi C_e} \tag{D-99}$$

According to the optimum criterion for the maximum power output at a given oscillator frequency, the negative real part of the output impedance Z_{out} has to be maximized. The possible optimal values of the feedback reactance, under which the negative value of R_{out} is maximized, are given by the following condition [179]:

$$\frac{\partial \text{Re}}{\partial X_1}[Z_{out}(I,\omega)] = 0 \Rightarrow \frac{\partial}{\partial X_1}[R_{out}] = 0 \tag{D-100}$$

$$\frac{\partial \text{Re}}{\partial X_2}[Z_{out}(I,\omega)] = 0 \Rightarrow \frac{\partial}{\partial X_2}[R_{out}] = 0 \tag{D-101}$$

The values of X_1 and X_2, which will satisfy the differential equations above, are given as X_1^* and X_2^*, which can be expressed in terms of a two-port parameter of the active device (bipolar) as

$$X_1^* = -X_{11} + \left[\frac{X_{12} + X_{21}}{2}\right] + \left[\frac{R_{21} - R_{12}}{X_{21} - X_{12}}\right]\left[\frac{R_{12} + R_{21}}{2} - R_{11} - R_1\right] \quad \text{(D-102)}$$

$$X_1^* = \frac{1}{2\omega C_c} - r_b \frac{\omega}{\omega_T} \quad \text{(D-103)}$$

$$X_2^* = -\left[\frac{X_{12} + X_{21}}{2}\right] - \left[\frac{(R_{21} - R_{12})(2R_2 + R_{12} + R_{21})}{2(X_{21} - X_{12})}\right] \Rightarrow -\frac{1}{2\omega C_c} - r_e \frac{\omega}{\omega_T}$$

$$\text{(D-104)}$$

$$X_2^* = -\frac{1}{2\omega C_c} - r_e \frac{\omega}{\omega_T} \quad \text{(D-105)}$$

By substituting the values of X_1^* and X_2^* into the equation above, the optimal real and imaginary parts of the output impedance Z_{out}^* can be expressed as

$$Z_{out}^* = R_{out}^* + X_{out}^* \quad \text{(D-106)}$$

$$R_{out}^* = R_2 + R_{22} - \left[\frac{(2R_2 + R_{21} + R_{12})^2 + (X_{21} - X_{12})^2}{4(R_{11} + R_2 + R_1)}\right] \quad \text{(D-107)}$$

$$R_{out}^* = r_c + \frac{r_b}{r_b + r_e + R_{11}}\left[r_e + R_{11} + \frac{a}{\omega_T C_e}\right] - \frac{a}{r_b + r_e + R_{11}}\left[\frac{1}{2\omega C_e}\right] \quad \text{(D-108)}$$

$$X_{out}^* = X_2^* + X_{22} - \left[\frac{R_{21} - R_{12}}{X_{21} - X_{12}}\right]\left[R_{out}^* - R_2 - R_{22}\right] \quad \text{(D-109)}$$

$$X_{out}^* = \frac{1}{2\omega C_e} - (R_{out}^* - r_c)\frac{\omega}{\omega_T} \quad \text{(D-110)}$$

Thus, in steady-state operation of the oscillator, the amplitude and phase balance conditions can be written as

$$R_{out}^* + R_L = 0 \quad \text{(D-111)}$$

$$X_{out}^* + X_L^* = 0 \quad \text{(D-112)}$$

The output power of the oscillator can be expressed in terms of load current and load impedance as

$$P_{out} = \frac{1}{2}I_{out}^2 \text{Re}[Z_L] \quad \text{(D-113)}$$

420   A COMPLETE ANALYTICAL APPROACH

I_{out} and V_{out} are the corresponding load current and drain voltage across the output.

$$I_{out} = \left[\frac{Z_{11} + Z_1 + Z_2}{Z_{11}Z_2 - Z_{12}(Z_1 + Z_2)}\right] V_{be} \tag{D-114}$$

$$V_{out} = V_c = \left[\frac{Z_{22}(Z_{11} + Z_1 + Z_2) - Z_{21}(Z_2 + Z_{12})}{Z_{12}(Z_1 + Z_2) - Z_{11}Z_2}\right] V_{be} \tag{D-115}$$

$$P_{out} = \frac{1}{2} I_{out}^2 \operatorname{Re}[Z_L] \tag{D-116}$$

$$P_{out} = aG_m^2(x) R_{out}^* \frac{(r_b + r_e + R_{11})}{(r_b + r_c - R_{out}^*)} \frac{V_1^2}{2} \tag{D-117}$$

V_1 is signal voltage, and x is the drive level across the base-emitter junction of the bipolar transistor. The large-signal transconductance $G_m(x)$ is given as

$$G_m(x) = \frac{qI_{dc}}{kTx}\left[\frac{2I_1(x)}{I_0(x)}\right]_{n=1} = \frac{g_m}{x}\left[\frac{2I_1(x)}{I_0(x)}\right]_{n=1} \tag{D-118}$$

$$V_1|_{peak} = \frac{kT}{q} x \tag{D-119}$$

$$g_m = \frac{I_{dc}}{kT/q} \tag{D-120}$$

where g_m is the small-signal transconductance.

D.4 PARALLEL FEEDBACK (BIPOLAR)

Figure D-4 shows the parallel feedback topology of the oscillator using a bipolar transistor in which the external feedback elements Y_1, Y_2, and Y_3 are shown outside the dotted line.

The optimum values of the feedback elements Y_1, Y_2, and Y_3 are given as

$$Y_1^{opt} = R_1^* + jX_1^* \tag{D-121}$$

$$Y_2^{opt} = R_2^* + jX_2^* \tag{D-122}$$

$$Y_3^{opt} = Z_L^{opt} = R_3^* + jX_3^* \tag{D-123}$$

$$Y_{out}^{opt} + Z_L^{opt} \Rightarrow 0 \tag{D-124}$$

$$Y_{out}^{opt} = R_{out}^* + jX_{out}^* \tag{D-125}$$

D.4 PARALLEL FEEDBACK (BIPOLAR)

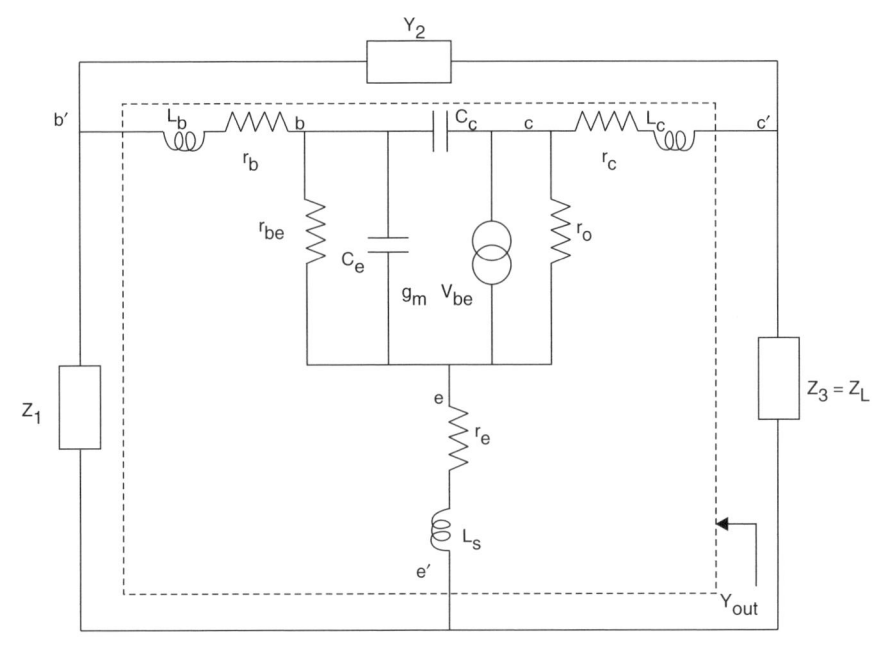

Figure D-4 A parallel feedback topology of the oscillator using a bipolar transistor.

The common source $[Y]$-parameters of the bipolar transistor are given as

$$[Y]_{cs} = \begin{bmatrix} Y_{11} & Y_{12} \\ Y_{21} & Y_{22} \end{bmatrix} \tag{D-126}$$

$$Y_{11} = G_{11} + jB_{11} \tag{D-127}$$

$$Y_{21} = G_{21} + jB_{21} \tag{D-128}$$

$$Y_{12} = G_{12} + jB_{12} \tag{D-129}$$

$$Y_{22} = G_{22} + jB_{22} \tag{D-130}$$

The optimum values of the output admittance Y_{out}^* and the feedback susceptance B_1^* and B_2^*, which can be expressed in terms of the two-port $[Y]$-parameter of the active device, are given as

$$B_1^* = -\left\{ B_{11} + \left[\frac{B_{12} + B_{21}}{2} \right] + \left[\frac{G_{21} - G_{12}}{B_{21} - B_{12}} \right] \left[\frac{G_{12} + G_{21}}{2} + G_{11} \right] \right\} \tag{D-131}$$

$$B_2^* = \left[\frac{B_{12} + B_{21}}{2} \right] + \left[\frac{(G_{12} + G_{21})(G_{21} - G_{12})}{2(B_{21} - B_{12})} \right] \tag{D-132}$$

The optimum values of the real and imaginary parts of the output admittance can be given as

$$Y^*_{out} = [G^*_{out} + jB^*_{out}] \tag{D-133}$$

where G^*_{out} and B^*_{out} are given as

$$G^*_{out} = G_{22} - \left[\frac{(G_{12} + G_{21})^2 (B_{21} - B_{12})^2}{4G_{11}}\right] \tag{D-134}$$

$$B^*_{out} = B_{22} + \left[\frac{G_{21} - G_{12}}{B_{21} - B_{12}}\right] - \left[\frac{(G_{12} + G_{21})}{2} + G_{22} - G^*_{out}\right] + \left[\frac{B_{21} + B_{12}}{2}\right] \tag{D-135}$$

The output power of the oscillator can be expressed in terms of the load current and the load impedance as

$$P_{out} = \frac{1}{2} I^2_{out} \text{Re}[Z_L] \tag{D-136}$$

where I_{out} and V_{out} are the corresponding load current and drain voltage across the output.

$$I_{out} = \left[\frac{Z_{11} + Z_1 + Z_2}{Z_{22}(Z_{11} + Z_1 + Z_2) - Z_{21}(Z_{12} + Z_2)}\right] V_{out} \tag{D-137}$$

D.5 AN FET EXAMPLE

Figure D-5 shows a 950 MHz MESFET oscillator circuit configuration [108] and the analytical approach used to determine optimum operating conditions for maximum

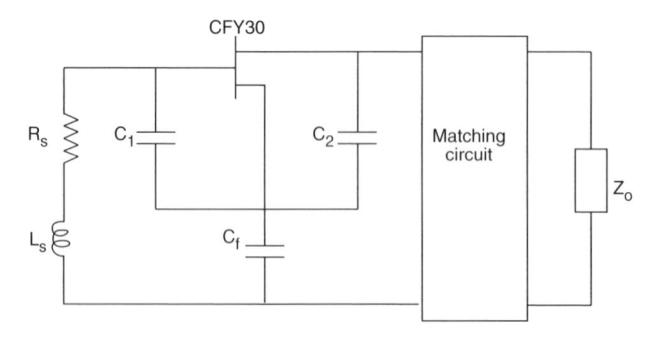

Figure D-5 A 950 MHz MESFET oscillator circuit configuration.

D.5 AN FET EXAMPLE

oscillator output power. The analysis is based on a quasi-linear approach and is experimentally supported with a conversion efficiency of 54%, the maximum conversion efficiency published for this topology. However, the publication does not emphasize the optimum phase noise, which is the key parameter for the oscillator design.

Power optimization of a GaAs 950 MHz MESFET oscillator:

The derivations of the analytical expressions are based on the open loop model of the oscillator. Figure D-6 shows an equivalent circuit of the oscillator shown in Figure 6-46.

Z_1 can be expressed as

$$Z_1 = \frac{\left[R_i + \frac{1}{j\omega C_{gs}}\right]\frac{1}{j\omega C_1}}{\left[R_i + \frac{1}{j\omega C_{gs}} + \frac{1}{j\omega C_1}\right]} = \frac{-\left[\frac{jR_i}{\omega C_1} + \frac{1}{\omega^2 C_{gs} C_1}\right]}{\left[R_i - j\left(\frac{1}{\omega C_{gs}} + \frac{1}{\omega C_1}\right)\right]} \quad \text{(D-138)}$$

Multiplying the numerator and the denominator by the conjugate yields

$$Z_1 = \frac{\left[\frac{-jR_i^2}{\omega C_1} - \frac{R_i}{\omega^2 C_{gs} C_1}\right] + \frac{R_i}{\omega^2 C_1}\left[\frac{1}{C_1} + \frac{1}{C_{gs}}\right] - \frac{j}{\omega^3 C_{gs} C_1}\left[\frac{1}{C_1} + \frac{1}{C_{gs}}\right]}{\left[R_i^2 + \left(\frac{1}{\omega C_{gs}} + \frac{1}{\omega C_1}\right)^2\right]} \quad \text{(D-139)}$$

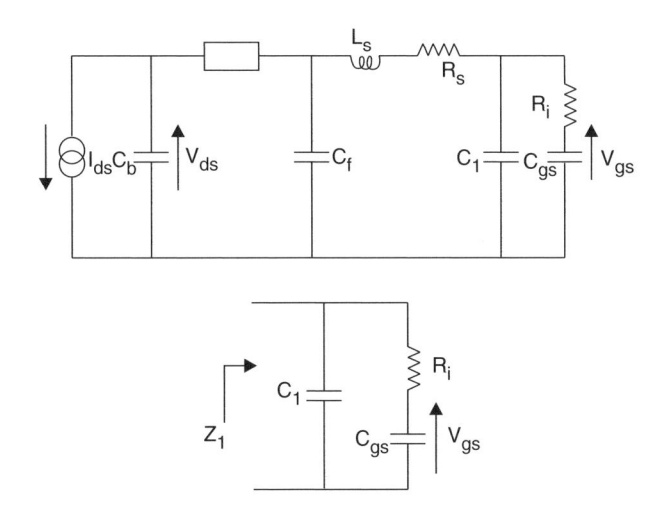

Figure D-6 An equivalent circuit of the open model MESFET oscillator.

The following assumptions are made for simplification purposes:

$$\frac{R_i}{\omega^2 C_{gs} C_1} \ll \frac{R_i}{\omega^2 C_1}\left[\frac{1}{C_1}+\frac{1}{C_{gs}}\right] \tag{D-140}$$

$$\frac{jR_i^2}{\omega C_1} \ll \frac{j}{\omega^3 C_{gs} C_1}\left[\frac{1}{C_1}+\frac{1}{C_{gs}}\right] \tag{D-141}$$

$$R_i \ll \left[\frac{1}{\omega C_{gs}}+\frac{1}{\omega C_1}\right] \tag{D-142}$$

Then modified Z_1 can be represented as

$$Z_1 = \left[\frac{\frac{R_i}{\omega^2 C_1}\left(\frac{1}{C_1}+\frac{1}{C_{gs}}\right) - \frac{j}{\omega^3 C_{gs} C_1}\left(\frac{1}{C_1}+\frac{1}{C_{gs}}\right)}{\left(\frac{1}{\omega C_{gs}}+\frac{1}{\omega C_1}\right)^2}\right] \tag{D-143}$$

$$Z_1 = \left[\frac{\left(\frac{R_i}{C_1}\right)}{\left(\frac{1}{C_1}+\frac{1}{C_{gs}}\right)} - \left(\frac{j}{\omega[C_1+C_{gs}]}\right)\right] \tag{D-144}$$

defining the three new variables as

$$C_a = C_1 + C_{gs} \tag{D-145}$$
$$C_b = C_2 + C_{ds} \tag{D-146}$$
$$R_a = R_s + \frac{C_{gs}^2}{C_a^2} R_i \tag{D-147}$$
$$X_a = \omega L_s - \frac{1}{\omega C_a} \tag{D-148}$$
$$\omega(X_a = 0) = \frac{1}{\sqrt{L_s C_a}} \tag{D-149}$$

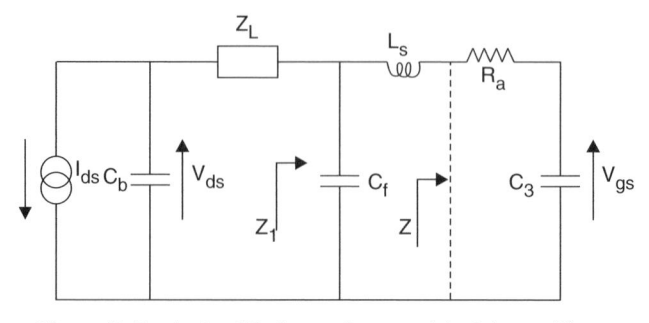

Figure D-7 A simplified open-loop model of the oscillator.

D.5 AN FET EXAMPLE

Figure D-7 shows a simplified open-loop model of the oscillator for easy analysis. In this model, the parasitic elements of the device are absorbed into the corresponding embedding impedances.

$$Z = R_a + \frac{1}{j\omega C_a} \Rightarrow R_s + \frac{C_{gs}^2}{C_a^2} R_i - \frac{j}{\omega C_a} \tag{D-150}$$

$$Z + j\omega L_s = R_s + \frac{C_{gs}^2}{C_a^2} R_i - \frac{j}{\omega C_a} + j\omega L_s \Rightarrow \left[R_s + \frac{C_{gs}^2}{C_a^2} R_i \right] + j\left[\omega L_s - \frac{1}{\omega C_a} \right]$$

$$\tag{D-151}$$

$$Z_a = Z + j\omega L_s \Rightarrow R_a + jX_a \tag{D-152}$$

$$R_a = R_s + \frac{C_{gs}^2}{C_a^2} R_i \tag{D-153}$$

$$X_a = \omega L_s - \frac{1}{\omega C_a} \tag{D-154}$$

$$Z_i = Z_a \| C_f \Rightarrow [Z + j\omega L_s] \| C_f \tag{D-155}$$

$$Z_i = \frac{-j\left[\frac{R_a + jX_a}{\omega C_f}\right]}{\left[R_a + jX_a - \frac{j}{\omega C_f}\right]} \Rightarrow \frac{[R_a + jX_a]}{[1 + jR_a\omega C_f - \omega C_f X_a]} = \frac{[R_a + jX_a]}{1 + j\omega C_f[R_a + jX_a]}$$

$$\tag{D-156}$$

The circuit model of the oscillator is shown in Figure D-8, where the output current through Z_L is given as

$$I = \frac{I_{ds}}{1 + j\omega C_b[Z_i + Z_L]} \tag{D-157}$$

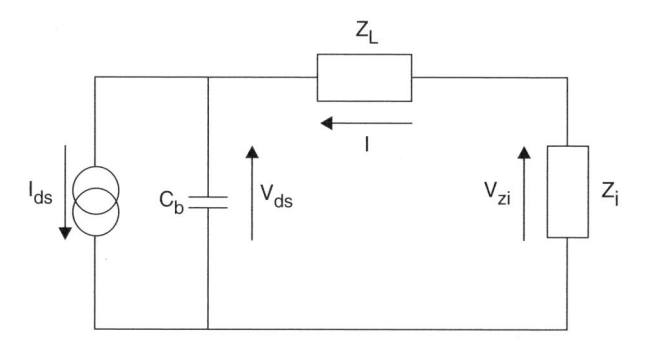

Figure D-8 A circuit model of an oscillator.

426 A COMPLETE ANALYTICAL APPROACH

The voltage across Z_i is given as

$$V_{zi} = IZ_i = -I_{ds}\left[\frac{[R_a + jX_a]}{1 + j\omega C_f[R_a + jX_a]}\right]\left[\frac{1}{1 + j\omega C_b[Z_i + Z_L]}\right] \qquad \text{(D-158)}$$

Applying the voltage divider in Figure D-8, V_{gs} can be expressed as

$$V_{gs} = -I_{ds}\left[\frac{1}{j\omega C_b[R_a + jX_a]}\right]\left[\frac{Z_i}{1 + j\omega C_b[Z_i + Z_L]}\right] \qquad \text{(D-159)}$$

Steady-state oscillation occurs when $I_{ds}(t) = I_1$ and $V_{gs} = V_p$. Consequently, the equation above can be written as

$$1 + j\omega C_b[Z_i + Z_L] = -\frac{I_{ds}}{V_{gs}}\frac{Z_i}{j\omega C_a(R_a + jX_a)} \qquad \text{(D-160)}$$

$$1 + j\omega C_b[Z_i + Z_L] = \frac{-g_{mc}Z_i}{j\omega C_a(R_a + jX_a)} \Rightarrow \frac{-g_{mc}[R_a + jX_a]}{j\omega C_a(R_a + jX_a)[1 + j\omega C_f(R_a + jX_a)]} \qquad \text{(D-161)}$$

$$1 + j\omega C_b[Z_i + Z_L] = \frac{-g_{mc}}{j\omega C_a[1 + jwC_f(R_a + jX_a)]} = \frac{g_{mc}}{\omega^2 C_f C_a - j[\omega C_a - \omega^2 C_f C_a X_a]} \qquad \text{(D-162)}$$

$$Z_L = Z_i\frac{g_{mc}}{\omega^2 C_b C_a[R_a + jX_a]} - Z_i - \frac{1}{j\omega C_b} \qquad \text{(D-163)}$$

$$Z_L = \frac{g_{mc}(R_a + jX_a)}{[1 + j\omega C_f(R_a + jX_a)][\omega^2 C_b C_a(R_a + jX_a)]} - \frac{(R_a + jX_a)}{[1 + j\omega C_f(R_a + jX_a)]} - \frac{1}{j\omega C_b} \qquad \text{(D-164)}$$

$$Z_L = \frac{g_{mc}}{\omega^2 C_b C_a[1 + j\omega C_f(R_a + jX_a)]} - \frac{(R_a + jX_a)}{[1 + j\omega C_f(R_a + jX_a)]} - \frac{1}{j\omega C_b} \qquad \text{(D-165)}$$

where

$$g_{mc} = \frac{I_1}{V_p} = \frac{I_{max}}{2V_p} \qquad \text{(D-166)}$$

In addition, V_{ds} can be determined by calculating I_{cb}, the current through C_b, with the help of Figure D-8.

$$I_{cb} = I_{ds}\frac{[Z_L + Z_i]}{Z_L + Z_i + 1/j\omega C_b} \qquad \text{(D-167)}$$

D.5 AN FET EXAMPLE

Based on the preceding result, we can conclude that

$$V_{ds} = I_{cb}\frac{j}{\omega C_b} = \frac{[Z_L + Z_i]}{Z_L + Z_i + 1/j\omega C_b}I_1 \qquad \text{(D-168)}$$

or in square magnitude form

$$V_{ds}^2 = \frac{[Z_L + Z_i]^2}{[1 + j\omega C_b(Z_L + Z_i)]^2}I_1^2 \qquad \text{(D-169)}$$

Also, Re$[Z_L]$ can be defined as follows:

$$\text{Re}[Z_L] = \frac{g_{mc}[1 - \omega C_f X_a - j\omega C_f R_a]}{\omega^2 C_b C_a[(1 - \omega C_f X_a)^2 + \omega^2 C_f^2 R_a^2]} - \frac{(R_a + jX_a)[1 - \omega C_f X_a - j\omega C_f R_a]}{[(1 - \omega C_f X_a)^2 + \omega^2 C_f^2 R_a^2]} \qquad \text{(D-170)}$$

$$\text{Re}[Z_L] = \frac{g_{mc}[1 - \omega C_f X_a - j\omega C_f R_a] - \omega^2 C_b C_a(R_a + jX_a)[1 - \omega C_f X_a - j\omega C_f R_a]}{\omega^2 C_b C_a[(1 - \omega C_f X_a)^2 + \omega^2 C_f^2 R_a^2]} \qquad \text{(D-171)}$$

$$\text{Re}[Z_L] = \frac{g_{mc}[1 - \omega C_f X_a] - \omega^2 C_b C_a R_a + \omega^3 C_b C_a C_f X_a - \omega^3 C_b C_a C_f R_a)}{\omega^2 C_b C_a[(1 - \omega C_f X_a)^2 + \omega^2 C_f^2 R_a^2]} \qquad \text{(D-172)}$$

The power delivered to the load Z_L and the magnitude of V_{ds} can be determined by

$$P_{out} = \frac{1}{2}I^2 \text{Re}[Z_L] \qquad \text{(D-173)}$$

$$P_{out} = \frac{1}{2}I_1^2 \frac{\text{Re}[Z_L]}{[1 + j\omega C_b(Z_L + Z_i)]^2} \qquad \text{(D-174)}$$

$$V_{ds}^2 = \frac{[Z_L + Z_i]^2}{[1 + j\omega C_b(Z_L + Z_i)]^2}I_1^2 \qquad \text{(D-175)}$$

$$[1 + j\omega C_b(Z_i + Z_L)]^2 = \frac{g_{mc}^2}{[\omega^2 C_f C_a R_a - j(\omega C_a - \omega^2 C_f C_a X_a)]} \qquad \text{(D-176)}$$
$$\times g_{mc}^2[\omega^2 C_f C_a R_a + j(\omega C_a - \omega^2 C_f C_a X_a)]$$

$$1 + j\omega C_b(Z_i + Z_L)]^2 = \frac{g_{mc}^2}{(\omega^2 C_f C_a R_a)^2 + (\omega C_a - \omega^2 C_f C_a X_a)^2}$$

$$= \frac{g_{mc}^2}{\omega^2 C_a^2[(1 - \omega^2 C_f C_a X_a)^2 + (\omega^2 C_f C_a R_a)^2]} \qquad \text{(D-177)}$$

A COMPLETE ANALYTICAL APPROACH

Based on the equations above, the output power can be estimated as

$$P_{out} = \frac{1}{2} I^2 \operatorname{Re}[Z_L] \tag{D-178}$$

$$P_{out} = \frac{1}{2} I_1^2 \frac{\operatorname{Re}[Z_L]}{[1 + j\omega C_b(Z_L + Z_i)]^2} \tag{D-179}$$

$$P_{out} = \frac{1}{2} I_1^2 \frac{\left\{ \dfrac{g_{mc}[1 - \omega C_f X_a] - \omega^2 C_b C_a R_a + \omega^3 C_b C_a C_f X_a - \omega^3 C_b C_a C_f R_a)}{\omega^2 C_b C_a [(1 - \omega C_f X_a)^2 + \omega^2 C_f^2 R_a^2]} \right\}}{\left\{ \dfrac{g_{mc}^2}{\omega^2 C_a^2 [(1 - \omega^2 C_f C_a X_a)^2 + (\omega^2 C_f C_a R_a)^2]} \right\}} \tag{D-180}$$

$$P_{out} = \frac{1}{2} I_1^2 C_a \frac{[1 - \omega C_f X_a] - \omega^2 C_b C_a R_a + \omega^3 C_b C_a C_f X_a - \omega^3 C_b C_a C_f R_a)}{g_{mc} C_b} \tag{D-181}$$

Below 5 GHz, it is valid to ignore some of the terms by assuming that

$$\omega^2 C_b C_a R_a \gg \omega^3 C_b C_a C_f X_a \tag{D-182}$$

$$\omega^2 C_b C_a R_a \gg \omega^3 C_b C_a C_f R_a \tag{D-183}$$

The power output is now expressed as

$$P_{out} = \frac{1}{2} I_1^2 C_a \frac{g_{mc}[1 - \omega C_f X_a] - \omega^2 C_b C_a C_f R_a}{g_{mc}^2 C_b} \tag{D-184}$$

$$P_{out} = \frac{1}{2} I_1^2 \left[C_a w \frac{[1 - \omega C_f X_a]}{\omega C_b} - \frac{\omega^2 C_a^2 R_a}{g_{mc}^2} \right] \tag{D-185}$$

$$P_{out} = \frac{1}{2} I_1^2 \left[\alpha \frac{[1 - \omega C_f X_a]}{\omega C_b} - \alpha^2 R_a \right] \tag{D-186}$$

$$\alpha = \frac{\omega C_a}{g_{mc}} \tag{D-187}$$

In a similar manner, V_{ds} is given by

$$V_{ds}^2 = I_1^2 \left[\frac{\alpha^2 [1 - \omega C_f X_a]^2 + [1 - \omega C_f R_a]^2}{\omega^2 C_b^2} \right] \tag{D-188}$$

Both the output power and V_{ds} depend on C_b if the other parameters are fixed. This is a limitation for the maximum value. However, there is a maximum value of the current and the voltage a transistor can take before it burns out. Therefore, setting $|V_{ds}| = V_{dsm}$ an optimal condition is given by

$$\frac{|V_{ds}|^2}{I_1^2} = \frac{|V_{dsm}|^2}{I_1^2} = \frac{\alpha^2[1-\omega C_f X_a]^2 + [1-\omega C_f R_a]^2}{\omega^2 C_b^2} \quad \text{(D-189)}$$

The optimum load impedance that the device requires to deliver the highest power is defined as

$$\frac{|V_{ds}|}{I_1} = \frac{2|V_{dsm}|}{I_{max}} = R_{opt} \quad \text{(D-190)}$$

leading to the following definition:

$$\omega C_b R_{opt} = \sqrt{\alpha^2[1-\omega C_f X_a]^2 + [1-\omega C_f R_a]^2} \quad \text{(D-191)}$$

Using the result above, the optimum P_{out} is, therefore, given by

$$P_{out} = \frac{V_{dsm} I_{dsm}}{4} \alpha \frac{[1-\omega C_f X_a]}{\sqrt{\alpha^2[1-\omega C_f X_a]^2 + [1-\omega C_f R_a]^2}} - (\omega C_a V_p)^2 \frac{R_a}{2} \quad \text{(D-192)}$$

The first term is the power available from the current source, and the second term is the power absorbed by R_a. This also indicates that a high-Q inductor minimizes the absorbed power, increasing the power available from the current source. P_{out} simplifies further at the oscillation frequency since $X_a \approx 0$.

$$P_{out} = \frac{V_{dsm} I_{dsm}}{4} \alpha \frac{1}{\sqrt{\alpha^2 + [1-\alpha\omega C_f R_a]^2}} - (\omega C_a V_p)^2 \frac{R_a}{2} \quad \text{(D-193)}$$

The above analysis gives the following important results:

1. Maximum output power is attained if we set

$$C_f = \frac{1}{\alpha \omega R_a} \quad \text{(D-194)}$$

and

$$P_{out}(\max) = \frac{V_{dsm} I_{max}}{4}\left[1 - \frac{1}{G}\right] \quad \text{(D-195)}$$

$$\frac{1}{G} = \frac{P_f}{P_{av}} = \omega^2 C_a^2 R_a \frac{2V_p^2}{V_{dsm} I_{max}} \quad \text{(D-196)}$$

Accordingly, DC/RF conversion efficiency is calculated by

$$P_{dc} = \frac{V_{DS}I_{max}}{\pi} \tag{D-197}$$

$$\eta_{max} = \frac{P_{out}(\max)}{P_{dc}} \tag{D-198}$$

$$\eta_{max} = \left[1 - \frac{1}{G}\right]\frac{V_{dsm}}{V_{DS}} \tag{D-199}$$

In order to maximize oscillator output power and efficiency, the loss resistance R_a of the input circuit has to be reduced (increasing G), and an optimal biasing condition V_{DS} has to be selected.

2.

$$C_b = \frac{[1 - \omega C_f R_a]C_a}{g_{mc}R_{opt}} \tag{D-200}$$

$$C_b(C_f = 0) = \frac{C_a}{g_{mc}R_{opt}} \tag{D-201}$$

3. Combining the above equations leads to expressions for Z_L in terms of

$$Z_L = \frac{1 + j\alpha}{1 + \alpha^2}R_{opt} \tag{D-202}$$

From these analytical calculations, the following results were achieved. The circuit simulation of the oscillator used a nonlinear Materka model.

Figure D-9 shows the schematic diagram of a practical oscillator operating at 950 MHz. A simple high-pass filter consisting of L_T and C_T is used to transfer the $Z_0/50\,\Omega$ load to the required Z_L value.

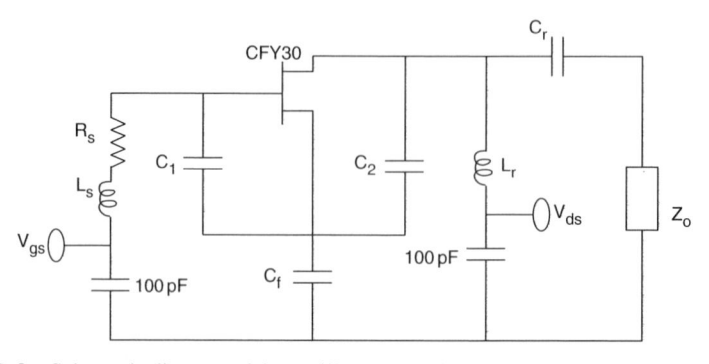

Figure D-9 Schematic diagram of the oscillator operating at 950 MHz as published in [108].

From the expressions above all the effective components of oscillator can be determined as follows:

1. Bias condition:
   $V_{DS} = 5\,\text{V}$
   $I_{DS} = 18\,\text{mA}$
2. Device parameters:
   $I_{max} = 45\,\text{mA}$
   $V_P = 1.25\,\text{V}$
   $V_K(\text{knee–voltage}) = 0.5\,\text{V}$
3. Device parasitics:
   $C_{gs} = 0.5\,\text{pF}$
   $C_{ds} = 0.2\,\text{pF}$
   $C_{gd} = 0.0089\,\text{pF}$
4. Oscillator parameters:
   $\omega = \dfrac{1}{\sqrt{L_s C_a}} \Rightarrow f = 950\,\text{MHz}$
   $C_1 = 6\,\text{pF}$
   $C_2 = 1.5\,\text{pF}$
   $C_f = 20\,\text{pF}$
   $L_s = 3.9\,\text{nH}$
   $C_a = C_1 + C_{gs} = 6.5\,\text{pF}$
   $C_b = C_2 + C_{ds} = 1.7\,\text{pF}$
   $L'_f = 18\,\text{nH}$
   $C'_f = 15\,\text{pF}$
   $R_a = R_s + (C_{gs}^2/C_a^2)R_i = 4\,\Omega$
5. Output matching circuit:
   $L_d(pacakge) = 0.7\,\text{nH}$
   $L_T = 8.9\,\text{nH}$
   $L'_T = 8.9\,\text{nH} - L_d = 8.7\,\text{nH}$
   $C_T = 1.91\,\text{pF}$
6. Calculation of R_{opt}:
   $I_{dc} = \dfrac{I_{max}}{\pi}$
   $I_1 = \dfrac{I_{max}}{2} = 22.5\,\text{mA}$
   $R_{opt} = \dfrac{V_{dsm}}{I_1} = \dfrac{V_{DS} - V_K}{I_1} = \dfrac{5V - 0.5V}{22.5\,\text{mA}} = 200\,\Omega$
7. Calculation of Z_L:
   $Z_L = \dfrac{1 + j\alpha}{1 + \alpha^2} R_{opt}$

$$g_{mc} = \frac{I_1}{V_p} = \frac{I_{max}}{2V_p} = \frac{45\,\text{mA}}{2*1.25} = 18.8\,\text{mS}$$

$$\alpha = \frac{w_0 C_a}{g_{mc}} = \frac{2*\pi*950\text{E}+6*6.5\text{E}-12}{0.0188} = 2.0$$

$$Z_L = \frac{1+j\alpha}{1+\alpha^2} R_{opt} = \frac{1+j2}{1+4} * 200 = 40 + j80\,\Omega$$

8. Output power:

$$P_{out}(\max) = \frac{V_{dsm} I_{max}}{4}\left[1 - \frac{1}{G}\right] = 16.6\,\text{dBm}$$

$$\frac{1}{G} = \frac{P_f}{P_{av}} = \omega^2 C_a^2 R_a \frac{2V_p^2}{V_{dsm} I_{max}}$$

9. DC/RF conversion efficiency:

$$P_{dc} = \frac{V_{DS} I_{max}}{\pi} = \frac{5*45\,\text{mA}}{\pi} = 71.62\,\text{mW}$$

$$\eta_{max} = \frac{P_{out}(\max)}{P_{dc}} = \frac{45.7\,\text{mW}}{71.62\,\text{mW}} = 0.64$$

$$\eta_{max} = 64\%$$

D.6 SIMULATED RESULTS

Figures D-10 to D-15 show the oscillator test circuit and its simulated results. After the oscillator circuit is analyzed in the HB program, the oscillator frequency is found

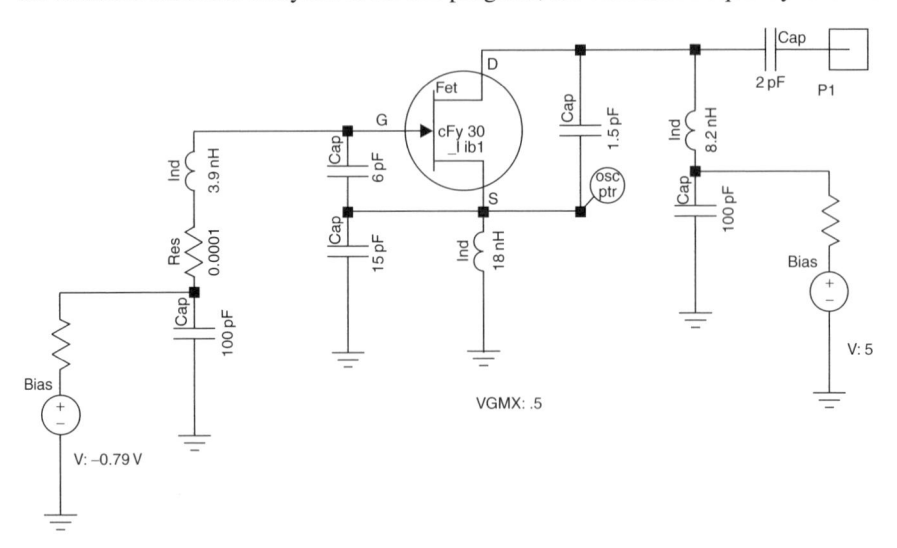

Figure D-10 Schematic of the MESFET test oscillator based on [108].

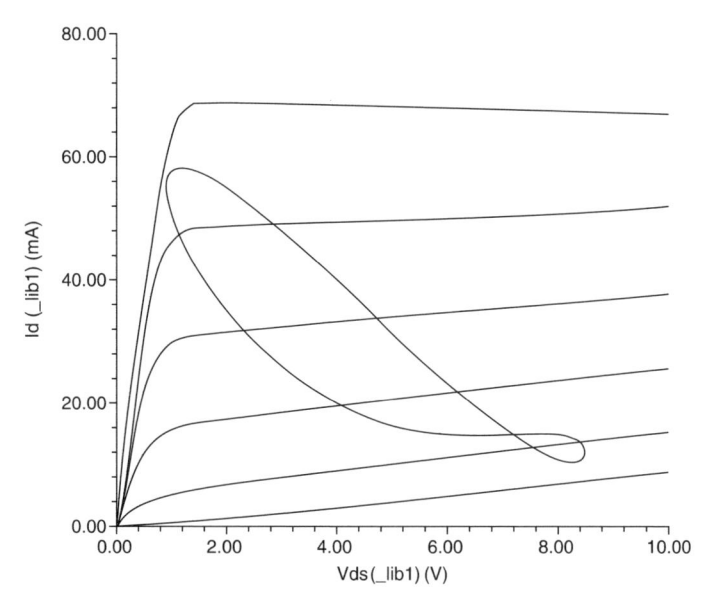

Figure D-11 Load line of the oscillator shown in Figure D-10. Because the load is a tuned circuit, the "load line" is a curve and not a straight line.

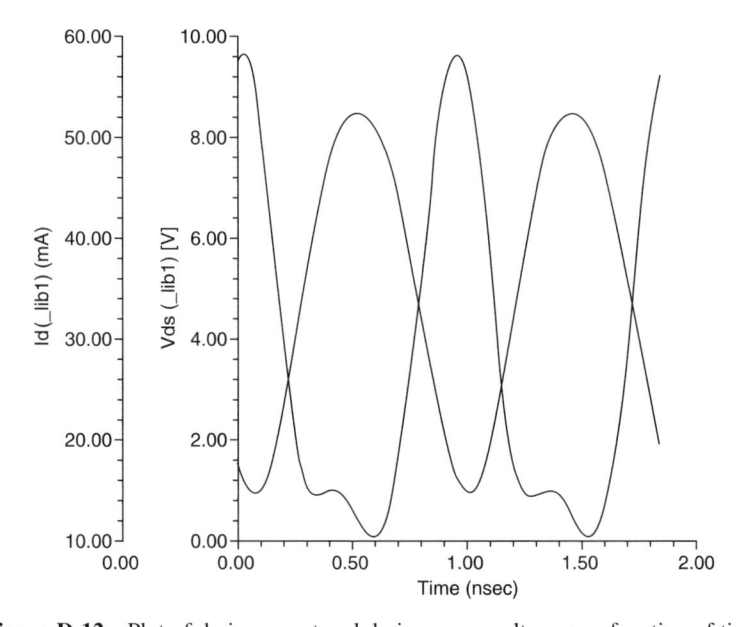

Figure D-12 Plot of drain current and drain source voltage as a function of time.

434   A COMPLETE ANALYTICAL APPROACH

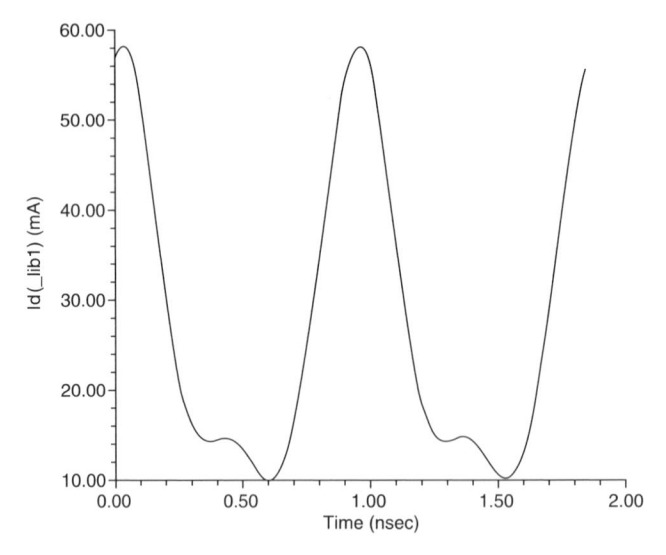

Figure D-13   AC drain current simulated for Figure D-10.

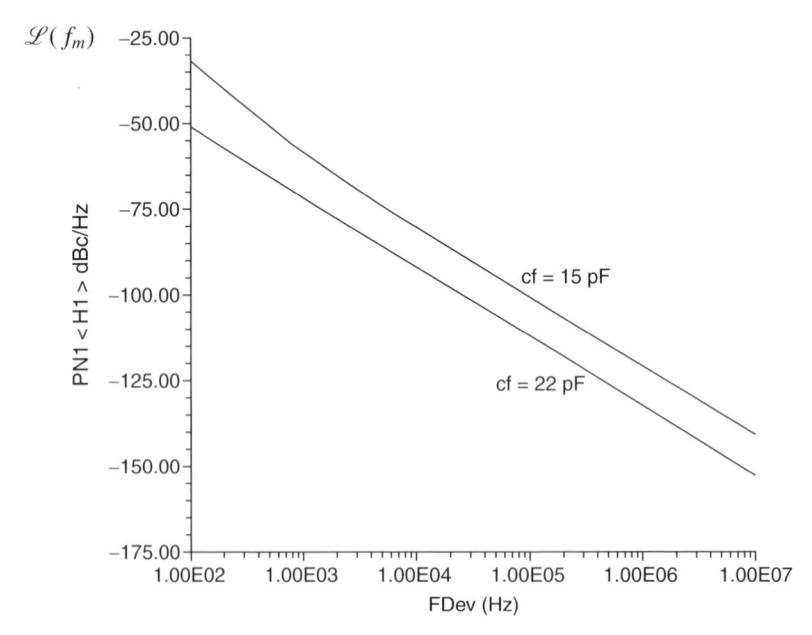

Figure D-14   Simulated noise of the circuit shown in Figure D-10. An increase in the feedback capacitor from 15 to 22 pF improves the phase noise.

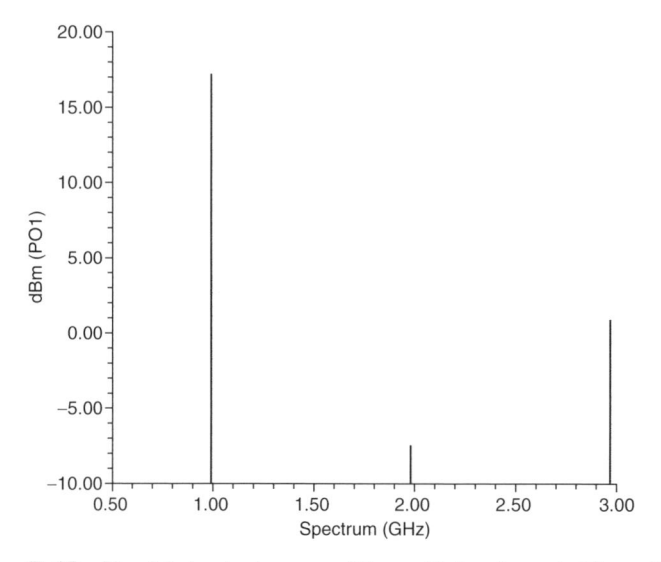

Figure D-15 Simulated output power of the oscillator shown in Figure D-10.

to be 1.08 GHz, and some tuning is required to bring the oscillator frequency back to the required value by changing L_s from 3.9 nH to 4.45 nH. The slight shift in oscillator frequency may be due to the device parasitic. The simulated power output is 17.04 dBm, which is about the same as the measured value in [108]. The DC/RF conversion efficiency at the fundamental frequency is 55%. The calculation in [108], as well as the calculation here, assumes an ideal transistor. When a better value between C_1 and C_2 was found, the efficiency was increased to 64% compared to the published result of 55%. This means that the circuit in [108] was not properly optimized.

Taking the published experimental results of [108] into consideration, the analytical expression gives excellent insight into the performance of the oscillator circuit.

The maximum achievable output power and efficiency for a given active device can be predicted by the closed-form expressions without the need for large-signal device characterization and an HB simulation. The publication [108] has not addressed power optimization and best phase noise, which are very important requirements for the oscillator. By proper selection of the feedback ratio at the optimum drive level, the noise is improved by 8 dB, keeping the output power approximately the same. In Chapter 8 we discuss adjustment of the optimum feedback ratio and the absolute values of the feedback capacitor, with consideration for the best possible phase noise.

APPENDIX E
CAD Solution for Calculating Phase Noise in Oscillators

This appendix address various important issues concerning noise. Noise and oscillators has already been discussed, but since many publications, which have been referenced, omit several important steps, the subject is difficult to follow. The discussion of noise has two aspects: physics-based and mathematics-based. In Chapter 8, noise is explained in terms of physics. In this appendix, all the necessary mathematical tools are presented. The mechanism that adds the noise, both close-in noise and far-out noise to the carrier, will be described mathematically. The resulting noise in the large-signal condition is an important issue. When modeling transistors, typically the noise correlation is incomplete. This correlation, which deals with the inner transistor, is discussed in Appendix D.

Two important linear noise models are necessary to understand SSB noise. One is the Leeson phase noise equation, and the other is based on the Lee and Hajimiri noise model. Noise theory can be divided into modulation noise and conversion noise. Both will be explained in detail.

E.1 GENERAL ANALYSIS OF NOISE DUE TO MODULATION AND CONVERSION IN OSCILLATORS

The degree to which an oscillator generates constant frequency for a specified period of time is defined as the *frequency stability* of the oscillator. Frequency instability is due to the presence of noise in the oscillator circuit that effectively modulates the signal, causing a change in frequency spectrum commonly known as *phase noise*.

The Design of Modern Microwave Oscillators for Wireless Applications: Theory and Optimization, by Ulrich L. Rohde, Ajay Kumar Poddar, Georg Böck
Copyright © 2005 John Wiley & Sons, Inc.

The unmodulated carrier signal is represented as

$$f(t) = A\cos(2\pi f_c t + \theta_0) \tag{E-1}$$

$$\theta = 2\pi f_c t + \theta_0 \tag{E-2}$$

$$f_c = \frac{1}{2\pi}\left(\frac{d\theta}{dt}\right) \tag{E-3}$$

In an unmodulated signal, f_c is constant and is expressed by the time derivative of the phase (angle θ). However in general, this derivative is not constant and can be represented as an instantaneous frequency, which can vary with time and is expressed as $f_i = 1/2\pi(d\theta_i/dt)$. The corresponding phase is determined as $\theta_i(t) = \int 2\pi f_i\, dt$. For the unmodulated carrier, $\theta_i(t) = (2\pi f_c t + \theta_0)$, where $\theta_0 = \theta_i(t)|_{t=0}$.

The phase/angle of the carrier can be varied linearly by the modulating signal $m(t)$, which results in phase modulation as $\theta_i(t)$.

$$\theta_i(t) = 2\pi f_c t + k_p m(t) \tag{E-4}$$

k_p is defined as phase sensitivity, and its dimension is given as radians per unit of the modulating signal. The instantaneous frequency ω_i of the carrier is modified by modulation with the modulating signal as

$$\omega_i = \frac{d\theta_i}{dt} \implies \omega_i = \omega_c + k_p \frac{dm}{dt} \tag{E-5}$$

The phase-modulated signal can be expressed in the time domain as

$$s(t) = A_c \cos[2\pi f_c t + k_p m(t)] \tag{E-6}$$

E.2 MODULATION BY A SINUSOIDAL SIGNAL

Consider a sinusoidal modulating signal given by $m(t) = A_m \cos(2\pi f_m t)$. The instantaneous frequency of the modulated signal is given as

$$f_i(t) = f_c + k_f A_m \cos(2\pi f_m t) \tag{E-7}$$

$$f_i(t) = f_c + \Delta f \cos(2\pi f_m t) \tag{E-8}$$

$$\Delta f = k_f A_m \tag{E-9}$$

Δf is defined as the frequency deviation corresponding to the maximum variation of the instantaneous frequency of the modulated signal from the carrier frequency.

E.2 MODULATION BY A SINUSOIDAL SIGNAL

The angle of the modulated signal is determined by integration as

$$\theta_i(t) = 2\pi \int_0^t f_i(t)\,dt \tag{E-10}$$

$$\theta_i(t) = 2\pi f_c + \frac{\Delta f}{f_m}\sin(2\pi f_m t) \tag{E-11}$$

The coefficient of the sine term is called the *modulation index* of the modulating signal and is denoted by $\beta = \Delta f/f_m$. The expression for the angle of the modulated signal can be written as $\theta_i(t) = 2\pi f_c + \Delta f/f_m \sin(2\pi f_m t)$, and the time representation of the modulated signal can be expressed as

$$s(t) = A_c \cos[\theta_i(t)] = A_c \cos[2\pi f_c t + \beta \sin(2\pi f_m t)] \tag{E-12}$$

$$s(t) = \text{Re}[A_c e^{j[2\pi f_c t + \beta \sin(2\pi f_m t)]}] \tag{E-13}$$

$$s(t) = \text{Re}[\sigma(t) A_c e^{j2\pi f_c t}] \tag{E-14}$$

$\sigma(t)$ is the complex envelope of the frequency-modulated signal and can be given as $\sigma(t) = A_{cn} e^{j\beta \sin(2\pi f_m t)}$. It is a periodic function of time with a fundamental frequency equal to the modulating frequency f_m and can be expressed as

$$\sigma(t) = \sum_{n=-\infty}^{\infty} C_n e^{j2\pi f_m t} \tag{E-15}$$

C_n is the Fourier coefficient given as

$$C_n = f_m \int_{-1/2 \pi f_m}^{1/2 \pi f_m} \sigma(t) e^{-j2\pi f_m t}\,dt \tag{E-16}$$

$$C_n = A_c f_m \int_{-1/2 \pi f_m}^{1/2 \pi f_m} e^{j[\beta \sin(2\pi f_m t) - j2\pi f_m t]}\,dt \tag{E-17}$$

$$C_n = A_c f_m \int_{-1/2 \pi f_m}^{1/2 \pi f_m} e^{j(\beta \sin x - nx)}\,dt \tag{E-18}$$

$x = 2\pi f_m t$, and the expression of coefficients may be rewritten as $C_n = (A_c/2\pi) \int_{-\pi}^{\pi} e^{j(\beta \sin x - nx)}\,dx$. This equation is called the *n*th *order Bessel function of the first kind with argument β*.

The expression for $\sigma(t)$ and $s(t)$ is given as

$$\sigma(t) = \sum_{n=-\infty}^{\infty} C_n e^{j2\pi f_m t} = A_c \sum_{n=-\infty}^{\infty} J_n(\beta) e^{j2\pi f_m t} \tag{E-19}$$

$$s(t) = A_c \text{Re}\left[\sum_{n=-\infty}^{\infty} J_n(\beta) e^{j2\pi(f_c+nf_m)t}\right] = A_c \sum_{n=-\infty}^{\infty} J_n(\beta) \cos[2\pi(f_c + nf_m)t] \tag{E-20}$$

Applying a Fourier transform to the time-domain signal $s(t)$ results in an expression for the discrete frequency spectrum of $s(t)$ as

$$s(f) = \frac{A_c}{2} \sum_{n=-\infty}^{\infty} J_n(\beta)[\delta(f - f_c - nf_m) + (f + f_c + nf_m)] \tag{E-21}$$

$$\sum_{n=-\infty}^{\infty} J_n(\beta)^2 = 1 \tag{E-22}$$

The spectrum of the frequency-modulated signal has an infinite number of symmetrically located sideband components spaced at frequencies of f_m, $2f_m$, $3f_m$ … nf_m around the carrier frequency. The amplitudes of the carrier component and the sideband components are the products of the carrier amplitude and a Bessel function.

E.3 MODULATION BY A NOISE SIGNAL

A noise signal is defined as $n(t) = r_n(t) \cos[2\pi f_c t + \theta + \Phi_n(t)]$ introduced into an oscillator circuit in a random fashion, and the desired oscillator output signal represented by $f(t) = A \cos(2\pi f_c t + \theta) \cdot r_n(t)$ is the coefficient of the noise signal having a Rayleigh distribution and functions of a noise signal. The phase $\Phi_n(t)$ is linearly distributed and is a distribution function of a noise signal. The output of the oscillator circuit is given as the superposition of the combined signal, which is expressed as

$$g(t) = f(t) + n(t) \tag{E-23}$$

$$g(t) = A \cos[(2\pi f_c t + \theta)] + r_n(t) \cos[2\pi f_c t + \theta + \Phi_n(t)] \tag{E-24}$$

$$g(t) = A \cos[(2\pi f_c t + \theta)] + r_n(t) \cos[\Phi_n(t)] \cos[2\pi f_c t + \theta]$$
$$- r_n(t) \sin[\Phi_n(t)] \sin[2\pi f_c t + \theta] \tag{E-25}$$

$$g(t) = \cos[(2\pi f_c t + \theta)]\{A + r_n(t) \cos[\Phi_n(t)]\}$$
$$- r_n(t) \sin[\Phi_n(t)] \sin[2\pi f_c t + \theta] \tag{E-26}$$

E.3 MODULATION BY A NOISE SIGNAL

$$g(t) = \sqrt{C_1^2 + C_2^2} \left[\frac{C_1}{\sqrt{C_1^2 + C_2^2}} \cos \Phi_e(t) - \frac{C_2}{\sqrt{C_1^2 + C_2^2}} \sin \Phi_e(t) \right] \quad \text{(E-27)}$$

$$g(t) = \left[\sqrt{C_1^2 + C_2^2} \right] \cos[\psi + \Phi_e(t)] \quad \text{(E-28)}$$

where

$$\sin[\Phi_e(t)] = \frac{C_2}{\sqrt{C_1^2 + C_2^2}} \quad \text{(E-29)}$$

$$\cos[\Phi_e(t)] = \frac{C_1}{\sqrt{C_1^2 + C_2^2}} \quad \text{(E-30)}$$

$$R(t) = \sqrt{C_1^2 + C_2^2} \quad \text{(E-31)}$$

$$g(t) = C_1 \cos \psi - C_2 \sin \psi \quad \text{(E-32)}$$

$$C_1 = A + r_n(t) \cos[\Phi_n(t)] \quad \text{(E-33)}$$

$$C_2 = r_n(t) \sin[\Phi_n(t)] \quad \text{(E-34)}$$

$$\psi = 2\pi f_c t + \theta \quad \text{(E-35)}$$

The phase term $\Phi_e(t)$ is a time-varying function and can be represented as

$$\Phi_e(t) = \tan^{-1}\left[\frac{C_2}{C_1}\right] = \tan^{-1}\left[\frac{r_n(t) \sin[\Phi_n(t)]}{A + r_n(t) \cos[\Phi_n(t)]}\right] \quad \text{(E-36)}$$

For a large signal-to-noise ratio (SNR), $\Phi_e(t)$ can be approximated as

$$\Phi_n(t) = \frac{r_n(t)}{A} \sin[\Phi_n(t)] \quad \text{(E-37)}$$

and the oscillator output signal can be expressed as

$$g(t) = f(t) + n(t) \implies g(t) = R(t) \cos[2\pi f_c t + \theta + \Phi_e(t)] \quad \text{(E-38)}$$

$$R(t) = \sqrt{C_1^2 + C_2^2} = \{[A + r_n(t) \cos[\Phi_n(t)]]^2 + r_n(t) \sin[\Phi_n(t)]^2\} \quad \text{(E-39)}$$

which is phase modulated due to the noise signal $n(t)$. The resultant oscillator output signal contains modulation sidebands due to noise present in the circuit, which is called *phase noise*.

The amplitude of a phase modulation sideband is given by the product of the carrier amplitude and a Bessel function of the first kind and can be expressed as

$$A_{SSB} = \frac{1}{2}A_c[J_n(\beta)]_{n=1} \tag{E-40}$$

$$\frac{A_{SSB}}{A_c} = \frac{1}{2}[J_1(\beta)] \tag{E-41}$$

$$\frac{A_{SSB}}{A_c} = \frac{P_{SSB}}{P_c} \tag{E-42}$$

$$L(f) = 10\log\left[\frac{P_{SSB}}{P_c}\right] - 10\log[BW_n] \tag{E-43}$$

$L(f)$ is phase noise due to noise modulation, A_{SSB} is the sideband amplitude of the phase modulation at offset Δf from the carrier, and BW_n is the noise bandwidth in hertz.

E.4 OSCILLATOR NOISE MODELS

At present, two separate but closely related models of oscillator phase noise exist. The first was proposed by Leeson [70], referred to as *Leeson's model*. The noise prediction using Leeson's model is based on time-invariant properties of the oscillator such as resonator Q, feedback gain, output power, and noise figure. This was discussed in Chapter 7.

Leeson introduced a linear approach for the calculation of oscillator phase noise. His noise formula was extended by Rohde et al., who added $2kTRK_0^2/f_m^2$ [75].

E.5 MODIFIED LEESON'S PHASE NOISE EQUATION

$$\mathscr{L}(f_m) = 10\log\left\{\left[1 + \frac{f_0^2}{(2f_m Q_L)^2(1-(Q_L/Q_0))^2}\right]\left(1+\frac{f_c}{f_m}\right)\frac{FkT}{2P_{sav}} + \frac{2kTRK_0^2}{f_m^2}\right\} \tag{E-44}$$

$\mathscr{L}(f_m)$ = SSB noise power spectral density defined as the ratio of sideband power in a 1 Hz bandwidth at f_m to total power in dB; the unit is dBc/Hz

f_m = frequency offset

f_0 = center frequency

f_c = flicker frequency—the region between $1/f^3$ and $1/f^2$

Q_L = loaded Q of the tuned circuit

Q_0 = unloaded Q of the tuned circuit
F = noise of the oscillator
$kT = 4.1 \times 10^{-21}$ at 300 K_0 (room temperature)
P_{sav} = average power at oscillator output
R = equivalent noise resistance of the tuning diode
K_0 = oscillator voltage gain

The last term of Leeson's phase noise equation is responsible for the modulation noise.

E.6 SHORTCOMINGS OF THE MODIFIED LEESON NOISE EQUATION

F is empirical, a priori, and difficult to calculate due to linear time-variant (LTV) characteristics of the noise.

Phase noise in the $1/f^3$ region is an empirical expression with fitting parameters.

E.7 LEE AND HAJIMIRI NOISE MODEL

The second noise model was proposed by Lee and Hajimiri [64]. It is based on the time-varying properties of the oscillator current waveform.

The phase noise equation for the $1/f^3$ region can be expressed as

$$\mathscr{L}(f_m) = 10 \log \left[\frac{C_0^2}{q_{max}^2} * \frac{i_n^2/\Delta f}{8 f_m^2} * \frac{w_{1/f}}{f_m} \right] \quad \text{(E-45)}$$

and the phase noise equation for the $1/f^2$ region can be expressed as

$$\mathscr{L}(f_m) = 10 \log \left[\frac{\Gamma_{rms}^2}{q_{max}^2} * \frac{i_n^2/\Delta f}{4 f_m^2} \right] \quad \text{(E-46)}$$

where

C_0 = coefficient of the Fourier series, 0th order of the ISF
i_n = noise current magnitude
Δf = noise bandwidth
$\omega_{1/f}$ = $1/f$ noise corner frequency of the device/transistor
q_{max} = maximum charge on the capacitors in the resonator
$\Gamma_{(rms)}$ = rms value of the ISF

E.8 SHORTCOMINGS OF THE LEE AND HAJIMIRI NOISE MODEL

The ISF function is tedious to obtain and depends upon the topology of the oscillator.

Although it is mathematical, the model lacks practicality.

$1/f$ noise conversion is not clearly specified.

E.9 MODULATION AND CONVERSION NOISE

Modulation noise of an oscillator is defined as the noise generated by modulating the diode. The noise associated with series loss resistance in the tuning diode will introduce frequency modulation, which is further translated into oscillator phase noise. This portion of the noise is responsible for the near-carrier noise. An additional phenomenon called *conversion noise* produces noise in a manner similar to that of the mixing process.

E.10 THE NONLINEAR APPROACH FOR THE COMPUTATION OF NOISE IN OSCILLATOR CIRCUITS

The mechanism of noise generation in autonomous circuits and oscillators combines the equivalent of modulation and frequency conversion (mixing) with the effect of AM-to-PM conversion [70, 109].

Traditional approaches relying on frequency conversion analysis are not sufficient to describe the complex physical behavior of a noisy oscillator. The accuracy of the nonlinear approach is based on the dynamic range of the HB simulator and the quality of the parameter extraction for the active device.

Figure E-1 shows a general noisy nonlinear network, which is subdivided into linear and nonlinear subnetworks and noise-free multiports. Noise generation is accounted for by connecting a set of noise voltage and noise current sources at the ports of the linear and nonlinear subnetworks. It is assumed that the circuit is forced by a DC source and a set of sinusoidal sources located at the carrier harmonics $k\omega_0$ and at the sideband $\omega + k\omega_0$.

The electrical regime in this condition of the autonomous circuit is quasi-periodic. The nonlinear system equations to be analyzed is expressed in terms of the HB error vector **E**, defined as the difference between linear and nonlinear current harmonics at the common ports of the circuits.

The solution of this nonlinear algebraic system can be expressed in the form

$$\mathbf{E}(\mathbf{X}_B, \mathbf{X}_H) = F \quad \text{(E-47)}$$

$$\mathbf{E}(\mathbf{X}_B, \mathbf{X}_H) \implies \mathbf{E}_B, \mathbf{E}_H \quad \text{(E-48)}$$

E.10 THE NONLINEAR APPROACH FOR THE COMPUTATION

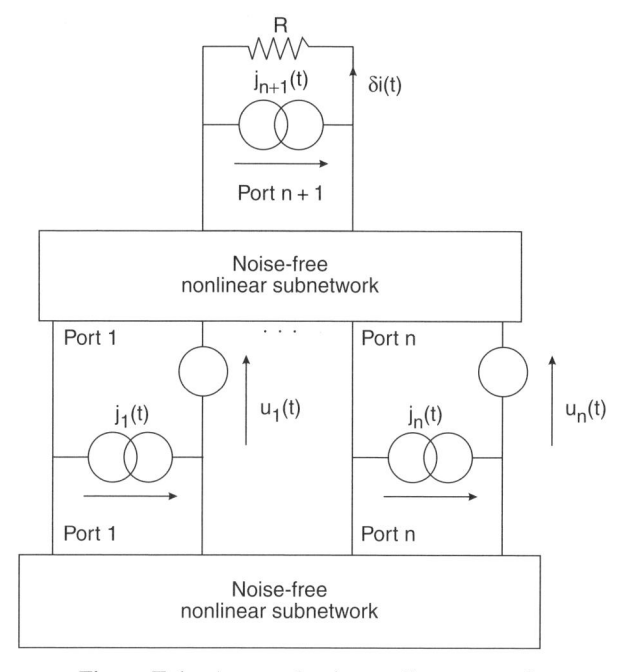

Figure E-1 A general noisy nonlinear network.

where

F = forcing term consisting of DC, harmonics, and sideband excitations
$\mathbf{X}_B$ = state-variable (SV) vectors consisting of components at the sideband
$\mathbf{X}_H$ = SV vectors consisting of components at the carrier harmonics
$\mathbf{E}$ = vector of real and imaginary parts of all (HB) errors
$\mathbf{E}_B$ = error subvector due to the sideband
$\mathbf{E}_H$ = error subvector due to carrier harmonics

Under autonomous (noiseless) steady-state conditions, the forcing term F contains only DC excitations and the possible solution for the nonlinear algebraic system $\mathbf{E}(\mathbf{X}_B, \mathbf{X}_H) = F$ has the form

$$\mathbf{X}_B = 0 \tag{E-49}$$

$$\mathbf{X}_H = \mathbf{X}_H^{SS} \longrightarrow \text{steady state} \tag{E-50}$$

Since the system operates under autonomous conditions, the phase of the steady state is arbitrary. The carrier frequency ω_0 represents one of the unknowns of the nonlinear algebraic system above $\mathbf{E}(\mathbf{X}_B, \mathbf{X}_H) = F$, so one of the harmonics of the vector $\mathbf{X}_H$ is replaced by ω_0.

Now let us assume that the steady-state condition of the autonomous (noiseless) circuit is perturbed by a set of small-signal noises generated inside the linear/nonlinear subnetwork ports of the circuit. This situation can be described by introducing a noise voltage and a noise current source at every interconnecting port, as shown in Figure E-1.

With small noise perturbations, the noise-induced deviation $[\partial \mathbf{X}_B, \partial \mathbf{X}_H]$ of the system from the autonomous (noiseless) steady state $[0, \mathbf{X}_H^{SS}]$ can be quantitatively expressed by the perturbing expression $\mathbf{E}(\mathbf{X}_B, \mathbf{X}_H) = F$ in the neighborhood of the steady state as

$$\left[\frac{\partial \mathbf{E}_B}{\partial \mathbf{X}_B}\right]_{SS} \partial \mathbf{X}_B + \left[\frac{\partial \mathbf{E}_B}{\partial \mathbf{X}_H}\right]_{SS} \partial \mathbf{X}_H = \mathbf{J}_B(\omega) \implies \mathbf{M}_{BB} \partial \mathbf{X}_B + \mathbf{M}_{BH} \partial \mathbf{X}_H = \mathbf{J}_B(\omega) \quad \text{(E-51)}$$

$$\left[\frac{\partial \mathbf{E}_H}{\partial \mathbf{X}_B}\right]_{SS} \partial \mathbf{X}_B + \left[\frac{\partial \mathbf{E}_H}{\partial \mathbf{X}_H}\right]_{SS} \partial \mathbf{X}_H = \mathbf{J}_H(\omega) \implies \mathbf{M}_{HB} \partial \mathbf{X}_B + \mathbf{M}_{HH} \partial \mathbf{X}_H = \mathbf{J}_H(\omega) \quad \text{(E-52)}$$

where

$$\mathbf{M}_{BB} = \left[\frac{\partial \mathbf{E}_B}{\partial \mathbf{X}_B}\right]_{SS} \quad \text{(E-53)}$$

$$\mathbf{M}_{BH} = \left[\frac{\partial \mathbf{E}_B}{\partial \mathbf{X}_H}\right]_{SS} \quad \text{(E-54)}$$

$$\mathbf{M}_{HB} = \left[\frac{\partial \mathbf{E}_H}{\partial \mathbf{X}_B}\right]_{SS} \quad \text{(E-55)}$$

$$\mathbf{M}_{HH} = \left[\frac{\partial \mathbf{E}_H}{\partial \mathbf{X}_H}\right]_{SS} \quad \text{(E-56)}$$

$\mathbf{M}$ is the Jacobian matrix of the HB errors and can be expressed as

$$\mathbf{M} = \begin{bmatrix} \dfrac{\partial \mathbf{E}_B}{\partial \mathbf{X}_B}\bigg|_{SS} & \dfrac{\partial \mathbf{E}_B}{\partial \mathbf{X}_H}\bigg|_{SS} \\ \dfrac{\partial \mathbf{E}_H}{\partial \mathbf{X}_B}\bigg|_{SS} & \dfrac{\partial \mathbf{E}_H}{\partial \mathbf{X}_H}\bigg|_{SS} \end{bmatrix} \quad \text{(E-57)}$$

In the steady-state condition, $\mathbf{X}_B = 0 \Rightarrow \mathbf{M}_{BH}$ and $\mathbf{M}_{HB} = 0$, and the system of equations will be reduced to the following uncoupled equations:

$$\left[\frac{\partial \mathbf{E}_B}{\partial \mathbf{X}_B}\right]_{SS} \partial \mathbf{X}_B = \mathbf{J}_B(\omega) \implies \mathbf{M}_{BB} \partial \mathbf{X}_B = \mathbf{J}_B(\omega) \quad \text{(E-58)}$$

$$\left[\frac{\partial \mathbf{E}_H}{\partial \mathbf{X}_H}\right]_{SS} \partial \mathbf{X}_H = \mathbf{J}_H(\omega) \implies \mathbf{M}_{HH} \partial \mathbf{X}_H = \mathbf{J}_H(\omega) \quad \text{(E-59)}$$

In equation (E-58), $\mathbf{M}_{BB}\partial\mathbf{X}_B = \mathbf{J}_B(\omega)$ is responsible for the mechanism of the conversion noise, which is generated by the exchange of power between the sidebands of the unperturbed large-signal steady state through frequency conversion in the nonlinear subnetwork/devices. The equation $\mathbf{M}_{HH}\partial\mathbf{X}_H = \mathbf{J}_H(\omega)$ describes the mechanism of modulation noise, which is considered a jitter of the oscillatory steady state.

E.11 NOISE GENERATION IN OSCILLATORS

The physical effects of random fluctuations in the circuit differ, depending on their spectral allocation with respect to the carrier.

Noise components at low-frequency deviations result in frequency modulation of the carrier through a mean square frequency fluctuation proportional to the available noise power.

Noise components at high-frequency deviations result in phase modulation of the carrier through a mean square phase fluctuation proportional to the available noise power.

E.12 FREQUENCY CONVERSION APPROACH

The circuit has a large-signal time-periodic steady state of fundamental angular frequency ω_0 (carrier). Noise signals are small perturbations superimposed on the steady state, represented by families of pseudo-sinusoids located at the sidebands of the carrier harmonics. The noise sources are modeled as pseudo-sinusoids having random amplitudes, phases, and deterministic frequencies corresponding to the noise sidebands.

Therefore, the noise performance of the circuit is determined by the exchange of power between the sidebands of the unperturbed steady state through frequency conversion in the nonlinear subnetwork. From the expression $\mathbf{M}_{BB}\partial\mathbf{X}_B = \mathbf{J}_B(\omega)$, it can be seen that oscillator noise is essentially additive noise that is superimposed on each harmonic of a lower and upper sideband at the same frequency offset.

E.13 CONVERSION NOISE ANALYSIS

Consider a set of noise currents and voltage sources connected to the linear/nonlinear subnetwork ports, as shown in Figure E-1. The vectors of the sideband phasor of such sources at the pth noise sideband $\omega + p\omega_0$ are represented by $\mathbf{J}_p(\omega)$ and $\mathbf{U}_p(\omega)$, respectively, where ω is the frequency offset from the carrier ($0 \leq \omega \leq \omega_0$). Due to the perturbative assumption, the nonlinear subnetwork can be replaced with a multifrequency linear multiport described by a conversion matrix. The flow of noise signals can be computed using the conventional linear circuit techniques.

Assuming that the noise perturbations are small, the kth sideband phasor of the noise current through a load resistance R may be expressed through frequency conversion analysis by the linear relationship as

$$\partial I_k(\omega) = \sum_{p=-nH}^{nH} \mathbf{T}_{k,p}^J(\omega) \mathbf{J}_p(\omega) = \sum_{p=-nH}^{nH} \mathbf{T}_{k,p}^U(\omega) \mathbf{U}_p(\omega) \qquad \text{(E-60)}$$

For $k = 0$, upper and lower sideband noise is $\partial I_0(\omega)$ and $\partial I_0(-\omega) = \partial I_0^\otimes(\omega)$. $\mathbf{T}_{k,p}^J(\omega)$ and $\mathbf{T}_{k,p}^U(\omega)$ are the conversion matrices, and n_H is the number of carrier harmonics taken into account in the analysis. From the equation above, the correlation coefficient of the kth and rth sidebands of the noise delivered to the load can be given as

$$C_{k,r}(\omega) = R\langle \partial I_k(\omega) \partial I_r^*(\omega) \rangle \qquad \text{(E-61)}$$

$$C_{k,r}(\omega) = R \sum_{p,q=-nH}^{nH} \left\{ \left[\mathbf{T}_{k,p}^J(\omega) \langle \mathbf{J}_p(\omega) \mathbf{J}_q^\otimes(\omega) \rangle \mathbf{T}_{r,q}^{J\otimes}(\omega) \right] \right. \qquad \text{(E-62)}$$

$$+ \left[\mathbf{T}_{k,p}^U(\omega) \langle \mathbf{U}_p(\omega) \mathbf{U}_q^\otimes(\omega) \rangle \mathbf{T}_{r,q}^{U\otimes}(\omega) \right] \qquad \text{(E-63)}$$

$$+ \left[\mathbf{T}_{k,p}^J(\omega) \langle \mathbf{J}_p(\omega) \mathbf{U}_q^\otimes(\omega) \rangle \mathbf{T}_{r,q}^{U\otimes}(\omega) \right] \qquad \text{(E-64)}$$

$$\left. + \left[\mathbf{T}_{k,p}^U(\omega) \langle \mathbf{U}_p(\omega) \mathbf{J}_q^\otimes(\omega) \rangle \mathbf{TJ}_{r,q}^{J\otimes}(\omega) \right] \right\} \qquad \text{(E-65)}$$

where $*$ denotes the complex conjugate, $\otimes$ denotes the conjugate transposed, and $\langle \bullet \rangle$ denotes the ensemble average. $\mathbf{J}_p(\omega)$ and $\mathbf{U}_p(\omega)$ are the sideband noise sources.

In the above expression of the correlation coefficients of the kth and rth sidebands $C_{k,r}(\omega)$, the power available from the noise sources is redistributed among all the sidebands through frequency conversion. This complex mechanism of interfrequency power flow is described by the family of sideband–sideband conversion matrices $\mathbf{T}_{k,p}^J(\omega)$ and $\mathbf{T}_{k,p}^U(\omega)$.

The noise power spectral density delivered to the load at $\omega + k\omega_0$ can be given as

$$N_k(\omega) = R\langle |\partial I_k(\omega)|^2 \rangle = C_{k,k}(\omega) \qquad \text{(E-66)}$$

E.14 NOISE PERFORMANCE INDEX DUE TO FREQUENCY CONVERSION

The PM noise due to frequency conversion, the AM noise to carrier ratio due to frequency conversion, and the PM-AM correlation coefficient due to frequency conversion can be expressed by simple algebraic combination of the equations above.

E.14 NOISE PERFORMANCE INDEX DUE TO FREQUENCY CONVERSION

PM noise for the kth harmonic can be expressed as

$$\langle |\delta\varphi_{ck}(\omega)|^2 \rangle = \left[\frac{[N_k(\omega) - N_{-k}(\omega)] - 2\mathrm{Re}[C^{\bullet}_{k,-k}(\omega)\exp(j2\varphi_k^{SS})]}{R|I_k^{SS}|^2} \right] \quad \text{(E-67)}$$

where

$\langle |\delta\varphi_{ck}(\omega)|^2 \rangle$ = PM noise at the kth harmonic; the subscript c in $\varphi_{ck}(\omega)$ stands for frequency conversion.

$N_k(\omega)$, $N_{-k}(\omega)$ = noise power spectral densities at the upper and lower sidebands of the kth harmonic

$C^{\bullet}_{k,-k}(\omega)$ = correlation coefficient of the upper and lower sidebands of the kth carrier harmonic

$|I_k^{SS}|\exp(j2\varphi_k^{SS})$ = kth harmonic of the steady-state current through the load

R = load resistance

$\langle \bullet \rangle$ = ensemble average

AM noise for the kth harmonic can be given as

$$\langle |\delta A_{ck}(\omega)|^2 \rangle = 2\left[\frac{[N_k(\omega) - N_{-k}(\omega)] + 2\mathrm{Re}[C^{\bullet}_{k,-k}(\omega)\exp(j2\varphi_k^{SS})]}{R|I_k^{SS}|^2} \right] \quad \text{(E-68)}$$

where

$\langle |\delta A_{ck}(\omega)|^2 \rangle$ = AM noise to carrier ratio at the kth harmonic; the subscript c in $A_{ck}(\omega)$ stands for frequency conversion

$N_k(\omega)$, $N_{-k}(\omega)$ = noise power spectral densities at the upper and lower sidebands of the kth harmonic

$C^{\bullet}_{k,-k}(\omega)$ = correlation coefficient of the upper and lower sidebands of the kth carrier harmonic

$|I_k^{SS}|\exp(j2\varphi_k^{SS})$ = kth harmonic of the steady-state current through the load

R = load resistance

$\langle \bullet \rangle$ = ensemble average

For $k = 0$, the expression for $\partial I_k(\omega)$ can be given as

$$\partial I_k(\omega) = \sum_{p=-nH}^{nH} \mathbf{T}^J_{k,p}(\omega)\mathbf{J}_p(\omega) = \sum_{p=-nH}^{nH} \mathbf{T}^U_{k,p}(\omega)\mathbf{U}_p(\omega) \quad \text{(E-69)}$$

$$\partial I_k(\omega)|_{k=0} = \partial I_0(\omega) \implies \partial I_0^{\otimes}(\omega) \quad \text{(E-70)}$$

450 CAD SOLUTION FOR CALCULATING PHASE NOISE IN OSCILLATORS

$$\partial I_{-k}(\omega)|_{k=0} \implies \partial I_0(-\omega) \implies \partial I_0^{\otimes}(\omega) \qquad \text{(E-71)}$$

$$\varphi_k^{SS}|_{k=0} = 0, \quad \pi \implies \langle|\delta\Phi_{c0}(\omega)|^2\rangle = 0 \qquad \text{(E-72)}$$

$N_k(\omega)|_{k=0} = N_0(\omega) \to$, which is pure AM noise.

The PM-AM correlation coefficient for the kth harmonic can be given as

$$\begin{aligned} C_{ck}^{PMAM}(\omega) &= \langle\delta\Phi_{ck}(\omega)\delta A_k(\omega)^*\rangle \\ &= -\sqrt{2}\left[\frac{2\text{Im}[C_{k,-k}^\bullet(\omega)\exp(j2\varphi_k^{SS})] + j[N_k(\omega) - N_{-k}(\omega)]}{R|I_k^{SS}|^2}\right] \end{aligned} \qquad \text{(E-73)}$$

where

$C_{ck}^{PMAM}(\omega) = $ PM-AM noise correlation coefficient for the kth harmonic; the subscript c in $C_{ck}^{PMAM}(\omega)$ stands for frequency conversion

$N_k(\omega), N_{-k}(\omega) = $ noise power spectral densities at the upper and lower sidebands of the kth harmonic

$C_{k,-k}^\bullet(\omega) = $ correlation coefficient of the upper and lower sidebands of the kth carrier harmonic

$|I_k^{SS}|\exp(j2\varphi_k^{SS}) = k$th harmonic of the steady-state current through the load

$R = $ load resistance

$\langle\bullet\rangle = $ ensemble average

The frequency conversion approach often used has the following limitations:

- It is not sufficient to predict the noise performance of an autonomous circuit. The spectral density of the output noise power, and consequently the PM noise computed by the conversion analysis, is proportional to the available power of the noise sources.
- In the presence of both thermal and flicker noise sources, PM noise due to frequency conversion rises as ω^{-1} and approaches a finite limit for $\omega \to \infty$ like kT.
- Frequency conversion analysis correctly predicts the far-carrier noise behavior of an oscillator, but the oscillator noise floor does not provide results consistent with the physical observations at low-frequency deviations from the carrier. This inconsistency can be eliminated by adding modulation noise analysis.

E.15 MODULATION NOISE ANALYSIS

The equation $[\partial \mathbf{E}_H/\partial \mathbf{X}_B]_{SS}\partial \mathbf{X}_B + [\partial \mathbf{E}_H/\partial \mathbf{X}_H]_{SS}\partial \mathbf{X}_H = \mathbf{J}_H(\omega)$ describes the noise-induced jitter of the oscillatory state, represented by the vector $\delta \mathbf{X}_H$. In this equation, PM noise is the result of direct frequency modulation by the noise sources present in the circuits.

E.15 MODULATION NOISE ANALYSIS

The noise sources in this approach are modeled as modulated sinusoids located at the carrier harmonics, with random pseudo-sinusoidal phase and amplitude modulation causing frequency fluctuations with a mean square value proportional to the available power of the noise sources. The associated mean square phase fluctuation is proportional to the available noise power divided by ω^2. This mechanism is referred as *modulation noise*.

One of the entries of $\delta \mathbf{X}_H$ is $\delta\omega_0$, where $\delta\omega_0(\omega)$ is the phasor of the pseudo-sinusoidal components of the fundamental frequency fluctuations in a 1 Hz band at frequency ω. Frequency jitter with a mean square value proportional to the available noise power is described by $\mathbf{M}_{HH} \partial \mathbf{X}_H = \mathbf{J}_H(\omega)$.

In the presence of both thermal and flicker noise, PM noise due to modulation rises as ω^{-3} for $\omega \to 0$ and tends to go to 0 for $\omega \to \infty$. Modulation noise analysis correctly describes the noise behavior of an oscillator at low deviations from the carrier but does not provide results consistent with physical observations at high deviations from the carrier. The combination of both phenomena explains the noise in the oscillator shown in Figure E-2, where near-carrier noise dominates below ω_X and far-carrier noise dominates above ω_X.

From a strict HB viewpoint, the forcing term $\mathbf{J}_H(\omega)$ of the uncoupled equation $\mathbf{M}_{HH} \partial \mathbf{X}_H = \mathbf{J}_H(\omega)$ represents a synchronous perturbation with time-independent spectral components at the carrier harmonics only. It can be expressed in terms of the sideband noise sources $\mathbf{J}_p(\omega)$ and $\mathbf{U}_p(\omega)$, whose correlation matrices are calculated from the following expressions:

$$C_{k,r}(\omega) = R \langle \partial I_k(\omega) \partial I_r^*(\omega) \rangle \tag{E-74}$$

$$C_{k,r}(\omega) = R \sum_{p,q=-nH}^{nH} \left\{ \left[\mathbf{T}_{k,p}^J(\omega) \langle \mathbf{J}_p(\omega) \mathbf{J}_q^\otimes(\omega) \rangle \mathbf{T}_{r,q}^{J\otimes}(\omega) \right] \right. \tag{E-75}$$

$$+ \left[\mathbf{T}_{k,p}^U(\omega) \langle \mathbf{U}_p(\omega) \mathbf{U}_q^\otimes(\omega) \rangle \mathbf{T}_{r,q}^{U\otimes}(\omega) \right] \tag{E-76}$$

$$+ \left[\mathbf{T}_{k,p}^J(\omega) \langle \mathbf{J}_p(\omega) \mathbf{U}_q^\otimes(\omega) \rangle \mathbf{T}_{r,q}^{U\otimes}(\omega) \right] \tag{E-77}$$

$$\left. + \left[\mathbf{T}_{k,p}^U(\omega) \langle \mathbf{U}_p(\omega) \mathbf{J}_q^\otimes(\omega) \rangle \mathbf{TJ}_{r,q}^{J\otimes}(\omega) \right] \right\} \tag{E-78}$$

In a conventional (deterministic) HB analysis, $\mathbf{J}_H(\omega)$ would contain real and imaginary parts of the synchronous perturbation phasor at $k\omega_0$. But, in reality, forcing the term $\mathbf{J}_H(\omega)$ at the kth harmonic occurs due to superposition of upper and lower sideband noise at $\omega + k\omega_0$. For noise analysis, the noise source waveforms may be viewed as sinusoidal signals at a frequency of $k\omega_0$ slowly modulated in both amplitude and phase at the rate of ω.

In the equation $\mathbf{M}_{HH} \partial \mathbf{X}_H = \mathbf{J}_H(\omega)$, the phasors of the deterministic perturbations are replaced by complex modulations laws, each generated by the superposition of an upper and a lower sideband contribution.

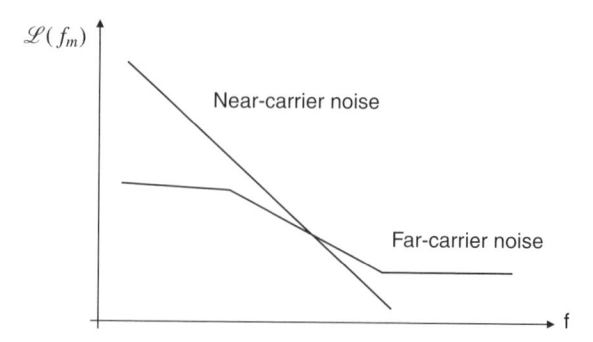

Figure E-2 Oscillator noise components.

From this quasi-stationary viewpoint, the real part of the constant synchronous perturbation is replaced by the phasor of the amplitude modulation law, and the imaginary part is replaced by the phasor of the phase modulation law.

Thus, the expression for the noising forcing term $\mathbf{J}_H(\omega)$ can be expressed as

$$\mathbf{J}_H(\omega) = \{[\mathbf{J}_0^T(\omega) \cdots [\mathbf{J}_k(\omega) + \mathbf{J}_{-k}(\omega)]^T \cdots [-j\mathbf{J}_K(\omega) + j\mathbf{J}_{-K}(\omega)]^T \cdots]^T\}$$
$$(1 \leq k \leq n_H) \quad \text{(E-79)}$$

T denotes the transpose operation, and from Figure E-1 the equivalent Norton phasor of the noise source sidebands is given as

$$\mathbf{J}_k(\omega) = -[\mathbf{J}_{Lk}(\omega) + \mathbf{J}_{Nk}(\omega) + Y(\omega + k\omega_0)\mathbf{U}_{Nk}(\omega)] \quad \text{(E-80)}$$

$Y(\omega + k\omega_0)$ is the linear subnetwork admittance matrix, and $\mathbf{J}_{Lk}(\omega)$ and $\mathbf{J}_{Nk}(\omega)$ are the forcing terms corresponding to the linear and nonlinear subnetworks.

$\mathbf{J}_k(\omega)$ and $\mathbf{U}_k(\omega)$ are the phasors of the pseudo-sinusoids representing the noise components in a 1 Hz bandwidth located in the neighborhood of the sidebands $\omega + k\omega_0$.

In phasor notation with a rotating vector $\exp(j\omega t)$, the forcing term $\mathbf{J}_H(\omega)$ can be given as

$$\mathbf{J}_H(\omega) = \begin{bmatrix} \mathbf{J}_0(\omega) \\ --- \\ \mathbf{J}_K(\omega) + \mathbf{J}_{-K}(\omega) \\ -j\mathbf{J}_K(\omega) + j\mathbf{J}_{-K}(\omega) \\ --- \end{bmatrix} \quad (1 \leq k \leq n_H) \quad \text{(E-81)}$$

After the forcing term $\mathbf{J}_H(\omega)$ in the earlier uncoupled equation $\mathbf{M}_{HH}\partial\mathbf{X}_H = \mathbf{J}_H(\omega)$ is replaced by $[\mathbf{J}_0^T(\omega) \cdots [\mathbf{J}_k(\omega) + \mathbf{J}_{-k}(\omega)]^T \cdots [-j\mathbf{J}_K(\omega) + j\mathbf{J}_{-K}(\omega)]^T \cdots]^T$, the entries of the perturbation vector $\partial\mathbf{X}_H$ become a complex phasor of the pseudo-sinusoidal fluctuations of the corresponding entries of the state vector $\delta \mathbf{X}_H$ at frequency ω.

The equations $\mathbf{M}_{HH}\partial \mathbf{X}_H = \mathbf{J}_H(\omega)$ for $\delta \mathbf{X}_H$ and for $\partial \omega_0$ can be solved by introducing the row matrix $\mathbf{S} = [000 \ldots 1 \ldots 0]$, where the nonzero element corresponds to the position of the entry $\partial \omega_0$ in the vector $\delta \mathbf{X}_H$ and we obtain

$$\partial \omega_0(\omega) = \mathbf{S}[\mathbf{M}_{HH}]^{-1}\mathbf{J}_H(\omega) = \mathbf{T}_F \mathbf{J}_H(\omega) \quad \text{(E-82)}$$

$\partial \omega_0(\omega)$ represents the phasor of the pseudo-sinusoidal component of the fundamental frequency fluctuations in a 1 Hz bandwidth at frequency ω, and $\mathbf{T}_F$ is a row matrix.

Furthermore, a straightforward perturbative analysis of the current of the linear subnetwork allows the perturbation of the current through the load resistor R to be linearly related to the perturbation on the state vector, $\delta \mathbf{X}_H$ is obtained from the equation $\mathbf{M}_{HH}\partial \mathbf{X}_H = \mathbf{J}_H(\omega)$, and the phasor of the pseudo-sinusoidal component of the load current fluctuations in a 1 Hz bandwidth at a deviation ω from $k\omega_0$ can be given as

$$\partial I_k(\omega) = \mathbf{T}_{AK}\mathbf{J}_H(\omega) \quad \text{(E-83)}$$

$\mathbf{T}_{AK}$ is a row matrix, and $\mathbf{J}_H(\omega)$ is a forcing term of the uncoupled equation.

E.16 NOISE PERFORMANCE INDEX DUE TO THE CONTRIBUTION OF NOISE MODULATION

PM noise due to noise modulation, AM noise due to noise modulation, and the PM-AM correlation coefficient due to noise modulation can be expressed in terms of a simple algebraic combination of the above equations.

PM noise for the kth harmonic due to the contribution of noise modulation is

$$\langle |\partial \Phi_k(\omega)|^2 \rangle = \frac{k^2}{\omega^2}\left[\mathbf{T}_F \langle |\mathbf{J}_H(\omega)\mathbf{J}_H^\otimes(\omega)| \rangle \mathbf{T}_F^\otimes \right] \quad \text{(E-84)}$$

where

$\langle |\delta \varphi_{mk}(\omega)|^2 \rangle = $ PM noise at the kth harmonic; the subscript m in $\varphi_{mk}(\omega)$ stands for the modulation mechanism

$\langle |\mathbf{J}_H(\omega)\mathbf{J}_H^\otimes(\omega)| \rangle = $ correlation matrix

$\mathbf{T}_F^\otimes = $ Conjugate transpose

$\mathbf{J}_H(\omega) = $ forcing term

$\langle \bullet \rangle = $ ensemble average

AM noise due to modulation contribution: the kth harmonic of the steady-state current through the load can be expressed as

$$I_k^{SS} = \mathbf{Y}_R(k\omega_0)\mathbf{V}_k(\mathbf{X}_H) \tag{E-85}$$

where

I_k^{SS} = kth harmonic of the steady-state current through the load
$\mathbf{Y}_R(k\omega_0)$ = transadmittance matrix
$\mathbf{V}_k$ = vector representation of the kth harmonics of the voltages at the nonlinear subnetwork ports

By perturbing I_k^{SS} in the neighborhood of the steady state, the phasor of the pseudo-sinusoidal component of the kth harmonic current fluctuations at frequency ω can be expressed as a linear combination of the elements of the perturbation vector $\delta \mathbf{X}_H$.

From the equation $[\partial \mathbf{E}_H/\partial \mathbf{X}_H]_{SS} \partial \mathbf{X}_H = \mathbf{J}_H(\omega) \Rightarrow \mathbf{M}_{HH} \partial \mathbf{X}_H = \mathbf{J}_H(\omega); \partial I_k(\omega) = \mathbf{T}_{AK} \mathbf{J}_H(w)$, the modulation contribution of the kth harmonic AM noise to carrier ratio at frequency ω can be expressed as

$$\langle \partial |A_{mk}(\omega)|^2 \rangle = \frac{2}{|I_k^{SS}|^2} \left[\mathbf{T}_{Ak} \langle |\mathbf{J}_H(\omega)\mathbf{J}_H^{\otimes}(\omega)| \rangle \mathbf{T}_{Ak}^{\otimes} \right] \tag{E-86}$$

where

$\langle |\delta A_{mk}(\omega)|^2 \rangle$ = AM noise to carrier ratio at the kth harmonic; the subscript m in $A_{mk}(\omega)$ stands for the modulation mechanism
$\langle |\mathbf{J}_H(\omega)\mathbf{J}_H^{\otimes}(\omega)| \rangle$ = correlation matrix
$\mathbf{J}_H(\omega)$ = forcing-term
$\mathbf{T}_{Ak}$ = row matrix
$\mathbf{T}_{Ak}^{\otimes}$ = Conjugate transpose

E.17 PM-AM CORRELATION COEFFICIENT

In the equation $[\partial \mathbf{E}_H/\partial \mathbf{X}_H]_{SS} \partial \mathbf{X}_H = \mathbf{J}_H(\omega) \Rightarrow \mathbf{M}_{HH} \partial \mathbf{X}_H = \mathbf{J}_H(\omega)$, the information on the RF phase is lost and it is not possible to calculate the phase of the PM-AM correlation coefficient from the expressions above. In order to calculate PM-AM correlation, the first-order approximation of the normalized kth harmonic PM-AM normalized correlation coefficient $C_{ck}(\omega)$ computed from frequency conversion analysis is given as

$$C_{ck}(\omega) = \left[\frac{C_{ck}^{PMAM}(\omega)}{\sqrt{\langle|\delta\Phi_{ck}(\omega)|^2\rangle\langle|\delta A_{ck}(\omega)|^2\rangle}} \right] \quad \text{(E-87)}$$

and can be correctly evaluated from frequency conversion analysis even for $\omega \to 0$, where $C_{ck}(\omega)$ is the normalized PM-AM correlation coefficient, which compensates for the incorrect dependency of $\langle|\delta\Phi_{ck}(\omega)|^2\rangle$ of the frequency at low-frequency offsets from the carrier. From the PM-AM correlation coefficient above due to modulation, the contribution to the kth harmonic can be given as

$$C_{mk}^{PMAM}(\omega) = C_{ck}(\omega)\left[\sqrt{\langle|\delta\Phi_{mk}(\omega)|^2\rangle\langle|\delta A_{mk}(\omega)|^2\rangle}\right] \quad \text{(E-88)}$$

$$C_{mk}^{PMAM}(\omega) \cong \langle\delta\Phi_k(\omega)\delta A_k(\omega)^*\rangle = \frac{k\sqrt{2}}{j\omega|I_k^{SS}|}\left[\mathbf{T}_F\langle\mathbf{J}_H(\omega)\mathbf{J}_H^\otimes(\omega)\rangle\mathbf{T}_{Ak}^\otimes\right] \quad \text{(E-89)}$$

Now the near-carrier noise power spectral density $N_k(\omega)$ of the oscillator, due to the modulation contribution at offset ω from $k\omega_0(-n_H \leq k \leq n_H)$, can be given as

$$N_k(\omega) = \frac{1}{4}R\left[\frac{k^2}{\omega^2}|I_k^{SS}|^2\mathbf{T}_F\langle\mathbf{J}_H(\omega)\mathbf{J}_H^\otimes(\omega)\rangle\mathbf{T}_F^\otimes\right] \quad \text{(E-90)}$$

$$+ \frac{1}{4}R[\mathbf{T}_{AK}[\mathbf{J}_H(\omega)\mathbf{J}_H^\otimes(\omega)]\mathbf{T}_{Ak}^\otimes] \quad \text{(E-91)}$$

$$+ \frac{kR}{2\omega}|I_k^{SS}|\text{Re}[\mathbf{T}_F\langle\mathbf{J}_H(\omega)\mathbf{J}_H^\otimes(\omega)\rangle\mathbf{T}_{Ak}^\otimes] \quad \text{(E-92)}$$

$$\Rightarrow N_k(\omega) = \frac{1}{4}R|I_k^{SS}|^2\left[\langle|\delta\Phi_{mk}(\omega)|^2\rangle + \frac{1}{2}\langle|\delta A_{mk}(\omega)|^2\rangle - \frac{k}{|k|}\sqrt{2}\,\text{Im}[C_{mk}^{PMAM}(\omega)]\right] \quad \text{(E-93)}$$

where

$\mathbf{J}_H(\omega)$ = vector of the Norton equivalent of the noise sources
$\mathbf{T}_F$ = frequency transfer matrix
R = load resistance
I_k^{SS} = kth harmonic of the steady-state current through the load.

APPENDIX F
General Noise Presentation

F.1 NOISE PARAMETERS AND THE NOISE CORRELATION MATRIX

The noise correlation matrices form a general technique for calculating noise in n-port networks. This method is useful because it forms a base from which we can rigorously calculate the noise of linear two-ports combined in arbitrary ways. For many representations, the method of combining the noise parameters is as simple as it is for combining the circuit element matrices.

Linear, noisy two-ports can be modeled as noise-free two-ports with two additional noise sources, as shown in Figures F-1a and F-1b.

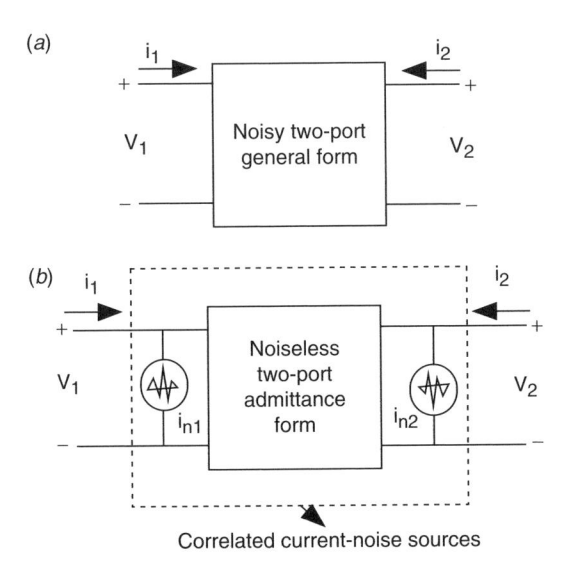

Figure F-1 (*a*) General representation of a noisy two-port. (*b*) General representation of a noiseless two-port with two current-noise sources at the input and output.

The Design of Modern Microwave Oscillators for Wireless Applications: Theory and Optimization,
by Ulrich L. Rohde, Ajay Kumar Poddar, Georg Böck
Copyright © 2005 John Wiley & Sons, Inc.

GENERAL NOISE PRESENTATION

The matrix representation is

$$\begin{bmatrix} I_1 \\ I_2 \end{bmatrix} = \begin{bmatrix} Y_{11} & Y_{12} \\ Y_{21} & Y_{22} \end{bmatrix} \begin{bmatrix} V_1 \\ V_2 \end{bmatrix} + \begin{bmatrix} i_{n1} \\ i_{n2} \end{bmatrix} \qquad \text{(F-1)}$$

where i_{n1} and i_{n2} are the noise sources at the input and output ports of the admittance form.

Since i_{n1} and i_{n2} (noise vectors) are random variables, it is convenient to work with the noise correlation matrix, which gives a deterministic number with which to calculate. The above two-port example can be extended to n-ports in a straightforward way as a matrix chain representation.

F.2 THE CORRELATION MATRIX

The *correlation matrix* is defined as the mean value of the outer product of the noise vector, which is equivalent to multiplying the noise vector by its adjoint (complex conjugate transpose: identical to the Hermitian matrix) and averaging the result.

Consider the $[Y]$-parameter noise correlation matrix as $[C_y]$; it can be given as

$$\langle \vec{i} \, \vec{i}^{+} \rangle = \begin{bmatrix} i_1 \\ i_2 \end{bmatrix} [i_1^* \ i_2^*] = \begin{bmatrix} \langle i_1 \ i_1^* \rangle & \langle i_1 \ i_2^* \rangle \\ \langle i_1^* \ i_2 \rangle & \langle i_2 \ i_2^* \rangle \end{bmatrix} = [C_y] \qquad \text{(F-2)}$$

The diagonal term represents the power spectrum of each noise source, and the off-diagonal terms are the cross-power spectrum of the noise source. Angular brackets denote the average value.

F.3 METHOD OF COMBINING TWO-PORT MATRIX

If we parallel the two matrices y and y', we have the same port voltages, and the terminal currents add as shown in Figure F-2.

$$\begin{aligned} I_1 &= y_{11}V_1 + y_{12}V_2 + y'_{11}V_1 + y'_{12}V_2 + i_1 + i'_1 \\ I_2 &= y_{21}V_1 + y_{22}V_2 + y'_{21}V_1 + y'_{22}V_2 + i_2 + i'_2 \end{aligned} \qquad \text{(F-3)}$$

In matrix form:

$$\begin{bmatrix} I_1 \\ I_2 \end{bmatrix} = \begin{bmatrix} y_{11} + y'_{11} & y_{12} + y'_{12} \\ y_{21} + y'_{21} & y_{22} + y'_{22} \end{bmatrix} \begin{bmatrix} V_1 \\ V_2 \end{bmatrix} + \begin{bmatrix} i_1 + i'_1 \\ i_2 + i'_2 \end{bmatrix} \qquad \text{(F-4)}$$

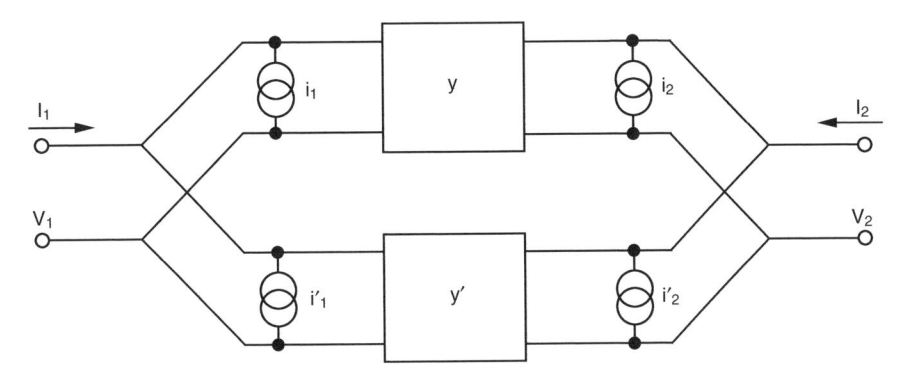

Figure F-2 Parallel combination of two-ports using y parameters.

Here we can see that the noise current vectors add just as the [Y] parameters add. Converting the new noise vector to a correlation matrix yields

$$\langle \bar{i}_{new} \bar{i}^+_{new} \rangle = \left\langle \begin{bmatrix} i_1 + i'_1 \\ i_2 + i'_2 \end{bmatrix} [i_1^* + i_1'^* \quad i_2 i_2'^*] \right\rangle \tag{F-5}$$

$$= \begin{bmatrix} \langle i_1 i_1^* \rangle + \langle i'_1 i_1'^* \rangle & \langle i_1 i_2^* \rangle + \langle i'_1 i_2'^* \rangle \\ \langle i_2 i_1^* \rangle + \langle i'_2 i_1'^* \rangle & \langle i_2 i_2^* \rangle + \langle i'_2 i_2'^* \rangle \end{bmatrix} \tag{F-6}$$

The noise sources from different two-ports must be uncorrelated, so there are no cross-products of different two-ports. By inspection, it can be seen that this is just the addition of the correlation matrices for the individual two-ports, which is given as

$$[C_{y\,new}] = [C_y] + [C'_y] \tag{F-7}$$

F.4 NOISE TRANSFORMATION USING [ABCD] NOISE CORRELATION MATRIX

Figure F-3 shows the noise transformation using the [ABCD] matrix, where $[C_A]$ and $[C'_A]$ are correlation matrices, respectively, for two noise-free two-port systems.

$$[C_A] = \begin{bmatrix} \overline{v_A v_A^*} & \overline{v_A i_A^*} \\ \overline{i_A v_A^*} & \overline{i_A i_A^*} \end{bmatrix} \tag{F-8}$$

$$[A_{new}] = [A][A]' \tag{F-9}$$

460   GENERAL NOISE PRESENTATION

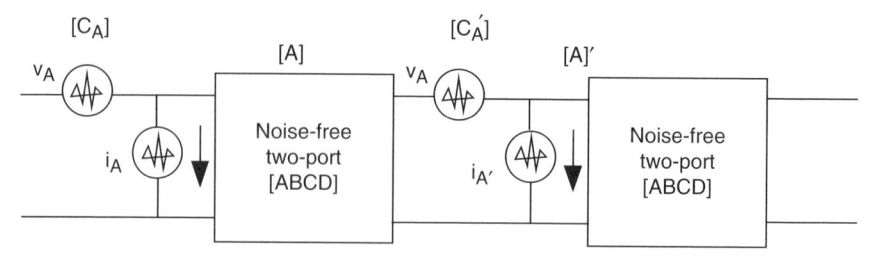

Figure F-3   Noise transformation using the [ABCD] matrix.

The corresponding correlation matrix is given as:

$$[C_{A_{new}}] = [C_A] + [A][C'_A][A]'\tag{F-10}$$

F.5   RELATION BETWEEN THE NOISE PARAMETER AND $[C_A]$

Figure F-4 shows the generator current and noise sources for the derivation of noise parameters.
where

I_G = generator current
Y_G = generator admittance
i_G, i_A, v_A = noise sources

Using the noise correlation matrix representation, the correlated noise voltage and current are located at the input of the circuit, supporting a direct relation with the noise parameters (R_n, Γ_{opt}, F_{min}).
From the matrix properties, i_A can be written as

$$i_A = Y_{cor}v_A + i_u \tag{F-11}$$

$$\overline{i_A} = \overline{(Y_{cor}v_A + i_u)} \tag{F-12}$$

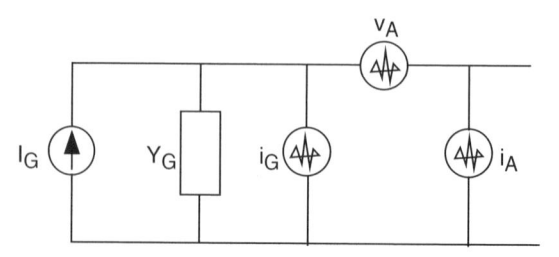

Figure F-4   Generator current with noise sources.

F.5 RELATION BETWEEN THE NOISE PARAMETER AND [C_A]

$$\overline{v_A i_A^*} = Y_{cor} v_A^2 \tag{F-13}$$

$$Y_{cor} = \frac{\overline{v_A i_A^*}}{v_A^2} \tag{F-14}$$

where i_u represents the uncorrelated port, Y_{cor} represents the correlation factor, and i_u and v_A are uncorrelated.

The noise factor F is now given as

$$F = \left|\frac{i_G}{i_G}\right|^2 + \left|\frac{(i_A + Y_G v_A)}{i_G}\right|^2 = 1 + \left|\frac{Y_{cor}v_A + i_u + Y_G v_A}{i_G}\right|^2$$

$$= 1 + \left|\frac{v_A(Y_{cor} + Y_G) + i_u}{i_G}\right|^2 \tag{F-15}$$

$$F = 1 + \left|\frac{i_u}{i_G}\right|^2 + \left|\frac{v_A(Y_{cor} + Y_G)}{i_G}\right|^2$$

$$= 1 + \frac{G_u}{G_G} + \frac{R_n}{G_G}[(G_G + G_{cor})^2 + (B_G + B_{cor})^2] \tag{F-16}$$

$$i_G^2 = 4kTBG_G \tag{F-17}$$

$$i_u^2 = 4kTBG_u \tag{F-18}$$

$$v_A^2 = 4kTBR_n \tag{F-19}$$

$$Y_{cor} = G_{cor} + jB_{cor} \tag{F-20}$$

where

$$G_u = \frac{1}{R_u}, \quad R_n = \frac{1}{G_n}, \quad Y_{cor} = \frac{1}{Z_{cor}} \tag{F-21}$$

$$F = 1 + \frac{R_u}{R_G} + \frac{G_n}{G_G}[(R_G + R_{cor})^2 + (X_G + X_{cor})^2] \tag{F-22}$$

The minimum noise factor F_{min} and corresponding optimum noise source impedance $Z_{opt} = R_{opt} + jX_{opt}$ are found by differentiating F with respect to the source resistance (R_G) and susceptance (X_G).

$$\left.\frac{dF}{dR_G}\right|_{Rg_{opt}} = 0 \tag{F-23}$$

$$\left.\frac{dF}{dX_G}\right|_{Xg_{opt}} = 0 \tag{F-24}$$

$$R_{opt} = \sqrt{\frac{R_u}{G_n} + R_{cor}^2} \qquad (F-25)$$

$$X_{opt} = -X_{cor} \qquad (F-26)$$

$$F_{min} = 1 + 2G_n R_{cor} + 2\sqrt{R_u G_n + (G_n R_{cor})^2} \qquad (F-27)$$

$$F = F_{min} + \frac{G_n}{R_G}|Z_G - Z_{opt}|^2 \qquad (F-28)$$

$$F = F_{min} + \frac{4r_n|\Gamma_G - \Gamma_{opt}|^2}{(1 - |\Gamma_G|^2)|1 + \Gamma_{opt}|^2} \qquad (F-29)$$

The noise factor of a linear two-port as a function of the source admittance can be expressed as

$$F = F_{min} + \frac{R_n}{G_g}[(G_{opt} - G_g)^2 + (B_{opt} - B_G)^2] \qquad (F-30)$$

where

Y_g (generator admittance) $= G_g + jB_G$
Y_{opt} (optimum noise admittance) $= G_{opt} + jB_{opt}$
F_{min} (minimum achievable noise factor) when $Y_{opt} = Y_g$
R_n (noise resistance) = sensitivity of the NF to the source admittance

F.6 REPRESENTATION OF THE [ABCD] CORRELATION MATRIX IN TERMS OF THE NOISE PARAMETERS

The following is the calculation of Y_{opt}, F_{min}, and R_n from the noise correlation matrix.

$$[C_A] = \begin{bmatrix} \overline{v_A v_A^\bullet} & \overline{v_A i_A^\bullet} \\ \overline{i_A v_A^\bullet} & \overline{i_A i_A^\bullet} \end{bmatrix} = \begin{bmatrix} C_{uu\bullet} & C_{ui\bullet} \\ C_{u\bullet i} & C_{ii\bullet} \end{bmatrix}$$

$$= 2kT \begin{bmatrix} R_n & \dfrac{F_{min} - 1}{2} - R_n Y_{opt}^\bullet \\ \dfrac{F_{min} - 1}{2} - R_n Y_{opt} & R_n |Y_{opt}|^2 \end{bmatrix} \qquad (F-31)$$

$$Y_{opt} = \sqrt{\frac{C_{ii\bullet}}{C_{uu\bullet}} - \left[\operatorname{Im}\left(\frac{C_{ui\bullet}}{C_{uu\bullet}}\right)\right]^2} + j\operatorname{Im}\left(\frac{C_{ui\bullet}}{C_{uu\bullet}}\right) \tag{F-32}$$

$$F_{min} = 1 + \frac{C_{ui\bullet} + C_{uu\bullet} Y_{opt}^{\bullet}}{kT} \tag{F-33}$$

$$R_n = \frac{C_{uu\bullet}}{2kT} \tag{F-34}$$

F.7 NOISE CORRELATION MATRIX TRANSFORMATIONS

For simplification of the analysis of the noise parameters of the correlation matrix, it is often necessary to transform from admittance to impedance and vice versa.

Two-port currents for admittance form can be written as

$$\begin{bmatrix} I_1 \\ I_2 \end{bmatrix} = [Y]\begin{bmatrix} V_1 \\ V_2 \end{bmatrix} + \begin{bmatrix} i_1 \\ i_2 \end{bmatrix} \tag{F-35}$$

Writing in terms of voltage (we can move the noise vector to the left side and invert y):

$$\begin{bmatrix} V_1 \\ V_2 \end{bmatrix} = [Y^{-1}]\begin{bmatrix} I_1 - i_1 \\ I_2 - i_2 \end{bmatrix} = [Y^{-1}]\begin{bmatrix} I_1 \\ I_2 \end{bmatrix} + [Y^{-1}]\begin{bmatrix} -i_1 \\ -i_2 \end{bmatrix} \tag{F-36}$$

Since $(Y)^{-1} = (Z)$, we have

$$\begin{bmatrix} V_1 \\ V_2 \end{bmatrix} = [Z]\begin{bmatrix} I_1 \\ I_2 \end{bmatrix} + [Z]\begin{bmatrix} -i_1 \\ -i_2 \end{bmatrix} \tag{F-37}$$

Considering only the noise source,

$$\begin{bmatrix} v_1 \\ v_2 \end{bmatrix} = [Z]\begin{bmatrix} -i_1 \\ -i_2 \end{bmatrix} = [T_{yz}]\begin{bmatrix} -i_1 \\ -i_2 \end{bmatrix} \tag{F-38}$$

Here the signs of i_1 and i_2 are superfluous since they will cancel when the correlation matrix is formed, and the transformation of the Y-noise current vector to the Z-noise voltage vector is done simply by multiplying $[Z]$. Other transformations are shown in Table F-1 for ready reference.

To form the noise correlation matrix, we again form the mean of the outer product

$$\langle vv^+ \rangle = \begin{bmatrix} \langle v_1 v_1^* \rangle & \langle v_1 v_2^* \rangle \\ \langle v_1^* v_2 \rangle & \langle v_2 v_2^* \rangle \end{bmatrix} = [Z]\left\langle \begin{bmatrix} i_1 \\ i_2 \end{bmatrix} [i_1^* \ i_2^*] \right\rangle [Z]^+ \tag{F-39}$$

TABLE F-1 Noise Correlation Matrix Transformations

		Original form (α form)					
		Y		Z		A	
Resulting form (β form)	Y	1	0	y_{11}	y_{12}	$-y_{11}$	1
		0	1	y_{21}	y_{22}	$-y_{21}$	0
	Z	z_{11}	z_{12}	1	0	1	$-z_{11}$
		z_{21}	z_{22}	0	1	0	$-z_{21}$
	A	0	A_{12}	1	$-A_{11}$	1	0
		1	A_{22}	0	$-A_{21}$	0	1

which is identical to

$$[C_z] = [Z][C_y][Z]^+ \tag{F-40}$$

$$v^+ = [i_1^* \ i_2^*][Z]^+ \tag{F-41}$$

This is called a *congruence transformation*. The key to all of these derivations is the construction of the correlation matrix from the noise vector. For passive circuit noise, a correlation matrix can be determined with only thermal noise sources. The $2kT$ factor comes from the double-sided spectrum of thermal noise.

$$[C_z] = 2kT\Delta f \text{Re}([Z]) \tag{F-42}$$

and

$$[C_y] = 2kT\Delta f \text{Re}([Y]) \tag{F-43}$$

For example, the transformation of the noise correlation matrix $[C_A]$ to $[C_Z]$ is

$$[C_A] = \begin{bmatrix} 1 & -Z_{11} \\ 0 & -Z_{22} \end{bmatrix} [C_A] = \begin{bmatrix} 1 & 0 \\ -Z_{11}^* & -Z_{22}^* \end{bmatrix} \tag{F-44}$$

F.8 MATRIX DEFINITIONS OF SERIES AND SHUNT ELEMENT

Figures F-5a and F-5b show series and shunt elements for the calculation of noise parameters.

$$[Z]_{series} = \begin{bmatrix} Z & Z \\ Z & Z \end{bmatrix} \tag{F-45}$$

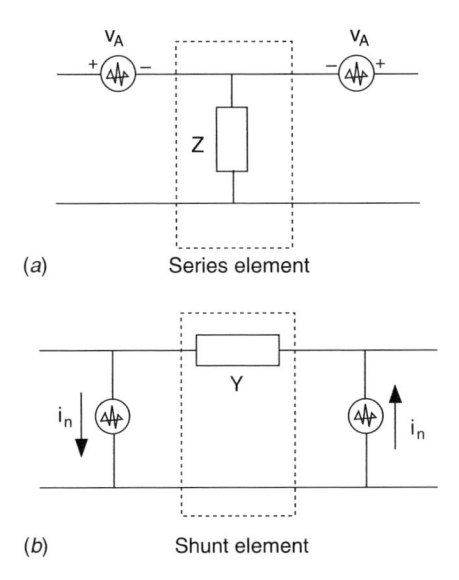

Figure F-5 (a) Series element for the calculation of noise parameters. (b) Shunt element for the calculation of noise parameters.

$$[C_Z]_{series} = 2kT\Delta f \mathrm{Re}([z]_{series}) \tag{F-46}$$

$$[Y]_{shunt} = \begin{bmatrix} Y & -Y \\ -Y & Y \end{bmatrix} \tag{F-47}$$

$$[C_Y]_{shunt} = 2kT\Delta f \mathrm{Re}([Y]_{shunt}) \tag{F-48}$$

F.9 TRANSFERRING ALL NOISE SOURCES TO THE INPUT

For easier calculation of noise parameters, all noise sources are transferred to the input with the help of the [ABCD] matrix.

Before a detailed analysis of the noise parameters is presented, some useful transformations using the [ABCD] matrix are discussed. They will be used later in forming the correlation matrices.

Figure F-6 shows the two-port [ABCD] parameter representation of a noise-free system.

The general expression of the two-port [ABCD] matrix is given as

$$V_i = AV_0 + BI_o \tag{F-49}$$

GENERAL NOISE PRESENTATION

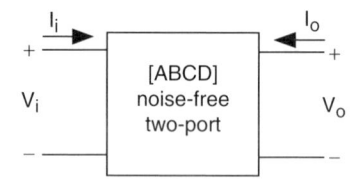

Figure F-6 Two-port [ABCD] parameter representation of a noise-free system.

$$I_i = CV_0 + DI_o \tag{F-50}$$

$$\begin{bmatrix} V_i \\ I_i \end{bmatrix} = \begin{bmatrix} A & B \\ C & D \end{bmatrix} = \begin{bmatrix} V_0 \\ I_o \end{bmatrix} \tag{F-51}$$

With the addition of the noise source at the output, as shown in Figure F-7, the matrix shown above can be expressed as

The expression of a two-port [ABCD] matrix with noise sources connected at the output can be given as

$$V_i = AV_0 + BI_o - AV_n - BI_n \tag{F-52}$$

$$I_i = CV_0 + DI_o - CV_n - DI_n \tag{F-53}$$

F.10 TRANSFORMATION OF NOISE SOURCES

Figure F-8 shows the noise sources transformed to the input, where the current and voltage noise sources are correlated.

The modified matrix is expressed as

$$V_i + AV_n + BI_n = AV_0 + BI_o \tag{F-54}$$

$$I_i + CV_n + DI_n = CV_0 + DI_o \tag{F-55}$$

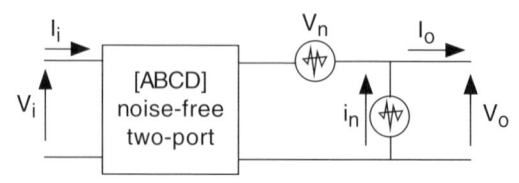

Figure F-7 Two-port [ABCD] parameter representation with noise source connected at the output.

F.10 TRANSFORMATION OF NOISE SOURCES 467

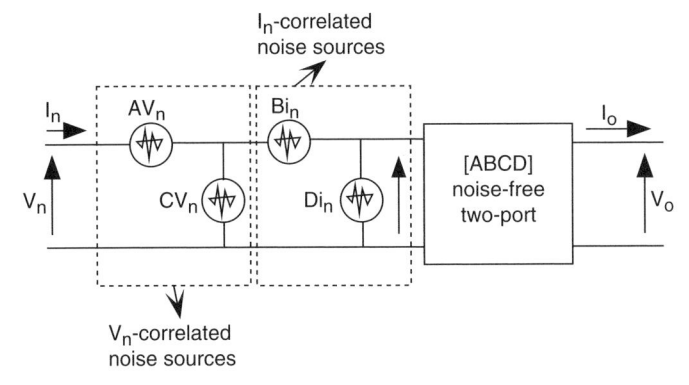

Figure F-8 Noise source transformed to the input.

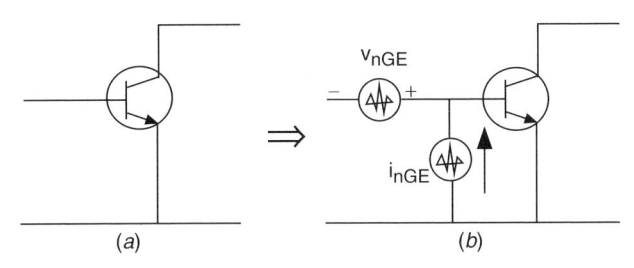

Figure F-9 (*a*) GE configuration of a noise-free transistor; (*b*) GE configuration with the noise sources at the input.

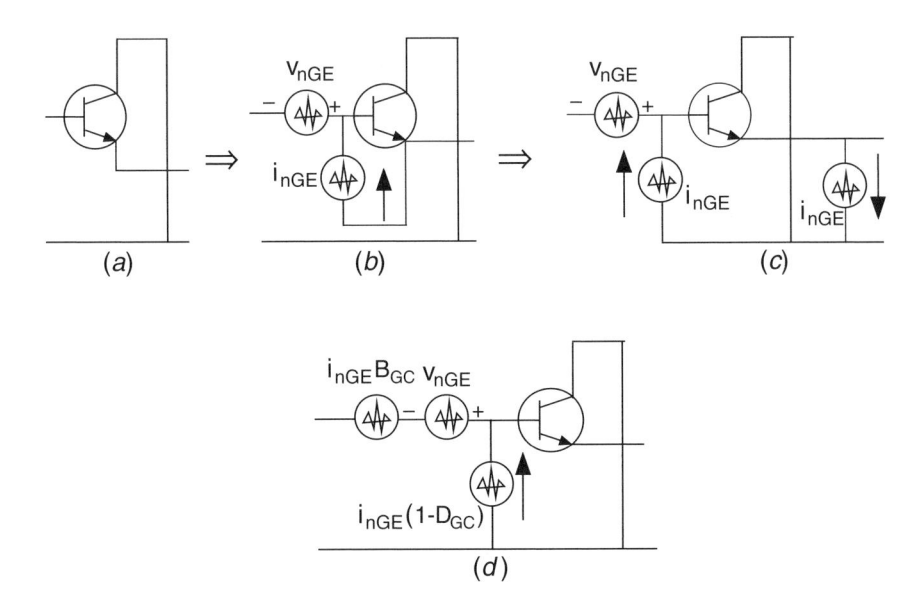

Figure F-10 (*a*) GC configuration; (*b*) GC with input noise sources; (*c*) GC noise with current splitting; (*d*) output current with noise source transferred to the ouput.

F.11 [ABCD] PARAMETERS FOR GE, GC, AND GB CONFIGURATIONS

Figures F-9a and F-9b show the grounded emitter (GE) configuration of the noise-free transistor and with the current and voltage noise sources at the input.

Common emitter:

$$[ABCD]_{GE} = \begin{bmatrix} A_{GE} & B_{GE} \\ C_{GE} & D_{GE} \end{bmatrix} \tag{F-56}$$

$$\begin{bmatrix} A_{GE} & B_{GE} \\ C_{GE} & D_{GE} \end{bmatrix} \Rightarrow \begin{matrix} v_{nGE} = v_{bn} + B_{GE}i_{cn} \\ i_{nGE} = i_{bn} + D_{GE}i_{cn} \end{matrix} \tag{F-57}$$

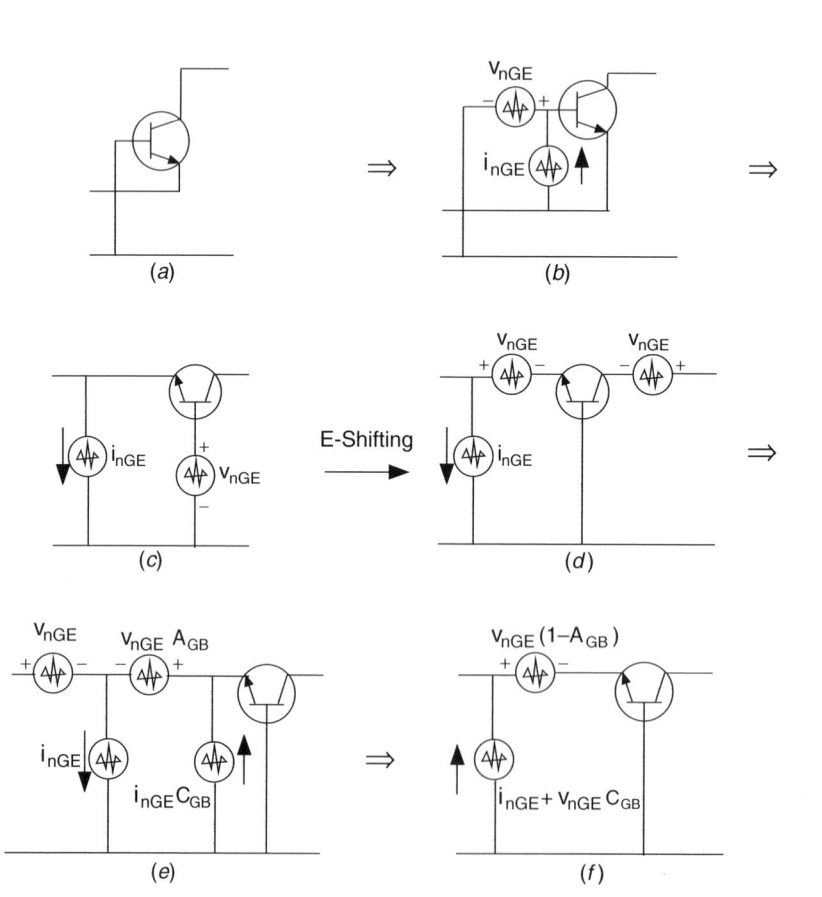

Figure F-11 Transformation of noise sources in GB configurations. (a) GB configuration; (b) GB with input noise sources; (c) GB orientation of noise sources; (d) noise sources with e-shift; (e) noise transformation from output to input; (f) orientation of noise sources at the input.

Common collector:

$$[ABCD]_{GC} = \begin{bmatrix} A_{GC} & B_{GC} \\ C_{GC} & D_{GC} \end{bmatrix} \quad \text{(F-58)}$$

Figures F-10a–d show the grounded collector (GC) configuration of the noise-free transistor and with the current and voltage noise sources at the input.

B_{CC} and D_{CC} are very small for the GC configuration because the transadmittance $1/B_{CC}$ and the current gain $1/D_{CC}$ are large. Therefore, the noise performance/parameters of the GC configuration are similar to those of the GE configuration.

Figures F-11a–f show the grounded base (GB) configuration of the noise-free transistor and the transformation of all noise sources to the input.

GB configuration:

$$[ABCD]_{CB} = \begin{bmatrix} A_{GB} & B_{GB} \\ C_{GB} & D_{GB} \end{bmatrix} \quad \text{(F-59)}$$

A_{GB} and C_{GB} are very small for the GB configuration because the transimpedance $(1/C_{GB} = (v_o/i_i)|_{io} = 0)$ and the voltage gain $(1/A_{GB} = v_o/v_i|_{io} = 0)$ are very large. With comparable bias conditions, the noise performance is similar to that of the GE configuration.

APPENDIX G
Calculation of Noise Properties of Bipolar Transistors and FETs

G.1 BIPOLAR TRANSISTOR NOISE MODELS

G.1.1 Hybrid-π Configuration

Figure G-1a shows the equivalent schematic of the bipolar transistor in a grounded emitter configuration (GE). The high frequency or microwave noise of a silicon bipolar transistor in a CE configuration can be modeled by using the three noise sources shown in the equivalent schematic (hybrid-π) in Figure G-1b. The emitter junction in this case is conductive, and this generates shot noise on the emitter. The emitter current is divided into base (I_b) current and collector current (I_c), both of which generate shot noise.

Collector reverse current (I_{cob}) also generates shot noise. The emitter, base, and collector are made of semiconductor material and have a finite resistance value associated with them, which generates thermal noise.

The value of the base resistance is relatively high in comparison to the resistance associated with the emitter and collector, so the noise contribution of these resistors can be neglected.

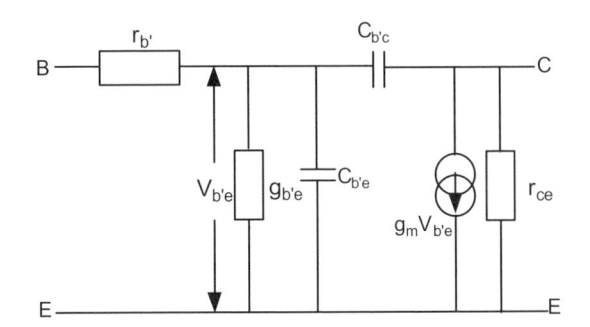

Figure G-1a π-Configuration of the GE bipolar transistor.

The Design of Modern Microwave Oscillators for Wireless Applications: Theory and Optimization,
by Ulrich L. Rohde, Ajay Kumar Poddar, Georg Böck
Copyright © 2005 John Wiley & Sons, Inc.

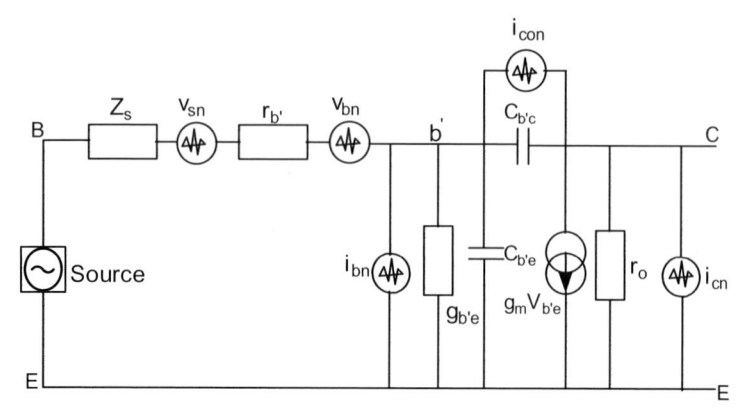

Figure G-1b π-Configuration of the CE bipolar transistor with noise sources.

For noise analysis, three sources are introduced in a noiseless transistor: noise due to fluctuation in the DC bias current (i_{bn}), the DC collector current (i_{cn}), and the thermal noise of the base resistance.

For evaluation of noise performance, the signal-driving source should also be taken into consideration because its internal conductance generates noise and its susceptance affects the noise level through noise tuning.

In silicon transistors, the collector reverse current (I_{cob}) is very small, so the noise (i_{con}) generated by this current can be neglected.

The mean square value of the above noise generator in a narrow frequency interval Δf is given by

$$\overline{i_{bn}^2} = 2qI_b\Delta f \tag{G-1}$$

$$\overline{i_{cn}^2} = 2qI_c\Delta f \tag{G-2}$$

$$\overline{i_{con}^2} = 2qI_{cob}\Delta f \tag{G-3}$$

$$\overline{v_{bn}^2} = 4kTR_b\Delta f \tag{G-4}$$

$$\overline{v_{sn}^2} = 4kTR_b\Delta f \tag{G-5}$$

I_b, I_c, and I_{cob} are average DC current over Δf noise bandwidth. The noise power spectral densities due to noise sources are

$$S(i_{cn}) = \frac{\overline{i_{cn}^2}}{\Delta f} = 2qI_c = 2KTg_m \tag{G-6}$$

$$S(i_{bn}) = \frac{\overline{i_{bn}^2}}{\Delta f} = 2qI_b = \frac{2KTg_m}{\beta} \tag{G-7}$$

$$S(v_{bn}) = \frac{\overline{v_{bn}^2}}{\Delta f} = 4KTR_b \qquad (G\text{-}8)$$

$$S(v_{sn}) = \frac{\overline{v_{sn}^2}}{\Delta f} = 4KTR_s \qquad (G\text{-}9)$$

r_b' and R_s are base and source resistance, and Z_s is the complex source impedance.

G.1.2 Transformation of the Noise Current Source to the Input of the GE Bipolar Transistor

Figure G-2a shows the GE π-configuration with the noise sources. In silicon transistors the collector reverse current (I_{cob}) is very small, and the noise (i_{con}) generated by this current can be neglected.

Figure G-2b shows the equivalent [ABCD] representation of the intrinsic transistor, which is given in terms of two-port noise-free parameters as A_{GE},

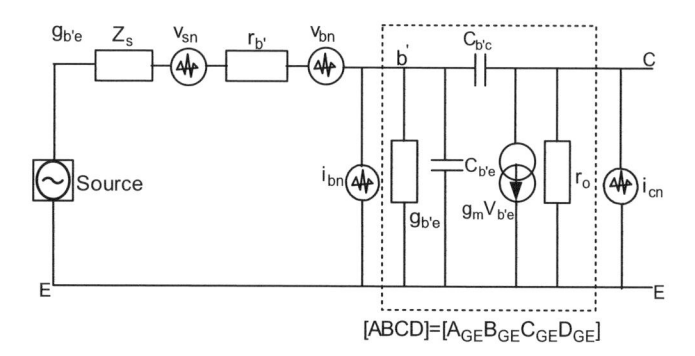

Figure G-2a π-Configuration of the bipolar transistor with noise sources.

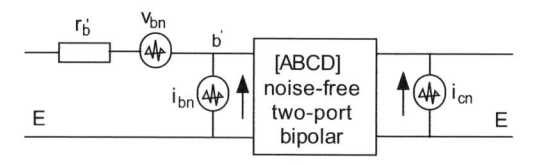

Figure G-2b Equivalent [ABCD] representation of the intrinsic transistor.

B_{GE}, C_{GE}, and D_{GE}.

$$[ABCD] = \begin{bmatrix} A_{GE} & B_{GE} \\ C_{GE} & D_{GE} \end{bmatrix} = \begin{bmatrix} 1 & 0 \\ g_{b'e} & 0 \end{bmatrix} \begin{bmatrix} 1 & 0 \\ sc_{b'e} & 0 \end{bmatrix} \begin{bmatrix} \dfrac{sc_{b'c}}{sc_{b'c} - g_m} & \dfrac{1}{sc_{b'c} - g_m} \\ \dfrac{g_m sc_{b'c}}{sc_{b'c} - g_m} & \dfrac{sc_{b'c}}{sc_{b'c} - g_m} \end{bmatrix} \quad (G\text{-}10)$$

$$\begin{bmatrix} A_{GE} & B_{GE} \\ C_{GE} & D_{GE} \end{bmatrix} = \begin{bmatrix} 1 & 0 \\ (g_{b'e} + sc_{b'e}) & 0 \end{bmatrix} \begin{bmatrix} \dfrac{sc_{b'c}}{sc_{b'c} - g_m} & \dfrac{1}{sc_{b'c} - g_m} \\ \dfrac{g_m sc_{b'c}}{sc_{b'c} - g_m} & \dfrac{sc_{b'c}}{sc_{b'c} - g_m} \end{bmatrix} \quad (G\text{-}11)$$

$$\begin{bmatrix} A_{GE} & B_{GE} \\ C_{GE} & D_{GE} \end{bmatrix} = \begin{bmatrix} \dfrac{sc_{b'c}}{sc_{b'c} - g_m} & \dfrac{1}{sc_{b'c} - g_m} \\ \dfrac{sc_{b'c}(g_m + g_{b'e} + sc_{b'e})}{sc_{b'c} - g_m} & \dfrac{(g_{b'e} + sc_{b'e} + sc_{b'c})}{sc_{b'c} - g_m} \end{bmatrix} \quad (G\text{-}12)$$

$$A_{GE} = \frac{sc_{b'c}}{sc_{b'c} - g_m} = \frac{1}{1 - \dfrac{g_m}{sc_{b'c}}} \quad (G\text{-}13)$$

$$B_{GE} = \frac{1}{sc_{b'c} - g_m} = \frac{-1}{g_m - jwc_{b'c}} = -r_e = -\frac{1}{g_m}; \quad (wr_e c_{b'c} \ll 1) \quad (G\text{-}14)$$

$$C_{GE} = \frac{sc_{b'c}(g_m + g_{b'e} + sc_{b'e})}{sc_{b'c} - g_m} \quad (G\text{-}15)$$

$$D_{GE} = \frac{g_{b'e} + sc_{b'e} + sc_{b'c}}{sc_{b'c} - g_m} = \frac{\left(\dfrac{1 + jwr_{b'e}c_{b'e}}{r_{b'e}} + jwc_{b'c}\right)}{g_m - jwc_{b'c}} \quad (G\text{-}16)$$

$$D_{GE} = -\left[\frac{1}{\beta} + j\frac{f}{f_T}\right] \quad \text{if } (wr_e c_{b'c} \ll 1) \quad (G\text{-}17)$$

where

$$\beta = \beta(f) = g_e(f) r_{b'e} \quad (G\text{-}18)$$

$$f_T = \left[\frac{g_e}{2\pi(C_{b'c} + C_{b'e})}\right] \quad (G\text{-}19)$$

r_o is normally very large and can be neglected for ease of analysis.

G.1.3 Noise Factor

Figures G-3a and G-3b show the two-port [ABCD] and the GE bipolar transistor presentation for calculation of noise.

G.1 BIPOLAR TRANSISTOR NOISE MODELS

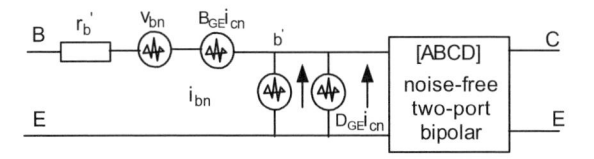

Figure G-3a Two-port [ABCD] with noise sources transferred to the input.

The resulting noise voltage $V_{n(network)}$, combining all the noise contributions, is expressed in terms of the chain parameters:

$$V_{n(network)} = V_{bn} + I_{cn}B_{ce} + (I_{bn} + D_{ce}I_{cn})(Z_s + r'_b) \tag{G-20}$$

$$V_{n(network)} = V_{bn} + I_{bn}(R_s + r'_b) + I_{cn}[B_{ce} + D_{ce}(R_s + r'_b)]$$
$$+ j(I_{bn}X_s + I_{cn}D_{ce}) \tag{G-21}$$

$$V_{n(network)} = V_{bn} + I_{bn}(R_s + r'_b) + I_{cn}(-r_e) + \left[\frac{1}{\beta} + j\frac{f}{f_T}\right](R_s + r'_b)$$
$$+ j(I_{bn}X_s) + j\left[\frac{1}{\beta} + j\frac{f}{f_T}\right]I_{cn} \tag{G-22}$$

$$V_{n(total)} = V_{sn} + V_{n(network)} \tag{G-23}$$

where

$V_{n(total)}$ = total noise voltage
V_{sn} = noise due to source
$V_{n(network)}$ = noise due to network

The noise factor F is the ratio of the total mean square noise current and the thermal noise generated from the source resistance.

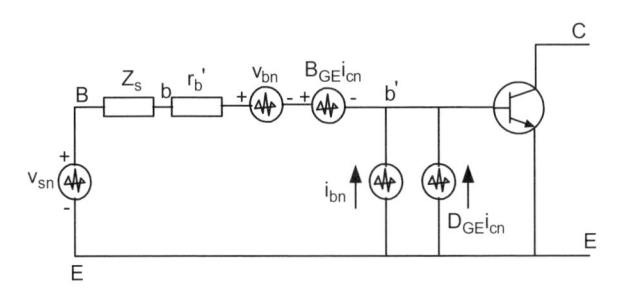

$$F = \frac{\overline{v^2_{n(total)}}}{\overline{v^2_{sn}}} = \frac{\overline{V^2_{sn}} + \overline{V^2_{network}}}{\overline{V^2_{sn}}} \tag{G-24}$$

Figure G-3b GE bipolar transistor with noise sources transferred to the input.

476 CALCULATION OF NOISE PROPERTIES

$$F = \frac{\overline{V_{sn}^2}}{\overline{V_{sn}^2}} + \frac{\overline{V_{network}^2}}{\overline{V_{sn}^2}} = 1 + \frac{\overline{V_{network}^2}}{\overline{V_{sn}^2}} \quad \text{(G-25)}$$

After substituting the value of V_{sn} and $V_{n(network)}$, the noise factor F can be expressed as

$$F = 1 + \frac{\overline{V_{network}^2}}{\overline{V_{ns}^2}} \quad \text{(G-26)}$$

$$= \frac{\overline{\left(V_{bn} + I_{bn}(R_s + r_b') - r_e I_{cn} - (\frac{1}{\beta} + j\frac{f}{f_T})(R_s + r_b')I_{cn} + jI_{bn}X_s + jI_{cn}(\frac{1}{\beta} + j\frac{f}{f_T})\right)^2}}{4kT\Delta f R_s} \quad \text{(G-27)}$$

$$F = 1 + \begin{bmatrix} \dfrac{[\overline{V_{bn}^2} + \overline{I_{bn}^2}(R_s + r_b')^2 + \overline{I_{cn}^2}r_e^2 + \overline{I_{cn}^2}(R_s + r_b')^2(\frac{f^2}{f_T^2}) + \overline{I_{cn}^2}(R_s + r_b')^2(\frac{1}{\beta^2})}{4kT\Delta f R_s} \\ \dfrac{+ \overline{I_{bn}^2}(R_s + r_b')(R_s + r_b' + 2r_e)]}{} \end{bmatrix}$$

$$+ \left[\frac{[\overline{I_{bn}^2}X_s^2 + \overline{I_{cn}^2}(\frac{1}{\beta^2}) - \overline{I_{cn}^2}(\frac{f^2}{f_T^2})]}{4kT\Delta f R_s}\right] \quad \text{(G-28)}$$

G.1.4 Real Source Impedance

In real source impedance, $X_s = 0$ and the noise factor F can be expressed as

$$F = 1 + \frac{r_b'}{R_s} + \frac{r_e}{2R_s} + \frac{(r_b' + R_s)(r_b' + R_s + 2r_e)}{2r_e \beta R_s} + \frac{(r_b' + R_s)^2}{2r_e R_s \beta^2} + \frac{(r_b' + R_s)^2}{2r_e R_s}\left(\frac{f}{f_T}\right)^2 \quad \text{(G-29)}$$

If $wr_e C_{b'c} \ll 1$ and $\beta \gg 1$, then F can be further simplified as

$$F = 1 + \frac{1}{R_s}\left[r_b' + \frac{(r_b' + R_s)^2}{2r_e \beta} + \frac{r_e}{2} + \frac{(r_b' + R_s)^2}{2r_e}\left(\frac{f^2}{f_T^2}\right)\right] \quad \text{(G-30)}$$

$$F = 1 + \frac{1}{R_s}\left[\langle r_b'\rangle + \left\langle\frac{(r_b' + R_s)^2}{2r_e \beta}\right\rangle + \left\langle\frac{r_e}{2} + \frac{(r_b' + R_s)^2}{2r_e}\left(\frac{f^2}{f_T^2}\right)\right\rangle\right] \quad \text{(G-31)}$$

where the contribution of the first term is due to the base resistance, that of the second term is due to the base current, and that of the last term is due to the collector current.

G.1.5 Formation of the Noise Correlation Matrix of the GE Bipolar Transistor

Figures G-4a and G-4b show the steps for the calculation of the noise correlation matrix.

Figure G-4c shows the noise transformation from the output to the input for the calculation of noise parameters.

$$\begin{bmatrix} A_{GE} & B_{GE} \\ C_{GE} & D_{GE} \end{bmatrix} = \begin{bmatrix} 1 & 0 \\ g_{b'e} & 0 \end{bmatrix} \begin{bmatrix} 1 & 0 \\ sC_{b'e} & 0 \end{bmatrix} \begin{bmatrix} \dfrac{sC_{b'c}}{sC_{b'c} - g_m} & \dfrac{1}{sC_{b'c} - g_m} \\ \dfrac{g_m sC_{b'c}}{sC_{b'c} - g_m} & \dfrac{sC_{b'c}}{sC_{b'c} - g_m} \end{bmatrix} \quad \text{(G-32)}$$

$$\begin{bmatrix} A_{GE} & B_{GE} \\ C_{GE} & D_{GE} \end{bmatrix} = \begin{bmatrix} 1 & 0 \\ (g_{b'e} + sC_{b'e}) & 0 \end{bmatrix} \begin{bmatrix} \dfrac{sC_{b'c}}{sC_{b'c} - g_m} & \dfrac{1}{sC_{b'c} - g_m} \\ \dfrac{g_m sC_{b'c}}{sC_{b'c} - g_m} & \dfrac{sC_{b'c}}{sC_{b'c} - g_m} \end{bmatrix} \quad \text{(G-33)}$$

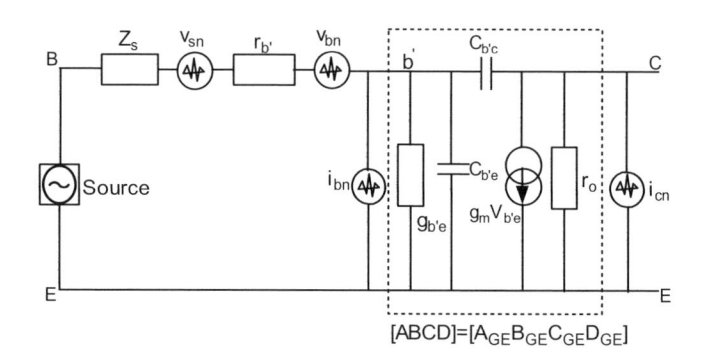

Figure G-4a π-Configuration of the bipolar transistor with noise sources.

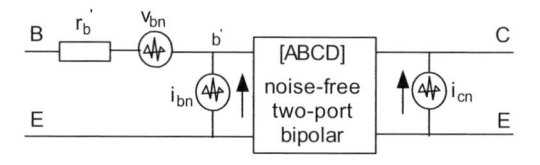

Figure G-4b Two-port [ABCD] with noise sources transferred to the input.

CALCULATION OF NOISE PROPERTIES

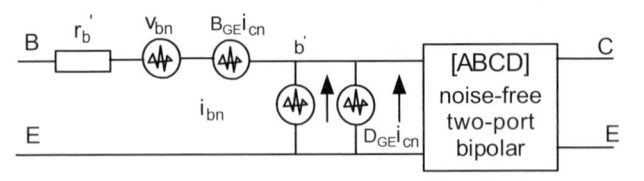

Figure G-4c Noise transformation from output to input.

$$\begin{bmatrix} A_{GE} & B_{GE} \\ C_{GE} & D_{GE} \end{bmatrix} = \begin{bmatrix} \dfrac{sC_{b'c}}{sC_{b'c} - g_m} & \dfrac{1}{sC_{b'c} - g_m} \\ \dfrac{sC_{b'c}(g_m + g_{b'e} + sC_{b'e})}{sC_{b'c} - g_m} & \dfrac{(g_{b'e} + sC_{b'e} + sC_{b'c})}{sC_{b'c} - g_m} \end{bmatrix} \quad (G\text{-}34)$$

$$A_{GE} = \frac{sC_{b'c}}{sC_{b'c} - g_m} = \frac{1}{1 - \dfrac{g_m}{sC_{b'c}}} \quad (G\text{-}35)$$

$$B_{GE} = \frac{1}{sC_{b'c} - g_m} = \frac{-1}{g_m - j\omega C_{b'c}} = -r_e = -\frac{1}{g_m}; \quad (\omega_e C_{b'c} \ll 1) \quad (G\text{-}36)$$

$$C_{GE} = \frac{sC_{b'c}(g_m + g_{b'e} + sC_{b'e})}{sC_{b'c} - g_m} \quad (G\text{-}37)$$

$$D_{GE} = \frac{(g_{b'e} + sC_{b'e} + sC_{b'c})}{sC_{b'c} - g_m} = \begin{bmatrix} \dfrac{\left(\dfrac{1 + j\omega r_{b'e} C_{b'e}}{r_{b'e}} + j\omega C_{b'c}\right)}{g_m - j\omega C_{b'c}} \end{bmatrix}$$

$$= -\left[\frac{1}{\beta} + j\frac{f}{f_T}\right]; \quad (\omega_e C_{b'c} \ll 1) \quad (G\text{-}38)$$

$$\beta = \beta(f) = g_e(f)r_{b'e}; \quad f_T = \left[\frac{g_e}{2\pi(C_{b'c} + C_{b'e})}\right];$$

$$r_o \gg (\text{neglected}) \quad (G\text{-}39)$$

The noise transformation to the input using the chain matrix is shown in Figure G-5. [ABCD] parameters of the noiseless transistor are given as A_{GE}, B_{GE},

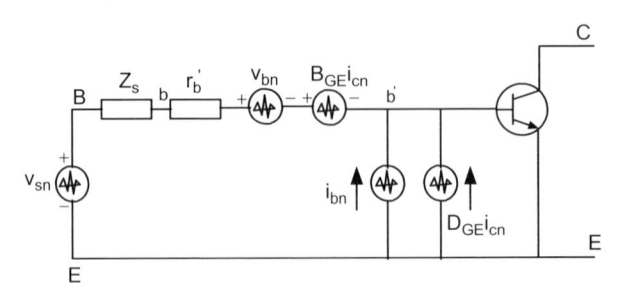

Figure G-5 GE bipolar transistor with noise sources transferred to the input.

C_{GE}, and D_{GE}. A_{GE} and C_{GE} can be ignored because only i_{cn} is being transformed from the output port to the input port of the noiseless transistor two-port.

G.1.6 Calculation of the Noise Parameter Ignoring the Base Resistance

Figures G-6a and G-6b show two uncorrelated noise sources located at the input (i_{bn}) and output (i_{cn}) terminals of the bipolar transistor. This equivalent circuit is analogous to y-representation, thereby converting the [ABCD]-parameter to the [Y]-parameter for the formation of the noise correlation matrix.

$$[Y]_{tr} = \begin{bmatrix} y_{11} & y_{12} \\ y_{21} & y_{22} \end{bmatrix} \tag{G-40}$$

$$y_{11} = \frac{D}{B} = g_{b'e} + s(C_{b'e} + C_{b'c}) \tag{G-41}$$

$$y_{12} = C - \frac{AD}{B} = -sC_{b'c} \tag{G-42}$$

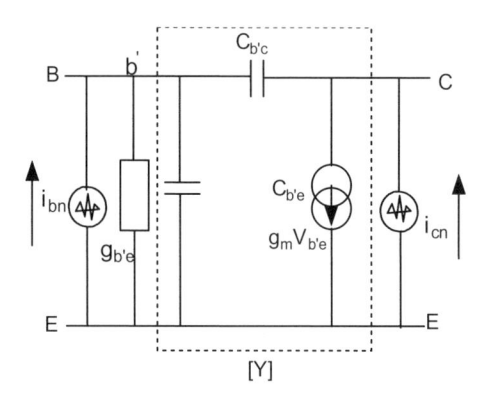

Figure G-6a [Y]-representation of the intrinsic bipolar transistor.

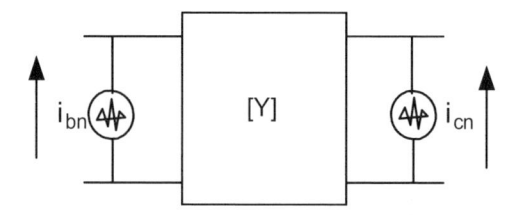

Figure G-6b Two-port [Y]-representation of the intrinsic bipolar transistor.

CALCULATION OF NOISE PROPERTIES

$$y_{21} = -\frac{1}{B} = g_m - sC_{b'c} \tag{G-43}$$

$$y_{22} = \frac{A}{B} = sC_{b'c} \tag{G-44}$$

$$[Y]_{tr} = \begin{bmatrix} y_{11} & y_{12} \\ y_{21} & y_{22} \end{bmatrix} \tag{G-45}$$

$$[Y]_{tr} = \begin{bmatrix} \{g_{b'e} + s(C_{b'e} + C_{b'c})\} & -sC_{b'c} \\ g_m - sC_{b'c} & sC_{b'c} \end{bmatrix} \tag{G-46}$$

$$[C_Y]_{tr} = [N]_{noise\ matrix} = \begin{bmatrix} \overline{i_{bn}i_{bn}^\bullet} & \overline{i_{cn}i_{bn}^\bullet} \\ \overline{i_{bn}i_{cn}^\bullet} & \overline{i_{cn}i_{cn}^\bullet} \end{bmatrix} \tag{G-47}$$

$$\overline{i_{bn}i_{cn}^\bullet} = 0 \tag{G-48}$$

$$\overline{i_{cn}i_{bn}^\bullet} = 0 \tag{G-49}$$

$$\overline{i_{bn}i_{bn}^\bullet} = kTg_m\Delta f \tag{G-50}$$

$$\overline{i_{cn}i_{cn}^\bullet} = \frac{kTg_m\Delta f}{\beta} \tag{G-51}$$

$$S(i_{cn}) = qI_c = kTg_m \tag{G-52}$$

$$S(i_{bn}) = qI_b = \frac{kTg_m}{\beta} \tag{G-53}$$

$$[C_Y]_{tr} = [N]_{noise\ matrix} = \begin{bmatrix} \overline{i_{bn}i_{bn}^\bullet} & \overline{i_{cn}i_{bn}^\bullet} \\ \overline{i_{bn}i_{cn}^\bullet} & \overline{i_{cn}i_{cn}^\bullet} \end{bmatrix} \tag{G-54}$$

$$[C_Y]_{tr} = kT \begin{bmatrix} \frac{g_m}{\beta} & 0 \\ 0 & g_m \end{bmatrix} \tag{G-55}$$

$$[C_a]_{tr} = [A][C_Y]_{tr}[A]^+ \tag{G-56}$$

$$[C_a]_{tr} = \begin{bmatrix} 0 & B_{CE} \\ 1 & D_{CE} \end{bmatrix}[C_Y]_{tr}\begin{bmatrix} 0 & 1 \\ B_{CE}^\bullet & D_{CE}^\bullet \end{bmatrix} \tag{G-57}$$

$$B_{ce} = \frac{-1}{g_m - jwc_{b'c}} = -r_e = -\frac{1}{g_m}; \quad (wr_eC_{b'c} \ll 1) \tag{G-58}$$

$$D_{ce} \rightarrow -\left[\frac{1}{\beta} + j\frac{f}{f_T}\right]; \quad (wr_{b'e}C_{b'e} \ll 1) \tag{G-59}$$

$$[C_Y]_{tr} = kT \begin{bmatrix} \frac{g_m}{\beta} & 0 \\ 0 & g_m \end{bmatrix} \tag{G-60}$$

G.1 BIPOLAR TRANSISTOR NOISE MODELS

$$[C_a]_{tr} = kT \begin{bmatrix} \dfrac{1}{g_m} & \left(\dfrac{1}{\beta} + j\dfrac{f}{f_T}\right) \\ \left(\dfrac{1}{\beta} - j\dfrac{f}{f_T}\right) & \left\{g_m\left(\dfrac{1}{\beta} + \dfrac{1}{\beta^2} + \dfrac{f^2}{f_T^2}\right)\right\} \end{bmatrix} \tag{G-61}$$

$$g_m = Y_{21} + sC_{b'c} \tag{G-62}$$

$$\beta \cong \dfrac{Y_{21}}{Y_{11}} \tag{G-63}$$

$$[C_a]_{tr} = kT \begin{bmatrix} \dfrac{1}{Y_{21} + sC_{b'c}} & \left(\dfrac{Y_{11}}{Y_{21}} + j\dfrac{f}{f_T}\right) \\ \left(\dfrac{Y_{11}}{Y_{21}} - j\dfrac{f}{f_T}\right) & Y_{21}\left\{\dfrac{Y_{11}}{Y_{21}} + \dfrac{Y_{11}^2}{Y_{21}^2} + \dfrac{f^2}{f_T^2}\right\} \end{bmatrix} \tag{G-64}$$

$$[C_a]_{tr} = \begin{bmatrix} 0 & -\dfrac{1}{g_m} \\ 1 & -\left(\dfrac{1}{\beta} + j\dfrac{f}{f_T}\right) \end{bmatrix} kT \begin{bmatrix} \dfrac{g_m}{\beta} & 0 \\ 0 & g_m \end{bmatrix} \begin{bmatrix} 0 & 1 \\ -\dfrac{1}{g_m} & -\left(\dfrac{1}{\beta} - j\dfrac{f}{f_T}\right) \end{bmatrix} \tag{G-65}$$

From $[C_a]_{tr}$, the noise parameters are given as

$$[C_a]_{tr} = \begin{bmatrix} C_{uu\bullet} & C_{ui\bullet} \\ C_{u\bullet i} & C_{ii\bullet} \end{bmatrix} = 2kT \begin{bmatrix} R_n & \dfrac{F_{\min} - 1}{2} - R_n Y_{opt}^{\bullet} \\ \dfrac{F_{\min} - 1}{2} - R_n Y_{opt} & R_n |Y_{opt}|^2 \end{bmatrix} \tag{G-66}$$

$$R_n = \dfrac{C_{uu\bullet}}{2kT} = \dfrac{1}{2g_m} \tag{G-67}$$

$$Y_{opt} = \sqrt{\dfrac{C_{ii\bullet}}{C_{uu\bullet}} - \left[\text{Im}\left(\dfrac{C_{ui\bullet}}{C_{uu\bullet}}\right)\right]^2} + j\,\text{Im}\left(\dfrac{C_{ui\bullet}}{C_{uu\bullet}}\right) \tag{G-68}$$

$$Y_{opt} = g_m\left(\dfrac{\sqrt{\beta+1}}{\beta} + j\dfrac{f}{f_T}\right) \Longrightarrow g_m\left(\sqrt{\dfrac{1}{\beta}} + j\dfrac{f}{f_T}\right); \quad \beta \gg 1 \tag{G-69}$$

$$F_{\min} = 1 + \dfrac{C_{ui\bullet} + C_{uu\bullet} Y_{opt}}{kT} \tag{G-70}$$

$$F_{\min} = 1 + \dfrac{1}{\beta} + \dfrac{\sqrt{\beta+1}}{\beta} \Longrightarrow 1 + \dfrac{1}{\sqrt{\beta}}; \quad \beta \gg 1 \tag{G-71}$$

CALCULATION OF NOISE PROPERTIES

The noise factor F is given as

$$F = 1 + \frac{1}{\beta} + \frac{G_s}{2g_m} + \frac{g_m}{2G_s\beta} + \frac{g_m}{2G_s\beta^2} + \frac{g_m w^2}{2G_s w_T^2} \tag{G-72}$$

$$F = F_{\min} + \frac{R_n}{G_g}[(G_{opt} - G_g)^2 + (B_{opt} - B_G)^2] \tag{G-73}$$

$$R_b \rightarrow 0 \tag{G-74}$$

where

Y_g (generator admittance) $= G_g + jB_G$
Y_{opt} (optimum noise admittance) $= G_{opt} + jB_{opt}$
$F_{\min}$ (minimum achievable noise figure) $\Rightarrow F = F_{\min}$ when $Y_{opt} = Y_g$
R_n (noise resistance) $=$ sensitivity of NF to the source admittance

For $r'_b > 0$, the resulting [ABCD] matrix is given as

$$\begin{bmatrix} A_{GE} & B_{GE} \\ C_{GE} & D_{GE} \end{bmatrix} = \begin{bmatrix} 1 & r'_b \\ 0 & 1 \end{bmatrix} \begin{bmatrix} 1 & 0 \\ g_{b'e} & 0 \end{bmatrix} \begin{bmatrix} 1 & 0 \\ sc_{b'e} & 0 \end{bmatrix} \begin{bmatrix} \dfrac{sc_{b'c}}{sc_{b'c} - g_m} & \dfrac{1}{sc_{b'c} - g_m} \\ \dfrac{g_m sc_{b'c}}{sc_{b'c} - g_m} & \dfrac{sc_{b'c}}{sc_{b'c} - g_m} \end{bmatrix} \tag{G-75}$$

$$\begin{bmatrix} A_{GE} & B_{GE} \\ C_{GE} & D_{GE} \end{bmatrix} = \begin{bmatrix} 1 & r'_b \\ 0 & 1 \end{bmatrix} \begin{bmatrix} 1 & 0 \\ (g_{b'e} + sc_{b'e}) & 0 \end{bmatrix} \begin{bmatrix} \dfrac{sc_{b'c}}{sc_{b'c} - g_m} & \dfrac{1}{sc_{b'c} - g_m} \\ \dfrac{g_m sc_{b'c}}{sc_{b'c} - g_m} & \dfrac{sc_{b'c}}{sc_{b'c} - g_m} \end{bmatrix} \tag{G-76}$$

$$\begin{bmatrix} A_{GE} & B_{GE} \\ C_{GE} & D_{GE} \end{bmatrix} = \begin{bmatrix} 1 & r'_b \\ 0 & 1 \end{bmatrix} \begin{bmatrix} \dfrac{sc_{b'c}}{sc_{b'c} - g_m} & \dfrac{1}{sc_{b'c} - g_m} \\ \dfrac{sc_{b'c}(g_m + g_{b'e} + sc_{b'e})}{sc_{b'c} - g_m} & \dfrac{(g_{b'e} + sc_{b'e} + sc_{b'c})}{sc_{b'c} - g_m} \end{bmatrix}$$

$$\tag{G-77}$$

$$\begin{bmatrix} A_{GE} & B_{GE} \\ C_{GE} & D_{GE} \end{bmatrix} =$$

$$\begin{bmatrix} \left\{\dfrac{sc_{b'c}}{(sc_{b'c} - g_m)} + \dfrac{r'_b sc_{b'c}(g_m + g_{b'e} + sc_{b'e})}{(sc_{b'c} - g_m)}\right\} & \left\{\dfrac{1}{(sc_{b'c} - g_m)} + \dfrac{r'_b(g_{b'e} + sc_{b'e} + sc_{b'c})}{(sc_{b'c} - g_m)}\right\} \\ \left\{\dfrac{sc_{b'c}(g_m + g_{b'e} + sc_{b'e})}{(sc_{b'c} - g_m)}\right\} & \left\{\dfrac{(g_{b'e} + sc_{b'e} + sc_{b'c})}{(sc_{b'c} - g_m)}\right\} \end{bmatrix}$$

$$\tag{G-78}$$

G.1 BIPOLAR TRANSISTOR NOISE MODELS

$$[C_Y]_{tr} = kT \begin{bmatrix} \frac{g_m}{\beta} & 0 \\ 0 & g_m \end{bmatrix} \tag{G-79}$$

$$[C_a]_{tr} = \begin{bmatrix} 0 & \left(\frac{1}{(sc_{b'c} - g_m)} + \frac{r'_b(g_{b'e} + sc_{b'e} + sc_{b'c})}{(sc_{b'c} - g_m)}\right) \\ 1 & \left(\frac{(g_{b'e} + sc_{b'e} + sc_{b'c})}{(sc_{b'c} - g_m)}\right) \end{bmatrix} kT \begin{bmatrix} \frac{g_m}{\beta} & 0 \\ 0 & g_m \end{bmatrix} * \tag{G-80}$$

$$\begin{bmatrix} 0 & 1 \\ \left(\frac{1}{(sc_{b'c} - g_m)} + \frac{r'_b(g_{b'e} + sc_{b'e} + sc_{b'c})}{(sc_{b'c} - g_m)}\right)^{\bullet} & \left(\frac{(g_{b'e} + sc_{b'e} + sc_{b'c})}{(sc_{b'c} - g_m)}\right)^{\bullet} \end{bmatrix} \tag{G-81}$$

$$[C_a]_{tr} = \begin{bmatrix} C_{uu^\bullet} & C_{ui^\bullet} \\ C_{u^\bullet i} & C_{ii^\bullet} \end{bmatrix} = 2kT \begin{bmatrix} R_n & \frac{F_{min}-1}{2} - R_n Y^\bullet_{opt} \\ \frac{F_{min}-1}{2} - R_n Y_{opt} & R_n |Y_{opt}|^2 \end{bmatrix} \tag{G-82}$$

$$C_{uu^\bullet} = kT\left[2r'_b\left(1 + \frac{1}{\beta}\right) + \frac{1}{g_m} + g_m(r'_b)^2\left(\frac{1}{\beta^2} + \frac{f^2}{f_T^2}\right)\right] \tag{G-83}$$

$$C_{ui^\bullet} = kT\left[\frac{1}{\beta} + g_m r'_b\left(\frac{1}{\beta^2} + \frac{f^2}{f_T^2}\right) - j\frac{f}{f_T}\right] \tag{G-84}$$

$$C_{u^\bullet i} = kT\left[\left\{1 + \frac{g_m r'_b}{\beta} + \frac{1}{g_m \beta} + r'_b\left(\frac{1}{\beta^2} + \frac{f^2}{f_T^2}\right)\right\} - j\left\{g_m r'_b \frac{f}{f_T} - \frac{f}{g_m f_T}\right\}\right] \tag{G-85}$$

$$C_{ii^\bullet} = kT\left[\frac{g_m}{\beta} + g_m\left(\frac{1}{\beta^2} + \frac{f^2}{f_T^2}\right)\right] \tag{G-86}$$

$$[C_a]_{tr} = [A][C_Y]_{tr}[A]^+ \tag{G-87}$$

$$[C_a]_{tr} = \begin{bmatrix} 0 & B_{CE} \\ 1 & D_{CE} \end{bmatrix} [C_Y]_{tr} \begin{bmatrix} 0 & 1 \\ B^\bullet_{CE} & D^\bullet_{CE} \end{bmatrix} \tag{G-88}$$

$$R_n = \frac{C_{uu^\bullet}}{2kT} \tag{G-89}$$

$$R_n = r'_b\left(1 + \frac{1}{\beta}\right) + \frac{1}{2g_m} + \frac{g_m(r'_b)^2}{2}\left(\frac{1}{\beta^2} + \frac{f^2}{f_T^2}\right)$$

$$= R_b\left(1 + \frac{1}{\beta}\right) + \frac{kT}{2qI_C} + \frac{qI_C(r'_b)^2}{2kT}\left(\frac{1}{\beta^2} + \frac{f^2}{f_T^2}\right) \tag{G-90}$$

$$Y_{opt} = \sqrt{\frac{C_{ii^\bullet}}{C_{uu^\bullet}} - \left[\text{Im}\left(\frac{C_{ui^\bullet}}{C_{uu^\bullet}}\right)\right]^2} + j\,\text{Im}\left(\frac{C_{ui^\bullet}}{C_{uu^\bullet}}\right) \tag{G-91}$$

$$Y_{opt} = G_{opt} + jB_{opt} \tag{G-92}$$

$$Y_{opt} = \frac{\left[\frac{1}{\beta^2} + 2g_m r'_b\left(1+\frac{1}{\beta}\right)\left\{\frac{1}{\beta^2}+\frac{f^2}{f_T^2}\right\} + (g_m r'_b)^2\left\{\frac{1}{\beta^2}+\frac{f^2}{f_T^2}\right\}^2\right]^{1/2} + j\frac{f}{f_T}}{\left[\frac{1}{g_m} + 2r'_b(1+\frac{1}{\beta}) + g_m(r'_b)^2\left\{\frac{1}{\beta^2}+\frac{f^2}{f_T^2}\right\}\right]} \quad \text{(G-93)}$$

$$G_{opt} = \frac{\left[\frac{1}{\beta^2} + 2g_m r'_b(1+\frac{1}{\beta})\left\{\frac{1}{\beta^2}+\frac{f^2}{f_T^2}\right\} + (g_m r'_b)^2\left\{\frac{1}{\beta^2}+\frac{f^2}{f_T^2}\right\}^2\right]^{\frac{1}{2}}}{\left[\frac{1}{g_m} + 2r'_b(1+\frac{1}{\beta}) + g_m(r'_b)^2\left\{\frac{1}{\beta^2}+\frac{f^2}{f_T^2}\right\}\right]} \quad \text{(G-94)}$$

$$B_{opt} = \frac{f/f_T}{\left[\frac{1}{g_m} + 2r'_b\left(1+\frac{1}{\beta}\right) + g_m(r'_b)^2\left\{\frac{1}{\beta^2}+\frac{f^2}{f_T^2}\right\}\right]} \quad \text{(G-95)}$$

$$Z_{opt} = \frac{1}{Y_{opt}} = \frac{\left[\frac{1}{g_m} + 2r'_b\left(1+\frac{1}{\beta}\right) + g_m(r'_b)^2\left\{\frac{1}{\beta^2}+\frac{f^2}{f_T^2}\right\}\right]}{\left[\frac{1}{\beta^2} + 2g_m r'_b\left(1+\frac{1}{\beta}\right)\left\{\frac{1}{\beta^2}+\frac{f^2}{f_T^2}\right\} + (g_m r'_b)^2\left\{\frac{1}{\beta^2}+\frac{f^2}{f_T^2}\right\}^2\right]^{1/2} + j\frac{f}{f_T}} \quad \text{(G-96)}$$

$$Z_{opt} = R_{opt} + jX_{opt} \quad \text{(G-97)}$$

$$Z_{opt} = \frac{\left[\frac{1}{g_m} + 2r'_b(1+\frac{1}{\beta}) + g_m(r'_b)^2\left(\frac{1}{\beta^2}+\frac{f^2}{f_T^2}\right)\right]\left\{\left[\frac{1}{\beta^2} + 2g_m r'_b(1+\frac{1}{\beta})\times\left(\frac{1}{\beta^2}+\frac{f^2}{f_T^2}\right) + (g_m r'_b)^2\left(\frac{1}{\beta^2}+\frac{f^2}{f_T^2}\right)^2\right]^{\frac{1}{2}} - j\frac{f}{f_T}\right\}}{\left[\frac{1}{\beta^2} + 2g_m r'_b(1+\frac{1}{\beta})\left(\frac{1}{\beta^2}+\frac{f^2}{f_T^2}\right) + (g_m r'_b)^2\left(\frac{1}{\beta^2}+\frac{f^2}{f_T^2}\right)^2 + \left(\frac{f}{f_T}\right)^2\right]} \quad \text{(G-98)}$$

$$R_{opt} = \frac{\left[\frac{1}{g_m} + 2r'_b\left(1+\frac{1}{\beta}\right) + \times g_m(r'_b)^2\left(\frac{1}{\beta^2}+\frac{f^2}{f_T^2}\right)\right]\left[\frac{1}{\beta^2} + 2g_m r'_b\left(1+\frac{1}{\beta}\right)\times\left(\frac{1}{\beta^2}+\frac{f^2}{f_T^2}\right) + (g_m r'_b)^2\left(\frac{1}{\beta^2}+\frac{f^2}{f_T^2}\right)^2\right]^{\frac{1}{2}}}{\left[\frac{1}{\beta^2} + 2g_m r'_b(1+\frac{1}{\beta})\left(\frac{1}{\beta^2}+\frac{f^2}{f_T^2}\right) + (g_m r'_b)^2\left(\frac{1}{\beta^2}+\frac{f^2}{f_T^2}\right)^2 + \left(\frac{f}{f_T}\right)^2\right]} \quad \text{(G-99)}$$

$$X_{opt} = \frac{\frac{f}{f_T}\left[\frac{1}{g_m} + 2r'_b(1+\frac{1}{\beta}) + g_m(r'_b)^2\left(\frac{1}{\beta^2}+\frac{f^2}{f_T^2}\right)\right]}{\left[\frac{1}{\beta^2} + 2g_m r'_b(1+\frac{1}{\beta})\left(\frac{1}{\beta^2}+\frac{f^2}{f_T^2}\right) + (g_m r'_b)^2\left(\frac{1}{\beta^2}+\frac{f^2}{f_T^2}\right)^2 + \left(\frac{f}{f_T}\right)^2\right]} \quad \text{(G-100)}$$

$$\Gamma_{opt} = \frac{Z_{opt} - Z_o}{Z_{opt} + Z_o}, \quad \Gamma_{opt} = \frac{Y_{opt} - Y_o}{Y_{opt} + Y_o} \quad \text{(G-101)}$$

$$F = F_{\min} + \frac{R_n}{G_g}[(G_{opt} - G_g)^2 + (B_{opt} - B_G)^2] \quad \text{(G-102)}$$

$$F_{\min} = 1 + \frac{C_{ui\bullet} + C_{uu}\bullet Y_{opt}}{kT} \quad \text{(G-103)}$$

$$F_{\min} = (1+\frac{1}{\beta})(1+r'_b G_s) + \frac{G_s}{2g_m} + \frac{g_m G_s}{2}(r'_b + \frac{1}{G_s})^2 \left\{ \frac{1}{\beta^2} + \frac{f^2}{f_T^2} \right\} \quad \text{(G-104)}$$

$$F = 1 + \frac{1}{R_s}\left[R_b + \frac{r_e}{2} + \frac{(R_b+R_s)(R_b+R_s+2r_e)}{2r_e\beta} \right.$$
$$\left. + \frac{(R_b+R_s)^2}{2r_e\beta^2} + \frac{(R_b+R_s)^2}{2r_e}\left(\frac{f^2}{f_T^2}\right) \right] \quad \text{(G-105)}$$

If $\omega r_e C_{b'c} \ll 1$ and $\beta \gg 1$, then the noise factor F can be further simplified to

$$F = 1 + \frac{1}{R_s}\left[R_b + \frac{(R_b+R_s)^2}{2r_e\beta} + \frac{r_e}{2} + \frac{(R_b+R_s)^2}{2r_e}\left(\frac{f^2}{f_T^2}\right) \right] \quad \text{(G-106)}$$

$$F_{\min} = \left(1+\frac{1}{\beta}\right)(1+r'_b G_s) + \frac{G_s}{2g_m} + \frac{g_m G_s}{2}(r'_b + \frac{1}{G_s})^2 \left\{ \frac{1}{\beta^2} + \frac{f^2}{f_T^2} \right\}$$
$$\implies 1 + \frac{1}{\sqrt{\beta}}; \quad \beta \gg 1 \quad \text{(G-107)}$$

where

$$\beta = \beta(f) = g_e(f) r_{b'e} \quad \text{(G-108)}$$

$$\beta = \frac{\alpha}{1-\alpha} = \frac{\alpha_0 \exp(-j\omega\tau)}{1+j\frac{f}{f_\alpha} - \alpha_0 \exp(-j\omega\tau)} = \frac{1}{\frac{1}{\alpha_0}\exp(j\omega\tau) + j\frac{f}{\alpha_0 f_\alpha}\exp(j\omega\tau) - 1} \quad \text{(G-109)}$$

$$\beta = \frac{\beta_o}{1+j\beta_o\left(\frac{f}{f_T}\right)} \quad \text{(G-110)}$$

$$\alpha = \frac{\alpha_0 \exp(-j\omega\tau)}{1+j\frac{f}{f_\alpha}}, \quad \beta_o = \frac{\alpha_o}{1-\alpha_o} \quad \text{(G-111)}$$

$$g_e(f \to low\ freq) = g_{eo}, \quad g_{eo} = \frac{1}{r_{eo}} \quad \text{(G-112)}$$

$$r_{eo} = \left(\frac{\partial I_E}{\partial V_{EB}}\right)^{-1} = \left(\frac{KT}{qI_E}\right)\left(\frac{\alpha_o}{\alpha_{DC}}\right) = \left(\frac{KT}{qI_C}\right)\alpha_o \quad \text{(G-113)}$$

$$f_T = \left[\frac{g_e}{2\pi(C_{b'c}+C_{b'e})}\right] \quad \text{(G-114)}$$

G.1.7. Bipolar Transistor Noise Model in T Configuration

Figure G-7 shows the T-equivalent circuit of the bipolar noise model, where C_{Te} is the emitter junction capacitance and Z_g is the complex source impedance.

486 CALCULATION OF NOISE PROPERTIES

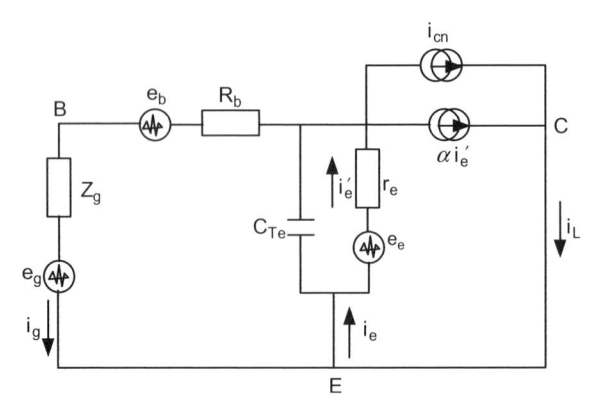

Figure G-7 T-equivalent circuit of the bipolar noise model.

For the calculation of minimum noise, the T-configuration is simpler than the hybrid-π. For the formation of the noise correlation matrix with the base collector capacitance C_{bc}, the hybrid-π topology is an easier approach for analysis. The noise of a silicon bipolar transistor can be modeled by the three noise sources, as shown in Figure G-1. The mean square values of noise sources in a narrow frequency range Δf are given as

$$\overline{e_e e_e^*} = \overline{e_e^2} = 2KTr_e\Delta f \tag{G-115}$$

$$\overline{e_g e_g^*} = \overline{e_g^2} = 4KTR_g\Delta f \tag{G-116}$$

$$\overline{e_b e_b^*} = \overline{e_b^2} = 4KTR_b\Delta f \tag{G-117}$$

$$\overline{i_{cp} e_e^*} = 0 \tag{G-118}$$

$$\alpha = \frac{\alpha_0}{1 + j\frac{f}{f_b}} \tag{G-119}$$

$$\beta = \frac{\alpha}{1-\alpha} \tag{G-120}$$

$$r_e = \frac{KT}{qI_e} \tag{G-121}$$

$$g_e = \frac{1}{r_e} \tag{G-122}$$

where the thermal noise voltage source due to base resistance is e_b, the shot noise voltage source e_e is generated by the forward-biased emitter-base junction r_e, and the collector noise current source i_{cp} comes from the collector partition, which is strongly correlated to the emitter-base shot noise.

G.1 BIPOLAR TRANSISTOR NOISE MODELS

The noise factor is defined as the SNR at the input of the transistor divided by the SNR of the output at the transistor or its ratio of the squared currents.

$$F = \frac{(S/N)_{in}}{(S/N)_{out}} = \frac{N_{out}}{N_{in}G} = \frac{N_{out}}{GkTB} \qquad \text{(G-123)}$$

The noise factor F is given by

$$F = \frac{\overline{i_L^2}}{\overline{i_{LO}^2}} \qquad \text{(G-124)}$$

where i_{LO} is the value of i_L due to the source generator e_g alone. From KVL, for the loop containing Z_g, r_b, and r_e, the loop equation can be expressed as

$$i_g(Z_g + r_b) + i'_e r_e = e_g + e_b + e_e \qquad \text{(G-125)}$$

$$i_L = \alpha i'_e + i_{cp} \qquad \text{(G-126)}$$

$$i'_e = \frac{i_L - i_{cp}}{\alpha} \qquad \text{(G-127)}$$

$$i_e = i'_e(1 + j\omega C_{Te}) - j\omega C_{Te} e_e \qquad \text{(G-128)}$$

$$i_g = i_e - i_L \qquad \text{(G-129)}$$

$$i_g = i'_e(1 + j\omega C_{Te} r_e) - j\omega C_{Te} e_e - i_L \qquad \text{(G-130)}$$

$$i_g = \frac{i_L - i_{cp}}{\alpha}(1 + j\omega C_{Te} r_e) - j\omega C_{Te} e_e - i_L \qquad \text{(G-131)}$$

$$i_g(Z_g + r_b) + i'_e r_e = e_g + e_b + e_e \qquad \text{(G-132)}$$

$$\left\{ \left(\frac{i_L - i_{cp}}{\alpha}(1 + j\omega C_{Te} r_e) - j\omega C_{Te} e_e - i_L \right)(Z_g + r_b) \right\}$$
$$+ \left(\frac{i_L - i_{cp}}{\alpha} \right) r_e = e_g + e_b + e_e \qquad \text{(G-133)}$$

$$\frac{i_L}{\alpha}\{(1 - \alpha + j\omega C_{Te} r_e)(Z_g + r_b) + r_e\} = e_g + e_b + e_e\{1 + j\omega C_{Te}(Z_g + r_b)\}$$
$$+ \frac{i_{cp}}{\alpha}\{(1 + j\omega C_{Te} r_e)(Z_g + r_b) + r_e\} \qquad \text{(G-134)}$$

$$i_L = \alpha \left[\frac{e_g + e_b + e_e\{1 + j\omega C_{Te}(Z_g + r_b)\} + \frac{i_{cp}}{\alpha}\{(1 + j\omega C_{Te} r_e)(Z_g + r_b) + r_e\}}{\{(1 - \alpha + j\omega C_{Te} r_e)(Z_g + r_b) + r_e\}} \right] \qquad \text{(G-135)}$$

i_L is the total load current or the collector current (AC short-circuited current) due to all the generators, such as e_e, e_b, e_g, and i_{cp}.

CALCULATION OF NOISE PROPERTIES

i_{Lo} is the value of i_L due to the source generator e_g alone. Other noise generators (e_e, e_b, e_g, and i_{cp}) are zero.

$$i_{LO} = \alpha \left[\frac{e_g}{\{(1-\alpha+j\omega C_{Te}r_e)(Z_g+r_b)+r_e\}} \right] \qquad \text{(G-136)}$$

$$F = \frac{\overline{i_L^2}}{\overline{i_{LO}^2}} = \frac{\overline{\left[e_g+e_b+e_e\{1+j\omega C_{Te}(Z_g+r_b)\ldots\}+\frac{i_{cp}}{\alpha}\{(1+j\omega C_{Te}r_e)(Z_g+r_b)+r_e\}\right]^2}}{\overline{e_g^2}} \qquad \text{(G-137)}$$

$$F = \frac{\overline{e_g^2}+\overline{e_b^2}+\overline{e_e^2\{1+j\omega C_{Te}(Z_g+r_b)\}^2}+\frac{\overline{i_{cp}^2}}{|\alpha|^2}\{(1+j\omega C_{Te}r_e)(Z_g+r_b)+r_e\}^2}{\overline{e_g^2}} \qquad \text{(G-138)}$$

$$F = \frac{4KTR_g + 4KTr_b + 2KTr_e\overline{\{1+j\omega C_{Te}(Z_g+r_b)\}^2} + \frac{2KT(\alpha_0-|\alpha|^2)}{|\alpha|^2 r_e}\{(1+j\omega C_{Te}r_e)(Z_g+r_b)+r_e\}^2}{4KTR_g} \qquad \text{(G-139)}$$

$$F = 1 + \frac{r_b}{R_g} + \frac{r_e}{2R_g}\overline{\{1+j\omega C_{Te}(Z_g+r_b)\}^2} + \frac{(\alpha_0-|\alpha|^2)}{2R_g|\alpha|^2 r_e}((1+j\omega C_{Te}r_e)(Z_g+r_b)+r_e)^2 \qquad \text{(G-140)}$$

$$F = 1 + \frac{r_b}{R_g} + \frac{r_e}{2R_g}|1+j\omega C_{Te}(R_g+r_b+jX_g)|^2$$
$$+ \left(\frac{\alpha_0}{|\alpha|^2}-1\right)\frac{|(1+jwC_{Te}r_e)(R_g+r_b+jX_g)+r_e|^2}{2R_g r_e} \qquad \text{(G-141)}$$

$$F = 1 + \frac{r_b}{R_g} + \frac{r_e}{2R_g}|1+j\omega C_{Te}(R_g+r_b)-\omega C_{Te}X_g|^2$$
$$+ \left(\frac{\alpha_0}{|\alpha|^2}-1\right)\frac{|R_g+r_b+r_e-\omega C_{Te}X_g r_e+j\omega C_{Te}r_e(R_g+r_b+X_g)|^2}{2R_g r_e} \qquad \text{(G-142)}$$

$$F = 1 + \frac{r_b}{R_g} + \frac{r_e}{2R_g}\left\{(1-\omega C_{Te}X_g)^2+\omega^2 C_{Te}^2(R_g+r_b)^2\right\}$$
$$+ \left(\frac{\alpha_0}{|\alpha|^2}-1\right)\frac{\left((R_g+r_b+r_e(1-\omega C_{Te}X_g))^2+(X_g+\omega C_{Te}e_e(R_g+r_b))^2\right)}{2R_g r_e} \qquad \text{(G-143)}$$

$$F = 1 + \frac{r_b}{R_g} + \frac{r_e}{2R_g} + \left(\frac{\alpha_0}{|\alpha|^2}-1\right)\frac{\left((R_g+r_b+r_e)^2+X_g^2\right)}{2R_g r_e}$$
$$+ \left(\frac{\alpha_0}{|\alpha|^2}\right)\left(\frac{r_e}{2R_g}\right)\left[\omega^2 C_{Te}^2 X_g^2 - 2\omega C_{Te}X_g + \omega^2 C_{Te}^2(R_g+r_b)^2\right] \qquad \text{(G-144)}$$

The noise terms and the generator thermal noise are given as ($\Delta f = 1$ Hz).

$$\overline{e_e^2} = 2KTr_e \tag{G-145}$$

$$\overline{e_g^2} = 4KTR_g \tag{G-146}$$

$$\overline{e_b^2} = 4KTR_b \tag{G-147}$$

$$\overline{i_{cp}^2} = \frac{2KT(\alpha_0 - |\alpha|^2)}{r_e} \tag{G-148}$$

$$r_e = \frac{KT}{qI_e} \tag{G-149}$$

$$\alpha = \frac{\alpha_0}{1 + j\dfrac{f}{f_b}} \tag{G-150}$$

G.1.8 Real Source Impedance

In the case of a real source impedance, rather than a complex, impedance, such as $R_g = 50$ ohms, $X_g = 0$, the equation of the noise factor becomes

$$F = 1 + \frac{r_b}{R_g} + \frac{r_e}{2R_g} + \left(\frac{\alpha_0}{|\alpha|^2} - 1\right)\frac{(R_g + r_b + r_e)^2}{2R_g r_e} + \frac{\alpha_0}{|\alpha|^2}\omega^2 C_{Te}^2 r_e^2 \frac{(R_g + r_b)^2}{2R_g r_e} \tag{G-151}$$

Substituting the value of α where f_b is the cutoff frequency of the base alone and introducing an emitter cutoff frequency $f_e = 1/2\pi C_{Te}r_e$,

$$F = 1 + \frac{r_b}{R_g} + \frac{r_e}{2R_g} + \left(1 - \alpha_0 + \frac{f^2}{f_b^2}\right)\frac{(R_g + r_b + r_e)^2}{2R_g r_e \alpha_0} + \left(1 + \frac{f^2}{f_b^2}\right)\frac{f^2}{f_e^2}\frac{(R_g + r_b)^2}{2R_g r_e \alpha_0} \tag{G-152}$$

Simplifying the equation above by f_e',

$$f_e' = f_e \frac{R_g + r_b + r_e}{R_g + r_b} = \frac{R_g + r_b + r_e}{2\pi C_{Te}r_e(R_g + r_b)} \tag{G-153}$$

The simplified equation of the noise factor for the real source impedance is given by

$$F = 1 + \frac{r_b}{R_g} + \frac{r_e}{2R_g} + \left\{\left(1 + \frac{f^2}{f_b^2}\right)\left(1 + \frac{f^2}{f_e'^2}\right) - \alpha_0\right\}\frac{(R_g + r_b + r_e)^2}{2R_g r_e \alpha_0} \tag{G-154}$$

G.1.9 Minimum Noise Factor

The minimum noise factor $F_{\min}$ and the corresponding optimum source impedance $Z_{opt} = R_{opt} + jX_{opt}$ are found by differentiating the general equation for the noise figure with respect to X_g and then R_g.

The equation of the noise factor can be expressed as

$$F = A + BX_g + CX_g^2 \qquad (G\text{-}155)$$

We determine

$$a = \left\{ 1 - \frac{|\alpha|^2}{\alpha_0} + \omega^2 C_{Te}^2 r_e^2 \right\} \frac{\alpha_0}{|\alpha|^2} \qquad (G\text{-}156)$$

The coefficients A, B, and C can be expressed as

$$A = a\frac{(R_g + r_b)^2}{2R_g r_e} + \frac{\alpha_0}{|\alpha|^2}\left\{ 1 + \frac{r_b}{R_g} + \frac{r_e}{2R_g} \right\} \qquad (G\text{-}157)$$

$$B = -\frac{\alpha_0}{|\alpha|^2}\frac{\omega C_{Te} r_e}{R_g} \qquad (G\text{-}158)$$

$$C = \frac{a}{2r_e R_g} \qquad (G\text{-}159)$$

Differentiating with respect to X_g for the optimum source reactance:

$$\left.\frac{dF}{dX_g}\right|_{X_{opt}} = B + 2CX_{opt} \qquad (G\text{-}160)$$

$$X_{opt} = \frac{-B}{2C} = \frac{\alpha_0}{|\alpha|^2}\frac{\omega C_{Te} r_e}{a} \qquad (G\text{-}161)$$

The corresponding noise factor is

$$F_{X_{opt}} = A - CX_{opt}^2 \qquad (G\text{-}162)$$

$$F_{X_{opt}} = a\frac{(R_g + r_b)^2}{2R_g r_e} + \frac{\alpha_0}{|\alpha|^2}\left\{ 1 + \frac{r_b}{R_g} + \frac{r_e}{2R_g} \right\} - \frac{aX_{opt}^2}{2r_e R_g} \qquad (G\text{-}163)$$

This must be further optimized with respect to the source resistance to derive $F_{\min}$.

G.1 BIPOLAR TRANSISTOR NOISE MODELS

Differentiating $F_{X_{opt}}$ with respect to source resistance, we get

$$F_{X_{opt}} = a\frac{(R_g + r_b)^2}{2R_g r_e} + \frac{\alpha_0}{|\alpha|^2}\left\{1 + \frac{r_b}{R_g} + \frac{r_e}{2R_g}\right\} - \frac{aX_{opt}^2}{2r_e R_g} \tag{G-164}$$

$$F_{X_{opt}} = A_1 + \frac{B_1}{R_g} + C_1 R_g \tag{G-165}$$

where

$$A_1 = a\frac{r_b}{r_e} + \frac{\alpha_0}{|\alpha|^2} \tag{G-166}$$

$$B_1 = a\frac{r_b^2 - X_{opt}^2}{2r_e} + \frac{\alpha_0}{|\alpha|^2}\left(r_b + \frac{r_e}{2}\right) \tag{G-167}$$

$$C_1 = \frac{a}{2r_e} \tag{G-168}$$

Differentiating the noise factor with respect to R_g to obtain the minimum noise figure, we get

$$\left.\frac{dF}{dR_g}\right|_{R_{opt}} = 0 = \frac{-B_1}{R_{opt}^2} + C_1 \tag{G-169}$$

$$R_{opt}^2 = \frac{B_1}{C_1} = r_b^2 - X_{opt}^2 \frac{\alpha_0}{|\alpha|^2}\frac{r_e(2r_b + r_e)}{a} \tag{G-170}$$

$$F_{\min} = A_1 + 2C_1 R_{opt} = a\frac{r_b + R_{opt}}{r_e} + \frac{\alpha_0}{|\alpha|^2} \tag{G-171}$$

Factor a can be simplified in terms of a simple symmetrical function of f_e and f_b:

$$a = \left\{1 - \frac{|\alpha|^2}{\alpha_0} + \omega^2 C_{Te}^2 r_e^2\right\}\frac{\alpha_0}{|\alpha|^2} \tag{G-172}$$

$$a = \left\{1 + \frac{f^2}{f_b^2} - \alpha_0 + \left(1 + \frac{f^2}{f_b^2}\right)\frac{f^2}{f_e^2}\right\}\frac{1}{\alpha_0} \tag{G-173}$$

$$a = \left\{\left(1 + \frac{f^2}{f_b^2}\right)\left(1 + \frac{f^2}{f_e^2}\right) - \alpha_0\right\}\frac{1}{\alpha_0} \tag{G-174}$$

If C_{Te} and X_{opt} are assumed to be zero, then the modified factor a can be given as

$$a = \left\{1 + \frac{f^2}{f_b^2} - \alpha_0\right\} \frac{1}{\alpha_0} \tag{G-175}$$

$$R_{opt}^2 = \frac{B_1}{C_1} = r_b^2 + \frac{1 + \frac{f^2}{f_b^2}}{1 + \frac{f^2}{f_b^2} - \alpha_0} \frac{r_e(2r_b + r_e)}{a} \tag{G-176}$$

$$F_{\min} = A_1 + 2C_1 R_{opt} = a \frac{r_b + R_{opt}}{r_e} + \frac{\alpha_0}{|\alpha|^2} \tag{G-177}$$

$$F_{\min} = \left(1 + \frac{f^2}{f_b^2} - \alpha_0\right) \frac{(r_b + R_{opt})}{\alpha_0 r_e} + \left(1 + \frac{f^2}{f_b^2}\right) \frac{1}{\alpha_0} \tag{G-178}$$

G.1.10 Noise Correlation Matrix of the Bipolar Transistor in the T-Equivalent Configuration

Figure G-8 shows the T-equivalent configuration of the CE circuit for the bipolar transistor. To use the noise correlation matrix approach, we transform the noise model to an equivalent one consisting of two noise sources a voltage source and a current source of a noiseless transistor.

The new bipolar transistor noise model from Figure G-2 now takes the form shown in Figure G-9. Since the system is linear, the two noise sources can be expressed in terms of three original noise sources by a linear transformation.

$$[Y]_{tr} = \begin{bmatrix} y_{11} & y_{12} \\ y_{21} & y_{22} \end{bmatrix} \tag{G-179}$$

$$[Y]_{tr} = \begin{bmatrix} \{(1-\alpha)g_e + j\omega C_e + Y_c\} & -Y_c \\ \alpha g_e - Y_c & Y_c \end{bmatrix} \tag{G-180}$$

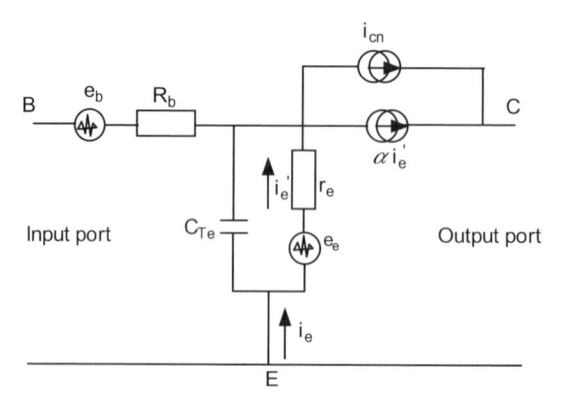

Figure G-8 T-equivalent configuration of a CE bipolar transistor.

G.1 BIPOLAR TRANSISTOR NOISE MODELS

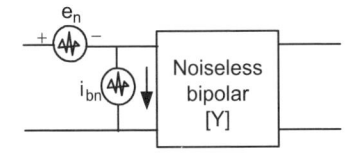

Figure G-9 Transformed bipolar transistor noise model represented as a two-port admittance matrix.

Now for easier representation, the matrix for the intrinsic device is defined as $[N]$ and for the transformed noise circuit as $[C]$.

$$[N]_{intrinsic} = \frac{1}{4KT\Delta f} \begin{bmatrix} \overline{e_e e_e^*} & \overline{e_e i_{cp}^*} \\ \overline{i_{cp} e_e^*} & \overline{i_{cp} i_{cp}^*} \end{bmatrix} = \begin{bmatrix} \frac{1}{2g_e} & 0 \\ 0 & \frac{g_e(\alpha_0 - |\alpha|^2)}{2} \end{bmatrix} \quad \text{(G-181)}$$

$$[C]_{transformed} = \frac{1}{4KT\Delta f} \begin{bmatrix} \overline{e_n e_n^*} & \overline{e_n i_n^*} \\ \overline{i_n e_n^*} & \overline{i_n i_n^*} \end{bmatrix} = \begin{bmatrix} C_{11} & C_{12} \\ C_{21} & C_{22} \end{bmatrix} \quad \text{(G-182)}$$

The noise correlation matrix C in terms of N can be obtained by a straightforward application of the steps outlined as

$$C = AZTN(AZT)^{\oplus} + ARA^{\oplus} \quad \text{(G-183)}$$

The sign $\oplus$ denotes the Hermitian conjugate.

Matrix Z is the inverse of the admittance matrix Y for the intrinsic portion of the model, and T is a transformation matrix which converts the noise sources e_e and i_{cp} to shunt current sources across the base-emitter and collector-emitter ports of the transistor, respectively.

$$T = \begin{bmatrix} -(1-\alpha)g_e & 1 \\ -\alpha g_e & -1 \end{bmatrix} \quad \text{(G-184)}$$

$$A = \begin{bmatrix} 1 & \frac{Z_{11}+r_b}{Z_{21}} \\ 0 & -\frac{1}{Z_{11}} \end{bmatrix} \quad \text{(G-185)}$$

$$R = \frac{1}{4KT\Delta f} \begin{bmatrix} \overline{e_b e_b^*} & 0 \\ 0 & 0 \end{bmatrix} = \begin{bmatrix} r_b & 0 \\ 0 & 0 \end{bmatrix} \quad \text{(G-186)}$$

Here Y_c is added as a fictitious admittance across the α-current generator to overcome the singularity of the actual Z matrix. However, in the final evaluation of C, Y_c is set to zero. Matrix A is a circuit transformation matrix, and matrix R is a noise correlation matrix representing thermal noise of the extrinsic base resistance.

$$C = \begin{bmatrix} C_{uu^\bullet} & C_{ui^\bullet} \\ C_{u^\bullet i} & C_{ii^\bullet} \end{bmatrix} = 2kT \begin{bmatrix} R_n & \dfrac{F_{\min}-1}{2} - R_n Y_{opt}^\bullet \\ \dfrac{F_{\min}-1}{2} - R_n Y_{opt} & R_n |Y_{opt}|^2 \end{bmatrix} \quad \text{(G-187)}$$

$$R_n = \frac{C_{uu^\bullet}}{2kT} \quad \text{(G-188)}$$

$$Y_{opt} = \sqrt{\frac{C_{ii^\bullet}}{C_{uu^\bullet}} - \left[\text{Im}\left(\frac{C_{ui^\bullet}}{C_{uu^\bullet}}\right)\right]^2} + j\,\text{Im}\left(\frac{C_{ui^\bullet}}{C_{uu^\bullet}}\right) \quad \text{(G-189)}$$

$$Y_{opt} = G_{opt} + jB_{opt} \quad \text{(G-190)}$$

The element of the noise correlation matrix C contains all the necessary information about the four extrinsic noise parameters $F_{\min}$, Rg_{opt}, Xg_{opt}, and R_n of the bipolar transistor. The expressions for $F_{\min}$, Rg_{opt}, and Xg_{opt} were derived above. The expression for R_n is

$$R_n = \frac{C_{uu^\bullet}}{2kT} \quad \text{(G-191)}$$

$$= r_b \left(\frac{1 + \left(\dfrac{f}{f_b}\right)^2}{\alpha_0^2} - \frac{1}{\beta_0} \right) + \frac{r_e}{2}\left[\frac{1 + \left(\dfrac{f}{f_b}\right)^2}{\alpha_0^2} + (g_e r_b)^2 \right.$$

$$\left. \times \left\{ 1 - \alpha_0 + \left(\frac{f}{f_b}\right)^2 + \left(\frac{f}{f_e}\right)^2 + \left(\frac{1}{\beta_0} - \left(\frac{f}{f_b}\right)\left(\frac{f}{f_e}\right)\right)^2 \right\} \right] \quad \text{(G-192)}$$

G.2 THE GaAs-FET Noise Model

G.2.1 The Model at Room Temperature

Figures G-10a and G-10b show a noise model of a grounded source FET with the noise sources at the input and output [110–113].

The mean square values of the noise sources in the narrow frequency range Δf are given by

$$\overline{i_d^2} = 4kT g_m P \Delta f \quad \text{(G-193)}$$

$$\overline{i_g^2} = \frac{4kT(\omega C_{gs})^2 R}{g_m} \Delta f \quad \text{(G-194)}$$

G.2 THE GaAs-FET Noise Model

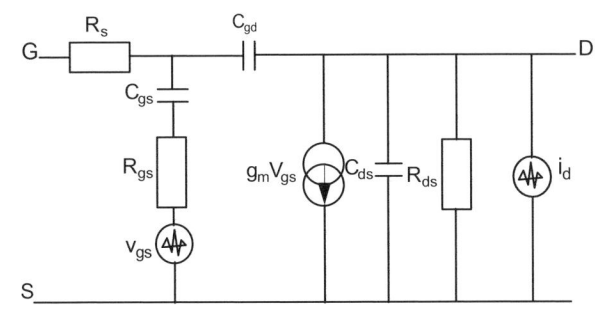

Figure G-10a Noise model of a FET with the voltage noise source at the input and the current noise source at the output.

$$\overline{i_g i_d^*} = -j\omega C_{gs} 4kTC\sqrt{PR}\Delta f \tag{G-195}$$

$$S(i_d) = \frac{\overline{i_d^2}}{\Delta f} = \left\langle \overline{|i_d^2|} \right\rangle = 4kTg_m P \tag{G-196}$$

$$S(i_g) = \frac{\overline{i_g^2}}{\Delta f} = \left\langle \overline{|i_g^2|} \right\rangle = \frac{4kT(\omega C_{gs})^2 R}{g_m} \tag{G-197}$$

$$S(i_g i_d^*) = \left\langle \overline{|i_g i_d^*|} \right\rangle = -j\omega C_{gs} 4kTC\sqrt{PR} \tag{G-198}$$

where P, R, and C are FET noise coefficients and can be given as

$$P = \left[\frac{1}{4kTg_m}\right]\overline{i_d^2}/Hz; \quad P = 0.67 \text{ for JFETs and } 1.2 \text{ for MESFETs}$$

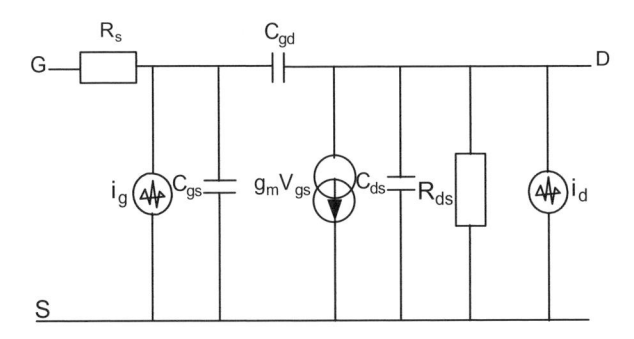

Figure G-10b Noise model of a FET with the current noise source at the input and output.

$$R = \left[\frac{g_m}{4kT\omega^2 C_{gs}^2}\right]\overline{i_g^2}/Hz; \quad R = 0.2 \text{ for JFETs and } 0.4 \text{ for MESFETs}$$

$$C = -j\left[\frac{\overline{i_g i_d^*}}{\sqrt{[\overline{i_d^2}\,\overline{i_g^2}]}}\right]; \quad C = 0.4 \text{ for JFETs and } 0.6-0.9 \text{ for MESFETs}$$

G.2.2 Calculation of the Noise Parameters

Figures G-11a and G-11b show the intrinsic FET with noise sources at the input and output.

$$[Y]_{Intrinsic\ FET} = \begin{bmatrix} y_{11} & y_{12} \\ y_{21} & y_{22} \end{bmatrix} \tag{G-199}$$

$$[C_Y] = [N]_{noise\ matrix} = \begin{bmatrix} \overline{i_g i_g^*} & \overline{i_g i_d^*} \\ \overline{i_d i_g^*} & \overline{i_d i_d^*} \end{bmatrix} \tag{G-200}$$

$$[C_Y]_{FET} = 4kT \begin{bmatrix} \dfrac{\omega^2 c_{gs}^2 R}{g_m} & -j\omega c_{gs} C\sqrt{PR} \\ j\xi c_{gs} C\sqrt{PR} & g_m P \end{bmatrix} \tag{G-201}$$

The noise transformation from output to input can be done to calculate the noise parameters. Figure G-12 shows the equivalent circuit of the FET with the noise

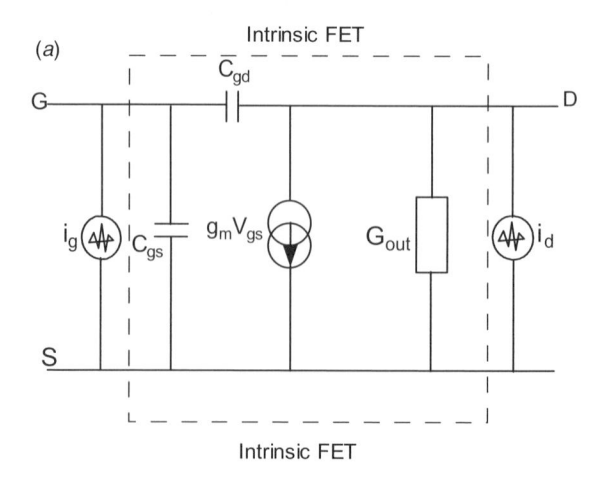

Figure G-11a Intrinsic FET with noise sources at the input and output.

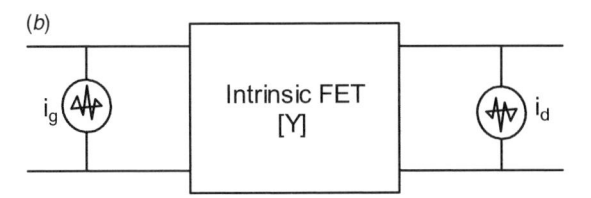

Figure G-11b Intrinsic FET with noise sources at the input and output.

source transferred to the input side.

$$[C_a]_{tr} = \begin{bmatrix} \overline{e_n e_n^*} & \overline{e_n i_n^*} \\ \overline{i_n e_n^*} & \overline{i_n i_n^*} \end{bmatrix} \tag{G-202}$$

$$[C_a]_{tr} = [T][C_Y]_{tr}[T]^+ \tag{G-203}$$

$$[T] = \begin{bmatrix} 0 & B_{CS} \\ 1 & D_{CS} \end{bmatrix} \tag{G-204}$$

$$[C_a]_{tr} = \begin{bmatrix} 0 & B_{CS} \\ 1 & D_{CS} \end{bmatrix} [C_Y]_{tr} \begin{bmatrix} 0 & 1 \\ B_{CS}^* & D_{CS}^* \end{bmatrix} \tag{G-205}$$

$$[ABCD]_{FET} = \begin{bmatrix} A_{CS} & B_{CS} \\ C_{CS} & D_{CS} \end{bmatrix} = \begin{bmatrix} 1 & R_s \\ 0 & 1 \end{bmatrix} \begin{bmatrix} 1 & 0 \\ sc_{gs} & 1 \end{bmatrix} \begin{bmatrix} \dfrac{sc_{gd}}{sc_{gd} - g_m} & \dfrac{1}{sc_{gd} - g_m} \\ \dfrac{g_m sc_{gd}}{sc_{gd} - g_m} & \dfrac{sc_{gd}}{sc_{gd} - g_m} \end{bmatrix}$$

$$\times \begin{bmatrix} 1 & 0 \\ g_{ds} & 1 \end{bmatrix} \begin{bmatrix} 1 & 0 \\ sc_{ds} & 1 \end{bmatrix} \tag{G-206}$$

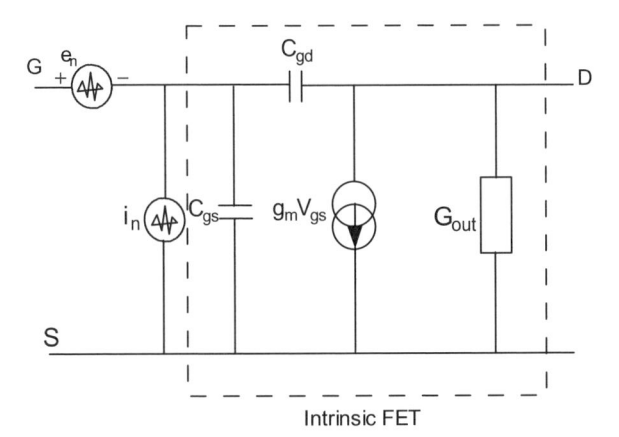

Figure G-12 Equivalent circuit representation of the FET with noise sources at the input.

CALCULATION OF NOISE PROPERTIES

$$[ABCD]_{FET} = \begin{bmatrix} 1 & R_s \\ 0 & 1 \end{bmatrix} \begin{bmatrix} \dfrac{sc_{gd}}{sc_{gd}-g_m} & \dfrac{1}{sc_{gd}-g_m} \\ \dfrac{sc_{gd}(g_m+g_{ds}+sc_{gs}+sc_{ds})}{sc_{gd}-g_m} & \dfrac{(sc_{gd}+g_{ds}+sc_{gs}+sc_{ds})}{sc_{gd}-g_m} \end{bmatrix} \quad (G\text{-}207)$$

$$[ABCD]_{FET} = \begin{bmatrix} \left(\dfrac{sc_{gd}}{sc_{gd}-g_m} + \dfrac{R_s sc_{gd}(g_m+g_{ds}+sc_{gs}+sc_{ds})}{sc_{gd}-g_m}\right) & \left(\dfrac{1}{sc_{gd}-g_m} + \dfrac{R_s(sc_{gd}+g_{ds}+sc_{gs}+sc_{ds})}{sc_{gd}-g_m}\right) \\ \left(\dfrac{sc_{gd}(g_m+g_{ds}+sc_{gs}+sc_{ds})}{sc_{gd}-g_m}\right) & \left(\dfrac{sc_{gd}+g_{ds}+sc_{gs}+sc_{ds}}{sc_{gd}-g_m}\right) \end{bmatrix}$$

$$(G\text{-}208)$$

$$[C_Y]_{FET} = 4kT \begin{bmatrix} \dfrac{\omega^2 c_{gs}^2 R}{g_m} & -j\omega c_{gs}C\sqrt{PR} \\ j\omega c_{gs}C\sqrt{PR} & g_m P \end{bmatrix} \quad (G\text{-}209)$$

$$[C_a]_{FET} = [T][C_Y]_{tr}[T]^+ \quad (G\text{-}210)$$

$$[C_a]_{FET} = \begin{bmatrix} 0 & \left(\dfrac{1}{sc_{gd}-g_m} + \dfrac{R_s(sc_{gd}+g_{ds}+sc_{gs}+sc_{ds})}{sc_{gd}-g_m}\right) \\ 1 & \left(\dfrac{sc_{gd}+g_{ds}+sc_{gs}+sc_{ds}}{sc_{gd}-g_m}\right) \end{bmatrix}$$

$$* \, 4kT \begin{bmatrix} \dfrac{\omega^2 c_{gs}^2 R}{g_m} & -j\omega c_{gs}C\sqrt{PR} \\ j\omega c_{gs}C\sqrt{PR} & g_m P \end{bmatrix} * K1 \quad (G\text{-}211)$$

$$K1 = \begin{bmatrix} 0 & 1 \\ \left(\dfrac{1}{sc_{gd}-g_m} + \dfrac{R_s(sc_{gd}+g_{ds}+sc_{gs}+sc_{ds})}{sc_{gd}-g_m}\right)^{\bullet} & \left(\dfrac{sc_{gd}+g_{ds}+sc_{gs}+sc_{ds}}{sc_{gd}-g_m}\right)^{\bullet} \end{bmatrix} \quad (G\text{-}212)$$

$$= 4kT \begin{bmatrix} \left(\dfrac{sc_{gs}C\sqrt{PR}}{sc_{gd}-g_m} + \dfrac{sc_{gs}R_sC\sqrt{PR}(sc_{gd}+g_{ds}+sc_{gs}+sc_{ds})}{sc_{gd}-g_m}\right) & \left(\dfrac{g_m P}{sc_{gd}-g_m} + \dfrac{g_m P R_s(sc_{gd}+g_{ds}+sc_{gs}+sc_{ds})}{sc_{gd}-g_m}\right) \\ \left(\dfrac{\omega^2 c_{gs}^2 R}{g_m} + \dfrac{(sc_{gd}+g_{ds}+sc_{gs}+sc_{ds})sc_{gs}C\sqrt{PR}}{sc_{gd}-g_m}\right) & \left(-j\omega c_{gs}C\sqrt{PR} + \dfrac{(sc_{gd}+g_{ds}+sc_{gs}+sc_{ds})g_m P}{sc_{gd}-g_m}\right) \end{bmatrix} * K2$$

$$(G\text{-}213)$$

$$K2 = \begin{bmatrix} 0 & 1 \\ \left(\dfrac{1}{sc_{gd}-g_m} + \dfrac{R_s(sc_{gd}+g_{ds}+sc_{gs}+sc_{ds})}{sc_{gd}-g_m}\right)^{\bullet} & \left(\dfrac{sc_{gd}+g_{ds}+sc_{gs}+sc_{ds}}{sc_{gd}-g_m}\right)^{\bullet} \end{bmatrix} \quad (G\text{-}214)$$

$$[C_a]_{FET} = \begin{bmatrix} C_{uu^{\bullet}} & C_{ui^{\bullet}} \\ C_{u^{\bullet}i} & C_{ii^{\bullet}} \end{bmatrix} = 4kT \begin{bmatrix} R_n & \dfrac{F_{\min}-1}{2} - R_n Y_{opt}^{\bullet} \\ \dfrac{F_{\min}-1}{2} - R_n Y_{opt} & R_n |Y_{opt}|^2 \end{bmatrix} \quad (G\text{-}215)$$

G.2 THE GaAs-FET Noise Model

$$C_{uu^\bullet} = 4kT\left[\left(\frac{g_m P}{sc_{gd}-g_m}+\frac{g_m P R_s(sc_{gd}+g_{ds}+sc_{gs}+sc_{ds})}{sc_{gd}-g_m}\right)\right.$$
$$\left.\times\left(\frac{1}{sc_{gd}-g_m}+\frac{R_s(sc_{gd}+g_{ds}+sc_{gs}+sc_{ds})}{sc_{gd}-g_m}\right)^\bullet\right] \quad \text{(G-216)}$$

$$C_{ui^\bullet} = 4kT\left[\left(\frac{sc_{gs}C\sqrt{PR}}{sc_{gd}-g_m}+\frac{sc_{gs}R_s C\sqrt{PR}(sc_{gd}+g_{ds}+sc_{gs}+sc_{ds})}{sc_{gd}-g_m}\right)\right]+A1 \quad \text{(G-217)}$$

$$A1 = \left[\left(\frac{g_m P}{sc_{gd}-g_m}+\frac{g_m P R_s(sc_{gd}+g_{ds}+sc_{gs}+sc_{ds})}{sc_{gd}-g_m}\right)\right.$$
$$\left.\times\left(\frac{(sc_{gd}+g_{ds}+sc_{gs}+sc_{ds})}{sc_{gd}-g_m}\right)^\bullet\right] \quad \text{(G-218)}$$

$$C_{u^\bullet i} = 4kT\left[\left(-j\omega c_{gs}C\sqrt{PR}+\frac{(sc_{gd}+g_{ds}+sc_{gs}+sc_{ds})g_m P}{sc_{gd}-g_m}\right)\right.$$
$$\left.\times\left(\frac{1}{sc_{gd}-g_m}+\frac{R_s(sc_{gd}+g_{ds}+sc_{gs}+sc_{ds})}{sc_{gd}-g_m}\right)^\bullet\right] \quad \text{(G-219)}$$

$$C_{ii^\bullet} = 4kT\left[\left(\frac{\omega^2 c_{gs}^2 R}{g_m}+\frac{(sc_{gd}+g_{ds}+sc_{gs}+sc_{ds})sc_{gs}C\sqrt{PR}}{sc_{gd}-g_m}\right)+B1\right] \quad \text{(G-220)}$$

$$B1 = \left(-j\omega c_{gs}C\sqrt{PR}+\frac{(sc_{gd}+g_{ds}+sc_{gs}+sc_{ds})g_m P}{sc_{gd}-g_m}\right)$$
$$\times\left(\frac{sc_{gd}+g_{ds}+sc_{gs}+sc_{ds}}{sc_{gd}-g_m}\right)^\bullet \quad \text{(G-221)}$$

After substituting the values of $C_{uu^\bullet}$, $C_{ui^\bullet}$, $C_{u^\bullet i}$, and $C_{ii^\bullet}$, we derive noise parameters as

$$R_n = \frac{C_{uu^\bullet}}{2kT} \quad \text{(G-222)}$$

$$F_{\min} = \left[1+\frac{C_{ui^\bullet}+C_{uu^\bullet}Y_{opt}}{kT}\right] \quad \text{(G-223)}$$

$$Y_{opt} = \sqrt{\frac{C_{ii^\bullet}}{C_{uu^\bullet}}-\left[\text{Im}\left(\frac{C_{ui^\bullet}}{C_{uu^\bullet}}\right)\right]^2}+j\,\text{Im}\left(\frac{C_{ui^\bullet}}{C_{uu^\bullet}}\right) \quad \text{(G-224)}$$

$$Y_{opt} = G_{opt}+jB_{opt} \quad \text{(G-225)}$$

$$\Gamma_{opt} = \frac{Z_{opt}-Z_o}{Z_{opt}+Z_o} \implies \frac{Y_{opt}-Y_o}{Y_{opt}+Y_o} \quad \text{(G-226)}$$

Neglecting the effect of gate leakage current I_{gd} and gate-to-drain capacitance C_{gd}, the models above will be further simplified as shown below.

500 CALCULATION OF NOISE PROPERTIES

Figures G-13a and G-13b show the equivalent configuration of the FET without gate-drain capacitance.

$$[C_a]_{tr} = \begin{bmatrix} \overline{e_n e_n^*} & \overline{e_n i_n^*} \\ \overline{i_n e_n^*} & \overline{i_n i_n^*} \end{bmatrix} \tag{G-227}$$

$$[C_a]_{FET} = [T][C_Y]_{FET}[T]^+ \tag{G-228}$$

$$[T] = \begin{bmatrix} 0 & B_{CS} \\ 1 & D_{CS} \end{bmatrix}_{FET} \tag{G-229}$$

$$[C_a]_{FET} = \begin{bmatrix} 0 & B_{CS} \\ 1 & D_{CS} \end{bmatrix}[C_Y]_{FET}\begin{bmatrix} 0 & 1 \\ B_{CS}^* & D_{CS}^* \end{bmatrix} \tag{G-230}$$

The transformation matrix T comes from the [ABCD] matrix of the intrinsic FET.

$$[Y]_{FET} = \begin{bmatrix} y_{11} & y_{12} \\ y_{21} & y_{22} \end{bmatrix} = \begin{bmatrix} sc_{gs} & 0 \\ g_m & G_{out} \end{bmatrix} \tag{G-231}$$

$$[ABCD]_{FET} = \begin{bmatrix} A_{CS} & B_{CS} \\ C_{CS} & D_{CS} \end{bmatrix} = \begin{bmatrix} -\dfrac{y_{22}}{y_{21}} & \dfrac{1}{-y_{21}} \\ \dfrac{\Delta}{y_{21}} & \dfrac{-y_{11}}{y_{21}} \end{bmatrix} = \begin{bmatrix} -\dfrac{G_{out}}{g_m} & \dfrac{-1}{g_m} \\ \dfrac{sc_{gs}G_{out}}{g_m} & \dfrac{-sc_{gs}}{g_m} \end{bmatrix} \tag{G-232}$$

$$[T] = \begin{bmatrix} 0 & B_{CS} \\ 1 & D_{CS} \end{bmatrix} = \begin{bmatrix} 0 & \dfrac{-1}{g_m} \\ 1 & \dfrac{-sc_{gs}}{g_m} \end{bmatrix} \tag{G-233}$$

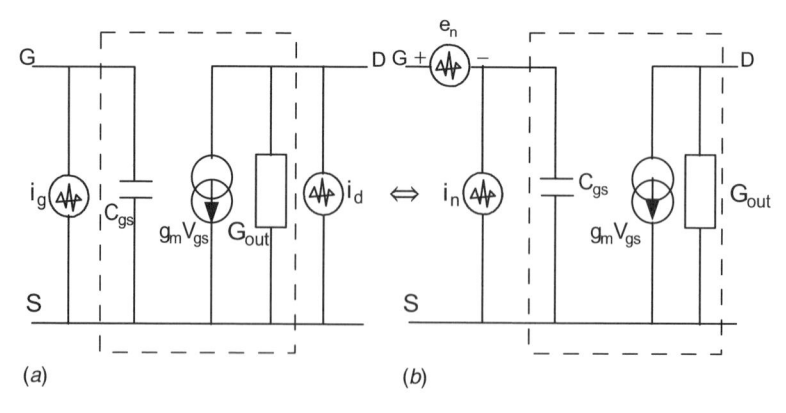

Figure G-13 Equivalent configuration of the FET without gate-drain capacitance. (a) Intrinsic FET with current noise sources at the input and output; (b) intrinsic FET with voltage and current noise sources at the input.

$$[T]^+ = \begin{bmatrix} 0 & 1 \\ B_{CS}^\bullet & D_{CS}^\bullet \end{bmatrix} = \begin{bmatrix} 0 & 1 \\ \dfrac{1}{-g_m} & \dfrac{sc_{gs}}{-g_m} \end{bmatrix} \tag{G-234}$$

$$[C_a]_{FET} = [T][C_Y]_{FET}[T]^+ \tag{G-235}$$

$$[C_a]_{FET} = \begin{bmatrix} 0 & \dfrac{1}{-g_m} \\ 1 & \dfrac{sc_{gs}}{-g_m} \end{bmatrix} * 4kT \begin{bmatrix} \dfrac{\omega^2 c_{gs}^2 R}{g_m} & -j\omega c_{gs} C\sqrt{PR} \\ j\omega c_{gs} C\sqrt{PR} & g_m P \end{bmatrix} * \begin{bmatrix} 0 & 1 \\ \dfrac{1}{-g_m} & \dfrac{sc_{gs}}{-g_m} \end{bmatrix} \tag{G-236}$$

$$[C_a]_{FET} = \dfrac{4kT}{g_m} \begin{bmatrix} P & -j\omega c_{gs}(P + C\sqrt{PR}) \\ j\omega c_{gs}(P + C\sqrt{PR}) & \omega^2 c_{gs}^2(P + R + 2C\sqrt{PR}) \end{bmatrix} \tag{G-237}$$

$$[C_a]_{FET} = \begin{bmatrix} C_{uu^\bullet} & C_{ui^\bullet} \\ C_{u^\bullet i} & C_{ii^\bullet} \end{bmatrix} = 4kT \begin{bmatrix} R_n & \dfrac{F_{\min} - 1}{2} - R_n Y_{opt}^\bullet \\ \dfrac{F_{\min} - 1}{2} - R_n Y_{opt} & R_n |Y_{opt}|^2 \end{bmatrix} \tag{G-238}$$

After substituting the values of $C_{uu^\bullet}$, $C_{ui^\bullet}$, $C_{u^\bullet i}$, $C_{ii^\bullet}$ we get the noise parameters as:

$$R_n = \dfrac{C_{uu^\bullet}}{4kT} = \dfrac{P}{g_m} \tag{G-239}$$

$$Y_{opt} = \sqrt{\dfrac{C_{ii^\bullet}}{C_{uu^\bullet}} - \left[\operatorname{Im}\left(\dfrac{C_{ui^\bullet}}{C_{uu^\bullet}}\right)\right]^2} + j\operatorname{Im}\left(\dfrac{C_{ui^\bullet}}{C_{uu^\bullet}}\right) \tag{G-240}$$

$$Y_{opt} = G_{opt} + jB_{opt} \tag{G-241}$$

$$G_{opt} = \dfrac{\omega c_{gs}}{P}\sqrt{PR(1 - C^2)} \tag{G-242}$$

$$B_{opt} = -\omega c_{gs}\left(1 + C\sqrt{\dfrac{R}{P}}\right) \tag{G-243}$$

$$F_{\min} = 1 + \dfrac{C_{ui^\bullet} + C_{uu^\bullet} Y_{opt}}{kT} = 1 + \dfrac{2\omega c_{gs}}{g_m}\sqrt{PR(1 - C^2)} \tag{G-244}$$

G.2.3 Influence of C_{gd}, R_{gs}, and R_s on Noise Parameters

The following is a calculation showing the values for the noise correlation matrix based on the FET parameters.

$$[C_a]_{FET} = \begin{bmatrix} \overline{e_n e_n^\bullet} & \overline{e_n i_n^\bullet} \\ \overline{i_n e_n^\bullet} & \overline{i_n i_n^\bullet} \end{bmatrix} \tag{G-245}$$

502 CALCULATION OF NOISE PROPERTIES

$$[C_a]_{FET} = \begin{bmatrix} C_{uu^\bullet} & C_{ui^\bullet} \\ C_{u^\bullet i} & C_{ii^\bullet} \end{bmatrix} \tag{G-246}$$

$$C_{uu^\bullet} = \left|\frac{g_m}{g_m - j\omega c_{gd}}\right|^2 \left(\frac{P + R - 2C_r\sqrt{PR}}{g_m}\right) + (R_s + R_{gs}) \tag{G-247}$$

$$C_{ui^\bullet} = \left|\frac{g_m}{g_m - j\omega c_{gd}}\right|^2 \left[\left(\frac{\omega^2 C_{gs}^2 C_{gd}^2}{g_m^2} + \frac{j\omega c_{gd}}{g_m}\right)(R - C\sqrt{PR}) - \frac{j\omega c_{gd}}{g_m}(P - C^\bullet \sqrt{PR})\right] \tag{G-248}$$

$$C_{ii^\bullet} = \left|\frac{g_m}{g_m - j\omega c_{gd}}\right|^2 \left[\left|\frac{\omega^2 C_{gs}^2 C_{gd}^2}{g_m^2} + \frac{j\omega c_{gd}}{g_m}\right|^2 \frac{R}{g_m} + \left|\frac{j\omega c_{gs}}{g_m}\right|^2 P g_m\right] \tag{G-249}$$

$$+ \left|\frac{g_m}{g_m - j\omega c_{gd}}\right|^2 \left[2\operatorname{Re}\left\{\left(\frac{\omega^2 C_{gs}^2 C_{gd}^2}{g_m^2} + \frac{j\omega c_{gd}}{g_m}\right)\left(\frac{j\omega c_{gs}}{g_m}\right) C\sqrt{PR}\right\}\right] \tag{G-250}$$

$$C_{u^\bullet i} = \left\{\left|\frac{g_m}{g_m - j\omega c_{gd}}\right|^2 \left[\left(\frac{\omega^2 C_{gs}^2 C_{gd}^2}{g_m^2} + \frac{j\omega c_{gd}}{g_m}\right)(R - C\sqrt{PR}) - \frac{j\omega c_{gd}}{g_m}(P - C^\bullet \sqrt{PR})\right]\right\}^\bullet \tag{G-251}$$

where

$$\frac{R}{g_m} = \overline{e_n e_n^\bullet} \tag{G-252}$$

$$P g_m = \overline{i_n i_n^\bullet} \tag{G-253}$$

$$C = \frac{\overline{|e_n i_n^\bullet|}}{\sqrt{\overline{(e_n e_n^\bullet)}\overline{(i_n i_n^\bullet)}}} = \frac{\overline{|e_n i_n^\bullet|}}{\sqrt{\overline{|e_n^2|}\overline{|i_n^2|}}} \tag{G-254}$$

$$[C_a]_{FET} = \begin{bmatrix} C_{uu^\bullet} & C_{ui^\bullet} \\ C_{u^\bullet i} & C_{ii^\bullet} \end{bmatrix} = \begin{bmatrix} R_n & \frac{F_{\min} - 1}{2} - R_n Y_{opt}^\bullet \\ \frac{F_{\min} - 1}{2} - R_n Y_{opt} & R_n |Y_{opt}|^2 \end{bmatrix} \tag{G-255}$$

The modified expressions for the noise parameters are now

$$R_n = C_{uu^\bullet} \tag{G-256}$$

$$Y_{opt} = \sqrt{\frac{C_{ii\bullet}}{C_{uu\bullet}} - \left[\text{Im}\left(\frac{C_{ui\bullet}}{C_{ii\bullet}}\right)\right]^2} + j\,\text{Im}\left(\frac{C_{ui\bullet}}{C_{uu\bullet}}\right) \quad \text{(G-257)}$$

$$Z_{opt} = \sqrt{\frac{C_{uu\bullet}}{C_{ii\bullet}} - \left(\frac{\text{Im}\,C_{ui\bullet}}{C_{ii\bullet}}\right)^2} - j\left(\frac{\text{Im}\,C_{ui\bullet}}{C_{ii\bullet}}\right) \quad \text{(G-258)}$$

$$Y_{opt} = G_{opt} + jB_{opt} \quad \text{(G-259)}$$

$$F_{\min} = 1 + 2\left[\text{Re}(C_{ui\bullet}) + C_{ii\bullet}\cdot\text{Re}(Z_{opt})\right] \quad \text{(G-260)}$$

$$F_{\min} = 1 + 2\left[\left(\frac{\omega^2 c_{gs}^2}{g_m^2}\right)(R_{gs} + R_s)Pg_m \right.$$
$$\left. + \sqrt{\frac{\omega^4 c_{gs}^4}{g_m^4}(R_{gs} + R_s)^2 P^2 g_m^2 + \left(\frac{\omega^2 c_{gs}^2}{g_m^2}\right)[PR(1-C^2) - Pg_m R_{gs}]}\right]$$

(G-261)

$$R_n = \left|\frac{g_m}{g_m - j\omega c_{gd}}\right|^2 \left(\frac{P + R - 2C_r\sqrt{RP}}{}\right) + (R_{gs} + R_s) \quad \text{(G-262)}$$

$$R_{opt} = \frac{1}{\omega c_{gs}}\sqrt{\frac{g_m(R_s + R_{gs}) + R(1 - C_r^2)}{P} + \omega^2 c_{gs}^2(R_s + R_{gs})^2} \quad \text{(G-263)}$$

$$X_{opt} = \frac{1}{\omega c_{gs}}\left(1 - C_r\sqrt{\frac{R}{P}}\right) \quad \text{(G-264)}$$

G.2.4 Temperature Dependence of the Noise Parameters of an FET

We now introduce a minimum noise temperature $T_{\min}$, and we will modify the noise parameters previously derived. This equation now will have temperature dependence factors.

For the use of FETs in amplifiers used in huge dishes scanning the universe, and in evaluating signals coming from outer space, extremely low-noise temperatures are needed. These amplifiers being cooled cryogenically to temperatures close to 0K. Since no measurements are available, the following method allows for the prediction of noise performance of FETs under "cooled" conditions. This derivation shows how to calculate the temperature-dependent noise performance.

Figures G-14a and G-14b are the familiar two-port noise representations of the intrinsic FET in admittance and ABCD matrix form.

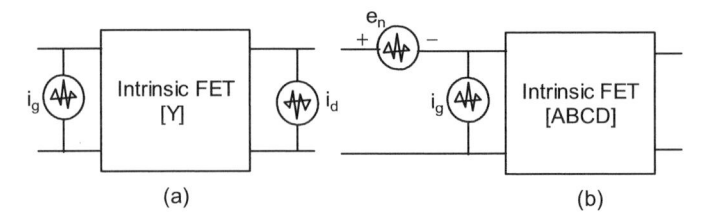

Figure G-14 Noise representation in the linear two-port. (*a*) Current noise sources at the input and output; (*b*) current and voltage noise sources at the output.

The admittance representation of the noise parameter of the intrinsic FET is expressed as

$$G_1 = \frac{\overline{|i_g^2|}}{4kT_0\Delta f} \tag{G-265}$$

$$G_2 = \frac{\overline{|i_d^2|}}{4kT_0\Delta f} \tag{G-266}$$

$$C_r = \frac{\overline{|i_g i_d^*|}}{\sqrt{\overline{|i_d^2|}\,\overline{|i_g^2|}}} \tag{G-267}$$

where k is Boltzman's constant ($k = 1.38\text{E-}23$), T_0 is the standard room temperature (290K), and Δf is the reference bandwidth.

The ABCD matrix representation and the corresponding noise parameters are

$$R_n = \frac{\overline{|e_n^2|}}{4kT_0\Delta f} \tag{G-268}$$

$$g_n = \frac{\overline{|i_n^2|}}{4kT_0\Delta f} \tag{G-269}$$

$$C_r = \frac{\overline{|e_n i_n^*|}}{\sqrt{\overline{|e_n^2|}\,\overline{|i_n^2|}}} \tag{G-270}$$

$$N = R_{opt} g_n \tag{G-271}$$

where g_n is noise conductance.

The noise temperature T_n and the noise measure M of a two-port driven by generator impedance Z_g are expressed as

$$T_n = T_{\min} + T_0 \frac{g_n}{R_g} |Z_g - Z_{opt}|^2 \tag{G-272}$$

$$T_n = T_{\min} + NT_0 \frac{|Z_g - Z_{opt}|^2}{R_g R_{opt}} \tag{G-273}$$

$$T_n = T_{\min} + 4NT_0 \frac{|T_g - T_{opt}|^2}{\left(1 - |T_{opt}|^2\right)\left(1 - |T_g|^2\right)} \tag{G-274}$$

$$T_{opt} = \frac{Z_{opt} - Z_o}{Z_{opt} + Z_o} \tag{G-275}$$

$$M = \frac{T_n}{T_0}\left(\frac{1}{1 - \frac{1}{G_a}}\right) \tag{G-276}$$

where Z_0 is the reference impedance and G_a is the available gain.

An extrinsic FET with parasitic resistances is shown in Figure G-15. These resistances contribute thermal noise, and their influence can be calculated based on the ambient temperature T_a.

$$G_1 = \frac{T_g}{T_o} \frac{R_{gs}(\omega C_{gs})^2}{\left(1 + \omega^2 C_{gs}^2 R_{gs}^2\right)} \tag{G-277}$$

$$G_2 = \frac{T_g}{T_o} \frac{g_m^2 R_{gs}}{\left(1 + \omega^2 C_{gs}^2 R_{gs}^2\right)} + \frac{T_d}{T_o} g_{gs} \tag{G-278}$$

$$C_{r_c} = C_r \frac{|i_g i_d^*|}{\sqrt{|i_d^2||i_g^2|}} = \frac{-j\omega g_m C_{gs} R_{gs}}{\left(1 + \omega^2 C_{gs}^2 R_{gs}^2\right)} \frac{T_g}{T_o} \tag{G-279}$$

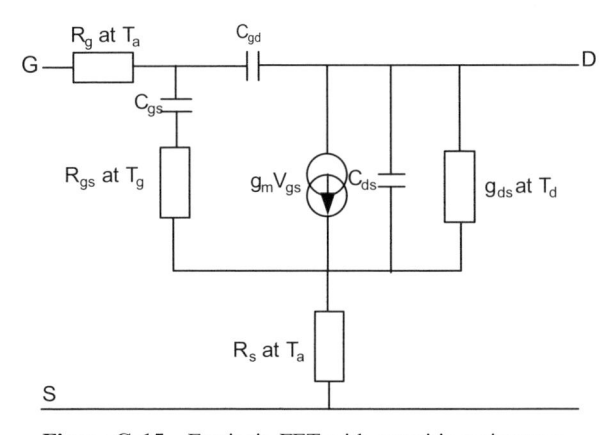

Figure G-15 Extrinsic FET with parasitic resistances.

CALCULATION OF NOISE PROPERTIES

The noise properties of the intrinsic FET are treated by assigning equivalent temperatures T_g and T_d to R_{gs} and g_{ds}.

No correlation is assumed between the noise sources represented by the equivalent temperatures T_g and T_d in Figure G-16.

The modified noise parameters are expressed as

$$Z_{opt} = R_{opt} + jX_{opt} \tag{G-280}$$

$$R_{opt} = \sqrt{\left(\frac{f_T}{f}\right)^2 \frac{R_{gs}}{R_{ds}} \frac{T_g}{T_d} + R_{gs}^2} \tag{G-281}$$

$$X_{opt} = \frac{1}{\omega C_{gs}} \tag{G-282}$$

$$T_{min} = 2\frac{f}{f_T}\sqrt{R_{gs}g_{ds}T_gT_d + \left(\frac{f_T}{f}\right)^2 R_{gs}^2 g_{ds}^2 T_d^2} + 2\left(\frac{f_T}{f}\right)^2 R_{gs}g_{ds}T_d \tag{G-283}$$

$$T_{min} = (F_{min} - 1)T_0 \tag{G-284}$$

$$g_n = \left(\frac{f_T}{f}\right)^2 \frac{g_{ds}T_d}{T_0} \tag{G-285}$$

$$f_T = \frac{g_m}{2\pi C_{gs}} \tag{G-286}$$

$$\frac{4NT_0}{T_{min}} = \frac{2}{1 + \dfrac{R_{gs}}{R_{opt}}} \tag{G-287}$$

$$R_n = \frac{T_g}{T_o}R_{gs} + \frac{T_d}{T_0}\frac{g_{ds}}{g_m^2}\left(1 + \omega^2 C_{gs}^2 R_{gs}^2\right) \tag{G-288}$$

$$C_r = C\sqrt{R_n g_n} = \frac{T_d}{T_0}\frac{g_{ds}}{g_m^2}\left(\omega^2 C_{gs}^2 R_{gs} + j\omega C_{gs}\right) \tag{G-289}$$

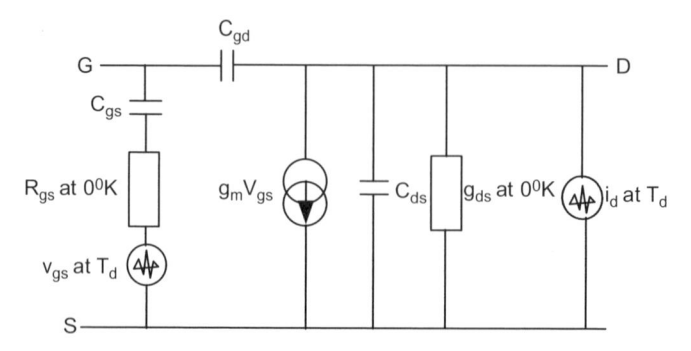

Figure G-16 Intrinsic FET with assigned equivalent temperatures.

G.2.5 Approximation and Discussion

With some reasonable approximation, the expression of the noise parameters becomes much simpler (by introducing the following approximation, the values obtained from the calculation typically vary less than 5% from the exact values):
If

$$\frac{f}{f_T} \leq \sqrt{\frac{R_{gs} T_g}{R_{ds} T_d}} \qquad (G\text{-}290)$$

and $R_{opt} \geq R_{gs}$, then

$$R_{opt} \cong \left(\frac{f_T}{f}\right) \sqrt{\frac{r_{gs} T_g}{r_{ds} T_d}} \qquad (G\text{-}291)$$

$$X_{opt} \cong \frac{1}{\omega C_{gs}} \qquad (G\text{-}292)$$

$$T_{min} \cong 2 \frac{f}{f_T} \sqrt{r_{gs} g_{ds} T_g T_d} \qquad (G\text{-}293)$$

$$g_n \cong \left(\frac{f_T}{f}\right)^2 \frac{g_{ds} T_d}{T_0} \qquad (G\text{-}294)$$

$$f_T = \frac{g_m}{2\pi C_{gs}} \qquad (G\text{-}295)$$

$$\frac{4NT_0}{T_{min}} \cong 2 \qquad (G\text{-}296)$$

Example: a linear FET model with the following intrinsic parameters is assumed:

R_{gs} = 2.5 ohms $\qquad C_{gd}$ = 0.042 pF
r_{ds} = 400 ohms $\qquad g_m$ = 57 mS
C_{gs} = 0.28 pF $\qquad f$ = 8.5 GHz
C_{ds} = 0.067 pF

The temperature-dependent noise parameters for an intrinsic FET are now calculated for two cases (room temperature).
Example 1: T_a = 297K, T_g = 304K, T_d = 5514K, V_{ds} = 2 V, I_{ds} = 10 mA

$$f_T = \frac{g_m}{2\pi C_{gs}} = 32.39 \text{ GHz} \qquad (G\text{-}297)$$

$$R_{opt} = \sqrt{\left(\frac{f_T}{f}\right)^2 \frac{r_{gs} T_g}{g_{ds} T_d} + r_{gs}^2} = 28.42 \, \Omega \qquad (G\text{-}298)$$

$$X_{opt} = \frac{1}{\omega C_{gs}} = 66.91\,\Omega \qquad \text{(G-299)}$$

$$T_{min} = 2\frac{f}{f_T}\sqrt{r_{gs}g_{ds}T_gT_d + \left(\frac{f_T}{f}\right)^2 r_{gs}^2 g_{ds}^2 T_d^2} + 2\left(\frac{f_T}{f}\right)^2 r_{gs}g_{ds}T_d = 58.74\,\text{K} \qquad \text{(G-300)}$$

$$g_n = \left(\frac{f_T}{f}\right)^2 \frac{g_{ds}T_d}{T_0} = 3.27\,\text{mS} \qquad \text{(G-301)}$$

$$F_{min} = \frac{T_{min}}{T_0} + 1 = \frac{58.7}{290} + 1 = 1.59\,\text{dB} \qquad \text{(G-302)}$$

$$R_n = \frac{T_g r_{gs}}{T_0} + \frac{g_{ds}T_d}{T_0 g_m^2}\left(1 + \omega^2 r_{gs}^2 c_{gs}^2\right) = 17.27\,\Omega \qquad \text{(G-303)}$$

Example 2: $T_a = 12.5\,\text{K}$, $T_g = 14.5\,\text{K}$, $T_d = 1406\,\text{K}$, $V_{ds} = 2\,\text{V}$, $I_{ds} = 5\,\text{mA}$ (cooled down to 14.5 K)

$$f_T = \frac{g_m}{2\pi C_{gs}} = 32.39\,\text{GHz} \qquad \text{(G-304)}$$

$$R_{opt} = \sqrt{\left(\frac{f_T}{f}\right)^2 \frac{r_{gs}}{g_{ds}} \frac{T_g}{T_d} + r_{gs}^2} = 12.34\,\Omega \qquad \text{(G-305)}$$

$$X_{opt} = \frac{1}{\omega C_{gs}} = 66.9\,\Omega \qquad \text{(G-306)}$$

$$T_{min} = 2\frac{f}{f_T}\sqrt{r_{gs}g_{ds}T_gT_d + \left(\frac{f_T}{f}\right)^2 r_{gs}^2 g_{ds}^2 T_d^2} + 2\left(\frac{f_T}{f}\right)^2 r_{gs}g_{ds}T_d = 7.4\,\text{K} \qquad \text{(G-307)}$$

$$g_n = \left(\frac{f_T}{f}\right)^2 \frac{g_{ds}T_d}{T_0} = 0.87\,\text{mS} \qquad \text{(G-308)}$$

$$F_{min} = \frac{T_{min}}{T_0} + 1 = \frac{7.4}{290} + 1 = 0.21\,\text{dB} \qquad \text{(G-309)}$$

$$R_n = \frac{T_g r_{gs}}{T_0} + \frac{g_{ds}T_d}{T_0 g_m^2}\left(1 + \omega^2 r_{gs}^2 c_{gs}^2\right) = 3.86\,\Omega \qquad \text{(G-310)}$$

These results are consistent with those published by Pucel [38] and Pospieszalski [40].

APPENDIX H
Noise Analysis of the N-Coupled Oscillator Coupled Through Different Coupling Topologies

The purpose of this appendix to find the analytical expression of the phase noise of N-coupled oscillators relative to a single free-running uncoupled oscillator for different coupling configurations. Three coupling topologies (global, bilateral, and unilateral) are described for the noise analysis of the N-coupled oscillator systems [187].

For an uncoupled oscillator coupling coefficient $\beta_{ij} \to 0$ and equations (10-191) and (10-192) can be expressed as

$$[\Delta A_i(\omega)]_{uncoupled} = \left[\frac{\mu \omega_i}{2Q\omega}\right][\alpha_i^2 - 3\dot{A}_i^2][\Delta A_i(\omega)]_{uncoupled} - \left[\frac{\omega_i}{2Q\omega}\right]\dot{A}_i G_{ni}(\omega);$$
$$i = 1, 2, 3, \ldots, N \quad \text{(H-1)}$$

$$[\Delta \theta_i(\omega)]_{uncoupled} = -\left[\frac{\omega_i}{2Q\omega}\right]B_{ni}(\omega); \quad i = 1, 2, 3, \ldots, N \quad \text{(H-2)}$$

From equation (H-1),

$$[\Delta A_i(\omega)]_{uncoupled} = \frac{-\dot{A}_i G_{ni}(\omega)}{\left[\frac{2\omega Q}{\omega_i}\right] + \mu\left[3\dot{A}_i^2 - \alpha_i^2\right]} \quad \text{(H-3)}$$

Noise spectral density due to amplitude fluctuation is given by

$$|\Delta A_i(\omega)|^2_{uncoupled} = \frac{\dot{A}_i^2 |G_{ni}(\omega)|^2}{\left[\frac{2\omega Q}{\omega_i}\right]^2 + \mu^2\left[3\dot{A}_i^2 - \alpha_i^2\right]^2} \quad \text{(H-4)}$$

The Design of Modern Microwave Oscillators for Wireless Applications: Theory and Optimization, by Ulrich L. Rohde, Ajay Kumar Poddar, Georg Böck
Copyright © 2005 John Wiley & Sons, Inc.

Noise spectral density due to phase fluctuation for a single uncoupled free-running oscillator is given from equation (H-2) as

$$|\Delta\theta_i(\omega)|^2_{uncoupled} = \frac{|B_{ni}(\omega)|^2}{\left[\dfrac{2Q\omega}{\omega_i}\right]^2} \tag{H-5}$$

For a series-tuned free-running oscillator, equations (H-4) and (H-5) can be rewritten as

$$|\Delta A_i(\omega)|^2_{uncoupled} = \frac{\dot{A}_i^2 |G_{ni}(\omega)|^2 R^2}{4L^2\omega^2 + (\mu R)^2[3\dot{A}_i^2 - \alpha_i^2]^2} \approx \frac{2|e|^2}{4L^2\omega^2 + (\gamma A_0)^2} \tag{H-6}$$

$$|\Delta\theta_i(\omega)|^2_{uncoupled} = \frac{|B_{ni}(\omega)|^2}{\left[\dfrac{\omega}{\omega_{3dB}}\right]^2} = \frac{|B_{ni}(\omega)|^2 \omega_i^2}{4\omega^2 Q^2} \approx \frac{2|e|^2}{4\omega^2 L^2 A_0^2} \tag{H-7}$$

where

$$Q = \frac{\omega L}{R}, \quad \gamma A_0 \approx (\mu R)[3\dot{A}_i^2 - \alpha_i^2], \quad |e|^2 \approx \dot{A}_i^2 |G_{ni}(\omega)|^2 R^2$$

Equations (H-6) and (H-7) represents the AM and PM noise for the uncoupled free-running series-tuned oscillator. They have the same forms given in equations (10-173) and (10-177).

H.1 GLOBALLY N-COUPLED OSCILLATOR SYSTEMS

Figure H-1 shows the globally N-coupled oscillator system.
From equation (10-194),

$$\left[\frac{2Q\omega}{\omega_i}\right][\Delta\theta_i(\omega)] = -\sum_{\substack{j=1 \\ j\neq i}}^{N} \beta_{ij}\big([\Delta\theta_i(\omega) - \Delta\theta_j(\omega)]\big)\cos[\dot{\theta}_i - \dot{\theta}_j] - B_{ni}(\omega);$$

$$i = 1, 2, 3, \ldots, N \tag{H-8}$$

For globally coupled topology, with coupling coefficient $\beta_{ij} = \beta$ for any ith and jth oscillators and with all oscillators in phase, equation (H-8) can be rewritten as

$$\left[\frac{2\omega Q}{\omega_i}\right][\Delta\theta_i(\omega)] = -\beta\sum_{\substack{j=1 \\ j\neq i}}^{N} \big([\Delta\theta_i(\omega) - \Delta\theta_j(\omega)]\big) - B_{ni}(\omega);$$

$$i = 1, 2, 3, \ldots, N \tag{H-9}$$

H.1 GLOBALLY N-COUPLED OSCILLATOR SYSTEMS

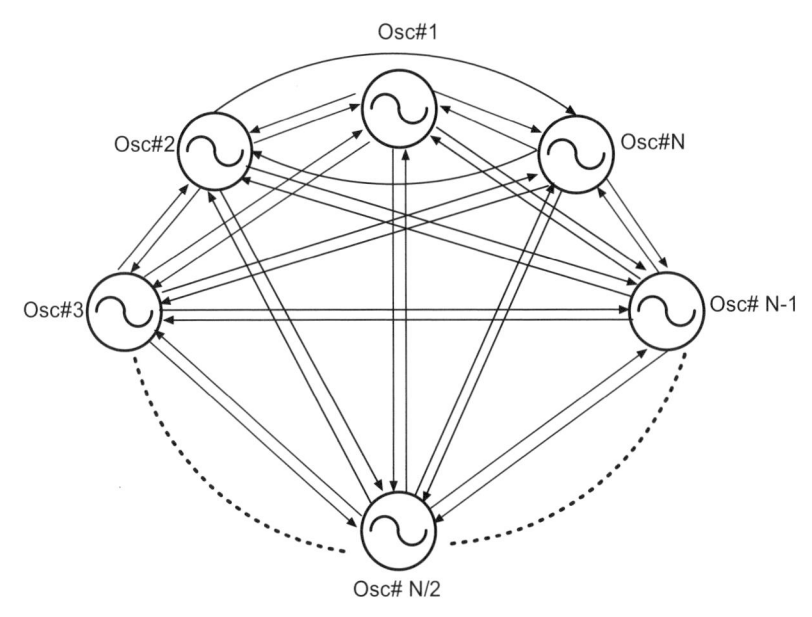

Figure H-1 Globally N-coupled oscillator system.

expanding the series of equation (H-9) as

$$\left[\frac{2\omega Q}{\beta\omega_i}\right][\Delta\theta_i(\omega)] = \{(N-1)[\Delta\theta_i(\omega)]\} + \{[\Delta\theta_1(\omega)] + [\Delta\theta_2(\omega)] + \cdots + [\Delta\theta_N(\omega)]\} - \frac{B_{ni}(\omega)}{\beta} \quad \text{(H-10)}$$

$$x[\Delta\theta_i(\omega)] = \left[\{(N-1)[\Delta\theta_i(\omega)]\} + \{[\Delta\theta_1(\omega)] + [\Delta\theta_2(\omega)] + \cdots + [\Delta\theta_N(\omega)]\} - \frac{B_{ni}(\omega)}{\beta}\right] \quad \text{(H-11)}$$

where

$$x = \frac{2\omega Q}{\beta\omega_i}$$

Following [138], from equations (10-195) and (H-11), $[\overline{C}]$ for global coupling can be described as

$$[\overline{C}] = \beta \begin{bmatrix} 1-N-x & 1 & 1 & \cdots & 1 \\ 1 & 1-N-x & 1 & \cdots & 1 \\ 1 & 1 & 1-N-x & \ddots & \vdots \\ \vdots & \vdots & \ddots & \ddots & 1 \\ 1 & 1 & \cdots & 1 & 1-N-x \end{bmatrix} \quad \text{(H-12)}$$

From equation (10-196),

$$[\bar{P}] = [\bar{C}]^{-1} = \frac{1}{-x\beta(N+x)} \begin{bmatrix} 1+x & 1 & 1 & \cdots & 1 \\ 1 & 1+x & 1 & \cdots & 1 \\ 1 & 1 & 1+x & \ddots & \vdots \\ \vdots & \vdots & \ddots & \ddots & 1 \\ 1 & 1 & \cdots & \cdots & 1+x \end{bmatrix} \quad \text{(H-13)}$$

From equation (10-197),

$$\sum_{j=1}^{N} p_{ij} = \frac{(N+x)}{-x\beta(N+x)} = \frac{-1}{x\beta} \quad \text{for all } i \quad \text{(H-14)}$$

From equation (10-205), the total output phase noise is given by

$$|\Delta\theta_{total}(\omega)|^2 = \frac{|B_n(\omega)|^2}{N^2} \sum_{j=1}^{N} \left|\sum_{i=1}^{N} p_{ij}\right|^2 = \frac{1}{N} \frac{|B_n(\omega)|^2}{\beta^2 x^2} = \frac{1}{N} \frac{|B_n(\omega)|^2}{\left[\dfrac{2\omega Q}{\omega_i}\right]^2} \quad \text{(H-15)}$$

Comparing equation (H-15) with the single-oscillator phase noise equation (H-5), we find

$$|\Delta\theta_{total}(\omega)|^2 = \frac{1}{N} |\Delta\theta_i(\omega)|^2_{uncoupled} \quad \text{(H-16)}$$

From equation (H-16), the total PM noise for globally N-coupled oscillators is reduced by the factor N of a single oscillator.

H.2 BILATERAL N-COUPLED OSCILLATOR SYSTEMS

Figure H-2 shows the nearest neighbor bilateral N-coupled oscillator system.
The coupling parameter β_{ij} for Figure H-2 is defined as [149]

$$\beta_{ij} = \begin{cases} \beta, & |i-j| = 1 \\ 0, & \text{otherwise} \end{cases} \quad \text{(H-17)}$$

assuming that the constant phase progression along the array of the N-coupled oscillator system is

$$\dot{\theta}_i - \dot{\theta}_{i+1} = \Delta\dot{\theta} \quad \text{(H-18)}$$

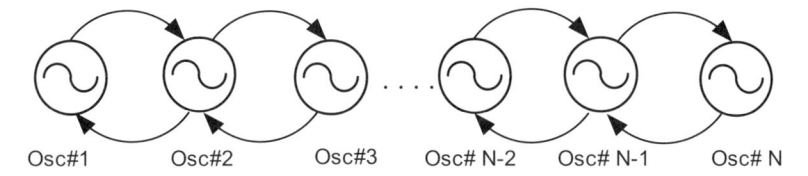

Figure H-2 Nearest neighbor bilateral N-coupled oscillator system.

From equation (10-195), $[\overline{C}]$ for bilateral coupling can be described as

$$[\overline{C}] = \beta\cos(\Delta\dot{\theta})\begin{bmatrix} -1-y & 1 & 0 & 0 & \cdots & 0 \\ 1 & -2-y & 1 & 0 & \cdots & 0 \\ 0 & 1 & -2-y & 1 & \cdots & \vdots \\ \vdots & \ddots & \ddots & \ddots & \ddots & 0 \\ 0 & \cdots & \ddots & 1 & -2-y & 1 \\ 0 & 0 & \cdots & 0 & 1 & -1-y \end{bmatrix} \quad \text{(H-19)}$$

where

$$y = \frac{\omega}{\left[\dfrac{\beta\omega_i}{2Q}\right]\cos(\Delta\dot{\theta})} \quad \text{(H-20)}$$

From equation (10-197)

$$\sum_{j=1}^{N} p_{ij} = \frac{-1}{\left[\dfrac{2Q\omega}{\omega_i}\right]} \quad \text{for all } i \quad \text{(H-21)}$$

From equation (10-205), the total output phase noise is given by

$$|\Delta\theta_{total}(\omega)|^2 = \frac{|B_n(\omega)|^2}{N^2}\sum_{j=1}^{N}\left|\sum_{i=1}^{N} p_{ij}\right|^2 = \frac{1}{N}\frac{|B_n(\omega)|^2}{\left[\dfrac{2\omega Q}{\omega_i}\right]^2} \quad \text{(H-22)}$$

Comparing equation (H-22) with the single-oscillator phase noise equation (H-5), we find

$$|\Delta\theta_{total}(\omega)|^2 = \frac{1}{N}|\Delta\theta_i(\omega)|^2_{uncoupled} \quad \text{(H-23)}$$

H.3 UNILATERAL N-COUPLED OSCILLATOR SYSTEMS

Figure H-3 shows the nearest neighboring unilateral N-coupled oscillator system.

514 NOISE ANALYSIS OF THE N-COUPLED OSCILLATOR

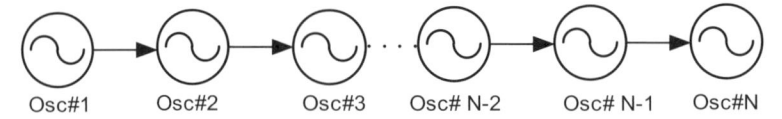

Figure H-3 Nearest neighbor unilateral N-coupled oscillator system.

In this topology, each successive oscillator in the array of an N-coupled oscillator system is slaved to the previous oscillator, and the first oscillator in the array is considered a master oscillator.

The coupling parameter β_{ij} for Figure H-3 is defined as [138]

$$\beta_{ij} = \begin{cases} \beta, & i-j = +1 \\ 0, & otherwise \end{cases} \tag{H-24}$$

From equation (10-195), $[\overline{C}]$ for unilateral coupling can be described as

$$[\overline{C}] = \beta \begin{bmatrix} -z & 0 & 0 & \cdots & 0 & 0 \\ 1 & -1-z & 0 & \cdots & \vdots & \vdots \\ 0 & 1 & -1-z & \cdots & \vdots & \vdots \\ \vdots & \ddots & \ddots & \ddots & 0 & \vdots \\ 0 & \cdots & \ddots & \ddots & -1-z & 0 \\ 0 & 0 & \cdots & 0 & 1 & -1-z \end{bmatrix} \tag{H-25}$$

$$[\overline{P}] = [\overline{C}]^{-1} = \frac{1}{\beta} \begin{bmatrix} \frac{-1}{z} & 0 & 0 & \cdots & 0 & 0 \\ \frac{-1}{z(1+z)} & \frac{-1}{(1+z)} & 0 & \cdots & 0 & 0 \\ \frac{-1}{z(1+z)^2} & \frac{-1}{(1+z)^2} & \frac{-1}{(1+z)} & \cdots & \vdots & 0 \\ \vdots & \vdots & \vdots & \ddots & 0 & \vdots \\ \frac{-1}{z(1+z)^{N-2}} & \frac{-1}{(1+z)^{N-2}} & \frac{-1}{(1+z)^{N-3}} & \ddots & \frac{-1}{(1+z)} & 0 \\ \frac{-1}{z(1+z)^{N-1}} & \frac{-1}{(1+z)^{N-1}} & \frac{-1}{(1+z)^{N-2}} & \cdots & \frac{-1}{(1+z)^2} & \frac{-1}{(1+z)} \end{bmatrix}$$

$$\tag{H-26}$$

H.3 UNILATERAL N-COUPLED OSCILLATOR SYSTEMS

where

$$z = \frac{2\omega Q}{\beta \omega_i}$$

The rows and columns of equation (H-26) form a geometric series; the expression of $\sum_{j=1}^{N} p_{ij}$ can be calculated analytically. For small values of z, the expression of $\sum_{j=1}^{N} p_{ij}$ can be given as

$$\sum_{j=1}^{N} p_{ij} = \begin{cases} \dfrac{-N}{\beta z}, & i = 1 \\ -\dfrac{N+1-i}{\beta}, & 2 \leq i \leq N \end{cases} \quad \text{(H-27)}$$

From equation (10-205), the total output phase noise is approximately given as [138]

$$|\Delta \theta_{total}(\omega)|^2 = \frac{|B_n(\omega)|^2}{N^2} \sum_{j=1}^{N} \left| \sum_{i=1}^{N} p_{ij} \right|^2 = \left[1 + z^2 \left(\frac{N}{3} - \frac{1}{2} + \frac{1}{6N} \right) |\Delta \theta_i(\omega)|^2_{uncoupled} \right]$$

(H-28)

$$|\Delta \theta_{total}(\omega)|^2 = \left[1 + \left(\frac{2\omega Q}{\beta \omega_i} \right)^2 \left(\frac{N}{3} - \frac{1}{2} + \frac{1}{6N} \right) |\Delta \theta_i(\omega)|^2_{uncoupled} \right] \quad \text{(H-29)}$$

From equation (H-29), it can be seen that there is noise degradation with respect to the single uncoupled free-running oscillator. The noise of the unilateral coupled oscillator increases quadratically away from the carrier and increases linearly with the number of oscillators in the array of the N-coupled oscillator system.

In general, $z \ll 1$, and close to the carrier frequency the noise is that of the first-stage oscillator. Thus, the total noise can be significantly reduced by making the first-stage oscillator the master oscillator, which has low noise.

INDEX

ABCD noise correlation matrix
  bipolar transistors
    GE configuration, 477–479
    noise current source transfer, 473–474
    noise factor calculations, 474–476
    noise parameters excluding base resistance, 479–485
  chain matrix, phase noise analysis, feedback models, 186–199
  resonator design, 191–193
  FET noise parameter calculation, 497–501
    temperature dependence, 503–506
  GE, GC, and GB configurations, 468–469
  input transfer, 465–466
  noise parameters and, 462–463
  noise transformation using, 459–460
Abrupt junctions, transistor design, tuning diodes, 56–60
Absolute values
  biasing networks, 230–232
  feedback capacitor phase noise and output power calculations, 401–405
AC equivalent circuits, negative resistance model, time domain noise signal calculations, 174
Active bias circuits
  1000 MHz CRO circuit validation, 233–237
  phase noise and temperature effects, 211–214
Active device systems
  metal semiconductor field-effect transistor, series feedback, 410–414
    ultra-low-noise wideband oscillators, 321–323
Admittance parameters
  FET noise parameter calculation, temperature dependence, 504–506
  mutually locked (synchronized) coupled oscillators, 255–257
  N-coupled (synchronized) oscillators, noise analysis, 282–300
  N-push coupled mode (synchronized) oscillators, mode analysis, 307–311
  ultra-low-noise wideband oscillators
    coupled resonator, 327–330
    inductor model, 338–339
Agilent ADS system, bias values, 232
Agilent HBFP-0405 transistor, biasing topology, 224–230
Almost periodic discrete Fourier transform (APDFT), historical background, 4–5
Amplitude measurements
  bipolar transistors, series feedback, 419–420
  coupled oscillator systems, 261–262
  large-signal oscillators, steady-state behavior, 100–101
  N-coupled (synchronized) oscillators, 263–266
  uncoupled oscillator noise, 272–276
Amplitude modulation (AM) noise
  coupled oscillator systems, 266–267
  mutually coupled oscillators, 278–282
  N-coupled oscillators, 285–300
  noise performance index
    frequency conversion, 448–450
    noise modulation, 453–454
    PM-AM correlation coefficient, 454–455

The Design of Modern Microwave Oscillators for Wireless Applications: Theory and Optimization, by Ulrich L. Rohde, Ajay Kumar Poddar, Georg Böck
Copyright © 2005 John Wiley & Sons, Inc.

517

518 INDEX

Ansoft circuit simulator
  coupled resonator design, 78–79
  field-effect transistors, 49–51
  large-signal oscillators, 99–100
  N-coupled oscillator noise analysis, push-pull configuration, 296–300
  oscillator conversion noise analysis, 144
  phase noise contributions
    bias values, 232
    CMOS transistors, 209–210
    ultra-low-noise wideband oscillators, coupled resonators, 330
    yttrium ion garnet resonators, 84–85
Attenuation effects, ultra-low-noise wideband oscillators, coupled resonator, 328–330
Autocorrelation function, negative resistance model, time domain noise signal calculations, 172–174

Bandwidth calculations
  coupled oscillator systems, 262–263
    noise analysis, 268
  negative resistance model, noise voltage calculations, 176–183
  1000–2000/2000–4000 MHz hybrid tuned VCO, 361–365
Barkhausen criteria
  large-signal oscillators, 95–100
  sinusoidal oscillator, 11
Base current, bipolar transistors, hybrid-π configuration, 471–473
Base emitter voltage
  biasing topololgy, Agilent HBFP-0405 transistor, 227–230
  bipolar transistors, series feedback, 420
  coupled oscillator noise sources, 268
  large-signal oscillators, time-domain behavior, 117–121
  phase noise, biasing and temperature effects, stability factors, 221–224
Base-lead inductance
  large-signal design, Bessel functions, 392–395
  N-coupled oscillator noise analysis, 289–292
  oscillator parasitics, 88–91
Base resistance
  bipolar transistor noise analysis, T-equivalent configuration, 486–489

bipolar transistors
  hybrid-π configuration, 471–473
  noise parameters excluding, 479–485
  feedback models, noise-shaping function, spectral density and transfer function, 196–199
BCR400W bias circuit, active bias circuits, phase noise and temperature effects, 212–214
Bessel functions
  large-signal oscillators
    design criteria, 389–395
    time-domain behavior, 102–121
  noise signal modulation, 442
  sinusoidal signal modulation, 439–440
β variation
  biasing topololgy, Agilent HBFP-0405 transistor, 226–230
  phase noise biasing and temperature effects, 210–211
  stability factors and, 218–224
BFP520 transistors
  large-signal design, time-domain behavior, 115–121
  NPN silicon RF transistor
    design principles, 41–47
    large-scale design, Bessel functions, 390–395
    parallel resonator oscillator design, large-signal S-parameters, 381–383
    S-parameter measurements, 67–69
BFP620 and 650 transistors, design principles, 43–47
Bias networks
  design topology, 224–230
  metal semiconductor field-effect transistor, 431
  negative resistance noise voltage, 181–183
  1000 MHz CRO circuit validation, 233–237
  1500–3000/3000–6000 MHz coupled resonator oscillator, circuit validation, 355–361
  phase noise effects
    calculation and optimization, 210–232
    active bias circuits, 211–214
    bias selection criteria, 230–232
    design criteria, 224–230

passive bias circuits, 214–218
  stability factors and
    temperature-dependent variables,
    218–224
  value selection criteria, 230–232
  output power, 399–405
Bilateral coupling, N-coupled system
  dynamics, 264–266
  noise analysis, 512–513
Bipolar transistors
  coupled oscillator systems,
    noise sources, 267–268
  design principles, 40–47
  large-signal oscillators, time-domain
    behavior, 101–121
  noise analysis
    base resistance parameters, exclusion
      of, 479–485
    current source transformation to
      input, GE configuration,
      473–474
    hybrid π configuration, 471–473
    minimum noise factor, 490–492
    noise correlation matrix
      GE configuration, 477–479
      T-equivalent configuration,
        492–494
    noise factor calculations, 474–476
    real source impedance,
      476–477, 489
    T configuration, 485–489
  phase noise analysis, feedback models,
    185–199
  series feedback, 417–420
  ultra-low-noise wideband oscillators,
    wideband VCO design technique,
    317–320
Blocking, oscillator phase noise effects,
  11–13
Boltzmann's constant
  FET noise parameter calculation,
    temperature dependence,
    504–506
  large-signal oscillators, time-domain
    behavior, 101–121
Broadband noise
  coupled oscillator systems,
    267–268
  1000–2000/2000–4000 MHz
    hybrid tuned VCO, 362–365
Bypass capacitor, phase noise,
  biasing and temperature effects,
  stability factors, 223–224

CAD solutions, phase noise effects
  conversion noise analysis,
    447–448
  frequency conversion approach, 447
  general analysis, 437–438
  Lee and Hajimiri noise model,
    443–444
  Leeson's equation modification,
    442–443
  modulation and conversion
    noise, 444
  modulation noise analysis,
    450–453
  noise generation, 447
  noise performance index
    frequency conversion, 448–450
    noise modulation contribution,
      453–454
  noise signal modulation,
    440–442
  nonlinear computation, 444–447
  oscillator noise models, 442
  PM-AM correlation coefficient,
    454–455
  sinusoidal signal, 438–440
Capacitance ratio, ultra-low-noise
  wideband oscillators, tuning
  characteristics, 324–326
Capacitor ratios, large-signal design,
  Bessel functions, 392–395
Carrier frequency
  N-coupled oscillator noise analysis,
    288–292
  ultra-low-noise wideband oscillators,
    active device, 323
  unilateral N-coupled oscillator noise,
    514–515
Cascade structure, N-push coupled mode
  (synchronized) oscillators,
  300–301
Ceramic resonators
  1000 MHz CRO circuit validation,
    235–237
  design principles, 79–81
CFY30 transistor model
  basic parameters, 49–51
  metal oxide semiconductors,
    51–56
Chalmers (Angelov) model, field-effect
  transistors, 47–51
Channel width/length ratio, 2400 MHz
  MOSFET-based push-pull
  oscillator, 204

520 INDEX

Circuit characteristics
   feedback models, spectral density and transfer function, 197–199
   metal semiconductor field-effect transistor, optimization, 425–432
   mutually locked (synchronized) oscillators, noise analysis, 277–282
   N-push coupled mode (synchronized) oscillators, mode analysis, 307–311
   oscillator noise analysis, 132–137
   ultra-low-noise wideband oscillators, inductor model, 336–339
   uncoupled oscillator noise, 272–276
Clapp oscillator, design techniques, 18–21
Closed-loop gain
   large-signal oscillators, time-domain behavior, 111–121
   oscillator noise analysis, Leeson equation, 126–132
   phase noise analysis, feedback models, resonator design, 191–193
CMOS device, phase noise effects, 207–210
Coarse tuning network, 1000–2000/2000–4000 MHz hybrid tuned VCO, 361–365
Collector current pulses
   biasing topololgy, Agilent HBFP-0405 transistor, 225–230
   bipolar transistor noise model, T-equivalent configuration, 487–489
   coupled oscillator noise sources, 268
   feedback models, noise-shaping function, spectral density and transfer function, 196–199
   large-signal oscillators, time-domain behavior, 117–121
   negative resistance models, phase noise analysis, 175
   1000–2000/2000–4000 MHz push-push oscillator, 349–355
   1500–3000/3000–6000 MHz coupled resonator oscillator, circuit validation, 357–361
   stability factors and, 218–224
Collector-emitter systems, stability factors, 218–224

Collector load, negative resistance models, phase noise analysis, 175
Collector resistor, stability factors, 219–224
Collector reverse current
   bipolar transistors, hybrid-π configuration, 471–473
   coupled oscillator noise sources, 268
Collector voltage, 1000 MHz CRO circuit validation, 234–237
Colpitts/Clapp oscillator, historical background, 2–5
Colpitts oscillator
   1000 MHz CRO circuit validation, 233–237
   basic components, 21–28
   design techniques, 18–21
   large-signal design, 99–100
      Bessel functions, 390–395
      time-domain behavior, 114–121
   N-coupled oscillator noise analysis, 289–292
   negative resistance model, time domain noise signal calculations, 174
   parallel resonator oscillator design, large-signal S-parameters, 382–383
   parallel tuned device, phase noise effects, 159–161
   parasitics, 88–91
   phase noise analysis
      feedback models, 186–199
      spectral density and transfer function, 197–199
      negative resistance model, 162–166
      output power and, 397–405
   series feedback configuration, 385–388
   77 GHz SiGe oscillator, circuit validation, 242–244
   Y-parameters, 28–32
Common-emitter (CE) amplifiers
   1000 MHz CRO circuit validation, 233–237
   ABCD parameters, 468–469
   bipolar transistors
      hybrid-π configuration, noise analysis, 471–473
      series feedback, 418–420
      T-equivalent configuration, 492–494
   coupled oscillator systems, noise sources, 267–268

INDEX **521**

Common mode (CM) pushing factor, N-coupled oscillator noise analysis, push-push configuration, 292–300
Conduction angle, large-signal oscillators, time-domain behavior, 105–121
Congruence transformation, noise correlation matrix, 463–464
Constant base currrent source bias circuit, phase noise and temperature effects, 215–216
Conversion noise analysis
    CAD solution, 444, 447–448
    oscillator design, 144
Correlation matrix
    ABCD matrix
        GE, GC, and GB configurations, 468–469
        input transfer, 465–466
        noise parameters and, 462–463
        noise transformation, 459–460
    bipolar transistors, T-equivalent configuration, 486–489, 492–494
    defined, 458
    GE bipolar transistor, 477–479
    generator current and noise parameter, 460–462
    noise parameters and, 457–458
    noise source transformation, 466–467
    series and shunt elements, 464–465
    transformations, 463–464
Coupled oscillator systems. *See also* Uncoupled oscillators
    amplitude and phase dynamics, 261–262
    classical pendulum analogy, mutal coupling, 247–254
    dynamics, 257–263
    locking bandwidth, 262–263
    mutually locked (synchronized) systems
        noise analysis, 276–282
        phase conditions, 254–257
    N-coupled (synchronized) oscillators
        dynamics, 263–266
        noise analysis, 282–300
            frequency doubler/multiplier configuration, 288–292
            push-push configuration, 292–300
            validation, 300
        noise analysis, 266–271
            Leeson's model, 270–271
            model characteristics, 269–270
            source identification, 267–268

N-push coupled mode (synchronized) oscillators
    design criteria, 311–315
    mode analysis, 307–311
    N-push subcircuits, 306–307
    push/push-push configuration, 303–304
    push-push/two-push design, 311–314
    quadruple-push configuration, 305–306
    system dynamics, 300–315
    triple-push configuration, 304–305
        design criteria, 314–315
    ultra-low-noise wideband oscillators
        active device, 321–323
        capacitance model, 338–339
        coupled resonator design, 326–330
        inductor model, 335–338
        optimum phase noise, loaded Q, 330–333
        passive device, 333–334
        resistor model, 339–340
        tuning characteristics and loaded Q, 324–326
        tuning network, 323–324
        voltage-controlled oscillator design, 320–321
        wideband VCO design, 315–320
    wideband coupled resonator VCOs, validation circuits
        1000–2000/2000–4000 MHz hybrid tuned VCO, 361–365
        1000–2000/2000–4000 MHz push-push oscillator, 346–355
        1500–3000/3000–6000 MHz coupled resonator oscillator, 355–361
        300-1100 MHz coupled resonator oscillator, 341–346
Coupled resonator
    three-dimensional design, 72, 74–79
    ultra-low-noise wideband oscillators, 326–330
Coupling capacitor values
    large-signal design, Bessel functions, 393–395
    phase noise and output power calculations, 402–405
    wideband coupled resonator VCOs, circuit validation, 300–1100 MHz coupled resonator oscillator, 342–346

522 INDEX

Coupling coefficient, N-coupled system dynamics, 263–266
Coupling topology, N-coupled oscillator noise analysis, 288
Cross-coupled oscillator circuits, 2400 MHz MOSFET-based push-pull oscillator
  design criteria, 206–207
  noise analysis, 199–205
Crystal oscillators, noise analysis, experimental variations, 145–148
Current amplitude
  mutually coupled oscillators, noise analysis, 279–282
  negative resistance model, time domain noise signal calculations, 171–174
Curtice-Ettenberg models
  field-effect transistors, 47–51
  transistor design, 39–40
Curtice-Quadratic model, field-effect transistors, 47–51
Curve-fitting procedures, negative resistance noise voltage, 182–183

DC bias points
  bipolar transistors, hybrid-π configuration, 472–473
  coupled oscillator noise sources, 268
  phase noise and output power, 398–405
  S-parameter measurements, 65–69
DC collector current, bipolar transistors, hybrid-π configuration, 472–473
DC offset voltage, large-signal oscillators, time-domain behavior, 103–121
DC/RF conversion efficiency, metal semiconductor field-effect transistor, optimization, 430–432
"Dead zones" of tuning, 1000–2000/2000–4000 MHz hybrid tuned VCO, 361–365
Device phase noise
  passive bias networks, temperature effects, 215–218
  synthesis procedure, 159–161
Device size
  phase noise analysis, 2400 MHz MOSFET-based push-pull oscillators, 204
  2400 MHz MOSFET-based push-pull oscillator, design criteria, 206

Device under test (DUT), S-parameter measurements, 65–69
Dielectric constant, N-coupled oscillator noise, push-push configuration, 295–300
Dielectric resonator
  ceramic design, 79–81
  design principles, 81–83
Differential equations
  metal semiconductor field-effect transistor, series feedback, 408–414
  mutually coupled oscillators
    noise analysis, 277–282
    pendulum analogy, 250–254
  N-coupled oscillator noise, push-push configuration, 293–300
  phase noise analysis, complete oscillator circuit, 164–166
  ultra-low-noise wideband oscillators, optimum phase noise, loaded Q, 332–333
  uncoupled oscillator noise, 271–276
Differential mode (DM) pushing factor, N-coupled oscillator noise analysis, push-push configuration, 292–300
Dipole construction, microwave oscillators, 1–5
Dirac delta function, negative resistance model, time domain noise signal calculations, 168–174
Discrete-device approach, 1000–2000/2000–4000 MHz hybrid tuned VCO, 362–365
Distributed/lumped resonators, design principles, 72–73
Drain current
  bipolar transistors, series feedback, 420
  metal semiconductor field-effect transistor
    series feedback, 411–414
    simulation, 433–435
  phase noise analysis
    CMOS transistors, 208–209
    2400 MHz MOSFET-based push-pull oscillators, 204
Drive levels
  large-signal oscillators, time-domain behavior, 117–121
  negative resistance noise voltage calculations, 179–183

INDEX   523

Driven voltage signals, large-signal oscillators, time-domain behavior, 101–121
Dynamic tuning, voltage-controlled oscillators, 35

Ebers-Moll equations, bipolar transistor models, 38–40
EKV3 model, field-effect transistors, 54–56
Electromotor force (EMF), negative resistance model, noise voltage calculations, 176–183
Emitter currents
  coupled oscillator noise sources, 267–268
  large-signal design, Bessel functions, 390–395
  large-signal oscillators, time-domain behavior, 106–121
  1500–3000/3000–6000 MHz coupled resonator oscillator, circuit validation, 358–361
  phase noise, biasing and temperature effects
    bias values, 230–232
    stability factors, 223–224
Emitter feedback-bias circuit, phase noise and temperature effects, 215, 217
Ensemble averaging, conversion noise analysis, C448AD solution
Equivalent circuit representation
  coupled oscillator systems, 257–263
  FET noise parameter calculation, 497–501
  metal semiconductor field-effect transistors, 423–432
  mutually locked (synchronized) oscillators, noise analysis, 277–282
  N-coupled oscillator noise, 290–292
  negative resistance models
    noise voltage, 176–183
    phase noise analysis, 163–166
  phase noise analysis, 2400 MHz MOSFET-based push-pull oscillators, 204–205
  triple-push configuration, 314–315
Equivalent series resistance (ESR), 1000 MHz CRO circuit validation, 235–237

Even-mode operation
  N-push coupled mode (synchronized) oscillators, 307–311
  push-push/two-push oscillator, 313–314
  triple-push configuration, 314–315
Extended resonance techniques, mutually locked (synchronized) coupled oscillators, 256–257
Extrinsic modeling, ultra-low-noise wideband oscillators, passive device, 333–334

Feedback capacitor ratio, phase noise and output power, 401–405
Feedback conditions
  large-signal oscillators, 94–100
  oscillator noise analysis
    Leeson equation, 127–132
    noise factors, 134–137
  parallel resonator oscillator design, large-signal S-parameters
    three-reactance oscillators, 27–32
  ultra-low-noise wideband oscillators, active devices, 321–323
Feedback model, phase noise calculation and optimization, 185–199
  noise sources, 190
  noise transfer function and spectral densities, 194–199
  nonunity gain, 193–194
  resonator noise-shaping function, 190–193
Ferrimagnetic resonance, yttrium ion garnet resonators, 84–85
Field-effect transistors (FETs). *See also* specific transistors, e.g. Gallium arsenide field-effect transistor
  active bias circuits, phase noise and temperature effects, 212–214
  design principles, 37–40
  GaAs FETs, 47–51
  large-signal oscillators, time-domain behavior, 101–121
  metal oxide semiconductors (BiCMOS), 51–56
  metal semiconductor field-effect transistor
    optimization and circuit configuration, 422–432
    parallel feedback, 414–417
    series feedback, 407–414
    test oscillator simulated results, 432–435

524 INDEX

Field-effect transistors (FETs) (*Continued*)
  microwave oscillator history,
    3–5, 18
  negative resistance noise voltage,
    180–183
  noise parameter calculation,
    496–501
  S-parameter measurements, 65–69
  temperature dependence of noise
    parameters, 503–506
  2000 MHz GaAs FET-based
    oscillator, circuit validation,
    241–242
Fine-tuning network, 1000–2000/
    2000–4000 MHz hybrid tuned
    VCO, 361–365
Flicker noise
  coupled oscillator systems, 267–268
  feedback models, spectral
    density and transfer function,
    195–199
  modulation noise analysis, 451–453
  negative resistance noise voltage,
    179–183
  oscillator noise analysis
    frequency conversion, 141–143
    Leeson equation, 126–132
  ultra-low-noise wideband oscillators,
    wideband VCO design
    technique, 320
Forcing terms, modulation noise
    analysis, 451–453
4100 MHz oscillator, transmission line
    resonators, circuit validation,
    237–241
Fourier coefficient, sinusoidal signal
    modulation, 439–440
Fourier series expansion, large-signal
    oscillators, time-domain
    behavior, 102–121
Fourier transform
  N-coupled (synchronized) oscillators,
    noise analysis, 284–300
  sinusoidal signal modulation, 440
  uncoupled oscillator noise, 272–276
Free-running oscillators
  mutually coupled (synchronized)
    systems, noise analysis, 281–282
  N-coupled oscillator noise analysis,
    topology characteristics, 510–513
  N-coupled system dynamics, 264–266
    noise analysis, 282–300
  uncoupled oscillator noise, 271–276

Frequency conversion
  oscillator noise analysis, 141–143
  phase noise effects, CAD solutions, 447
  noise performance index, 448–450
Frequency-dependent feedback network
  large-signal oscillators, 94–100
  sinusoidal oscillator, 10–11
Frequency-dependent forward loop gain
  large-signal oscillators, 94–100
  sinusoidal oscillator, 10–11
Frequency-dependent measurements,
    S-parameters, 66–69
Frequency deviation, sinusoidal signal
    modulation, 438–440
Frequency doubler/multiplier
  N-coupled oscillator noise analysis,
    288–292
  validation, 300
  N-push coupled mode (synchronized)
    oscillators, 300–301
Frequency errors, oscillator phase
    noise model, synthesis
    procedure, 161
Frequency modulated continuous wave
    (FMCW) radars, ultra-low-noise
    wideband oscillators, 323–324
Frequency pushing, voltage-controlled
    oscillator, 16
Frequency response
  1000 MHz CRO circuit validation,
    235–237
  coupled microstrip line resonator,
    75–79
  mutually coupled oscillators, pendulum
    analogy, 252–254
  N-coupled (synchronized)
    oscillators, 266
  oscillator noise analysis, 134–137
  300–1100 MHz coupled resonator
    oscillator, 344–347
  2400 MHz MOSFET-based push-pull
    oscillator, design criteria,
    206–207
  uncoupled oscillator noise, 275–276
  voltage-controlled oscillator, 13
Frequency stability/instability
  coupled oscillator system noise,
    269–270
  phase noise analysis, 437–438
Fundamental frequency ratio
  large-signal oscillators, time-domain
    behavior, 107–121
  modulation noise analysis, 452–453

INDEX 525

N-coupled oscillator noise
  frequency doubler/multiplier,
    291–292
  push-push configuration, 294–300
N-push coupled mode (synchronized)
  oscillators, quadruple-push
  configuration, 305–306
  1000–2000/2000–4000 MHz
    push-push oscillator, 349–355
  1500–3000/3000–6000 MHz coupled
    resonator oscillator, circuit
    validation, 359–361
Fundamental oscillator phase noise,
  N-coupled oscillator noise
  analysis, 288–292

Gallium arsenide (GaAs) technology
  field-effect transistor
    design criteria, 47–51
    noise analysis
      approximation and discussion,
        507–508
      noise correlation matrix values,
        501–503
      parameter extraction, 496–501
      room temperature, 494–496
      temperature dependence, 503–507
  microwave oscillator history, 18
  transistor design, 39–40
  2000 MHz GaAs FET-based
    oscillator, circuit validation,
    241–242
Gate current, CMOS phase noise,
  207–208
Gate drain capacitance
  FET noise parameter calculation,
    499–501
  metal semiconductor field-effect
    transistor, series feedback,
    411–414
Gate leakage current, FET noise
  parameter calculation,
  499–501
Gate-oxide capacitance, phase noise
  analysis, 2400 MHz
  MOSFET-based push-pull
  oscillators, 203
Gate resistance, CMOS phase noise,
  207–208
Globally N-coupled oscillators, noise
  analysis, 510–512
Grounded base (GB) oscillator
  ABCD parameters, 468–469

basic components, 21–28
ultra-low-noise wideband oscillators
  optimum phase noise, loaded Q,
    332–333
wideband VCO design technique,
  316–320
Grounded collector (GC)
  ABCD parameters, 468–469
  noise parameters and, 460–462
  noise source transformation,
    466–467
Grounded-emitter (GE) configuration
  ABCD parameters, 468–469
  bipolar transistors
    hybrid-π configuration, noise
      analysis, 471–473
    noise correlation matrix, 477–479
    noise current source, input transfer,
      473–474
    noise factor calculations,
      474–476
    phase noise analysis, feedback
      models, 185–199
  coupled oscillator noise sources,
    267–268
  noise source transformation,
    466–467
Gummel-Poon bipolar
  transistor model
    design principles, 38–40
    historical background, 3–5

Half-butterfly resonators
  active bias circuits, phase noise and
    temperature effects, 211–214
  900–1800 MHz half-butterfly
    resonator-based oscillator, circuit
    validation, 245–246
Harmonic generation/suppression.
  See also Second-order
  harmonics
  large-signal design, Bessel functions,
    394–395
  large-signal oscillators, time-domain
    behavior, 102–121
  modulation noise analysis, 451–453
  mutually coupled oscillators, noise
    analysis, 277–282
  noise performance index
    frequency conversion, 448–450
    noise modulation, 454
  N-push coupled mode (synchronized)
    oscillators, 300–307

526 INDEX

Harmonic generation/suppression
(*Continued*)
  push/push-push configuration,
    303–304
  quadruple-push configuration,
    305–306
  triple-push configuration, 304–305
  1000–2000/2000–4000 MHz
    hybrid tuned VCO, 363–365
  oscillator conversion noise
    analysis, 144
  300–1100 MHz coupled resonator
    oscillator, 344–346
  ultra-low-noise wideband oscillators,
    wideband VCO design technique,
    319–320
Harmonic suppression, voltage-controlled
  oscillator, 14–15
Hermitian conjugate
  bipolar transistors, T-equivalent
    configuration, noise correlation
    matrix, 493–494
  noise correlation matrix, 458
Heterojunction bipolar transistor (HBT)
  model
  basic principles, 37–40
  historical background, 3–5
High-frequency deviations
  oscillator noise analysis, 140
  phase noise analysis, CAD
    solutions, 447
High-pass configuration, oscillator
  design, 27–28
Hybrid π configuration, bipolar transistor
  noise analysis, 471–473
  T-equivalent configuration, 486–489
Hybrid-π configuration, phase noise
  analysis, feedback models,
  186–199
Hybrid tuned voltage-controlled
  oscillators, 1000–2000/
  2000–4000 MHz hybrid tuned
  VCO, 361–365
Hyperabrupt junction, transistor design,
  tuning diodes, 56–60

IAF (Berroth) model, field-effect
  transistors, 47–51
Impedance transformation
  bipolar transistors, real source
    impedance, 476–477, 489
  large-signal oscillators, time-domain
    behavior, 109–121

push-push/two-push oscillator,
  312–314
ultra-low-noise wideband oscillators
  coupled resonator, 326–330
  inductor model, 339
uncoupled oscillator noise, 272–276
Impulse sensitivity function (ISF)
  historical background, 3–5
  oscillator noise analysis, Lee and
    Hajimiri noise model, 138–139
Inductor model
  large-signal design, Bessel functions,
    392–395
  1500–3000/3000–6000 MHz coupled
    resonator oscillator, circuit
    validation, 357–361
  ultra-low-noise wideband oscillators,
    335–339
Infineon transistors
  active bias circuits, phase noise and
    temperature effects, 212–214
  bipolar transistor design, 41–47
  S-parameter measurements, 67–69
Input impedance
  N-coupled oscillator noise analysis,
    289–292
    push-push configuration, 294–300
  oscillator parasitics, 87–91
  ultra-low-noise wideband oscillators,
    passive device, 334
Input stability circle, 300–1100 MHz
  coupled resonator oscillator,
  343–346
Input transfer
  ABCD matrix, noise transfer using,
    465–466
  bipolar transistors, noise current source
    transfer, 473–474
  noise source transformation,
    466–467
Insertion loss parameter, ultra-low-noise
  wideband oscillators, active
  device, 323
Integrated resonators, design principles,
  72–79
ISF function, Lee and Hajimiri noise
  model, 444
ITT PFET/TFET models, field-effect
  transistors, 47–51

Jacobian matrices, phase noise analysis,
  CAD solutions, 446–447
Jitter, modulation noise analysis, 450–453

INDEX 527

Junction field-effect transistors (JFETs)
  design principles, 38–40
  noise analysis, room temperature conditions, 495–496

Kirchoff's voltage law (KVL)
  bipolar transistor noise model, T-equivalent configuration, 487–489
  mutually coupled oscillators, pendulum analogy, 249–254
  negative resistance noise voltage calculations, 177–183
  S-parameter measurements, 65–69
  uncoupled oscillator noise, 271–276

Large-signal oscillator design
  Bessel functions, 389–395
  bipolar transistors, series feedback, 420
  design principles, 94–121
    startup condition, 94–100
    steady-state behavior, 100–101
    time-domain behavior, 101–121
  metal semiconductor field-effect transistor, series feedback, 413–414
  S-parameters, 63–69
    parallel resonator oscillator, 381–383
    series feedback configuration, 384–388
  2400 MHz MOSFET-based push-pull oscillator, transconductance measurements, 206
  ultra-low-noise wideband oscillators passive device, 333–334
    wideband VCO design technique, 319–320
Large-signal transconductance, phase noise and output power, 401–405
Lateral defused MOS (LDMOS), field-effect transistors, 54–56
Lead inductance, oscillator parasitics, 87–91
Lecher lines, microwave oscillators, 17–18
Lee and Hajimiri noise model
  coupled oscillator system noise, 269–270
  oscillator noise analysis, 137–139
  phase noise analysis, CAD solution, 443–444

Leeson equation
  basic principles, 6
  coupled oscillator system noise, 269–271
  oscillator noise analysis, 124–132
  oscillator phase noise model, synthesis procedure, 161
  phase noise effects, 12–13
    CAD solutions, 442–443
Leeson's noise model, coupled oscillator systems, 269–271
Linearity
  large-signal oscillators, 94–100
  oscillator design, 23–28
  voltage-controlled oscillator tuning, 17
Linear representation, phase noise analysis, feedback models, 187–199
Linear theory equations
  oscillator design, 87
  oscillator noise, 123–137
    Leeson's technique, 124–132
Linear time invariant (LTIV) properties
  coupled oscillator system noise, 269–270
    Leeson's model, 271
  large-signal oscillators, 96–100
  Leeson noise equation, 443
  ultra-low-noise wideband oscillators, passive device, 334
Linear tuning response, 1000–2000/2000–4000 MHz hybrid tuned VCO, 363–365
Load change sensitivity, voltage-controlled oscillator, 16
Load conductance
  mutually locked (synchronized) coupled oscillators, 255–257
  phase noise analysis, feedback models, 186–199
Load current amplitude
  bipolar transistor noise model, T-equivalent configuration, 487–489
  bipolar transistors, series feedback, 420
  metal semiconductor field-effect transistor
    parallel feedback, 416–417
    series feedback, 408–414
Load impedance, metal semiconductor field-effect transistor
  optimization, 428–432
  parallel feedback, 416–417
  series feedback, 408–414

528 INDEX

Load lines, metal semiconductor
field-effect transistor simulation,
432–434
Load pulling, S-parameter measurements,
66–69
Locking bandwidth, coupled oscillator
systems, 262–263
Loop approach
bipolar transistor noise model,
T-equivalent configuration,
487–489
noise analysis, 6
Loop gain, phase noise and output power,
401–405
Loop transfer function, large-signal
oscillators, 95–100
Loss resistances
metal semiconductor field-effect
transistor, optimization, 430–432
mutually coupled oscillators, pendulum
analogy, 248–254
Lossy resonators, phase noise and output
power, 404–405
Low-frequency deviations
coupled oscillator system noise,
267–268
N-coupled oscillator noise analysis,
288–292
oscillator noise analysis, 140
phase noise analysis,
CAD solutions, 447
Low-pass configuration, oscillator design,
21–28
Lumped capacitor (LC) resonator, design
principles, 71–72

Materka-Kacprzak intrinsic model
bipolar transistor design, 44–47
field-effect transistors, 48–51
Materka model, metal semiconductor
field-effect transistor,
optimization, 430–432
MATHCAD equations, S-parameter
measurements, 67–69
Maximum value limits, metal
semiconductor field-effect
transistor, optimization, 428–432
Mean square noise voltage
bipolar transistors, hybrid-π
configuration, 472–473
coupled oscillator noise, 268
GaAs-FET noise analysis, room
temperature conditions, 494–496

negative resistance model, time
domain noise signal calculations,
168–174
phase noise analysis, feedback models,
185–199
Measurement techniques, S-parameters,
63–69
Mechanical post tuning, dielectric
resonator design, 82–83
Meissen oscillator
basic components, 21–28
Y-parameters, 28–32
MESFET model, field-effect transistors,
48–51
Metal oxide semiconductor
(MOS) models
field-effect transistors, 51–56
model level 3, 55–56
historical background, 3–5
transistor design, 38–40
Metal-oxide silicon field-effect transistor
(MOSFET)
tuning diodes, 57–60
2400 MHz MOSFET-based push-pull
oscillator, phase noise
calculation and optimization,
199–210
design equations, 201–205
device size, 206
large-signal transconductance, 206
oscillation frequency, 206–207
phase noise design calculations,
207–210
resonator parallel loss resistance, 206
Metal semiconductor field-effect transistor
(MESFET)
noise analysis, room temperature
conditions, 495–496
optimization and circuit configuration,
422–432
parallel feedback, 414–417
series feedback, 407–414
test oscillator simulated results,
432–435
Microstrip resonators, design
principles, 72–79
distributed/lumped resonators, 72
integrated resonators, 72–79
Microwave oscillators
design criteria, 18–21
historical background, 1–5, 17–18
metal semiconductor field-effect
transistor (MESFET)

optimization and circuit
configuration, 422–432
parallel feedback, 414–417
series feedback, 407–414
test oscillator simulated results,
432–435
Minimum noise factor
bipolar transistor noise analysis,
490–492
generator current and, 461–462
Modulation frequency, mutually coupled
oscillators, pendulum analogy,
252–254
Modulation index, sinusoidal signal
modulation, 439–440
Modulation noise
CAD solution, 444, 450–453
oscillator noise analysis
Leeson equation, 130–132
nonlinear calculations, 144–145
MOS BISM Model 3v3, field-effect
transistors, 54–56
Multidimensional fast Fourier transform
(MFFT), historical
background, 4–5
Multioctave band capability, 1000–2000/
2000–4000 MHz push-push
oscillator, 349–355
μ factor, N-coupled system
dynamics, 264–266
Mutually coupled (synchronized)
oscillators
noise analysis, 276–282
phase condition, 254–257

Narrowband noise signal
negative resistance model,
time domain noise signal
calculations, 170–174
uncoupled oscillator noise, 274–276
N-coupled (synchronized) oscillators
coupling parameters, 265–266
dynamics, 263–266
frequency parameters, 266
noise analysis, 282–300
bilateral systems, 512–513
frequency doubler/multiplier
configuration, 288–292
globally coupled systems, 510–512
push-push configuration, 292–300
topology characteristics, 509–513
unilateral systems, 513–515
validation, 300

NE68830 system
1000 MHz CRO circuit validation,
233–237
nonlinear parameters, 403–405
oscillator noise analysis, 133–137
ultra-low-noise wideband oscillators,
328–330
Nearest neighbor unilateral coupling,
N-coupled (synchronized)
oscillators, noise analysis,
283–300, 513–515
Negative feedback, phase noise,
biasing and temperature effects,
stability factors, 219–224
Negative gate voltage MOS transistor
(NMOS), transconductance
equations, 201–203
Negative resistance/capacitance
large-signal oscillators, 94–100
steady-state behavior, 100–101
time-domain behavior, 117–121
oscillator parasitics, 89–91
Negative resistance model
1000 MHz CRO circuit validation,
235–237
coupled oscillator systems, 258–263
mutually coupled oscillators, noise
analysis, 278–282
1500–3000/3000–6000 MHz coupled
resonator oscillator, circuit
validation, 355–361
phase noise analysis, 162–184
collector current equations, 175
noise voltage calculation, 176–183
nonlinear negative resistance region,
166–167
oscillator design parameters,
183–184
time domain noise signal, 167–174
2400 MHz MOSFET-based
push-pull oscillators, 200–210
push-push/two-push oscillator, 314
series feedback oscillator, 386–388
300–1100 MHz coupled resonator
oscillator, circuit validation,
342–346
ultra-low-noise wideband oscillators,
tuning characteristics,
324–326
uncoupled oscillator noise, 271–276
900–1800 MHz half-butterfly
resonator-based oscillator,
circuit validation, 245–246

530 INDEX

Noise analysis. *See also*
  Phase noise effects
 ABCD parameters, GE, GC, and GB
    configurations, 468–469
 bipolar transistor design, 43–47
    base resistance parameters, exclusion
      of, 479–485
    current source transformation
      to input, GE configuration,
      473–474
    hybride π configuration, 471–473
    minimum noise factor, 490–492
    noise correlation matrix
      GE configuration, 477–479
      T-equivalent configuration,
        492–494
    noise factor calculations, 474–476
    real source impedance,
      476–477, 489
    T configuration, 485–489
 correlation matrix
   ABCD matrix
     parameter extraction, 462–463
     transformation using, 459–460
   n-port networks, 457–458
   transformations, 463–464
 coupled oscillator systems, 266–271
   Leeson's model, 270–271
   model characteristics, 269–270
   source identification, 267–268
 GaAs FET model
   approximation and discussion,
     507–508
   noise correlation matrix values,
     501–503
   parameter extraction, 496–501
   room temperature, 494–496
   temperature dependence,
     503–507
 generator current and noise parameters,
   460–462
 large-signal oscillators, time-domain
   behavior, 120–121
 metal oxide field-effect transistor,
   tuning diodes, 57–60
 metal oxide semiconductor field-effect
   transistors, 55–56
 microwave oscillators, 2–5
 modulation rates
   oscillator design, 131–132
   phase noise, synthesis, 161
 mutually locked (synchronized)
   oscillators, 276–282

 N-coupled (synchronized) oscillators,
   282–300
   bilateral systems, 512–513
   frequency doubler/multiplier
     configuration, 288–292
   globally coupled systems, 510–512
   push-push configuration, 292–300
   topology characteristics, 509–513
   unilateral systems, 513–515
   validation, 300
 offset frequency, N-coupled
   (synchronized) oscillators, noise
   analysis, 285–300
 oscillators
   Lee and Hajimiri model, 137–139
   linear techniques, 123–137
     Leeson's approach, 124–132
   nonlinear phase noise calculations,
     139–148
   phase noise measurements,
     148–153
     spectrum analyzer, 149
     test protocols, 149–153
   support circuits, 153–157
     comb generator, 156–157
     crystal oscillators, 153–157
 series and shunt matrix definitions,
   464–465
 signal modulation, CAD solutions,
   440–442
 source transfer to input, 465–466
 source transformation, 466–467
 two-port matrix combination,
   458–459
 ultra-low-noise wideband oscillators,
   active devices, 321–323
 uncoupled oscillators, 271–276
Noise factors
 bipolar transistor noise analysis,
   474–476
   minimum noise factor, 490–492
   noise parameters excluding base
     resistance, 482–485
   real source impedance, 489
   T-equivalent configuration,
     487–489
 generator current and, 461–462
Noiseless oscillators
 noise parameters and correlation
   matrix, 457–458
 phase noise analysis, oscillator circuit
   equivalent representation,
   163–166

INDEX   531

Noise performance index, frequency conversion, 448–450
Noise transfer function, phase noise analysis, feedback models
  resonator design, 192–193
  spectral density, 194–199
Noise voltage
  bipolar transistors, noise factor calculations, 475–476
  negative resistance model
    equivalent representations, 176–183
    time domain noise signal calculations, 168–174
  phase noise analysis
    feedback models, 188–199
    oscillator circuit equivalent representation, 163–166
Noisy negative resistance, basic principles, 162–166
Nonlinear circuits
  negative resistance calculations, 166–167
  noise computation, 444–447
  oscillator noise analysis, equivalent representations, 140
  phase noise and output power, 403–405
Nonlinearity
  large-signal oscillators, 94–100
    time-domain oscillators, 101–121
  oscillator noise analysis, 139–148
    circuit equivalent representation, 140
    conversion noise, 144–145
    experimental variations, 145–148
    frequency conversion, 141–143
    noise generation, 140
Nonlinear time variant (NLTV) properties
  large-signal oscillators, 96–100
  oscillator noise analysis, 137–139
Nonstabilized bias circuits, phase nosie and temperature effects, 215–218
Nonunity gain, phase noise analysis, feedback models, 193–194
Normalized fundamental currents
  large-signal design, Bessel functions, 390–395
  large-signal oscillators, time-domain behavior, 104–121
  phase noise and output power, 398–405
Norton equivalent, oscillator noise analysis, nonlinear calculations, 145

Norton phasor, modulation noise analysis, 452–453
NPN transistors
  4100 MHz oscillator, circuit validation, 237–241
  silicon germanium, design principles, 43–47
N-push coupled mode (synchronized) oscillators
  design criteria, 311–315
  mode analysis, 307–311
  N-push subcircuits, 306–307
  push/push-push configuration, 303–304
  push-push/two-push design, 311–314
  quadruple-push configuration, 305–306
  system dynamics, 300–315
  triple-push configuration, 304–305
    design criteria, 314–315
Nyquist criterion, sinusoidal oscillator, 10–11

Odd-mode operation
  N-push coupled mode (synchronized) oscillators, 307–311
  push-push/two-push oscillator, 311–314
  triple-push configuration, 314–315
One-port oscillators
  design techniques, 18–21
  large-signal oscillators, 96–100
  parallel feedback, 92–93
  1000–2000/2000–4000 MHz hybrid tuned VCO, circuit validation, 361–365
  1000–2000/2000–4000 MHz push-push oscillator, circuit validation, 346–355
  1000 MHz CRO, circuit validation, 233–237
  1500–3000/3000–6000 MHz coupled resonator oscillator, circuit validation, 355–361
Open-circuit noise voltage, negative resistance model, 176–183
Open-loop gain
  metal semiconductor field-effect transistors, 424–432
  phase noise analysis, feedback models, 189–199
Operating point calculation, phase noise and output power, 398–405

532 INDEX

Orthogonal equations
  negative resistance model, time domain noise signal calculations, 167–174
N-push coupled mode (synchronized) oscillators, 309–311
uncoupled oscillator noise, 275–276
Oscillation frequency
  metal semiconductor field-effect transistor
    optimization, 429–432
    series feedback, 408–414
  1000–2000/2000–4000 MHz push-push oscillator, 349–355
  1500–3000/3000–6000 MHz coupled resonator oscillator, circuit validation, 357–361
  phase noise and output power calculations, 402–405
  2400 MHz MOSFET-based push-pull oscillator, design criteria, 206–207
Oscillator circuit equivalent representation
  negative resistance model, noise voltage calculations, 176–183
  phase noise analysis, negative resistance model, 163–166
Oscillator noise models, CAD solutions, 442
Oscillators
  basic properties, 9
  Colpitts, 21–28
  equations, 87–93
    large-scale time-domain behavior, 111–121
    linear theory, 87
    parallel feedback, 91–93
    parasitics, 87–91
  grounded-base design, 21–28
  large-signal design, 94–121
    S-parameters, 381–388
    startup condition, 94–100
    steady-state behavior, 100–101
    time-domain behavior, 101–121
  Meissen oscillator, 21–28
  microwave oscillators
    design techniques, 18–21
    history, 17–18
  noise analysis
    Lee and Hajimiri model, 137–139
    linear techniques, 123–137
      Leeson's approach, 124–132
      nonlinear phase noise calculations, 139–148
    phase noise measurements, 148–153
      spectrum analyzer, 149
      test protocols, 149–153
      support circuits, 153–157
        comb generator, 156–157
        crystal oscillators, 153–157
  phase noise effects, 11–13
  sinusoidal, 10–11
  three-reactance Y-parameters, 28–32
  voltage-controlled oscillators, 13–17
    basic components, 32–35
    frequency pushing, 16
    frequency range, 13
    harmonic suppression, 14–15
    load change sensitivity, 16
    output power, 14
    phase noise effect, 14
    post-tuning drift, 16
    power consumption, 17
    spurious response, 15–16
    temperature-dependent output power, 15
    tuning characteristics, 17
Output impedance
  bipolar transistors, series feedback, 418–420
  metal semiconductor field-effect transistor, series feedback, 408–414
  series feedback oscillators, 384–388
Output power
  bipolar transistors
    parallel feedback, 421–422
    series feedback, 418–420
  coupled oscillator systems, 260–263
  design examples, 397–405
  large-signal oscillators, time-domain behavior, 106–121
  Meissen oscillators, 26–28
  metal semiconductor field-effect transistor
    optimization techniques, 422–432
    parallel feedback, 415–417
    series feedback, 408–414
  N-coupled system dynamics, 264–266
  push-push configuration, 296–300
  1000–2000/2000–4000 MHz push-push oscillator, 348–355
  parallel feedback oscillator, 92–93
  parallel resonator oscillator design, large-signal S-parameters, 382–383
  phase noise analysis

INDEX **533**

CMOS oscillators, 209–210
  feedback models, resonator design, 191–193
  2400 MHz MOSFET-based push-pull oscillators, 201–203
  300–1100 MHz coupled resonator oscillator, 344–346
  voltage-controlled oscillator, 14
  temperature effects, 15
Output voltage, sinusoidal oscillator, 10–11

Package capacitance
  N-coupled oscillator noise analysis, 289–292
  oscillator parasitics, 87–91
  phase noise and output power, 404–405
Parallel admittance topology, large-signal oscillators, 99–100
Parallel feedback models
  bipolar transistors, 420–422
  metal semiconductor field-effect transistor, 414–417
  N-coupled system dynamics, 264–266
  oscillator theory, 91–93
  ultra-low-noise wideband oscillators
    coupled resonators, 328–330
    tuning characteristics, 325–326
    wideband VCO design technique, 316–320
Parallel loss resistance, 2400 MHz MOSFET-based push-pull oscillator, 206
Parallel negative conductance, N-coupled (synchronized) oscillators, noise analysis, 282–300
Parallel resonator oscillator design, large-signal S-parameters, 381–383
Parallel-tuned circuits
  integrated resonators, 72, 74–79
  Meissen oscillators, 24–28
  phase noise and output power, 397–405
Parallel two-port oscillators, combined matrices, 458–459
Parameter extraction
  ABCD correlation matrix, 462–463
  bipolar transistors, base resistance exclusion, 479–485
  coupled oscillator systems, 265–266
  FET noise parameters, 501–503
    approximation techniques, 507–508
  field-effect transistors, 49–51
  GaAs FET noise analysis, 496–501

negative resistance phase noise models, 183–184
  noise correlation matrix, 457–458
  radio frequency integrated circuits, 57–60
  transistor design, 39–40
Parasitics
  FET noise parameter calculation, temperature dependence, 505–506
  large-scale series feedback oscillator, 386–388
  metal semiconductor field-effect transistor, optimization, 425–432
  oscillator equations, 87–91
  oscillator noise analysis, 133–137
  ultra-low-noise wideband oscillators
    coupled resonators, 328–330
    passive device, 333–334
    tuning characteristics, 324–326
Passive bias circuits, phase noise and temperature effects, 214–218
Passive device, ultra-low-noise wideband oscillators, 333–334
Pendulum analogy, mutually coupled oscillators, 247–254
Phase dynamics
  bipolar transistors, series feedback, 419–420
  coupled oscillator systems, 261–262
  mutually coupled oscillators, noise analysis, 278–282
  mutually locked (synchronized) coupled oscillators, 254–257
  N-coupled oscillators, 263–266
  uncoupled oscillator noise, 272–276
Phase-locked loop (PLL)-based frequency synthesizer, ultra-low-noise wideband oscillators
  tuning network, 323–324
  wideband VCO design technique, 315–320
Phase modulation (PM) noise
  coupled oscillator systems, 266–267
  globally N-coupled oscillators, 510–512
  mutually coupled oscillators, 280–282
  N-coupled oscillators, 285–300
  noise performance index
    frequency conversion, 448–450
    noise modulation, 453–454
  PM-AM correlation coefficient, 454–455

Phase noise effects. *See also* Noise analysis
  1000 MHz CRO circuit validation, 236–237
  CAD solution
    conversion noise analysis, 447–448
    frequency conversion approach, 447
    general analysis, 437–438
    Lee and Hajimiri noise model, 443–444
    Leeson's equation modification, 442–443
    modulation and conversion noise, 444
    modulation noise analysis, 450–453
    noise generation, 447
    noise performance index
      frequency conversion, 448–450
      noise modulation contribution, 453–454
    noise signal modulation, 440–442
    nonlinear computation, 444–447
    oscillator noise models, 442
    PM-AM correlation coefficient, 454–455
    sinusoidal signal, 438–440
  calculation and optimization
    biasing and temperature effects, 210–232
      active bias circuits, 211–214
      bias selection criteria, 230–232
      design criteria, 224–230
      passive bias circuits, 214–218
      stability factors and temperature-dependent variables, 218–224
    feedback model, 185–199
      noise sources, 190
      noise transfer function and spectral densities, 194–199
      nonunity gain, 193–194
      resonator noise-shaping function, 190–193
    negative resistance model, 162–184
      collector current equations, 175
      noise voltage calculation, 176–183
      nonlinear negative resistance region, 166–167
      oscillator design parameters, 183–184
      time domain noise signal, 167–174
      oscillator configurations, 159–160
      synthesis model, 159–161
  2400 MHz MOSFET-based push-pull oscillator, 199–210
    design equations, 201–205
    device size, 206
    large-signal transconductance, 206
    oscillation frequency, 206–207
    phase noise design calculations, 207–210
    resonator parallel loss resistance, 206
  Colpitts series feedback oscillator, 385–388
  coupled oscillator systems, VCO models, 269–270
  coupled resonator design, 77–79
    design examples, 397–405
    dielectric resonator design, 81–83
  4100 MHz oscillator, circuit validation, 239–241
  large-signal design, Bessel functions, 394–395
  large-signal oscillators, time-domain behavior, 112–121
  Meissen oscillators, 26–28
  metal semiconductor field-effect transistor, 433–435
  1000–2000/2000–4000 MHz hybrid tuned VCO, 362–365
  1000–2000/2000–4000 MHz push-push oscillator, 349–355
  1500–3000/3000–6000 MHz coupled resonator oscillator, circuit validation, 358–361
  oscillator design, 11–13
  oscillator noise analysis
    Leeson equation, 124–132
    measurement techniques, 148–153
    nonlinear approaches, 139–148
  output power and, 401–405
  77 GHz SiGe oscillator, circuit validation, 242–244
  300–1100 MHz coupled resonator oscillator, 343–346
  ultra-low-noise wideband oscillators, 330–333
  VCO design criteria, 320–321
  voltage-controlled oscillator, 14
  yttrium ion garnet resonators, 83–85
Phase sensitivity, phase noise analysis, 438

INDEX **535**

PIN diodes, voltage-controlled oscillators, 32–35
π-configuration
  bipolar transistors
    noise correlation matrix, GE configuration, 477–479
    noise current source transfer, 473–474
  coupled oscillator noise sources, 267–268
  hybrid-π configuration
    bipolar transistor noise analysis, 471–473
    phase noise analysis, feedback models, 186–199
    1500–3000/3000–6000 MHz coupled resonator oscillator, circuit validation, 358–361
PM-AM correlation coefficient
  oscillator conversion noise analysis, 144
  phase noise analysis, CAD solution, 454–455
PN junction
  biasing topololgy, Agilent HBFP-0405 transistor, 227–230
  transistor design, tuning diodes, 56–60
Polynomial coefficients, oscillator parasitics, 90–91
Port-coupling networks, N-coupled (synchronized) oscillators, noise analysis, 282–300
Positive gate voltage MOS transistor (PMOS), transconductance equations, 201–203
Post-tuning drift, voltage-controlled oscillator, 16
Power consumption, voltage-controlled oscillator, 17
Pushing factor, N-coupled oscillator noise analysis
  frequency doubler/multiplier, 288–292
  push-push configuration, 292–300
Push-pull configuration, phase noise analysis, 2400 MHz MOSFET-based push-pull oscillators, 199–210
Push-push configuration
  N-coupled oscillator noise, 292–300
  validation, 300

N-push coupled mode (synchronized) oscillators
  design limitations, 302
  mode analysis, 307–311
  1000–2000/2000–4000 MHz push-push oscillator, circuit validation, 346–355
Push/push-push oscillator
  N-push coupled mode (synchronized) configuration, 303–304
  ultra-low-noise wideband oscillators, VCO design criteria, 321
Push-push/two-push oscillator, design analysis, 311–314
p values, negative resistance noise voltage, 182–183

Q factor
  ceramic resonator design, 79–81
  coupled oscillator system noise, 269–271
  dielectric resonator design, 81–83
  distributed/lumped resonators, 72–73
  feedback models, noise-shaping function, spectral density and transfer function, 197–199
  integrated resonator design, 72–79
  large-scale series feedback oscillator, 386–388
  large-signal oscillators, 95–100
  lumped capacitor resonator, 71–72
  N-coupled system dynamics, 264–266
  noise analysis, 283–300
  negative resistance model, time domain noise signal calculations, 173–174
  N-push coupled mode (synchronized) oscillators, push/push-push configuration, 303–304
  1500–3000/3000–6000 MHz coupled resonator oscillator, circuit validation, 358–361
  oscillator noise analysis, Leeson equation, 127–132
  phase noise analysis
    negative resistance model, 162–166
    2400 MHz MOSFET-based push-pull oscillators, 199–210
    300–1100 MHz coupled resonator oscillator, circuit validation, 341–346
  transistor design, tuning diodes, 59–60
  ultra-low-noise wideband oscillators
    active device, 322–323

Q factor (*Continued*)
  coupled resonator, 326–330
  loaded tuning characteristics, 324–326
  optimum phase noise, loaded Q, 330–333
  yttrium ion garnet resonators, 83–85
Quadratic equations, large-signal design, Bessel functions, 391–395
Quadruple-push oscillator, N-push coupled mode (synchronized) oscillators, 305–306
Quality factor
  large-signal design, Bessel functions, 392–395
  ultra-low-noise wideband oscillators
    coupled resonator, 328–330
    passive device, 334
Quasi-linear analysis, metal semiconductor field-effect transistor, series feedback, 412–414
q values, negative resistance noise voltage, 182–183

Radio frequency integrated circuits (RFICs)
  4100 MHz oscillator, circuit validation, 239–241
  large-signal oscillators, time-domain behavior, 101–121
  microwave oscillator history, 18
  parameter extraction, 56–60
  2400 MHz MOSFET-based push-pull oscillators, 199–210
Radio frequency (RF) measurements
  microwave oscillator history, 1–5
  mutually locked (synchronized) oscillators, noise analysis, 277–282
  negative resistance model, time domain noise signal calculations, 171–174
  nonlinear negative resistance calculation, 166–167
  1000–2000/2000–4000 MHz push-push oscillator, 349–355
  S-parameters, 65–69
  ultra-low-noise wideband oscillators
    inductor model, 335–339
    passive device, 333–334
Rayleigh distribution, noise signal modulation, 440–442
Raytheon (Statz) model, field-effect transistors, 48–51

Real source impedance, bipolar transistors
  noise analysis, 476–477
  noise factor calculations, 489
Reciprocal mixing, oscillator phase noise effects, 11–13
Reciprocity, triple-push configuration, 310–311
Rectangular pulse amplitude, negative resistance model, time domain noise signal calculations, 168–174
Reflected wave measurements, S-parameters, 63–69
Resistor model, ultra-low-noise wideband oscillators, 339–340
Resistor values
  biasing topololgy, Agilent HBFP-0405 transistor, 225–230
  coupled oscillator noise sources, 268
Resonance condition
  large-signal oscillators, 97–100
  oscillator parasitics, 90–91
Resonance frequency
  parallel resonator oscillator design, large-signal S-parameters, 383
  2400 MHz MOSFET-based push-pull oscillator, 204–205
  uncoupled oscillator noise, 274–276
Resonator design
  ceramic resonators, 79–81
  CMOS phase noise, 207–208
  Colpitts, grounded base, and Meissen oscillators, 23–28
  dielectric resonators, 81–83
  feedback models, noise-shaping function, 190–193
    spectral density and transfer function, 195–199
  historical background, 4–5
  LC resonators, 71–72
  loss resistance, negative resistance model, noise voltage calculations, 176–183
  microstrip resonators, 72–79
    distributed/lumped resonators, 72
    integrated resonators, 72–79
  N-coupled oscillator noise, push-push configuration, 295–300
  N-push coupled mode (synchronized) oscillators, 306–307
  2400 MHz MOSFET-based push-pull oscillator, parallel loss resistance, 206

ultra-low-noise wideband oscillators,
    active devices, 321–323
wideband coupled resonator VCOs,
    circuit validation
  1000–2000/2000–4000 MHz hybrid
      tuned VCO, 361–365
  1000–2000/2000–4000 MHz
      push-push oscillator, 346–355
  1500–3000/3000–6000 MHz coupled
      resonator oscillator, 355–361
  300–1100 MHz coupled resonator
      oscillator, 341–346
  yttrium ion garnet resonators, 83–85
RLC resonator
  N-coupled system dynamics, push-push
      configuration, 296–300
  phase noise analysis, feedback models,
      nonunity gain, 194
Roger substrate, N-coupled oscillator
    noise, push-push configuration,
    295–300
Rohde oscillator model
  coupled oscillator system noise, 269–270
  uncoupled oscillator noise, 271–276
Rohde & Schwarz ZVR network
    analyzer, S-parameter
    measurements, 65–69
Room temperature conditions, GaAs-FET
    noise analysis, 494–496

Second-order harmonics. *See also* Harmonic
    generation/suppression
  1000–2000/2000–4000 MHz push-push
      oscillator, 348–355
  1500–3000/3000–6000 MHz coupled
      resonator oscillator, circuit
      validation, 359–361
  300–1100 MHz coupled resonator
      oscillator, 344–347
Second-order homogeneous equations,
    mutually coupled oscillators,
    pendulum analogy, 250–254
Selectivity measurement, phase noise
    effects, 12–13
Self-resonant frequency (SRF)
  1500–3000/3000–6000 MHz coupled
      resonator oscillator, circuit
      validation, 355–361
  ultra-low-noise wideband oscillators,
      passive device, 333–334
Series capacitance
  coupled resonator, ultra-low-noise
      wideband oscillators, 326–330

ultra-low-noise wideband oscillators,
    inductor model, 335–339
Series elements, noise correlation matrix
    and, 464–465
Series feedback configuration
  bipolar transistors, 417–420
  large-signal S-parameters, 384–388
  metal semiconductor field-effect
      transistor, 407–414
  ultra-low-noise wideband oscillators,
      wideband VCO design technique,
      316–320
Series loss resistance
  mutually coupled oscillators, pendulum
      analogy, 248–254
  push-push/two-push oscillator,
      312–314
  ultra-low-noise wideband oscillators,
      inductor model, 335–339
Series resonator circuits
  coupled oscillator systems, 259–263
  4100 MHz oscillator, circuit validation,
      237–241
  N-coupled oscillator noise analysis,
      topology characteristics,
      510–513
  negative resistance model, time domain
      noise signal calculations, 174
  77 GHz SiGe oscillator, circuit validation,
      242–244
Shot noise
  coupled oscillator systems,
      267–268
  feedback model phase noise, 190
    spectral density and transfer
        function, 195–199
  N-coupled oscillators, 290–292
  negative resistance noise voltage,
      180–183
  synthesis procedure, 160–161
  ultra-low-noise wideband oscillators,
      wideband VCO design
      technique, 320
Shunt element, noise correlation matrix
    and, 464–465
Signal-driving source, coupled oscillator
    noise, 268
Signal-to-noise ratio (SNR)
  bipolar transistor noise analysis,
      T-equivalent configuration,
      487–489
  oscillator noise analysis, Leeson
      equation, 130–132

Silicon germanium (SiGe) technology
  microwave oscillator history, 18
  77 GHz SiGe oscillator, circuit
    validation, 242–244
Silicon transistors, noise analysis,
  hybrid-π configuration,
  472–473
Simulation results, metal semiconductor
  field-effect transistor,
  432–433
Single noise pulse, negative resistance
  model, time domain noise signal
  calculations, 168–174
Single sideband (SSB) phase noise,
  N-coupled oscillator noise
  analysis, 289–292
Sinusoidal oscillator
  basic properties, 10–11
  phase noise modulation, 438–440
Small-signal conditions
  bipolar transistors, series feedback,
    418–420
  large-signal oscillators, time-domain
    behavior, 109–121
  metal semiconductor field-effect
    transistor, series feedback,
    413–414
  negative resistance model, time domain
    noise signal calculations, 174
  phase noise analysis, CAD solutions,
    446–447
  S-parameter measurements, 64–69
  ultra-low-noise wideband oscillators
    optimum phase noise, loaded Q,
      332–333
    wideband VCO design technique,
      317–320
SMD inductor, ultra-low-noise wideband
  oscillators, 335–339
S-parameters
  bipolar transistor design, 41–47
  coupled resonator design, 78–79
  large-signal parameters
    defined, 61–63
    measurements, 63–69
    oscillator design, 381–388
  parallel feedback oscillator, 91–93
  ultra-low-noise wideband oscillators
    inductor model, 336–339
    resistor model, 339–340
Spectral density calculations
  conversion noise analysis, C448AD
    solution, 448

coupled oscillator noise, 268
N-coupled oscillator noise analysis
  frequency doubler/modifier,
    287–292
  push-push configuration, 293–300
  topology characteristics, 509–513
negative resistance model, time
  domain noise signal calculations,
  171–174
oscillator phase noise model, synthesis
  procedure, 160–161
phase noise analysis, feedback models,
  186–199
  noise transfer function, 194–199
uncoupled oscillator noise, 275–276
SPICE parameters
  CFY30 field effect transistors, 51–56
  oscillator noise analysis, 133–137
  phase noise contributions, CMOS
    transistors, 209–210
  transistor design, 38–40
  tuning diodes, 59–60
  tuning diodes, 59–60
  ultra-low-noise wideband oscillators,
    coupled resonators, 328–330
Spurious signals, voltage-controlled
  oscillator, 15–16
Stability properties
  biasing topololgy, Agilent HBFP-0405
    transistor, 227–230
  of oscillators, 9
  passive bias circuits, phase noise and
    temperature effects, 214–218
  phase noise, biasing, and temperature
    deffects, 218–224
  wideband coupled resonator VCOs,
    circuit validation, 300–1100 MHz
    coupled resonator oscillator,
    343–346
Startup condition
  large-signal oscillators, 94–100
  2400 MHz MOSFET-based push-pull
    oscillator, 205
Steady-state oscillation
  bipolar transistors, series feedback,
    419–420
  metal semiconductor field-effect transistor
    optimization, 426–432
    series feedback, 413–414
  negative resistance noise voltage, 184
  nonlinear negative resistance calculation,
    166–167
  series feedback configuration, 384–388

INDEX **539**

Steady-state stationarity
  large-signal oscillators, 97–101
  parallel feedback oscillator, 92–93
Steady-state synchronized frequency
  mutually coupled oscillators, noise analysis, 280–282
  N-coupled (synchronized) oscillators, 266
  noise analysis, 284–300
  N-push coupled mode (synchronized) oscillators, 307–311
  push-push/two-push oscillator, 312–314
Stop-band design, ceramic resonators, 79–81
Substrate loss, ultra-low-noise wideband oscillators, passive device, 334
Surface acoustical wave (SAW) resonator, oscillator design, 20–21
Switched capacitor system, ultra-low-noise wideband oscillators, VCO design criteria, 321
Symmetry
  1000–2000/2000–4000 MHz push-push oscillator, 348–355
  1500–3000/3000–6000 MHz coupled resonator oscillator, circuit validation, 358–361
  triple-push configuration, 310–311
Synchronized oscillators
  modulation noise analysis, 452–453
  mutually locked systems
    noise analysis, 276–282
    phase condition, 254–257
  N-coupled system dynamics, 263–266
  push-push configuration, 295–300
  N-push coupled mode (synchronized) oscillators
    design criteria, 311–315
    mode analysis, 307–311
    N-push subcircuits, 306–307
    push/push-push configuration, 303–304
    push-push/two-push design, 311–314
    quadruple-push configuration, 305–306
    system dynamics, 300–315
    triple-push configuration, 304–305
    design criteria, 314–315
Synthesis procedure, oscillator phase noise model, 159–161

Taylor-series expansion, metal semiconductor field-effect transistor, series feedback, 413–414
Temperature effects
  dielectric resonator design, 82–83
  FET noise parameters, 503–506
  phase noise calculation and optimization, 210–232
    active bias circuits, 211–214
    bias selection criteria, 230–232
    design criteria, 224–230
    passive bias circuits, 214–218
    stability factors and temperature-dependent variables, 218–224
  voltage-controlled oscillator output, 15
T-equivalent configuration, bipolar transistors
  noise correlation matrix, 486–489, 492–494
  noise model, 485–489
Thermal noise
  bipolar transistors
    hybrid-π configuration, 472–473
    noise analysis, T-equivalent configuration, 486–489
  coupled oscillator systems, 267–268
  feedback model phase noise, 190
    spectral density and transfer function, 195–199
  modulation noise analysis, 451–453
  N-coupled oscillators, 290–292
  negative resistance noise voltage, 180–183
  ultra-low-noise wideband oscillators, wideband VCO design technique, 320
300–1100 MHz coupled resonator oscillator, circuit validation, 341–346
Three-port oscillators
  design techniques, 20–21
  triple-push configuration, 310–311
Three-reactance oscillators, Y-parameters, 28–32
Time-dependent calculations
  coupled oscillator systems, 257–263
  phase noise analysis, negative resistance model, 162–166

540　INDEX

Time-domain behavior
　coupled oscillator system noise, 269–270
　large-signal oscillators, 101–121
　negative resistance model, noise signal calculations, 167–174
　uncoupled oscillator noise, 272–276
Time-varying negative resistance noise analysis, 6
N-push coupled mode (synchronized) oscillators
　push/push-push configuration, 303–304
　quadruple-push configuration, 305–306
　triple-push configuration, 304–305
Topological analysis
　biasing networks, 224–230
　bipolar transistors
　　parallel feedback, 420–422
　　series feedback, 417–420
　metal semiconductor field-effect transistor
　　parallel feedback, 414–417
　　series feedback, 407–414
　N-coupled oscillator noise, 288–292, 509–513
　　bilateral systems, 512–513
　　globally coupled systems, 510–512
　　push-push configuration, 293–300
　　unilateral systems, 513–515
　N-push coupled mode (synchronized) oscillators, 302–315
　1000–2000/2000–4000 MHz hybrid tuned VCO, 361–365
　1000–2000/2000–4000 MHz push-push oscillator, 346–355
　parallel admittance, large-signal oscillators, 99–100
　300–1100 MHz coupled resonator oscillator, circuit validation, 341–346
　ultra-low-noise wideband oscillators, wideband VCO design technique, 316–320
Transconductance equations
　bipolar transistors, series feedback, 420
　large-signal design
　　Bessel functions, 391–395
　　transconductance, phase noise and output power, 401–405
　　metal semiconductor field-effect transistor, series feedback, 412–414
　phase noise analysis, 2400 MHz MOSFET-based push-pull oscillators, 201–203
　2400 MHz MOSFET-based push-pull oscillator, large-signal transconductance, 206
Transfer function (TF)
　large-signal oscillators, 95–100
　phase noise analysis, feedback models
　　nonunity gain, 193–194
　　resonator design, 192–193
　　spectral densities and, 194–199
　ultra-low-noise wideband oscillators, active devices, 321–323
Transformation factor (n)
　ABCD noise correlation matrix, 459–460
　feedback models, spectral density and transfer function, 197–199
　large-signal design, Bessel functions, 391–395
　negative resistance noise voltage, 182–183
　noise correlation matrix, 463–464
　noise sources, 466–467
Transformation ratio, large-signal oscillators, time-domain behavior, 110–121
Transient analysis, large-signal oscillators, 99–100
Transistor model properties
　basic principles, 37–40
　bipolar transistors, 40–47
　field-effect transistors, 47–56
　　GaAs models, 47–51
　　MOS (BiCMOS) models, 51–55
　　MOS Model Level 3 model, 55–56
　tuning diodes, 56–60
Transit time, metal semiconductor field-effect transistor, series feedback, 411–414
Transmission line resonators
　4100 MHz oscillator, circuit validation, 237–241
　mutually locked (synchronized) coupled oscillators, 255–257
Transverse electromagnetic (TEM) transmission lines, ceramic resonator design, 79–81

INDEX 541

Triple-push oscillator
  design analysis, 314–315
  mode analysis, 310–311
  N-push coupled mode (synchronized) configuration, 304–305
TriQuint models, field-effect transistors, 48–51
Tuned circuits
  1000–2000/2000–4000 MHz hybrid tuned VCO, 361–365
  1500–3000/3000–6000 MHz coupled resonator oscillator, circuit validation, 357–361
  oscillator noise analysis, 135–137
    Lee and Hajimiri noise model, 137–139
    phase noise contributions, CMOS transistors, 209–210
  ultra-low-noise wideband configuration, 323–326
Tuning diodes
  dielectric resonator design, 82–83
  oscillator noise analysis, Leeson equation, 129–132
  transistor design, 56–60
  ultra-low-noise wideband oscillators
    tuning characteristics, 324–326
    VCO design criteria, 321
  voltage-controlled oscillator, 17
Two-port oscillators
  ABCD matrix, noise transfer, 465–466
  bipolar transistors
    GE configuration, noise correlation matrix, 477–479
    noise factor calculations, 475–476
    noise parameters excluding base resistance, 479–485
    parallel feedback, 421–422
    series feedback, 419–420
    T-equivalent configuration, noise correlation matrix, 493–494
  combined matrices, 458–459
  design techniques, 19–21
  metal semiconductor field-effect transistor
    parallel feedback, 415–417
    series feedback, 410–414
  noise parameters and correlation matrix, 457–458
    transformation, 463–464
  parallel feedback, 92–93

  phase noise analysis, feedback models, 186–199
  series feedback configuration, 384–388
  S-parameters, 61–63
  ultra-low-noise wideband configuration, resistor model, 340
2000 MHz GaAs FET-based oscillator, circuit validation, 241–242
2400 MHz MOSFET-based push-pull oscillator, phase noise calculation and optimization, 199–210
  design equations, 201–205
  device size, 206
  large-signal transconductance, 206
  oscillation frequency, 206–207
  phase noise design calculations, 207–210
  resonator parallel loss resistance, 206

Ultra-low-noise wideband oscillators
  active device, 321–323
  capacitance model, 338–339
  coupled resonator design, 326–330
  inductor model, 335–338
  optimum phase noise, loaded Q, 330–333
  passive device, 333–334
  resistor model, 339–340
  tuning characteristics and loaded Q, 324–326
  tuning network, 323–324
  voltage-controlled oscillator design, 320–321
  wideband VCO design, 315–320
Uncoupled oscillators
  mutually coupled oscillators, noise analysis, 279–282
  N-coupled oscillator noise analysis, push-push configuration, 296–300
  noise analysis, 271–276
  topology characteristics, 509–513
Unilateral N-coupled oscillator systems, noise analysis, 513–515
Unity-gain system, phase noise analysis, feedback models, resonator design, 191–193
Unmodulated signals, phase noise analysis, 438

Validation circuits
  4100 MHz oscillator, transmission line resonators, 237–241
  N-coupled oscillators, noise analysis, 300
  900–1800 MHz half-butterfly resonator-based oscillator, 245–246
  1 MHz CRO, 233–237
  77 GHz SiGe oscillator, 242–244
  2000 MHz GaAs FET-based oscillator, 241–242
  wideband coupled resonator VCOs
    1000–2000/2000–4000 MHz hybrid tuned VCO, 361–365
    1000–2000/2000–4000 MHz push-push oscillator, 346–355
    1500–3000/3000–6000 MHz coupled resonator oscillator, 355–361
    300–1100 MHz coupled resonator oscillator, 341–346
Van der Pol (VDP) model, coupled oscillator systems, 257–263
  negative resistance, 261
Varactor-tuned VOC oscillators, ultra-low-noise wideband configuration, 323–324
  tuning characteristics, 324–326
Variable coupling capacitor, 300-1100 MHz coupled resonator oscillator, circuit validation, 342–346
Vector presentation
  negative resistance model, time domain noise signal calculations, 169–174
  noise correlation matrix transformation, 463–464
  uncoupled oscillator noise, 274–276
Voltage-controlled oscillator (VCO)
  active bias circuits, phase noise and temperature effects, 212–214
  basic properties, 13–17
    basic components, 32–35
    frequency pushing, 16
    frequency range, 13
    harmonic suppression, 14–15
    load change sensitivity, 16
    output power, 14
    phase noise effect, 14
    post-tuning drift, 16
    power consumption, 17
    spurious response, 15–16
    temperature-dependent output power, 15
    tuning characteristics, 17
  bias values, 230–232
  coupled oscillator systems, noise analysis, 269–270
  coupled resonator design, 75–79
  historical background, 1–5
  locking bandwidth, 263
  mutually coupled systems, classical pendulum analogy, 247–254
  noise analysis
    experimental variations, 147–148
    Leeson equation, 129–132
  N-push coupled mode (synchronized) topology, 302–307
  phase noise effects, 12–13
  push-push/two-push design, 311–314
  transistor design, tuning diodes, 56–60
  triple-push configuration, 314–315
  ultra-low-noise wideband oscillators
    basic design criteria, 320–321
    passive device, 333–334
    tuning characteristics, 324–326
    tuning network, 323–324
    wideband design technique, 315–320
  wideband coupled resonator validation circuits
    1000–2000/2000–4000 MHz hybrid tuned VCO, 361–365
    1000–2000/2000–4000 MHz push-push oscillator, 346–355
    1500–3000/3000–6000 MHz coupled resonator oscillator, 355–361
    300–1100 MHz coupled resonator oscillator, 341–346
Voltage division
  large-signal oscillators, time-domain behavior, 109–121
  phase noise, biasing and temperature effects, stability factors, 220–224
Voltage feedback bias networks
  metal semiconductor field-effect transistor, parallel feedback, 416–417
  phase noise and temperature effects, 215–218
  stability factors, 218–224

Voltage gain, 1500–3000/3000–6000 MHz coupled resonator oscillator, circuit validation, 358–361
Voltage-source bias circuits
  large-signal design, Bessel functions, 390–395
  phase noise and temperature effects, 215, 217

White additive noise, synthesis procedure, 159–161
Wideband coupled resonator VCOs
  ultra-low-noise wideband oscillators, 315–320
  validation circuits
    1000–2000/2000–4000 MHz hybrid tuned VCO, 361–365
    1000–2000/2000–4000 MHz push-push oscillator, 346–355
    1500–3000/3000–6000 MHz coupled resonator oscillator, 355–361
    300–1100 MHz coupled resonator oscillator, 341–346

Y-parameters
  bipolar transistors
    noise parameters excluding base resistance, 479–485
    parallel feedback, 420–422
    T-equivalent configuration, 493–494
    historical background, 61
    large-signal oscillators, time-domain behavior, 107–121
    metal semiconductor field-effect transistor, parallel feedback, 414–417
    mutually locked (synchronized) coupled oscillators, 256–257

N-coupled oscillator noise analysis, 290–292
noise correlation matrix transformation, 463–464
parallel feedback oscillator, 91–93
three-reactance oscillator design, 28–32
ultra-low-noise wideband oscillators, coupled resonator, 327–330
Yttrium ion garnet (YIG) resonator
  design principles, 83–85
  ultra-low-noise wideband configuration, 323–324

Zero reactance, phase noise analysis, 2400 MHz MOSFET-based push-pull oscillators, 200–210
Z-parameters
  bipolar transistors
    series feedback, 417–420
    T-equivalent configuration, 493–494
  historical background, 61
  metal semiconductor field-effect transistor
    optimization, 422–432
    series feedback, 407–414
  noise correlation matrix transformation, 463–464
  N-push coupled mode (synchronized) oscillators, 309–311
  push-push/two-push oscillator, 311–314
  series feedback configuration, 384–388
  ultra-low-noise wideband oscillators
    inductor model, 336–339
    wideband VCO design technique, 316–320